Lecture Notes on Data Engineering and Communications Technologies

Volume 54

Series Editor

Fatos Xhafa, Technical University of Catalonia, Barcelona, Spain

The aim of the book series is to present cutting edge engineering approaches to data technologies and communications. It will publish latest advances on the engineering task of building and deploying distributed, scalable and reliable data infrastructures and communication systems.

The series will have a prominent applied focus on data technologies and communications with aim to promote the bridging from fundamental research on data science and networking to data engineering and communications that lead to industry products, business knowledge and standardisation.

Indexed by SCOPUS, INSPEC.

All books published in the series are submitted for consideration in Web of Science.

More information about this series at http://www.springer.com/series/15362

Ashish Khanna · Deepak Gupta ·
Zdzisław Pólkowski · Siddhartha Bhattacharyya ·
Oscar Castillo
Editors

Data Analytics and Management

Proceedings of ICDAM

Springer

Editors
Ashish Khanna
Maharaja Agrasen Institute of Technology
New Delhi, India

Deepak Gupta
Maharaja Agrasen Institute of Technology
New Delhi, India

Zdzisław Pólkowski
Jan Wyzykowski University
Polkowice, Poland

Siddhartha Bhattacharyya
CHRIST (Deemed to be University)
Bengaluru, India

Oscar Castillo
Tijuana Institute of Technology
Tijuana, Mexico

ISSN 2367-4512 ISSN 2367-4520 (electronic)
Lecture Notes on Data Engineering and Communications Technologies
ISBN 978-981-15-8334-6 ISBN 978-981-15-8335-3 (eBook)
https://doi.org/10.1007/978-981-15-8335-3

This Springer imprint is published by the registered company Springer Nature Singapore Pte Ltd.
The registered company address is: 152 Beach Road, #21-01/04 Gateway East, Singapore 189721, Singapore

Dr. Ashish Khanna would like to dedicate this book to his mentors Dr. A. K. Singh and Dr. Abhishek Swaroop for their constant encouragement and guidance and his family members including his mother, wife and kids. He would also like to dedicate this work to his (Late) father Sh. R. C. Khanna with folded hands for his constant blessings.

Dr. Deepak Gupta would like to dedicate this book to his father Sh. R. K. Gupta, his mother Smt. Geeta Gupta for their constant encouragement, his family members including his wife, brothers, sisters, kids and to his students who are close to his heart.

Dr. Zdzisław Pólkowski would like to dedicate this book to his wife, daughter and parents.

Prof. (Dr.) Siddhartha Bhattacharyya would like to dedicate this book to his father Late Ajit Kumar Bhattacharyya, his mother Late Hashi Bhattacharyya, his beloved wife Rashni and his colleagues Jayanta Biswas and Debabrata Samanta.

ICDAM-2020 Committees

Steering Committee Members

Patrons

Dr. Tadeusz Kierzyk, Professor of UJW, Rector of Jan Wyzykowski University, Polkowice, Poland
Lukasz Puzniecki, Mayor of Polkowice, Poland
Mr. Rajeev Jain, Chairman, B. M. Institute of Engineering and Technology, India
Mr. Rakesh Kuchhal, Manager BOG, B. M. Institute of Engineering and Technology, India

General Chairs

Prof. Dr. Janusz Kacprzyk, Polish Academy of Sciences, Systems Research Institute, Poland
Prof. Dr. Cesare Alippi, Polytechnic University of Milan, Italy
Prof. Dr. Siddhartha Bhattacharyya, CHRIST (Deemed to be University), Bengaluru, India

Honorary Chairs

Prof. Dr. Aboul Ella Hassanien, Cairo University, Egypt
Prof. Dr. Vaclav Snasel, Rector, VSB-Technical University of Ostrava, Czech Republic

Conference Chairs

Dr. Zdzisław Pólkowski, Professor of UJW, Jan Wyzykowski University, Polkowice, Poland
Prof. Dr. Harish Mittal, Principal, B. M. Institute of Engineering and Technology, India
Prof. Dr. Joel J. P. C. Rodrigues, Universidade Estadual do Piau Teresina, Brazil

Prof. Dr. Abhishek Swaroop, Bhagwan Parshuram Institute of Technology, Delhi, India
Prof. Dr. Anil K. Ahlawat, KIET Group of Institutes, Ghaziabad, India

Technical Programme Chairs

Dr. Stanislaw Piesiak, Professor of UJW, Jan Wyzykowski University, Polkowice, Poland
Dr. Jan Walczak, Jan Wyzykowski University, Polkowice, Poland
Dr. Anna Wojciechowicz, Jan Wyzykowski University, Lubin, Poland
Dr. Abhinav Juneja, Associate Director, B. M. Institute of Engineering and Technology, India
Dr. Vishal Jain, Dean Academics, B. M. Institute of Engineering and Technology, India

Technical Programme Co-chairs

Prof. Dr. Victor Hugo C. de Albuquerque, Universidade de Fortaleza, Brazil
Dr. Gulshan Shrivastava, National Institute of Technology Patna, India

Conveners

Dr. Ashish Khanna, Maharaja Agrasen Institute of Technology (GGSIPU), New Delhi, India
Dr. Deepak Gupta, Maharaja Agrasen Institute of Technology (GGSIPU), New Delhi, India
Dr. Pradeep Kumar Mallick, KIIT Deemed to be University Bhubaneswar, Odisha, India
Dr. Akash Kumar Bhoi, Sikkim Manipal University, India

Publication Chair

Dr. Jerzy Widerski, Professor of UJW, V-Rector of Jan Wyzykowski University, Polkowice, Poland

Publicity Chairs

Dr. Pawel Gren, Professor of UJW, V-Rector of Jan Wyzykowski University, Polkowice, Poland
Dr. Aditya Khamparia, Lovely Professional University, Punjab, India

Co-convener

Mr. Moolchand Sharma, Maharaja Agrasen Institute of Technology, India

Local Committee

Leszek Cybulski, Director of Jan Wyzykowski University, Polkowice, Poland

Advisory Committee

Dr. Tadeusz Kierzyk, Professor of UJW, Rector of Jan Wyzykowski University, Polkowice, Poland
Prof. Vincenzo Piuri, University of Milan, Italy
Prof. Aboul Ella Hassanien, Cario University, Egypt
Prof. Marcin Paprzycki, Polish Academy of Science, Poland
Prof. Valentina Emilia Balas, Aurel Vlaicu University of Arad, Romania
Prof. Marius Balas, Aurel Vlaicu University of Arad, Romania
Prof. Mohamed Salim Bouhlel, Sfax University, Tunisia
Prof. Cenap Ozel, King Abdulaziz University, Saudi Arabia
Prof. Ashiq Anjum, University of Derby, Bristol, UK
Prof. Mischa Dohler, King's College London, UK
Prof. David Camacho, Universidad Autonoma de Madrid, Spain
Prof. Parmanand, Dean, Galgotias University, Uttar Pradesh, India
Prof. Maryna Yena, Medical University of Kiev, Ukraine
Prof. Giorgos Karagiannidis, Aristotle University of Thessaloniki, Greece
Prof. Tanuja Srivastava, Department of Mathematics, IIT Roorkee
Dr. D. Jude Hemanth, Karunya University, Coimbatore
Prof. Tiziana Catarci, Sapienza University, Rome, Italy
Prof. Salvatore Gaglio, University Degli Studi di Palermo, Italy
Prof. Bozidar Klicek, University of Zagreb, Croatia
Prof. A. K. Singh, NIT Kurukshetra, India
Prof. Anil Kumar, KIET Group of Institutes, India
Prof. Chang-Shing Lee, National University of Tainan, Taiwan
Dr. Paolo Bellavista, Alma Mater Studiorum–Universitȧ di Bologna
Prof. Sanjay Misra, Covenant University, Nigeria
Prof. Benatiallah Ali, Adrar University, Algeria
Prof. Suresh Chandra Satapathy, PVPSIT, Vijayawada, India
Prof. Marylene Saldon-Eder, Mindanao University of Science and Technology
Prof. Özlem Onay, Anadolu University, Eskisehir, Turkey
Prof. Kei Eguchi, Department of Information Electronics, Fukuoka Institute of Technology
Prof. Zoltan Horvath, Kasetsart University
Dr. A. K. M. Matiul Alam Vancouver British Columbia, Canada
Prof. Joong Hoon Jay Kim, Korea University
Prof. Sheng-Lung Peng, National Dong Swa Uinversity, Taiwan
Dr. Dusanka Boskovic, University of Sarajevo, Sarajevo
Dr. Periklis Chat Zimisios, Alexander TEI of Thessaloniki, Greece
Dr. Nhu Gia Nguyen, Duy Tan University, Vietnam
Dr. Ahmed Faheem Zobaa, Brunel University, London
Prof. Ladjel Bellatreche, Poitiers University, France
Prof. Victor C. M. Leung, The University of British Columbia, Canada
Prof. Huseyin Irmak, Cankiri Karatekin University, Turkey

Preface

We hereby are delighted to announce that Jan Wyzykowski University, Polkowice, Poland, and B.M. Institute of Engineering and Technology, Haryana, India, have hosted the eagerly awaited and much coveted International Conference on Data Analytics and Management (ICDAM-2020). The first version of the conference was able to attract a diverse range of engineering practitioners, academicians, scholars and industry delegates, with the reception of abstracts including more than 1540 authors from different parts of the world. The committee of professionals dedicated towards the conference is striving to achieve a high-quality technical programme with tracks on data analytics, data management, big data, computational intelligence and communication networks. All the tracks chosen in the conference are interrelated and are very famous amongst present-day research community. Therefore, a lot of research is happening in the above-mentioned tracks and their related sub-areas. More than 380 full-length papers have been received, among which the contributions are focused on theoretical, computer simulation-based research and laboratory-scale experiments. Amongst these manuscripts, 70 papers have been included in the Springer proceedings after a thorough two-stage review and editing process. All the manuscripts submitted to the ICDAM-2020 were peer reviewed by at least two independent reviewers, who were provided with a detailed review proforma. The comments from the reviewers were communicated to the authors, who incorporated the suggestions in their revised manuscripts. The recommendations from two reviewers were taken into consideration while selecting a manuscript for inclusion in the proceedings. The exhaustiveness of the review process is evident, given the large number of articles received addressing a wide range of research areas. The stringent review process ensured that each published manuscript met the rigorous academic and scientific standards. It is an exalting experience to finally see these elite contributions materialize into a book volume as ICDAM proceedings by Springer entitled "Data Analytics and Management: Proceedings of ICDAM".

ICDAM-2020 invited three keynote speakers, who are eminent researchers in the field of computer science and engineering, from different parts of the world. In addition to the plenary sessions on each day of the conference, six concurrent technical sessions are held every day to assure the oral presentation of around 70 accepted papers. Keynote speakers and session chair(s) for each of the concurrent sessions have been leading researchers from the thematic area of the session. A technical exhibition is held during the conference, which has put on display the latest technologies, expositions, ideas and presentations. The delegates were provided with a book of extended abstracts to quickly browse through the contents, participate in the presentations and provide access to a broad audience of the audience. The research part of the conference was organized in a total of six special sessions. These special sessions provided the opportunity for researchers conducting research in specific areas to present their results in a more focused environment.

An international conference of such magnitude and release of the ICDAM-2020 proceedings by Springer has been the remarkable outcome of the untiring efforts of the entire organizing team. The success of an event undoubtedly involves the painstaking efforts of several contributors at different stages, dictated by their devotion and sincerity. Fortunately, since the beginning of its journey, ICDAM-2020 has received support and contributions from every corner. We thank them all who have wished the best for ICDAM-2020 and contributed by any means towards its success. The edited proceedings volumes by Springer would not have been possible without the perseverance of all the steering, advisory and technical programme committee members.

All the contributing authors owe thanks from the organizers of ICDAM-2020 for their interest and exceptional articles. We would also like to thank the authors of the papers for adhering to the time schedule and for incorporating the review comments. We wish to extend our heartfelt acknowledgment to the authors, peer reviewers, committee members and production staff whose diligent work put shape to the ICDAM-2020 proceedings. We especially want to thank our dedicated team of peer reviewers who volunteered for the arduous and tedious step of quality checking and critique on the submitted manuscripts. We wish to thank our faculty colleague Mr. Moolchand Sharma for extending their enormous assistance during the conference. The time spent by them and the midnight oil burnt is greatly appreciated, for which we will ever remain indebted. The management, faculties, administrative and support staff of the college have always been extending their services whenever needed, for which we remain thankful to them.

Lastly, we would like to thank Springer for accepting our proposal for publishing the ICDAM-2020 conference proceedings. Help received from Mr. Aninda Bose, the acquisition senior editor, in the process has been very useful.

New Delhi, India

Ashish Khanna
Deepak Gupta
Organizers, ICDAM-2020

About This Book

International Conference on Data Analytics and Management (ICDAM-2020) was held on 18 June via virtual mode and jointly organized by Jan Wyzykowski University, Polkowice, Poland, and B. M. Institute of Engineering and Technology, Haryana, India. This conference was able to attract a diverse range of engineering practitioners, academicians, scholars and industry delegates, with the reception of papers including more than 1540 authors from different parts of the world. Only 70 papers have been accepted and registered with an acceptance ratio of 18% to be published in one volumes of prestigious Springer Lecture Notes on Data Engineering and Communications Technologies series.

Contents

About the Editors

Dr. Ashish Khanna has expertise in Teaching, Entrepreneurship, and Research & Development of 16 years. He received his Ph.D. degree from National Institute of Technology, Kurukshetra, in March 2017. He has completed his M.Tech. and B.Tech. from GGSIPU, Delhi. He has completed his PDF from the Internet of Things Lab at Inatel, Brazil. He has around 100 research papers along with book chapters including more than 40 papers in SCI Indexed Journals with cumulative impact factor of above 100 to his credit. Additionally, he has authored and edited 19 books. Furthermore, he has served the research field as a Keynote Speaker/Session Chair/Reviewer/TPC member/ Guest Editor, and many more positions in various conferences and journals. His research interest includes image processing, distributed systems and its variants, and machine learning. He is currently working at the CSE, Maharaja Agrasen Institute of Technology, Delhi. He is Convener and Organizer of ICICC Springer conference series.

Dr. Deepak Gupta is an eminent academician and plays versatile roles and responsibilities juggling between lectures, research, publications, consultancy, community service, Ph.D. and postdoctorate supervision, etc. With 12 years of rich expertise in teaching and two years in industry, he focuses on rational and practical learning. He has contributed massive literature in the fields of–computer interaction, intelligent data analysis, nature-inspired computing, machine learning, and soft computing. He is working as an Assistant Professor at Maharaja Agrasen Institute of Technology (GGSIPU), Delhi, India. He has served as Editor-in-Chief, Guest Editor, and Associate Editor in SCI and various other reputed journals. He has authored/edited 33 books with national/international level publisher (Elsevier, Springer, Wiley, Katson). He has published 96 scientific research publications in reputed international journals and conferences including 46 SCI Indexed Journals of IEEE, Elsevier, Springer, Wiley, and many more.

Dr. Zdzisław Pólkowski is an Adjunct Professor at Faculty of Technical Sciences at the Jan Wyzykowski University, Poland. He is also the Rector's Representative for International Cooperation and Erasmus Programme and Former Dean of the

Technical Sciences Faculty during the period of 2009–2012. His area of research includes management information systems, business informatics, IT in business and administration, IT security, small medium enterprises, CC, IoT, big data, business intelligence, and block chain. He has published around 60 research articles. He has served the research community in the capacity of Author, Professor, Reviewer, Keynote Speaker, and Co-Editor. He has attended several international conferences in the various parts of the world. He is also playing the role Principal Investigator.

Dr. Siddhartha Bhattacharyya [FIEI, FIETE, LFOSI, SMIEEE, SMACM, SMIETI, LMCSI, LMISTE, MIET (UK), MIAENG, MIRSS, MIAASSE, MCSTA, MIDES, MISSIP, MSDIWC] is currently the Principal of RCC Institute of Information Technology, Kolkata, India. In addition, he is also serving as the Professor of Computer Application and Dean (Research and Development and Academic Affairs) of the institute. He is the recipient of several coveted awards like Adarsh Vidya Saraswati Rashtriya Puraskar, Distinguished HoD Award, Distinguished Professor Award, Bhartiya Shiksha Ratan Award, Best Faculty for Research, and Rashtriya Shiksha Gaurav Puraskar. He received the Honorary Doctorate Award (D. Litt.) from The University of South America and the SEARCC International Digital Award ICT Educator of the Year in 2017. He has been appointed as the ACM Distinguished Speaker for the tenure 2018–2020. He is a co-author of 4 books and the co-editor of 16 books and has more than 200 research publications in international journals and conference proceedings to his credit.

Dr. Oscar Castillo holds the Doctor in Science degree (Doctor Habilitatus) in Computer Science from the Polish Academy of Sciences (with the Dissertation "Soft Computing and Fractal Theory for Intelligent Manufacturing"). He is a Professor of Computer Science in the Graduate Division, Tijuana Institute of Technology, Tijuana, Mexico. Currently, he is President of HAFSA (Hispanic American Fuzzy Systems Association) and Past President of IFSA (International Fuzzy Systems Association). Prof. Castillo is also Chair of the Mexican Chapter of the Computational Intelligence Society (IEEE). His research interests are in Type-2 Fuzzy Logic, Fuzzy Control, Neuro-Fuzzy, and Genetic-Fuzzy hybrid approaches. He has published over 300 journal papers, 10 authored books, 40 edited books, 200 papers in conference proceedings, and more than 300 chapters in edited books, in total 865 publications according to Scopus (H index=60), and more than 1000 publications according to Research Gate (H index=72 in Google Scholar).

A Forecasting-Based DLP Approach for Data Security

Kishu Gupta and Ashwani Kush

Abstract Sensitive data leakage is the major growing problem being faced by enterprises in this technical era. Data leakage causes severe threats for organization of data safety which badly affects the reputation of organizations. Data leakage is the flow of sensitive data/information from any data holder to an unauthorized destination. Data leak prevention (DLP) is set of techniques that try to alleviate the threats which may hinder data security. DLP unveils guilty user responsible for data leakage and ensures that user without appropriate permission cannot access sensitive data and also provides protection to sensitive data if sensitive data is shared accidentally. In this paper, data leakage prevention (DLP) model is used to restrict/grant data access permission to user, based on the forecast of their access to data. This study provides a DLP solution using data statistical analysis to forecast the data access possibilities of any user in future based on the access to data in the past. The proposed approach makes use of renowned simple piecewise linear function for learning/training to model. The results show that the proposed DLP approach with high level of precision can correctly classify between users even in cases of extreme data access.

Keywords Data leakage · Data leakage prevention · Forecast · Guilty agent · Statistical analysis

1 Introduction

The NIST explains computer security as "protection afforded to an automated information system in order to attain the applicable objectives of preserving the integrity, availability, and confidentiality of information system resources (includes hardware,

K. Gupta (✉)
Department of Computer Science and Applications, Kurukshetra University, Kurukshetra, India
e-mail: kishugupta2@gmail.com

A. Kush
Institute of Integrated and Hons. Studies, Kurukshetra University, Kurukshetra, India
e-mail: akush20@gmail.com

A. Khanna et al. (eds.), *Data Analytics and Management*, Lecture Notes on Data Engineering and Communications Technologies 54,
https://doi.org/10.1007/978-981-15-8335-3_1

software, firmware, information/data, and telecommunications)" [1, 2]. Advancement in technology allows easy and speedy transfer of data. Data is the key to conduct business activities nowadays, and hence, a need arises to share data among various stakeholders/third parties like human resources working from outside the site (e.g., on laptops), business colleague, and clients [3]. For example, service provider requires access to the company intellectual property and other confidential information to carry out their services [4, 5].

Data loss/leakage has emerged as the biggest threat that organizations are facing today. In the present scenario, almost all business activities depend on extensive sharing of sensitive/confidential data, within or outside the organization [6, 7]. Data leakage is an event that may occur either accidentally or maliciously that permits data access to unauthorized user. Sensitive data loss/leakage rigorously hampers reputation of organization, confidence/faith of customers in company which may ultimately lead to shut down company or even may lead to severe political crisis like WikiLeak leaks [8]. Leakage is thus a subset of data loss with a spotlight on the data security goal.

To minimize the risk of data loss, organizations usually make use of DLP solutions as a protection/defense mechanism. Prior to DLP security, mechanisms/technologies like firewalls and IDS were in use [9]. DLPSs are used to protect all kind of data, that is, data in use, data at rest and data in transit. DLPSs use the statistical/ analytical approach, data fingerprinting, regular expressions on context and content of data to identify and avoid unauthorized access to data [10]. DLP approach performs deep content analysis and observes the data access by users to discover improper usage [11–13]. DLP systems employ a model using either knowledge of an expert or may train/learn from available past records (Fig. 1).

This scenario provides ample space to produce requirement of a mechanism that can identify leakage with more precision for greater data security. The proposed DLP model tries to provide data security by observing users' trend to access the date, uses learning-based approach to highlight the user whoever is performing different

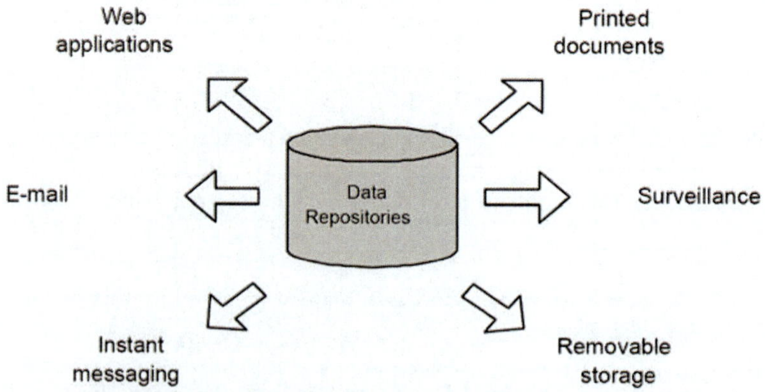

Fig. 1 Possible leak channels [14]

to trend observed. This enables organization to take suitable action like imposing access restriction on sensitive data for data security.

The paper layout is as follows: in Sect. 2, an overview of DLP solutions related to work has been presented. Furthermore, an overview of the proposed data fitting model framework has been discussed in Sect. 3. Section 4 deals with the experimental results, and Sect. 5 represents conclusion of paper, respectively.

2 Overview of DLP Approach

This section specifies major benefits of DLP solution. DLP model generally distinguishes suspicious activity from normal activity and performs either detection, i.e., raise alert if doubtful activity happens, or prevention, i.e., block nasty activity. DLP model can be characterized by various dimensions like model construction, filtering approach, network-based, host-based, etc. Model building approach to describe how the model is constructed is most relevant and best suited for the proposed work. Specification and learning-based are two approaches for model construction. Specification-based approach uses expert's knowledge and hence more precise/accurate, while learning-based model automatically learns using statistical techniques, i.e., machine learning.

The proposed framework reflects numerous benefits over existing solutions for DLP. First, learning-based framework tailors itself to the user's behavior and hence makes feasible to detect unknown and insider attacks. Additionally proposed DLP approach provides better control on data from being misused along with providing flexibility to access the data simultaneously. Moreover, this approach integrates data protection with user identity, thus making organization capable to implement data protection policy based on user identity and their role. The proposed approach tries to forecast guilty user based on available records of user access to the organization data. Finally, this forecasting-based framework proves to be more practical and efficient (Fig. 2).

3 Data Training Model

The proposed DLP model fulfills the objective of data security by employing machine learning-based approach and provides forecast for further action. The proposed model considers multiple agents (m) which have accessibility to organization data anytime and for any number of times. Each and every access to organization data by any user is entered in the form of a user's accessibility dataset containing many important details like date to access data, duration for which organization data was in use by particular user, i.e., y. The user's accessibility dataset continues to grow with time and is fitted and trained by machine learning techniques to obtain trend. Trend in this study determines the pattern of data accessibility by various users along

Fig. 2 DLP overview

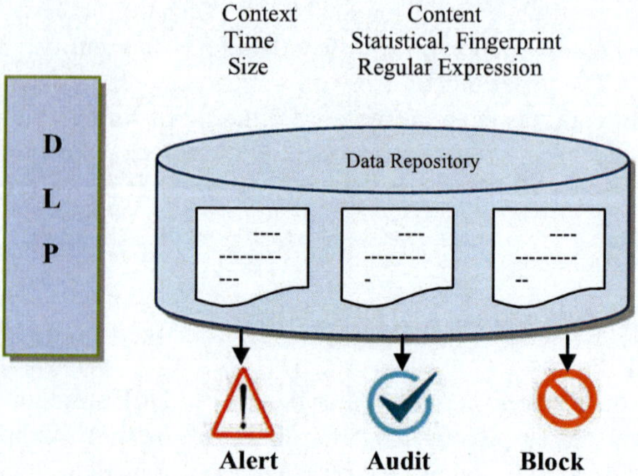

a period of time. Thus, the proposed model provides future insight by studying the past events. This section uses dataset of 2014 to 2018 to train the model and then predicts the trend of particular user after 2018. If the trend exceeds the defined upper limit, then it raises alert to prevent or restrict the user access to organization data. The discussed approach is explained through Fig. 3.

3.1 *Model Equations*

The model uses a simple piecewise linear model-based function as shown in Eq. (1) named as model fitting equation.

$$g(t) = (k + \alpha(t)'\delta)t + (m + \alpha(t)'\gamma)\ . \tag{1}$$

The above model fitting equation is used to fit and train the user's accessibility dataset to evaluate the predicted time value to be spent by user $\hat{y}$ along with many other parameters required for study. The model executes all entries in user accessibility database, i.e., $i = 1$ to n.

3.2 *Model Accuracy*

After obtaining predicted time value, the model calculates the error existing between actual and predicted value to determine the accuracy of approach. Here, ϵ_i shown in Eq. (2A) is error existing between actual and predicted values of time spend by user to access data. $£_i$ is percentage error represented by Eq. (2B).

Fig. 3 Data training model

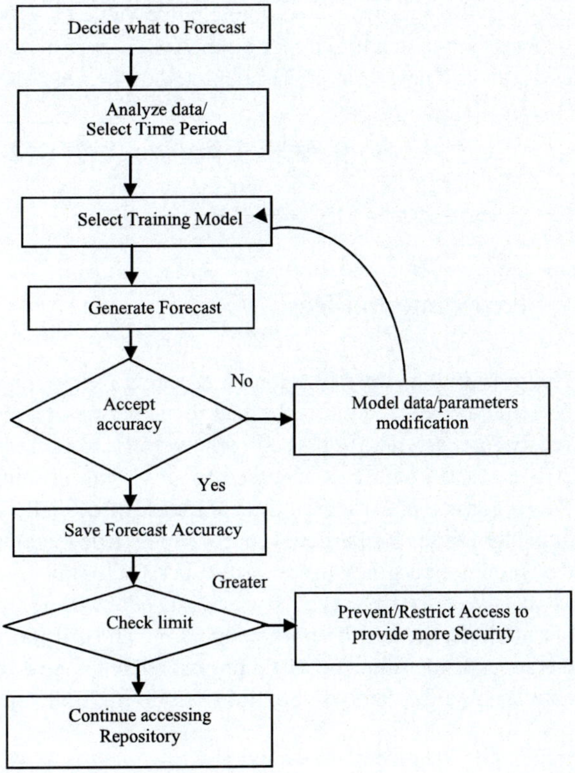

$$\epsilon_i = \left((y_i + \alpha) - (\widehat{y}_i - \alpha)\right) \tag{2A}$$

$$\epsilon_i = 100 * \frac{\epsilon_i}{(y_i + \alpha)}. \tag{2B}$$

3.3 *Calculating Limit*

To define the upper bound and lower bound for a user to access the data, the study computes $Æ_i$, L_i and υ_i in Eqs. (3A, 3B and 3C), respectively.

$$Æi = |\varepsilon i|. \tag{3A}$$

$$\mathcal{L}i = \left(\hat{y}i - (\mu * \varsigma * \sigma)\right). \tag{3B}$$

$$\mho i = \left(\hat{y}i + (\mu * \varsigma * \sigma)\right). \tag{3C}$$

$Æ_i$ is absolute error, i.e., nonnegative value of error as calculated in Eq. (2A). L_i is lower bound and υ_i is upper bound for users to access data. Here, μ and σ are mean and standard deviation of Æ calculated as shown below.

$$\mu = \frac{\sum_{i=1}^{n} Æi}{n} \text{ and } \sigma = \sqrt{\frac{1}{n}\sum_{i=1}^{n}(Æi - \overline{Æ})^2}$$

4 Experimental Result

The proposed model is implemented using Python on Jupyter Notebook platform in Anaconda environment to conduct the experiments. Figure 4a shows the overall half yearly forecast, and Fig. 4b shows overall annual forecast; it is observed that from 2014 to 2018, data was accessed for less than 80 min, given a year or half yearly. These graphs show access time forecast of overall system, i.e., all users together. Separate graphs for particular users can also be generated. Forecasting results depict that some user is going to access the data for longer duration in 2019 which is unusual as compared to the previous years. Based on trend, it can be a scenario/possibility of data leakage; hence, more restriction can be imposed to the user by checking on the access time limit for particular user. This is how this model will help to prevent data leakage for database being in use by multiple users and multiple repositories.

5 Conclusion

The biggest challenge in the present era is to shield sensitive data from leakage, which imposes a big threat for organization's growth/security/health. The paper highlights DLP approach with better control on data to protect data from being misused and also provide flexibility to access the data simultaneously. Moreover, this approach integrates data protection with user identity, hence enabling organization to enforce data protection policy based on user identity and their role. The proposed approach tries to forecast seems to be guilty user based on available records of user access to the organization data. On basis of model prediction output; access rights of any particular user can be restricted or blocked completely, hence proposed model provides enhanced data security for organization data. Thus, the proposed model provides future insight by studying the past events. Conclusion drawn from this study is that the system based on forecasting approach to identify guilty user is more practical and efficient.

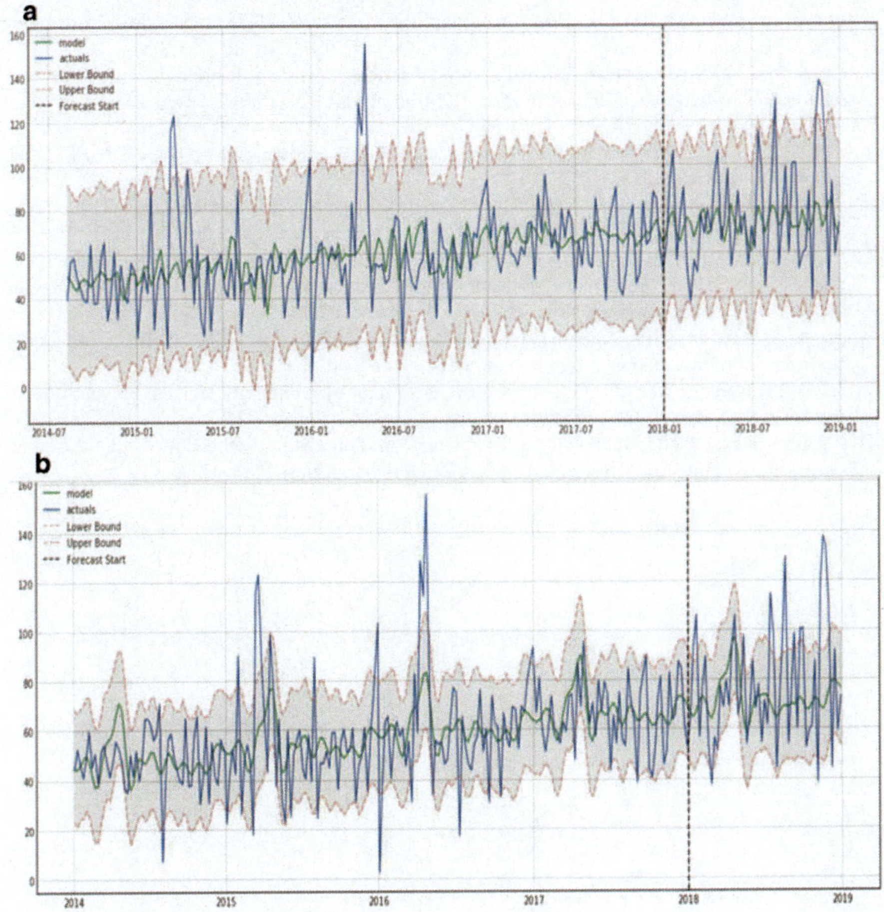

Fig. 4 **a** Overall half yearly forecast, **b** overall annual forecast

References

1. Gupta K, Kush A (2017) A review on data leakage detection for secure. Int J Eng Adv Technol (IJEAT) 7(1):153–159
2. Gupta K, Kush A (2018) Performance evaluation on data leakage detection for secure communication. In: 5th international conference on computing for sustainable global development: INDIACom, New Delhi, India
3. Alneyadi S, Sithirasenan E, Muthukkumarasamy V (2016) A survey on data leakage prevention systems. Elsevier J Netw Comput Appl 62(1):137–152
4. Papadimitriou P, Molina H (2011) Data leakage detection. IEEE Trans Knowl Data Eng 23(1):51–63
5. Sodagudi S, Kurra RR (2016) An approach to identify data leakage in secure communication. In: 2nd international conference on intelligent computing and applications, Singapore
6. Shabtai A, Elovici Y, Rokach L (2012) A survey of data leakage detection and prevention solutions. Springer, pp 1–87

7. Dattana V, Gupta K, Kush A (2019) A probability based model for big data security in smart city. In: International conference on big data and smart city (ICBDSC), Muscat
8. Liu F, Shu X, Yao D, Butt A (2015) Privacy-preserving scanning of big content for sensitive data exposure with MapReduce. In: 5th ACM conference data application security, privacy (CODASPY), Texas, USA
9. Shu X, Yao D (2015) Privacy-preserving detection of sensitive data exposure. IEEE Trans Inform Forensics Sec 10(5):1092–1103
10. Shu X, Zhang J, Yao D, Feng WC (2016) Fast detection of transformed data leaks. IEEE Trans Inform Forensics Sec 11(3):528–542
11. Fin Y, Lina W, Rongwei Y, Xiaoyan M (2013) A distribution model for data leakage prevention. In: International conference on mechatronic sciences, electric engineering and computer (MEC), IEEE, Shenyang, China
12. Peneti S, Rani B (2016) Data leakage prevention system with time stamp. In: International conference on information communication and embedded system (ICICES)
13. Kadu RS, Gadicha VB (2017) Review on seuring data by using data leakage prevention and detection. Int J Recent Innov Trends Comput Commun 5(5):731–735
14. Raman P, Kayacık HG, Somayaji A (2011) Understanding data leak prevention. In: 6th annual symposium on information assurance (ASIA'11), Albany, New York, USA

Detection of Anterior Cruciate Ligament Tear Using Deep Learning and Machine Learning Techniques

Vansh Kapoor, Nakul Tyagi, Bhumika Manocha, Ansh Arora, Shivangi Roy, and Preeti Nagrath

Abstract Magnetic resonance imaging (MRI) is used by surgeons to analyse different tears in the knee part of the body. This technique has demonstrated significant accuracy for the diagnosis of injuries like meniscus tears, ligament injuries, ACL tears, etc. However, studying these MRI manually is very time-consuming and has high chances of wrong prediction. It can take more than thirty minutes to properly examine a knee MRI and come to a result. It is a time-intensive process and has high error probability. Therefore, an automated model for examining knee images to predict the tears would cut the time cost and also the human errors. Generally, deep learning models work best with large amounts of dataset. But, there is not much data present out there of knee injuries which can help to train the model properly. For this purpose, we decided to analyse different deep learning and machine learning algorithms to compare and find the most efficient method. These models were applied on knee MRI dataset for the prediction of anterior cruciate ligament (ACL) tear from sagittal plane MRI scans which provide scans of totally ruptured, partially injured and healthy knees. After our analysis of various models, it was found that the best results were given by Support Vector Machine (SVM) followed by Convolutional Neural Network (CNN) on the masked dataset.

V. Kapoor · N. Tyagi · B. Manocha (✉) · A. Arora · S. Roy · P. Nagrath
Department of Computer Science and Engineering, Bharati Vidyapeeth's College of Engineering, Delhi, India
e-mail: bhumika.manocha@gmail.com

V. Kapoor
e-mail: vanshkapoorvk7@gmail.com

N. Tyagi
e-mail: nakul1tyagi@gmail.com

A. Arora
e-mail: ansh1999arora@gmail.com

S. Roy
e-mail: shivangi22.sr@gmail.com

P. Nagrath
e-mail: preeti.nagrath@bharatividyapeeth.edu

A. Khanna et al. (eds.), *Data Analytics and Management*, Lecture Notes on Data Engineering and Communications Technologies 54,
https://doi.org/10.1007/978-981-15-8335-3_2

Keywords Anterior cruciate ligament tear · MRI · Algorithm analysis on knee injuries · CNN · SVM

1 Introduction

Deep learning methods are being widely used in health care for better accuracy and medical decision making. These techniques can be very effectively used and can be trained on MRI data. There are numerous techniques that have been developed to improve MRI processing [1]. These new innovative techniques enhance the analysis process. CNN has been greatly used for automatic segmentation in MRI [2, 3]. Here, we use them to classify knee injuries by training them on knee MRI dataset [4].

Knee is the largest joint that is present in the human body. It holds up the maximum weight of our body and hence can be injured easily [5]. Our knee has four main parts: cartilage, bones, tendons and ligaments. There are different ligaments, and any of these could be torn apart due to sudden twisted movement usually while playing a sport or while performing any other physical activity. The ligament tears are very common and require medical imaging to diagnose. There are some very common injuries: posterior cruciate ligament (PCL) tear, ACL tear, meniscus tear [6] and general abnormalities. The ACL is the longest ligament connecting the thigh bone to the shin bone. There could be a strain or tear in the ligament due to twisting motion of the knee. These injuries are very common and require proper diagnosis before their surgery. MRI is the most common and safest method for the diagnosis, but identifying the tear from MRI by a medical professional has many chances of error. They are wrongfully diagnosed and lead to unnecessary surgeries. Automating this procedure of identifying the tear in MRI exams will help the medical professionals [7]. It will reduce their time in identifying and will reduce the human error in prediction, thereby reducing the number of unnecessary surgeries.

We applied Logistic Regression, Support Vector Machine (SVM), Convolutional Neural Network (CNN), Recurrent Neural Network (RNN) and Deep Neural Network (DNN) on the data containing MRI scans [4] of the patients diagnosed concerning the condition of ACL. Tables 1 and 2 show the results of above models.

Table 1 Results obtained with CNN, DNN and RNN

S. No.	Exam name	Neural network model used	Dataset (masked/unmasked)	Accuracy (%)	Model loss percentage (%)
1.	ACL tear	CNN	Masked	82.00	4.20
			Unmasked	74.30	5.55
2.	ACL tear	DNN	Masked	82.00	4.30
			Unmasked	82.00	4.52
3.	ACL tear	RNN	Masked	81.80	4.55
			Unmasked	81.80	4.59

Table 2 Results obtained with Logistic Regression and SVM

S. No.	Exam name	Machine learning model used	Dataset (masked/unmasked)	Accuracy (%)
1.	ACL tear	SVM	Masked	88.23
2.	ACL tear	Logistic regression	Masked	67.87

The MRI dataset consists of sagittal plane 736 exams each constituting 32 × 320 × 320 images, where 32 represents the count of images in the exam and 320 × 320 is the size ratio of images. The scans represent healthy, partially injured or completely ruptured ACL tears. So, to sum up our model mainly proposes:

- The prediction of ACL tear from sagittal plane MRI scans which provide scans of totally ruptured, partially injured and healthy knees.
- The study of different DL techniques like CNN, DNN and RNN on masked as well as unmasked dataset of knee MR images.
- The study of various ML techniques like SVM and Logistic Regression on a dataset of knee MR images.

Subsequently, the advantages and weaknesses of each algorithm are discussed and objective assessments of these algorithms are provided.

Our paper is organized in six sections. Section 2 comprises the analysis of different papers related to the research. Section 3 contains detailed discussions about the methodologies like MRI, deep learning models and machine learning models. Section 4 comprises the solution we are proposing and Sect. 5 consists of its results and analysis. At the end, Sect. 6 contains the scope of our paper in future and how we can achieve more by implementing advanced technologies at later stages.

2 Related Work

2.1 CNN Analysis

3D CNN is an advanced algorithm that is applied in the MR images of the knee. It evidently helped in segmentation of knee cartilage to accurately analyse osteoarthritis [8]. Osteoarthritis can be predicted by analysing the knee by segmenting it in parts and further analysing the part that contributes most in the prediction using the 3D CNN model.

A segmentation model developed using CNN and augmentation on brain MRI dataset [3] gives results showing it can be used for preprocessing steps like bias field corrections over existing methods.

MRNet model can be used to map 3D MRI images to a series of probability using CNN as it reduces the time taken to identify tears and improve accuracy, thereby reducing unwanted surgeries of knee [7]. This technique can help to reduce wrong diagnosis of ACL tears. Deep learning has also been used for pneumonia prediction using X-ray images [9, 10] and neuro disorders [11].

Segmentation of knee cartilage [12] can be studied using triplanar CNN by comparing voxel classification technique with standard voxel classification technique. The classification of voxel with integration of three 2D CNN will be obtained as a result using triplanar CNN.

2.2 *Classification Algorithms Analysis*

The papers [2, 13, 14] use algorithms like SVM and Naïve Bayes techniques for classification. These techniques have been applied to the MRI of articular cartilage, brain and liver. Naïve Bayes is an algorithm based on the Bayer's theorem in which the features have Naïve (strong) independence assumptions. It predicts the probability of how much a member can belong to each class. The class which gets the highest probability is then considered to be the most likely one. SVM is a supervised machine learning algorithm. It is mostly used for classification but can also be used for regression. The data items are plotted in an n-D space. The coordinates are the features of the data items. The articular tissue can be interpreted and visualized using MR imaging. Cartilage segmentation has been implemented using Naïve Bayes algorithm. The methodology used shows good consistency with the manual results. SVM has been applied to brain MRI in order to diagnose PD and PSP. The accuracy obtained using this algorithm was above 90%. SVM and Naïve Bayes are used for the classification of various liver diseases. In comparing the two techniques, SVM has given better accuracy but the execution time of Naïve Bayes is lesser.

2.3 *Visualization Analysis*

The papers [15, 16] are using computer vision and image processing techniques for visualization of knee MRI. Visualization helps to understand the complex relationships within the dataset. Image processing can be used for pattern recognition, localization, measurement, etc. It can be used to extract some useful information from an image or to get an enhanced image by performing some operations on the image. Computer vision enables the computer to understand digital images or videos. This involves extracting information from the image like a text, an object, some 3D model, etc. Computer vision is able to automate the task of human vision. Image processing is used to calculate the cartilage thickness to compare the normal and OA subjects. It is also used for the visualization of the meniscal tear. Here, the mask generation is based on ROI polygons. Computer vision has been used for the knee articular cartilage segmentation. It has been noticed that the proposed approach has the potential to replace or even complement the preprocessing steps which results in the improvement of the segmentation performance.

Fig. 1 Axial view

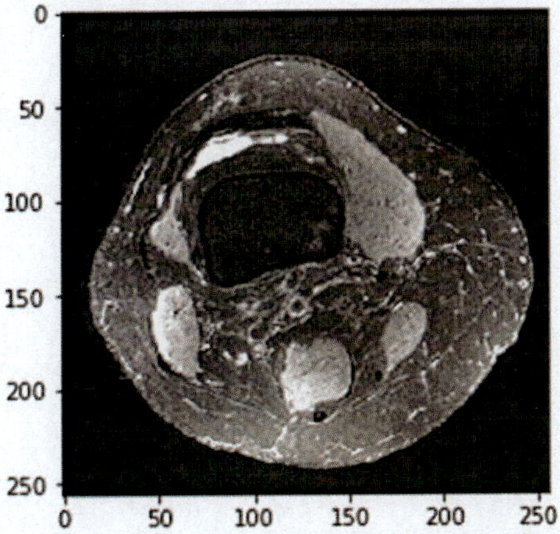

3 Methodology

3.1 Magnetic Resonance Imaging (MRI)

Magnetic resonance imaging is used to produce scans of muscles, tissues and bones which can be used to diagnose several medical conditions. Radio waves and magnetic fields are used to produce these scans. It is a non-invasive test to function. Strong magnetic fields are used instead of ionizing radiations. MRI scans can help diagnose many conditions. All major internal injuries are diagnosed using MRI scans like knee injuries, brain tumours, etc. It is not a single image of an organ. It is actually a set of different images or slices stacked together in a volume which is used to view the organ through different depths easily. There are 3 major kinds of knee MRI exams based on the plane of images that are axial exam, coronal exam and sagittal exam. Figures 1, 2 and 3 show different view planes. The MRI dataset used consists of a sagittal plane.

3.2 Deep Learning Techniques

3.2.1 Convolutional Neural Network (CNN)

Convolutional Neural Network is a popular technique used in artificial intelligence and deep learning. It is almost the first solution that comes to mind for the machine vision projects as it works great in image detection tasks. The most basic component of a CNN is its layers. A layer is like a filter whose purpose is to spot different patterns at different spots like contrast or colour. A number of layers are out on top of each other to produce a 3D network.

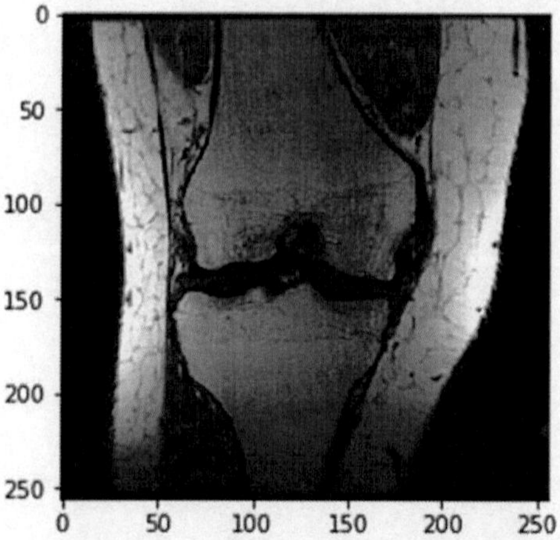

Fig. 2 Coronal view

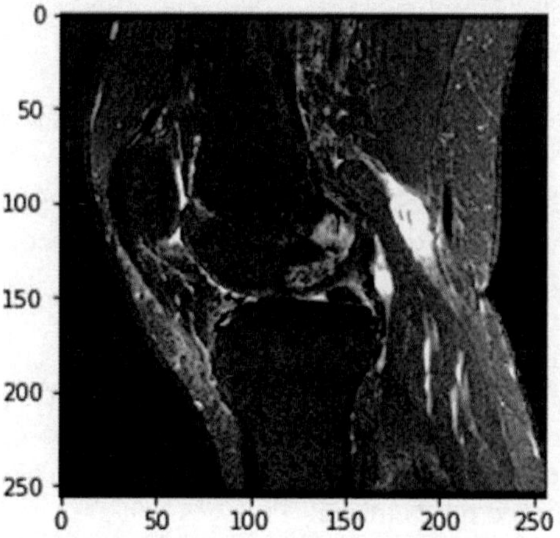

Fig. 3 Sagittal view

Each layer does the same task as the other, i.e. filtering through the image for a specific pattern. Each layer is made up of neurons which can be assigned different weights and biases. In CNN, the neurons are not connected to all the neurons of the adjacent layer, rather to a fixed amount of neurons only. Bias and weight values are the same for the neurons present in a layer. Therefore, the same pattern is filtered across the whole layer in different sections. This is the first set of layers called convolutional layers. This is followed by two sets of other layers called activation layer and pooling layer. These assist in building a simple pattern on the basis of what the convolutional layer discovers. Activation layer is used for proper training and the latter for dimensionality reduction. These 3 layers are repeated several times. These help in discovering complex patterns but with no ability to understand these patterns. Therefore, to understand the pattern, a fully connected layer is used at the end of this net. This gives the functionality to classify the data samples.

3.2.2 Deep Neural Network (DNN)

DNN is a subcategory of Artificial Neural Network. A DNN consists of various layers including the input and output layers. The probability of each layer is calculated as the network is parsed through the layers. In DNN, every neuron in a particular layer gets an output from every neuron of the previous layer.

3.2.3 Recurrent Neural Network (RNN)

RNN is another type of neural network. Here, the output of a layer is passed on to the next layer which acts as an input to that layer. RNNs have memory which helps them to process a sequence of inputs. It can remember patterns from the previous inputs which help to predict the outcome for the next input. The main feature is the hidden state in RNN. Its function is to remember some information in a sequence. RNNs can be used for speech recognition, next-text prediction, etc.

3.3 Machine Learning Techniques

3.3.1 Logistic Regression

Logistic regression is a technique that predicts whether something is true or false using a logistic function. In regression analysis, it is used for predicting the parameters of a logistic model. It is used for modelling the probability of the default class. It works by modelling the probability of output in terms of input without performing statistical classification. A classifier can also be made using Logistic Regression. We can choose a cut-off value and classify objects in 2 classes, one which has probability greater than cut-off and another which has lower value.

3.3.2 Support Vector Machine (SVM)

SVM, a supervised machine learning method, is trained on a labelled dataset. At a basic level, SVM finds a hyperplane between two classes of data. It does complex data transformation and separates the classes using an optimal boundary. Non-linear SVM uses kernel trick. There are functions which take low-dimension input space and map it to a high-dimensional space where the data points can be separated. We take this solution back to low-dimension input space and we have a non-linear separation. SVM is highly useful as we can get complex relationships by applying the kernel trick and not perform difficult steps on our own. Although a key thing to remember is that SVM is computationally intensive and therefore training time can be large.

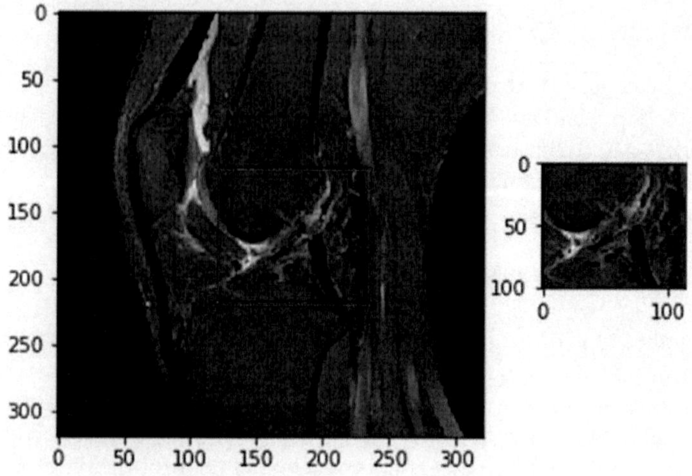

Fig. 4 ROI view

3.4 Region of Interest (ROI)

These are a set of coordinates for a given exam and for a given image that contains the particular tear for the case of knee MRI. The image for which we know its ROI is generally specified as a masked image. The ROI points are 5 that are X-axis, Y-axis, Z-axis, depth and height. Figure 4 shows the ROI area for a knee MRI exam.

4 Proposed Solution

Various deep learning and machine learning techniques were applied on masked as well as unmasked dataset of knee MR images. The masked images are the one which has a region of interest (ROI coordinates) which determines the area where the tear is present in the overall image while the unmasked dataset is without the ROI coordinates. Deep learning models like DNN, CNN and RNN have been used for the detection of ACL tear. Figure 5 represents the data flow diagram for the deep learning models. In DNN and RNN, 6 dense layers have been used. The activation function is ReLU. After the third layer, Flatten () is used to obtain a 1D array from the data. Dropout of 30% has been applied to the fourth and fifth layers. In RNN, the kernel initializer used is Glorot Uniform. In CNN, though the architecture is similar to DNN, max pooling has been applied in the first three layers which reduces the dimensions of the data of the window size (5.5) and down-samples it to (2.2). The activation function used in the last layer is softmax in the case of all the three models. The loss function used is Binary-Cross Entropy, and the optimizer used is Adam.

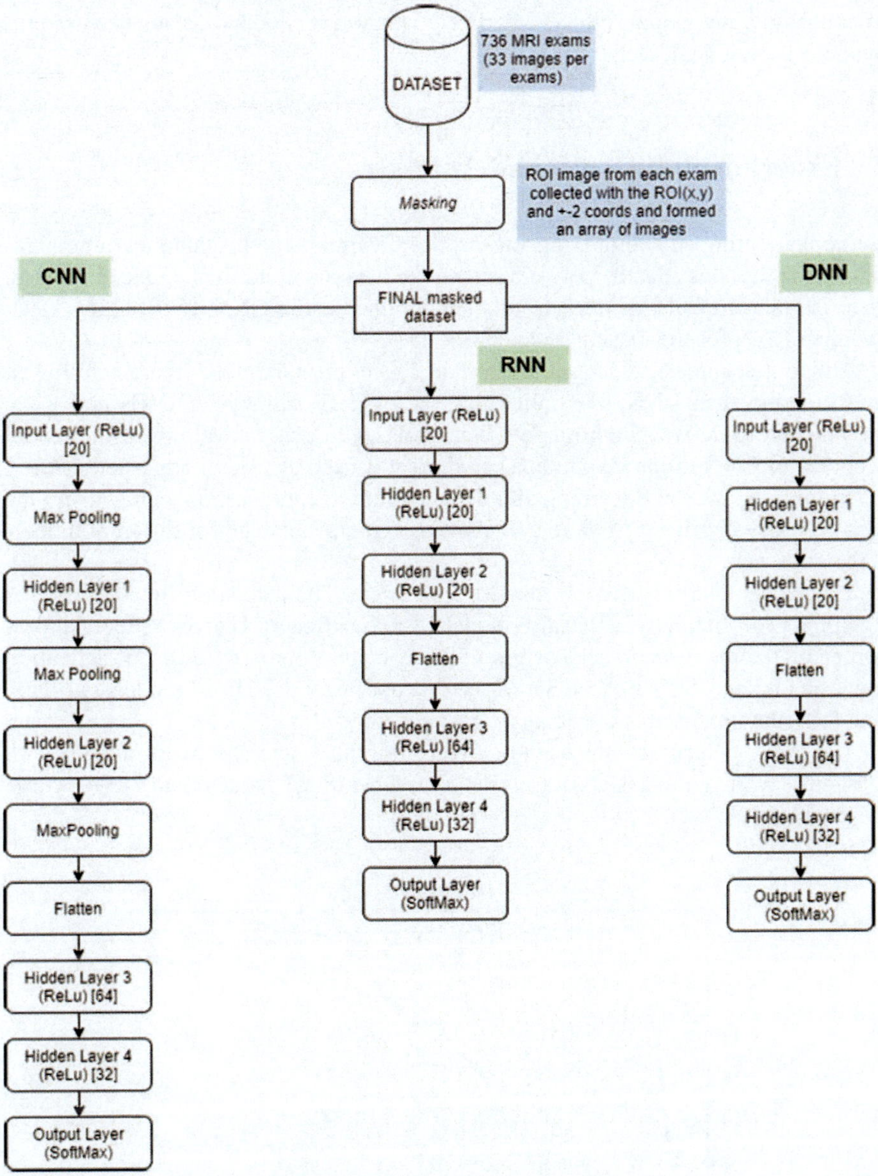

Fig. 5 Data flow diagram for deep learning models

Machine learning models like SVM and Logistic Regression have also been used for the detection of ACL tear.

5 Experimental Results and Analysis

By implementing and comparing various deep learning and machine learning algorithms for the identification of ACL tears, we have come to the conclusion that in deep learning models, the best results are obtained using CNN, followed by DNN and then RNN for the masked images.

And for the unmasked dataset, DNN model gives the maximum accuracy followed by RNN and then CNN. We notice that the model accuracy for DNN and RNN remains the same with both masked and unmasked datasets. Moreover, the model accuracy of CNN drops if an unmasked dataset is used instead of a masked dataset. On comparing machine learning and deep learning techniques, SVM has given the best accuracy. Figures 6, 7, 9 and 10 show the graphs obtained for different models (Fig. 8).

The 3 deep learning models consist of 6 layers. The activation function used is Rectified Linear Unit (reLU) in all 3 models. The function reLU gives better results as compared to other functions like sigmoid, tanh, etc. Softmax is used as the activation function in the last layer of the 3 models. The optimizer used in all models is Adam. The loss function used is the Binary-Cross Entropy.

The training accuracy has been plotted for all the 3 neural network models with masked as well as unmasked dataset. Receiver Operating Characteristic (ROC) curve has been plotted for SVM.

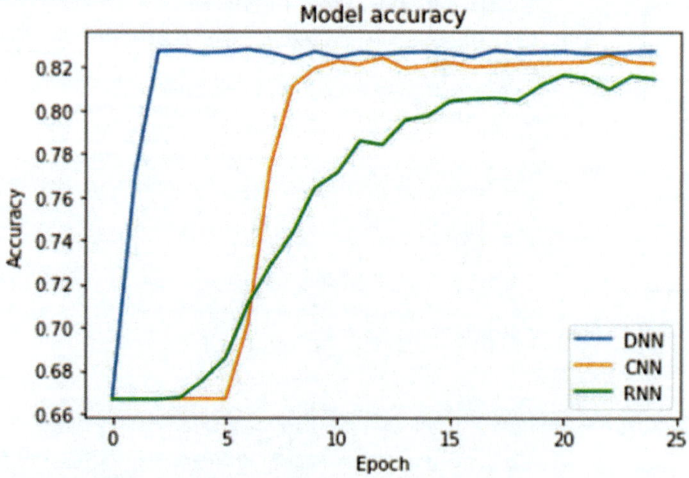

Fig. 6 Training accuracy with masked dataset

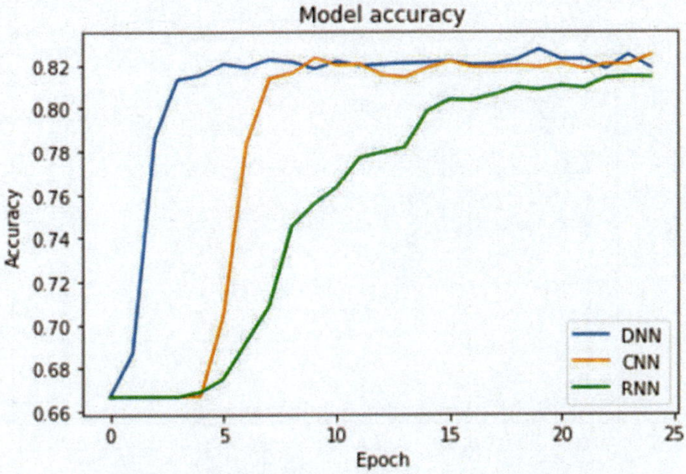

Fig. 7 Training accuracy with unmasked dataset

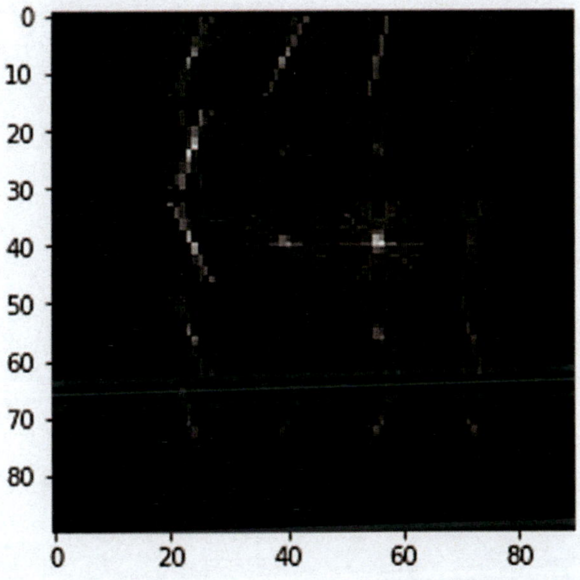

Fig. 8 HOG plot

Using machine learning techniques, the best results are obtained with SVM while the accuracy with Logistic Regression comes out to be lowest among all models, for masked dataset. SVM provides the best test case accuracy among both machine learning and deep learning models for ACL tear detection with an accuracy of 88.23% for the masked dataset. It was interesting to note that the traditional computer vision

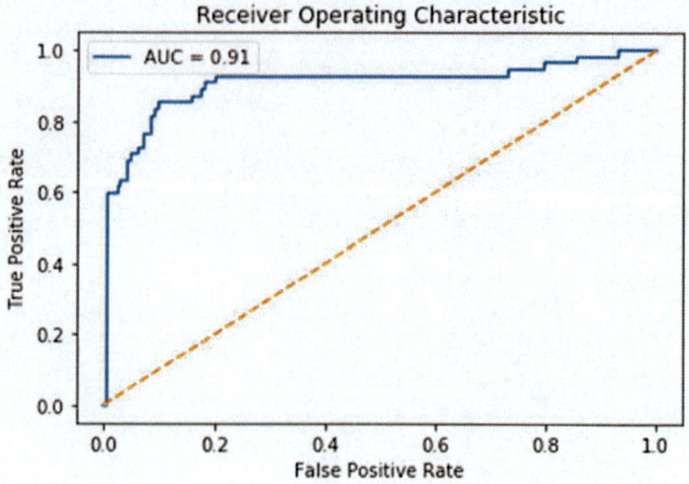

Fig. 9 ROC curve for SVM

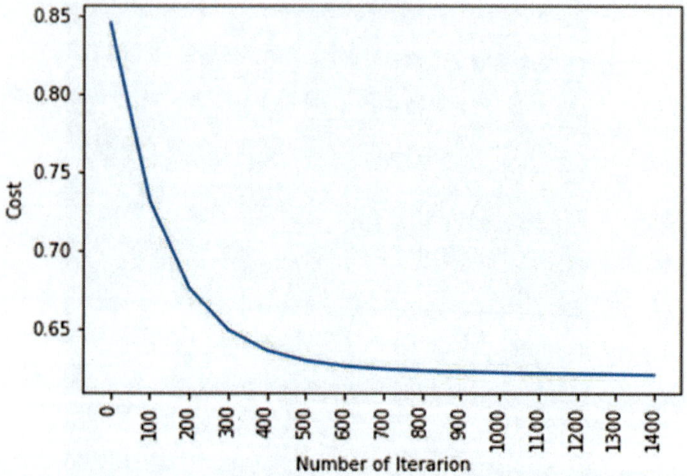

Fig. 10 Plot of cost vs number of iteration

model provided better results compared to the neural networks used for classification on knee MR images.

Histogram of oriented gradients (HOG) which is a feature detector was used for the purpose of object detection. The SVM model was trained on the HOG features of the knee image corresponding to ROI-Z from the exam of 33 images. The following is a HOG plot for a knee ROI-Z image.

We also altered our dataset; we intentionally added more labels for ACL tears to some images which were not marked as one and then added these images too in training the model. With this, we achieved an accuracy of 91.28% and AUC of 0.95 with SVM. This proves that even by a slight increase in the dataset, there is a good amount of increase in accuracy and AUC.

6 Future Scope

The analysis of the various models is implemented on a moderate dataset with predictions of only ACL tear and thereby could not analyse the exams for other abnormalities and tears. With a more clinically suitable dataset, these models could show acceptable results for practical use. With a detailed dataset, the models could analyse the exams for any abnormalities that are not known. To overcome the lack of dataset problem, generative adversarial network (GAN) can be used to provide better results for the following dataset.

References

1. Liu J, Pan Y, Li M, Chen Z, Tang L, Lu C, Wang J (2018) Applications of deep learning to MRI images: a survey. Big Data Min Anal 1(1):1–18. https://doi.org/10.26599/bdma.2018.9020001
2. Pang J, Li P, Qiu M, Chen W, Qiao L (2015) Automatic articular cartilage segmentation based on pattern recognition from knee MRI images. J Dig Imag 28(6):695–703. https://doi.org/10.1007/s10278-015-9780-x
3. Khalili N, Lessmann N, Turk E, Claessens N, de Heus R, Kolk T, Išgum I et al (2019) Automatic brain tissue segmentation in fetal MRI using convolutional neural networks. Magn Resonan Imag. doi: 10.1016/j.mri.2019.05.020
4. Štajduhar I, Mamula M, Miletić D, Unal G Semi-automated detection of anterior cruciate ligament injury from MRI. Comput Methods Programs Biomed. https://doi.org/10.1016/j.cmpb.2016.12.006
5. Naraghi AM, White LM (2016) Imaging of athletic injuries of knee ligaments and menisci: sports imaging series. Radiology 281(1):23–40. https://doi.org/10.1148/radiol.2016152320
6. Ramakrishna B, Liu Weimin, Saiprasad G, Safdar N, Chang Chein-I, Siddiqui K, Lee San-Kan et al (2009) An automatic computer-aided detection system for meniscal tears on magnetic resonance images. IEEE Trans Med Imag 28(8):1308–1316. https://doi.org/10.1109/tmi.2009.2014864
7. Bien N, Rajpurkar P, Ball RL, Irvin J, Park A, Jones E, Lungren MP et al (2018) Deep-learning-assisted diagnosis for knee magnetic resonance imaging: development and retrospective validation of MRNet. PLOS Med 15(11):e1002699. https://doi.org/10.1371/journal.pmed.1002699
8. Raj A, Vishwanathan S, Ajani B, Krishnan K, Agarwal H (2018) Automatic knee cartilage segmentation using fully volumetric convolutional neural networks for evaluation of osteoarthritis. In: 2018 IEEE 15th international symposium on biomedical imaging (ISBI 2018). https://doi.org/10.1109/isbi.2018.8363705
9. Jaiswal AK, Tiwari P, Kumar S, Gupta D, Khanna A, Rodrigues JJPC (2019) Identifying pneumonia in chest X-rays: a deep learning approach. Measurement (Elsevier) 145:511–518
10. Chouhan V, Singh SK, Khamparia A, Gupta D, Tiwari P, Moreira C, Damaševičius R, Victor H.C. de Albuquerque (2020) A novel transfer learning based approach for pneumonia detection in chest X-ray images. Appl Sci (MDPI) 10(2):559

11. Khamparia A, Singh A, Anand D, Gupta D, Khanna A, Arun Kumar N, Tan J (2018) A novel deep learning based multi-model ensemble methods for prediction of neuromuscular disorders. Neural Comput Appl (Springer). https://doi.org/10.1007/s00521-018-3896-0
12. Prasoon A, Petersen K, Igel C, Lauze F, Dam E, Nielsen M (2013) Deep feature learning for knee cartilage segmentation using a triplanar convolutional neural network. In: Lecture notes in computer science, pp 246–253. https://doi.org/10.1007/978-3-642-40763-5_31
13. Salvatore C, Cerasa A, Castiglioni I, Gallivanone F, Augimeri A, Lopez M, Quattrone A et al (2014) Machine learning on brain MRI data for differential diagnosis of Parkinson's disease and progressive supranuclear palsy. J Neurosci Methods 222:230–237. https://doi.org/10.1016/j.jneumeth.2013.11.016
14. Vijayarani S, Dayanand S (2015) Liver disease prediction using SVM and naïve bayes algorithms 4(4)
15. Patel J, Modi H, Patel H (n.d) Measurement of cartilage thickness in osteoarthritis and visualization of meniscus tear of knee MRI image processing. IJCSMC 5(1):39–52)
16. Kumar D, Gandhamal A, Talbar S, Hani AFM (2018) Knee articular cartilage segmentation from MR images. ACM Comput Surv 51(5):1–29. https://doi.org/10.1145/3230631

A Study on Image Analysis and Recognition Using Learning Methods: CNN as the Best Image Learner

Vidushi and Manisha Agarwal

Abstract To ensure the correct image, recognition is a notable and limelight task which called for efficient measure. Learning algorithms in the area of pattern recognition have attracted attention of researchers owing to their extraordinary, exciting, and astounding performance. To alleviate the analysis, recognition, and detection problems such as speedy operations with powerful computing precision from huge dataset, learning algorithms are showing the paramount and remarkable success. However, to work with images using machine learning still remains an open issue. Therefore, this paper intends to perform learning approaches study in detail. The paper emphasized on image recognition using learning models. The study shows the learning techniques subtle comparison and finds out the best way to detect the images. Furthermore, the document conducts experiment and the results indicate that convolutional neural network shows the significant performance as the best image analyzer or learner over well-known state-of-the-art learning approaches. However, the machine learning approaches failed to achieve the required and comparatively better result.

Keywords Image recognition · Handwriting recognition · MNIST dataset · CIFAR 10 dataset · Machine learning · Deep learning · Convolutional neural network

1 Introduction

Today, many fields like diabetic retina detection, cyber security need expensive and advance technologies but machine learning adoption accomplish the tasks more accurately and less expensively [1]. Due to high popularity of Internet, the traffic also increases rapidly day by day, and this makes cyber security an enticing goal

Vidushi (✉) · M. Agarwal
Banasthali Vidyapith, Rajasthan 304022, India
e-mail: vidushi.mtech@gmail.com

M. Agarwal
e-mail: manishaagarwal18@yahoo.co.in

A. Khanna et al. (eds.), *Data Analytics and Management*, Lecture Notes on Data Engineering and Communications Technologies 54,
https://doi.org/10.1007/978-981-15-8335-3_3

[2]. Human authentication is an upcoming and essential area, mainly with security purpose, that needs crucial research analysis [3]. The connection of heterogeneous devices using IoT technology is the current research demand like in real projects smart city [4]. Machine learning is the extraordinary application of artificial intelligence that provides a way of self-learning to machines on the basis of history and also be able to improve as experience increases. Machine learning techniques are applied almost everywhere like speech recognition and in business. Image matching [5] is a difficult task because of advance technologies like remote sensing that makes acquired image larger and more complex, deep learning can be used in analyzing such critical issues [5]. Handwritten digit recognition is the very limelight problem and needs attention because of different handwriting patterns; deep learning models can be used in resolving such issues [6]. To monitor and analyze the water quality for the treatment of waste water in urban areas is present big challenges that can be solving using IoT sensors and artificial neural network [7]. The subset of this learning is deep learning where the concept of artificial neural network comes. The idea of this neural network comes from and is vaguely motivated by the human brain [8]. Health-related complex diseases like neuromuscular disorder, whose prediction could be made using deep learning [9]. The nation's economy growth directly depends on agriculture, especially in country like India; it makes the prediction of diseases like seasonal crop disorder that can be performed by convolutional neural model [10]. The repetition of same work enhances the model accuracy until reached the saturation point [11]. Learning rate is the essential parameter for any model which directly affects the model's performance [12]. To predict the diseases different learning models can be used [13, 32].

The document elaborates the survey of literature work that is already done in Sect. 2. As in the continuous fashion, a research is going on in this field, and the already work performed has vital significance for the upcoming research. Next, learning approaches are described in Sect. 3 of this study. This section throws the light on heterogeneous and various models of machine learning along with some prominent approaches of deep learning. Succeeding Sect. 4 shows pictorially and discussed the experimental outcomes. In addition, paper conclusion and future work presented in Sect. 5.

2 Background

Javier et al. [14] present a novel method that is software-based used to detect the fake one and for ensuring the presence of actual legitimate trait. The proposed system objective is toward enhancing the security for frameworks related to biometric recognition. The authors [15] present a survey related to deep hashing comprehensively to accomplish the task of retrieving the image having multiple labels. In this, the study categorizes the methods as per the ways of input image treatment and the relationship between them. The treatment of image can be pointwise, tripletwise and listwise, and pairwise. Additionally, the study discusses the space cost, described models efficiency as well as search quality and also opens the issue with the opportunities of future work.

The research authors [16] developed GTDA, which stands for general tensor discriminant analysis, as a linear discriminant analysis (LDA) processing step. Authors develop GTDA by getting the motivation due to the success of 2LDA, called two-dimensional LDA. The research conducts the experiment and gets good performance as compared with 9 state-of-the-art methods for classification by using proposed method to recognize gait that has ground on sequences of image from HumanID database from USF, i.e., the University of South Florida.

The applications of various image deformation models have nonlinear nature which are presented to perform image recognition task by the authors of this research [17]. The experiments performed by authors showed that the performance of model is good with different four tasks of handwritten digit recognition, as well as for the medical image classification. In particular, 0.54 percent error rate is achieved with MNIST dataset, along with, it also gets error rate of 12.6 percent in 2005 with ImageCLEF evaluation categorization of medical image.

Ling et al. [18] perform the survey of different transfer learning state-of-the-art algorithms in applications of visual categorization such as recognition of image, object, and human action. In this paper, [19] authors purpose is providing the overview of the processing with analysis of MRI images comprehensively that is based on deep learning. The paper briefly introduces the deep learning as well as given the MRI images for imaging modalities. The research also presents the common architecture of deep learning. The document also addressed the future aspects of MRI images related to deep learning.

Weilong et al. [20] investigate the way of how to perform the image visual quality blind evaluation by using the learning rules from the descriptions regarding linguistic. A model called blind IQA is proposed by the authors of this study, and the model directly learns the qualitative evaluations with outputs scores in numerical for fair comparison and general utilization. The author's [21] of this context provides the survey of series for four models of deep learning including CNN, ELM-RVFL, GAN, and other series to understand comprehensively the analytical techniques in the field of image processing. Besides, the researcher's described the open issues with the future research promising direction for image processing using deep learning new generation.

Fuyong et al. [22] introduce the famous neural network called deep neural network plus summarizes the current achievements of deep learning in variety of microscopy image examination and analysis tasks such as detection, classification, as well as segmentation. In particular, this paper explains the principles along with architecture of different neural networks and also interprets their respective formulations on microscopic images. In addition, open challenges along the future research potential trends by deep learning for analysis of the microscopic images are discussed. The paper authors [23] proposed MSRL-DehazeNet, a novel architecture that is based on deep learning for the removal of image haze particularly single relying on image decomposition and MSRL, where MSRL is for multi-scale residual learning. The authors of this paper [24] present the survey about the threats along with attacks in information security with various categories including network, application, as well as host. Further, it also depicted the relationship between attacks in system and information. This paper [25] includes the detailed survey on the blockchain, working related to blockchain, analyze the security and privacy threat on blockchain, and various applications on blockchain.

3 Learning Algorithms

The breakthrough achievement gained through machine learning and its subset called deep learning attracts the scientists earthly to apply them in different areas [26]. There are a number of different learning algorithms that show the good result. Some of the well-known algorithms are as follows:

3.1 Random Forest

This algorithm is supervised learning which can be applied for both crucial tasks called regression and classification like security analysis [27]. It is ensemble method, which means it is a collection of more than one decision trees. From each of the trees, prediction is taken and finally by voting sort out the best one [27].

3.2 K-Nearest Neighbor

This is also supervised learning and can be applied for regression and classification. It is based on the distance between the items means; it assumes similar things remain in the close proximity [28]. K in the algorithm refers to the nearest neighbors for voting to classify the new item in one of the present class [28]. To find out the closeness, different distance measurement functions can be used.

3.3 Logistic Regression

It is the supervised classification algorithm and basically used for assigning the observations to classes of discrete sets [29]. It uses the function called sigmoid function with the purpose to transform the output by returning the probability value. It relies on probability concept.

3.4 Support Vector Machine

This algorithm is one of the supervised learning, and its objective is finding the hyperplane that is able to classify distinctly the data points in the space having N-dimensions [30].

3.5 Artificial Neural Network and Convolutional Neural Network

The human intelligence has been adapted to the computerized platform due to the new advancement in technology for analysis [31]. It is very useful even in handling nonlinear dataset. The complete network is divided into number of layers, mainly into three layers, called one input, one or more hidden/s, and output layer. These layers are consisting of neurons, and these artificial neurons are connected with each other to transmit and receive the information. Convolutional Network is deep learning neural network and commonly used in analyzing the imagery visual. Its basic working is same as artificial neural network, but in addition it has more layers. The network basic layers are input layer, convolutional layers, pooling layers, fully connected, and output.

4 Experiment and Result Discussion

The study uses MNIST and CIFAR 10 image datasets to prove the results. Table 1

Table 1 Learning algorithms accuracy using MNIST and CIFAR-10 dataset

	MNIST dataset		CIFAR-10 dataset
Learning model	Training accuracy	Testing accuracy	Testing accuracy
Random forest	0.9596	0.96028	0.4135
K-nearest neighbor	0.9638	0.96557	0.1446
Logistic regression	0.8789	0.887	0.3982
Support vector machine	0.9808	0.981	0.48409
Artificial neural network	0.9164	0.95971	0.4021
Convolution network	0.9903	0.99028	82.24

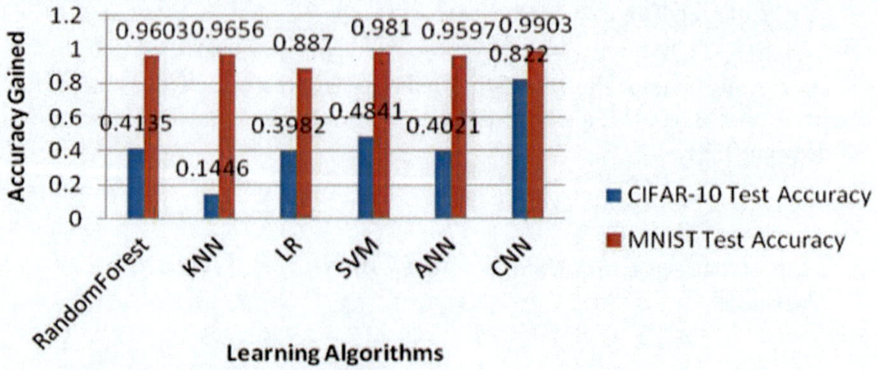

Fig. 1 Learning algorithm versus accuracy

shows different learning models results in terms of accuracy during training and testing phase using MNIST dataset.

In Table 1, it can be clearly observed that convolutional neural network gain the highest accuracy as compared, and after it, in machine learning models support vector machine shows the better result. Table 1 also shows the result during testing phase in terms of accuracy using CIFAR 10 dataset. Figure 1 pictorially compares the accuracy using MNIST dataset, and CIFAR 10 dataset.

5 Conclusion and Future Work

To recognize the images using learning approaches, this research compares the algorithms using two well-known datasets, and on the basis of the result in terms of accuracy concluded that convolutional neural network is the best classifier than others. In both cases either MNIST dataset or CIFAR 10 dataset, this deep learning neural network gains 0.9903 and 0.822 accuracy, which is much better than the other ones. It can be concluded that this neural network as the best learning approach toward image analysis and recognition and hence can be used in further image recognition tasks. In the upcoming image-related research, this result would be useful for researchers and further work can be done using this neural network.

References

1. Gaur J, Goel AK, Rose A, Bhushan B (2019) Emerging trends in machine learning. In: 2019 2nd international conference on intelligent computing, instrumentation and control Technologies (ICICICT). https://doi.org/10.1109/icicict46008.2019.8993192

2. Soni S, Bhushan B (2019) Use of machine learning algorithms for designing efficient cyber security solutions. In: 2019 2nd international conference on intelligent computing, instrumentation and control technologies (ICICICT). https://doi.org/10.1109/icicict46008.2019.8993253
3. Ferro M, Pioggia G, Tognetti A, Carbonaro N, Rossi DD (2009) A sensing seat for human authentication. IEEE Trans Inform Forensics Sec 4(3):451–459. https://doi.org/10.1109/tifs.2009.2019156
4. Jindal M, Gupta J, Bhushan B (2019) Machine learning methods for IoT and their future applications. In: 2019 international conference on computing, communication, and intelligent systems (ICCCIS). https://doi.org/10.1109/icccis48478.2019.8974551
5. Zhu H, Jiao L, Ma W, Liu F, Zhao W (2019) A novel neural network for remote sensing image matching. IEEE Trans Neural Netw Learn Syst 30(9):2853–2865
6. Agarwal M (2019) Intelligent handwritten digit recognition based on multiple parameters using CNN. Int J Comput Sci Eng 7(5):636–641. doi: 10.26438/ijcse/v7i5.636641
7. Aggarwal S, Gulati R, Bhushan B (2019) Monitoring of input and output water quality in treatment of urban waste water using IoT and artificial neural network. In: 2019 2nd international conference on intelligent computing, instrumentation and control technologies (ICICICT). https://doi.org/10.1109/icicict46008.2019.8993244
8. Mishra MV, Agarwal S\M, Puri N (2018) Comprehensive and comparative analysis of neural network. Int J Comput Appl 2(8). https://doi.org/10.26808/rs.ca.i8v2.15
9. Khamparia A, Singh A, Anand D, Gupta D, Khanna A, Kumar NA, Tan J (2018) A novel deep learning-based multi-model ensemble method for the prediction of neuromuscular disorders. Neural Comput Appl. doi: 10.1007/s00521-018-3896-0
10. Khamparia A, Saini G, Gupta D, Khanna A, Tiwari S, Albuquerque VHCD (2019) Seasonal crops disease prediction and classification using deep convolutional encoder network. Circ Syst Sig Process 39(2):818–836. https://doi.org/10.1007/s00034-019-01041-0
11. Mishra V, Agarwal M (2019) Effect of redundant work on machine. SSRN Electron J. https://doi.org/10.2139/ssrn.3444784
12. Agarwal M (2020) Performance analysis on the basis of learning rate. Innov Comput Sci Eng Lecture Notes Netw Syst 41–46. doi: 10.1007/978-981-15-2043-3_6
13. Mishra V, Agarwal M, Rajoria R (2019) Analysis of various classification methods. In: 2019 international conference on issues and challenges in intelligent computing techniques (ICICT). https://doi.org/10.1109/icict46931.2019.8977688
14. Galbally J, Marcel S, Fierrez J (2014) Image quality assessment for fake bio-metric detection: application to iris, fingerprint, and face recognition. IEEE Trans Image Process 23(2):710–724. https://doi.org/10.1109/tip.2013.2292332
15. Rodrigues J, Cristo M, Colonna JG (2020) Deep hashing for multi-label image retrieval: a survey. Artif Intell Rev. doi: 10.1007/s10462-020-09820-x
16. Tao D, Li X, Wu X, Maybank SJ (2007) General tensor discriminant analysis and gabor features for gait recognition. IEEE Trans Pattern Anal Mach Intell 29(10):1700–1715. https://doi.org/10.1109/tpami.2007.1096
17. Keysers D, Deselaers T, Gollan C, Ney H (2007) Deformation models for image recognition. IEEE Trans Pattern Anal Mach Intell 29(8):1422–1435. https://doi.org/10.1109/tpami.2007.1153
18. Shao L, Zhu F, Li X (2015) Transfer learning for visual categorization: a survey. IEEE Trans Neural Netw Learn Syst 26(5):1019–1034. https://doi.org/10.1109/tnnls.2014.2330900
19. Liu J, Pan Y, Li M, Chen Z, Tang L, Lu C, Wang J (2018) Applications of deep learning to MRI images: a survey. Big Data Min Anal 1(1):1–18
20. Hou W, Gao X, Tao D, Li X (2014) Blind image quality assessment via deep learning. IEEE Trans Neural Netw Learn Syst 26(6):1275–1286
21. Jiao L, Zhao J (2019) A survey on the new generation of deep learning in image processing. IEEE Access 7:172231–172263
22. Xing F, Xie Y, Su H, Liu F, Yang L (2017) Deep learning in microscopy im-age analysis: a survey. IEEE Trans Neural Netw Learn Syst 29(10):4550–4568

23. Yeh CH, Huang CH, Kang LW (2019) Multi-scale deep residual learning-based single image haze removal via image decomposition. IEEE Trans Image Process
24. Sinha P, Rai AK, Bhushan B (2019) Information security threats and attacks with conceivable counteraction. In: 2019 2nd international conference on intelligent computing, instrumentation and control technologies (ICICICT). https://doi.org/10.1109/icicict46008.2019.8993384
25. Soni S, Bhushan B (2019) A comprehensive survey on blockchain: working, security analysis, privacy threats and potential applications. In: 2019 2nd international conference on intelligent computing, instrumentation and control technologies (ICICICT). https://doi.org/10.1109/icicict46008.2019.8993210
26. Khamparia A, Singh KM (2019) A systematic review on deep learning architectures and applications. Expert Syst 36(3). https://doi.org/10.1111/exsy.12400
27. Pal M (2005) Random forest classifier for remote sensing classification. Int J Remote Sens 26(1):217–222
28. Yi X, Paulet R, Bertino E, Varadharajan V (2014) Practical k nearest neighbor queries with location privacy. In 2014 IEEE 30th international conference on data engineering. IEEE, pp. 640–651
29. Kleinbaum DG, Dietz K, Gail M, Klein M, Klein M (2002) Logistic regres-sion. Springer, New York
30. Zhang L, Zhou W, Jiao L (2004) Wavelet support vector machine. IEEE Trans Syst Man Cybern Part B (Cybern) 34(1):34–39
31. Sil R, Roy A, Bhushan B, Mazumdar A (2019) Artificial intelligence and machine learning based legal application: the state-of-the-art and future research trends. In: 2019 international conference on computing, communication, and intelligent systems (ICCCIS). https://doi.org/10.1109/icccis48478.2019.8974479
32. Jaiswal AK, Tiwari P, Kumar A, Gupta D, Khanna A, Rodrigues JJPC (2019) Identifying pneumonia in chest X-rays: a deep learning approach. Measurement (Elsevier) 145:511–518

ANN Model for Forest Cover Classification

Ayush Chauhan and Deepali Kamthania

Abstract In order to develop efficient ecosystem management, an ANN model has been developed to present a factual way to capture futuristic dynamics on forest cover. The forest cover-type Kaggle dataset provided by US Forest Service has been considered for the study. The different activation functions are used to perform varied calculation between the layers of MLP architecture to predict different forest types. The ANN model predicts projection by giving accuracy of 100% with ReLU (ReLU on internal node and Softmax at output node) in comparison with TanH activation function giving 61.21% of accuracy. The results illustrate the toughness and efficiencies of the ANN representation with the combination of ReLU and Softmax. This work offers a consistent means for projecting forest cover and farming yields under provided prospective circumstances, supporting administrative management in consistent land development, management, and protection.

Keywords Multiple layer perceptron (MLP) · Rectifier linear unit (ReLU) · Softmax · TanH · Forest cover · Artificial neural network (ANN) · Activation function (AF) · US Forest Service (USFS) · US Geological Survey (USGS) · Sustainable forest management (SFM)

1 Introduction

The forest cover change and exhaustion are of worldwide concern because of its role in global warming [1]. SFM requires information regarding forest land. Forest cover information is extensive with a various type of data and features influencing a particular activity. There are different approaches to deal with danger that forests are facing all over the world. ANN approach can be used for generalization and

A. Chauhan · D. Kamthania (✉)
School of Information Technology, Vivekanada Institute of Professional Studies, Delhi 110034, India
e-mail: deepali102@gmail.com

A. Chauhan
e-mail: ayush96cse@gmail.com

A. Khanna et al. (eds.), *Data Analytics and Management*, Lecture Notes on Data Engineering and Communications Technologies 54,
https://doi.org/10.1007/978-981-15-8335-3_4

predictions on unseen data. Blackardet developed ANN model to study land cover classification for remotely sensed Colorado state forest having 71.1% accuracy [2]. The logistic regression model (LRM) was applied for prediction of forest dynamics of Bhanupratappur in India using forest cover for 10 years with 73% accuracy [3]. Gupta et al. observed accuracy of different ML algorithms as randomized decision trees (85%), gradient boosting model (86%), support vector machines (77%), voting classifier (86%), and deep learning techniques (71%) [4]. The examination on the impact of forest cover on the extent of flood damage controlling various social, economic, and demographic aspects in India is still going on by the researchers [5]. In another study, four different techniques regression, random forest, decision tree, and GBM are applied. It has been observed that random forest gives better prediction with 74.8% accuracy [6]. A recent study on forest cover in Australia has been presented by applying long short-term memory (LSTM) deep learning neural networks. It has been observed that performance is improved by 44% RMSE and 12% (pretend R-squared) [7]. In this paper, an attempt has been made to predict forest cover type using ANN MLP architecture with different AFs between the hidden layers and the output layers.

2 Proposed Methodology

Figure 1 shows the methodology applied for forest cover-type prediction using ANN model. The details of explanation of each block are as follows.

2.1 Getting Data

The open-source Kaggle dataset forest cover type prediction from USGS and USFS of four rough country areas located in northern Colorado at Roosevelt National Forest has been considered for the study. The data is in unrefined form and contains binary columns for qualitative non-dependent variables such as rough country areas and soil type [8].

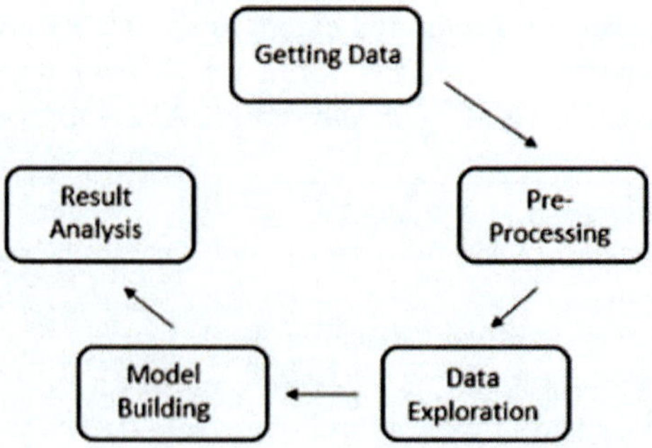

Fig. 1 Workflow design for data analysis and model building

2.2 Preprocessing

In this stage, raw, incomplete or inconsistent or noisy un-processed data is dealt in the following way:

2.2.1 Data Cleaning

In this step, data smoothing has been performed by filling the missing values and identifying removing of the outlier, and outliners were identified through inter-quartile range (IQR), scatter plot and box plot and eliminated. Some columns in dataset do not help in any prediction because they do not have any specific distinct values to make them considerable [9] and just creating some noises here determined by calculating standard deviation of each column through statistical description. In the study, no column had standard deviation equal to zero, so column was removed. The dataset has large number of outliners which are detected through scatter plot. The plots are overlapped by the class distribution. Hill shade patterns give nice ellipsoid patterns with each other, aspect and hill shade attributes form a sigmoid pattern, and horizontal and vertical distance to hydrology gives an almost linear pattern. The features like Wilderness_Area and Soil_Type are ignored as they are not continuous but binary. The Pearson correlation coefficient has been used to determine the statistical association, between two continuous variables with the threshold up to 0.5 to select only highly correlated attributes, as: Aspect and Hillshade_3pm = 0.65, Horizontal _Distance_To_Hydrology and Vertical_Distance_To_Hydrology = 0.61, Hillshade_Noon and Hillshade_3pm = 0.59.

2.2.2 Data Transformation

In this stage, normalization was carried out to obtain aggregation. 04 wilderness areas and 40 types of soils were normalized to form single dataset of forest cover type. The standard scalar was used for transforming the data to obtain its distribution for achieving a mean value zero and standard deviation of one. The dataset has high range of integers to bring feature values at same-scale normalization in the range of 0 to 1 which has been performed.

2.2.3 Data Reduction

The data reduction is necessary to obtain optimum model performance. The attribute soil types seven and fifteen are eliminated as these were containing zero values. In this paper, through PCA, the dimensionality of data is reduced by reducing dimensionality of space feature by feature extraction from 54 features to 2 features, while maintaining the most of data variance. Before PCA, the mean of the data has been zeroed out and each coordinate for one of the data has been set to obtain unit variance for evenly treatment of each attribute. Two of the features have the variance ratio of 56.39% and 42.91%, respectively, which combines to 99.30% which is very good variance.

3 Exploratory Study on Feature Selection

The Python functions can be used to explore the data. It has been observed that the dataset has 581012 instances having 54 attributes with no null values having total memory usage 243.8 Mb. The negative values are present in Vertical_Distance_To_Hydrology attribute, so chisq tests cannot be used. Wilderness_Area and Soil_type are one hot encoded so can be converted back for analysis. All attributes have some distinct values with no constant attributes, so there is no sense in removing any columns. As the scales were not same, rescaling and standardization have been applied. Skewness can be computed to describe the level of distribution which is different from a normal distribution [10]. Several attributes in Soil_Type show a large skew MLP may benefit if skewness between variables is good. Class distribution determines whether all classes have same number of instances or not. Inequitable allocation of classes, it is observed that the one class belonged by number of observations is notably lesser than those belonging to the other classes [11].

Figure 2 shows the target class distribution of seven cover types, namely: spruce/fir, lodgepole pine, ponderosa pine, cottonwood/willow, aspen, Douglas fir, and krummholz having totally 581,102 observations. Class 4 is having lowest appearances of a class in 'Cover_type' having value 2747 with less than 1 percentage of presence that is 0.47. Classes 3, 4, 5, 6, 7 are having very low presence in comparison with Classes 1 and 2 in 'Cover_type'.

Fig. 2 Class distribution

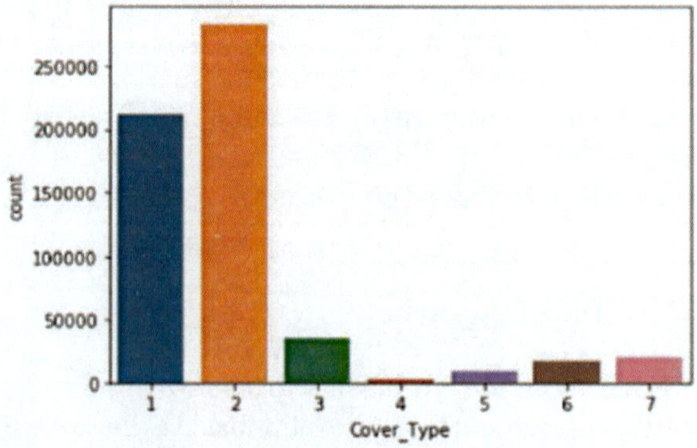

4 Data Visualization

The forest cover dataset has complex and large number of information, features that combined together. The data visualization is used to identify new patterns.

The heat map helps determine correlation between the variables in the database. Figure 3 helps in visualizing the heat map to depict their high dimensionality with every other feature in dataset. The box with black color suggests that the correlation between features is extremely poor, and the one with white and light orange color shows good correlation between the features.

Figure 4 shows the quartile, median, min–max, and outliners through box plot between slope and cover_type. Cover type 1 has 30 as maximum, whereas cover types 3 and 4 have 45 degrees of slope. Cover type 2 has maximum number of outliners, whereas cover type 4 has no outliners. Cover types 3 and 4 have same number of maximum data points with respect to slope degree, but cover type 3 has higher 3rd quartile in comparison with cover type 4. Another observation that this visualization gives, outliners of cover types 2 and 7 are continuous whereas cover type 6 is having discrete outliners which makes it difficult to remove. Every cover

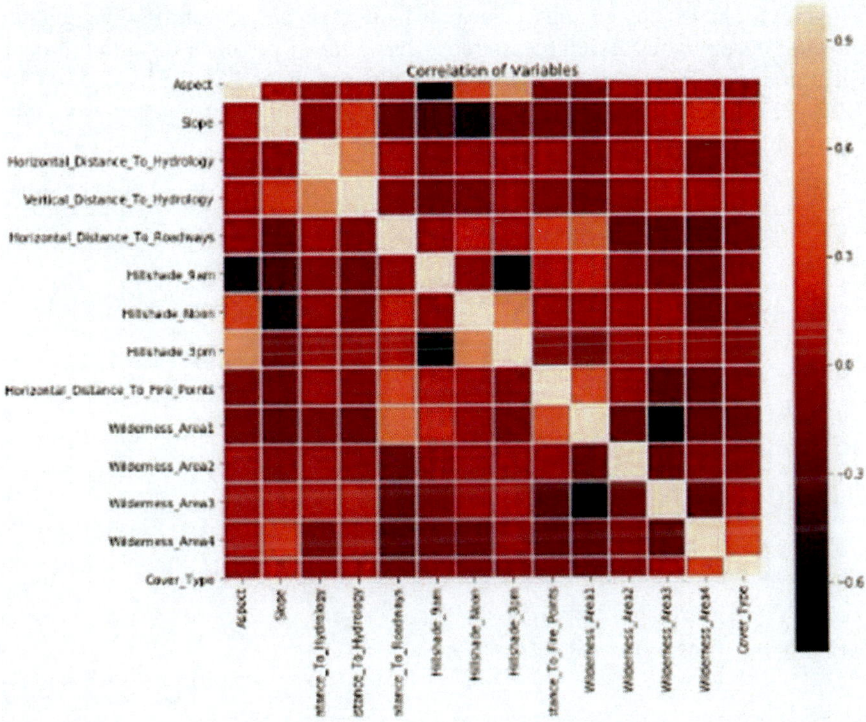

Fig. 3 Heat map

Fig. 4 Box plot of cover type

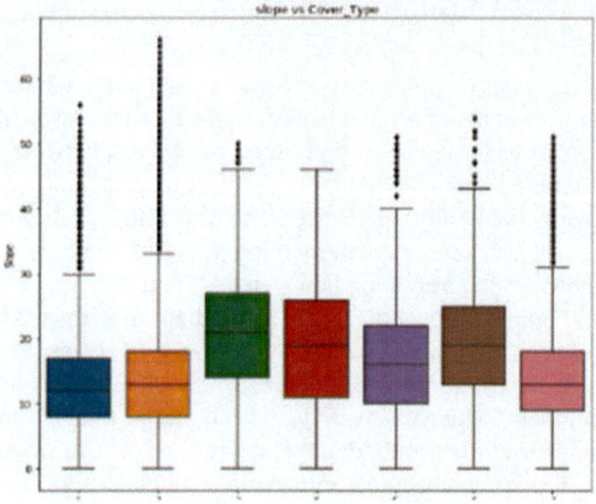

type points their minimum to zero, which means every cover has some patches which are totally flat and having zero-degree slope.

Figure 5 shows hot encoding between soil_type and cover_type having the following observations: WildernessArea_4 has a lot of presence for cover_type 4. WildernessArea_3 does not have good class distinction, and Soil_Type 1-6, 10-14, 17, 22-23, 29-33, 35, 38-40 offer lot of class distinctions as counts for some soil_type are very high.

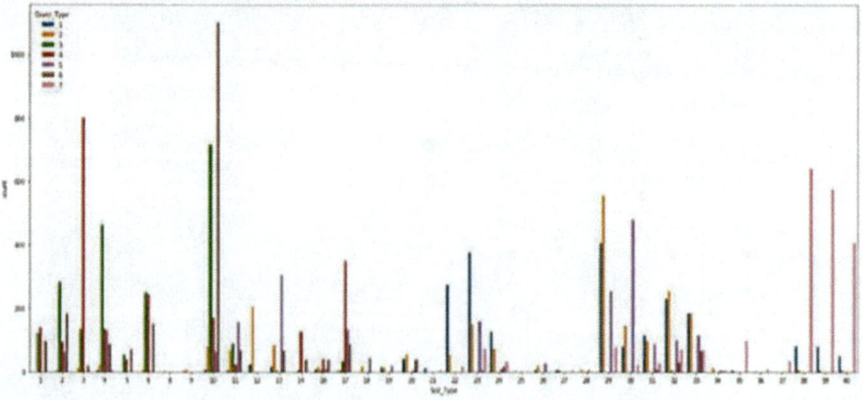

Fig. 5 Hot encoding between soil and cover type

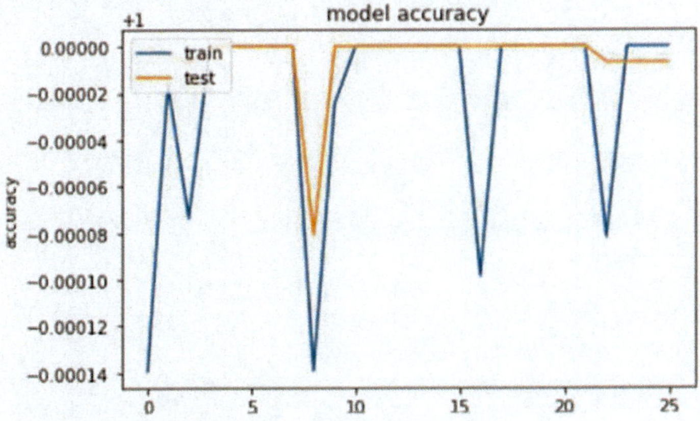

Fig. 6 Model accuracy graph for ReLU

5 Evaluation, Prediction, and Analysis

The MLP architecture is designed using 2 hidden layers with 64 neurons in each hidden layer and 7 neurons on output layer. The back-propagation algorithm has been applied where the weighted inputs of each neuron are mapped to output by the linear function. The output layer has 7 different classes. Due to the vanishing gradient problem, the sigmoid and hyperbolic tangent activation functions cannot be used in networks. Hence, the study explores impact of different variants of the ReLU activation function on hidden and output layers. In neural networks, the Softmax activation is used in the output layer and ReLU is used in the middle layer. For this model, a batch size of 60 has been considered. The performance of 26 epochs has been taken for all the training examples. Figures 6, 7, and 8 show training and validation model accuracy with ReLU Tan and TanH, respectively.

The mapping of negative inputs with strongly negative and mapping zero inputs with near zero in TanH graph is an advantage of the function. Applying 54 neurons on input layer as the total columns are of same size, exactly 100 neurons are placed with 4 hidden layers and Softmax function is used in output layer. To deal regularization in MLP model dropout function with a value of 0.2 has been used.

6 Conclusions

In this study, MLP architecture applies three usual differentiable and monotonic activation functions and generalized delta rule learning. TanH, ReLU, and Softmax activation functions are used in this study. On comparing their performances, it has been observed that ReLU function has 100% classification accuracy and Tanh–Softmax combination gives good results instead of 'Tanh–Tanh' combination. The

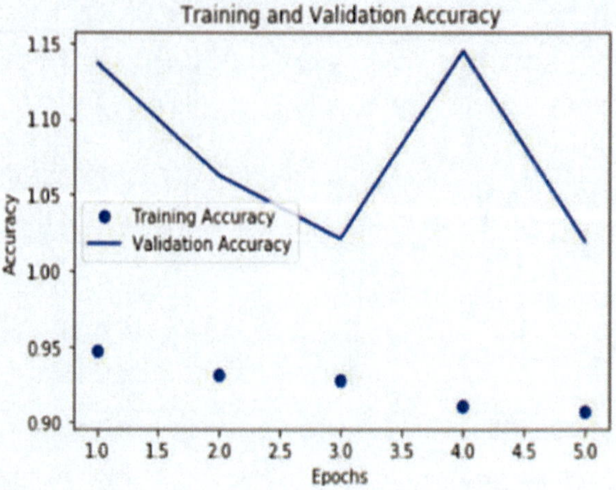

Fig. 7 Accuracy with Tan

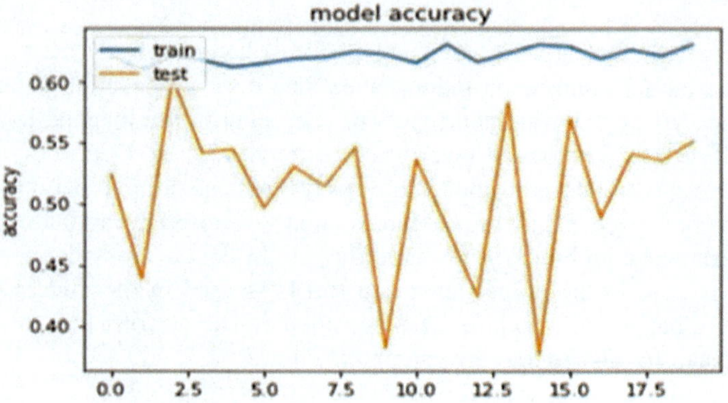

Fig. 8 Model accuracy for TanH

'Tanh–Tanh' combination has accuracy of 61%. Hence, it can be concluded when neural network uses ReLU–Softmax grouping of activation functions for neurons of hidden and output layers and good results are obtained.

References

1. Freitasab SR, Hawbaker TJ, Metzgera JP (2010) Effects of roads, topography, and land use on forest cover dynamics in the Brazilian Atlantic Forest 259(3):410-417
2. Blackard JA, Dean DJ (1999) Comparative accuracies of artificial neural networks and discriminant analysis in predicting forest cover types from cartographic variables. Comput Electron Agric 24:131–151

3. Kumar R, Nandy S, Agarwal R, Kushwaha SPS (2014) Forest cover dynamics analysis and prediction modelling using logistic regression model. Ecol Ind 45:444–455
4. Gupta A, Jagadeesh R, Sahwney H, Zalani Z (2015) Classifying forest categories using cartographic variables. In: IITK/CS771A/2015. Indian Institute of Technology, Kanpur
5. Bhattacharjee K, Behera B (2017) Forest cover change and flood hazards in India. Land Use Policy 67:436–448
6. Wagh TA, Bhargavi R, Wagh TA, Samant RM (2018) Forest cover type prediction using cartographic variables. Int J Comput Appl 182(30). (0975-8887)
7. Ye L, Gao L, Marcos-Martinez R, Mallants D, Bryan BA (2019) Projecting Australia's forest cover dynamics and exploring influential factors using deep learning. Environ Model Softw 119:407–417
8. Kaggle Forest cover type. https://www.kaggle.com/c/forest-cover-type-prediction
9. Abbott D (2017) Applied predictive analytics, principles and techniques for the professional data analayst, Chapter 3, 4, 6 and 8
10. Hastie T, Tibshirani R, Friedman J (2009) The elements of statistical learning: data mining, inference, and prediction. Springer, New York, NY
11. Class imbalance problem. Posted Aug 30, 2013, http://www.chioka.in/class-imbalance-problem/

An Approach to Detect Sarcasm in Tweets

Jyoti Godara and Rajni Aron

Abstract When academic and business ventures are discussed, electronic documents form the crucial part of receiving and transferring information. There is no use of online information if we cannot extract it and use it to cater our ventures. In order to frame up any overview, it is required to find the relevant text with complete omission of unnecessary information while keeping the focus on details and compile them into a document. The sentiment analysis is the approach used to evaluate users' sentiments on websites, forums, comments, feedback as negative, positive or neutral. But, sometimes, people express their negative sentiment in a positive manner. This flips the polarity of the sentence and sentiment analysis performance is affected. Thus, automatic detection of this sarcastic text is an essential aspect of sentiment analysis. Motivation for this paper includes to provide overview of Sentiment Analysis and Sarcasm. Research studies in the area of sentiment analysis and Sarcasm are also discussed. There are multiple existing techniques for automatic detection of Sarcasm, according to the literature studied ensemble learning has performed better than others. In this paper, designing and development of an ensemble classification method is proposed. Base classifiers selected are Decision Tree, Naive Bayes and K-nearest Neighbor. It will thus improve various parametric values for the sarcasm detection.

Keywords Sentiment analysis · Sarcasm · Ensemble classifier · Machine learning

1 Introduction

Internet era has changed the way individuals communicate their views and opinions. Millions of people use social networking sites to share thoughts, emotions, and views about their daily life [15]. There are around 823 new tweets posted on Twitter

J. Godara (✉) · R. Aron
Lovely Professional University, Phagwara, Punjab, India
e-mail: jyotipoonia6@gmail.com

R. Aron
e-mail: rajni3387@gmail.com

A. Khanna et al. (eds.), *Data Analytics and Management*, Lecture Notes on Data Engineering and Communications Technologies 54,
https://doi.org/10.1007/978-981-15-8335-3_5

every second, 520 comments every minute, 294,000 statuses and 138,000 snapshots posted on Facebook and more than 1 million customer transactions per day are manged by Walmart [5]. Such huge data can be valuable for various applications such as election outcome forecast, analysis of market trend, e-governance, business intelligence as customers rely heavily on the online content for the decision making. This voluminous content is too high to be processed by human, thus it needs to be automated. This task of creating a system for gathering and analysing the sentiment about an entity's sentiment in the forum, comments, feed backs or tweets is called sentiment analysis.

It analyzes people's views, thoughts, emotions and perceptions towards things such as goods, services, organisations, person, issues, events, topics and their attributes. It can be applied at three different levels- document level, sentence level and aspect level [16]. From the sentiments expressed, we can easily express what the speaker has in his mind. Sentiments may be like- dislike, agree-disagree, good-bad, positive or negative sentiments or multi-class sentiments-strong positive, positive, neutral, negative, strong negative. Sometimes, analysis of sentiment can be misled by the use of terms which are of strong polarity but are used sarcastically, implying that the opposite polarity was intended.

Sarcasm is a specific type of sentiment where an individual conveys his negative feelings using positive exaggerated words in the text [17]. Example—"I love music, it can take you to another Place, for example, Meghan Trainor is playing in this cafe so now I am going to a different café." "Oh! He is out on a duck, what a legendry batsman." [22]. In the example, person is expressing positive emotion but overall tweet is reflecting negativity towards the batsman. It is a very elegant way to indirectly convey message which makes it difficult to detect [7].

From business perspective, it is very crucial to understand product reviews, movie popularity and social opinions, as they may be suffered, if considered in the wrong category [18]. Detection of sarcastic text will refine the performance of sentiment analysis system to make it more reliable and accurate as it will help to eliminate the intentional ambiguity raised in the text [20]. Detection of sarcasm while analysing the sentiment in the given text is a type of classification task. Various approaches for detection have been categorized into machine learning and lexicon based approach [15]. Machine learning approach initiates with the collection of training data sample. Then, a classifier on the training information is trained. After the selection of a classification method, a significant conclusion is made. Machine learning technique engages both supervised learning for labelled dataset and unsupervised learning for the unlabelled dataset. Few classifiers using supervised learning approach are Decision Tree, Neural Network, Support Vector Machine, Naïve Bayes, Maximum Entropy, while the unsupervised approach are clustering methods. Lexicon-based approach [10] uses the dictionary of sentiments with the terms of opinion and combines them with the data for polarity determinations. This assigns sentiment scores to the words of opinion explaining whether the terms in the dictionary are optimistic, negative and objective. Dictionary based lexical approach includes looking for synonyms or antonyms of the dictionary words while Corpus based approach provides dictionaries related to specific domains.

Ensemble approaches in machine learning puts together the effect of several machine learning methods on the problem to obtain a greater predictive ability in solitude than its comprising algorithms [14]. The overarching objective of this approach is to combine outputs of base classifiers, in order to produce a single aggregated output which exceeds any of the isolated base classifiers. The first stage in the classifier ensemble process is to produce various base classifiers. One strategy is to apply N learning technique, with a single collection of training data, to obtain N different models of classification. An alternative approach is to build N datasets from the training data and use single learning algorithm for each set [13]. Bagging is the technique of development of multiple training datasets from a single dataset. Boosting applies different datasets and same learning method. Stacking is used to combine various models which were built using different learning models on single data.

In proposed method, ensemble classifier will be generated from the machine learning classifiers to improve the accuracy for detecting sarcasm.

Major contribution of this paper includes:

- Understanding basic concepts of Sentiment Analysis concepts.
- Literature review of existing techniques for analysing sentiment and detection of sarcastic text.
- Proposing an ensemble learning technique to automatically detect Sarcasm thus improving the performance of Sentiment Analysis.

The following sections are organized as follows: Sect. 2 discusses existing research studies related to Sentiment Analysis and Sarcasm Detection. In Sect. 3, an approach to automatically detect Sarcasm is proposed. Section 4 concludes the paper.

2 Literature Review

This section includes existing literature related to the sentiment analysis, ensemble classifier and detection of sarcasm in the textual data.

Ren et al. (2019) integrated sentiment analysis using SVM classifier into a machine learning framework [21]. In addition, more accurate and practical sentiment indexes were generated by taking into account the day-to-week impact. It was possible to improve the prediction accuracy of movement direction of the SSE 50 Index up to 89.93% as per the obtained experimental results. A growth of 18.6% was seen in the accuracy rate after including sentiment variables. In the meantime, the proposed model also provided support to the shareholders in efficient decision making.

Iqbal et al. (2019) proposed a new feature reduction approach based on Genetic Algorithm (GA) [2]. This hybrid technique had the ability of reducing the dimension of feature-set up to 42% without bothering accuracy. The proposed feature reduction approach was compared with more extensively used feature lessening algorithms based on Principal Component Analysis (PCA) and Latent Semantic Analysis (LSA). The proposed approach showed improved accuracy rate of 15.4% and 40.2% over PCA and LSA based approaches respectively.

l-Twairesh et al. (2019) made an attempt to discover about sentiment-particular word embeddings from Arabic tweets [1]. These word embeddings were later used for classifying the sentiments of Arabic Tweets. Furthermore, a novel feature which was an ensemble model of surface and deep features was recommended in this work. The surface features were retrieved physically. On the other hand, the deep features were generic word embeddings and sentiment-specific word embeddings. A lot of tests were carried out for testing the efficiency of ensembling surface and deep features. These tests also examined pooling functions, embeddings dimension, and cross-dataset models.

Fu et al. (2018) recommended a lexicon-enhanced LSTM model. Initially, This model used sentiment lexicon as additional pre-training information for a word sentiment classifier [9]. Afterward, this classifier got the sentiment embeddings of words. These embeddings included non-lexicon words. It was possible to represent words more accurately by combining the sentiment embedding and its word embedding. Also, a novel technique was described in this work for finding the attention vector in wide-ranging sentiment analysis devoid of any target. This phenomenon could enhance the LSTM ability for getting inclusive information of sentiment. The tested outcomes depicted that the recommended models outperformed the other accessible models.

Yu et al. (2018) recommended a word vector refinement model. Refining existing pretrained word vectors was the main aim of this model [23]. For this purpose, this model used real-valued sentiment intensity scores given by sentiment lexicons. The concept of the recommended model was to improve every word vector for bringing it nearer to both semantically and sentimentally alike words and more far from sentimentally unlike words in lexicon. The tested outcomes revealed that the suggested model could improve both traditional word embeddings and already suggested sentiment embeddings for classifying binary, ternary, and fine-grained sentiments. Two data sets called SemEval and Stanford Sentiment Treebank were used in this work to perform tests.

Hayran et al. (2017) suggested a new scheme to classify the sentiments microblog posts in automatic manner [12]. The recommended scheme was based on strong feature demonstration and combination. This work used word embedding method as the feature depiction. On the other hand, Support Vector Machine was used as a classification model. The tested outcomes depicted that that the proposed technique efficiently reduced the size of tweet depiction and enhanced the precision of sentiment classification. The suggested approach showed accuracy rate of 80.05% in the classification of sentiments and proved its supremacy over other existing techniques.

Da Silva et al. (2014) generated a single classifier from multiple classifiers—Naïve Bayes, Maximum Entropy and Support Vector Machine to improve the accuracy of tweet classification [6].

Perikos et al. (2016) proposed a sentiment analysis method for automated emotion detection in a text using classifier ensembles [19].Ensemble classifier was based on three key classifiers-a learner from Naïve Bayes, a maximum entropy and knowledge-based system. Results showed that ensemble schema performed better in recognizing emotion present in the text and determining text polarity than the sole classifiers.

Gidhe et al. (2017) stated that a lot of researches were conducted on dataset including # sarcasm tag. These researchers mostly used structural and sentiment attributes. However, some cases did not include # sarcasm tags [11]. Emotional and semantic similarity features were needed to transcribe these types of cases. In this work, a novel scheme had been recommended for the classification of sarcastic and non-sarcastic sentences. For this purpose, structural, emotional and semantic similarity features with MLP-BP approach were used in this work.

Dharwal et al. (2017) stated that it was a very complex task to detect sarcasm in automatic manner in Sentiment Analysis [8]. The detection task included composite linguistic analysis and machine learning algorithms. In this work, a lot of sarcasm analysis techniques were reviewed. These techniques were used to filter the sarcastic statements from a text. These techniques also analyzed the application of Automatic sarcasm detection in the classification of product review texts and tweets.

Bharti et al. (2017) stated that several sarcasm detection approaches had been recommended for detecting sarcastic within text [3]. Detecting sarcasm in English text was the main aim of almost all techniques. A novel context-based pattern had been recommended in this work to detect sarcasm within Hindi tweets. In the proposed pattern, sarcasm was represented as a conflict between a tweet and the reference of its relevant news. The suggested technique used Hindi news as the reference of a tweet within the similar timestamp and achieved 87% accuracy rate.

Mukherjee et al. (2017) considered different text independent feature sets. This kind of features included n-grams and function words [18]. In this work, Naïve Bayes and fuzzy clustering algorithms were used for the testing of several feature sets. The tested results depicted that it was advantageous to add some features capturing the blogging style of the microblog writers to detect sarcasm. Accuracy rate of about 65% was achieved by the recommended approach in this work.

Bouazizi et al. (2016) proposed a novel scheme based on pattern for twitter sarcasm detection [4]. In this work, four feature sets had been utilized. These feature sets covered the different sorts of identified sarcasm. These feature sets were utilized for classifying tweets as sarcastic and non-sarcastic. The recommended scheme showed accuracy and precision rate of 83.1% and 91.1% respectively. This work studied the significance of every suggested feature set. This work also evaluated its additional value to the classification.

As per the existing research work, we have analysed that hybrid classifier is performing better than standalone classifiers, thus an ensemble classifier is proposed to increase the accuracy to detect sarcasm.

3 Proposed Methodology

This article proposes a prototype for detecting sarcasm during sentiment analysis of tweets. The basic steps are also described in Fig. 1.

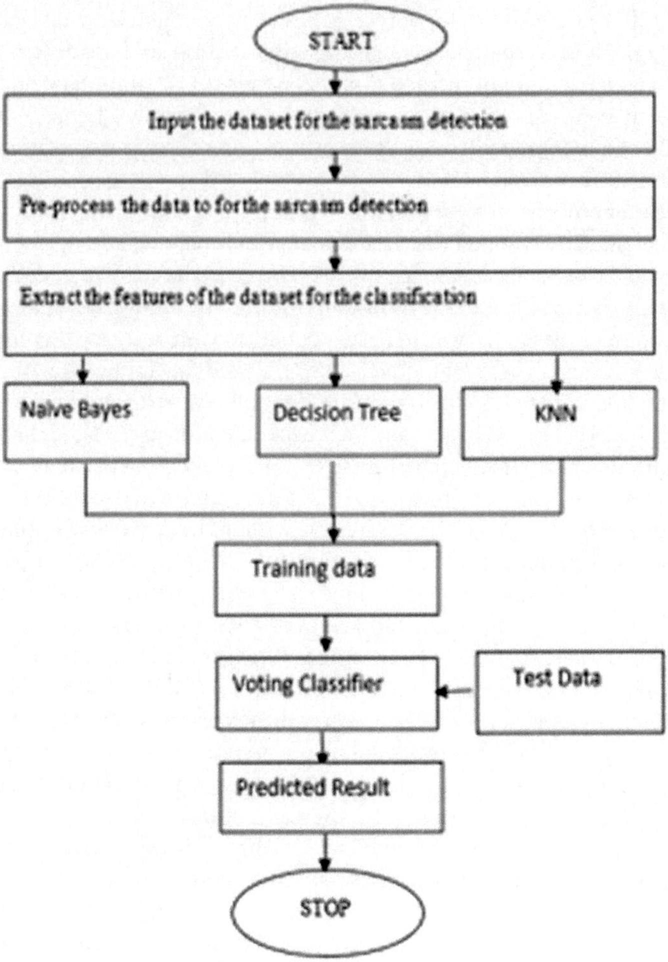

Fig. 1 Proposed methodology

3.1 Data Collection

The data sets needed for the experiments will be obtained from Twitter's official website using appropriate APIs and programs in the form of tweets—a micro blogging network that allows its users to write up to 140 characters. Tweets of #sarcasm or #sarcastic or #not of a certain length are taken as sarcastic datasets. It also extracts an appropriate number of regular tweets that have positive or negative sentiment scores. These corpora would then be used at a later stage to train the systems.

3.2 Data Pre-processing

The collected dataset will be then pre-processed to be converted into computer readable format. Following steps will be used:

- DataCleaning: URLs, @mention, retweets, hashtags, ampersands, extra white spaces, hexacharacters, double quotations, emoticons, numbers, and lines are eliminated. All uppercase letters are transformed to lowercase cases. Redundant tweets are removed. Tweets of fewer than three words are also deleted.
- Tokenization: Word sequences are broken down to symbols, phrases, words, or other elements that are useful and are known as tokens.
- POS (Parts Of Speech) tagging: It is used to extract important words which have a greater impact on the classification of emotions, such as adjective, adverb, verb, noun. Each word is be assigned a POS tagger.
- Removal of stop words: Stop Words such as the, in, at, that, which, and on are removed as they will increase the work required by the software to parse them, while providing minimal benefit.
- Stemming and Lemmatization: Stemming reduces the derived word to its original word stem. E.g. Attending is stemmed to attend. The main objective of this technique is to eliminate included suffixes and amount of words. Lemmatization removes the affixes and also adds the missing character to the root to make the complete word. E.g. Stemming: decided- decid; Lemmatization: decide

3.3 Feature Extraction

- Point-wise Mutual Information: The measure of mutual information provides a traditional way of modeling the mutual information between the features and classes.
- Chi-square and PMI: These are two distinct ways to measure the association between words and groups.
- Punctuation marks: Punctuation includes, ; : ? !, emoticons, quotes, capitalized words, etc. In certain cases, Sarcastic sentences show a certain pattern. Punctuation marks play a significant role in understanding such sarcastic sentences. For example, Wednesday is latin for "almost Friday."
- Lexical features: These are significant when sarcastic statements are delivered and identified. These can be based on factors such as counter-factuality (e.g., yet), temporal compression (e.g., suddenly), repeated sequences (e.g., n sequences in a sentence-grams, grams skip).
- Semantic characteristics: While some attention has been given to semantic characteristics (e.g., ADV + ADJ + N) and temporal inconsistency (e.g., dislike, not). They are of vital importance in recent investigations to demonstrate inconsistency in the sense of sentences. Sarcasm requires the background to be reliably known.

Those features are selected which are highly co-related to the class.

3.4 Classification

In the proposed methodology, ensemble classifier will be generated by selecting Naïve Bayes, Decision Tree and K nearest Neighbor as baseline classifiers.

- Naive Bayes is a probabilistic classification model. It makes assumption about the strong independence among the features. This classifier needs less number of training data for computing the parameters to carry out prediction task [8]. This is the major benefit of this approach. Due to the independence of features, this classifier computes merely the variance of the feature rather than computing the complete covariance matrix. The conditional probability for each class given a review is $P(c)$ for a text review received 'd' and for a class 'c' (positive, negative); It is expressed as:

$$P(c/d) = \frac{(P(d/c) * P(c))}{P(d)} \tag{1}$$

- A decision tree resembles a flowchart-like tree structure. In this approach, every interior node depicts a test on a feature and every branch demonstrates the test result. Every leaf node or terminal node represents a class label. The testing of attribute values of the tuple is performed against the decision tree for a given tuple X. The tracing of a route is done from the root to a leaf node. This node holds the class prediction for the tuple [9].
- K-Nearest Neighbors is a very fundamental and important classifier. This classification algorithm is related to the supervised learning approach. This classifier is extremely useful in pattern recognition, data mining and attack revealing. Some prior data known as training data is used to classify coordinate into classes.

Ensemble classifier will be generated by combining the base classifiers using stacking approach. Initially all instances from the validation set are classified by all the base classifiers. A model is built by combining the decisions of base classifier through majority voting technique.

3.5 Evaluation

The ensemble model will be classifying the new test data in the sarcastic or non-sarcastic data. Performance will be evaluated by the confusion matrix. The parameters of accuracy, precision, recall, and F-measures are based on the classification's true positive, true negative, false positive, and false negative outcomes. True Positive: The correctly defined values are considered true positive ones. True Negatives: The attributes that are wrongly placed into a separate class are True negatives. False positives: False positives are the values that do not belong to a certain class but are mistakenly listed in that class. False Negatives: False negatives are the values that neither belonged to a certain class nor were included in that class.

$$\text{Precision} = \frac{(\text{TruePositive})}{(\text{TruePositive} + \text{FalsePositive})} \tag{2}$$

$$\text{Recall} = \frac{(\text{TruePositive})}{(\text{TruePositive} + \text{FalseNegative})} \tag{3}$$

$$\text{Fmeasure} = \frac{(2 * \text{recall} * \text{precision})}{(\text{precision} + \text{recall})} \tag{4}$$

$$\text{Accuracy} = \frac{(\text{TruePositive} + \text{TrueNegative})}{(\text{TruePositive} + \text{FalsePositive} + \text{TrueNegative} + \text{FalseNegative})} \tag{5}$$

4 Conclusion

Sentiment analysis examines the polarity of the sentiments from the data. Due to presence of sarcastic text, the accuracy is affected. Thus, detection of sarcasm in the text is very crucial during analysis of sentiments. In this paper, an approach to detect sarcasm in tweets is proposed by generating an ensemble classifier combining outputs of Naive Bayes, K-nearest Neighbor and Decision Tree using majority voting approach. Performance will be improved using this approach. In further studies, this model can be implemented on product review tweets or other social networking websites. It can also be extended to stock markets, news articles and political discussions.

References

1. Al-Twairesh N, Al-Negheimish H (2019) Surface and deep features ensemble for sentiment analysis of arabic tweets. IEEE Access 7:84122–84131
2. Anuprathibha T, KanimozhiSelvi CS Penguin search optimization based feature selection for automated opinion mining
3. Bharti SK, Babu KS, Raman R (2017) Context-based sarcasm detection in hindi tweets. In: 2017 Ninth international conference on advances in pattern recognition (ICAPR). IEEE, pp 1–6
4. Bouazizi M, Ohtsuki TO (2016) A pattern-based approach for sarcasm detection on twitter. IEEE Access 4:5477–5488
5. Chidananda HT, Das D, Sagnika S (2018) Sentiment analysis using n-gram technique. Progress in computing, analytics and networking. Springer, Berlin, pp 359–367

6. Da Silva NF, Hruschka ER, Hruschka ER Jr (2014) Tweet sentiment analysis with classifier ensembles. Decis Support Syst 66:170–179
7. Dave AD, Desai NP (2016) A comprehensive study of classification techniques for sarcasm detection on textual data. In: 2016 International conference on electrical, electronics, and optimization techniques (ICEEOT). IEEE, pp. 1985–1991
8. Dharwal P, Choudhury T, Mittal R, Kumar P (2017) Automatic sarcasm detection using feature selection. In: 2017 3rd International conference on applied and theoretical computing and communication technology (iCATccT). IEEE, pp 29–34
9. Fu X, Yang J, Li J, Fang M, Wang H (2018) Lexicon-enhanced lstm with attention for general sentiment analysis. IEEE Access 6:71884–71891
10. Ge X, Jin X, Miao B, Liu C, Wu X (2018) Research on the key technology of chinese text sentiment analysis. In: 2018 IEEE 9th International conference on software engineering and service science (ICSESS). IEEE, pp 395–398
11. Gidhe P, Ragha L (2017) Sarcasm detection of non# tagged statements using mlp-bp. In: 2017 International conference on advances in computing, communication and control (ICAC3). IEEE, pp 1–4
12. Hayran A, Sert M (2017) Sentiment analysis on microblog data based on word embedding and fusion techniques. In: 2017 25th Signal processing and communications applications conference (SIU). IEEE, pp 1–4
13. Jurek A, Bi Y, Wu S, Nugent C (2014) A survey of commonly used ensemble-based classification techniques. Knowl Eng Rev 29(5):551–581
14. Kanakaraj M, Guddeti RMR (2015) Performance analysis of ensemble methods on twitter sentiment analysis using nlp techniques. In: Proceedings of the 2015 IEEE 9th international conference on semantic computing (IEEE ICSC 2015). IEEE, pp 169–170
15. Kharde V, Sonawane P et al (2016) Sentiment analysis of twitter data: a survey of techniques. arXiv preprint arXiv:1601.06971
16. Kiruthika M, Woonna S, Giri P (2016) Sentiment analysis of twitter data. Int J Innov Eng Technol 6(4):264–273
17. Kumar A, Sangwan SR, Arora A, Nayyar A, Abdel-Basset M et al (2019) Sarcasm detection using soft attention-based bidirectional long short-term memory model with convolution network. IEEE Access 7:23319–23328
18. Mukherjee S, Bala PK (2017) Sarcasm detection in microblogs using naïve bayes and fuzzy clustering. Technol Soc 48:19–27
19. Perikos I, Hatzilygeroudis Ioannis (2016) Recognizing emotions in text using ensemble of classifiers. Eng Appl Artif Intell 51:191–201
20. Porwal S, Ostwal G, Phadtare A, Pandey M, Marathe MV (2018) Sarcasm detection using recurrent neural network. In: 2018 Second international conference on intelligent computing and control systems (ICICCS). IEEE, pp 746–748
21. Ren R, Wu DD, Liu T (2018) Forecasting stock market movement direction using sentiment analysis and support vector machine. IEEE Syst J 13(1):760–770
22. Rendalkar S, Chandankhede C (2018) Sarcasm detection of online comments using emotion detection. In: 2018 International conference on inventive research in computing applications (ICIRCA). IEEE, pp 1244–1249
23. Yu L-C, Wang J, Lai KR, Zhang X (2017) Refining word embeddings using intensity scores for sentiment analysis. IEEE/ACM Trans Audio Speech Lang Process 26(3):671–681

A Trust-Based Approach to Extract Social Relationships for Recommendation

Jyoti Shokeen, Chhavi Rana, and Poonam Rani

Abstract Recommender systems are gaining momentum with the increasing set of user choices in recent years. Also, growing proliferation of social networking sites is using these systems to accelerate their usage. User preferences are likely to be similar or influenced by their friends on social networks. Due to this, social relations can help the recommender systems in solving cold-start and data sparsity problems. This paper presents a trust-based approach to extract social relations for recommendation in social networks. We use matrix factorization approach to factorize the user-item rating matrix. Both rating matrix and trusted relations play a crucial role in calculating the weighted trust between users in our proposed approach. Based on the weighted trust value, the algorithm recommends top-N items to the user. The motivation behind the extraction of social relationships is to use the recommendations of friends.

Keywords Social networks · Recommender system · Trust · Similarity

1 Introduction

Recently, a tremendous use of social media sites has generated huge amount of data. A recommender system (RS) is a computer system that gives suggestions. In the era of digital laziness, RSs are alternatives to search engines. Netflix suggests movies, Amazon suggests products to purchase and Facebook suggests friends (people you may know). The aspect of RS in social networks is essential to keep them running,

J. Shokeen (✉) · C. Rana
CSE, University Institute of Engineering and Technology,
Maharshi Dayanand University, Rohtak, Haryana, India
e-mail: jyotishokeen12@gmail.com

C. Rana
e-mail: chhavi1jan@yahoo.com

P. Rani
COE, Netaji Subhas University of Technology, Dwarka, Delhi, India
e-mail: poonamrani2017.nsit@gmail.com

A. Khanna et al. (eds.), *Data Analytics and Management*, Lecture Notes on Data Engineering and Communications Technologies 54,
https://doi.org/10.1007/978-981-15-8335-3_6

alive and effective. The different aspects and challenges of social networks are discussed by Rani et al. [13].

The traditional RS techniques are usually classified into three categories. Content-based filtering (CBF) involves the features of items. In this, the explicit user feedback plays an important role in adjusting the corresponding weights for each item for a particular user. In contrast, collaborative filtering (CF) does not need the information of items. Instead, CF demands a large amount of data to draw the similarities between users for their taste prediction. It is based on the idea that people with analogous preferences in the past will have same liking in the future. Most of the existing recommendation systems use CF approach to find similar users. Model-based and memory-based are the two forms of CF techniques [24]. Despite its unparalleled performance, this approach has some inherent drawbacks. This is due to the fact that this approach uses similar users from the user-item rating matrix. However, a large proportion of such rating matrix is generally sparse due to which it performs poorly for cold-start users and cold-start items. Alternatively, hybrid techniques combine the above techniques to use the best of both worlds. Different techniques have their own advantages and disadvantages. Therefore, the hybrid of these techniques gives best results, on the contrary, when executed alone. Apart from these techniques, deep learning [20], fuzzy clustering [21], ontology-based are some other techniques to build RS.

Social networks analysis is important to predict relationships [14], compare social networks [11] and quantify relationships quanitatively and qualitatively [10]. The exploding use of social networks allows people to make social relationships over Internet. These relationships serve as important information in improving the performance of RS. The system that makes use of social relationships to recommend items to users is called social RS. A social RS is the system that integrates social information into recommendation engine [22]. Social information is also beneficial in education and e-learning [12, 17]. Social data aids in improving user experience [15]. Various parameters define the quality of improving the social recommendation performance. Shokeen et al. [18] highlight these parameters of social RS. The major contributions of this paper are as follows:

- Aiming at the limitations of using user-item rating matrix alone in the recommendation models, this paper combines trusted relations of users in the rating matrix to improve recommendations.
- We present an algorithm that extracts social relationships for recommendation.
- The approach proposed in this paper aims to remove cold-start problem, data sparsity problem and issues related to malicious attacks. The reason beyond this is that relations from social networks are more trustworthy and reasonable.

In this paper, Sect. 2 reviews the related work in this sphere. Section 3 describes the proposed work and Sect. 4 concludes the paper.

2 Related Work

A number of techniques to build social RS are given by Shokeen and Rana [21]. Various researchers [3, 8, 23] attempted to combine trust awareness in recommendation technique to improve the performance. Authors in [16] experimentally demonstrated the importance of social relationships in affecting user tastes and consequently their similarities. Ma et al. [7] proposed SoRec as a trust-based model which is based on probabilistic matrix factorization. This model reduces data sparsity problems in recommendation framework. RSTE proposed by [6] is another trust-based model that uses matrix factorization and social information to recommend items to users. Most of the works in literature use memory-based approaches to employ the use of trust propagation in recommendation. Very few of them have used model approaches to propagate trust in social networks and use the results in recommendation. Matrix factorization is a model-based approach that factorizes the user-item matrix. TidalTrust [2] used shortest path distance to calculate the trust between raters and the source user. Dang et al. [1] used deep learning approach to propose a social recommendation approach called *dTrust*.

A web RS is proposed by [4] that considers sequential information based on the user's usage patterns of web pages. To create soft clusters of users, they employ fuzzy c-means clustering approach. In our previous paper [19], we have designed an architecture for trust-aware social recommender system in which rating is predicted using both rating matrix and trust from social relationships. We also discussed that there are two types of trust in social networks, i.e., direct trust and indirect trust. When two users are connected in a social network, then we define it as direct trust. On the other hand, when two users are similar in many perspectives and also possess some common neighbors but these users are not connected, then we can compute indirect trust between them. Li et al. [5] employed user interaction to propose a social recommendation model. They used social similarity tendency to predict ratings for items to users. Recently, trust-based recommendation models are also applied in the field of Internet of Things for secure and authentic transmission [9].

3 Proposed Work

We state our recommendation problem as follows: Given a set of users denoted by U, a set of items denoted as I and R represents the user-item rating matrix. The aim is to predict the ratings in R when it encounters an unknown rating by a user for an item. One of the methodologies to cope with data sparsity in RS is to implement both rating information and social relationships into RS. In social networks, social ties play a crucial role in dictating user's behavior. In this paper, we use matrix factorization approach to incorporate hidden features of items and users. This approach is used to extract the hidden features of users and items. Let $R_{m \times n}$ is the rating matrix consisting of m users and n items. If there are k latent features and X denotes the latent user

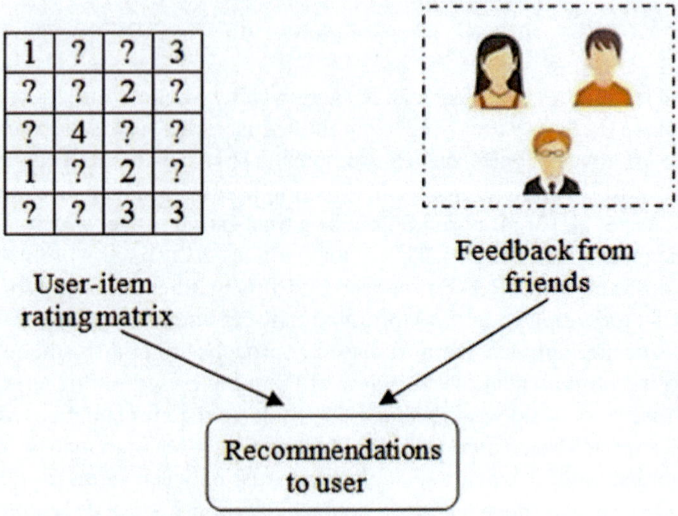

Fig. 1 Factors affecting recommendations

vector and Y denotes the latent item vector, then we can represent $R_{m\times n}$ as follows:

$$R_{m\times n} = X.Y^{\mathrm{T}} \tag{1}$$

where $X \in R_{m\times k}$ and $Y \in R_{n\times k}$. We define the conditional probability of the observed ratings as follows:

$$p(R \mid X, Y, \sigma_R^2) = \prod_{u=1}^{m}\prod_{i=1}^{n}[\mathfrak{N}(R_{u,i} \mid h(X.Y^{\mathrm{T}}), \sigma_r^2)]^{I_{u,i}^R} \tag{2}$$

where $\mathfrak{N}(x \mid \mu, \sigma^2)$ denotes normal distribution with variance σ^2 and mean μ. The function $h(x)$ is the logistic function that bounds the range of $X.Y^{\mathrm{T}}$ within [0, 1] and $h(x) = \frac{1}{1+\mathrm{e}^{-x}}$. If a user rates an item i, then indicator function $I_{u,i}^R$ becomes 1, otherwise its value is 0.

In the real-world scenario, users prefer to trust their friends for recommendation. In this paper, we design a trust-based approach that aims to reduce cold-start and data sparsity problems of RS. When a user is new to a recommendation system, then the system does not possess any information regarding that user. In such a case, social networking information and relations can assist the recommendation engine in finding similar users to the target user. Consequently, social relationships reduce the issues of data sparsity occurring in user-item rating matrix. Figure 1 depicts that both user-item rating matrix and feedback from social relations affect the recommendations. There exist a number of methods for trust computation in literature. Our work aims at finding the top-N similar items for a given user in social networks.

When any two users give common ratings to some items, then it is easy to predict the similarity between such users in the user-item rating matrix using the cosine formula as follows:

$$\mathrm{Sim}(u_x, u_y) = \frac{\Sigma u_x.u_y}{\sqrt{\Sigma (u_x)^2.(u_y)^2}} \quad (3)$$

In this paper, we use the cosine similarity formula to measure the similarity between any two users in the user-item rating matrix.

3.1 *Trust-Based Approach*

Many a times, users meet virtually while rating same item or items. The rater satisfaction during the encounter is reflected clearly by their ratings. At the end of virtual encounter, the rating of user i for user j gives us $S_{u_i}^{u_j}$ which can be described by computing the difference between their ratings for any item I_p as follows:

$$S_{u_i}^{u_j} = \begin{cases} \mathrm{VHT} : 0.0 \leq | r_{u_j,I_p} - r_{u_i,I_p} | \leq 0.5 \\ \mathrm{HT} \;\; : 0.5 < | r_{u_j,I_p} - r_{u_i,I_p} | \leq 1.0 \\ \mathrm{AT} \;\; : 1.0 < | r_{u_j,I_p} - r_{u_i,I_p} | \leq 2.0 \\ \mathrm{LT} \;\; : 2.0 < | r_{u_j,I_p} - r_{u_i,I_p} | \leq 3.0 \\ \mathrm{VLT} \qquad\qquad\qquad : \text{otherwise} \end{cases}$$

When the difference between the ratings for item I_p by user i and user j is less than or equal to 0.5, then it shows very high trust (VHT) between user i and user j. When this difference lies between 0.5 and 1.0, then it shows high trust (HT) between these users. For the difference between the ratings is greater than 1.0 but less than equal to 2.0, we assume that there exists average trust (AT) between the users. We assume low trust (LT) between user i and user j when this difference goes to 3. A further variation in the ratings represents very low trust (VLT).

In our trust-based social approach, we first choose a user from the given trust file and discover the first friends of this user based on the given trust value. We add the first friends of this user in the set of influencing users. We also find the second and third friends of the user by computing the direct friends of the first friend and the direct friends of the second friend, respectively. We then set the trust of the second-level friends as 0.5 and friends at third level as 0.25 and hence calculated the similarity between the users by the following method. We chose a user from the trust file, and corresponding to each of his second and third friends, we found out the similar items that the user has with each of them. This gives us the similarity value $T_{u,v}$.

For example, a user 3 has a second friend user 5, and these users have 4 movies in common. For computation of similarities between any two users, u and v, we use the common movies rated by the users. For each common movie between these two users, we calculate the difference in their movie ratings and append the differences in a ratingdiff array. Now, we calculate the cosine of each of the rating differences. After calculating the cosine values, we sum up all the cosine values of the rating differences and divide the sum by the length of the movies which the initial user rated. This gives us the similarity value S_{u_i,u_j}. When any two users are directly connected on social networks, then we assume the trust $T_{u,v}$ between them as 1. However, if these users are not connected directly but have many common friends between them, then we compute the similarities between them using user-item rating matrix.

The pseudocode of our proposed algorithm is given below:

Algorithm 1: Proposed Algorithm for Trust-based Recommendation

---

Require: Graph G(U,E), social dataset DS
Ensure: Top-N item recommendation
1: Initial user set = u_i
2: $getFriendList(u_i)$;
3: **for** each item $i_p \in DS$, do:;
    while *(isUserNode(i_p) == true and i_p = NextAdjacentUser(u_j));*
    do
        if u_i *and* u_j *are friends* **then**
            Compute T_{u_i,u_j};
        else
            Calculate S_{u_i,u_j} using similarity matrix;
            Use $S_{u_i}^{u_j}$ to determine the degree of trust ;
4: Calculate: $WT(u,v) = T_{u,v} \times S_{u_i}^{u_j} + S_{u_i,u_j}$
5: Recommend item list to user u_i

This proposed algorithm aims to recommend top-N items to any user based on the similarities between their ratings and the social connections existing between them on social networks. In the algorithm, we initialize a user from the user set and retrieve the friends of this user. For any item i_p that belongs to dataset DS, the algorithm checks whether u_i is an adjacent user to any user in the given set U. If the adjacent user u_j is a friend of user u_i, then the algorithm computes the trust between them, otherwise, similarity is calculated between them based on the cosine similarity formula as given in Eq. 3. On calculating similarity between the users, we compute the differences between their ratings and then, compute the degree of trust ranging from very high trust to very low trust which gives us $S_{u_i}^{u_j}$. If both similarity and trust exist between them, then the algorithm computes the weighted trust between them to recommend top-N items to the user. We compute the weighted trust value between them by the formula as follows:

$$\mathrm{WT}(u,v) = T_{u,v} \times S_{u_i}^{u_j} + S_{u_i,u_j} \tag{4}$$

4 Conclusion

The social relationships as extra information have been used in literature to enhance the recommendations. In this paper, we attempt to improve the recommendation quality by combining the user-item rating data with user relations. We present a trust-based approach that combines the similarity between users and trust to calculate the weighted trust between the users. The direct connections between users are used as the trust between them. Due to the space issue, we target to implement our proposed work on real-world social networking dataset as our future work. To further our research, we would like to explore implicit relations and distrust to get better recommendation performance and accuracy.

References

1. Dang QV, Ignat CL (2017) dTrust: A simple deep learning approach for social recommendation. In 3rd IEEE international conference on collaboration and internet computing, pp 209–218. https://doi.org/10.1109/CIC.2017.00036
2. Golbeck J (2006) Combining provenance with trust in social networks for semantic web content filtering. In: International provenance and annotation workshop. Springer, pp 101–108
3. Jamali M, Ester M (2010) A matrix factorization technique with trust propagation for recommendation in social networks. In: Proceedings of the 4th ACM conference on recommender systems, pp 135–142. https://doi.org/10.1145/1864708.1864736
4. Katarya R, Verma OP (2017) An effective web page recommender system with fuzzy c-mean clustering. Multimedia Tools Appl 76(20):21481–21496
5. Li Y, Liu J, Ren J (2019) Social recommendation model based on user interaction in complex social networks. PloS one 14(7):e0218957. https://doi.org/10.1371/journal.pone.0218957
6. Ma H, Lyu M, King I (2009) Learning to recommend with trust and distrust relationships. In: Proceedings of the third ACM conference on Recommender systems, pp 189–196. https://doi.org/10.1145/1639714.1639746
7. Ma H, Yang H, Lyu MR, King I (2008) SoRec: Social recommendation using probabilistic matrix factorization. In: Proceedings of the 17th ACM conference on information and knowledge management, CIKM '08. Association for Computing Machinery, New York, NY, USA, pp 931–940. https://doi.org/10.1145/1458082.1458205
8. Mei JP, Yu H, Shen Z, Miao C (2017) A social influence based trust model for recommender systems. Intell Data Anal 21(2):263–277
9. Mohammadi V, Rahmani AM, Darwesh AM, Sahafi A (2019) Trust-based recommendation systems in Internet of Things: a systematic literature review. Hum-centric Comput Inf Sci 9(1):21
10. Rani P, Bhatia M, Tayal D (2019) A comparative study of qualitative and quantitative SNA. In: 2019 6th International conference on computing for sustainable global development (INDIACom). IEEE, pp 500–504
11. Rani P, Bhatia M, Tayal DK (2018) An astute SNA with OWA operator to compare the social networks. Int J Inf Technol Comput Sci 3:71–80
12. Rani P, Shokeen J (2019) Designing a project leader recommender system using twitter feed analysis. In: Proceedings of the international conference on machine learning, big data, cloud and parallel computing: trends, prespectives and prospects, COMITCon 2019. IEEE, pp 56–60. https://doi.org/10.1109/COMITCon.2019.8862261

13. Rani P, Tayal DK, Bhatia M (2018) Different aspects, challenges, and impact of social networks with a mathematical analysis of teaching learning process. J Adv Res Dyn Control Syst 10(14):1576–1590
14. Rani P, Tayal DK, Bhatia M (2019) Predicting facebook group relationship. Int J Innov Technol Exploring Eng 8(11):1862–1869
15. Rani P, Tayal DK, Bhatia M (2019) SNA using user experience. In: 2019 International conference on machine learning, big data, cloud and parallel computing (COMITCon). IEEE, pp 125–128
16. Said A, De Luca EW, Albayrak S (2010) How social relationships affect user similarities. In: Proceedings of the 2010 workshop on social recommender systems, pp 1–4
17. Shokeen J (2018) On measuring the role of social networks in project recommendation. Int J Comput Sci Eng 6(4):215–219
18. Shokeen J, Rana C (2018) A review on the dynamics of social recommender systems. Int J Web Eng Technol 13(3):255–276. https://doi.org/10.1504/IJWET.2018.095184
19. Shokeen J, Rana C (2018) A study on trust-aware social recommender systems. In: Proceedings of the 12th INDIACom and 5th international conference on computing for sustainable global development. IEEE, pp 4268–4272
20. Shokeen J, Rana C (2019) An application-oriented review of deep learning in recommender systems. Int J Intell Syst Appl 11(5):46. https://doi.org/10.5815/ijisa.2019.05.06
21. Shokeen J, Rana C (2020) Social recommender systems: techniques, domains, metrics, datasets and future scope. J Intell Inf Syst 54(3):633–667. https://doi.org/10.1007/s10844-019-00578-5
22. Shokeen J, Rana C (2020) A study on features of social recommender systems. Artif Intelligence Rev 53(2):965–988. https://doi.org/10.1007/s10462-019-09684-w
23. Wang M, Ma J (2016) A novel recommendation approach based on users' weighted trust relations and the rating similarities. Soft Comput 20(10):3981–3990. https://doi.org/10.1007/s00500-015-1734-1
24. Yang X, Guo Y, Liu Y, Steck H (2014) A survey of collaborative filtering based social recommender systems. Comput Commun 41:1–10. https://doi.org/10.1016/j.comcom.2013.06.009

A Green 6G Network Era: Architecture and Propitious Technologies

Sukriti Goyal, Nikhil Sharma, Ila Kaushik, Bharat Bhushan, and Nitin Kumar

Abstract With the frequent growth in smartphones and media tablets, the industry of computer systems captured leadership from the industry of telecommunication in carry forwarding technological evolution. However, the principal component impacting the growth of the future will be the extent to which spectrum policy as well as management can rise spectrum capacity and give the necessary radio spectrum frequency efficiency. This entry describes why this is the bounding component for 5G mobile communication evolution so, around the world the research organizations begun to look beyond fifth-generation network, and it is expected from 6G to develop into green networks which supplies high capacity of energy and quality of service (QoS). With the goal of fulfill the requirements of forthcoming applications, considerable changes require to be made in structure of mobile communication networks. This paper gives a survey in detail on wireless development toward green 6G networks with an aim to show a pathway for further research works in the field of green 6G networks.

Keywords 6G · Blockchain · Green network · IoT · Energy harvesting

S. Goyal · N. Sharma (✉) · B. Bhushan
HMR Institute of Technology and Management, Delhi, India
e-mail: nikhilsharma1694@gmail.com

S. Goyal
e-mail: sukriti.s9900@gmail.com

B. Bhushan
e-mail: bharat_bhushan1989@yahoo.com

I. Kaushik · N. Kumar
Krishna Institute of Engineering and Technology, Ghaziabad, U.P, India
e-mail: ila.kaushik.8.10@gmail.com

N. Kumar
e-mail: nitin.kumar@kiet.edu

A. Khanna et al. (eds.), *Data Analytics and Management*, Lecture Notes on Data Engineering and Communications Technologies 54,
https://doi.org/10.1007/978-981-15-8335-3_7

1 Introduction

In today's time of different–different technologies, the coloration used for mobile phone interactions is becoming over-subscribed. Existing systems cannot continually meet the requirements of consumers for data. During the time of huge usage of data, consumers may suffer slow speed of data, loss or delay of services, and unsteady connections which results in continuous growth of the demand for the data as the number of gadgets linked to the Internet increases and here comes, 5G which will fulfill the demands of the consumers. Industries and consumers will depend upon the network of 5G to power the gadgets as well as send the data that run their day-to-day workings. The fifth generation of wireless systems, known as 5G, is in the process of structured and disseminated. The wireless network 5G marks the starting of an actual digital community and instates considerable blowouts in reference of delay, rates of data, maneuverability, and number of linked gadgets in opposition to trailing descents [1]. As the wireless network 5th-generation is in the introductory phases of industrialization, so, the present is the correct moment to start the investigation on inheritor (generation) of 5G.

In the previous few seasons, for the growth of 6G [2], many vital schemes are issued by some countries. The 6th-Genesis Flagship Program was proclaimed by Finland in the year 2018 which is a program of eight-years with the complete expense of $290 million to evolve an overall surrounding of 6G network [3]. 6G (the sixth-generation wireless network) is an inheritor of fifth-generation cellular technology. The networks of 6G will be potentially viable to use frequencies much higher than the networks of 5G and will supply much lesser delay and practicably higher capacity. For 6G namely quantum technology, the governments of Germany and USA have invested in some efficient technologies, and research on the mobile network 6G based on terahertz was begun by the USA.

In the digital economy, big data-based intelligence will become the real exhortation for innovation, and networks of 6G will not only be thoroughfares for sending the info but also will combine the technologies core and edge computing much more seamlessly as a section of an integrated interactions and computation architecture structure. 6th-generation network is looking forward to endorse speeding of a TB/s (terabyte per second) which is an unexampled degree of delay as well as efficiency, which will enhance the execution of applications of 5G apart from enlarging the scope of efficiencies in reference of growing new and innovative applications across the fields of wireless sensing, imaging, and cognition.

In the coming sections, the evolution of mobile communication networks will be introduced in Sect. 2. Then, in Sect. 3, a detailed architecture of the green 6G network is explained. Further, Sect. 4 gives an oversight of promising technologies of 6G in brief, and finally, at last, a conclusion is provided in Sect. 5.

2 Development of Mobile Connection Networks

In the future world, there will soon be more mobile phones than human beings. The prodigious development of mobile phones has been structured by the convergence of expeditiously developing technologies in the industries of computer and telecommunication. This development is not just a one-step process, but comprises of many inheritors which have various criterions, efficiencies, and technologies. The development of mobile networks is illustrated in Fig. 1.

2.1 *Evolution of Mobile Connection Networks from 1G to 3G*

In the world, on December 1, 1979, the first industrial 1G mobile network was started by Nippon Telephone and Telegraph (NTT) in Tokyo, Japan, which was structured for services of data with the rate of data up to 2.4 kbps. This cellular technology employed many, concentrated maintained cell sites or base stations, each giving service to a cell (small area). Analog signals were used by it to send data and because of the absence of any ubiquitous wireless medium, it led to several disadvantages like the absence of surveillance, problematic hand-off and the low transfer capacity [4]. In comparison, of 1G (first generation) networks, 2G phone systems used digital transmission rather than analog transmission, which made it to rely upon digital modulation technologies that are, time-division multiple access (TDMA) and code-division multiple access (CDMA) having speed of data up to 64 kbps, providing not only superior services of voice than the previous generation, but also services of short message service (SMS) [5]. In the mid-period of 2000s, the development of 3G network technology begun to be executed, viz. high-speed downlink packet access (HSDPA). The primary technological dissimilarity that differs from the 3G network from the 2G network is the usage of packet switching instead of circuit switching for transfer of information.

2.2 *4G*

Consequentially, the industries began viewing to data reformed technologies of the 4G (4th-generation), with the parole of advancements of speed up to tenfold over current technologies of 3rd-generation. Generally, it is the expansion of 3G network with increase in bandwidth and services proposals in the existing 3G network. More better attributes of audio or video streaming over end-to-end Internet protocol (IP) is the primary expectance for the technology of 4G network. In the late of 2000s, an all IP relied on network, that is, 4G was established which has the capability of giving data rates of the high speed of 1G bits/s downlink and 500 M bits/s uplink. The Wireless Interoperability for Microwave Access (WiMax) standard and the Long Term Evolution-Advanced (LTE-A) standard were the first two technologies of 4G,

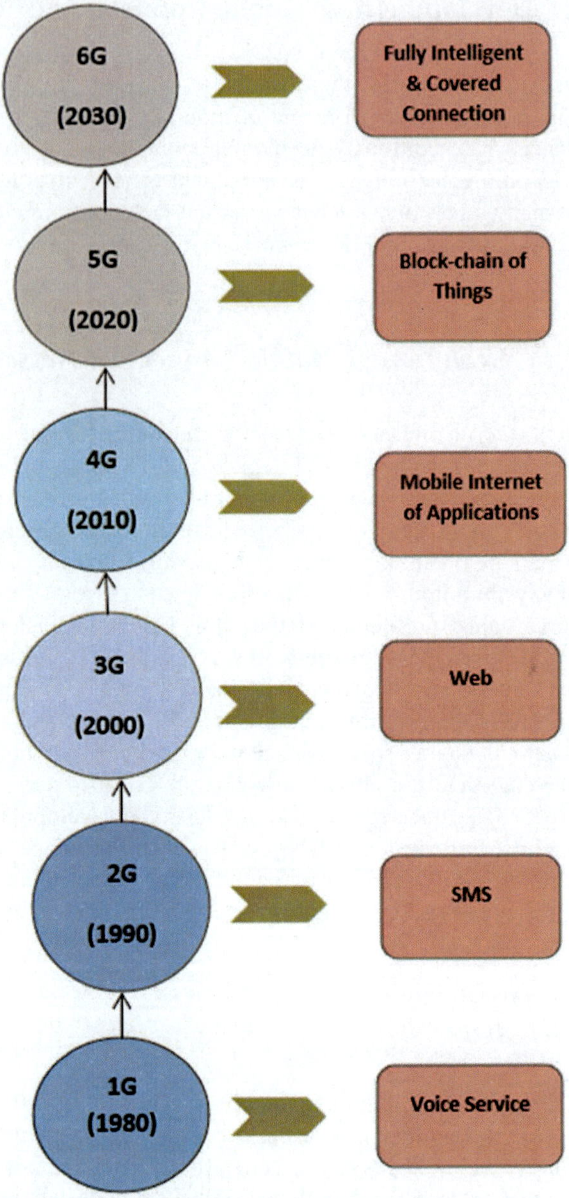

Fig. 1 Development of mobile connection networks

that were industrially accessible and first proposed by Telia Sonera in Scandinavia. The primary reason for which network of 4G is distinct technically from 3G network was the removal of circuit switching [6]. Table 1 shows comparison among 4 mobile phone generations.

Table 1 Comparison among 4 mobile phone generations

Parameters	1G	2G	3G	4G
Standards	TACS, AMPS	EDGE, GSM, CDMA, GPRS	HSPDA, UTMS, EVDO, CDMA2000	LTE advanced, IEEE 802.16
Technology	Analog	Digital	Digital	Digital
Speed of data	<2.4 kbps	<64 kbps	2–100 mbps	<2 gbps
Voice services	Circuit	Circuit	Circuit	Packet
SMS	No	Yes	Yes	Yes
Data services	Circuit	Circuit	Packet	Packet

2.3 5G

Frequently, the world's connectivity requirements are substituting. Before the termination of 2024, it is expected to amplify the traffic of global mobile data by 5. That is where a new G comes into the game. With industrial networks of 5G being switched on, the first usages are the development of mobile broadband, which results in better experiences for the users of smartphones, and availability of fixed wireless network by providing fiber speeds without the use of fiber to homes. Basically, 5G is a technology to digitalize the industries and a pathway to serve consumers. The motive of 5G is to make huge advancements in delay, speed of rates of data, capacity of energy, huge connectivity, and credibility of network [7]. 5G is a game-changer among the game of 4 mobile phone generations; it will enhance the network connection and consumers won't have to face disturbances when transferring videos from mobbed areas. Filter bank multicarrier (FBMC) and beam-division multiple access (BDMA) are developed available technologies applied by 5G. To enhance the implementation of network, several rising technologies are unified into 5G [8] that are as follow:

- Information Centralized Networking (ICN) for the decrease in slicing and traffic of network for fast expansion of many services.
- Software Clarified (defined) Networking (SDN) for pliability in the network.
- Massive MIMO for increase in efficiency [9].
- Device-to-device (D2D) for spectral capacity.

The technology of the 5G is enabling a new gateway of innovation. It has the capability for changing the world, moreover nerving the hottest trends in technical world now: augmented reality (AR), Internet of things (IoT), artificial intelligence (AI), and many others.

2.4 Motive of Green 6G

As the network of 5th-Generation is getting in the rising stage of industries, throughout the countries, research organizations have started to focus on 6G network, which is taken into account to be unfurled in nearly 2030. It is expectance from green 6G to advance the implementation of the transfer of data with 1 Tbps peak data speed and very low delay in time period micro-seconds [10]. In comparison, of network of 5G, it comprises of communication of terahertz frequency and spatial multiplying, giving 1000 times huge efficiency. By unifying networks of communication as well as underwater communication, to give oecumenical distribution [11] is one of the main motives of 6G to instate universal connectivity. There are three new services classes of 6G network that are illustrated in [12]:

- ultrahigh Data Density (uHDD),
- universal Mobile Ultra-Broadband (uMUB) and
- ultrahigh speed with low delay communications (uHSLLC).

Therefore, it is rightly said by Roberto Saracco, EIT Digital that, "The 6G network is likely to be a concept, a virtual one, and not a 'real' network you can put a boundary around."

3 Green 6G Network: Architecture

A new G into the game of mobile generations for the smart society of information of 2030, i.e., 6th-generation, it is expected from the network of 6G to give implementation preferential than the network of 5G and indulge rising applications and services. Hence, it is essential to research on forthcoming frameworks of the network. To delineate the structure of Network 2030 (6G), FG NET-2030 has installed Sub-Group 3. Still, it is impossible to exactly describe what will be the forthcoming structure of the network. The structure of the network from distinct dimensions instead of describing an integrated base is interpreted by Sub-Group 3. In this part, according to Fig. 2, the structural changes related with 6G from three different dimensions are introduced.

3.1 From Terrestrial to Universal 3D Distribution

One of the aims of the forthcoming inheritor's structure of the network is to extend the broadness and deepness of communication distribution. The structure of existing network relied on legacy terrestrial cellular structure has the below shown disadvantages:

- It has incapability to fulfill the deep-sea and high-altitude communication cases which leads to the neglection of needs for coming period services.

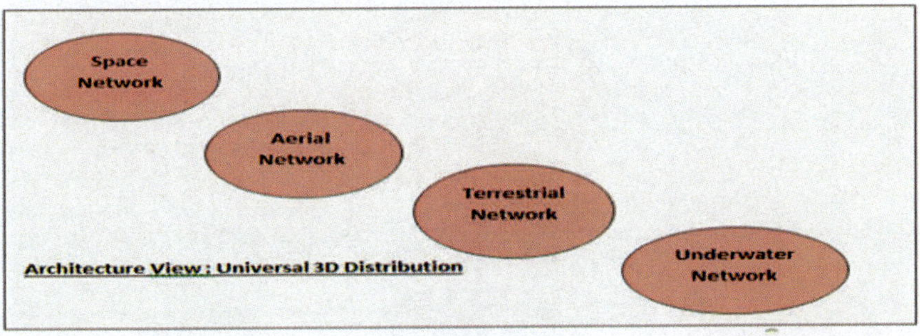

a. Architecture View

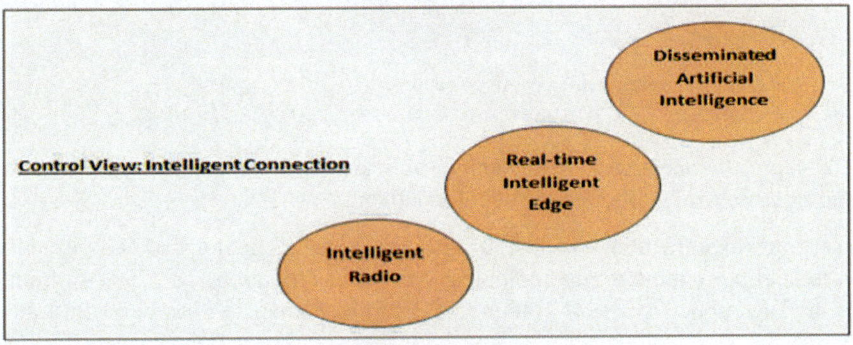

b. Control View

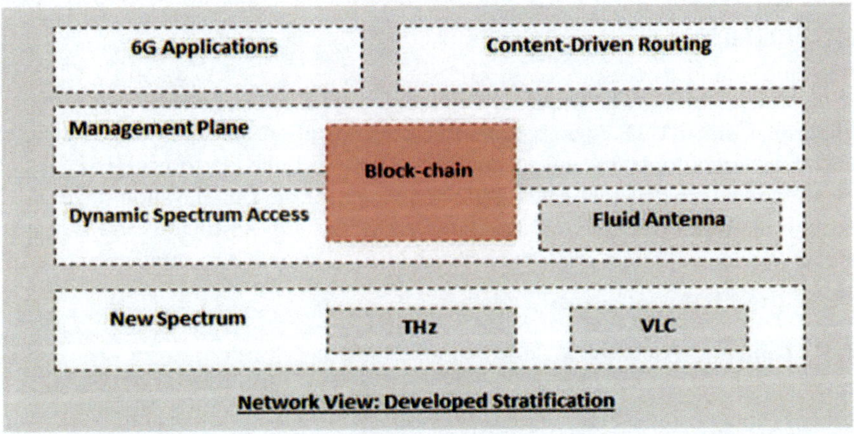

c. Network View

Fig. 2 Structure of distinct dimensions of green 6G network

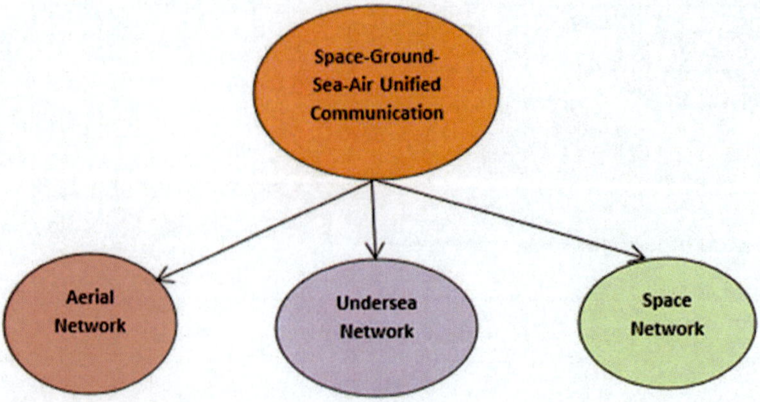

Fig. 3 Space-ground-sea-air unified communication

- To supply connectivity in the universal scale, it has repressively sumptuous provisioning value of concrete cellular networks.

So, to resolve the above-shown disadvantages, the technology of 6G will unify non-terrestrial systems to give complete wireless distribution [13]. In [14], initiatory depicted about Air-Sea-Ground-Space unified connection has been illustrated in Fig. 3.

3.1.1 Aerial Network

The aerial networks can be widely sorted into two classes: first one is low or inferior altitude platforms (LAP) which generally executes at an elevation of less than or equal to several kilometers and second one is high-altitude platforms (HAP) which typically execute in the stratosphere. In comparison, of inferior elevation stages, high elevation platforms are able to cover large areas and long-lasting durability, but its benefits overlap with network low Earth orbit (LEO) satellite to some extent.

3.1.2 Undersea Network

Acoustic, optical, and RF communication are involved in undersea communication. The juxtaposition among the mentioned three communications is shown in Table 2 [15]. Sometimes, incalculable and complicated surrounding of underwater leads to drastic attenuation of signals, complicate deployment of network and damage of component physically [16], renouncing many problems to be resolved.

Table 2 Comparison of distinct undersea communication technologies

Communication	Power consumption	Delay	Attenuation	Data rates	Distance of transmission
Acoustic	High	High	Lowest	~Kbps	<100 km
Optical	Low	Low	Turbidity	~Gbps	<100 m
RF	Moderate	Moderate	High	~Mbps	<10 m

3.1.3 Space Network

The networks of high-throughput satellite (HTS) are able to Internet available performance equivalent to terrestrial performances in reference of bandwidth as well as price. A capacity structure of the space-terrestrial integrated/unified network (STIN) is illustrated in [17], consists of the Terrestrial Networks (TN), space-based backbone network (SBN) of Geostationary satellites and, also their inter-satellite links (ISLs) linking them and space base access networks (SAN) of low Earth orbit (LEO) and medium Earth orbit (MEO) satellites.

3.2 Toward Intelligent Network

In current years, a lot of consideration has been considered by artificial intelligence (AI) but more particularly machine learning (ML) from industries and organizations. Also, initial intelligence has been employed to several facets of mobile network of fifth generation [18]. To further develop the "intelligence" level of the unit of investigation, artificial intelligence and machine learning algorithms are introduced to provide the ability of self-learning in the network. Moreover, the integration of technologies such as edge computing and artificial intelligence endorses to decrease the prices and enhance the quality of experience (QoE) [19]. For the performance of various consumptions, like health maintenance [20], fresh probabilities are provided by edge learning. Still, in the networks of 5G, the usage of artificial intelligence is finite to the consumption of the structure of the conventional network [21]. The conformation of 6th-generation structure should take into account probabilities of AI in system pervasively and adhere an AI-based viewpoint where intelligence definitely will be an interior feature structure of 6G, to achieve the motive of intelligent networks [22]. According to Fig. 4, the absolute expectance form intelligent networks are the autonomous development of networks. For intelligent network, here are two key enablers which are illustrated below.

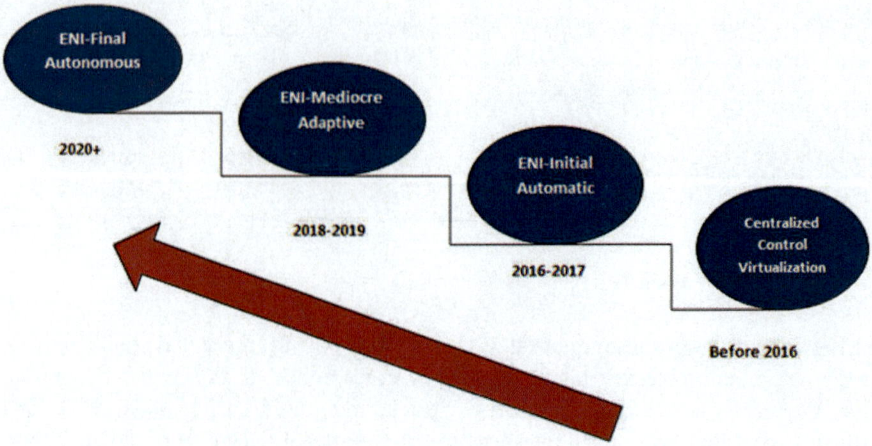

Fig. 4 Autonomous development of networks

3.2.1 Intelligent Radio (IR)

Radios that have the capability of self-learning are considered as intelligent radios which are cognitive radios as shown in Fig. 5. These self-learning radios give the end-user with the probability of elongating execution, even in addition executing self-adaptation to fulfill cognitive aims and by making decisions according to experience. Hardware and transceiver algorithms are separated by a wider as well as

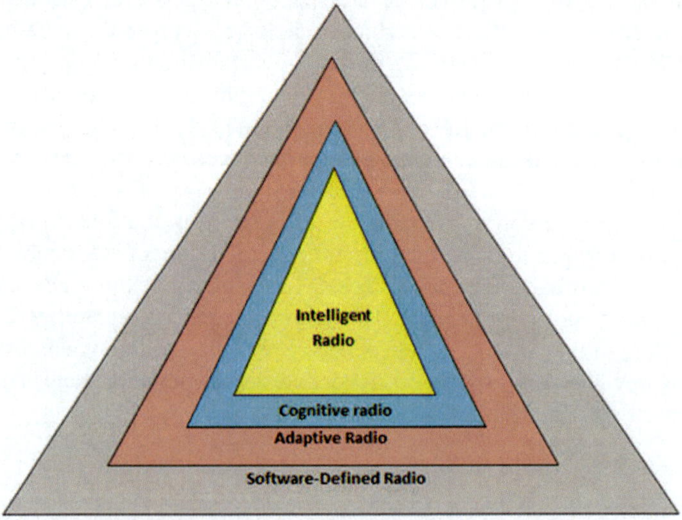

Fig. 5 Relationship of IR with CR, AR, and SDR

rooted concept that is, intelligent radio (IR) in comparison with deployed physical layer (PHY) with initial intelligence. By means of the viewpoint of a physical layer, IR has the ability to access the feasible spectrum, maintain transfer power, and control TP with the aid of artificial intelligence [23]. The active adaptation to multitudinous and upgradable hardware is allowed by a modern reformation illustration by disconnecting transceiver from hardware.

3.2.2 Real-Time Intelligent Edge

Networks of forthcoming inheritor will need the endorsement of communicative artificial intelligence-enabled performances, and some other performances such as automobiles are susceptive to react delay, which requires to communicate smartly with their surroundings in actual time. As it is observed that, concentrated cloud artificial intelligence with static information is unable of fulfilling such services as well as there is an importunate requirement of real-time intelligent edge where intelligent presumption, prognostication, and conclusion are formed on live data. Many primary organizations as well as industries have started to evolve technologies and software parts that fulfill the actual time needs in associative investigation laboratories like the Berkeley RISELab [24]. RTIE has an other working feature of high-implementation hardware.

3.3 Modern Structure of Network Protocol Stack

The world's most widely used non-possessory protocol suite is TCP/IP because it makes computer systems able to use different software and hardware platforms, on distinct types of network to interact [25].

As illustrated in Fig. 6, the layers of TCP/IP network are:

- Application Layer
- Transport Layer
- Internet Layer
- Data Link Layer
- Physical Network Layer.

Basically, TCP/IP is a current architecture of the Internet protocol stack [26], which was generally reformed for transmitting of data and has fulfilled great success over the last forty years. However, many remarkably issues are confronted by recent Internet networks, but these networks cannot give warranty to advance application distribution shortages for example, delay. But on the other side, layer packets of recent network adhere a particular *"Header + Payload"* resource, which is separated from the needs of applications included in it.

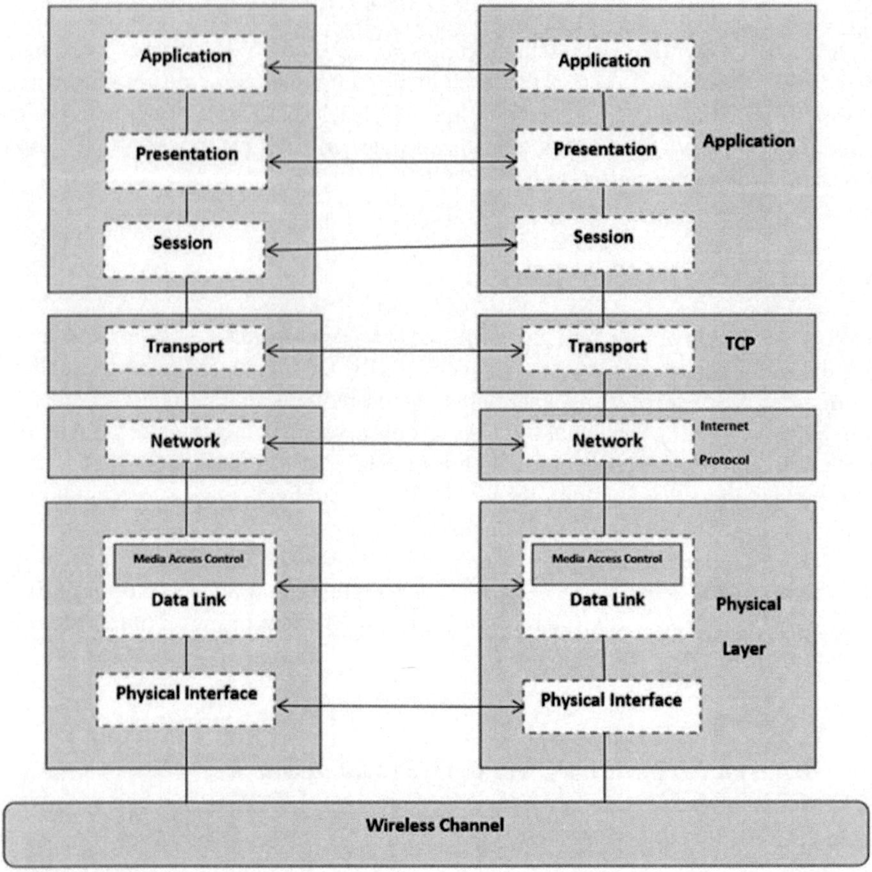

Fig. 6 Structure of network protocol stack

4 Propitious Technologies of Green 6G

Ultimately, we find various plighting technologies for the surrounding of 6G which are introduced below one by one.

4.1 New Communication Paradigm

The technology of new communication paradigm includes, basically, two techniques that are:

- Quantum Communication
- Molecular Communication.

4.1.1 Quantum Communication

One of the plighting exemplifications of communication with unlimited safety is quantum communication (QC). The basic distinction between recognized binary relied on communication and quantum communication are whether intruding can be identified on-location or not [27]. Because of the superposition feature of qubits, quantum communication can enhance rates of information. After the 10 years of investigation, there are many offshoots of quantum communication such as [28];

- Quantum Teleportation
- Quantum Safe Direct Communication
- Quantum Key Dissemination
- Quantum Secret Sharing.

In [29], the abilities and characteristics of channels of quantum communication are described. Its efficiency at large extent in long-distance communication is one of the fascinating characteristics. It is observed in a current experiment that intercontinental quantum communication with a maximum distance of around 7600 km depending on low Earth orbit satellites.

4.1.2 Molecular Communication

The capability of engineered biological nanomachines to interact with biological systems at the molecular stage is expected to empower future applications like regeneration of biological organs and tissues and many other functions. Different from the recent paradigm of network communication, molecules are used by the molecular communication, where molecules act as the carriers of info; after that, information on molecules is encoded by the sender biological nanomachines which then free the molecules in the atmosphere [30]. Further, those molecules spread in the atmosphere to receiver biological nanomachines, and then, to decode the info, the molecules are reacted biochemically with the receiver biological nanomachines. The list of the fundamental processes involved in each stage of molecular communication is listed below:

- Encoding
- Sending
- Spreading/Propagation
- Receiving
- Decoding.

4.2 Fundamental Techniques

The fundamental techniques are explained as follows:

4.2.1 Blockchain for Decentralized Protection

After the Internet, the technology of blockchain is taken into account to be the next huge trendy and attractive technology as it is renovating the way human works and lives. Blockchain is a disseminated ledger-based database where the transaction is safely updated and recorded without any interference by third party [31]. The salient characteristics of blockchain, including invariableness, legality, transparency, security, and decentralization make it an example for several other applications [32]. Blockchain-based decentralized control procedures empower direct interaction links to be installed between entities of the network which decreases the prices of administration [33]. Rather than concentrated database, the unification of the technology of blockchain in network of spectrum sharing can increment the spectral potential. Moreover, blockchain accelerates the unification of each single network evolved by distinct operator by giving an integrated permission and verification process as well as billing system [34].

4.2.2 Energy Harvesting and Management

A very plighting technology, energy harvesting is the storage of small amounts of ambient energy to empower wireless gadgets. For processing of artificial intelligence and rising dissemination of IoT gadgets [35], the continuous requisitions of computation are determining considerable defiance's to the energy capacity of communication component. Over the last 10 years, there have been several endeavors spent on the concept of energy harvesting and management researches. With an aim of dynamically improving the moderation between energy demand and supply, here is a plighting process of smart energy management [36].

4.3 Spectrum Communication Techniques

The technique of spectrum communication is the base of mobile connections and in the period of 1980s, since the growth of communication network of mobile, due to the everlasting target for data rates we have seen terrific elaboration of sources of spectrum in every fresh descent [37]. To supply Tbps, agglomerated bitrate is one of the major aims of 6G and it is imperative to execute operations at higher frequencies for accessible bandwidth as well as spectrum. The two main spectrums are visible light and terahertz (THz). To fulfill the double objectives high-speed data communication and lightning, visible light communication (VLC) takes complete benefit of LED. For communication networks, Terahertz (THz) communication is a complementary wireless technology, which permits high-speed wireless explication of the optical fibers for beyond 5G.

5 Conclusion

An exponential growth in wireless network, specifically, multimedia information and a fast elaboration of all types of smart gadgets are fixing the platform for coming wireless development toward the technology 6G. 6G networks are plighting a considerable growth in the feasible coming time period. A survey on wireless development toward the green 6G networks is detailed in this paper. Initially, the growth of wireless mobile networks from 1G to 5G is illustrated, which is indicating the growth in the trend of 6G network. Then, the architecture of green 6G is described which has three different dimensions: unification of non-terrestrial and terrestrial networks, AI powered intelligent networks and also developed architecture of network protocol stack. Moving forward, finally, the attention is focused on rising technologies, i.e., new communication techniques like quantum and molecular communication. Also, renaissances in the fundamental techniques and spectrum communication are also described. So, overall, this paper provides some guidelines for coming time period research works in the field of green 6G wireless network communications.

References

1. Soldani D, Manzalini A (2015) Horizon 2020 and beyond: on the 5G operating system for a true digital society. IEEE Veh Technol Mag 10(1):32–42
2. Yang P, Xiao Y, Xiao M, Li S (2019) 6G Wireless communications: vision and potential techniques. IEEE Netw 33(4):70–75
3. Katz M, Matinmikko-Blue M, Latva-Aho M (2018) 6G enesis flagship program: building the bridges towards 6G-enabled wireless smart society and ecosystem. In: Proceedings of 10th IEEE Latin-American conference on communications (LATINCOM), Nov 2018, pp 1–9
4. Gupta A, Jha ERK (2015) A survey of 5G network: architecture and emerging technologies. IEEE Access 3:1206–1232
5. David K, Berndt H (2018) 6G vision and requirements: Is there any need for beyond 5 G? IEEE Veh Technol Mag 13(3):72–80
6. Parikh J, Basu A (2011) LTE advanced: The 4G mobile broadband technology. Int J Comput Appl 13(5):17–21
7. Shafi M, Molisch AF, Smith PJ, Haustein T, Zhu P, De Silva P, Tufvesson F, Benjebbour A, Wunder G (2017) 5G: A tutorial overview of standards, trials, challenges, deployment, and practice. IEEE J Sel Areas Commun 35(6):1201–1221
8. Zhou Z, Dong M, Ota K, Wang G, Yang LT (2016) Energy efficient resource allocation for D2D communications underlaying cloud RAN-based LTE-A networks. IEEE Internet Things J 3(3):428–438

9. Wu J, Dong M, Ota K, Li J, Yang W, Wang M (2019) Fog-computing enabled cognitive network function virtualization for an information centric future internet. IEEE Commun Mag 57(7):48–54
10. Wu J, Dong M, Ota K, Li J, Guan Z (2018) Big data analysis-based secure cluster management for optimized control plane in software-defined networks. IEEE Trans Netw Serv Manag 15(1):27–38
11. Yastrebova A, Kirichek R, Koucheryavy Y, Borodin A, Koucheryavy A (2018) Future networks 2030: architecture and requirements. In Proceedings of 10th international conference on ultra modern telecommunications and workshops (ICUMT), Nov 2018, pp 1–8
12. Zong B, Fan C, Wang X, Duan X, Wang B, Wang J (2019) 6G technologies: Key drivers, core requirements, system architectures, and enabling technologies. IEEE Veh Technol Mag 14(3):18–27
13. Zhang Z, Xiao Y, Ma Z, Xiao M, Ding Z, Lei X, Karagiannidis GK, Fan P (2019) 6G wireless networks: vision, requirements, architecture, and key technologies. IEEE Veh Technol Mag 14(3):28–41
14. Zhao Y, Yu G, Xu H (2019) 6G mobile communication network: vision, challenges and key technologies. arXiv:1905.04983. [Online]. Available https://arxiv.org/abs/1905.04983
15. Kaushal H, Kaddoum G (2016) Underwater optical wireless communication. IEEE Access 4:1518–1547
16. Felemban E, Shaikh FK, Qureshi UM, Sheikh AA, Qaisar SB (2015) Underwater sensor network applications: a comprehensive survey. Int J Distrib Sens Netw 2015(11):896832
17. Yao H, Wang L, Wang X, Lu Z, Liu Y (2018) The space-terrestrial integrated network: an overview. IEEE Commun Mag 56(9):178–185
18. Li R, Zhao Z, Xuan Z, Ding G, Yan C, Wang Z, Zhang H (2017) Intelligent 5G: ehen cellular networks meet artificial intelligence. IEEE Wirel Commun 24(5):175–183
19. Sharma N, Kaushik I, Rathi R, Kumar S (2020) Evaluation of accidental death records using hybrid genetic algorithm. SSRN Electron J. doi:10.2139/ssrn.3563084
20. Harjani M, Grover M, Sharma N, Kaushik I (2019) Analysis of various machine learning algorithm for cardiac pulse prediction. In: 2019 international conference on computing, communication, and intelligent systems (ICCCIS). https://doi.org/10.1109/icccis48478.2019.8974519
21. Li G, Xu G, Sangaiah AK, Wu J, Li J EdgeLaaS: edge learning as a service for knowledge-centric connected healthcare. IEEE Netw
22. Jiang C, Zhang H, Ren Y, Han Z, Chen K-C, Hanzo L (2017) Machine learning paradigms for next-generation wireless networks. IEEE Wirel Commun 24(2):98–105
23. Fowers J, Ovtcharov K, Papamichael M, Massengill T, Liu M, Lo D, Alkalay S, Haselman M, Adams L, Ghandi M, Heil S, Patel P, Sapek A, Weisz G, Woods L, Lanka S, Reinhardt SK, Caulfield AM, Chung ES, Burger D (2018) A configurable cloud-scale DNN processor for real-time AI. In: Proceedings of ACM/IEEE international symposium computer architecture, June 2018, pp 1–14
24. RISELAB. Real-time intelligent secure explainable systems. [Online]. Available https://rise.cs.berkeley.edu
25. Hu J-Y, Yu B, Jing M-Y, Xiao L-T, Jia S-T, Qin G-Q, Long G-L (2016) Experimental quantum secure direct communication with single photons. Light Sci Appl 5(9):e16144
26. Kaushik I, Sharma N, Singh N (2019) Intrusion detection and security system for blackhole attack. In: 2019 2nd international conference on signal processing and communication (ICSPC). https://doi.org/10.1109/icspc46172.2019.8976797
27. Zhang W, Ding D-S, Sheng Y-B, Zhou L, Shi B-S, Guo G-C (2017) Quantum secure direct communication with quantum memory. Phys Rev Lett 118(22):220501
28. Gyongyosi L, Imre S, Nguyen HV (2018) 'A survey on quantum channel capacities. IEEE Commun Surv Tuts 20(2):1149–1205
29. Lin X, Li J, Wu J, Liang H, Yang W (2019) Making knowledge tradable in edge-ai enabled IoT: a consortium blockchain-based efficient and incentive approach. IEEE Trans Ind Inform 15(12):6367–6378. https://doi.org/10.1109/tii.2019.2917307

30. Liang H, Wu J, Mumtaz S, Li J, Lin X, Wen M (2019) MBID: micro-blockchain-based geographical dynamic intrusion detection for V2X. IEEE Commun Mag 57(10):77–83
31. Ling X, Wang J, Bouchoucha T, Levy BC, Ding Z (2019) Blockchain radio access network (B-RAN): towards decentralized secure radio access paradigm. IEEE Access 7:9714–9723
32. Varshney T, Sharma N, Kaushik I, Bhushan B (2019) Authentication and encryption based security services in blockchain technology. In: 2019 international conference on computing, communication, and intelligent systems (ICCCIS). https://doi.org/10.1109/icccis48478.2019.8974500
33. Jaidka H, Sharma N, Singh R (2020) Evolution of IoT to IIoT: applications and challenges. SSRN Electron J. doi:10.2139/ssrn.3603739
34. Jadon S, Choudhary A, Saini H, Dua U, Sharma N, Kaushik I (2020) Comfy smart home using IoT. SSRN Electron J. https://doi.org/10.2139/ssrn.3565908
35. Varshney T, Sharma N, Kaushik I, Bhushan B (2019). Architectural model of security threats and their countermeasures in IoT. In: 2019 international conference on computing, communication, and intelligent systems (ICCCIS). https://doi.org/10.1109/icccis48478.2019.8974544
36. Zhou Z, Gong J, He Y, Zhang Y (2017) Software defined machine-tomachine communication for smart energy management. IEEE Commun Mag 55(10):52–60
37. Strinati EC, Barbarossa S, Gonzalez-Jimenez JL, Ktenas D, Cassiau N, Maret L, Dehos C (2019) 6G: the next frontier: from holographic messaging to artificial intelligence using subterahertz and visible lightcommunication. IEEE Veh Technol Mag 14(3):42–50

Onion Price Prediction for the Market of Kayamkulam

Anubha, Kaustubh Tripathi, Kshitiz Kumar, and Gopesh Khandelwal

Abstract Time series data have been a routine part of our lives, be it weather forecast or stock prices prediction of companies. ARIMA models have proved themselves to be in competent class of models to measure the occurrence of random variations present in any economic (price-based) time series. The same model, thus developed for the use in stock prices can be used for prediction of rates of daily agricultural commodities which vary in pattern, seasonally. Therefore, the prime purpose of this research is to apply the popular ARIMA model on annual prices of Onion for the Market of Kayamkulam in Kerala.

Keyword ARIMA

1 Introduction

Forecasting refers to a practice in which the future trends of a dynamical event are estimated based on the understanding and characterisation of the past system. Considering the instability, the dynamic system, which we are scrutinising, the initial circumstances become one of the most significant constraints of the time series response. This situation is called as sensitive dependence on initial conditions and is related to chaotic behaviour for the dynamical system. The fluctuations in prices are a matter of apprehension among patrons, farmers and policymakers, and hence, forecasting is

Anubha · K. Tripathi (✉) · K. Kumar · G. Khandelwal
Maharaja Agrasen Institute of Technology, New Delhi, India
e-mail: ktkaustubh4@gmail.com
URL: https://mait.ac.in

Anubha
e-mail: anubha@mait.ac.in

K. Kumar
e-mail: kkrajput09@gmail.com

G. Khandelwal
e-mail: gopeshkk@gmail.com

A. Khanna et al. (eds.), *Data Analytics and Management*, Lecture Notes on Data Engineering and Communications Technologies 54,
https://doi.org/10.1007/978-981-15-8335-3_8

extremely important for efficient planning and monitoring [1]. Numerous attempts have been made in past and in recent times, to come up with a price forecasting model for various agricultural commodities [2]. The fluctuation in onion prices over the years can be seen due to the variations in production and market arrival time at places. Thus, the practical importance of modelling and forecasting the periodic behaviour of prices surge [3]. The unpredictability of the prices of such agricultural commodity makes it more challenging to predict future trends. The genesis of this practice has its roots embedded in the dependencies of the stakeholders. Crop failure in the mainland of India is not an unfamiliar phenomenon, as it is frequent in nature. The failure of the crop at one place in India affects the prices of the same commodity in the neighbouring region as well. Hence, the volatility of crop prices is extreme in India, especially of principal crops which includes onion. India is one of the prime producers of onions in the world after China. The production is often outnumbered by the consumption of onion in India. India has witnessed an onion crisis in 2010 [4].

2 Related Work

There has been a hefty sum of publications on forecasting and prediction of vegetable price. This unit presents a brief appraisal of the related and recent studies [5], Alionue Dieng investigated the execution of parametric models for predicting vegetable prices and to make a sanction to the probable stakeholder. The author of the research implemented two forecasting methodologies. The prediction methods implemented entails three alternative parameter models and a nonparametric model. Parametric models consist of the exponential-smoothing-models, naïve-based model and Box–Jenkins interacted moving average model (ARIMA) [6]. The author accumulated monthly mean rates of onion, tomato and potato for duration between 1980 and 2003. Both the parametric model and the non-parametric model were implemented to produce a prediction of onion, tomato and potato rates. According to the outcomes, the parametric models were suggested for estimating vegetable prices.

Among this, ARIMA methodology has the highest recognition. Amegbeto [7] elicited research that scrutinised the undercurrents of vegetable rates and the amount provided to the chief vegetables and fruits market hub in Kabul, Afghanistan. Prediction models were created to facilitate marketing strategies of vegetables using data gathered from Aug 2004 to Dec 2005. The outcomes show that rates and arrival of some commodities were unpredictable and inversely related in many situations, and vegetables supplied to Kabul market during the time of November to April, which usually relate to low arrival of commodities and surge rates. Manish Shukla and Sanjay Jharkharia [8] examined the capability of this method in the wholesale market of vegetables and fruits; models were constructed for selling data of a specific vegetable for Ahmadabad wholesales vegetable market in India.

The aim of this study is the application of these models on wholesale vegetable data and predicting the upcoming requests with precision. In [9] offered a forecasting

Table 1 Dataset information

S. No.	Data columns	Entries
1	Index	209 non-null int64
2	State	209 non-null object
3	District	209 non-null object
4	Market	209 non-null object
5	Variety	209 non-null object
6	Arrival	209 non-null object
7	Min_price	209 non-null int64
8	Max_price	209 non-null int64
9	Modal_price	209 non-null float64

model of vegetable rates, by applying the neural network-based genetic algorithms. Taking the quantity of mushrooms as an example [10] presents a combined technique to predict monthly vegetable rates in the market. In this research paper, four distinct models were built for the forecasting. The authors accumulated Lentinus edodes rates between the time frame from 2003 to 2009 for Beijing Xinfudi vegetable market. The outcomes displayed that the integrated prediction model was a better position than all the others. This effort aims to develop an ARIMA model to predict the rates of onion in Kayamkulam market, further to study forecast results (Table 1).

3 Materials and Methods

3.1 Dataset

The National Informatics Centre (NIC), Ministry of Electronics and Information Technology, Government of India, maintains a website called data.gov.in, which is also developed and hosted by them. It makes publicly available the daily data on mandi prices and arrival volumes of many commodities, including onion and potato, from across various mandis and markets in the country. The data is available for research purposes. The data provided by the abovementioned website is monitored and reliable. The data available has 2 types of granularity: one is weekly, and the other one is daily, from which daily granularity is chosen. The data is furnished in the form of comma-separated values, which can be easily interpreted by python3. This data contained many missing values, though, and therefore, our analysis is restricted to only a few retail centres and mandis for which we had enough data. The selection was made to ensure that there was no missing data for more than 60 consecutive days, and at least 65% of data for all the days were available (Fig. 1).

The dataset has various fields such as the name of the state, the name of the district, the name of the market, variety of the onion sold, arrival date, min price, max price and modal price. The data in hand consider two types of Onions, which are ‘big’ and

	state	district	market	commodity	variety	arrival_date	min_price	max_price	modal_price
0	Andaman and Nicobar	Nicobar	Car Nicobar	Onion	Other	07/06/2019	6000	10000	8000.0
1	Andaman and Nicobar	Nicobar	Car Nicobar	Onion	Other	11/06/2019	6000	10000	8000.0
2	Andaman and Nicobar	Nicobar	Car Nicobar	Onion	Other	12/06/2019	6000	10000	8000.0
3	Andaman and Nicobar	Nicobar	Car Nicobar	Onion	Other	13/06/2019	6000	10000	8000.0
4	Andaman and Nicobar	Nicobar	Car Nicobar	Onion	Other	14/06/2019	4000	8000	6000.0

Fig. 1 Head of the dataset in Jupyter Notebook

'small.' The arrival date shows when was the lot arrived. The min price indicates the lowest price for which the lot has been sold for that particular day and similarly, the max rate indicates the maximum amount for which the lot has been sold that specific day. The modal price basically shows the most frequent price at which a lot has been sold. All the prices incorporated, i.e., min price, max price and the modal price, are all for prices per Kg. The states involved may not be the states where the production is happening, so prices of one state's onion may be dependent on the costs of some other state/s. A similar notion is observed for the districts. The prices are given from the first week of January till August.

The data includes all the major states and their major markets, but since the task is to predict prices for a particular district and one specific variety of the commodity; the data was refined to get the modal prices for 'big' variety for the Kayamkulam of Kerala. The final data has 209 instances/rows of data, each row having the onion's modal prices.

3.2 Methodology

Autoregressive integrated moving average, commonly known as ARIMA, is a generalisation of an autoregressive moving average (ARMA) [11]. Both of these models are used in time series either to understand the data or to forecast the future values. Also, the ARIMA model is used to eliminate the non-stationarity of data, which can be removed by initial differencing as an integrated part of the model, further repeating it multiple times to reduce it. The prevalent domain where ARIMA is most commonly used is to give a prediction of an existing time series based on its past values, that is, its lags per its lagged forecast errors.

ARIMA is used to model the time series [12] that are non-seasonal, exhibiting a pattern, and also, it should not be any random white noise. ARIMA in its generalised form is denoted as ARIMA (p, d, q) where the parameters p, d and q are the non-negative integers. Here, p denotes the order or the number of time lags of the AR (autoregressive) model, d indicates the degree of differencing or the number of times the past values have been subtracted in the data. Finally, q means the order of the moving average (MA) model. Further, moving forward to the terminologies that are autoregression (AR) and moving average (MA), discussing their significance in an ARIMA model.

ARIMA is specially made up of two parts, namely the autoregression (AR) and the moving average (MA). A pure AR model represents an evolving variable that is regressed in its lags or the prior values. Going into the mathematical representation of this model, consider a variable Y_t and below is an equation which shows the dependence of Y_t on its lags as follows:

$$Y_t = \alpha + \beta_1 Y_{t-1} + \beta_2 Y_{t-2} + \ldots + \beta_p Y_{t-p} + \epsilon_1$$

where Y_{t-1} denotes the lag_1 of the series, and accordingly, Y_{t-2} represents the lag_2 and so on. The model and alpha predict the Beta terms that can be seen as the coefficients of the lags represents the intercept term which is also predicted by the model.

Now coming to the second part of ARIMA that is moving average (MA) is used to remove the non-determinism from a time series. It forms a dependency between observation and residual error after a moving average model is applied to a lagged observation.

Mathematically, Y_t here is a variable that depends on the lags of its forecast errors:

$$Y_t = \alpha + \epsilon_1 + \phi_1 \epsilon_{t-1} + \phi_2 \epsilon_{t-2} + \ldots + \phi_p \epsilon_{t-p}$$

Generally, a time series is non-stationary, and stationarity is one of the conditions to be fulfilled while applying ARIMA models. Therefore, the differencing of time series is done at least once to make it stationary.

After combining both the AR and MA terms, the equation becomes:

$$Y_t = \beta_1 Y_{t-1} + \beta_2 Y_{t-2} + \ldots + \beta_0 Y_0 + \epsilon_1$$
$$Y_{t-1} = \beta_1 Y_{t-2} + \beta_2 Y_{t-3} + \ldots + \beta_0 Y_0 + \epsilon_{t-1}$$

At last, *I* (for Integrated) here represents that the data values have been replaced by the difference between their values and the previous values.

Further easing the understanding of the ARIMA model, it can be expressed in words as:

Predicted Y_t = Constant + Linear combination Lags of Y (up to p lags) + Linear Combination of Lagged forecast errors (up to q lags) (Fig. 2).

Autoregressive integrated moving average (ARIMA) has found its usage in a variety of domains, and one of such domains is to study market variations, that too of agricultural goods. ARIMA models have proved themselves to be the best class of models to measure the occurrence of random variations present in any economic (price-based) time series. As a result of this capability shown by the ARIMA models, they were subjected to the differencing and eventually the prediction of the future prices of the onion by analysing the economic time series particular to some selected markets. Stationarity of time series is one of the basic requirements before applying

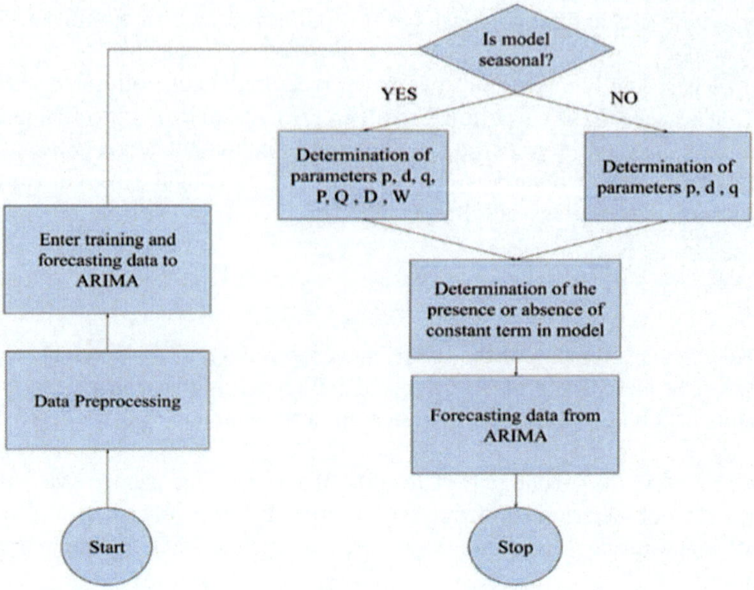

Fig. 2 Flow diagram of the methodology practice

the time series; therefore, a suitable order of differencing is applied to it to make it into a stable set. The absence of trend component in the series is confirmed by the computation of the autocorrelation and partial autocorrelation coefficient of the time series.

To predict y (y can be any variable or an economic time series), ARIMA models are the best to predict the future values based on the prior one. The real-world events cited as (φ_t), to predict future values of y. Additional parameters such as ω_t^k denoting the weight of each event $k \in [K]$ at time t and also the time lag δ are also taken into the record. The new equation is as follows:

$$y_t = \epsilon_t + \alpha_1 y_{t-1} + \ldots + \alpha_p y_{t-p} - \beta_1 e_{t-1} \ldots - \beta_q e_{t-q} \\ + \sum_{k=1}^{K} \omega_t^k \phi_{tk} + \sum_{k=1}^{K} \omega_{t-1}^k \phi_{(t-1)k} + \ldots + \sum_{k=1}^{K} \omega_{t-\delta}^k \phi_{(t-\delta)k}$$

where e_t and t are the moving average components and the error of the ARIMA model, respectively.

Trend, seasonality, cyclic and error made up the components of any time series. Autoregressive integrated moving average (ARIMA) stands out as the most common model for the time series forecasting. For a regression or the linear equation, the time series stands out to be stationary, and in the equation, the lags and forecast errors are common. The projected value in a non-seasonal ARIMA model is computed by:

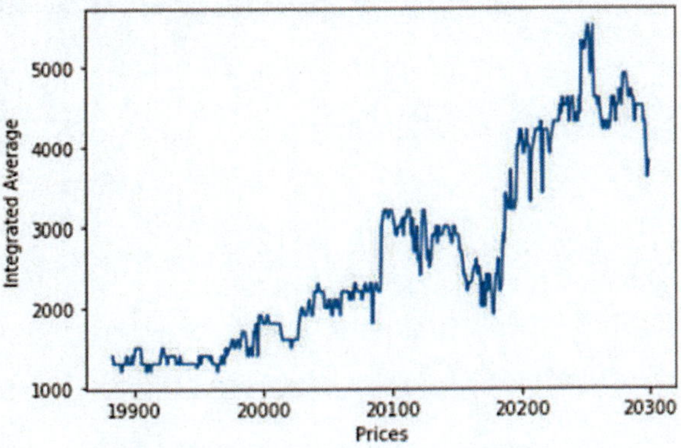

Fig. 3 Integrated average variation with prices

$$y'_t = \epsilon + \alpha_1 y'_{t-1} + \ldots + \alpha_p y'_{t-p} - \beta_1 e_{t-1} - \ldots - \beta_q e_{t-q}$$

here the differenced time series is denoted by Y_t with differencing d of the first degree, p denotes the order or the number of time lags of the autoregressive (AR) model (the parameters are α), and finally, q means the order of the moving average (MA) model (the parameters are β).

4 Results

The result drawn by applying the Arima model on the dataset was by far close to the actual forecast values. The rapid changes in the prices were following no noticeable trend. Hence, the use of ARIMA model showcased the hidden moving average pattern, which is the trademark of this algorithm (Fig. 3), sharply displays the variation of the integrated average with the sum of prices of Onion.

The graph (Fig. 4) shows the close relationship between the predicted values of the costs over the course of and 120 days. As 63% of the day was taken to train the model, and the rest was employed to compare the results. The measurability of the values of the difference between the predictions and the actual amount was taken using mean squared error. The MSE value for the model was equal to 64,394.03. The visual outlook of the graph also confirms that the model has predicted values near about to the actual ones.

5 Conclusion

As the same algorithm has sought its profound use in varied industries, the same can be applied for field of agriculture. The predictability of prices of an agricultural commodity will put the stakeholders in a better position to estimate the productivity of

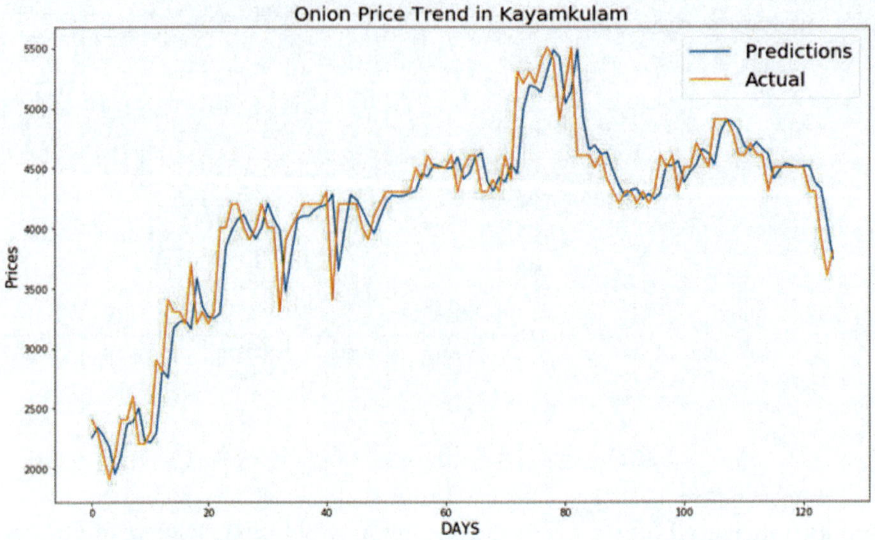

Fig. 4 Prediction results

a crop. This methodology has not yet reached to the agriculture domain because of the schism, between information technology and agriculture. Also, as the model is highly dependent on the accuracy and reliability of the data, the data fed to the algorithm must be reliable. Hence, if authenticity and abundance of data is the situation for an agricultural commodity, the ARIMA model can display pleasing and substantial results.

6 Discussion

The data that we have is a time series data, and only the past values of the entity to predict the new ones. There is no meta-data that runs along our data that can be used to learn some mathematical functional model that predicts the future values on the basis of learnt parameters. ARMA is one of the most traditional and widely known statistical models which is used for the purpose of predicting future values on the basis of past values, but it has its own shortcomings, that is mainly the stationarity of the dataset. It fails to work for non-stationary data, and that is where the recent model, ARIMA comes into the picture, that has a ‘integrate’ component, that makes the data stationary (to a great extent) by differencing the series with its delayed copy and so on, a number of times.

Now while this ARIMA outperforms ARMA in all aspects, we still cannot deny the presence of deep learning models such as LSTM [13], that have specialised memory units that are defined specifically to handle the time-variant nature of time series data. But in order to train any deep learning model, a requirement of a lot of data is presumed. LSTM for a fact requires relatively a lot of data, and considering the niche data that has to be forecast, ARIMA outshines as it learns the statistical features of the data and hence can do well in a small amount of data.

References

1. Qamar W, Younas M, Waseem M (2019) Price fluctuations of rice crop in district Sheikhupura. J Agric Stud Macrothink Inst VII(3):227–239
2. Bhattacharya D, Dhakre DS (2016) Price behaviour of potato in Agra Market-a statistical analysis. Indian Res J Extension Edu
3. Singh DK, Pynbianglang K, Pandey NK (2017) Market arrival and price behaviour of potato in Agra District of Uttar Pradesh. Economic Affairs, New Delhi
4. Ahmed M, Singla N (2017) Market integration and price transmission in major onion. Economic Affairs 62(3):405–417
5. Dieng A (2008) Alternative forecasting techniques and vegetable prices in Senegal. Revue senegalaise des recherches agricoies et agroalimentaires 1(3)
6. Ansari Saleh A, Suryo G, Abdurakhman (2016) Modeling data containing outliers using ARIMA additive. In: Joint workshop of KO2PI 2017 & ICMSTEA, 2016
7. Amegbeto KN (2006) Vegetable price analysis and marketing windows for protected agriculture technology adopters in Afghanistan. In: Protected agriculture project in Afghanistan
8. Manish Shukla SJ (2011) Applicability of ARIMA models in wholesale vegetable market, an investigation. In: International conference on industrial engineering and operations management, KualaLumour, Malasia
9. Guo Q, Luo C-S, Wei Q-F (2011) Prediction and research on vegetable price based on genetic algorithm and Neural network model. Asia Agric Res 148–150
10. Chapgshou L, Qingfeng W, Liying Z, Jungeng Z, Suien S (2011) Prediction of vegetable price based on neural network and genetic algorithm. IFIP AICT, Springer, pp 672–681
11. Kalpakis K, Gada D, Puttagunta V (2001) Distance measures for effective clustering of ARIMA time-series. In: Proceedings 2001 IEEE international conference on data mining
12. de Santos Júnior DS, de Oliveira JFL, de Mattos Netoa PSG (2018) An intelligent hybridization of ARIMA with machine learning models. Knowl Based Syst 175:72–86
13. Siami-Namini S, Tavakoli N, Siami Namin AR (2018) A comparison of ARIMA and LSTM in forecasting time series. In: 17th IEEE international conference on machine learning and applications, Texas, USA

A Technical Review Report on Deep Learning Approach for Skin Cancer Detection and Segmentation

Keerthana Duggani and Malaya Kumar Nath

Abstract Skin cancer is growth of abnormal cells, mainly caused due to exposure of ultraviolet rays from sun. There are various types of skin cancer, among them melanoma is the most hazardous. Manual detection of skin cancer is a time-consuming task. In order to reduce the time constraint, various computer aided diagnosis are introduced. Among them deep learning is more advantageous because it can be performed on large amount of data. Deep learning is a subset of machine learning, which extracts features from raw input. This paper deals with various deep learning architectures proposed by several researchers for detection and segmentation of skin cancer. The commonly used architectures for detection of skin cancer are convolutional neural network (CNN), *k*-nearest neighbor (*k*-NN), artificial neural network (ANN), deep convolutional neural network (DCNN) and you only look once (YOLO). For skin cancer segmentation DermoNet, U-net, GrabCut, saliency based and fully convolutional network (FCN) are used. For the abovementioned architectures, the authors have used different datasets such as ISIC 2017, ISIC 2018, ISBI 2016, ISBI 2017 and PH2 for detection and segmentation of skin cancer. To access the correctness of segmentation and classification various performance measures such as: accuracy, sensitivity, specificity, Jaccard coefficient, Dice similarity coefficient, Hammoude distance, XOR and area under curve are computed for different architecture. A detailed comparison of various methods based on their performances is discussed in this paper. Among them the method using CNN has attained highest accuracy of 97.49%.

Keywords Deep learning · Melanoma · CNN and Accuracy

K. Duggani (✉) · M. K. Nath
Department of ECE, NIT Puducherry, Karaikal, India
e-mail: keerthanaduggani@gmail.com

M. K. Nath
e-mail: malaya.nath@nitpy.ac.in

A. Khanna et al. (eds.), *Data Analytics and Management*, Lecture Notes on Data Engineering and Communications Technologies 54,
https://doi.org/10.1007/978-981-15-8335-3_9

1 Introduction

Abnormal growth of skin cells leads to skin cancer, which is one of the most widespread type of cancers in the world [15]. This growth of cells can lead to two types of tumors, namely benign and malignant [3]. Benign tumors are normal cells that divide and grow but do not disturb the functions of normal cells surrounding them. Malignant tumors are excessive growth of abnormal cells that are cancerous and divide without control [8]. There are three different types of skin cancer, namely basal cell carcinomas, squamous cell carcinomas and melanoma [7]. The most lethal type of skin cancer found in the humans is melanoma, which causes pigmented moles on the skin [2]. Basal cell carcinoma affects basal cells and appears as tiny transparent bump on the skin. Squamous cell carcinoma is slow growing skin cancer and spreads to tissues, bones and lymph nodes. Melanoma can be caused by weak immune system, fair skin, hereditary factors and excessive exposure to ultraviolet (UV) light [2]. Melanoma can be further divided into four types based on its appearance on the body region; they are superficial spreading melanoma, nodular melanoma, lentigo maligna melanoma and acral lentiginous melanoma. Classification of skin cancer is shown in Fig. 1.

Approximately, 91,270 new cases of melanoma are identified in the year 2019. White people are 20 times more likely to be diagnosed with melanoma than black people. Generally, 2.6% of white people suffer from melanoma, while the life time risk of the black people is 0.1% (1 in every thousand). Melanoma is common in people younger than 30 years of age. Skin cancer is diagnosed visually, initiating with a clinical screening followed by dermoscopic analysis, histopathological assessment and a biopsy [2]. Doctors follow ABCDE rule to check weather it is skin cancer or

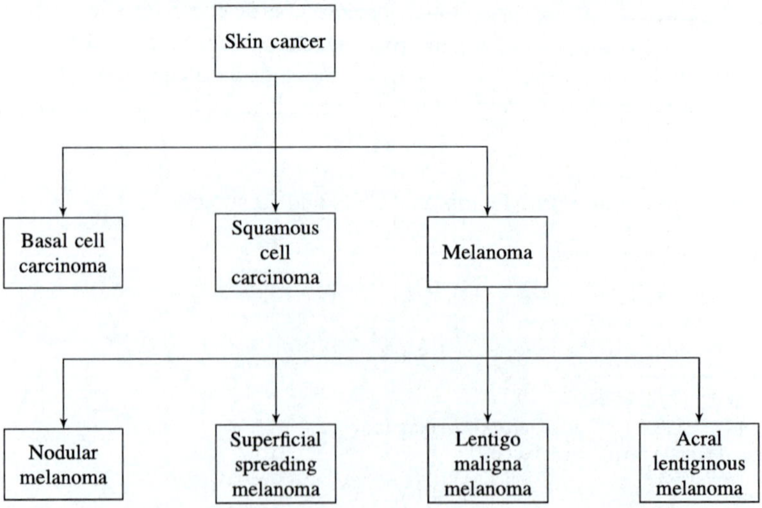

Fig. 1 Different types of skin cancer

not. The ABCDE rule stands for asymmetry, border irregularity, color that may not be uniform, diameter >6 mm and evolving size, shape or color [12].

Early diagnosis of skin cancer is very important for increasing the survival rate [13]. But its identification is challenging by naked eye. The medical images are extracted from computer-aided investigation analysis through various imaging systems such as magnetic resonance imaging (MRI), computer tomography (CT) scan and ultrasound B-scan [14]. Over the years, several different imaging techniques have been proposed for classification [10] and segmentation which can be broadly divided into conventional and deep learning techiques. In this paper, a comparative study of different techniques for skin cancer detection and segmentation has been discussed along with their performance measures. This analysis provides the information about the datasets publicly available for skin cancer studies. It will help the readers to understand the detailed information about the methods for skin cancer detection and segmentation using deep learning and analyze them. The rest part of the paper is organized as follows: Section 2 discuses about the various architecture used for skin cancer detection. Section 3 discuses about datasets and performance measures. Conclusion is discussed in Sect. 4.

2 Discussion

Different methods have been reported for classification and segmentation of skin cancer using deep learning. Brief discussion is given below.

Automatic melanoma detection using multi-scale lesion-biased representation (MLR) and joint reverse classification (JRC) was proposed by Bi et al. [6] which consists of three blocks, namely hair removal, MLR block and JRC block. The block diagram is represented in Fig. 2. The hair pixels are detected with the help of different templates and are replaced with non-hair pixels. MLR block finds multiple histograms along 12 directions and Gaussian blur of four scales. For every direction and scale, scale invariant feature transform (SIFT) features are extracted. Bag of words (BoW) encoding approach is used by the author to represent histograms of visual words, which are obtained by quantizing local descriptors. JRC block uses dual sampling and support vector machine with linear kernel to differentiate between

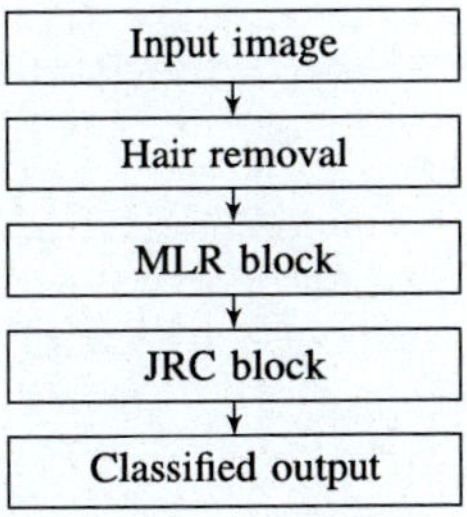

Fig. 2 Automatic melanoma detection using multi-scale lesion-biased representation (MLR) and joint reverse classification (JRC) proposed by Bi et al. [6]

melanoma and normal cells. Performance measures such as accuracy, sensitivity and specificty for PH2 dataset have been reported to be 92%, 87.5% and 93.13%, respectively. The lesion detection can be further improved with optimized feature selection [6].

Begum and Asra have detected melanoma skin cancer using fuzzy C means (FCM) and *k*-NN classifier. The method consists of preprocessing, feature extraction and classification block as shown in Fig. 3. Fuzzy C means clustering is used for preprocessing to identify group of information. The clustered output obtained from the preprocessing block is fed to the feature extraction block. In feature extraction block, fractal feature and color correleogram features are extracted and used for classification by ANN and *k*-NN. *k*-NN is a non-parametric technique used for characterization and relapse [5].

Yuan et al. [16] have come up with a method, which consists of convolution, maxpooling, dropout, deconvolution and upsampling for segmentation of melanoma skin cancer from ISBI 2016 and PH2 dataset and achieved a Jaccard index of 84.7% and Dice index of 93.8%, respectively. The systematic representation of the steps carried out by this method is given in Fig. 4. The whole network consists of 19 layers with two set (290, 129) of trainable parameters. A stride of 1 and ReLU activation function is used for all convolution and deconvolution layers. Sigmoid activation

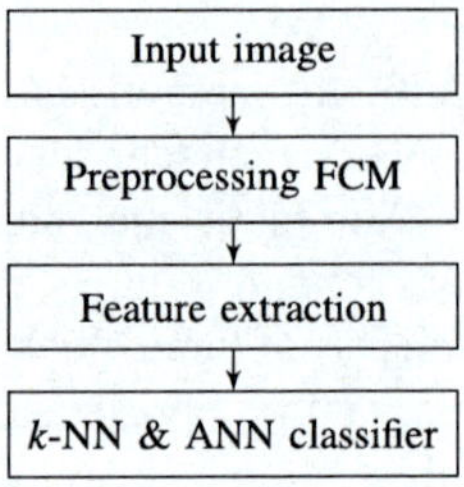

Fig. 3 Architecture for detection of melanoma skin cancer proposed by Begum and Asra [5]

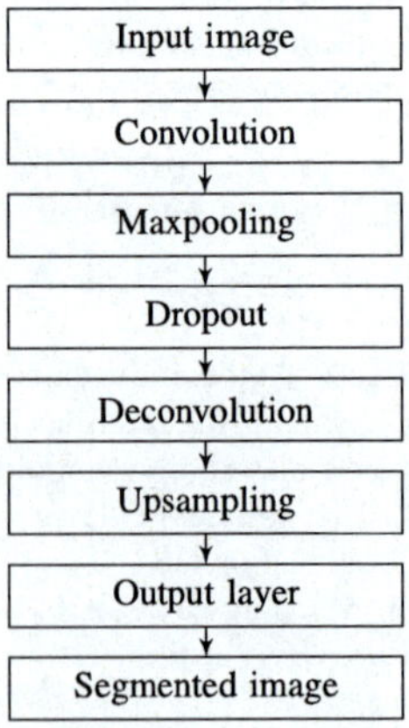

Fig. 4 Systematic representation of segmentation of skin cancer proposed by Yuan et al. [16]

function is used in output layer. Convolution and pooling layers give the contextual information. Image resolution is given by deconvolution and upsampling layers [16].

Ahn et al. [1] developed a saliency-based skin lesion segmentation for automatic detection of melanoma cancer from indistinct lesion border and low contrast images. It consists of a hair removal step in the preprocessing block. This is performed by gray scale morphological closing operation with three structuring elements at different directions. The preprocessed output is partitioned into *n* segments by simple linear iterative clustering method to create multiscale background templates. The background templates are extracted at various scales to identify the structural information represented by each scale. This is followed by sparse reconstruction and error calculation, context-based error propagation, pixel-level sparse reconstruction, saliency map creation via Bayesian framework. Bayesian model is used for the final saliency map, which uses Bayes formula to measure the saliency value of individual pixel by aposterior probability. Aperior probability of the lesion and the background are computed by binarizing an initial saliency map created from pixel-level sparse reconstruction error proposed by Huang et al. [11]. The likelihood probabilities for the lesion is computed from the color histogram. The background probability is computed from the color distribution histogram by using a threshold proposed by Lei et al. [6]. The lesion is segmented from the final saliency map. This is followed by a morphological dilation operation having a disk radius of 5 pixel, to feel the tiny holes and remove the isolated single pixels. This method was experimented on ISIC and PH2 dataset. A mean average Dice similarity coefficient (DSC), Hammound distance (HM) and XOR are found to be 91.05%, 15.49% and 16.45%, respectively on PH2 dataset. For ISIC dataset DSC, HM, XOR are found to be 83.41%, 25.67% and 36.21%, respectively. The block diagram for the abovementioned method is given in Fig. 5.

A deep convolution neural network (DCNN) has been used by Hosny et al. [9] to detect melanoma, atypical nevus and common nevus from PH2 skin cancer dataset. The semantic block diagram for this method is given in Fig. 6. The authors are inspired by the various applications of DCNN architecture (VGGNet, ResNet, AlexNet, ZFNet, LeNet and GoogLeNet) and have used the AlexNet to classify

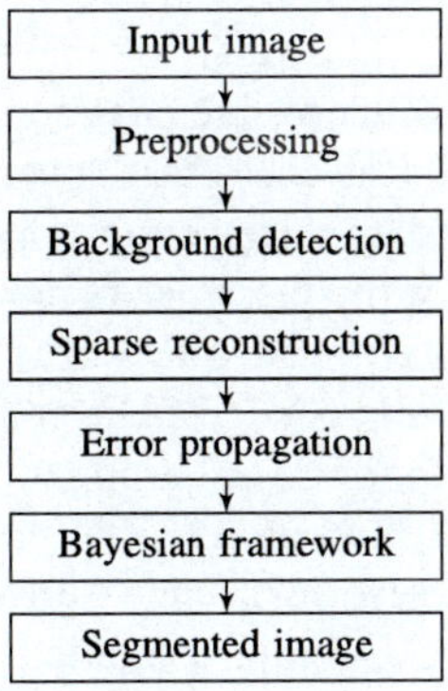

Fig. 5 Saliency based skin lesion segmentation proposed by Ahn et al. [1]

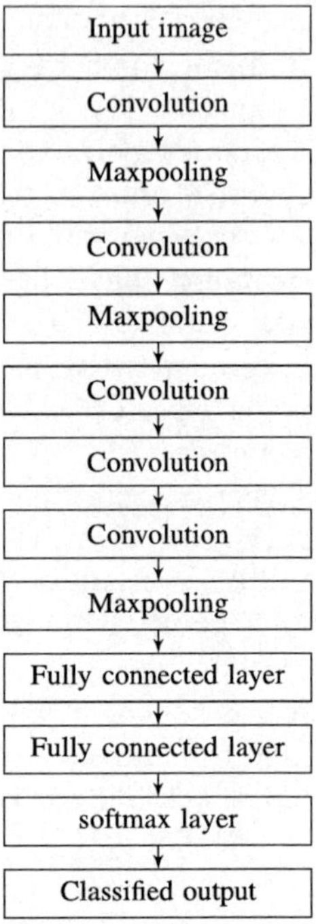

Fig. 6 Semantic block diagram for detection of melanoma proposed by Hosny et al. [9]

different skin cancers from PH2 dataset. AlexNet was initially used for visual recognition of Imagenet, which consists of five convolution layers that are followed by maxpooling layer and three fully connected layers. The third and fourth convolution layers do not have pooling layer. In this case, the last layer of the AlexNet has been replaced with softmax layer to classify the abovementioned skin lesions. Back propagation is used for fine tuning the weights and stochastic gradient is used for weight updation. The experiment was conducted on MATLAB 2017, 64 bit environment by an IBM computer having a core i5 processor with 8 GB DDRAM and NVDIA Gforce 920m GPU. The accuracy, sensitivity, specificity and precesion reported on PH2 dataset are 98.61%, 98.33%, 98.93% and 97.73%, respectively.

Marwan Ali Albahar proposed a method for skin lesion classification using CNN with novel regularizer, which is based on the standard deviation of the weight matrix of the classifier. Novel regularizer is embedded in the convolution layer because the filter values correspond to the weight matrix. By limiting its values, the complexity of the classifier is reduced. This method consists of two convolutional layers (one having 32 filters of size 3×3 and other having 64 filters), pooling layer, dropout, fully connected layer and output. An 1D array is formed by flattening the 2D outputs followed by dropout. These are fully connected with the next layer consisting of 128 neurons. This method was experimented over ISIC archive dataset. Various performance measures such as accuracy, sensitivity, specificity and area under curve (AUC) are computed for ISIC dataset and have been found to be 97.49%, 94.3%,

Fig. 7 Skin lesion classification using CNN and Novel Regularizer proposed by Albahar [2]

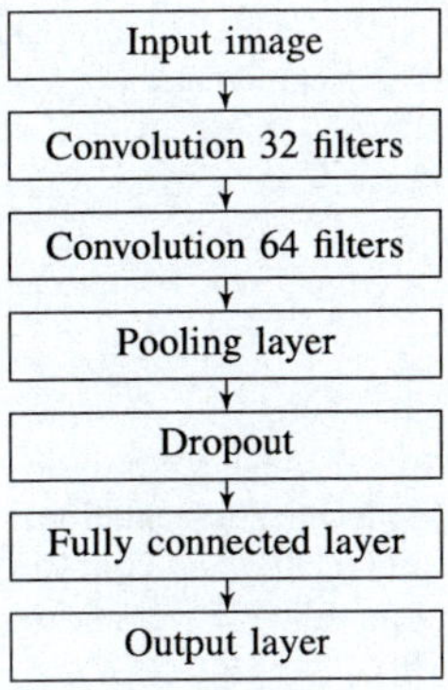

Fig. 8 DermoNet architecture for skin cancer segmentation proposed by Baghersalimi et al. [4]

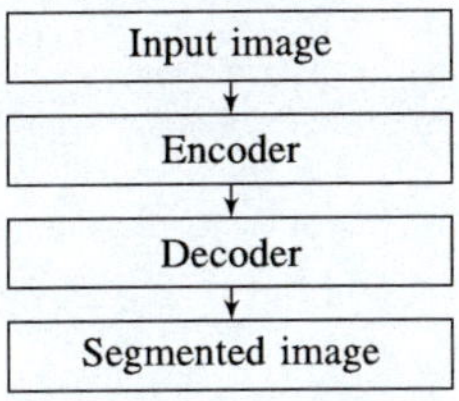

93.6% and 98%, respectively. This method may not be suitable for feature selection or feature reduction [2]. The systematic diagram for this method is given in Fig. 7.

Baghersalimi et al. [4] have proposed DermoNet, consists encoder and decoder block architecture for skin cancer segmentation. The encoder block performs convolution on an input image with a kernel of size 7×7 and stride 2 followed by maxpooling with stride 2. The output feature dimension of each layer consists of k feature maps and they are connected to the input. This process is repeated four times and the final output of the encoder block is the sum of all the outputs obtained in the previous layers. The block helps the network to process high resolution features from early layers and high semantic features from the deeper layers. The decoder block is subdivided into four blocks; each consists of three layers such as a convolutional layer with a kernel 1×1, full convolution layer with a kernel of 3×3 (followed by upsampling with a factor 2) and a convolutional layer with kernel 1×1. The whole network is connected to three convolutional layers and two bilinear upsampling layers of factor 2. It generates a segmented image. The method is tested over ISBI 2017, ISBI 2016 and PH2 datasets. Authors reported an average Jaccard coefficient of 78.3%, 82.5% and 85.3% on PH2, ISBI 2016 and ISBI 2017 datasets, respectively [4]. The block diagram for DermoNet architecture proposed by Baghersalimi et al. is shown in Fig. 8.

A 46-layer U-net architecture (consists of encoder and decoder) has been proposed by Hasan et al. [8] for skin lesion segmentation (shown in Fig. 9). It consists of preprocessing and segmentation phase. Detection and removal of artifacts takes place in preprocessing phase. At first, the images are converted into gray scale using Gaussian blur and borders are drawn around artifacts using sobel and canny filters,

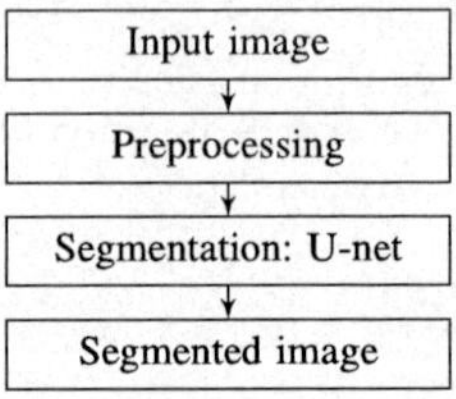

Fig. 9 Architecture for skin lesion segmentation proposed by Hasan et al. [8]

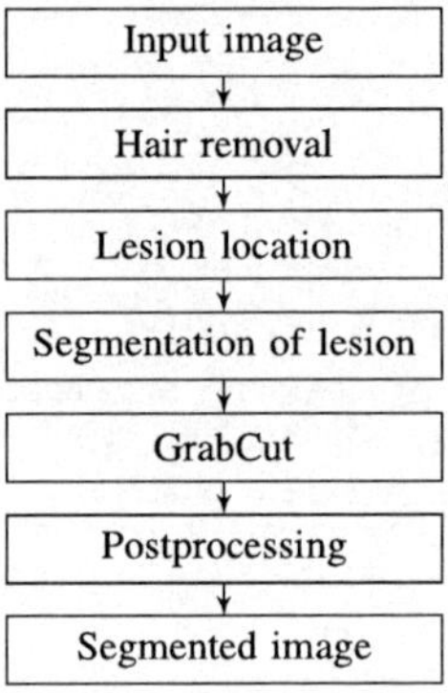

Fig. 10 Architecture for skin lesion segmentation using YOLO and GrabCut proposed by Unver and Ayan [15]

followed by binary thresholding. Image restoration takes place prior to the outlining of the edges by morphological dilate. These images are converted to gray scale and resized to 448×448 pixels. The preprocessed images are fed into the U-net architecture, which consists of convolutions, activation functions, downsampling and upsampling. The encoder is splitted into six blocks each comprising of three layers, such as two non-padded convolutional layer with exponential linear unit (ELU) activation function and a maxpooling layer. A six blocks decoder each consists of four layers is used for decoding. Each block's fourth layer is a 2×2 upsampling layer. Authors have tested this method over ISIC 2018 dataset and found an accuracy, sensitivity, specificity to be 93%, 91% and 97%, respectively [8].

Unver and Ayan have come up with a method for melanoma segmentation by combining a deep convolutional neural network (You only look once (YOLO) and GrabCut algorithm). It determines lesion location and information about the segmented lesion. The various steps used in this method for segmentation of lesion are hair removal, detection of lesion location, segmentation of lesion area from the background, GrabCut and a postprocessing step (morphological operation on segmented binary image). The architecture is shown in Fig. 10. The hair locations are determined by a morphological closing operation followed by bilinear interpolation at the location of hair pixels. This follows an adaptive median filtering to remove the hair pixels. YOLO is a deep learning technique, which splits the input image into non-overlapped grid cells to predict the bounding boxes and the confidence score. The confidence score is used to determine the presence or absence of the object in the bounding box. It is computed by the probability of the object and the intersection over union. The YOLO architecture has 24 convolution layers and two fully connected layers. The last layer uses linear activation function, whereas LeakyReLU activation function is used for the rest of layers. GrabCut is a semi-automatic image segmentation method represented by a graph (which is built by specifying a minimum cost reduction function to produce the segmented output). In this case, a minicut maxflow technique is used for segmenting the graph. Gaussian mixture model is used by GrabCut to determine the region information present in the color image. GrabCut basically finds whether the pixel is belonging to the background or lesion. This follows a morphological operation on the binary image to remove noise. Authors used PH2 and ISBI 2017 dataset to check the correctness of lesion segmentation. Various performance measures such as IOU, sensitivity, specificity, Dice coefficient and Jaccard coefficient are reported to be 90%, 83.63%, 94.02%, 88.13%, 79.94% and 94.92% for PH2 dataset and 86%, 90.82%, 92.68%, 84.26%, 74.81% and 93.39% for ISBI 2017 dataset [15].

Various methods are compared based on their performance measures such as accuracy, sensitivity, specificity, Jaccard coefficient, Dice similarity coefficient, Hammoude distance, XOR and area under curve. Detailed description is given in Table 1. It may be noticed that researchers mainly focused on CNN architecture for the segmentation.

3 Datasets and Performance Measures

Researchers have used publicly available datasets such as PH2, ISIC 2017, ISIC 2018, ISBI 2016 and ISBI 2017 for skin cancer detection and segmentation. These datasets are available in RGB format (8 bit for each channel) and contain the information about various skin lesions along with the ground truth for the lesion. Description about these datasets (represented in Table 2) are mentioned below:

3.1 PH2 Dataset

PH2 dataset is given by University of Porto collected from Hospital Pedro Hispano, Portugal. It consists of 200 lesion images out of which 80 are atypical nevi, 40 are melanoma and 80 are common nevi. It has a resolution of 768 × 560. The images are taken using a lens with a magnification of 20×.

3.2 ISIC 2017 and ISIC 2018

ISIC stands for International skin imaging collaboration. ISIC 2017 dataset consists of 23,906 dermoscopic images with resolution varying from 540 × 722 to 4499 × 6748. The images present in the dataset are distinguished as seborrheic keratosis,

Table 1 Comparision of various architectures

Method	Architecture	Datasets	Performance	Remarks
Bi et al. [6]	Multi-scale lesion-biased representation and joint reverse classification	PH2	Acc—92% (PH2) Sen—87.50% (PH2) Spe—93.13% (PH2)	Optimized feature selection
Begum and Asra [5]	FCM, KNN & ANN	–	–	–
Yuan et al. [16]	Fully convolution network	ISBI 2016 and PH2	JC—84.7% (ISBI 2016) DSC—93.8% (PH2)	–
Ahn et al. [1]	Saliency based lesion segmentation	ISIC and PH2	DSC—91.05% (PH2), 83.41% (ISIC) HM—15.49% (PH2), 25.67% (ISIC) XOR—16.45% (PH2), 36.21% (ISIC)	–
Hosny et al. [9]	Deep convolution neural network	PH2	Acc - 98.61% (PH2) Sen—98.33% (PH2) Spe—98.93% (PH2) Pre—97.73% (PH2)	
Marwan Ali Albahar [2]	CNN using novel regularizer	ISIC 2017	Acc—97.49% (ISIC 2017) Sen—94.3% (ISIC 2017) Spe—93.6% (ISIC 2017) AUC—98% (ISIC 2017)	Cannot be used for feature selection and reduction
Baghersalimi et al. [4], 2019	Dermonet	ISBI 2016, ISBI 2017 and PH2	JC—82.5% (ISBI 2016), 78.3% (ISBI 2017), 85.3% (PH2)	
Hasan et al. [8]	U-net	ISIC 2018	Acc—93% (ISIC 2018) Sen—91% (ISIC 2018) Spe—97% (ISIC 2018)	–
Unver and Ayan [15]	YOLO and GrabCut	ISBI 2017 and PH2	Acc—92.99% (PH2), 93.39% (ISBI 2017) Sen—83.63% (PH2), 90.82% (ISBI 2017) Spe—94.02% (PH2), 92.68% (ISBI 2017) JC—94.02% (PH2), 92.68% (ISBI 2017) DSC—94.02% (PH2), 92.68% (ISBI 2017)	–

Table 2 Various datasets used for skin cancer detection and segmentation

Dataset	Information	Remarks
PH2 [4]	200 images 8-bit RGB 768 × 560	Ground truth is available
ISIC 2017 [2]	23,906 images 8-bit RGB 540 × 722 to 4499 × 6748	Ground truth is available
ISIC 2018 [8]	2594 training and 1000 test images 8-bit RGB	Ground truth is available
ISBI 2016 [4]	900 training and 379 test images 8-bit RGB 1022 × 767 to 4288 × 2848	Ground truth is available
ISBI 2017 [4]	2000 training and 600 test images 8-bit RGB 771 × 750 to 6748 × 4499	Ground truth is available

melanoma and benign. ISIC 2018 dataset consists of 2594 training images and 1000 test images.

3.3 ISBI 2016 and ISBI 2017

ISBI is a subset of ISIC dataset. ISBI 2016 dataset consists of 900 training images and 379 test images with resolution varying between 1022 × 767 and 4288 × 2848. ISBI 2017 dataset consists of 2000 training images and 600 test images with resolution varying between 771 × 750 and 6748 × 4499.

Various performance measures such as accuracy, sensitivity, specificity, Jaccard coefficient, Dice similarity coefficient, Hammoude distance, XOR, area under curve and precision are computed for preciseness of segmentation and classification of skin lesion. The measures are tabulated in Table 3. These are computed by measuring the true positive (TP), false positive (FP), false negative (FN) and true negative (TN) values. Among the various performance measures, researchers mainly computed accuracy, sensitivity and specificity for their computational efficiency.

Table 3 Performance measures

Name	Representation	Range
Accuracy (Acc) [9]	$\frac{TP+TN}{TP+FP+FN+TN}$	[0–100]
Sensitivity (Sen) [9]	$\frac{TP}{TP+FN}$	[0–100]
Specificity (Spe) [9]	$\frac{TN}{FP+TN}$	[0–100]
Jaccard coefficient (JC) [4]	$\frac{TP}{TP+FN+FP}$	[0–100]
Dice similarity coefficient (DSC) [4]	$\frac{2*TP}{2*TP+FN+FP}$	[0–100]
Hammmoude distance (HM) [1]	$\frac{FP+FN}{TN}$	[0–100]
Area under curve (AUC) [2]	–	[0–100]
XOR [1]	–	[0–100]
Precision (Pre) [9]	$\frac{TP}{TP+FP}$	[0–100]

where, TP—true positive, FP—false positive, TN—true negative, FN—false negative

4 Conclusion

In this paper, a detailed comparision of various methods for skin cancer detection and segmentation is discussed along with their performance. Among the various performance measures, accuracy, sensitivity and specificity are widely used. It is observed that the method proposed by Marwan using CNN has achieved an highest accuracy of 97.49% on ISIC 2017 dataset. Detailed description of various dataset is provided for clarity. Among the available dataset, ISIC 2017 contains highest, 23,906 number of images. Some of the authors have used preprocessing and postprocessing step along with the segmentation or classification architecture. It is observed that the performance metrics is improved by the preprocessing and postprocessing step. Generally, preprocessing is used to remove artifacts and postprocessing is used to remove noise. Based on the CNN architecture the performance may be varied. This information may help the readers to provide an insight for skin cancer segmentation and classification. This will help the researchers to develop and modify the various stages along with CNN architecture for efficient and robust segmentation and classification.

References

1. Ahn E, Kim J, Bi L, Kumar A, Li C, Fulham M, Feng DD (2017) Saliency-based lesion segmentation via background detection in dermoscopic images. IEEE J Biomed Health Inf 21:1685–1693. https://doi.org/10.1109/JBHI.2017.2653179
2. Albahar MA (2019) Skin lesion classification using convolutional neural network with novel regularizer. IEEE Access 7:38306–38313. https://doi.org/10.1109/ACCESS.2019.2906241
3. Alfed N, Khelifi F, Bouridane A, Seker H (2015) Pigment network-based skin cancer detection. In: 37th Annual international conference of the IEEE engineering in medicine and biology society (EMBC), Milan, pp 7214–7217. https://doi.org/10.1109/EMBC.2015.7320056

4. Baghersalimi S, Bozorgtabar B, Schmid-Saugeon P, Ekenel HK, Thiran JP (2019) Dermonet: densely linked convolutional neural network for efficient skin lesion segmentation. EURASIP J Image Video Process 7. https://doi.org/10.1186/s13640-019-0467-y
5. Begum S, Asra S (2017) Extraction of skin lesions from non dermoscopic images using deep learning. Int J Sci Res Comput Sci 2:591–596
6. Bi L, Kim J, Ahn E, Feng D, Fulham M (2016) Automatic melanoma dtection via multi-scale lesion-biased representation and joint reverse classification, pp 1055–1058. https://doi.org/10.1109/ISBI.2016.7493447
7. Farooq MA, Azhar MAM, Raza RH (2016) Automatic lesion detection system (alds) for skin cancer classification using svm and neural classifiers. In: IEEE 16th international conference on bioinformatics and bioengineering (BIBE), Taichung, pp 301–308. https://doi.org/10.1109/BIBE.2016.53
8. Hasan SN, Gezer M, Azeez RA, Gulsecen S (2019) Skin lesion segmentation by using deep learning techniques, pp 1–4. https://doi.org/10.1109/TIPTEKNO.2019.8895078
9. Hosny KM, Kassem MA, Foaud MM (2018) In: 9th Cairo international biomedical engineering conference (cibec), pp 90–93. https://doi.org/10.1109/CIBEC.2018.8641762
10. Khamparia A, Singh A, Anand D, Gupta D, Khanna A, Kumar NA, Tan J (2018) A novel deep learning-based multi-model ensemble method for the prediction of neuromuscular disorders. In: Neural computing and applications. https://doi.org/10.1007/s00521-018-3896-0
11. Huang LK, Wang MJJ (1995) Image thresholding by minimizing the measures of fuzziness. Pattern Recogn 28:41–51. https://doi.org/10.1016/0031-3203(94)E0043-K January
12. Jafari MH, Karimi N, Esfahani E, Samavi S, Soroushmehr SMR, Ward K, Najarian K (2016) Skin lesion segmentation in clinical images using deep learning. In: 23rd International conference on pattern recognition (ICPR), Cancun, pp 337–342. https://doi.org/10.1109/ICPR.2016.7899656
13. Mishra R, Daescu O (2017) Deep learning for skin lesion segmentation. In: IEEE international conference on bioinformatics and biomedicine (BIBM), Kansas City, pp 1189–1194. https://doi.org/10.1109/BIBM.2017.8217826
14. Raj RJS, Shobana SJ, Pustokhina IV, Pustokhin DA, Gupta D, Shankar K (2020) Optimal feature selection-based medical image classification using deep learning model in internet of medical things. IEEE Access 8:58006–58017. https://doi.org/10.1109/ACCESS.2020.2981337
15. Unver HM, Ayan E (2019) Skin lesion segmentation in dermoscopic images with combination of yolo and grabcut algorithm. Diagnostics 9(72):97–114. https://doi.org/10.3390/diagnostics9030072 July
16. Yuan Y, Chao M, Lo YC (2017) Automatic skin lesion segmention using deep fully convolutional networks with jaccard distance. IEEE Trans Med Imaging 36:1876–1886. https://doi.org/10.1109/TMI.2017.2695227

A Secure Epidemic Routing Using Blockchain in Opportunistic Internet of Things

Poonam Rani, Pushpinder Pal Singh, Arnav Balyan, Jyoti Shokeen, Vibha Jain, and Devesh Sangwan

Abstract With expanding applications of the opportunistic Internet of things, there has been an ever-increasing enthusiasm in integrating blockchain with the IoT for enhancing privacy and security. However, due to high resource consumption by the blockchain, its implementation becomes a potential task for IoT networks. The main purpose of the paper is to develop a lightweight and secure communication protocol using blockchain architecture for decentralized IoT networks. Blockchain keeps a track of the messages, which are transmitted through the network. A traditional opportunistic IoT network sends multiple copies of messages for communicating information from source to destination, which gives rise to the risk of network security. Therefore, this paper proposes a secure routing protocol based on epidemic nature by integrating blockchain to ensure security and privacy. The main idea behind using blockchain is to keep a distributed ledger of all the messages (transactions), which are being transmitted through the network. For the simulation of our proposed network, we use the random waypoint model which we believe perfectly simulates the mobility of opportunistic devices.

Keywords Blockchain · Epidemic routing · IoT · Opportunistic IoT

P. Rani (✉) · P. P. Singh · A. Balyan · V. Jain · D. Sangwan
Netaji Subhas University of Technology, Dwarka, Delhi, India
e-mail: poonamrani2017.nsit@gmail.com; poonam.rani@nsut.ac.in

P. P. Singh
e-mail: pushpinderpal19@gmail.com

A. Balyan
e-mail: arnavbalyan1@gmail.com

V. Jain
e-mail: vibha.jain.cs19@nsut.ac.in

D. Sangwan
e-mail: dev.sangwan2001@gmail.com

J. Shokeen
CSE, UIET, Maharshi Dayanand University, Rohtak, Haryana, India
e-mail: jyotishokeen12@gmail.com

A. Khanna et al. (eds.), *Data Analytics and Management*, Lecture Notes on Data Engineering and Communications Technologies 54,
https://doi.org/10.1007/978-981-15-8335-3_10

1 Introduction

Traditional IoT uses several connected smart physical devices with an infrastructure-based network. Opportunistic IoTs [1] (OIoTs) work on ad hoc policies with opportunistic network protocols using short-range communication technologies like Zigbee, Bluetooth, near-field communication (NFC), etc. [2]. These smart devices consist of smart cell phones, wearable devices, laptops, smart gateways, etc. With the increasing popularity of IoT devices, the security of smart devices becomes an important task [3]. The security of the IoT network has been a weak point due to its constraint and cost-driven resources [4]. However, most of the IoT devices are deployed in a remote location, which makes them inaccessible and exposes them to various attacks easily [5, 6]. In ad hoc networks, each node acts as a carrier for the message. There are no central servers for an ad hoc network. Ad hoc networks rely on the peer-to-peer (P2P) network protocol [7]. An opportunistic network is a wireless ad hoc network where the nodes can be static or mobile; that is, infrastructure is not fixed. Each node has a fixed finite range in which they can operate and send messages. A message can be transmitted only when a node comes within a range of another node. In this network, internal nodes forward the message from the source node to the destination node.

Traditional routing protocols are unable to work in intermittent connectivity scenarios. To overcome these issues and increase the message delivery rate over large distances with intermittent connectivity, we use epidemic routing, which is introduced by Amin Vahdat and David Becker [8]. A mobility model is used to simulate the movement of users in the mobile ad hoc network (MANET). It describes how the location, velocity, and acceleration of nodes vary to time. These models simulate performance, evaluate various routing strategies, and check potential problems. This paper uses the random waypoint model and integrates blockchain into regular ad hoc networks. Novelties of our work are given by these major contributions:

- We propose an epidemic routing environment in opportunistic IoT for message delivery from source to destination.
- To simulate the real-world atmosphere, a random way mobility model is used to replicate the mobility for IoT devices.
- Security is a major issue while flooding information in the entire network. This issue is handled by implementing blockchain distributive among the IoT nodes.

The paper is designed as follows: Sect. 2 describes the related work. Section 3 covers system architecture and design goals. Section 4 contains the simulation and performance results of the proposed network. Section 5 concludes the work.

2 Related Work

In this section, we discuss the related work to epidemic routing and random waypoint model. Epidemic routing gives an unbeatable performance in routing when working with short-range communication techniques and low-energy mobile opportunistic IoT nodes. Amin Vahdat and David Becker [8] first proposed epidemic routing. They proposed a routing technique in an ad hoc network where a host node floods the system with a message, which would be picked by a carrier node, and then it floods the system to be picked by another carrier node. This process continues until the message reaches its destination. The message in this routing spreads like an epidemic of a disease. The main aim of employing epidemic routing in our work is to maximize the message delivery rate and reduce message delivery latency. The simulation of our work is based on the random waypoint model where nodes come in contact with another node randomly. The random waypoint model is one of the simple and widely used mobility models [9]. In this model, users are free to move anywhere randomly without any restrictions. The speed, destination, and direction of a user vary randomly over the simulation. In the simulation, every node will pause and move to a new destination randomly with random velocity; after this, it will then pause for a fixed time and repeat this for the duration of the simulation. However, delivering messages to the destination using more than one hop nodes threatens network security. While transmitting the message from one source node to destination, the same packet encounters many relay nodes, which can perform malicious activities by modification of messages.

Authors in [9–11] have explored the epidemic routing and its modified versions, but none of these papers in the literature have explored security and privacy aspect with them. This is a big research gap in the literature. Moreover, OIoT networks have IoT nodes, where this problem becomes more potential. This paper tries to fill this research gap by incorporating blockchain technology. Blockchain [12, 13] is a safe and integrated, distributed ledger technique, where it has several blocks chained together. It is having an immutable database for transferring messages very securely and privately among nodes. It has been proved very useful and impenetrable to all types of attacks [12]. In other words, blockchain technology is a highly trustworthy distributed storage system with encryption to deal with distrusted behavior of client nodes. So, this paper explores the epidemic routing with blockchain in OIoT. The use of blockchain in the ad hoc network will provide improved security of the network. We use the random waypoint model for the simulation of the opportunistic network, as we believe this model very closely replicates our potential users. Blockchain is a way to transmit data from one point to another, which is highly secure and fully automated [13, 14]. To transmit a message, a node creates a block that is verified by every other node on that network. If there is a conflict with the contents of a block, the version which majority nodes possess will be accepted as the correct version. Blockchain provides a better way to secure our IoT devices [15].

3 Proposed Work

3.1 System Model

The proposed model integrates the blockchain technique into communication via ad hoc networks using epidemic routing. Each node has an ID, energy level, current condition (dead or alive), array of current neighbors; a message vector contains blockchain data (previous hash, time stamp, data, and current hash of the block) [16]. In our simulation, whenever a node comes in contact with any other node (carrier), the host node shares the summary vector with the carrier node, which will then be compared with the data in the message vector block, which is already buffered. If there exists some data on the host, which is not present in the carrier, a new block will be generated with new data, time stamp, and the previous hash of the message vector present on the carrier node. On every hop, i.e., data transfer between host and carrier, the hop count will be reduced by 1. This process will continue until either the message vector reaches its destination or the hop count drops to 0.

3.2 Consensus Algorithms

A consensus algorithm is a mechanism, which is used to bring all nodes in agreement, and in confidence with each other, in an environment, these nodes do not trust any other. We make consensus algorithms to achieve reliability in a network having numerous unreliable nodes. We mainly use two types of algorithms—proof of work (PoW) and proof of stake (PoS). In proof of work, any node may get involved in adding blocks to the chain, by proving that it has performed a computationally expensive task or work. Adding new blocks using the PoW is termed mining. In PoS, the primary objective is to reduce the high power consumption by the PoW concept in blockchain networks. Hence as a substitute to avoid computationally expensive puzzle solving, the proof of stake aims to stake peer economic share within the network. Here, validators substitute miners, and analogous to the PoW concept, a validator is selected to publish a block in the blockchain.

Blockchains that use PoW concept are slow and highly inefficient [17]. PoW has high-energy expenditure. However, PoS lacks randomization, which may cause peers with large stakes to continuously keep on winning the consensus and hinder the trust value updating of peers with low stakes. In contrast, we use a hybrid of PoW and PoS consensus mechanism, which takes the sum of absolute values of offsets (offset of trust) in the candidate block as the stake. The offset of trust is calculated for all the nodes involved in the network based on the credit rating of the node. All nodes assess the credibility of received messages/data and then generate ratings for them.

Algorithm 1: Proposed Algorithm for Epidemic Blockchain

Require: Simulation parameters as given in Table 1, Model: Random Way-point model

Ensure: Data within a Blockchain at the receiver

1: **for** i_i =1 to n **do**;
  Check for neighbors within range;
  2: **for** i_2=1 to n **do**
    3: **for** every neighbor i_2
    4: Compare Summary vectors of the nodes
    5: Create blocks of the missing data
    6: Send these blocks to the other node
    7: Proceed to a new destination at a random speed

4 Simulation and Results

Our simulation aims primarily on the operation of the routing protocol used and chain generation and not on the underlying layers. We use a random waypoint mobility model to simulate the environment of the nodes. The nodes choose a destination and speed randomly. We perform our simulation with MATLAB on the Intel i5 Quad-Core@2.5 GHz processor with 16 GB RAM.

Upon arrival, the node pauses for a randomly chosen duration during which it communicates data and then choose a new destination. Our scenario consists of 1500×300 m^2 area where 50 nodes are placed randomly. The transmission range is taken to be 50 m, and the simulation is then run for around 2000s. Table 1 depicts various parameters and external factors that are taken into account for the simulation. Using 50 nodes in an area of 1500×300 m^2, we have successfully simulated the feasibility of our network. In our simulation, we use the random waypoint model, which makes the nodes to move randomly throughout the simulation. Every node is

Table 1 Parameters used in the simulation

Simulation parameters	Values
Simulation area	1500×300 m^2
Number of nodes	50
Transmission range	50 m
Simulation time	2000s
Speed	0–20 m/s
Direction	Random
Max hop count	50

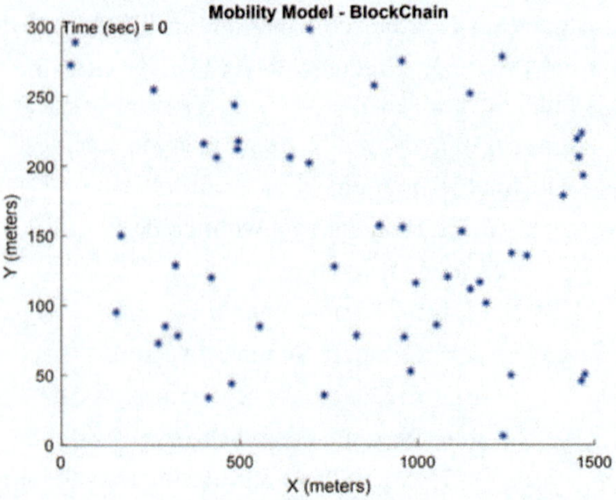

Fig. 1 Visual distribution of nodes after a 2000s simulation

placed randomly at the beginning of the simulation and moves after a fixed period by which nodes communicate their data. We run our simulation for 2000s with every node transmission range of 50 m. Figure 1 shows the visual distribution of simulation of nodes after the 2000s. We received high successful delivery rates where about 68% of messages are successfully delivered after the 2000s simulation.

We formulate the energy of different nodes using the following equations:

$$E_{TX} = E_{\text{elec}} \times k + E_{\text{amp}} \times k \times \text{tran\_range} \tag{1}$$

$$E_{RX} = (E_{\text{elec}} + \text{EDA}) \times k \tag{2}$$

where $E_{T\ X}$ represents the energy required for transmission of a node and E_{RX} represents the energy required for receiving a node. We initialize all nodes with an initial energy of 2 J. In above equations, *tran_range* is transmission range of each node, E_{elec} is energy being dissipated to run the transceiver circuit, E_{amp} is the amount of energy spent by the amplifier to transmit energy in bits per square meter, *EDA* is the data aggregation energy used to reduce the energy consumption, and k is the size of the data packet in bits which is taken to be 4000.

Figure 2 shows a comparison of the energy level of nodes with different transmission ranges (50 m, 75 m, 100 m). Figure 3 depicts the time difference between the use of blockchain and without blockchain. This time difference is found to be around 3 s. We argue that the use of blockchain provides massive gains in terms of security. On integrating blockchain, we guarantee that the sent messages do not tamper during the transmission of such messages to the host. Hence, our current simulation verifies the successful implementation of blockchain on decentralized opportunistic networks and guarantees very high security and safeguards against attacks and vulnerabilities in the network. Figure 4 compares the successful deliveries of nodes for 50 m, 75 m, and 100 m transmission ranges. Further, Fig. 5 gives us insights into each node's

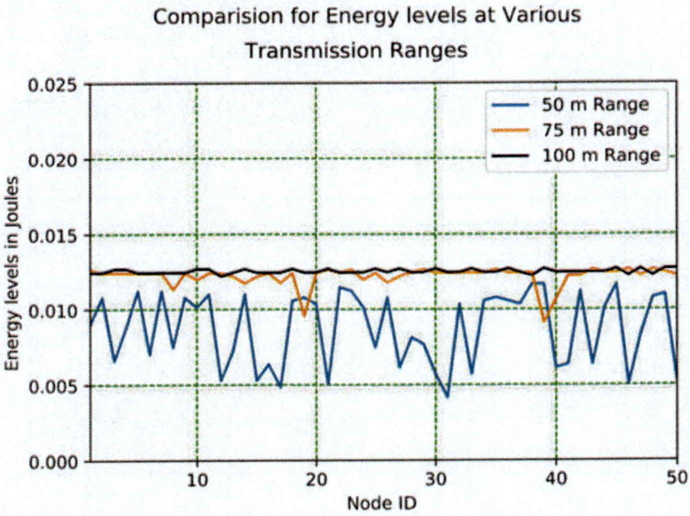

Fig. 2 Energy levels of various nodes in joules simulated at different transmission ranges

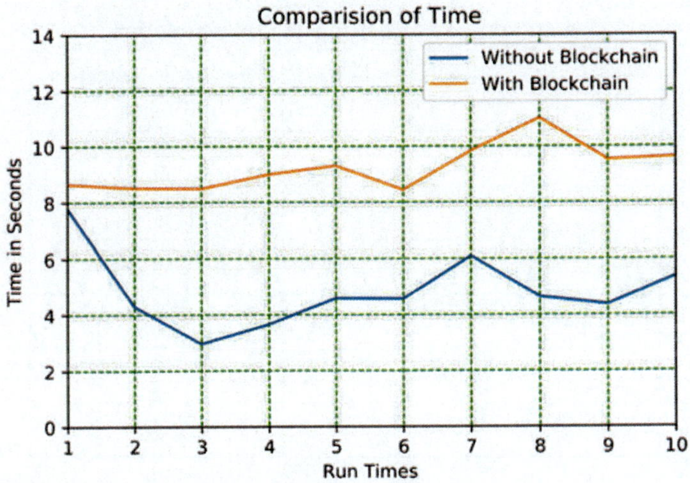

Fig. 3 Comparison of time in seconds with and without blockchain

buffer size after a 2000s simulation. We see an average of around 30 as the buffer size of nodes after the simulation. Figure 6 shows the remaining energy levels of each node after the simulation. We see an average of 1.99 joules energy remaining at the end of the simulation.

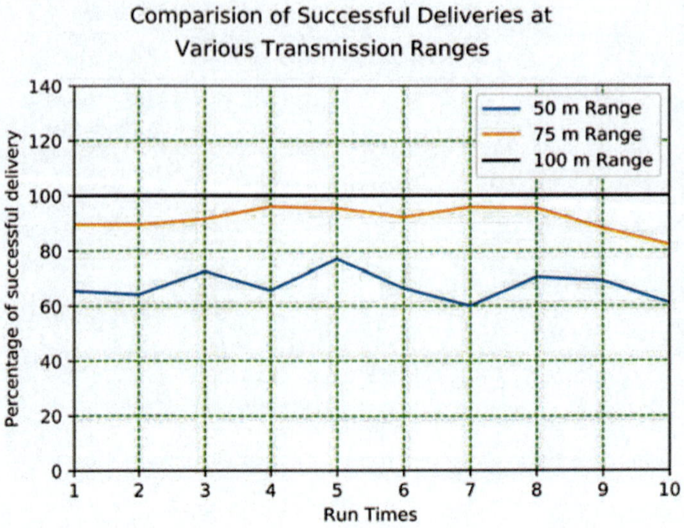

Fig. 4 Comparison of successful deliveries for 50, 75, and 100 m transmission range of the nodes

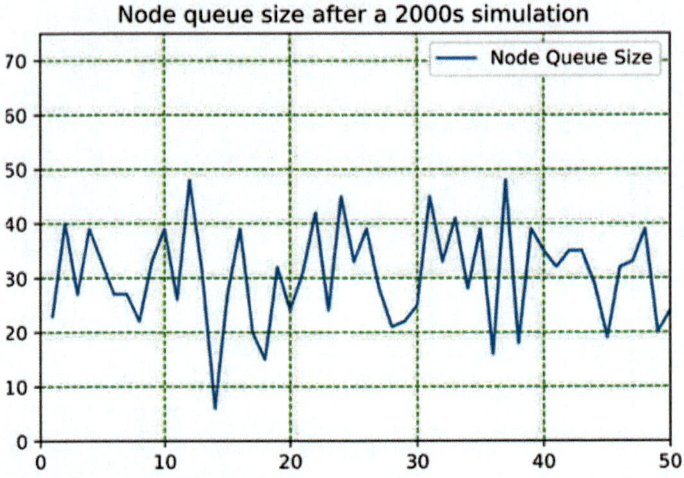

Fig. 5 Node queue size after a 2000s simulation

5 Conclusion

In this paper, we successfully introduced a blockchain-based ad hoc system in IoT networks that ensures the credibility of the message. With the help of this proposed system, IoT devices can communicate securely with the credibility of received messages. Several simulations are performed to assess the performance and reliability of the whole system. The achieved simulation results reveal that a small

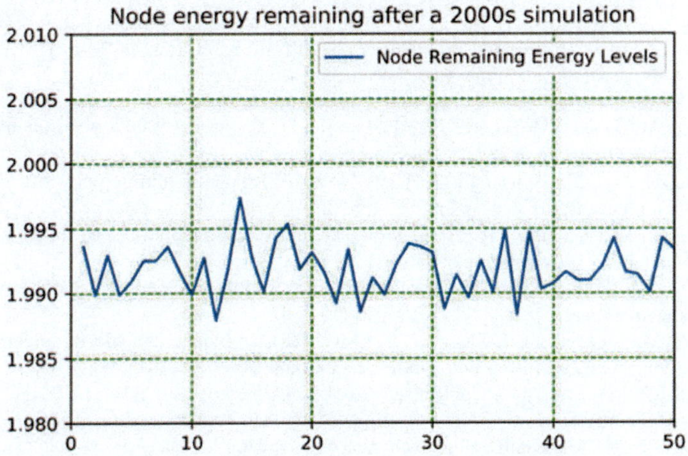

Fig. 6 Remaining energy levels of all the nodes after a 2000s simulation

increase in processing time improves security drastically. Simulation results prove that the proposed system is feasible and effective for secure communication in a decentralized IoT network. This system can greatly help IoT devices evaluate the credibility of neighboring nodes and create a secure and efficient IoT network. We achieved a successful delivery rate of about 68%, which is not pretty impressive in real life. A delivery rate of 68% would put a huge strain on the system in terms of redelivering the message, which may bog down our network a fair bit over time if the generation of messages is constant. For future work, we will use probabilistic routing for better performance [18]. We can also apply this approach to make secure social networks as discussed in different aspects, issues, and challenges explored in paper [19].

References

1. Guo B Opportunistic IoT : exploring the harmonious interaction between human and the internet of things, pp 1–21
2. Ammar M, Russello G, Crispo B (2018) Journal of information security and applications internet of things: a survey on the security of IoT frameworks. J Inf Secur Appl 38:8–27
3. Sujitha B, Subbiah V, Laxmi PE, Rani P, Polkowski Z (2020) Optimal deep learning based image compression technique for data transmission on industrial internet of things applications. Trans Emerg Tel Tech 1–13
4. Khamparia A, Singh PK, Rani P, Samanta D, Khanna A, Bharat B (2020) An internet of health things-driven deep learning framework for detection and classification of skin cancer using transfer learning. Trans Emerg Tel Tech 1–11
5. Lakshmanaprabu SK et al (2018) Effective features to classify big data using social internet of things 6
6. Gochhayat SP, Kaliyar P, Conti M, Tiwari P, Prasath VBS, Gupta D, Khanna A (2019) LISA: lightweight context-aware iot service architecture. J Cleaner Prod 212:1345–1356

7. Vergata T, Loreti P, Bracciale L (2019) Optimized neighbor discovery for opportunistic networks of energy constrained IoT devices. IEEE Trans Mob Comput PP(c):1
8. Vahdat A, Becker D Epidemic routing for partially-connected ad hoc networks
9. Yoon J, Liu M, Noble B Random waypoint considered harmful
10. Rashidi L, Towsley D, Mohseni-kabir A, Movaghar A, Member S On the performance analysis of epidemic routing in non-sparse delay tolerant networks, pp 1–16
11. Wang R, Wang Z, Ma W, Deng SU, Huang H (2019) Epidemic routing performance in DTN with selfish nodes. IEEE Access 7:65560–65568
12. Banerjee M, Lee J, Choo KR (2018) A blockchain future for internet of things security: a position paper. Digit Commun Netw 4(3):149–160
13. Cai C, Yuan X, Wang C (2017) Towards trustworthy and private keyword search in encrypted decentralized storage
14. Reyna A, Martín C, Chen J, Soler E, Díaz M (2018) On blockchain and its integration with IoT. Challenges and opportunities. Futur Gener Comput Syst 88(2018):173–190
15. Jurdak R (2017) Towards an optimized blockchain for IoT. In: 2017 IEEE/ACM second international conference on internet-of-things design and implementation (IoTDI)
16. Singh M, Kim S Blockchain : a game changer for securing IoT data, pp 51–55
17. Nandan S et al (202) An efficient Lightweight integrated Blockchain (ELIB) model for IoT security and privacy. Futur Gener Comput Syst 102:1027–1037
18. Lindgren A, Doria A Probabilistic routing in intermittently connected networks, pp 1–8
19. Rani P, Tayal DK, Bhatia MPS (2018) Different aspects, challenges, and impact of social networks with a mathematical analysis of teaching learning process. JARDCS 14:1576–1590

Improved MAC Design-based Dynamic Duty Cycle for Vehicular Communications over M2M System

Mariyam Ouaissa, Mariya Ouaissa, Meriem Houmer, and Abdallah Rhattoy

Abstract With the growing proliferation of machine to machine (M2M) applications, several terminals will be deployed in many types of applications such as health, surveillance, industrial automation and especially transportation. This work concentrates on M2M networks in the intelligent transportation system (ITS) and considers areas of enhanced vehicular networking for M2M concepts. In M2M communications, an efficient medium access control (MAC) is important to avoid such collisions by helping devices to decide when and how to access support. This paper intends to develop robust MAC protocol-based contention to be used in intra-cluster for vehicular communication over M2M system. Our proposed is a multi-layers MAC protocol with an adaptive duty cycle that reduces the amount of energy wasted on idle listening via sending all messages in a variable active period according to a load of traffic and return to sleep mode. Simulation results illustrate that our proposed outperforms other traditional MAC protocols in terms of energy consumption, average delay and collisions probability.

Keywords M2M · ITS · VANET · MAC · Energy consumption · Duty cycle

M. Ouaissa (✉) · M. Ouaissa · M. Houmer
Laboratory of Modelization Mathematic and Computer Science, ISIC Team, ENSAM, Moulay-Ismail University, Meknes, Morocco
e-mail: mariyam.ouaissa@edu.umi.ac.ma

M. Ouaissa
e-mail: mariya.ouaissa@edu.umi.ac.ma

M. Houmer
e-mail: houmer.m@gmail.com

A. Rhattoy
High School of Technology, Department of Computer Science, ISIC Team, Moulay-Ismail University, Meknes, Morocco
e-mail: rhattoy@gmail.com

A. Khanna et al. (eds.), *Data Analytics and Management*, Lecture Notes on Data Engineering and Communications Technologies 54,
https://doi.org/10.1007/978-981-15-8335-3_11

1 Introduction

Internet of things (IoT) [1] has become a new concept referring to the increasing advance of infrastructures and the continuous growth of everyday intelligent devices. Machine to machine (M2M) [2] communication is seen as an integral part of the IoT movement. M2M system allows one or more autonomous machines to communicate directly with each other without human intervention. These communications are constantly growing with new technologies and have just become part of our daily life as they can be used in almost all areas, among them we find the transportation applications [3].

In this context, vehicular ad hoc network (VANET) is the most crucial aspect of the intelligent transportation system (ITS). In this network, vehicles are equipped with a wireless short and medium range communication. Actually, the vehicles can communicate with each other with two methods, either vehicle to vehicle (V2V) or vehicle to infrastructure (V2I), where vehicle communicates with the equipment next to the road named road side units (RSU). Since connected vehicles are mainly a network of autonomic machines, vehicle networks may benefit greatly from support for M2M communications [4].

In this paper, we investigate an enhanced medium access control (MAC) design for vehicle-based M2M communications; our scheme combines with contention and reservation-based MAC protocols by applying the concept of cluster. Then we present and analyze an improved energy efficient MAC protocol-based contention for intra-cluster communication to reduce the energy consumption of communication and the number of collisions. Our proposed protocol is a multi-layers MAC protocol based on contention and uses two modes, active and sleep, to save the power consumption with an adaptive and variable duty cycle.

The remainder of this paper is organized as follows: the next section presents the system model for M2M system in vehicular context. A detailed description of our considered medium access design is provided in Sect. 3. In Sect. 4, we propose an improved multi-layers MAC protocol in intra-cluster communication. Section 5 evaluates and analyzes the performances of the existing protocols as well as our proposition. Finally, we draw our conclusion in Sect. 6.

2 System Model

This article considers a topology for the ITS application in M2M communication, which may be an appropriate solution for merging the characteristics of both wireless sensor networks (WSN) and vehicular ad hoc network [5]. Figure 1 presents a vehicular network over M2M system.

For the lower layer, this is a vehicle domain made up of sensors nodes deployed alongside the RSS highway and on-board sensors that are carried in vehicles. These nodes can be transmitted via short-range communication with each other or the

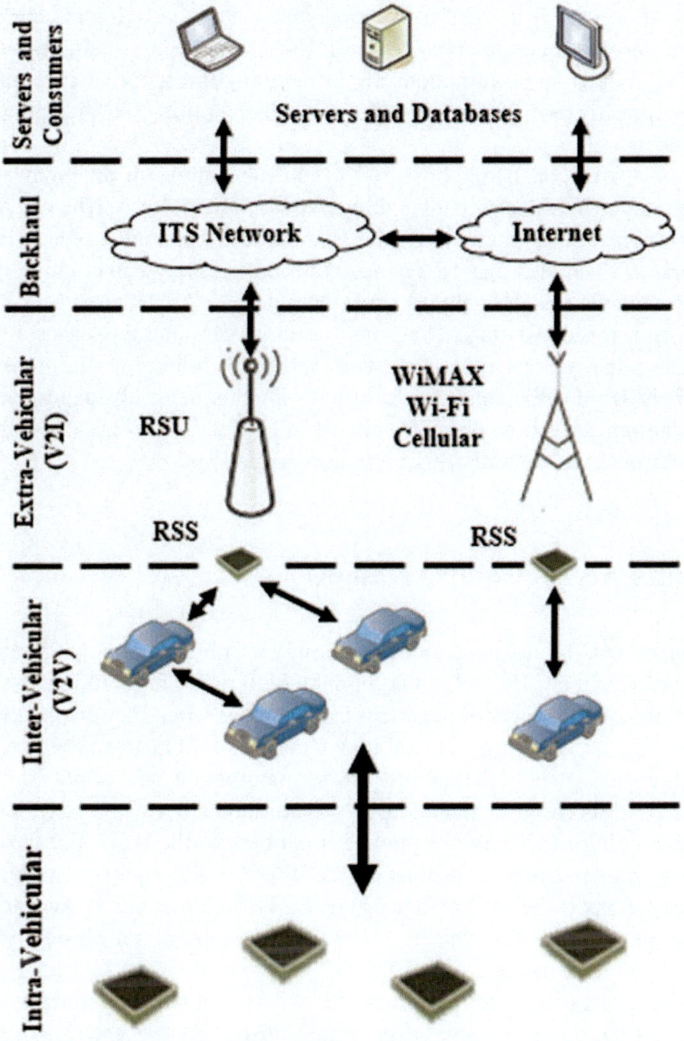

Fig. 1 M2M system model in a vehicular application

roadside sensor. The sensor readings must be transmitted through RSS nodes to RSU or base stations (BS). [6]. The components are:

Vehicular Sensor Nodes: are physical devices installed on each vehicle. They allow the measurement and the collection of information concerning the state of the vehicle as well as that of its environment. These sensors open up wider possibilities for vehicle network applications.

Road Side Sensors (RSS): are positioned next to the road at a set pace. For vehicle nodes, RSSs serve as cluster heads. RSS nodes accept mobile node data and retransmit it to the RSU or BSs.

Road Side Unit (RSU): is an infrastructure located near roads. It plays the role of the router, which provides connectivity between V2V or between a vehicle and another infrastructure V2I. Its main functions are broadening the scope of communication, providing connectivity between vehicles and other entities, and running security applications.

A regional layer consisting of a traffic control station on all highways is the intermediate layer. Wireless broadband base stations (cellular, WIFI, WiMAX, etc.) and RSU stations act as gateways on the Internet and as connection services on an ITS network. All vehicles that are connected to the gateway can communicate with the main servers via the ITS network and internet.

The upper layer, consisting of internet based servers and databases, allows end users to access the various vehicle network resources remotely. The transportation authority may handle the infrastructure and vehicles centrally using servers and databases that are linked to the ITS network. ITS details can also be downloaded from the Internet to make end users more accessible [7].

3 Medium Access Control Design

The proposed MAC model is a hybrid solution that minimizes and saves energy and at the same time extends the lives of equipment batteries. This solution also includes clustering because the power of the nodes can decrease when they are put in a cluster. Clustering is the act of putting several nodes in different clusters in which a node is chosen as the cluster head (CH). While cluster members (CM) cannot communicate with the RSU directly, it is mandatory to communicate with CH, which has the role of transmitting messages from cluster members to the RSU. For the proposed MAC design frame structure, it must divide into two large parts in order to avoid interference between the cluster head and the cluster members, one intra-cluster between the member clusters, the cluster head, and another inter-cluster between the cluster head and the infrastructure (Fig. 2).

In CSMA, energy consumption increases with the increasing numbers of nodes increases; it is not due to the increasing data transmitted, but because of the traffic density that increases the back off and collisions. For the high traffic of cluster members, it can cause congestion at the CH level. In the case of a small cell, within each cluster, traffic load is too low to provoke idle listening and collisions; it is for this reason that the technique can be used inside the cluster is CSMA. The CSMA period is divided into different layers to reduce passive listening. Therefore, for communication inside a cluster, we can choose the multi-layers CSMA protocol [8].

The TDMA technique is used for communication between infrastructures because it consumes less energy between RSU and CH. Therefore, for cluster head and unclustered nodes, use protocol with TDMA to communicate directly with RSU. The packet path is limited in two hops, which means that the node that can only speak to the RSU is the unclustered node and the CH.

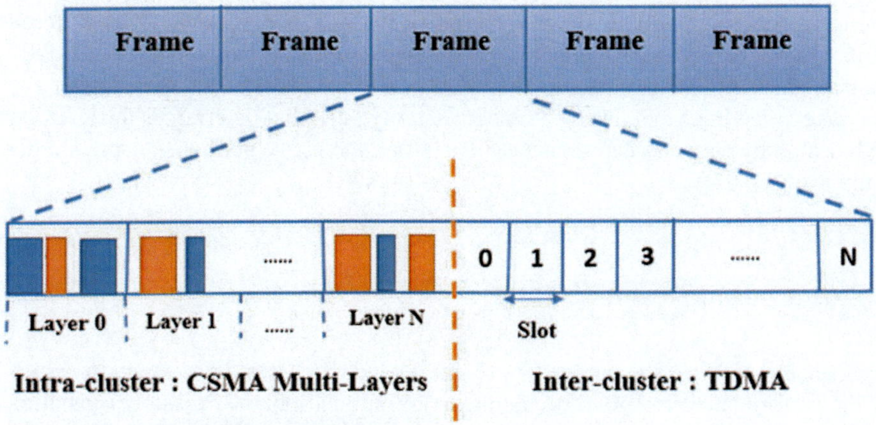

Fig. 2 Frame structure of proposed MAC design

4 Proposed CSMA Multi-layers Protocol for Intra-Cluster Vehicular Communication

Contention-based protocols cannot prevent collisions and idle listening if the load of the network increases. In the following, we propose an improvement of CSMA multi-layers protocol in intra-cluster communication in order to reduce idle listening and the number of collisions and further improve energy efficiency.

An effective MAC protocol has a huge effect on consumption; in our work, we propose an improved multi-layers protocol-based CSMA more precisely the non-persistent CSMA for M2M communication that contains a small packet, which is divided into n phases, wherein each phase, a part of the cluster members using the CSMA protocol to send their packets, in order to avoid idle listening and collisions. Idle listening is the primary source of energy waste that is why different MAC protocols are modeled to save energy by putting the radio on sleep mode. In the design of our proposed, sensors return to sleep mode that will minimize the energy consumed in the listening phase.

The MAC protocol proposed is a protocol that has been designed for variable traffic load to address the weaknesses of the multi-layers CSMA protocol existing. The protocol proposed is characterized by a low value of duty cycle and the number of collisions. In our protocol, sensor nodes reduce the energy consumed during the communication with each other; this is due to their short listening time. In addition, the number of collisions is minimized where several sensor nodes attempt to transmit data at the same time. The idea of our proposed is to minimize the phenomenon of the passive period in the listening mode by putting a variable duty cycle dependent on the traffic load, contrary to the existing multi-layers protocol, which characterizes by a fixed duty cycle. Each sensor node wakes up periodically to communicate with its neighbors and then returns to its sleep mode until the beginning of the next sequence. In the meantime, the messages destined for their receiver will be stored in its buffer. During the active period, a node will listen for support and potentially transmit packets during this time. The active period ends when there is no longer any message-receiving event during a timeout activity (TA) that determines the duration of listening by active asleep sequence. Therefore, the active period is adapted to the volume of the traffics.

5 Simulation Results

This section presents the parameters (Table 1) and discusses the simulation results of the performances for MAC protocols. In this simulation, the Poisson distribution model was chosen to illustrate the generation of traffic. This concept for traffic assumes that nodes generate statistically traffic based on an exponentially distributed time of inter-arrival. This model was developed to test the efficiency of the protocol at various arrival rates [9]. We consider that the vehicular sensor nodes are moved at the speed 30 m/s and the road side sensors are fixed [10–12].

The total energy consumption of a node is composed of three types of energy; transmission energy, listen energy and sleep energy. Figure 3 shows that our proposed consumes less energy compared to other protocols which are; S-MAC, non-persistent CSMA and CSMA multi-layers, because the listening periods of our proposed are variable and adapted to the traffic load, unlike CSMA multi-layers and S-MAC who have a constant duty cycle, which means a reduction of the energy consumption.

Figure 4 shows that the average delay of the proposed has a longer period than the CSMA multi-layers, S-MAC and CSMA, because the nodes spend significant time in sleep mode for the proposed protocol, then the packets will meet further delay. This delay is the latency that is able to follow a packet as it is stored in the buffer node before the packet is delivered to its destination without collision. The delay consists of two elements: transmission delay and queuing delay.

Figure 5 illustrates the energy consumed by a node according to the number of

Table 1 Simulation parameters

Parameters	Values
Number of nodes	200
Arrival distribution	Poisson distribution
Speed of vehicular nodes	30 m/s
Node transmitting power	24.75 mW
Node listening power	13.5 Mw
Node sleeping power	15 Uw
Node transmission data rate	6 Mbps
Average packet length	200 Bytes
Simulation time	120 s

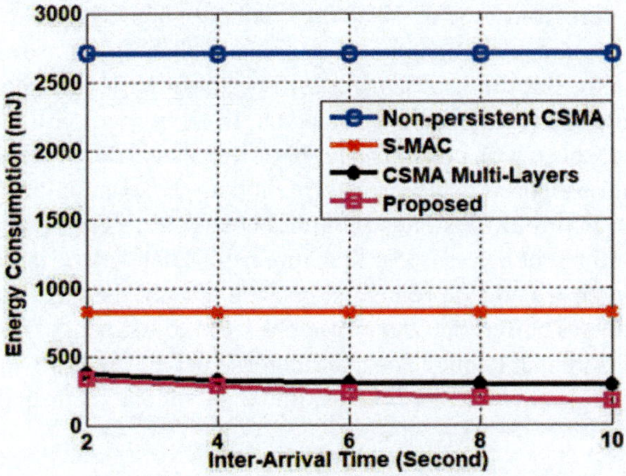

Fig. 3 Total energy consumption

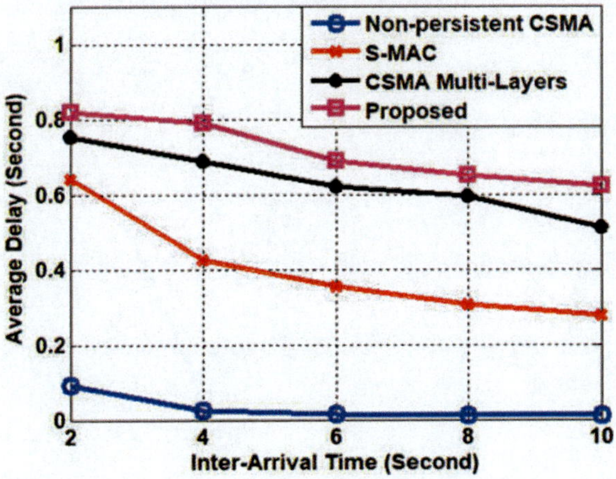

Fig. 4 Average delay

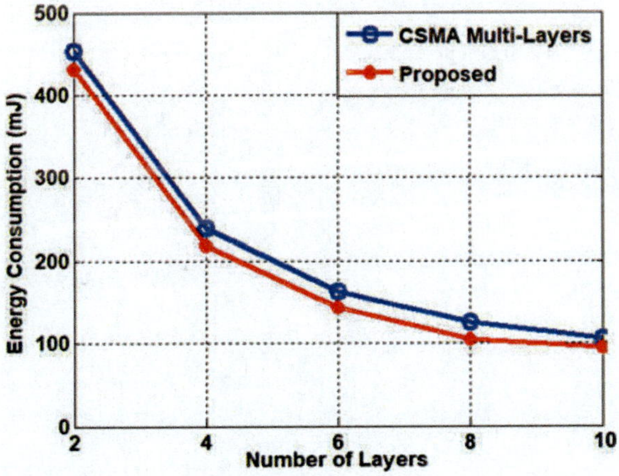

Fig. 5 Total energy consumption

layers from 1 to 10. When the number of layers is less than 6, the energy consumption decreases significantly. After six layers, energy consumption is not reduced as previously since most packages will be for other layers and nodes will spend more time to wake up to different calendars. In addition, it will increase the number of control packets which consumes more power. According to this figure, we can conclude that our proposed is more performant than the existing one.

Figure 6 illustrates the influence of adding several phases over the delay. If the number of layers is less than three, the delay will be increased significantly. However, when we add more layers, packets are not delayed because the next frame loop is always buffered. For our proposed, the adjustment of duty cycle is allowed in order to ensure that latency does not increase before the limit on system resources is reached.

Figure 7 indicates the number of layers according to collision probability for

Fig. 6 Average delay per packet

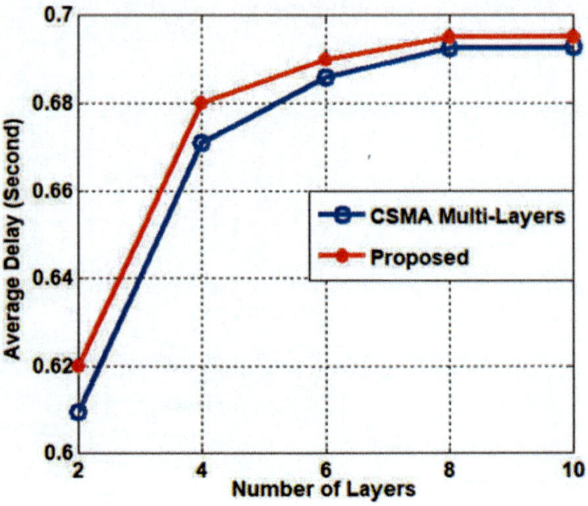

Fig. 7 Collision probability

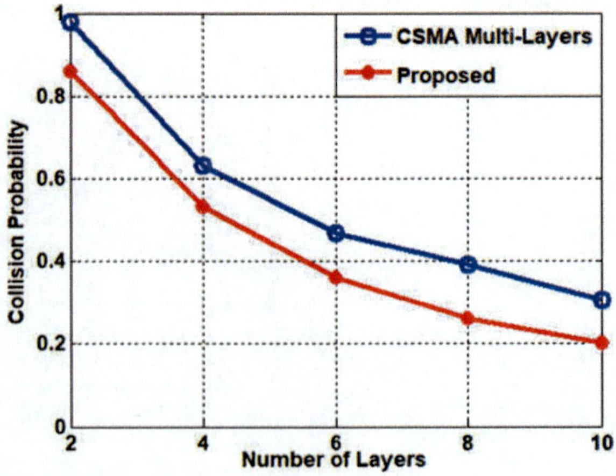

CSMA multi-layers and our proposed, when the number of layers increases, the collisions decreases significantly, because the request packets per phase spread sufficiently so that the risk of collision is minimized to this kind of traffic.

6 Conclusion

In this article, an enhanced CSMA multi-layers protocol for intra-cluster vehicular communication over M2M system has been proposed, according to the requirements that we needed to design a MAC protocol for M2M communication in order to satisfy the needs of M2M applications. Our protocol is based on non-persistent CSMA with an adaptive and variable duty cycle to reduce the phenomenon of passive listening. The active period is adapted to the volume of traffics, to avoid idle listening, reduce energy and the number collisions. The results show that our proposed is performant than the other existing protocols in terms of reducing energy consumption, average delay and collisions.

References

1. Ouaissa M, Ouaissa M, Rhattoy A (2019) An efficient and secure authentication and key agreement protocol of LTE mobile network for an IoT system. Int J Intell Eng Syst (IJIES) 12(4)
2. Ouaissa M, Rhattoy A, Lahmer M (2018) New method to control congestion for machine to machine applications in long term evolution system. Int J Commun Antenna Propag (I.Re.C.A.P.) 8(4)
3. Chavhan S, Gupta D, Chandana BN, Khanna A, Rodrigues JJ (2019) IoT-based context-aware intelligent public transport system in a metropolitan area.
4. Sheikh MS, Liang J, Wang W (2019) A survey of security services, attacks, and applications for vehicular ad hoc networks (VANETs). Sensors 19(16)
5. Booysen MJ, Gilmore JS, Zeadally S, van Rooyen GJ (2012) Machine-to-machine (M2M) communications in vehicular networks. KSII Trans Internet Inform Syst 6(2)
6. Piran MJ, Cho Y, Yun J, Ali A, Suh DY (2014) Cognitive radio-based vehicular ad hoc and sensor networks. Int J Distrib Sens Netw
7. Losilla F, Garcia-Sanchez AJ, Garcia-Sanchez F, Garcia-Haro J, Haas ZJ (2011) A comprehensive approach to WSN-based ITS applications: a survey. Sensors 11(11)
8. Azari A, Miao G (2014) Energy efficient MAC for cellular-cased M2M communications. In: IEEE global conference on signal and information processing (GlobalSIP)
9. 3rd Generation Partnership Project (2011) 3GPP TR 37.868 V11.0.0: study on RAN improvements for machine-type communications. Tech Rep
10. Deshpande KV, Rajesh A (2017) Investigation on IMCP based clustering in LTE-M communication for smart metering applications. Eng Sci Technol Int J 20(3)

11. Malandra F, Rochefort S, Potvin P, Sansò B (2017) A case study for M2M traffic characterization in a smart city environment. In: The 1st international conference on internet of things and machine learning
12. Yang F, Zou S, Tang Y, Du X (2016) A multi-channel cooperative clustering-based MAC protocol for V2V communications. Wirel Commun Mo Comput 16(18)

Ruggedizing LTE Security Using Hybridization of AES and RSA to Provide Double Layer Security

Anu Ahlawat and Vikas Nandal

Abstract Achieving better security for the transmission of data over cellular networks has always being a hard nut to crack. Security of long-term evolution (LTE) network is prime concern because of existence of malicious attackers. The security of network can be enhanced by utilizing cryptography. However, each cryptographic algorithm has its own pros and cons. So, hybridization (merging two or more algorithms) is one of the solutions to achieve better security. In this paper, double layer protection is implemented in order to ruggedize the security mechanism of LTE networks. The proposed algorithm utilizes better speed of AES and robust security of Rivest–Shamir–Adleman (RSA) together to achieve a hybrid system with improved performance compared to the traditional algorithm. The cryptanalysis has been intricated by increasing the number of keys to six. Further, the time required for encryption of keys is less when compared to decryption making the network more rugged and secure. In order to further enhance the speed, the concept of parallel processing is utilized. Then, proposed framework is analysed based on various parameters, and it is experimentally concluded that it provides approximately 22.7% increment in acceleration when compared to encryption rate and about 22.3% increment in acceleration when compared to decryption rate. 10% enhancement in percentage of avalanche effect has been achieved thereby increasing confusion in ciphertext and reducing chances of decoding. 5.39% increment in PSNR values of the proposed algorithm has been achieved when compared to existing algorithm.

Keywords Cryptosystem · AES · RSA algorithm · Private key · Public key · Encryption · Decryption · LTE

A. Ahlawat (✉) · V. Nandal
Department of Electronics and Communication, UIET, MDU, Rohtak, India
e-mail: ahlawat.anu@gmail.com

V. Nandal
e-mail: nandalvikas@gmail.com

A. Khanna et al. (eds.), *Data Analytics and Management*, Lecture Notes on Data Engineering and Communications Technologies 54,
https://doi.org/10.1007/978-981-15-8335-3_12

1 Introduction

The concern of security of data transmission through cellular network has become more crucial these days because cellular phones are widely used for various tasks that involve sensitive data transmission like credit card numbers, Internet banking details, social security numbers [1]. Much of the information shared over the air is highly confidential and should not be disclosed to adversaries. Hence, the effects of data security breach can be catastrophic. So, the following factors should be taken into consideration in order to achieve better security services for the cellular networks:

1. Authentication
2. Confidentiality
3. Integrity
4. Availability
5. Non-repudiation.

The introduction of all IP-based networks has been caused by a transition from circuit-switched network to packet-switched network [2]. This transition has made the network open and easily accessible but has introduced whole new set of vulnerabilities and threats. LTE is truly a fourth-generation network which provides high data rates, scalability, and quality of service and also supports secure wireless services. However, various possible threats and vulnerabilities in LTE network can be categorized into four types, namely

1. Attacks against Authentication
2. Attacks against Availability
3. Attacks against Privacy
4. Attacks against Integrity.

Authentication and privacy-preserving security mechanisms of LTE utilize various countermeasures to overcome or lessen the possibility of attacks. Figure 1 presents the classification of countermeasures used by LTE networks.

As it is evident from Fig. 1 that to protect the data from unauthorized access and prevent the discloser of confidential information, the concept of cryptography came into existence. Cryptography (a Greek word) is a technique of secret writing or it may be defined as an art which helps in encoding and decoding the information in order to prevent the disclosure of information among adversaries. It helps to achieve data security by providing authenticity, availability, integrity, and confidentiality. In LTE networks, AES algorithm is utilized for providing encryption and integrity protection in order to provide better authentication and privacy to the users [3]. However, AES has drawback of key-distribution. Thus, into order this drawback, hybridization is proposed.

Here in this present research work:

- Symmetric-key (AES) and asymmetric-key (RSA) algorithms are implemented to obtain a hybrid algorithm in order to improve efficiency and achieve better security.

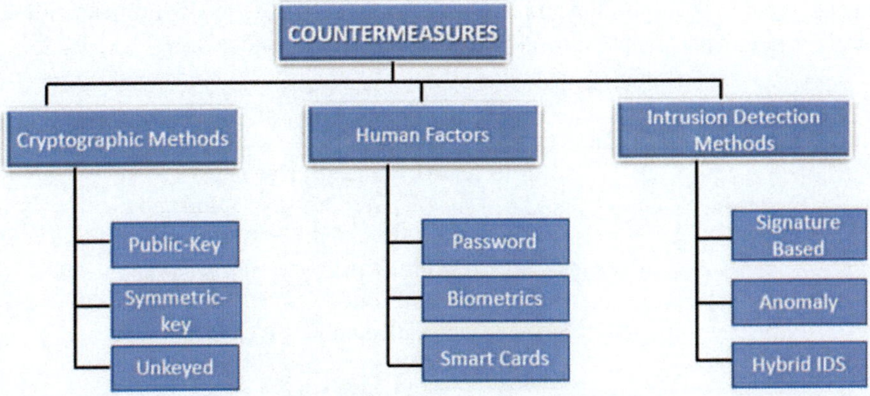

Fig. 1 Countermeasures used by LTE networks

- The keys used by AES-128 are encrypted before transmission using RSA to achieve double layer protection. The number of keys used in the proposed work is six making cryptanalysis more difficult, thereby enhancing the security and making the network more rugged.
- The confusion in ciphertext is intensified which improves the percentage of avalanche effect and decreases chances of decoding.
- Higher PSNR value is observed in the proposed framework which reduces distortion and increases the integrity of LTE network.

1.1 Symmetric-Key Algorithm

AES is a symmetric-key encryption standard. AES uses variable block and key sizes. In AES generally, the block size is fixed to 128 bits and variable key sizes of 128, 192, 256 bits are used [4]. Unlike DES, it is not based on Feistel structure rather it uses four different functions for each round. The length of the key determines the number of rounds in AES as depicted by Table 1.

Each round of AES-128 bit consists of four different functions. All the rounds consist of four stages except for the last round which consists of three stages except mix column [5]. The 128-bit input is at first copied into state matrix that is modified

Table 1 Key-round combinations

Length of key	Number of rounds
128-bits	10
192-bits	12
256-bits	14

at each round of the algorithm and finally after ten rounds of the algorithm an output matrix is obtained. The pseudocode for AES can be represented as:

```
{
AddRoundKey (0);
        for (round=1; round<9; round++)
         {
                SubByte ( );
                ShiftRows ( );
                MixColumn ( );
                AddRoundKey ( );
         }
                SubByte ( );
                ShiftRows ( );
                AddRoundKey ( );
}
```

1.2 Asymmetric-Key Algorithm

RSA is the most popular and widely used asymmetric cryptosystem. It is that public-key algorithm which performs encryption and decryption based on number theory. It is a block cipher which uses variable key sizes depending upon the requirements. Long keys are used for enhanced security, and short keys are used for better efficiency (typically size is kept 512 bytes) [6]. The entire RSA algorithm is divided into three sections:

1. Key Generation
2. Encryption
3. Decryption.

The aforementioned symmetric-key and asymmetric-key algorithms have their own set of strengths and weakness. AES is more efficient in terms of speed as compared to RSA, whereas RSA since uses two keys for encryption provides better security. There is problem of key-sharing in AES since the same key is being used for encryption and decryption; however, there is no such issue with RSA [7].

At present, researchers are continuously working on new cryptographic algorithm. In order to increase robustness of cryptographic algorithm, various factors are paid attention to, including number of keys used, complexity of algorithm, memory usage, time complexity, etc. Hybridization is one of the techniques which enhance the security of algorithms by utilizing the concept of merging two or more algorithms. Since each algorithm has its own set of strengths and drawbacks. Hybridization uses the strengths of one algorithm in order to overcome the drawbacks of another algorithm, and thereby the security achieved by hybridization is more robust and enhanced as compared to the traditional algorithm [8].

Protection of core network signalling protocols of LTE network is of utmost importance. Thus, adaptation of a hybrid model will help in enhancing user-to-network and end-to-end security mechanisms. Thus, to address security scheme of LTE, RSA and AES were together implemented in order to achieve better security for encryption and integrity protection in order to provide better authentication to the users.

The rest of the paper is organized as Sect. 2 provides current related research survey, and Sect. 3 describes the proposed framework adopted for hybrid implementation in order to enhance the robustness of security algorithm. Section 4 analyses and validates the results obtained. Finally, the paper is concluded in last section.

2 Current Related Research Survey

The authors in [9] proposed an efficient implementation of RSA algorithm by using two public keys instead of one as used in basic RSA algorithm so as to enhance the security. Then the proposed algorithm was compared with the basic RSA algorithm, and it was concluded that the proposed algorithm was less vulnerable to brute-force attacks and provided high communication overheads. Moreover, the two public keys were transmitted separately which further enhances the system security. The authors in [10] presented a new approach to enhance the efficiency and reliability of data transmission by using three Mersenne prime numbers to construct a modified RSA algorithm. The strength of modified RSA depends upon the unbreakable large prime numbers. The proposed approach of using three large prime numbers which makes factorization more difficult and hence system more secure. In [11], a novel hybrid technique which can be implemented for both wired and ad hoc networks was proposed. The hybrid technique used AES-128 and ECC-128 for encryption of data. Also, in order to achieve security of key ECDH algorithm was used. It was experimentally concluded that proposed framework provided higher security with low costs and complexity. The authors in [12] proposed a hybrid algorithm by using combination of RSA algorithm and block cipher so as to enhance the system security. However, the security strength of RSA algorithm depends upon the exponential process and the factorial number which is very difficult to factorize till now. However, by using the combination of aforementioned algorithms the authentication of plaintext was made more secured. The authors in [13] presented an enhancement in RSA algorithm by making use of optimized dynamic key management techniques in order to achieve better confidentiality, authenticity, and integrity of data transmitted. The proposed algorithm used five prime numbers making the factorization task difficult. The results were simulated using MATLAB12a. It was concluded that the security of proposed algorithm was increased but there was decreased in computation time. The authors in [14] proposed an enhanced version of RSA algorithm which provides more secure path for data transmission than conventional RSA algorithm. The value of 'n' is not transmitted over the air in the proposed algorithm due to which the factorization of 'n' becomes more difficult for the intruder. Instead the authors proposed a new number, i.e. 'f' which can be used for calculations of both private key and public key. The substitution of 'n' with 'f' makes the factorization of 'n' into two prime numbers 'p' and 'q' very time consuming, and difficult task hence increases the complexity of algorithm making it more secure. In this paper [15], authors proposed some modifications to basic AES and RSA in order to optimize their process of encryption and decryption. In order to reduce encryption and decryption time of AES algorithm, the number of computations of mix column and inverse mix column were reduced. Also, to further optimize the performance of AES the key expansion mechanism was enhanced. Chinese remainder theorem together with Montgomery modular multiplication was implemented for better performance of RSA algorithm. Then implemented modified algorithms in order to obtain mixed encryption system which provided higher speed, proper key management, and higher efficiency. In [16], authors proposed a novel method theta uses dynamic features of chaotic system and encryption techniques of modern S-AES. Due to the usage of chaos-based hybrid technique, the algorithm so obtained was more efficient in terms of memory usage and faster as compared to some other proposed hybrid algorithms. In [17], with the aim of preventing unauthorized access to channel a hybrid encryption algorithm was proposed by the authors. Hybridization was achieved by combining substitution and transposition techniques in order to enhance security of data or plaintext. The number of keys implemented was not predefined rather depended on the size of message to be encrypted. To evaluate the performance of proposed algorithm, possible keys combinations were analysed and it was concluded that with 2.560 X 106 possible combinations the cryptanalysis of the proposed algorithm was difficult. In [18], authors surveyed various traditional and modern hybrid encryption techniques including blowfish, DES-RSA, RSA based on ECC with AVK, etc., in order to provide comparative analysis of their performances. Blowfish was found to be more secure than AES, DES, and IDEA. Also, it was better to compress the data prior to encryption to enhance performance of security. However, quantum encryption techniques outperformed all other techniques in terms of ruggedness, requires less resources and also key-distribution problem is solved. In [19], authors used the concept of both steganography and cryptographic techniques to obtain a hybrid method. Random nature of chaotic neural network (CNN) is utilized for encrypting the message to be transmitted. The access to secret key is given only to authorized users to further enhance the security. The experimental results showed a higher value of peak signal-to-noise ratio (PSNR) which indicates higher security with less distortion.

3 Proposed Hybrid Algorithm

In this section, a novel hybrid technique which merges both the algorithms (AES and RSA), i.e. symmetric and asymmetric techniques, is proposed. Hybridization is proposed along with parallel processing in order to speed up the execution process with ruggedized security of the cellular networks. The proposed work is divided into two phases. Both the phases operate in parallel in order to achieve better speed with optimum efficiency of the network.

4 Encryption Process

The proposed algorithm uses AES and RSA for encryption process. At first step, the message to be encrypted (MG) is divided into two blocks, namely BL_1 and BL_2. The values of these blocks are given by Eqs. (1) and (2):

$$BL_1 = \sum_{i=1}^{(n-1)/2} MG_i \tag{1}$$

where n = total length of message (MG).

$$BL_2 = \sum_{i=(n+1)/2}^{n} MG_i \tag{2}$$

The values in two blocks BL_1 and BL_2 are converted into binary form and then groups of 128-bits are created, and the concept zero padding is utilized in case if any part of any block is not equal to 128-bits. Using a random number generator two keys, i.e. K_1 and K_2, with each being equal to 128-bits are generated.

Using K_1 key generated by random number generator and the values in BL_1 as input to AES-128-bit algorithm the encryption is carried out. The Encrypted Message1 (EMG_1) is given by Eq. 3:

$$EMG_1 = AES\_Enc(BL_1, K_1) \tag{3}$$

After getting EMG_1, the values are converted into binary form and bit-wise ex-ored with values of Key (K_1) as given by Eq. 4 and then back converted into hexadecimal values before transmission.

$$EMG_1' = Bitxor(EMG_1, K_1) \tag{4}$$

Using public key of RSA algorithm, the value of K_1 is also encrypted before transmission. The encryption is specified by Eq. 5:

$$K_1' = RSA\_Encrypt(K_1, publickey) \tag{5}$$

At the same time using the concept of parallel processing, the values in BL_2 are encrypted using AES and randomly generated key (K_2). The Encrypted Message2(EMG_2) is given by Eq. 6:

$$EMG_2 = AES\_Enc(BL_2, K_2) \tag{6}$$

After getting EMG_2, the values are converted into binary form and bit-wise ex-ored with values of Key (K_2) as given by Eq. 7 and then back converted into hexadecimal values before transmission.

$$EMG_2' = Bitxor(EMG_2, K_2) \tag{7}$$

Using public key of RSA algorithm, the value of K_1 is also encrypted before transmission. The encryption is specified by Eq. 8:

$$K_2' = RSA\_Encrypt(K_2, publickey) \tag{8}$$

After the above steps, the values of EMG_1' and EMG_2' are concatenated to get the final ciphertext and transmitted along with the encrypted keys, i.e.K_1' and K_2'.

5 Decryption Process

At the phase, the first step is to divide the concatenated message back in two parts to get values of EMG_1' and EMG_2'. After achieving this, the values of keys obtained are decrypted using private key component of RSA algorithm as given by Eqs. 9 and 10.

$$K_1 = RSA\_Decrypt\left(K_1', privatekey\right). \tag{9}$$

$$K_2 = RSA\_Decrypt\left(K_2', privatekey\right) \tag{10}$$

When the values of both the keys are decrypted, the parallel processing is again implemented at the receiver's side in order to speed up the decryption phase. After getting EMG_1', the values are converted into binary form and bit-wise ex-ored with values of Key (K_1) as given by Eq. 11 and then back converted into hexadecimal values before decryption.

$$EMG_1 = Bitxor\left(EMG_1', K_1\right) \tag{11}$$

Using K_1 as Key and EMG_1 as input to AES-128-bit algorithm the decryption is carried out. The decrypted Message (BL_1) is given by Eq. 12:

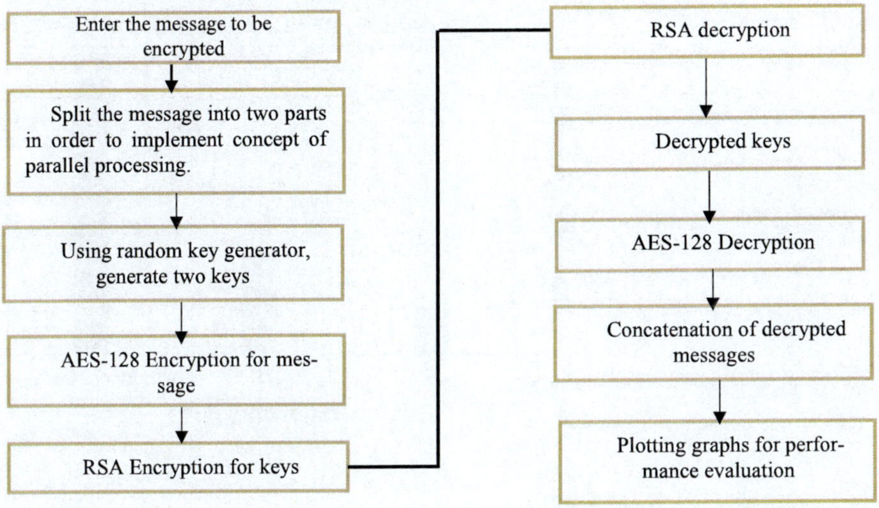

Fig. 2 Flow diagram of proposed hybrid implementation (AES and RSA)

$$BL_1 = AES\_Dec(EMG_1, K_1) \tag{12}$$

At the same time using the concept of parallel processing, the values in EMG_2 are decrypted using MAES and key (K_2). The decrypted Message2(BL_2) is given by Eq. 13:

$$BL_2 = AES\_Dec(EMG_2, K_2) \tag{13}$$

Finally, the values of both the decrypted values are concatenated in order to obtain the original message which was transmitted. The flow diagram given below explains the proposed hybrid model (Fig. 2).

6 Results and Discussions

The hybrid implementation has been proposed in order to achieve better security without compromising system complexity and speed of cellular networks. The following section clearly analysis that the requirements are meet by taking into consideration various parameters including number of keys used for algorithm, time taken for encryption and decryption of keys, total time taken for encryption and decryption of string, PSNR, and diffusion and confusion tests.

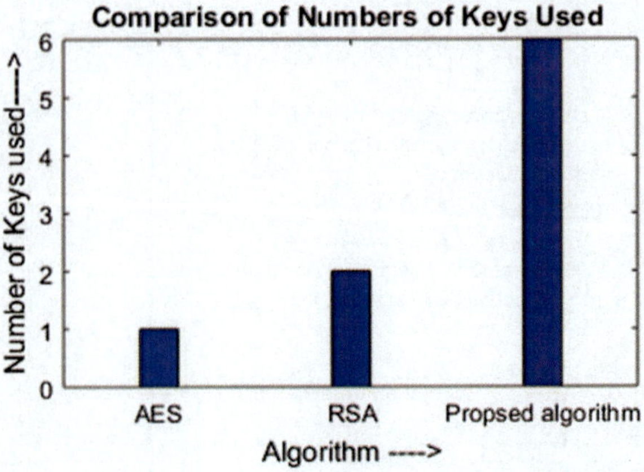

Fig. 3 Comparison of number of keys used

6.1 Number of Keys Used

The number of keys used in hybrid implementation of AES with RSA is compared to the traditional AES and RSA is depicted in Fig. 3. The AES algorithm uses single key for encryption and decryption as a result of which key-distribution is one of the major problems faced in such algorithms. RSA algorithm uses two keys, i.e. public and private keys. However, the proposed framework uses six keys making the cryptanalysis more difficult and thereby enhancing security and making the network more rugged.

6.2 Encryption and Decryption Time for Keys (K_1 and K_2)

The keys used for encryption using AES-128 are encrypted before transmission using RSA algorithm in order to achieve double layer protection. The time required for encryption and decryption of key using RSA is shown in Figs. 4 and 5. The time required for encryption is less when compared to decryption which indicates an enhancement in security. Because of higher decryption time, it can be concluded that performing cryptanalysis of key will be a difficult task for adversaries, and hence, data will be more securely transmitted.

6.3 Time Taken for Encryption and Decryption Phases

The total time taken to convert the given message string into ciphertext is defined as encryption time and the total time taken for obtaining original message string from ciphertext is termed as decryption time.

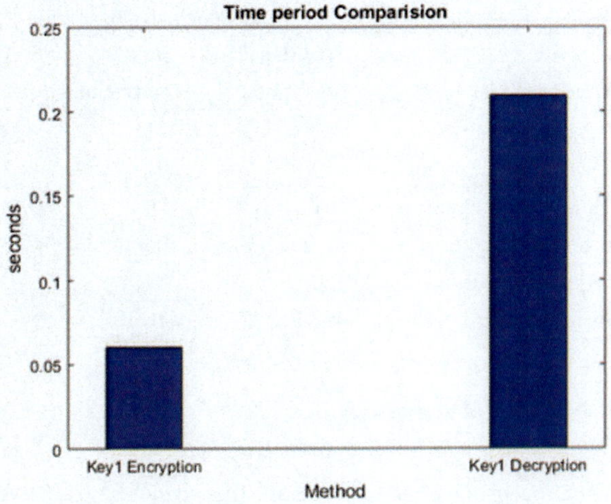

Fig. 4 Time period for encryption and decryption of K_1

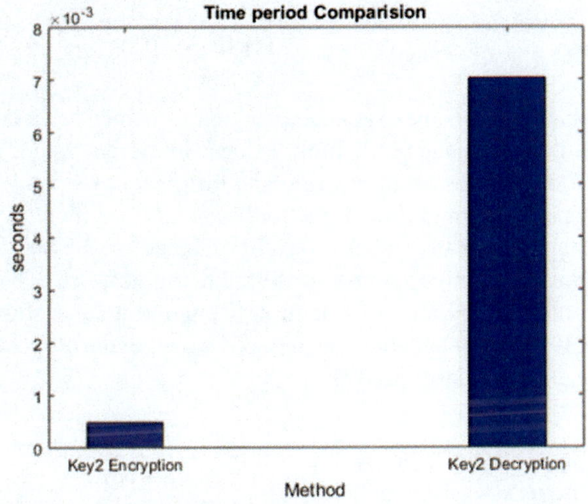

Fig. 5 Time period for encryption and decryption of K_2

Table 2 provides comparison of the encryption and decryption time of proposed algorithm with traditional AES algorithm. In order to validate the experimental results obtained, process of encryption and decryption is performed 10000 times on five different groups of strings.

Table 2 depicts an increment of acceleration of about 22.7% when compared to encryption rate and about 22.3% increment of acceleration when compared to decryption rate. Hence, the proposed algorithm provides better speed when compared to traditional AES without comprising security. This increment is achieved by parallel processing of message string.

Table 2 Encryption and decryption time comparison

Algorithm	Traditional AES algorithm		Proposed algorithm	
Time (ms)	Encryption	Decryption	Encryption	Decryption
First sample string	470	491	365	377
Second sample string	450	501	348	380
Third sample string	461	490	353	375
Fourth sample string	471	500	368	390
Fifth sample string	460	481	350	365
Average value	462.1	486.3	356.8	377.4

6.4 PSNR Analysis

After the process of encryption, the proposed framework is tested by determining PSNR values. The PSNR value is calculated using the formula given below:

$$\mathrm{PSNR} = 10\log_{10}\frac{R^2}{\mathrm{MSE}} \tag{14}$$

where MSE is mean square error and is defined as error between the encrypted data and cover image used. Table 3 depicts different values of PSNR obtained by taking different sizes of input string and different cover images. Since the value of PSNR is high so amount of distortion is less, it is difficult to detect the change in cover image. Since the values of PSNR obtained are higher, therefore the integrity of LTE network is more compared to traditional network. The value of PSNR obtained is compared to values obtained in [7] and plotted as shown in Fig. 6. There is about 5.39% increment obtained in PSNR values of the proposed algorithm when compared to existing algorithm [7].

Table 3 PSNR values for different sizes of ciphertext

Size and type of cover image used	Size of encrypted message	PSNR value
Peppers.png (384 × 512 × 3)	2000 bytes	72.1775
	4000 bytes	67.0500
	7000 bytes	65.4277
Cameraman.tif (256 × 256)	2000 bytes	62.8301
	4000 bytes	58.6153
	7000 bytes	55.8314

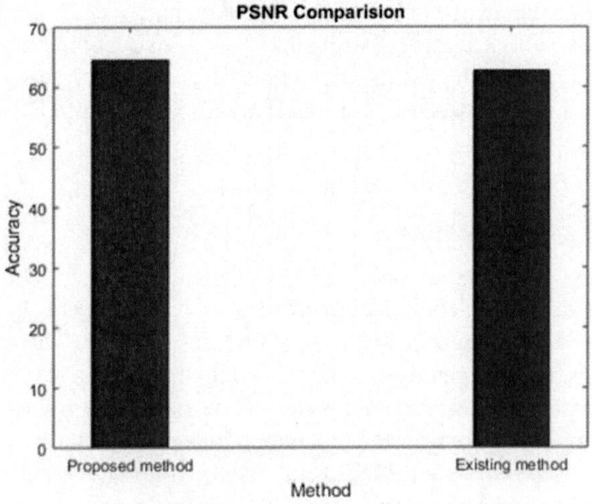

Fig. 6 Comparison of PSNR obtained with Existing algorithm

6.5 *Validation Using Confusion and Diffusion Tests*

In order to reduce the chances of statistical analysis of transmitted message by adversaries the concept of diffusion and confusion proposed by Shannon are taken into consideration. At first, the number of bits is varied in plaintext without variation in number of bits of keys and corresponding variations in ciphertext are recorded and tabulated in Table 4.

Now, the confusion of the proposed algorithm is tested using variations in key keeping the plaintext constant. The recorded variations are given in Table 5.

The changes obtained in ciphertext when the bits in plaintext and ciphertext are varied are approximately 10% more than the original algorithm. This indicates an

Table 4 Number of bits variation in ciphertext when key remains unchanged

No of bits changed in Plaintext	No of bits variations in ciphertext	
	Traditional AES	Proposed algorithm
1-bit	61 ± 6	64 ± 4
2-bit	63 ± 4	65 ± 6
3-bit	60 ± 6	63 ± 6

Table 5 Number of bits variation in ciphertext when plaintext remains unchanged

No of bits changed in Plaintext	No of bits variations in ciphertext	
	Traditional AES	Proposed algorithm
1-bit	63 ± 4	66 ± 6
2-bit	62 ± 6	64 ± 6
3-bit	64 ± 4	65 ± 4

enhancement in percentage of avalanche effect, thereby increasing the confusion in the ciphertext and reducing the chances of decoding the message transmitted. Hence, due to the improved avalanche effect, the encryption mechanism of LTE networks will be more secure and robust when compared to basic AES.

7 Conclusion

In this paper, double layer protection achieved by the hybridization of AES with RSA provides enhanced security with less distortion. The concept of parallel processing has been efficiently combined with hybridization resulting in about 22.7% increment in acceleration rate. To make the system more robust in terms of key-locking, the number of keys used is six which makes cryptanalysis task cumbersome. To further add, the values of PSNR obtained are high which indicates less distortion and more efficiency. Hence, the concept hybridization has helped to overcome the cons of the traditional algorithms and make system more secure and robust. Thus, the encryption and integrity protection mechanism of LTE which provides better authentication as well as privacy to the users which were earlier based on AES algorithm, when implemented using the proposed hybrid algorithm provides ruggedized security and better efficiency.

References

1. Viswanath G, Krishna PV (2020) Hybrid encryption framework for securing big data storage in multi-cloud environment. Evol Intel. https://doi.org/10.1007/s12065-020-00404-w
2. Mohanty SN, Ramya KC, Rani SS, Gupta D, Shankar K, Lakshmanaprabu SK, Khanna A (2020) An efficient lightweight integrated blockchain (ELIB) model for IoT security and privacy. Future Gener Comput Syst 12:1027–1037. https://doi.org/10.1016/j.future.2019.09.050
3. Shankar K, Lakshmanaprabu SK, Gupta D, Khanna A, de Albuquerque VHC (2018) Adaptive optimal multi key based encryption for digital image security. Wiley Online Library. https://doi.org/10.1002/cpe.5122
4. Tarigan SY, Ginting DS, Gaol ML, Sitompul KL (2017) The combination of RSA and block chiper algorithms to maintain message authentication. In: International conference on information and communication technology (IconICT). https://doi.org/10.1088/1742-6596/930/1/012009
5. Bhargavi I, Veeraiah D, Maruthi Padmaja T (2017) Securing BIG data: a comparative study across RSA, AES, DES, EC and ECDH. In: Satapathy S, Bhateja V, Raju K, Janakiramaiah B. (eds) Computer communication, networking and internet security, Lecture notes in networks and systems, vol 5. Springer, Singapore. Online ISBN 978-981-10-3226-4, https://doi.org/10.1007/978-981-10-3226-4_36
6. Suri S, Vijay R (2017) An AES–chaos-based hybrid approach to encrypt multiple images. In: Patnaik S, Popentiu-Vladicescu F (eds) Recent developments in intelligent computing, communication and devices, Advances in intelligent systems and computing, vol 555. Springer, Singapore. Online ISBN 978-981-10-3779-5, https://doi.org/10.1007/978-981-10-3779-5_6

7. Kumar B, Hussain M, Kumar V (2018) BRRC: a hybrid approach using block cipher and stream cipher. In: Saeed K, Chaki N, Pati B, Bakshi S, Mohapatra D (eds) Progress in advanced computing and intelligent engineering, Advances in intelligent systems and computing, vol 563. Springer, Singapore. Online ISBN 978-981-10-6872-0, https://doi.org/10.1007/978-981-10-6872-0_21
8. Dash S, Das M, Das M (2018) Implementation of chaotic-based hybrid method to provide double layer protection. In: Pattnaik P, Rautaray S, Das H, Nayak J (eds) Progress in computing, analytics and networking, Advances in intelligent systems and computing, vol 710. Springer, Singapore. Online ISBN 978-981-10-7871-2, https://doi.org/10.1007/978-981-10-7871-2_34
9. Ghosh Ramkrishna (2016) An efficient and robust modified RSA based security algorithm in modern cryptography. J Comput Sci Eng Inform Technol Res (JCSEITR) 6(2):15–22
10. Padmaja ChJL, Bhagavan VS, Srinivas B (2016) RSA encryption using three Mersenne primes. Int J Chem Sci 14(4):2273-2278. ISSN 0972-768X
11. Chourasia S, Singh KN (2016) An efficient hybrid encryption technique based on DES and RSA for textual data. In: Satapathy S, Mandal J, Udgata S, Bhateja V (eds) Information systems design and intelligent applications, Advances in intelligent systems and computing, vol 433. Springer, New Delhi. Online ISBN 978-81-322-2755-7, https://doi.org/10.1007/978-81-322-2755-7_9
12. Dutta SC, Singh S, Singh DK (2016) Enhancement of mobile ad hoc network security using improved RSA algorithm. In: Pant M, Deep K, Bansal J, Nagar A, Das K (eds) Proceedings of fifth international conference on soft computing for problem solving, Advances in intelligent systems and computing, vol 436. Springer, Singapore, pp 1027-1036. Online ISBN 978-981-10-0448-3, https://doi.org/10.1007/978-981-10-0448-3_86
13. Talukdar D, Saikia LP (2017) Simulation and analysis of modified RSA cryptographic algorithm using five prime numbers. Int J Recent Innov Trends Comput Commun 5(6):224–228
14. Sahu J, Singh V, Sahu V, Chopra A (2017) An enhanced version of RSA to increase the security. J Netw Commun Emerg Technol (JNCET) 7(4):1-4
15. Liu J, Fan C, Tian X, Ding Q (2018) Optimization of AES and RSA algorithm and its mixed encryption system. Adv Intell Inform Hiding Multimedia Sig Process Smart Innov Syst Technol 82:393-403. https://doi.org/10.1007/978-3-319-63859-1-48
16. ÜnalÇavu S, Kaçar S, Zengin A, Pehlivan I (2018) A novel hybrid encryption algorithm based on chaos and S-AES algorithm. Nonlinear Dyn 92(4):1745-1759. ISSN 1573-269X, https://doi.org/10.1007/s11071-018-4159-4
17. Ghosh P, Thakor V (2018) Optimization of hybrid encryption algorithm for secure communication system. In: Third international congress on information and communication technology, pp 973-981. Online ISBN 978-981-13-1165-9, https://doi.org/10.1007/978-981-13-1165-9_89
18. Dixit P, Gupta AK, Trivedi MC, Yadav VK (2018) Traditional and hybrid encryption techniques: a survey. In: Networking communication and data knowledge engineering, pp 239-248. Online ISBN 978-981-10-4600-1, https://doi.org/10.1007/978-981-10-4600-1_22
19. Farooq S, Prashar D, Jyoti K (2018) Hybrid encryption algorithm in wireless body area networks (WBAN). In: Singh R, Choudhury S, Gehlot A (eds) Intelligent communication, control and devices, Advances in intelligent systems and computing, vol 624. Springer, Singapore. Online ISBN 978-981-10-5903-2, https://doi.org/10.1007/978-981-10-5903-2_41

Performance Evaluation of Merging Techniques for Handling Small Size Files in HDFS

Vijay Shankar Sharma and N. C. Barwar

Abstract When dealing with the storage of large files, HDFS is one of the good choices as a distributed storage. Processing a large number of small files results in the performance bottleneck of HDFS. A massive number of small files will produce excessive metadata that leads to inefficient utilization of the Name Node memory, and frequent function calls will consume all over more time to process; therefore, it can be concluded that HDFS degrades when handling with small files. A detailed performance evaluation is being conducted to understand the impact of increasing small files in Hadoop for processing. This paper mainly evaluates sequential files, CombineFileInputFormat, HAR and Hadoop streaming techniques to deal with small file problem in HDFS. Empirical evaluation conducted in this paper shows that HAR and CombineFileInputFormat perform better and have consistent and stable results when increasing number of files for processing.

Keywords Hadoop · MapReduce · HAR · Hadoop streaming · Sequential file · CombineFileInputFormat · Small files · HDFS

1 Introduction

The function of Hadoop is identified by its two major core components, i.e., Hadoop Distributed File System (HDFS) and MapReduce. The HDFS stores huge data into the computing nodes. This huge data cannot be stored directly to the computing nodes; HDFS divides the large file into 128 MB data chunks, and these data chunks

---

This Research work is done under the Project "The Optimization of Storage and Access Efficiency in Hadoop Framework for small file applications." The project is sponsored by the TEQIP-III at M.B.M Engg. College, Jodhpur (Rajasthan), India.

V. S. Sharma (✉) · N. C. Barwar
Department of Computer Sc. & Engg., MBM Engineering College, Jodhpur, India
e-mail: vijay.mbmit09@gmail.com

N. C. Barwar
e-mail: ncbarwar@gmail.com

A. Khanna et al. (eds.), *Data Analytics and Management*, Lecture Notes on Data Engineering and Communications Technologies 54,
https://doi.org/10.1007/978-981-15-8335-3_13

are executed in parallel by the MapReduce. The power of Hadoop is such that it can handle the large file applications in a much efficient way but Hadoop does not perform well in case of the small file applications, i.e., file produced by weather sensors, powerpoint and pdf files, files produced by social media network servers, bioinformatics files and geographic information files. When handling with the small files, the performance of the Hadoop is impacted badly. Although Hadoop provides a variety of alternative MapReduce techniques to solve the problem of small files, these strategies are not directly designed for the small files. To understand the problem of small files, it should be clear that the system will identify a file as small file if its size is less than 50% of the HDFS block size. Name Node of the Hadoop is responsible for storing metadata and data block information over the network, and each metadata occupies a certain amount of memory.

Processing millions of small files will lead to a lot of issues like there will be the requirement of more space for Name Node in RAM, network traffic will increase that results in consumption of more time to store the data, MapReduce will take longer times to process the requests, etc.; therefore, there is the requirement of the efficient data management technique that can solve the problem of handling small files efficiently. Several solutions are proposed to the problem of handling small files in the Hadoop, i.e., Hadoop Archives (HAR), sequential file, CombineFileInputFormat, Hadoop streaming, etc. In this paper, performance evaluation of main merging techniques is being conducted on the different number of files, whereas the size of the dataset is the same in all the cases. The parameters considered for the performance evaluation are execution time of the merging techniques, HDFS_Byte_Read, HDFS_Byte_Written, HDFS_Read_Operations and HDFS_Write_Operations. **Experimental evaluation conducted in this paper shows that HAR and CombineFileInputFormat perform better in comparison with the sequential file and Hadoop streaming and have consistent and stable results when increasing number of files for processing, whereas the performance of the Hadoop streaming and sequential file will degrade as the numbers of files is increased.**

The other sections of the research paper are organized as follows. Section 2 explains the small file problem in detail. Section 3 briefs various approaches for the solution of the small file storage in HDFS. Section 4 presents merging techniques, their merits and demerits for managing small files in Hadoop. Section 5 elaborates detailed empirical evaluation of the merging techniques. The last Sect. 6 of the research papers presents the overall conclusion.

2 Problem of Small File in HDFS

The problem of small files can easily be explained by the fact that native Hadoop is known especially for fault-tolerant, consistent storage, efficient processing and management of the large/massive files. There is a server in the Hadoop called "Name Node" that will be responsible for managing all the files in HDFS for that Name Node keeps the metadata of each file that will be stored in the HDFS in its main memory; in this situation, when there are massive small files to process, the overall performance of the HDFS degrades severely because these small files impose an excess burden on the Name Node performance in terms of handling smaller data blocks that are less than the predefined size of the data block and small files will require more space to manage and process these files. In HDFS, the I/O performance of the files is poor because HDFS does not provide any prefetching technique and also does not consider correlations among the files to improve the access efficiency. Due to the increasing popularity of small file applications in the Web world, it attracts the attention of the researchers across the globe that the Hadoop framework should be modified in such a way that it can efficiently handle the small files. As a consequence, a number of solutions are proposed to the small file problem, i.e., HAR [1], SequenceFile [2], CombineFileInputFormat [3] and Hadoop Streaming [4].

3 Related Work: A Brief Survey

Hadoop Perfect File (HPF) is a type of new archive file that is proposed by the Tchaye-Kondi et al. [5]; in this archive file, there is no need to process the whole file, and metadata of a particular file can be accessed directly with the help of its index file. The distribution of the metadata is accomplished through index files with the help of dynamic hash function, and the position of each file's metadata in the index file will be maintained by a special-order preserving hash function. Jing et al. [6] proposed dynamic queue-based solution—Dynamic Queue of Small Files (DQSF). This solution uses an analytical hierarchal process which classifies the small files into several different categories. Peng et al. [7] proposed SHDFS. It is based on the native Hadoop, and there are two main modules—caching module and merging module. Merging module is responsible for reducing the total number of files; for that, user-oriented collaborative filtering is used to identify co-related files, and then, these correlated files are merged; thus, files get reduced. In the caching module, frequently accessed data can search out using the log-linear model, and a special memory subsystem is proposed by the author to cache the frequently accessed data. In this way, by the use of merging and caching modules, SHDFS provides better performance in comparison with the original HDFS. Cai et al. [8] consider the distribution and correlation of the files in the merging algorithm. The correlation between files strengthens the identification of relevant files and reduces the required HDFS block to process the small files. A comparison has been carried out in this paper between native HDFS access method and PS algorithm-based access method; experimental results show that access efficiency and storage efficiency are improved at a significant level. Kim et al. [9] presented a digital archive that is based on the access patterns and data characteristics of the application. Massive small files can easily be handled by this digital archive. A modification to the on-disk in-node structure of the existing file system was proposed by the author, which leads to reduced memory I/O operations and hence improve the overall performance.

Lyu et al. [10] introduced an optimized algorithm; in this algorithm, size of small files is considered while merging the massive small files and map record is created for each small file. Experimental results show that the use of prefetching and caching technique improves the access efficiency. To enhance the file accessing in the cloud environment, Fu et al. [11] introduced a block replica placement algorithm. Based on the relevance among the small files, they are characterized in four dimensions, and then, merging will be carried out according to the types of the file. Experimental result shows that D2I can efficiently reduce the read/write for small files. Mu et al. [12] presented an efficient method that makes use of the secondary indexes and enhances the storage architecture. Experimental results show that the proposed scheme in the paper will improve storage efficiency and reduce the Name Node memory space. Wang et al. [13] presented modified PLSA that will be used to construct correlations among access files, applications and access tasks. Files are merged based on access tasks, and prefetching targets are selected based on transition probability of the tasks. He et al. [14] presented a merging technique that is based on the balance of data blocks. This algorithm optimizes the volume distribution of the merged file that results in reducing the data blocks of HDFS. The reduced data blocks lower the memory overheads of major nodes of the cluster and improve the overall performance. The author showed that the use of balancing data block in the merging algorithm will improve the performance significantly. Fu et al. [15] presented flat storage architecture (iFlatLFS) that is based on the simple metadata management strategy. The proposed iFlatLFS can handle the millions of small files effectively and efficiently. In this paper, metadata occupies lesser space in comparison with the traditional file system. The author showed that iFlatLFS along with a new metadata strategy will greatly improve the results in terms of accessing the small files and storage efficiency. Mao et al. [16] presented an optimized merging technique SIFM; in this technique, multi-level file indexes are used. The results show that in this algorithm, caching and prefetching techniques are used to accelerate the access efficiency, and structured metadata storage is used to the I/O operations. In this paper, the author compares the proposed SIFM algorithm with the original Hadoop and Hadoop Archive. The experiment shows that storing and accessing of the small files in HDFS become easy and cost-effective, The SIFM can effectively manage the millions of small files in Hadoop framework. Prasad et al. [17] presented an efficient indexing and merging technique. The experimental results show that there is a big enhancement in the performance of processing small files. The proposed techniques improve the overall performance for managing the millions of small files significantly and also reduce the name node memory utilization. Dong et al. [18] proposed a new technique for managing the small files. In this technique, small files are identified in two categories that are logically oriented small files and structurally oriented small files. The experimental result shows that this technique improves the storage and access efficiencies when file prefetching and merging scheme are applied for structurally oriented small files, and prefetching and file grouping technique are used for logically oriented small files. Ahad et al. [19] presented a dynamic merging technique (DM-SFS) for efficient processing of the small files. In this technique, filtering criteria, that is based on the file size and type, will be applied over the incoming data/files. Security to the data is also provided using the Twofish cryptographic technique.

4 Merging Techniques for Handling Small Size Files

4.1 HAR

Small files can be archived using the HAR. HAR reduces the Name Node memory usage. HAR is the general-purpose solution for handling the small file issues in Hadoop. HAR contains three files; two of them are index file and another one is a data file; due to two index files, the access efficiency of HAR is poor, and Hadoop archives are created by combining small files into large files; this will reduce the map operations and lessen the storage overhead on Name Node that results in the increased performance of the Hadoop. "Hadoop Archive" command is used to create the HAR file:

hadoop archive—archiveName [Archive Name]—p <Parent Dir> <Source Path> <Destination Path>

Example: hadoop archive—archiveName smallfiles.har—p/hduser/hadoop/sourcedir1 hduser/hadoop/outdir.

4.1.1 Benefits of HAR Files

- HAR can effectively address the namespace issues related to storing various small files.
- To achieve high-level transparency while accessing original files, small files are merged into large files
- HAR reduce the memory usage of the Name Node and namespace; therefore, it will increase the overall scalability of the system,
- Memory optimization is done in orthogonal fashion in the Name Node and distributing namespace management across multiple Name Nodes.
- HAR provides parallel access to the original files by MapReduce and compatible with MapReduce.

4.1.2 Limitation of HAR Files

- HAR files are immutable. One cannot add or remove files after creation of the archive.
- HAR does not support the compression of files.
- Small files are processed separately in HAR by the mapper that is inefficient.

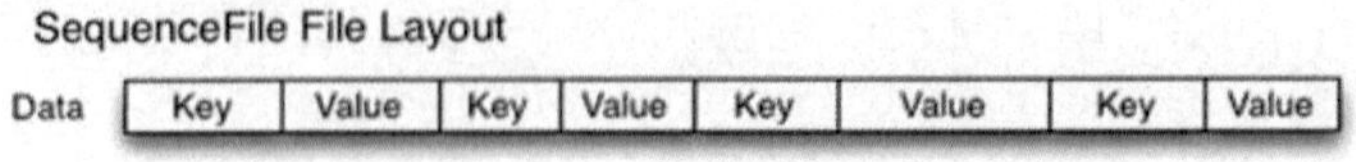

Fig. 1 Sequential file layout

4.2 Sequence File

Hadoop API provides a binary file format called "Sequential File"; by combining several small files, sequential files are created. The structure of sequential files is based on the binary key–value pair. Different compression techniques are used for compression of file blocks in small files. Poor access efficiency is the main disadvantage of the sequential files because HDFS reads the content of the files sequentially (Fig. 1).

4.2.1 Benefits of Sequential Files

- Sequential files are more compact than text files as they use binary files.
- Sequential files provide optional support for compression at a record and block level.
- Sequential files can be executed in a parallel fashion using the concept of file splitting.
- Sequential files pretend as large files for HDFS and MapReduce by serving as a container for small files and avail all the benefits of MapReduce for small files.
- The intermediate results of mappers are also stored using sequence file format.

4.2.2 Limitations of Sequential Files

- Sequence files are appended only.
- Interaction with sequential files can only be done using Java API. It does not provide multi-language support.

4.3 CombineFileInputFormat

CombineFileInputFormat is a new type of I/P format that accepts multiple small files as input. In this format, a mapper can process multiple data; therefore, map and reduce operations are faster in comparison with other techniques. The modification to the input format class can be achieved by combining multiple files in a single spilt; hence, each map operation will receive more inputs to execute. When mappers are receiving more input, the time required to execute the millions of small files get reduced significantly. In HDFS, generally, there is a provision of the more than one mapper and single reducer; to achieve the parallelism, CombineFileInputFormat allows multiple reducers. Name Node is accountable for the combining small files in single input split, and this input split given to the map task that generates intermediate results. Now, these intermediate results are available as input for the multiple reducers. Multiple reducers provide sorted merge output that will be used further for processing. In such a way, CombineFileInputFormat reduces the processing time and achieves more than average performance improvement [20].

4.4 Hadoop Streaming

Hadoop provides a utility called “Hadoop Streaming” for creating and running MapReduce tasks with any script or executable as the reducer and/or the mapper. Reducer and mapper are executables that read the input from standard input and emit the output to standard output. In our case, small files are merged using Hadoop streaming by setting mapper and reducer as the command “cat” (Fig. 2).

Example: hadoop jar/usr/local/hadoop/share/Hadoop/tools/lib/hadoop-streaming-3.1.3 (Path to Hadoop—Streaming file in Hadoop Distribution)—Dmapred.reduce.tasks = 1 (Number of Reducers)—input “Path to Input Directory”—output “Path to Output Directory”—mapper cat—reducer cat.

5 Empirical Evaluation

The following experimental setup is being adopted to evaluate the performance of the merging strategies.

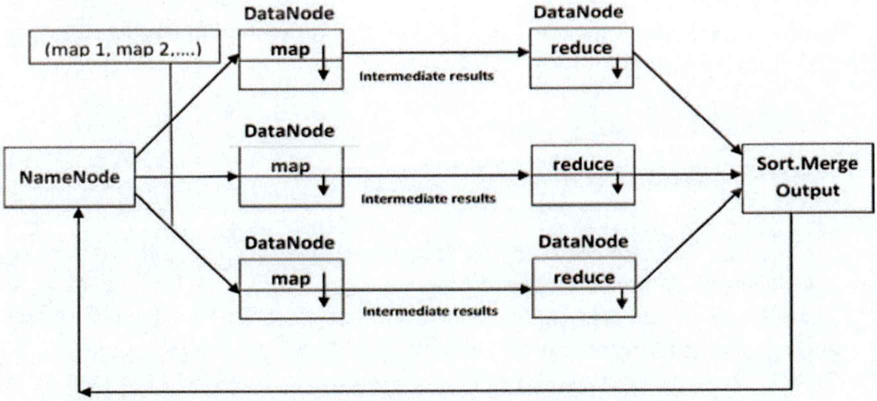

Fig. 2 Combined file input format work flow [21]

Table 1 Execution time (Sec.) of merging strategies

S. No.	No. of files	CombineFileInputFormat	Execution time (Sec.)		
			Sequential file	HAR	Hadoop streaming
1	41	232	205	168	367
2	1500	258	280	182	401
3	3741	262	338	175	402

- **Processor**: Intel® core™ i5-7500 CPU@3.40 GHz, 64-bit Operating System
- **Installed RAM**: 4.00 GB
- **Java**: Open JDK-11.0.4
- **Operating System**: 18.04.1 LTS
- **Hadoop**: Hadoop-3.1.3 with 4 Data Nodes and 1 Name Node (Replication Factor:04)
- **Ethernet**: 1 GBPS (Backbone)/100 MBPS Network
- **Input Dataset**: 1 GB size text files are considered (Netflix Prize Data) for the evaluation. To explore the effect of number of files on the performance of the merging strategies, the same dataset is splitted into 41, 1500 and 3741 files.
- **Job Execution**: All the merging strategies, i.e., CombineFileInputformat, sequential file, Hadoop streaming and HAR, are executed on 41, 1500 and 3741 text files.
- **Parameters for Evaluation**: Execution Time (Sec), HDFS_Byte_Read, HDFS_Byte_Written, HDFS_Read_Operation, HDFS_Write_Operation.

5.1 Execution Time

Table 1 presents the processing time of the merging strategies over different number of files.

Figure 3 displays the comparative analysis of execution time of merging strategies over the different number of files.

5.2 HDFS_Bytes_Read

Table 2 presents the number of bytes read by the map/reduce task for merging strategies over the different number of files.

Figure 4 displays the comparative analysis of bytes read by the map/reduce task for merging strategies over the different number of files.

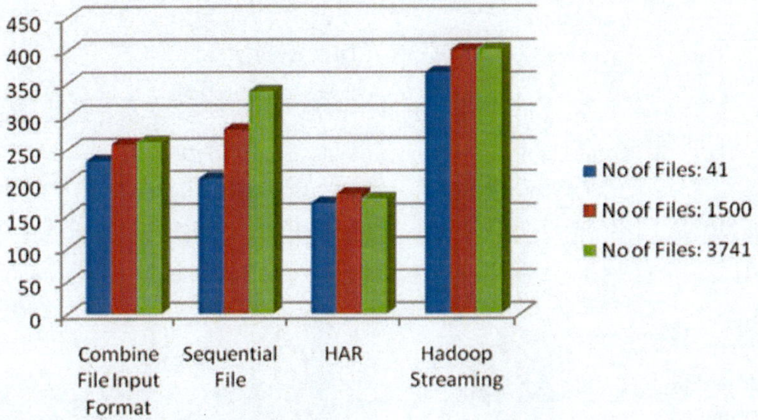

Fig. 3 Execution time (Sec.) of merging strategies

Table 2 Map/reduce byte read

S. No.	No. of files	CombineFileInputFormat	HDFS_Bytes_Read		
			Sequential file	HAR	Hadoop streaming
1	41	18412248118	30653727951	2095195122	36158751951
2	1500	22828721648	826665121884	2102424766	826665121884
3	3741	19191670550	3264159578760	1834320282	3264159578760

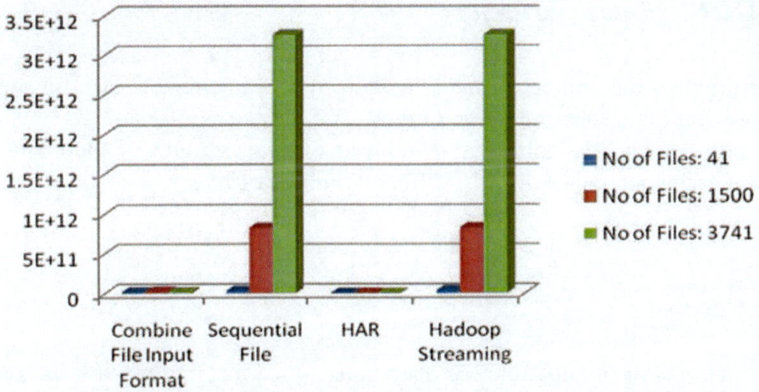

Fig. 4 Map/reduce byte read

Table 3 Map/reduce byte written

S. No.	No. of files	CombineFileInputFormat	HDFS_Bytes_Read		
			Sequential file	HAR	Hadoop streaming
1	41	2926832491	6442606362	2095201631	7694356162
2	1500	3149019860	7387808469	2102618346	7720784988
3	3741	2168431703	6470007374	1834826967	6609345138

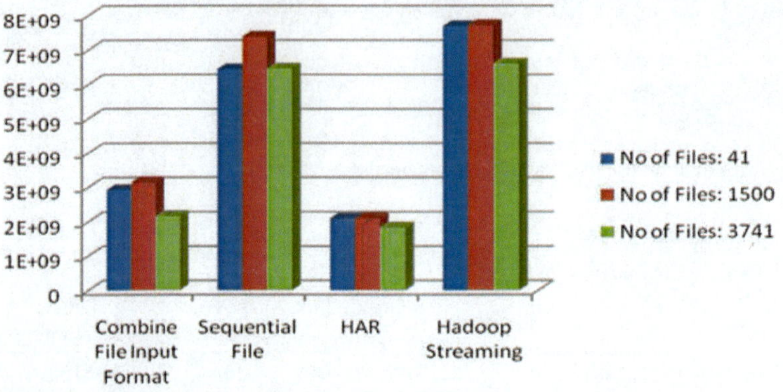

Fig. 5 Map/reduce byte written

5.3 *HDFS_Bytes_Written*

Table 3 presents the number of bytes written by the map/reduce task for merging strategies over the different number of files.

Figure 5 displays the comparative analysis of bytes written by the map/reduce task for merging strategies over the different number of files.

5.4 *HDFS_Read_Operation*

Table 4 presents the number of read operations by the map/reduce task for merging strategies over the different number of files.

Figure 6 displays the comparative analysis of read operations by the map/reduce task for merging strategies over the different number of files.

Table 4 Read operations by map/reduce task

S. No.	No. of files	CombineFileInputFormat	HDFS_Read_Operation		
			Sequential file	HAR	Hadoop streaming
1	41	1491	2973	200	3405
2	1500	33620	2307618	6050	2307618
3	3741	77785	14149099	15006	14149069

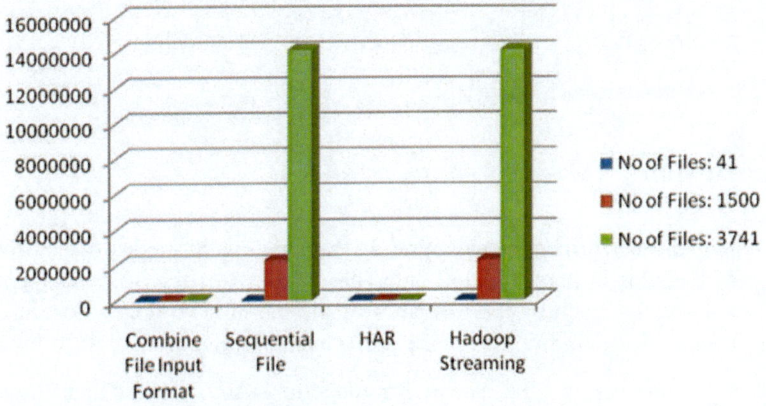

Fig. 6 Read operations by map/reduce task

5.5 *HDFS_Write_Operation*

Table 5 presents the number of write operations by the map/reduce task for merging strategies over different number of files.

Figure 7 displays the comparative analysis of write operations by the map/reduce task for merging strategies over the different number of files.

Table 5 Write operations by map/reduce task

S. No.	No. of files	CombineFileInputFormat	HDFS_Write_Operation		
			Sequential file	HAR	Hadoop streaming
1	41	207	232	18	236
2	1500	211	1697	18	1697
3	3741	77785	3936	18	3931

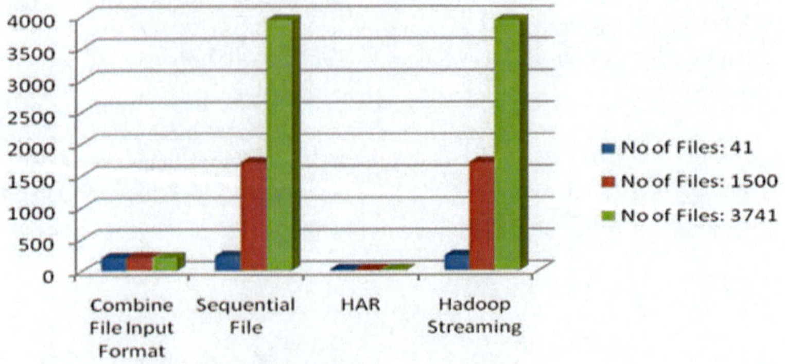

Fig. 7 Write operations by map/reduce task

6 Conclusion

In this paper, the performance evaluation of the merging strategies over different numbers of the files is being carried out. The objective of this research paper is to identify the variation in the performance of the merging strategies for different number of files. The main findings of the performance evaluation are as follows:

- In terms of processing time, the performance of HAR is better than the other merging strategies. Sequential file strategy has a direct impact on the processing time, and adding more files to the input dataset will proportionally increase the processing time. The maximum processing time is consumed by the Hadoop streaming.
- In terms of byte read by the map/reduce task, both HAR and CombineFileInputFormat will perform well and the impact of adding more files to the input dataset is avoidable, whereas Hadoop streaming and sequential file have the proportional impact of adding more files to the input dataset.
- In terms of byte written by the map/reduce task, both HAR and CombineFileInputFormat perform comparatively better. The impact of adding more files to the input dataset is reverse in the nature which means lesser numbers of bytes are required.
- In terms of read operation, the performance of HAR is better than other merging strategies. CombineFileInputFormat performs better in comparison with the sequential file and Hadoop streaming. The performance of the sequential file and Hadoop streaming is almost identical. All the merging strategy has the proportional impact of adding more files to the input dataset.
- In terms of write operations, both HAR and CombineFileInputFormat will perform well and the impact of adding more files to the input dataset is avoidable, whereas Hadoop streaming and sequential file have the proportional impact of adding more files to the input dataset.

At last the conclusion of this paper is that HAR and CombineFileInputFormat will perform better than Hadoop streaming and sequential files, while increasing the number of files as input, HAR and CombineFileInputFormat results consistent performance. The performance of the Hadoop streaming and sequential file will degrade as the numbers of files is increased.

References

1. HAR [Online]. Available https://hadoop.apache.org/docs/r1.2.1/hadoop_archives.html
2. SequenceFile [Online]. Available https://examples.javacodegeeks.com/enterprise-java/apache-hadoop/hadoop-sequence-file-example/
3. CombineFileInputFormat [Online]. Available https://hadoop.apache.org/docs/r2.4.1/api/org/apache/hadoop/mapreduce/lib/input/CombineFileInputFormat.html
4. Hadoop Streaming [Online]. Available https://hadoop.apache.org/docs/r1.2.1/streaming.html
5. Tchaye-Kondi et al (2019) Hadoop perfect file: a fast access container for small files with direct in disc metadata access, 26 Apr. 2019. https://arxiv.org/abs/1903.05838
6. Jing W et al (2018) An optimized method of HDFS for massive small files storage. Comput Sci Inf Syst 15(3):533–548. https://doi.org/10.2298/csis171015021j
7. Peng J, Wei W, Zhao H, Dai Q, Xie G, Cai J, He K (2018) Hadoop massive small file merging technology based on visiting hot-spot and associated file optimization. In: Proceedings of 9th international conference, BICS 2018, Xi'an, China, July 7–8 2018. https://doi.org/10.1007/978-3-030-00563-4_50
8. Cai X et al (2018) An optimization strategy of massive small files storage based on HDFS. In: Proceedings of the 2018 joint international advanced engineering and technology research conference (JIAET 2018). https://doi.org/10.2991/jiaet-18.2018.40
9. Kim H, Yeom H (2017). Improving small file I/O performance for massive digital archives. In: 2017 IEEE 13th international conference on e-science (e-Science). https://doi.org/10.1109/escience.2017.39
10. Lyu Y, Fan X, Liu K (2017) An optimized strategy for small files storing and accessing in HDFS. In: 22017 IEEE international conference on computational science and engineering (CSE) and IEEE international conference on embedded and ubiquitous computing (EUC). https://doi.org/10.1109/cse-euc.2017.112
11. Fu X, Liu W, Cang Y, Gong X, Deng S (2016) Optimized data replication for small files in cloud storage systems. Math Probl Eng 2016:1–8. https://doi.org/10.1155/2016/4837894
12. Mu Q, Jia Y, Luo B (2015) The optimization scheme research of small files storage based on HDFS. In: 2015 8th international symposium on computational intelligence and design (ISCID). https://doi.org/10.1109/iscid.2015.285
13. Wang T, Yao S, Xu Z, Xiong L, Gu X, Yang X (2015) An effective strategy for improving small file problem in distributed file system. In: 2015 2nd international conference on information science and control engineering. https://doi.org/10.1109/icisce.2015.35
14. He H, Du Z, Zhang W, Chen A (2015) Optimization strategy of Hadoop small file storage for big data in healthcare. J Supercomput 72(10):3696–3707. https://doi.org/10.1007/s11227-015-1462-4
15. Fu S, He L, Huang C, Liao X, Li K (2015) Performance optimization for managing massive numbers of small files in distributed file systems. IEEE Trans Parallel Distrib Syst 26(12):3433–3448. https://doi.org/10.1109/tpds.2014.2377720
16. Mao Y et al (2015) Optimization scheme for small files storage based on hadoop distributed file system. Int J Database Theor Appl 8(5):241–254.https://doi.org/10.14257/ijdta.2015.8.5.21
17. Improving the performance of processing for small files in hadoop: a case study of weather data analytics. CiteSeerX, https://citeseerx.ist.psu.edu/viewdoc/summary?doi=10.1.1.659.7461

18. Dong B, Zheng Q, Tian F, Chao K-M, Ma R, Anane R (2012) An optimized approach for storing and accessing small files on cloud storage. J Netw Comput Appl 35(6):1847–1862. https://doi.org/10.1016/j.jnca.2012.07.009
19. Ahad MA, Biswas R (2018) Dynamic merging based small file storage (DM-SFS) architecture for efficiently storing small size files in hadoop. Proc Comput Sci 132:1626–1635. https://doi.org/10.1016/j.procs.2018.05.128
20. Sharma VS, Barwar NC (2019) Data management techniques in hadoop framework for handling small files: a survey. Springer AIS Series (ISSN: 2524-7565)
21. Raut S, Phakade P (2014) An innovative strategy for improved processing of small files in hadoop. Int J Appl Innov Eng Manage 3:278–280

Analysis, Visualization and Forecasting of COVID-19 Outbreak Using LSTM Model

Suhas Harbola, Priyanka Jain, and Deepak Gupta

Abstract The COVID-19 outbreak has been treated as a pandemic disease by the World Health Organization (WHO). Severe diseases like Middle East respiratory syndrome (MERS) and severe acute respiratory syndrome (SARS) are caused by members of a large family of viruses called coronavirus (CoV). A new strain was identified in humans in December 2019, named coronavirus (COVID-19). The highest affected countries are unable to predict the pace of the outbreak of COVID-19. So, AI is helpful to analyze the COVID-19 outbreak in the world. We have used the LSTM model to predict the outbreak of COVID-19 all over the world with limited epidemiological data. A variant of recurrent neural network (RNN) which has the capability of learning long-term dependencies with feedback connections, also known as long short-term memory (LSTM), is used in resolving the problems related to time series in deep learning. LSTM is capable of processing a single data point and an entire sequence of data related to any field. We observe that the LSTM model is useful to predict the ongoing outbreak so that authorities can take preventive action earlier. The LSTM model result shows that the growth rate of infected cases of COVID-19 increased exponentially every week.

Keywords COVID-19 · LSTM model · LSTM prediction

S. Harbola (✉) · P. Jain
University School of Information, Communication and Technology, GGSIPU, New Delhi, India
e-mail: suhas.harbola@gmail.com

P. Jain
e-mail: priyankajain.cse17@gmail.com

D. Gupta
Maharaja Agrasen Institute of Technology, New Delhi, Delhi, India
e-mail: deepakgupta@mait.ac.in

S. Harbola
National Informatics Centre, New Delhi, India

A. Khanna et al. (eds.), *Data Analytics and Management*, Lecture Notes on Data Engineering and Communications Technologies 54,
https://doi.org/10.1007/978-981-15-8335-3_14

1 Introduction

On December 31, 2019, the Chinese government reported WHO about an unknown virus causing pneumonia to some people in the city of Wuhan, China. The virus has spreaded in most countries due to the travel of passengers through international flights from China to other parts of the world. On January 30, 2020, WHO declared this virus as Public Health Emergency for the entire world due to infection in around 10 thousand people within China and 125 cases in other countries. COVID-19 was the name given by WHO to this new disease on February 11, 2020 [1]. On March 11, 2020, WHO upgraded it as pandemic due to a significant increase in several cases around the entire world due to a lack of preventive measures taken by various countries. COVID-19 has been identified as the new family member of severe acute respiratory syndrome (SARS) which spreaded in the year 2002–03 and took 774 lives in South China.COVID-19 has spread across 179 countries causing 7807 deaths and 191,127 positive/infected cases till March 18, 2020, as reported to WHO [2].

COVID-19 is transmittable from one person to another one through microdroplets of an infected person via the nose or mouth whenever he/she coughs or exhales. These droplets enter directly to the body of another person through nose or mouth, if they are close to each other and come in contact with the objects around the infected person. It is very important to stay at least 1 m (3 ft) distance or more away from each other to prevent the spread of COVID-19.

Worldwide, the governments are taking various measures to stop or control the virus's transmission, such as inhabitants are warned to stay inside their homes and leave only in the case of urgency, restricting or suspending public transport, and even suggesting quarantine to the suspects and the foreigners landing in the country. Some nations also suspended aviation services. These precautionary steps are expected to mitigate the spread of the virus [3, 4].

- In this paper, we observed that the highest infected developed countries like the USA, Italy, Spain, France, etc., are affected due to misinterpreting the rate of the outbreak of COVID-19.
- AI is very helpful for tracking and predicting the future effect which helps to take a preventive measure to reduce the speed of COVID-19 outbreak.
- LSTM is best to predict the time series sequential problems like stock, sales, and weather, electricity bill, and epidemic spread. So, we have used the LSTM model for future prediction of the number the infected and death cases of COVID-19.

In this study, the data from "Johns Hopkins Center for Systems Science and Engineering" was taken and analyzed with long short-term memory (LSTM). This result of the LSTM model shows that every week, the growth rate of the infected case of COVID-19 increased exponentially. We believe that this study may help an authority to take timely preventive actions, especially in the contaminated zone.

2 Related Work

Artificial intelligence (AI) is playing a critical role for researchers for the prediction of an epidemic or outbreak of disease. The forecasting of any pandemic is of great help to the authorities to prepare themselves to control the situation and take preventive measures. Forna et al. [5] used a machine learning model—boosted regression trees—to predict the characteristics of the Ebola virus in West African. AI models like SVM were helpful in the prediction and classification of diseases as shown for SARS [6] and brain tumor [7]. Machine learning and AI algorithm are being used to analyze or forecast the growth of the disease based on historical data. GAN has been already applied to predict COVID-19 [8].

The outbreak of COVID-19 has inspired several researchers to help the community in resolving the situation. A deep learning-based model has been proposed to identify the COVID-19 patient by analyzing data of chest X-ray reports [9]. Jia et al. [10] have proposed three models, namely logistic, Gompertz, and Bertalanffy, to predict the coronavirus disease. A simple mathematical model is proposed by Zhong and authors [11].

Time series model is used in the stock market [12, 13] and many businesses to forecast the sales demand [14] so that they can take requisite actions like in case a forecast shows the demand of any product might decrease then the company may reduce the scale of production to save raw material and storage cost. Many AI algorithms like a deep neural network [15], recurrent neural network (RNN) [16], auto-regressive integrated moving average (ARIMA) [17] and LSTM [18] have been explored for time series problem to forecasts in different fields.

3 Materials and Methods

3.1 Data

The COVID-19 data is taken from the "GitHub account of Johns Hopkins Coronavirus Resource Center" [19]. They consolidated the data from various sources, namely CDC, NHC,1point3 acres, ECDC, WHO, DXY, Worldometers.info, BNO, national and state government health department, and local media reports. We have taken data from date January 22, 2020, to March 20, 2020, for training the model and March 21–27, 2020, for comparing the accuracy.

3.2 LSTM

LSTM has a recursive structure like RNN which is used in a time prediction model. There are several variants of the model, but no single model is capable of improving

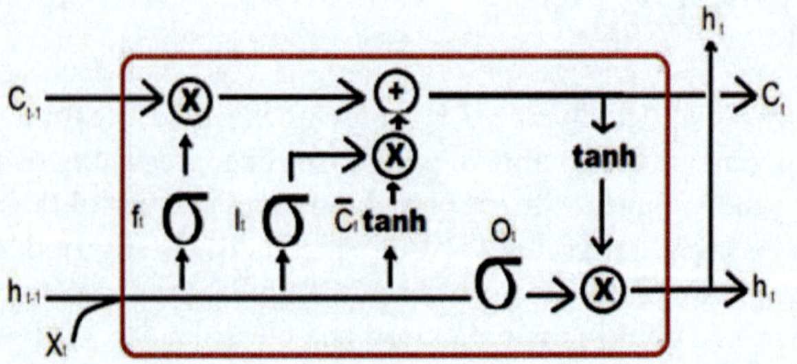

Fig. 1 LSTM network node structure

the architecture of standard LSTM mode [20]. So, we choose the standard LSTM model architecture to predict the COVID-19. It contains four modules that interact with each other in a specific way to ensure long-term dependencies [21, 22]. Standard LSTM node architecture is shown in Fig. 1.

A simple LSTM cell has three gates:

(1) Forget Gate
After getting the output from the previous cell ($h_t - 1$), the forget gate (f_t) can decide which information is needed to discard from the previous state. It uses a sigmoid function layer which helps to crush the input between [0,1]. It is represented as:

$$f_t = \sigma(W_f * [X_t, h_{t-1}] + b_f)$$

where

f_t	output of the forget gate,
σ	sigmoid function,
W_f	weight vectors for forget gate,
X_t	input of the cell,
h_{t-1}	input of the previous cell,
B_f	bias.

(2) Input gate
In the next stage, the main role of the input gate is to determine what information is needed to be stored in the LSTM cell. Initially, the values to be updated are decided by a sigmoid layer, and the tanh layer creates a new candidate vector that is added to the current state of the cell.

$$i_t = \sigma(W_i[h_{t-1}, X_t] + b_i)$$
$$C'_t = \tanh(W_c[h_{t-1}, X_t] + b_c)$$

Now, after calculating the present state of the cell, the output of ($i_t * C'_t$) and ($f_t * C_{t-1}$) is added as shown below.

$$C_t = f_t * C_{t-1} + i_t * C'_t$$

where

i_t	output of input gate,
W_i	weight vector of input gate,
W_c	weight vector of candidate,
b_i	bias of input gate,
b_c	bias of candidate,
C_{t-1}	previous cell state,
C_t	new cell state.

(3) Output gate
The output gate updates the status of the LSTM cell, which will be done by a sigmoid function. Then tanh processes the cell state to obtain a value in the range −1 to 1, and finally, it is multiplied with the output generated by a sigmoid function (O_t) to obtain the output value.

$$O_t = \sigma(W_o[h_{t-1}, X_t] + b_o)$$
$$h_t = O_t * \tanh(C_t)$$

where

O_t	output of the sigmoid function,
W_o	weight vector of output gate,
b_o	bias,
h_t	output of LSTM cell.

In the training phase, LSTM uses a back-propagation algorithm to determine the accurate vectors.

3.3 Model and Architecture

We have used a multi-layered long short-term memory neural network to predict the last values of a sequence. We gave input of cases registered from January 22, 2020, to March 20, 2020, for prediction of an early outbreak of COVID-19. Figure 2 shows the method and architecture used for making predictions.

Data Preprocessing
Figure 3 shows the COVID-19 dataset which has date-wise confirmed and death cases for each country. This dataset is used for training and validation of LSTM models.

Before giving the input to the LSTM model, data preprocessing steps are applied on Fig. 3 COVID-19 dataset mentioned below need to be conducted:

- Group the case_type (Confirmed, death) per date and country
- Create the dataset and normalize the features.

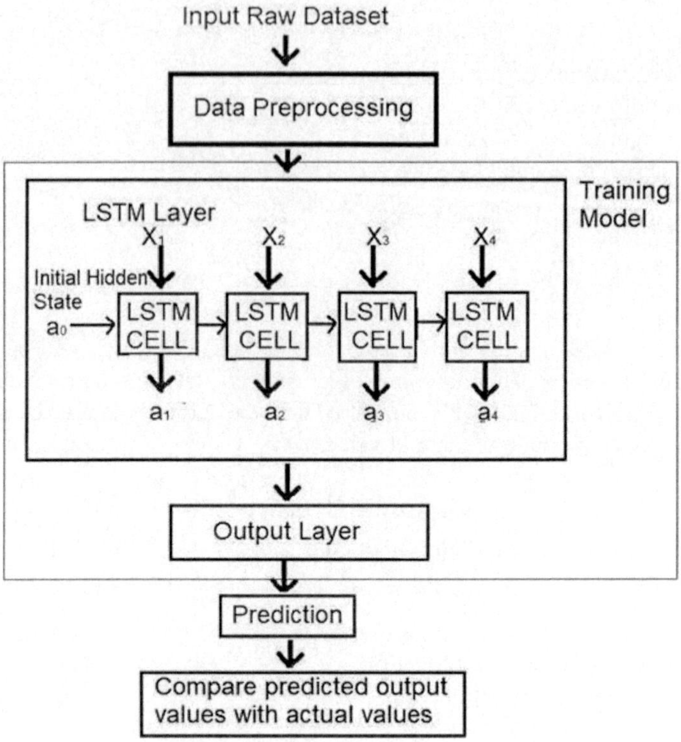

Fig. 2 Method and architecture followed

- The dataset is bifurcated into training and testing sets.
- The total size of the dataset:49 (From January 22, 2020, to March 20, 2020).
- Size of training dataset: 44 (From January 22, 2020, to March 15, 2020).
- Size of test dataset: 5 (from March 16, 2020, to March 20, 2020).
- The dataset matrix is created from an array of values, and then, it is reshaped into $X = t$ and $Y = t + 1$.
- Reshape train and test input into 3D, i.e., number of samples, number of time steps, number of features.

3.4 Steps for Training the LSTM

- Construct the LSTM model with a four-cell hidden layer and one neuron as an output layer that returns a single vector for predicting the confirmed and death cases of COVID-19.

$$\text{LSTMmodel.Add(LSTM(4,input\_shape} = ()))$$

	Date	Country_Region	Confirmed	Deaths
0	1/22/2020	Korea, South	1	0
1	1/22/2020	Afghanistan	0	0
2	1/22/2020	Albania	0	0
3	1/22/2020	Algeria	0	0
4	1/22/2020	Andorra	0	0
...	...	...	...	...
10379	3/20/2020	Venezuela	42	0
10380	3/20/2020	Vietnam	91	0
10381	3/20/2020	West Bank and Gaza	47	0
10382	3/20/2020	Zambia	2	0
10383	3/20/2020	Zimbabwe	1	0

10384 rows × 4 columns

Fig. 3 COVID-19 dataset

- The input shape is assigned to the LSTM model is a one-time step with 120 features.

$$\text{Input\_ shape} = (1, \text{look\_ back})$$

- Use "mean square error" as a loss function and "Adam" for stochastic gradient descent for training deep learning models.

$$\text{LSTMmodel.compile(loss} = \text{'mean\_squared\_error','optimizer'} = \text{'adam')}$$

- The model fit for 300 training epochs with the value of batch size is 1 and verbose is 2.

$$\text{LSTMmodel.fit(trainX, trainY, epochs} = 300, \text{batch\_ size} = 1, \text{verbose} = 2)$$

4 Result and Discussion

We have analyzed the result with a different perspective in terms of training loss and also analyzed the trend of disease outbreak in the world. We have also observed the prediction of the LSTM model and compared it with the actual value of confirmed and death cases. Finally, we have concluded all the findings in the end.

4.1 Plot Model Loss for Confirmed and Death Cases

Figure 4a, b shows training loss coverage of confirmed and death cases of COVID-

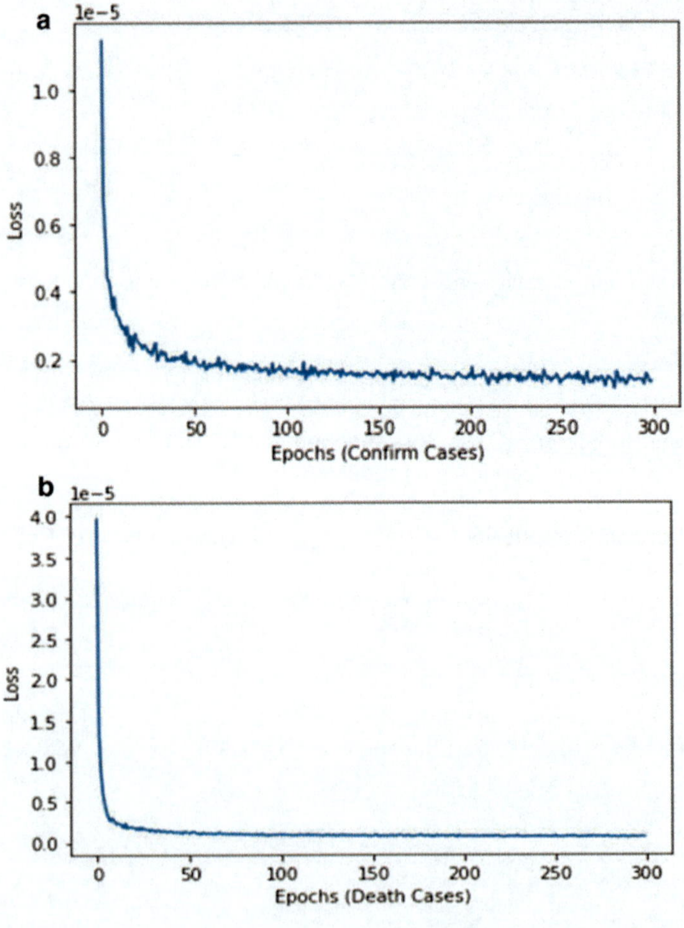

Fig. 4 **a** Training loss coverage of confirmed cases, **b** training loss coverage of death cases

19, respectively. We can observe that training loss convergence of confirmed cases is constant after 200 epochs. However, after 50 epochs, training loss coverage of death cases remains the same.

4.2 Diseases Trend Analysis

The analysis shows the trend of disease outbreak in the world. Figure 5 shows a comparison between the cases registered in China and other countries varies with time. Initially, the cases registered for COVID-19 were more in China with respect to the other countries. After 18th February, the rate of cases registered in China is shown constant, and this may be due to the strict measures, such as warning citizens from going outdoors, suspending public transport, and locking down several cities, taken by the Chinese government which resulted in reducing the spread of the virus. After the 15th March, the rate of cases registered in other countries tremendously increased and crossed the rate of cases registered in China. It also shows how the COVID-19 virus initially spread in China and further to the rest of the world within 2 months.

In the initial 15 days, the death rate and confirmed rate was high. As the authorities identified the symptoms and took necessary measures to stop the spread of the virus, the rates of spread remain steady. It is also seen that in almost every 10–15 days, there is a spike in rate showing the increase in the confirmed infected cases and deaths. We have a COVID-19 dataset starting from January 22, 2020, to March 27, 2020. We used data between January 22, 2020, and March 20, 2020, from COVID-19 cases dataset for training the LSTM model, and the remaining dataset is used to check the correctness of the model. LSTM calculated the number of deaths and confirmed cases for the next 15 days. The prediction is based only on the initial data provided. There might be several cases that are not reported and persist in the symptoms. After identification of those cases, the count will drastically increase and the situation might get out of control.

Figure 6 represents a pie chart diagram of the topmost affected countries from COVID-19 except for Mainland China. The only USA has 19.4% confirmed cases

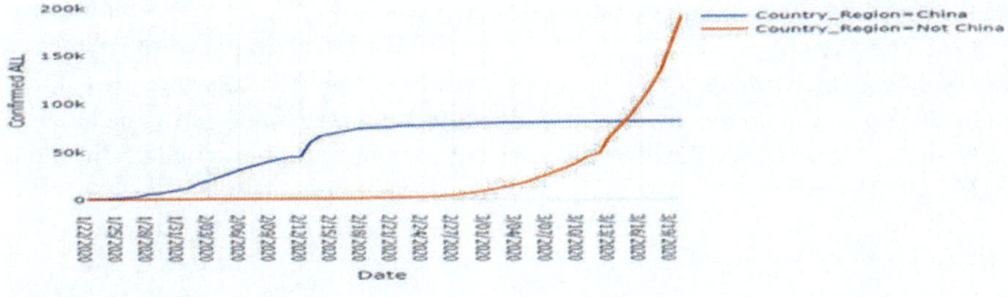

Fig. 5 Shows a trend between the outbreak of the disease in China and the rest of the world

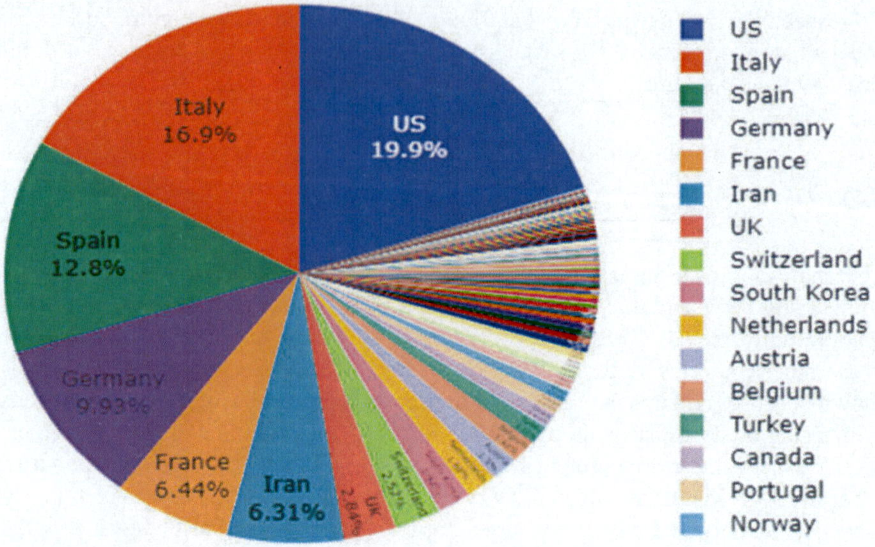

Fig. 6 Top highest infected countries [not Mainland China]

infected by COVID-19. After that, Italy, Spain, Germany, and France share the major part of the pie graph with the percentage value of 16.9%, 12.8%, 9.93%, and 6.44%, respectively.

We observe that the list of highest infected countries is mainly USA, Italy, Spain, Germany, and France which also has the highest number of tourist visitors [23] and international students [24], which may be responsible for the transmission of COVID-19. In Italy, various fashion brands have outsourced their manufacturing to Wuhan, Mainland China. And Wuhan is the first place where the outbreak was reported. The business is so intricate that there are direct flights between Wuhan and Milan, and Wuhan and Rome [25].

4.3 Comparison of Prediction to Actual Values

As shown in Fig. 7 we compared the actual and prediction infected values for the next 7 days (after March 21, 2020, to March 27, 2020). We can observe that the predicted values closely match with actual values of confirmed cases which seem great at first sight. In Fig. 8, comparisons of actual death to predicted death are depicted. We observe that the predicted death values flow almost parallel to the actual death values with less difference.

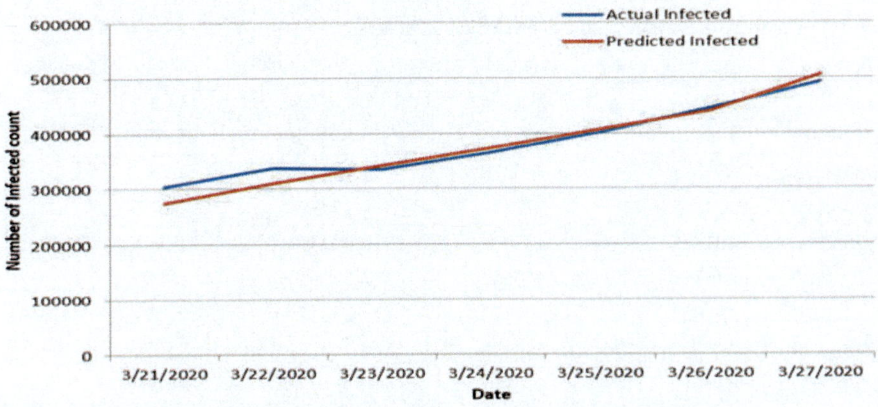

Fig. 7 Comparison of prediction and actual values for infected cases

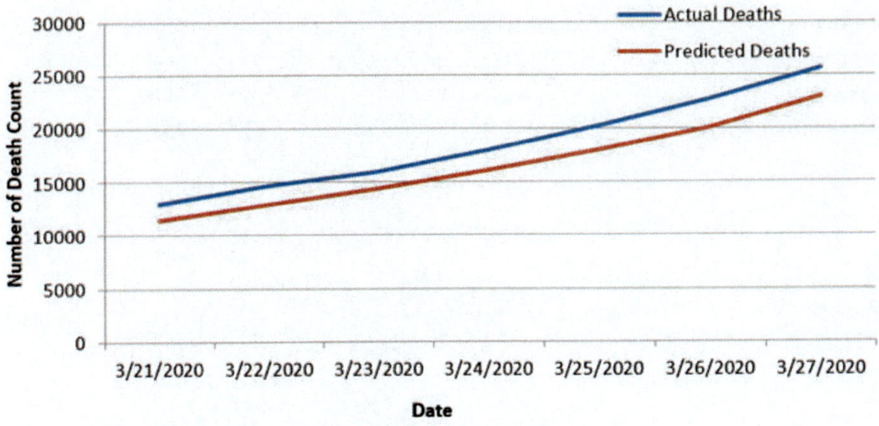

Fig. 8 Comparison of prediction and actual values for death cases

4.4 *Tracking and Prediction of Epidemic*

Table 1 represents the value of predicted, confirmed, and death cases from March 21, 2020, to April 5, 2020, generated by the LSTM Model. We validate that the predicted values (confirmed and death cases) are almost close to the actual values of confirmed and death cases from March 21, 2020, to March 27, 2020.

As shown in Fig. 9, the result of the LSTM model shows a very close value to actual data. The result of this work is that the prediction of the model mimics the actual values of the sequence. We observed that the total infected people as on March 23, 2020, is 0.34 Million, which in turn doubled to 0.68 million within the next 7 days till March 30, 2020; after that, these infected cases again doubled in the next 6 days to 1.29 million till April 5, 2020. This shows that the growth rate of infected cases

Table 1 Comparison of predicted values and actual values

Date	Actual infected	Predicted infected	Actual deaths	Predicted deaths
3/21/2020	304324	273334	12973	11348
3/22/2020	336870	309161	14650	12876
3/23/2020	334388	340280	15948	14343
3/24/2020	364305	372210	17919	16055
3/25/2020	401875	407004	20239	17906
3/26/2020	445755	440552	22761	19999
3/27/2020	491634	504573	25617	22977
3/28/2020		559078		25770
3/29/2020		616311		29060
3/30/2020		679582		32690
3/31/2020		740885		36792
04/01/2020		854771		42525
04/02/2020		953593		48000
04/03/2020		1059882		54491
04/04/2020		1178451		61780
04/05/2020		1293193		70021

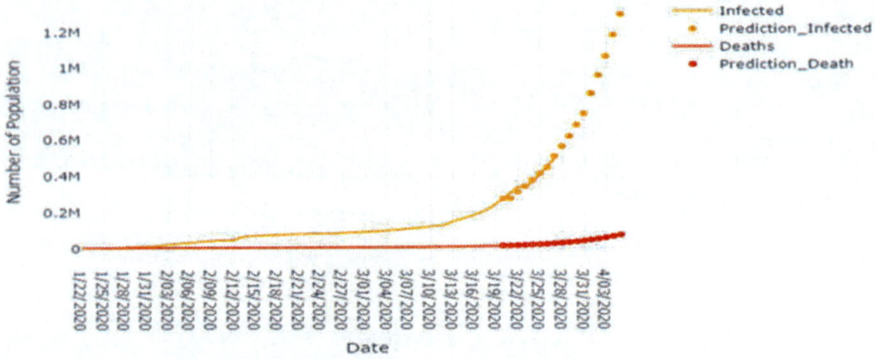

Fig. 9 Plot comparison of predicted infected value with actual infected

increased exponentially within 1 week. This prediction is supposed to be helpful in taking preventive action to control the ongoing 2019-nCoV transmission in the world. Therefore, based on the trend predicted by the LSTM model, the decision-making authority should pay more attention to the medical service.

5 Conclusion

In this paper, we discussed the prediction model for COVID-19 infected and death cases along with related technologies. The standard LSTM model is used for time series and sequential problems. This approach is best if we have a large dataset and also suitable for prediction of the stock market, electricity bill, etc. It might be too early to predict the epidemic evolution outbreak due to the availability of limited data for our analysis. LSTM prediction shown both in Table 1 and graphically represented in Fig. 8 depicts that by April 5, 2020, about 1.29 million of the world population would be infected by the virus and around 70 thousand people would lose their lives due to it.

This prediction is based only on the cases registered by various countries. There are several other factors such as climate condition on different longitude and latitude, the transmission rate of the virus, immunity level of each individual, resistance level of COVID-19 virus on different surfaces, and some other biological factors that are not considered in this observation, while these factors also largely affect the number of infection cases of COVID-19.

References

1. World Health Organization (2005) WHO statement on the second meeting of the international health regulations. In: Emergency committee regarding the outbreak of novel coronavirus (2019-nCoV), Geneva, Swiss [online]. Available https://www.who.int/news-room/detail/30-01-2020-statement-on-the-second-meeting-of-the-international-health-regulations-(2005)-emergency-committee-regarding-the-outbreak-of-novel-coronavirus-(2019-ncov) (2020)
2. World Health Organization (2005) WHO statement on the second meeting of the international health regulations. In: Emergency committee regarding the outbreak of novel coronavirus (2019-nCoV) [online]. Available https://www.who.int/health-topics/coronavirus (2020)
3. World Health Organization (2005) WHO Statement on the second meeting of the international health regulations. In: Emergency committee regarding the outbreak of novel coronavirus (2019-nCoV) [online]. Available https://www.who.int/emergencies/diseases/novel-coronavirus-2019
4. World Health Organization (2005) WHO statement on the second meeting of the international health regulations. In: Emergency committee [online]. Avalaible https://www.who.int/emergencies/diseases/novel-coronavirus-2019/events-as-they-happen
5. Forna A, Nouvellet P, Dorigatti I, Donnelly C (2019) Case fatality ratio estimates for the 2013–2016 west African Ebola epidemic: application of boosted regression trees for imputation. Int J Infect Dis 79:128
6. Hu B, Gong J (2010) Support vector machine-based classification analysis of SARS spatial distribution. In: 2010 sixth international conference on natural computation, Yantai, 2010, pp 924–927
7. Parveen, Singh A (2015) Detection of a brain tumor in MRI images, using a combination of fuzzy c-means and SVM. In: 2015 2nd international conference on signal processing and integrated networks (SPIN), Noida, 2015, pp 98–102
8. Waheed A, Goyal M, Gupta D, Khanna A, Al-Turjman F, Pinheiro FR CovidGAN: data augmentation using auxiliary classifier GAN for improved covid-19 detection. IEEE Access. https://doi.org/10.1109/ACCESS.2020.2994762

9. Luz E, Silva P, Pedrosa Silva R, Moreira G (2020). Towards an efficient deep learning model for COVID-19 patterns detection in X-ray Images. https://arxiv.org/pdf/2004.05717.pdf
10. Jia L, Li K, Jiang Y, Guo X et al (2019) Prediction and analysis of coronavirus disease 2019. https://arXiv.org/2003.05447
11. Zhong L, Mu L, Li J, Wang J, Yin Z, Liu D (2020) Early prediction of the 2019 novel coronavirus outbreak in the Mainland China based on simple mathematical model. IEEE Access 8:51761–51769
12. Thomas K (2019) Time series prediction for the stock price and opioid incident location. In: Ph.D. dissertation, Arizona State University, 2019
13. Lorenzo A, Olivas JA (2019) Some considerations on the use of ai techniques for prediction and forecasting in political elections and the stock market. In: Proceedings on the international conference on artificial intelligence (ICAI). The steering committee of the world congress in computer science, pp 403–407
14. Bandara K, Shi P, Bergmeir C, Hewamalage H, Tran Q, Seaman B (2019) Sales demand forecast in e-commerce using a long short-term memory neural network methodology. In: International conference on neural information processing. Springer, pp 462–474
15. Hasan A, Kalıpsız O, Akyokuş S (2017) Predicting financial market in big data: deep learning. In: 2017 international conference on computer science and engineering (UBMK), Antalya, 2017, pp 510–515
16. Dong Y, Wen R, Li Z, Zhang K, Zhang L (2019) Clu-RNN: a new RNN based approach to diabetic blood glucose prediction. In: 2019 IEEE 7th international conference on bioinformatics and computational biology (ICBCB), Hangzhou, China, 2019, pp 50–55
17. Putra JA, Basbeth F, Bukhori S (2019) Sugar production forecasting system in PTPN XI semborojember using autoregressive integrated moving average (ARIMA) method. In: 2019 6th international conference on electrical engineering, computer science and informatics (EECSI), Bandung, Indonesia, 2019, pp 448–453
18. Sugiyarto AW, Abadi AM (2019) Prediction of Indonesian palm oil production using long short-term memory recurrent neural network (LSTM-RNN). In: 2019 1st international conference on artificial intelligence and data sciences (AiDAS), Ipoh, Perak, Malaysia, 2019, pp 53–55
19. Johns Hopkins University and Medicide Coronavirus Resource Center [online]. Available https://github.com/CSSEGISandData/COVID-19/
20. Gers FA, Schmidhuber J, Cummins F (2000) Learning to forget: continual prediction with LSTM. Neural Comput 12(10):2451–2471
21. Sherstinsky A (2018) Fundamentals of recurrent neural network (RNN) and long short-term memory (LSTM) network
22. Chalvatzaki G, Koutras P, Hadfield J, Papageorgiou XS, Tzafestas C, Maragos P (2018) Online human gait stability prediction using LSTMs for the fusion of deep-based pose estimation and LRF-based augmented gait state estimation in an intelligent robotic rollator
23. Tourist Data [online] Available https://worldpopulationreview.com/countries/most-visited-countries/
24. International Students Data [online] Available https://www.easyuni.com/advice/top-countries-with-most-international-students-1184/
25. New Source [online] Available https://www.dailyo.in/variety/coronavirus-coronavirus-pandemic-wuhan-covid-19-quarantine/story/1/32587.html

Virtual Machine Replication in the Cloud Computing System Using Fuzzy Inference System

Priti Kumari and Parmeet Kaur

Abstract Cloud computing provides various services and scalable computing resources through the internet. Due to the features that allow sharing and multiplexing computing resources across numerous tenants, cloud reliability has gained an extensive foothold in recent times. However, in a cloud computing-based environment, it is critical to enhance the reliability of cloud services such as the virtual machine (VM)-based services. To ensure the reliability of VMs in the Infrastructure as a Service (IaaS) cloud computing model, this paper proposes two fuzzy inference systems (FIS), namely Physical Machine FIS1 (PMFIS1) and PMFIS2 for fault tolerance of VMs in cloud computing. The proposed inference systems use replication as a fault tolerance mechanism for VMs. These systems aid in the selection of optimal physical machines (PMs) to place the replicas of virtual machines (VMs). Implementation of proposed FIS is performed in MATLAB to compare the FISs with each other in terms of complexity, flexibility, and better selection of PMs. However, from the simulation result, it is observed that the PMFIS1 is less complex than PMFIS2, but the PMFIS2 is more flexible and makes a better selection of PM than PMFIS1.

Keywords Cloud computing · Virtual machines · Fault tolerance · Replication · Fuzzy inference system

1 Introduction

The immense adoption of the cloud computing paradigm by many enterprises is a result of its features such as high scalability, on-demand services, and maintenance with a low cost [1]. Infrastructure as a service (IaaS) cloud computing-based model virtualizes PMs into VMs to execute different types of tasks submitted by users.

P. Kumari (✉) · P. Kaur
Jaypee Institute of Information Technology, Noida, India
e-mail: priti9sep@gmail.com

P. Kaur
e-mail: parmeet.kaur@jiit.ac.in

A. Khanna et al. (eds.), *Data Analytics and Management*, Lecture Notes on Data Engineering and Communications Technologies 54,
https://doi.org/10.1007/978-981-15-8335-3_15

These VMs can be provisioned, started, and stopped to meet service requirements on-demand [2, 3]. However, in many cases, VMs can stop working or fail while executing the user request because of hardware or software faults. The VM failures lead to the boundless impact on performance, scalability, revenue as well as consumer trust. To deal with VM failure, fault tolerance (FT) approaches are used where a cloud system continues to work successfully in the presence of a fault [4, 5].

The most frequently used FT methods are replication and checkpointing. In replication, multiple copies of tasks or VMs are placed at a different location to handle the failure [6]. In checkpointing, the error-free execution state of tasks or VMs is saved periodically at the storage system [5]. A VM can be restarted from its saved checkpoint upon recovery.

This paper addresses the issue of FT in the cloud system by proposing a fuzzy logic-based VM replication scheme. For this, two fuzzy rule-based systems, namely PMFIS1 and PMFIS2, have been proposed to select the optimal candidate PMs to place the replicas of VMs in the cloud computing system, which is defined as follows:

- The PMFIS1 uses two input attributes: PM storage (PMST) and PM CPU speed (PMCS). The output parameter is the PM selection decision (PMSD).
- For PMFIS2, we have considered three input attributes adding PM temperature (PMTE) as a new parameter.
- The implementation of both fuzzy systems is done in MATLAB. The performance evaluation of the proposed FISs has demonstrated that although PMFIS1 is less complex than PMFIS2, but PMFIS2 is more flexible and makes a better selection of PMs than PMFIS1.

The rest of the paper is planned as follows: Sect. 2 deliberates the related work in the area of FT, FIS, and cloud computing. Section 3 presents the proposed fuzzy inference systems. Section 4 discusses the performance evaluation of both FIS and its results of the simulation. In Sect. 5, we have concluded our presentation.

2 Related Work

Fault tolerance (FT) is crucial for the cloud system to offer the needed services even in the presence of one or more failed components. According to literature, FT methods are generally of two types: reactive and proactive. The most commonly used reactive FT methods are replication, checkpointing, resubmission, etc. Of all the reactive FT methods, the most frequently employed FT method is replication [5, 6].

Abdelfattah et al. [7] have proposed a new method using replication as well as resubmission FT methods to enhance the reliability cloud system. Zhou et al. [6], have projected an optimal redundant VM placement model using the replication FT method to enhance the consistency of server-based cloud computing services.

Soft computing (SC)-based approaches can play an important role in making cloud systems more reliable. SC is tolerant of imprecision, uncertainty, partial truth as well as provides an approximation to achieve robustness, tractability, and low-cost near-optimal solutions. The most frequently used SC methods are artificial neural networks and fuzzy logic (FL), etc. [8, 9]. The term FL deals with imprecise as well as vague information. Monil et al. [10] have developed a fuzzy-based method for selecting and migrating virtual machines in the cloud system. A fuzzy-based method has been proposed by Bui et al. [11] for early diagnosis of the faults that occur in the IaaS cloud layer. The proposed approach can enhance the performance of the cloud system in terms of accuracy as well as the reaction rate. The fuzzy can also be applied for usability prediction [12, 13].

The present work makes use of a fuzzy rule base system in determining a suitable PM to place replicas of VM. Subsequently, an optimal host is selected based on its storage, CPU speed, and temperature.

3 Proposed Method

Each PM in the cloud environment contains one or more VMs. The PM that hosts a VM (primary VM) is known as the host PM. In order to tolerate the failures of VMs, they are replicated on PMs other than their own host. In case a primary VM fails, its replica can be used to successfully execute the submitted tasks and hence prevent any loss due to the VM's failure. The PMs where replicas of VMs (backup VM) may be placed is referred to as the candidate PMs and are chosen from available PMs in the cloud. Appropriate selection of candidate PMs from the available PMs is important since those PMs should be selected which are less vulnerable to failures.

This paper proposes two fuzzy rule base systems, namely PMFIS1 and PMFIS2, to select the optimal candidate PMs in a cloud environment. The input and output variables of proposed fuzzy systems are demonstrated in Fig. 1.

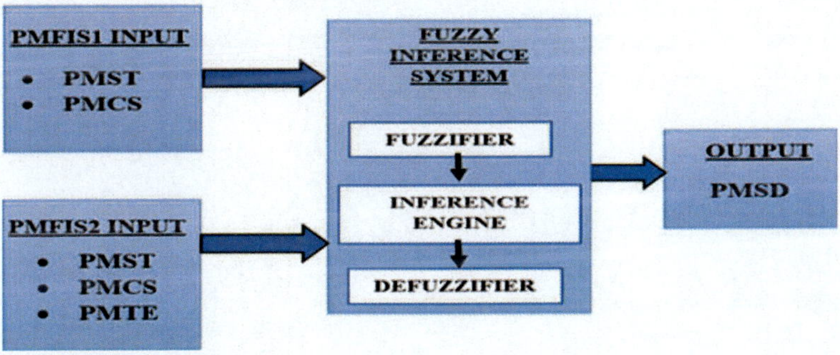

Fig. 1 Proposed fuzzy system model

3.1 Description and Membership Function (MF) of Input and Output Variables

To diminish the complexities, this paper uses the combination of trapezoid and triangle MF to describe the input as well as output attributes. The detail description and fuzzy set membership for the attributes PMST, PMCS, PMTE, and PMSD are defined as follows:

Physical Machine Storage (PMST): This input parameter is used to specify the amount of storage available at the candidate PM. The PMST is defined on a scale of 0–100. Three fuzzy sets are used to describe this attribute, namely less, medium, and high. The fuzzy set diagram for the PMST attribute is demonstrated in Fig. 2. The MF of the PMST is defined in Eq. 1.

$$\mu_{\text{less}}(x) = \begin{cases} \frac{(30-x)}{10}, & \text{if } 20 \le x \le 30, \end{cases}$$

$$\mu_{\text{medium}}(x) = \begin{cases} \frac{(x-20)}{30}, & \text{if } 20 \le x \le 50 \\ \frac{(80-x)}{30}, & \text{if } 50 \le x \le 80 \end{cases},$$

$$\mu_{\text{high}}(x) = \begin{cases} \frac{(x-40)}{30}, & \text{if } 40 \le x \le 70 \end{cases} \quad (1)$$

Physical Machine CPU Speed (PMCS): This input variable is used to specify the CPU speed of candidate PM. The PMCS is defined on a scale of 0–100. Three fuzzy sets are used to describe this attribute, namely less, medium, and high. The fuzzy set diagram for the PMCS attribute is demonstrated in Fig. 3. The MF of the PMCS is defined in Eq. 2.

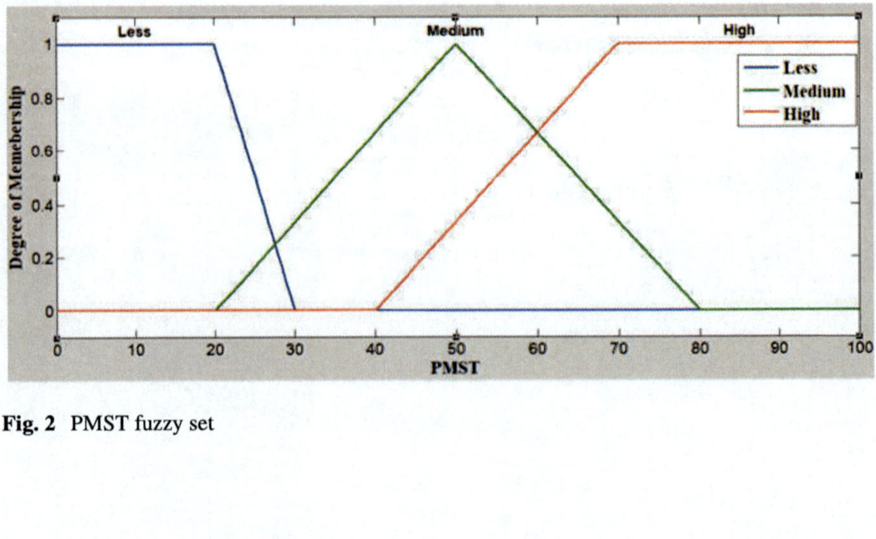

Fig. 2 PMST fuzzy set

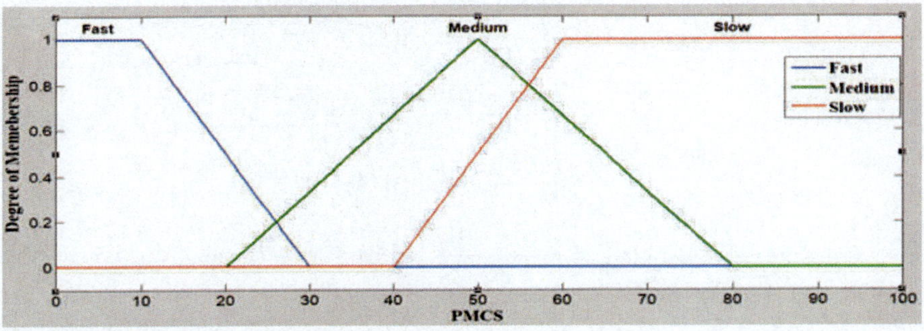

Fig. 3 PMCS fuzzy set

$$\mu_{\text{fast}}(x) = \left\{ \frac{(30 - x)}{20}, \quad \text{if } 10 \le x \le 30, \right.$$

$$\mu_{\text{medium}}(x) = \begin{cases} \dfrac{(x - 20)}{30}, & \text{if } 20 \le x \le 50 \\ \dfrac{(80 - x)}{30}, & \text{if } 50 \le x \le 80 \end{cases},$$

$$\mu_{\text{slow}}(x) = \left\{ \frac{(x - 40)}{20}, \quad \text{if } 40 \le x \le 60 \right. \tag{2}$$

Physical Machine Temperature (PMTE): This input parameter denotes the temperature value of the candidate PM. If the temperature of candidate PM is high, then it is more prone to failure. The PMTE is defined on a scale of 0–100. This attribute is described using three fuzzy sets, namely low, medium, and high. The fuzzy set diagram for the PMTE attribute is demonstrated in Fig. 4. The MF of the PMTE variable is defined in Eq. 3.

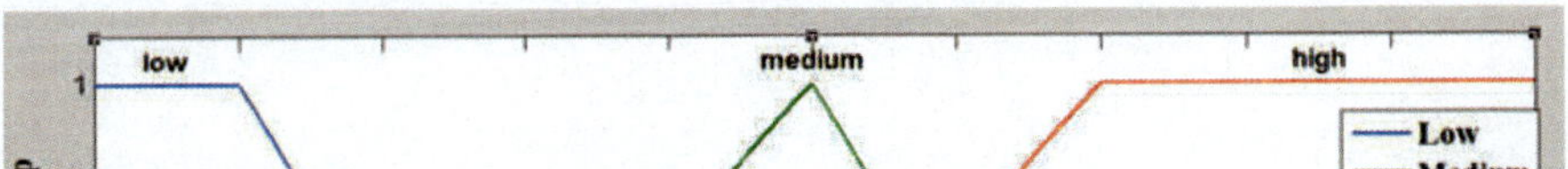

Fig. 4 PMTE fuzzy set

$$\mu_{\text{low}}(x) = \left\{ \frac{(30-x)}{20}, \quad \text{if } 10 \le x \le 30, \right.$$

$$\mu_{\text{medium}}(x) = \begin{cases} \frac{(x-20)}{30}, & \text{if } 20 \le x \le 50 \\ \frac{(70-x)}{20}, & \text{if } 50 \le x \le 70 \end{cases},$$

$$\mu_{\text{high}}(x) = \left\{ \frac{(x-40)}{30}, \quad \text{if } 40 \le x \le 70 \right. \tag{3}$$

Physical Machine Selection Decision (PMSD): This output variable is defined on a scale of 0–100. This attribute deliberates the following levels of candidate PM selection:

Low selection probability (LSP): The candidate PM will have less possibility to be selected.

Medium selection probability (MSP): The candidate PM will have an average possibility to be selected to carry out the replicated VMs.

High selection probability (HSP): The candidate PM has all the essential information and potential to be selected to carry out the replica of VMs.

The fuzzy set diagram for the PMSD attribute is demonstrated in Fig. 5. The MF of the PMSD variable is defined in Eq. 4.

$$\mu_{\text{LSP}}(x) = \begin{cases} \frac{x}{25}, & \text{if } 0 \le x \le 25 \\ \frac{(40-x)}{15}, & \text{if } 25 \le x \le 40 \end{cases},$$

$$\mu_{\text{MSP}}(x) = \begin{cases} \frac{(x-30)}{20}, & \text{if } 30 \le x \le 50 \\ \frac{(70-x)}{20}, & \text{if } 50 \le x \le 70 \end{cases},$$

$$\mu_{\text{HSP}}(x) = \begin{cases} \frac{(x-50)}{25}, & \text{if } 50 \le x \le 75 \\ 1, & \text{if } x \ge 75 \end{cases} \tag{4}$$

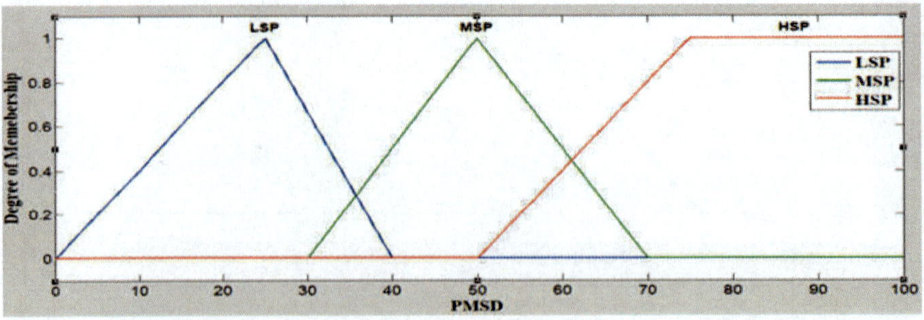

Fig. 5 PMSD fuzzy set

Table 1 Proposed rules for PMFIS1

No.	PMST	PMCS	PMSD
1	Less	Fast	MSP
2	Less	Medium	LSP
3	Less	Slow	LSP
4	Medium	Fast	HSP
5	Medium	Medium	MSP
6	Medium	Slow	MSP
7	High	Fast	HSP
8	High	Medium	HSP
9	High	Slow	MSP

3.2 Proposed Fuzzy Rules for PMFIS1 and PMFIS2

The PMFIS1 is composed of 9 fuzzy rules for defining the value of the PMSD variable (output variable) based on the value of two input attributes such that PMST and PMCS. The proposed rules for PMFIS1 are defined in Table 1. The PMFIS2 is composed of 27 fuzzy rules for defining the value of the PMSD variable based on the value of three input attributes such that PMST, PMCS, and PMTE. The proposed rules for PMFIS2 are defined in Table 2.

4 Performance Evaluation

In this section, we evaluate the performance of the proposed fuzzy systems by simulation experiments. The performance of PMFIS1 and PMFIS2 is compared with each other. The simulation of PMFIS1 and PMFIS2 is carried out using MATLAB.

The simulation result of the PMFIS1, presented in Fig. 6, shows how the output attribute PMSD is affected by the values of PMST and PMCS. From the result, it can be observed that as the value of a PM's storage (PMST) increases and the value of (PMCS) decreases then the possibility of the PM's selection as a candidate (PMSD) increases. However, a candidate PM with less storage capacity (PMST) and with slow CPU speed (PMCS) has a low probability to be selected.

The simulation result of the PMFIS2 is presented in Fig. 7a–c. We show how the output attribute PMSD is affected by the value of PMST, for different values of PMSC and PMTE. Then, we increase the value of PMTE and repeat the simulations.

From the result, it can be said that when the value of PMCS decreases, PMST increases and PMTE decreases then the possibility of PMSD increases. However, a PM with less storage capacity, slow CPU speed, and high temperature has a low probability to be selected.

Hence, from the simulation result, it can be observed that both the proposed FIS aids to make a better selection of candidate PM. However, PMFIS1 is less complex

Table 2 Proposed rules for PMFIS2

Rule No.	PMST	PMCS	PMTE	PMSD
1	Less	Fast	Low	MSP
2	Less	Fast	Medium	MSP
3	Less	Fast	High	LSP
4	Less	Medium	Low	MSP
5	Less	Medium	Medium	LSP
6	Less	Medium	High	LSP
7	Less	Slow	Low	LSP
8	Less	Slow	Medium	LSP
9	Less	Slow	High	LSP
10	Medium	Fast	Low	HSP
11	Medium	Fast	Medium	HSP
12	Medium	Fast	High	MSP
13	Medium	Medium	Low	MSP
14	Medium	Medium	Medium	MSP
15	Medium	Medium	High	LSP
16	Medium	Slow	Low	MSP
17	Medium	Slow	Medium	LSP
18	Medium	Slow	High	LSP
19	High	Fast	Low	HSP
20	High	Fast	Medium	HSP
21	High	Fast	High	MSP
22	High	Medium	Low	HSP
23	High	Medium	Medium	HSP
24	High	Medium	High	MSP
25	High	Slow	Low	MSP
26	High	Slow	Medium	MSP
27	High	Slow	High	LSP

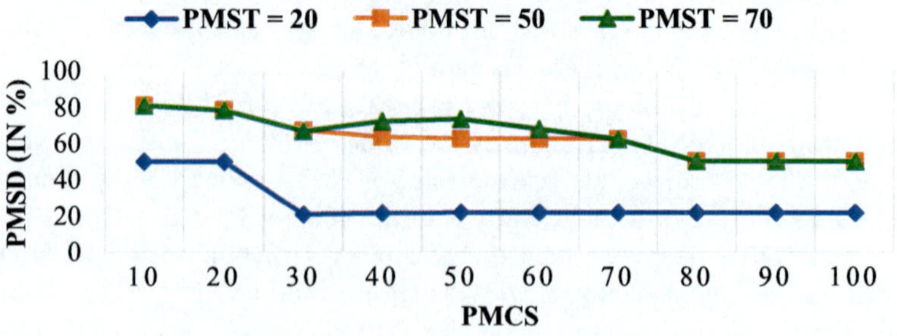

Fig. 6 Simulation result of PMFIS1

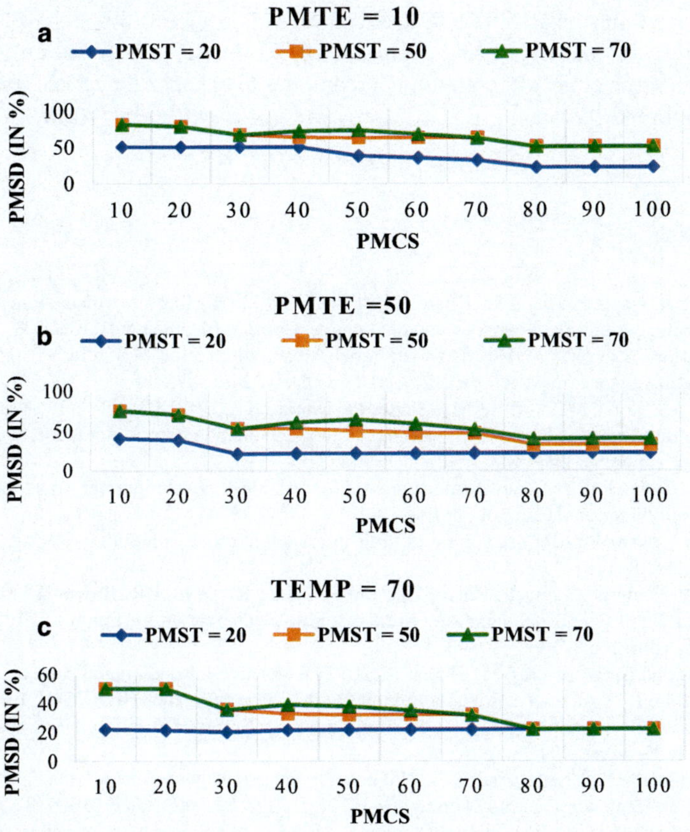

Fig. 7 **a** Simulation result of PMFIS2 for PMTE = 10, **b** Simulation result of PMFIS2 for PMTE = 50, **c** Simulation result of PMFIS2 for PMTE = 50

than PMFIS2. But, PMFIS2 is more flexible and makes a good selection of candidate PM than PMFIS1.

5 Conclusion

This paper proposes a fuzzy inference system to implement a replication fault tolerance scheme to enhance the reliability of cloud systems. The proposed approach consists of two fuzzy inference system, namely PMFIS1 and PMFIS2. For PMFIS1, we have used two input attributes: PMST and PMCS. The output parameter is the PMSD. For PMFIS2, we have considered three input attributes such that PMST, PMCS, and PMTE. The implementation of each fuzzy system is done in MATLAB. The performance evaluation of each proposed FIS is done by comparing it with each other. As a result, the PMFIS1 is less complex than PMFIS2 but the PMFIS2 is more flexible and makes a better selection of PM than PMFIS1. Future work will focus to use some other soft computing techniques to enhance the reliability of cloud computing-based services.

References

1. Mustafa S, Nazir B, Hayat A, Khan AR, Madani SA (2015) Resource management in cloud computing: taxonomy, prospects, and challenges. Comput Electr Eng 47:186–203
2. Guo J, Liu F, Lui JCS, Jin H (2015) Fair network bandwidth allocation in IaaS datacenters via a cooperative game approach. IEEE/ACM Trans Netw 24(2):873–886
3. Parvez AS, Robel MRA, Rouf MA, Podder P, Bharati S (2019) Effect of fault tolerance in the field of cloud computing. In: International conference on inventive computation technologies. Springer, Cham, pp 297–305
4. Gill SS, Buyya R (2018) Failure management for reliable cloud computing: a taxonomy, model and future directions. Comput Sci Eng: 1–1
5. Kumari P, Kaur P (2018) A survey of fault tolerance in cloud computing. J King Saud Univ Comput Inf Sci
6. Zhou A, Wang S, Cheng B, Zheng Z, Yang F, Chang RN, Lyu MR, Buyya R (2017) Cloud service reliability enhancement via virtual machine placement optimization. IEEE Trans Serv Comput 10(6):902–913
7. AbdElfattah E, Elkawkagy M, El-Sisi A (2017) A reactive fault tolerance approach for cloud computing. In: 13th international computer engineering conference (ICENCO). IEEE
8. Srivastava NP, Srivastava RK (2014) Soft Computing Approaches to Fault Tolerant Systems. Int J Adv Netw Appl 5(6):2096
9. Cheraghlou MN, Khademzadeh A, Haghparast M (2019) New fuzzy-based fault tolerance evaluation framework for cloud computing. J Netw Syst Manage 27(4):930–948
10. Monil MAH, Rahman RM (2016) VM consolidation approach based on heuristics, fuzzy logic, and migration control. J Cloud Comput 5(1):8
11. Bui DM, Lee S (2018) Early fault detection in IaaS cloud computing based on fuzzy logic and prediction technique. J Supercomput 74(11):5730–5745
12. Gupta D, Ahlawat A (2017) Usability prediction of live auction using multistage fuzzy system. Int J Artif Intell Appl Smart Dev 5(1)
13. Gupta D, Ahlawat A (2016) Usability determination using multistage fuzzy system. In: 2016 Procedia Computer Science ELSEVIER, SCOPUS.https://doi.org/10.1016/j.procs.2016.02.042

A Dimensional Representation of Depressive Text

Tara Rawat and Shikha Jain

Abstract Depression is presently one of society's main psychological disorders. An intensified public mental health concern has been prompted by recent experiences with the emergence of corona virus disease 2019 (COVID-19). At present, the emphasis of research on human emotional state representation has changed from basic emotions to a large number of emotions in continuous three-dimensional space owing to the complexity of describing and evaluating a vast number of emotions within a single framework. Significant considerations of 3D continuous valence, arousal and dominance space while overseeing mental health issues are important as they relate to the expression of emotion and behavioural reactions. The goal of this research is to design a machine learning regressor modal to estimate the continuous valence, arousal and dominance score which results from the process of emotional intelligence via text interpretation. In the pursuit of goal, EmoBank dataset, which contains text information as well as valence–arousal–dominance values and for validation ISEAR, a labelled corpus of categorical emotions datasets is used. We learn an embedding using three pre-trained word embeddings: word2vec, Doc2vec and BERT, and find that BERT significantly outperforms the result. In a future study, the regressor model will be adopted in depression detection by distributing the categorical negative emotions in terms of VAD.

Keywords Depression · Mental health · COVID-19 · Emotion · Emotional state · Valence · Arousal · Dominance · Textual data · Machine learning · Word embedding

---

T. Rawat (✉) · S. Jain
Department of Computer Engineering and Information Technology, Jaypee Institute of Information Technology, Noida, India
e-mail: tara.rawat@gmail.com

S. Jain
e-mail: shi_81@rediffmail.com

A. Khanna et al. (eds.), *Data Analytics and Management*, Lecture Notes on Data Engineering and Communications Technologies 54,
https://doi.org/10.1007/978-981-15-8335-3_16

1 Introduction

Globally, the World Health Organization (WHO) [17] is reporting that hundreds of millions of people struggle with depression. Even though medical science has advanced, there is still an inadequate diagnosis and care for a significant proportion of the sufferers. The emerging pandemic corona virus (COVID-19) is a specific and unusual phenomenon. This is likely to result in increased risk of depression and suicide. Depression is faced by a burden that interferes with thoughts, perceptions and various facets of life. Nowadays, depression has become a very common mental illness and a serious mood disorder that has an adverse effect on how one feels, think and behave. Luckily, depression is preventable and can be treated. It is a disease which requires professional care.

Adverse feelings or negative words are depressive symptoms. An automatic detection via machine learning techniques may be the most significant and feasible step in the identification of depression. Social media offers a means to capture behavioural characteristics relevant to the thinking, mood interaction, and socialization of an individual [7], analyse linguistic and behavioural patterns of users via their public twitter profile. Author used various distinctive characteristics to construct an SVM classifier that could predict the probability of an individual's depression. With 70% classification precision, the model yielded promising outcomes. Lin et al. [15] explores social media blogging sites to detect psychological stress among people. The tweets labelled with stressed and non-stressed are used in the study. To discover set of features in text and images, several statistical aspects of tweets such as behavioural, social, linguistic patterns are analysed. Lee et al. [14] develops an smartphone app and used wearable activity trackers as well as some standard clinical assessment. The recorded passive information from patients is used to develop and verify mood state prediction algorithms by using random forest. Havigerová et al. [10] focuses on the automatic analysis of the linguistic characteristics of the composed content along with the emotional state of the writer. To measure emotional states of affective disorder, DASS-21 (Depression, Anxiety, and stress scale 21 items) [9] is used.

DASS-21 is based on dimensional conception of psychological disorders. Salimath and Thomas [20] designs a metric for the depression levels and classify the participant appropriately. They use an intensity-based approach and duration-based approach to detect the depression level from the text. As human emotions are not limited to discrete groups, but can be defined suitably by psychology models in three-dimensional space: valence, arousal and dominance, respectively. Machine learning approaches have been commonly used to classify the affective states of the individual since the development of social media. However, different psychological theories describe several ways to understand the human affective state, which helps in yielding accurate, reliable, evidence based and promising results. Different psychologists have created their own theories as to why somebody develops depressive symptoms. Hankin et al. [10] has researched that a specific risk factor for future depression is the interaction of cognitive vulnerability with adverse occurrences. Since the relationships between human behaviour, mood and environmental impact is very complex. Huu Son [11] have proposed a dynamic computational mood model, which is based on psychological theories: appraisal theory (OCC) [6], theory of depression (Aaron Beck) [1] and theory of Thayer (2-D model) [23]. The author has assessed stress event with two parameters: coping and resilience. The results show that the model's computation is reasonable for the detection of stress based on daily data. However the limitations are that the model equations are built up by the subjective relationship between variables not mathematically. Therefore, the precision of the model must be proved mathematically. Since most of the current methods can only detect valence and fail to capture user's emotional state. In this context [12], propose a sentiment analyses tool called DEVA, which is intended for software engineering text. They also capture the user's emotional state with the assistance of the 2D model. Based on recent observations of adverse effects of the COVID-19 pandemic, there is an immediate need to consider mental health problems, which are closely related to depression and lifelong attempts for suicide. To counter this, we are in urgent need of identification and reporting of levels of depression, self-harm, suicide and other mental health problems, both to understand causes and crucially to advise treatments. The ultimate goal is to create a model that helps to develop a greater understanding (as opposed to diagnosis) of the disorder such as depression, which allows experimental evaluations, and ultimately can contribute to improved psychological approaches for depression treatment.

Recently, automation of textual data analysis has enabled the use of quantitative research strategies. There are two emotional models available: the categorical model and the dimensional model. The categorical model uses discrete class of emotion such as disgust, joy, fear, happy, sadness, anger and surprise [13]. In each particular category, the types of emotions consist of distinct elements and a wide variety of emotions. However, not all emotions are included in each since they are classified by a single category. A categorical model therefore has the limitations of an interpretation task in order to distinguish the exact emotional states that people experience. The dimensional model is another way of interpreting emotions. There are different emotional states in this model in a common set of dimensions. They can be defined in two dimensional spaces (valence and arousal) or three-dimensional spaces (valence, arousal and dominance) [19]. The aspect of valence determines the positive or negative emotion and ranges from bad to friendly (feeling of happiness). The arousal aspect refers to the degree of enthusiasm; the emotion reflects and varies from creeping to wild excitement. The aspect of dominance refers to the degree of power, such as a sense of emotion regulation.

As previously mentioned, both the categorical model and the dimensional model of emotions have a large body of work supporting them and offer different perspectives that help our understanding of emotions. However, very little research is being done to identify depressive state using both models. Most of the previous research on textual expressions were focused on precisely one model or the other, but not both. Several studies in depression detection utilize only the polarity tag as a measure for the overall emotion, which makes it impossible for the model to make fine-grained decisions. The detailed model of emotion that represents an emotional state as a point in the emotion space using VAD values provides detailed insights on human behaviour. We believe that by learning the three-dimension emotion model and extracting emotions from it, we can predict the emotional state more precisely than the use of the typical classification model.

1.1 Need for the Study

Depression is a significant under-addressed and top-most public health issue. No treatments are in particular accessible for it. Medications are prescribed through clinicians instead of understanding the category of disorder. Medical experts as of now interview patients, inquire specific questions about their past life experience, lifestyle, temperament/mood. The specialist tries to differentiate the disease based on the patient's response. But these techniques have a tendency to anticipate patient's condition based on the person's particular answers to particular questions. However, due to the specified symptoms, the difficulty in diagnosing particular types of mental health problem is often overlapped. Therefore, the lack of all quality research dealing with mood disorder is noteworthy and contributes to the slow advancement in finding new ways of treatment. Several researchers have studied how psychological dimensional model influences the emotions of human in different modalities (text, image, video, audio, etc.). This paper describes latest progress to use viable and cost effective machine learning based tools with psychological theories to improve standardized result assessment.

1.2 Contributions

The key contributions of our work with respect to previous studies are:

- Regressor model: We presented dimensional representation of text in 3D VAD space on a given text using EmoBank dataset instead of predicting categorical emotion labels directly as traditional emotion classifiers. For consistency, all emotional VAD values are linearly scaled in the range $[-1, 1]$.
- The proposed solution outperforms the simple regressor model-SVR with BERT word embedding, $R2$ score and on mean square error.
- The trained regressor model is used on ISEAR dataset to learn VAD score from the text with categorical emotions.
- Finally, the model will reduce the complexity of expressing and evaluating vast quantities of emotions within one framework.

The remainder of this paper is organized as follows. We, first, provide a brief description of dataset used in this study and then present the experimental approach in Sect. 2 for predicting VAD values using machine learning-based tools with psychological theories to improve standardized result assessment, which may be used in identifying emotional state of the user. We report our results and provide a discussion in Sects. 3 and 4, respectively. Finally, the paper is concluded with future scope in Sect. 5.

2 Method

The various related work for emotion recognition using categorical model has been studied, and the limitations related to intense sadness has been determined in order to identify depression. We therefore, introduce a dimensional model, which further will be mapped with categorical model using machine learning techniques to add value in emotion detection specifically for depression which a rarity in emotion detection.

2.1 Dataset

EmoBank [3] is a large-scale text corpus manually annotated with emotion according to the psychological system of valence–arousal–dominance. It was designed in the JULIE Building, Jena University. Since the dataset offers a meaning of valence, anticipation and superiority (VAD) for a given sentence, this dataset is ideal for dimensional text emotional recognition. The VAD score range is from 1 to 5 using the 5-scale Self-Assessment Manikin (SAM) [2]. Table 1 shows excerpt of EmoBank dataset and its dimensional VAD values.

We decided to focus our attention on text data for two main reason. First, they have typically a high load of emotional content, as it describes the mood and attitude of the users. Second, the communications between users are the best source of knowing their behaviour and their activity routine. In the dataset, for each sentence, VAD values are given. We then use various machine learning models to train the dataset.

Table 1 Excerpt of EmoBank dataset with its VAD values

Text	V	A	D
Remember what she say in my last letter	3	3	3.2
If was not working here	2.8	3.1	2.8
Goodwill help people get off of public assistance	3.44	3	3.22
Sherry learn through our future work class that she could rise out of the mire of the	3.55	3.27	3.46
Come to goodwill be the first step towards my become totally independent	3.6	3.6	3.6

2.2 Experimental Approach

ISEAR [21] is selected for the validation experiments; it contains different emotion categories for each text. The emotion identification system had several steps to ensure promising results. The emotion identification system had several steps to ensure promising results. In order to relate emotions, we use valence–arousal–dominance space (VAD) because it allows us to obtain emotion intensity values in terms of pleasure, excitement and dominance. The proposed approach is shown as below in Fig. 1. Here, the focus will be on predicting VAD-dimensional value. These VAD values corresponding to text can further be used in identifying emotional state of the user.

As a first step, we preprocessed the data to have a clean and noise-free text. The sentences S can be represented as.

$\mathbf{S} = \{\mathbf{w}_1, \mathbf{w}_2, \ldots \mathbf{w}_n\}$ and after preprocessing, $\mathbf{S_p} = \{\mathbf{w}_{p1}, \mathbf{w}_{p2}, \ldots \mathbf{w}_{pn}\}$. Then, we used a word embedding approach for representing text data in vectorize form to map with some emotional state as shown in the equation:

$$\mathbf{f}_{\mathbf{em}:}\mathbf{S_p}_{\rightarrow}\mathbf{S_{DE}}: \mathbf{VS_{DEi}}[\mathbf{W_{ei}}], \quad \text{where } \mathbf{W_{ei}} \text{ belongs to [word2vec, Doc2VEc, BERT].}$$

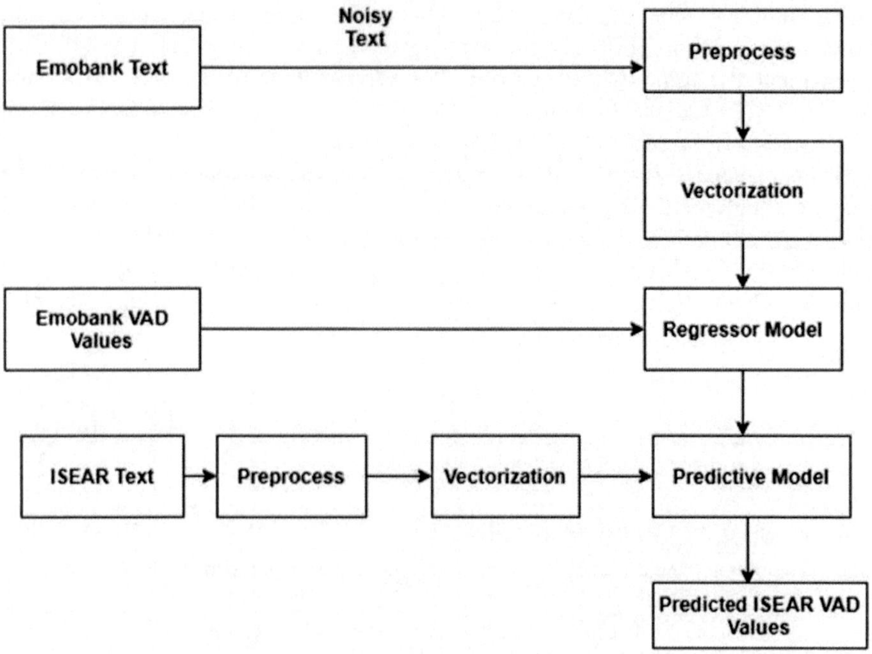

Fig. 1 Block diagram for predicting VAD values

To expand our study of the use of text functionality, we analyse the use of word embedding weighting derived from text data. Three pre-trained models word2vec [5], Doc2vec [8] and BERT [25] are tested without initial weighting conditions. This weighting factor is described in the layer of embedding. The regression process for dimensional values is performed on vectorized data using various machine learning models. The VAD values are represented as:

$$\mathbf{V_i} = \mathbf{f_v}(\mathbf{S_{DEi}})$$
$$\mathbf{A_i} = \mathbf{f_a}(\mathbf{S_{DEi}})$$
$$\mathbf{D}_i = \mathbf{f_d}(S_{\mathbf{DEi}})$$

The data in the text corpus (EmoBank) consists of two parts, the text data which a set of features will be extracted from, and the VAD dimensional values, which is fed to a machine learning model to build a predictive model. The feature extraction process converts the sentences in the text corpus to vector form by word embedding technique. This word vector is fed into model to produce the output, as three dimensions of valence, arousal and dominance. Once the predictive model is built with EmoBank Dataset, the regression task has performed for predicting dimensional VAD values from the ISEAR dataset. We have experimented with different machine learning models and evaluate their $R2$ score and mean squared error (MSE). The following Table 2 demonstrate the models along with word embeddings used in this research with $R2$ score and mean squared error (MSE).

Experimental result shows that support vector regressor with Bert embedding performs best than other embeddings and random forest performs better with BERT embedding, while with other embeddings, it gives poor results. As it is shown in Table 2, models are build with word2vec, Doc2vec and BERT embeddings and random forest regressor performs better with all the word embeddings with $R2$ score of around 80%, 79% and 82%, respectively, though support vector regressor gives best results with Bert embedding, where $R2$ score is 0.96 and MSE is 0.001296. Continue with our best model support vector regressor with BERT embedding, we train our ISEAR dataset text and predict VAD values for each row as shown in Table 3. The $R2$ score and MSE for ISEAR dataset with support vector regressor model along with the BERT embedding is 0.78 and 0.0079, respectively. The predicted VAD values for ISEAR dataset are shown in Table 3.

3 Results

The first evaluation on this research is to evaluate the system on emotion dimensional values. Two datasets are used for this purpose: EmoBank with VAD values, ISEAR datasets with seven categories and Russell's emotion-PAD values [18]. We utilize different machine learning techniques to assess the execution learned by word2vec, doc2vec and BERT embedding. Based on these results as Figs. 2 and 3 demon-

Table 2 Models with *R*2 score and mean squared error (MSE)

	Random forest								Support vector	
	Regressor		Linear regressor		MLP regressor		XGB regressor		Regressor	
	*R*2 score	MSE	*R*2 score	MSE	*R*2 score	MSE	*R*2 score	MSE	*R*2 score	MSE
Word2Vec	0.803	0.007	0.232	0.028	0.226	0.029	0.275	0.027	−3.112	0.154
Doc2Vec	0.796	0.007	0.030	0.036	0.0005	0.038	0.118	0.033	−0.016	0.038
Bert embedding	0.821	0.006	0.307	0.025	0.642	0.013	0.353	0.024	0.966	0.001

Table 3 Excerpt of ISEAR dataset with its predicted VAD values

ISEAR_text	PRED_V	PRED_A	PRED_D
On days when I feel close to my partner and other friends. When I feel at peace with myself and also experience a close contact with people whom I regard greatly	0.0714	−0.04	0.2136
Every time I imagine that someone I love or I could contact a serious illness, even death	−0.0476	0.04	−0.1282
When I had been obviously unjustly treated and had no possibility of elucidating this	0.0714	−0.04	0.0427
When I think about the short time that we live and relate it to the periods of my life when I think that I did not use this short time	0.3333	−0.04	0.2307
At a gathering I found myself involuntarily sitting next to two people who expressed opinions that I considered very low and discriminating	0.3988	0.176	0.4358

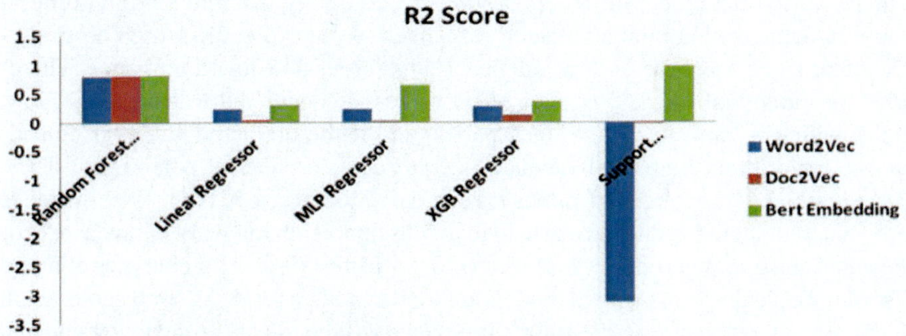

Fig. 2 *R*2 Score for each regressor model

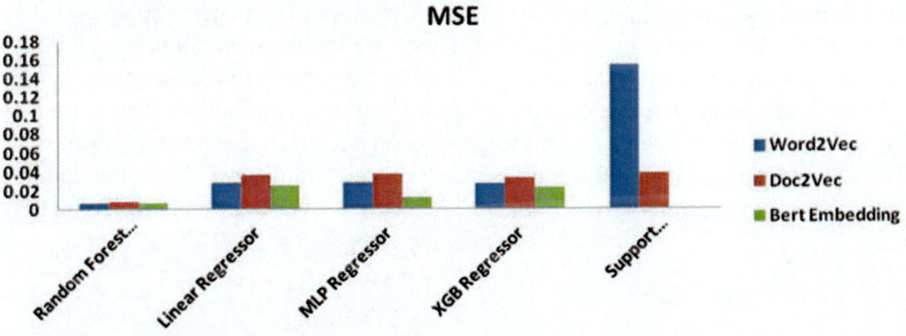

Fig. 3 MSE for each regressor model

strate the regression performance of various models with the respective embeddings, support vector regressor model along with the BERT embedding with $R2$ score 0.96 and MSE 0.0012, respectively, was chosen.

The system is trained to learn the value (valence, arousal, dominance) of the given word embeddings on this dimension-based text emotion-recognition task. The resultant predictive model is used on ISEAR dataset to predict VAD values, which will be extended in further research, in order to add value to emotion detection specifically for depression which a rarity in emotion detection.

As per our proposed model, the sentence is believed to have an emotional VAD value, which reflects the degree to which a situation is perceived and describe the state of mind of people. A sentence's VAD values show a person's perception of the situation. For discrete-dimensional annotation, researchers tend to use various levels of intensity such as here 5-point scale (e.g. ≤3-Low VAD and >3-High VAD). High VAD values reflects that the thoughts or feelings of the person are more positive than average. Negative words or sentence will cause VAD values at the lower end, which is often the case with a depression. In [22], every discrete emotion reflects a multidimensional combination such as: Anger $\rightarrow$ low valence + high arousal + low dominance and Fear $\rightarrow$ low valence + high arousal + low dominance. Therefore, for individual, If the predicted VAD values is considered to have (high, low, high), which is equivalent to relaxed person and if the predicted values is considered to have (high, high, high), which is equivalent to cheerful person. Similarly, for individual, if the predicted values is considered to have(low, high, low), which is equivalent to intense sadness, may lead to depression. High intensity of any aspect of the mood can be dangerous because it can be transformed into different types of mood disorder. Basically, dimensional models are used to study mood. As per researches, in depressive mood disorders, pleasure, dominance will be mostly at the lower end and arousal is very high. In [16], Mehrabian and Bemath analysed depression aspects in PAD's personality by examining a variety of commonly used depression scales and describing the behaviour components of the Mehrabian Depression Scale as −0.42 Pleasure, +0.09 Arousal and −0.37 Dominance. It indicates that depression consists of aggressive, submissive and slightly excitable features of the personality. On the basis of these considerations, we can outline some general predictions regarding how the basic emotion should be related to depression detection diagnoses. Future work can build on these data to develop a more complete framework for understanding the complex associations between three-dimensional VAD values and the depressive state.

4 Discussion

In this paper, we proposed a regressor model based on best-performing support vector regressor with BERT using VAD values of texts obtained from the Emobank. The VAD values will be used in extended paper as a reference to consider how people feel, which helps an automatic depression system to recognize the feeling/emotion of the user in depth. The performance of the VAD regressor model depends on kernel functions as it is a nonparametric technique. Here, the default kernel RBF is used. The advantage of SVR is that it helps the creation of a nonlinear model without modifying the explanatory variables, thereby allowing the resultant model to be properly understood. Although, there were some limitations of the study, such as the corpus was not sufficiently large enough to understand the general sense of word's or contextual relations and lacking comparison with deep learning methods. Furthermore, we compare our proposed framework with various machine learning approaches such as random forest, linear regressor, MLP and XGB regressor along with with three word embedding techniques, word2vec, doc2vec and BERT, respectively. The comparative results based on RMSE and $R2$ score are reported in Table 2 for EmoBank dataset. The obtained results suggest that the proposed SVR approach with BERT yields much better predictions that any of the other considered machine learning approaches. Cho [4] has used EmoBank dataset to learn the emotion embedding and sentiment embedding in order to see its generality on the other dataset. Wang et al. [24] proposed tree-structured CNN-LSTM model in order to predict text valence–arousal score, that is used in explore related context and favourable topics. EmoBank is relative a new dataset, and limited studies are available. We found no existing system evaluating RMSE and $R2$ Score for the regressor models and also how these psychologically predicted VAD values can be used in predicting mental illness.

5 Conclusion and Future Work

We propose learning from the text to forecast VAD ratings. Our framework predicts distributions of VAD score for a given text by using word embeddings, rather can classify probabilities for each class. Learning VAD values further enables simultaneous estimation of categorical emotions, where distinct emotion represents a multidimensional combination of some emotional state such as low pleasure, high arousal and low dominance values represent intense sadness state. And if sadness lasts for a long period, depression can occur.

Therefore, this paper contributes to the field of mental health issues by raising our understanding of how emotions play a role in mood disorder, in particular related to depression. Our study makes contributions:

- To oul approach in identifyingr knowledge, this is the first attempt to use the dimensiona users mental state of depression. We do this by EmoBank datasets and applying it to a very large repository of textual information collected from different sources.
- We studied that VAD score can be used to explain the emotional state of users. There is substantial literature showing that positive/negative response in terms of VAD, correlate with depression state.

Future research would benefit from finding emotional values based on predicted VAD values from textual dataset, which provides new ways to deal with depressive mood disorders and may also helpful in providing guiding principles for emergency clinical crisis strategies to reduce COVID-19 epidemic psychosocial impacts. We would also like to evaluate the proposed model on other modalities also such as facial images, audio and video. These modalities help us shed some light on how psychological systems work with analytical and computational methods and find potential solutions for various psychological mood disorders such as depression before attempting to perform patient trials.

References

1. Beck AT, Alford BA (2009) Depression: causes and treatment, 2nd edn. Depression: causes and treatment
2. Bradley MM, Lang PJ (1994) Measuring emotion: the self-assessment manikin and the semantic differential. J Behav Ther Exp Psychiatry 25(1):49–59. https://doi.org/10.1016/0005-7916(94)90063-9
3. Buechel S, Hahn U (2017). EmoBank: studying the impact of annotation perspective and representation format on dimensional emotion analysis, vol 2, pp 578–585. https://doi.org/10.18653/v1/e17-2092
4. Cho SC (2017) Exploring emotion embedding and the dimensional model of emotion's predictive power in sentiment
5. Church KW (2017) Word2Vec. Nat Lang Eng 23(1). https://doi.org/10.1017/s1351324916000334
6. Colby BN, Ortony A, Clore GL, Collins A (1989) The cognitive structure of emotions. Contemp Sociol 18(6):957. https://doi.org/10.2307/2074241
7. De Choudhury M, Gamon M, Counts S, Horvitz E (2013) Predicting depression via social media. In: Proceedings of the international AAAI conference on weblogs and social media (ICWSM), 2
8. Djuric N, Zhou J, Morris R, Grbovic M, Radosavljevic V, Bhamidipati N (2015) Doc2Vec. In: Proceedings of the 24th international conference on World Wide Web—WWW '15 Companion, 32. https://doi.org/10.1145/2740908.2742760
9. Gloster AT, Rhoades H, Novy D, Klotsche J, Senior A, Kunik M, Wilson N, Stanley M (2008) Psychometric properties of the depressional anxiety. Journal of Affective Disorders, 110(3), 248–259. https://doi.org/10.1016/j.jad.2008.01.023.Psychometric
10. Hankin BL, Abramson LY, Miller N, Haeffel GJ (2004) Cognitive vulnerability-stress theories of depression: examining affective specificity in the prediction of depression versus anxiety in three prospective studies. Cogn Ther Res. https://doi.org/10.1023/B:COTR.0000031805.60529.0d
11. Havigerová JM, Haviger J, Dalibor Kucera PH (2019) Text-based detection of the risk of depression. Front Psychol 10:1–11. https://doi.org/10.3389/fpsyg.2019.00513
12. Huu Son H (2017) Toward a computational model of mood. Proc Comput Sci.https://doi.org/10.1016/j.procs.2017.06.085
13. Islam R, Zibran MF (2018) DEVA: sensing emotions in the valence arousal space in software engineering text, pp 1536–1543
14. Lee H, Cho C, Lee T, Kim M, In HP (2019) Mood prediction of patients with mood disorder by machine learning using passive digital phenotypes based on circadian rhythm: a prospective observational cohort study. J Med Internet Res 21(4):e11029
15. Lin H, Jia J, Guo Q, Xue Y, Li Q, Huang J (2014) User-level psychological stress detection from social media using deep neural network. In ACM MM: 507–516.

16. Mehrabian A, Bernath MS (1991) Factorial composition of commonly used self-report depression inventories: relationships with basic dimensions of temperament. J Res Pers 25(3):262–275. https://doi.org/10.1016/0092-6566(91)90019-M
17. Mental Health Atlas 2017 (2018) Geneva: World Health Organization. Licence: CC BY-NC-SA 3.0 IGO
18. Russell JA (1980) A circumplex model of affect. J Pers Soc Psychol. https://doi.org/10.1037/h0077714
19. Russell JA, Mehrabian A (1977) Evidence for a three-factor theory of emotions. J Res Pers. https://doi.org/10.1016/0092-6566(77)90037-X
20. Salimath AK, Thomas RK (2019) Detecting levels of depression in text based on metrics. https://ArXiv.org/1807.03397
21. Scherer KR, Wallbott HG (1994) Evidence for universality and cultrual variation 66(2): 310–328
22. Smith CA, Ellsworth PC (1985) Patterns of cognitive appraisal in emotion. J Pers Soc Psychol. https://doi.org/10.1037/0022-3514.48.4.813
23. Thayer R (2000) Mood regulation and general arousal systems. psychological inquiry
24. Wang J, Yu LC, Lai KR, Zhang X (2020) Tree-structured regional CNN-LSTM model for dimensional sentiment analysis. IEEE/ACM Trans Audio Speech Lang Process. https://doi.org/10.1109/TASLP.2019.2959251
25. Wang Q, Liu P, Zhu Z, Yin H, Zhang Q, Zhang L (2019) A text abstraction summary model based on BERT word embedding and reinforcement learning. Appl Sci 9(21). https://doi.org/10.3390/app9214701

Machine Learning Algorithms to Predict Potential Dropout in High School

Vaibhav Singh Makhloga, Kartikay Raheja, Rishabh Jain, and Orijit Bhattacharya

Abstract In a developing country like India, the growth of its citizens and consequently the advancement of the nation depend on the education provided to them. However, the process of delivering education has been hindered by considerable dropout rates which have multiple social and economic consequences. Hence, it is crucial to find out ways to overcome this problem. The advent of machine learning and the availability of an immense amount of data have enabled the development of data science and consequently, its application in educational institutions. Educational data mining enables the educator/teacher to monitor student requirement and provides the necessary response and counselling. In this paper, we use advance machine learning algorithms like logistic regression, decision trees and K-nearest neighbours to predict whether a student will drop out or continue his/her education. The accuracy of such models is calculated and studied. On the basis of the results, it was found that ML techniques prove to be useful in this domain with random forest being the most accurate classifier for predicting dropout rate. Educational institutions can analyse which students may need more attention using this research as it is base, thus modifying teaching methods to achieve the end goal of 0% dropout rate.

Keywords Educational data mining · Classification · KNN · Decision tree · Logistic regression predictive analysis · Dropout detection

V. S. Makhloga · K. Raheja · R. Jain (✉) · O. Bhattacharya
Dr. Akhilesh Das Gupta Institute of Technology and Management, New Delhi, Delhi 110053, India
e-mail: rishab1300@gmail.com

V. S. Makhloga
e-mail: makhlogavaibhav@gmail.com

K. Raheja
e-mail: rahejakartikay99@gmail.com

O. Bhattacharya
e-mail: orijit98@gmail.com

A. Khanna et al. (eds.), *Data Analytics and Management*, Lecture Notes on Data Engineering and Communications Technologies 54,
https://doi.org/10.1007/978-981-15-8335-3_17

1 Introduction

The growth in educational sector in India has accelerated over the past few years. There are around 1,250,775 government schools in India and an estimated 339,000 private schools. The education sector is faced with a major concern and challenge in the form of high and increasing dropout rate in the senior secondary Level [1]. Transition from middle school to high school, i.e. classes 10th and 11th grade, is considered to be the most critical period when students show warning signs. Courses become intellectually demanding, and peer groups are larger and students experience personal freedom [2]. Other factors such as social and economic background, demographic and family background also influence a child's schooling [3]. Negative consequences of these student dropouts are significant both to individual and society. There will also be an impact on the educational institution's reputation which may lead to lower ranking and reduced government funding. A lack of formal education in case of dropout students could restrict their economic well-being later in life. The nation as well as society could suffer as dropouts are frequent recipients of welfare schemes and unemployed subsidies [4]. Policymakers, educators and researchers have long considered student dropouts as a serious education problem that needs to be tackled.

The base of this research paper is built upon the emerging research field of EDM. The analysis of educational data using statistics and exploration has proven to be insightful in understanding student behaviour [5]. The rapid development of machine learning models and their accuracy at predicting future trends which makes it a promising tool to tackle a wide range of real-life problems. The scope of this research is to combine the above two fields and determine its accuracy in predicting potential dropouts early so that schools can intervene at the right time.

In this paper, we apply machine learning and build:

- Predictive models to calculate the precision and accuracy of classification algorithms such as logistic regression (along with gradient descent), KNN, decision trees in predicting school dropouts. The educational data is formed by combining data sets collected from National Informatics Centre (NIC) and Ministry of Electronics.
- EDM methodologies are used to find correlation between various factors that lead to student dropout based on graphs generated by algorithms.
- The result of the classifiers is analysed using the precision metrics, and to prevent unbalanced classification, ROC curve is used

The selection of the data set and its features plays a crucial role in our research. They will take into account various factors like

- Behaviour (e.g. number of suspensions).
- Performance (e.g. examination and test marks).
- Facilities (e.g. Internet access, number of toilets, teaching staff) among others.
- Family background (e.g. caste and guardians).

The result has shown that decision tree algorithm has been able to predict dropout with high accuracy. This shows that the usage of machine learning to the increasing dropout problem could be useful. A warning system can be developed based on this research built upon machine learning algorithm which can inform the school authorities about potential dropout behaviour so that there can be an early intervention and special attention can be given to those at risk.

The following section describes in brief the background and related work. A brief description of machine learning techniques used in our study is Sect. 3. Section 4 presents details about the data set. The experiment along with training and testing phases is mentioned in Sect. 5. Section 6 discusses the result and has further analysis. Finally, Sect. 7 concludes our research.

2 Background and Related Work

Kotsiantis et al. [6] used machine learning techniques to deal with the problem of student dropout in universities providing distance education. Attempts have been made by the researchers to develop an appropriate learning algorithm for prediction of student's dropout.

Yukselturk et al. [7] examined the prediction of dropouts using data mining approaches in an online education program. Data was collected through online questionnaires which included variables such as gender, age, educational level, self-efficacy, previous online experience and online learning readiness. To classify data, four data mining approaches were used based on k-nearest neighbour, decision tree, naive Bayes and neural network. Moreover, the genetic algorithm was used to find the significant determinant factor(s) amongst the factors mentioned above.

Aulck et al. [8] described initial efforts to model student dropout using the most extensive data set on higher education attrition that tracks over 32,500 students' demographics and transcripts. Their model highlights several early indicators and accurate dropout prediction, even based on a single-term academic transcript data.

Suh et al. [9] attempted to identify the critical factors of at-risk students: those with low-grade point averages, those who had been suspended and those from low socioeconomic backgrounds. The author carried out a logistic regression analysis of the data, obtained from the National Longitudinal Survey of Youth-1997, which indicated that student dropout rates were affected differently by students' membership in the three at-risk categories.

3 Machine Learning Techniques

3.1 *Logistic Regression*

As per Klein [10] and Menard [11], it is a model based on statistics that uses logistic functions in its basic form to model a binary-dependent variable, even though many more complex extensions are existing. It estimates the parameters of the logistic model only, which means a form of binary regression. We have used logistic regression to classify the pattern of students into two different classes where one is dropping out from any particular course, and the another continues the course initiated.

3.2 *KNN Algorithm*

K-nearest neighbours algorithm is easy to implement a supervised machine learning algorithm. KNN is used to solve regression and classification problems. It has no model rather than storing the entire data set. It relies on labelled input data to learn a function that produces an appropriate output when given new unlabeled data. Soucy and Mineau [12] and Yigit [13] proposed KNN algorithms for text categorization. When KNN is used for classification, the output can be calculated as the class with the highest frequency from the K-most similar instances.

3.3 *Random Forest*

Random forest model includes a simple decision tree which is a weak learner and gives us the output at average, or the output is the majority vote of the decision tree. Random forest model is known for various advantages such as robustness against noise, quick learning and easy setting of hyperparameters. Liaw and Weiner [14] used random forest for implementing classification and regression. We have used random forest to classify our given set of data into a more straightforward form. Logistic regression and random forest both work as classifier but at different accuracies together gives us a trained data which is sent to the Scikit-learn for the construction of prediction model.

3.4 *ROC Curve*

ROC curve is a graphical display of sensitivity (TPR) on the y-axis and (1—specificity) (FPR) on the x-axis for varying cut-off points of test values. This is generally depicted in a square box for convenience, and its both axes are from 0 to 1. The area under the curve (AUC) is an effective and combined measure of sensitivity and specificity for assessing inherent validity of a diagnostic test. Maximum AUC = 1 and it means diagnostic test is perfect in differentiating diseased with non-diseased subjects. The investigation done by Bradley [15] supported the use of ROC in our research.

4 Data Set

The data was collected by the National Informatics Centre (NIC), Ministry of Electronics and Information Technology, Government of India. The National Informatics Centre (NIC) is an attached office under the Ministry of Electronics and Information Technology (MeitY) in the Indian Government, established in 1976 under the Planning Commission of the Indian government.

It is a representative of the quality of high school education provided in our country in relation to student dropouts in 2011–12. The data contains some features about the conditions in the school and the performance of the student to determine whether he/she drops out or not. It features conditions provided at the school in the following categories total toilets, availability of science and language teachers, Internet availability and establishment year of the school. Student performance is judged using marks in science, mathematics and language; also, features like caste, gender and present guardian of the student are also taken into account to determine whether he/she drops out. In this paper, we considered factors other than the student's performance, such as socio-economic conditions, age, infrastructure barriers of the students while extracting the data from NIC portals.

Student id	Gender	Caste	Mathematics marks	English marks	Science marks	Science teacher
s04566	F	BC	0.408	0.798	0.408	9
s00939	F	BC	0.266	0.623	0.266	7
s00470	F	BC	0.347	0.538	0.347	4
s15504	M	OC	0.646	0.317	0.646	6

Language teacher	Guardian	Internet	School id	Total students	Total toilets	Establishment year	Continue drop
5	Mother	True	322	179	8	1955	Drop
6	Other	True	326	177	17	1986	Continue
5	Father	False	341	430	44	1959	Continue
7	Mother	True	339	245	14	1840	Drop

4.1 Data Processing

Due to high skewness in data towards the number of students continuing the education, bias was introduced in the data for the prediction model. Initially, the data contained information about 19,000 students out of which we sampled 1800 values as a representative of the data to create asymmetry between dropouts and students continuing the education. After some data cleaning process, 1763 data values were finalized due to the presence of some null values.

4.2 Numerical Data

Other than cleaning the data and dropping the null values, some feature scaling was also done to bring certain features values on the same scale.

$$z = x - u/r \tag{1}$$

where μ = Mean Value σ or r = Standard Deviation.

We scaled two features total_students and total_toilets to get them to the same scale so that the value of one feature should not dominate over the others and hinder the performance of the learning algorithm.

4.3 Categorical Data

Features like gender, caste, guardian and Internet had categorical values, so they were encoded using a label encoder to integer values.

Caste	Encoding	Gender	Encoding	Guardian	Encoding	Internet	Encoding
BC	0	F	0	Father	0	False	0
OC	1	M	1	Mixed	1	True	1
SC	2			Mother	2		
ST	3			Other	3		

Encoding will enable these columns to be used as features in our classifier.

5 Experiment and Result

5.1 Training Phase

Data were divided into training and testing parts with training data size of 1188 values with 11 features and one label. While testing data size was 586 values with 11 features and one label. This split was done by shuffling the data to get rid of any patterns in the data format that would bring bias or cost us in reduced accuracy. This training data was fed into various classification models for training the respective models, whereas the purpose of testing data set was to test the model for overfitting or any bias during the prediction phase.

Logistic regression model gave a training accuracy of 66.07% and a testing accuracy of 64.86% using L2 regularization for a penalty. Regularization is a method of avoiding overfitting of classification models by penalizing high-valued regression coefficients. L2 regularization adds penalty equal to the square of the magnitude of coefficients.

Confusion matrices using heat maps were plotted for each model to obtain following values:

- True Positive (TP): Observation is positive and prediction is also positive.
- False Negative (FN): Observation is positive but prediction is negative.
- True Negative (TN): Observation is negative and prediction is negative.
- False Positive (FP): Observation is negative but prediction is positive (Fig. 1).

Regression Coefficients: (1.23971592, 0.0410947, −0.2475107, 0.60797969, −0.10031828, 0.03504581, −0.01807157, 0.31297533, −0.8502106, 2.8747373, −0.8502106).

In decision tree, the feature science_teacher has the maximum gain ratio and which has made it the starting node and the most effective attribute for the initial split. For a student to be classified as a dropout, the essential features according to the generated decision tree are science marks, English marks, number of language teachers and caste. Gini index or coefficient was used for the statistical analysis (Fig. 2).

KNN classifier gave us an accuracy of 92.2% the training set and an accuracy of 88.39% on the test set. The prediction was based on five nearest neighbours and a Minkowski metric. The Minkowski distance is a metric in a normed vector space which can be considered as a generalization of both the Euclidean distance and the Manhattan distance (Fig. 3).

5.2 Method of Analysis

The entire data was divided into training and testing sets in the ratio 4:1, and a classification metric was generated to get the prediction result on the test data. Precision rate, recall rate, F-measure and the overall accuracy rate were used as indicators to get the effectiveness of the predictions.

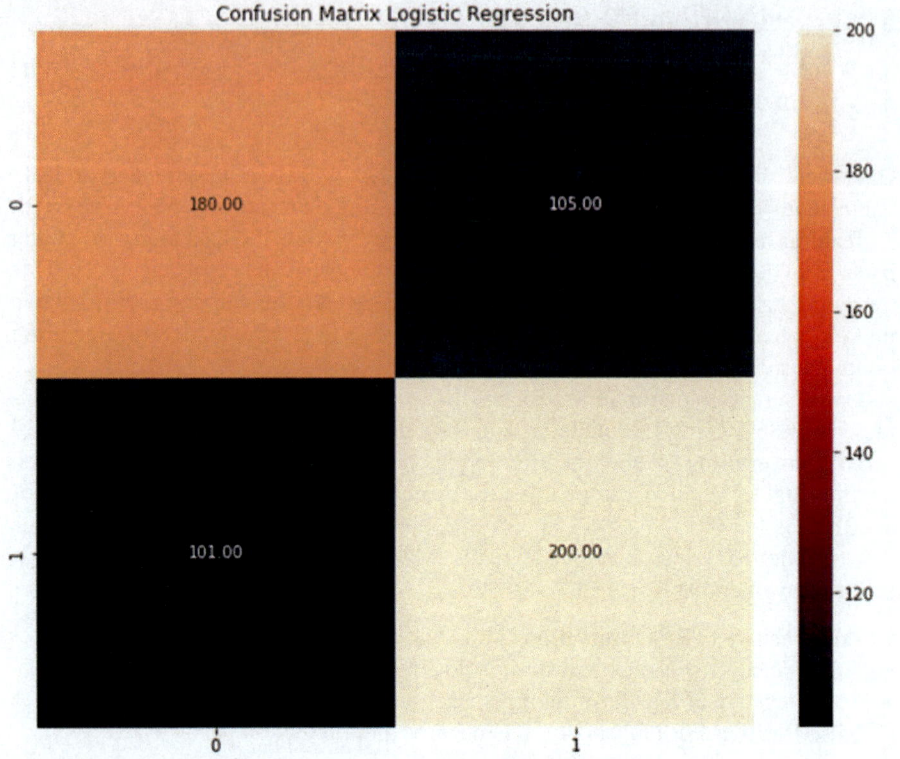

Fig. 1 Using Logistic Regression we get, TP = 180, FP = 105, TN = 200, FN = 101

Precision defines the percentage of samples with a specific predicted class label belonging to that class label.

$$\text{precision} = \text{TP}/(\text{TP} + \text{FP}) \tag{2}$$

Recall defines the percentage of samples of a particular class which were correctly predicted as belonging to that class. The $f1$ score is defined as the harmonic mean of precision and recall and is a far better indicator of model performance than precision and recall (usually).

$$\text{recall} = \text{TP}/(\text{TP} + \text{FN}) \tag{3}$$

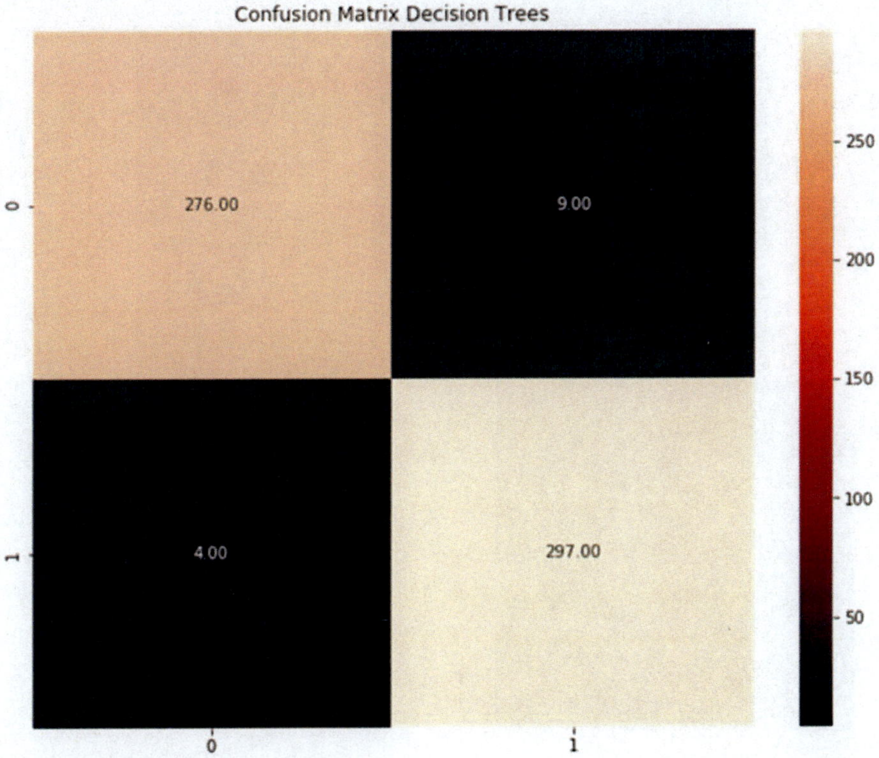

Fig. 2 Using decision trees classifier, we get TP = 276, FP = 9, FN = 4 and FP = 297

6 Discussion

Three classification techniques were applied to the data set to build the perfect model. These techniques are KNN, logistic regression and decision forest. After the preprocessing and preparation of raw data into a usable format, we applied the three algorithms on it and got the accuracy score. Decision tree was successfully able to predict the correct *Y*-Label, i.e. dropout approximately 98% of the time compared to basic algorithms such as logistic regression which was correctly able to predict only 66% of the time. KNN showed some promise with selecting 5 as nearest neighbour. Some fine-tuning of the algorithm and its parameters may lead to improved results. Being a complex classifier, decision tree was able to perform better. It was seen that it took 16 levels of splits to get the desired accuracy. Levels were reduced so as to not overfit the data. We try to analyse the ROC curve visually. ROC curve is used here as a diagnostic tool to check if there are any imbalanced classification (i.e. a negative case with the majority of examples and a positive case with a minority of examples). It will also help us to compare the performance of the used classifiers using AUC (area under the curve) (Fig. 4).

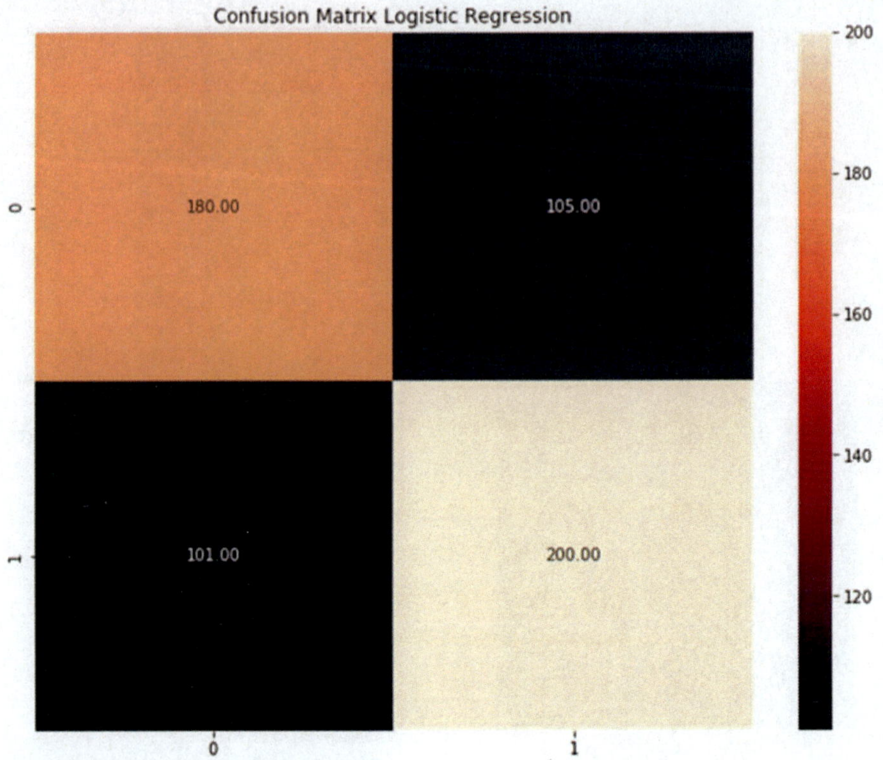

Fig. 3 Using KNN classifier, we get TP = 220, FP = 65, FN = 3 and TN = 298

After studying the combined ROC curve, it is clear that the random forest gives the maximum accuracy in predicting dropouts while logistic regression has the least accuracy. AUC of KNN, logistic regression and decision tree is as follows: 0.97, 0.71, 0.98.

The following tables show the classification metrics for each logistic regression, decision tree and KNN. They have precision values like 0.664, 0.987 and 0.847, respectively.

Logistic Regression Classification metrics:

	0	1	Accuracy	Macro avg	Weighted avg
$f1$-score	0.67012	0.677	0.674061	0.674015	0.674095
Precision	0.66438	0.683	0.674061	0.674029	0.674226
Recall	0.67595	0.672	0.674061	0.674099	0.674061
Support	287.000	299.0	0.674061	586.0000	586.000000

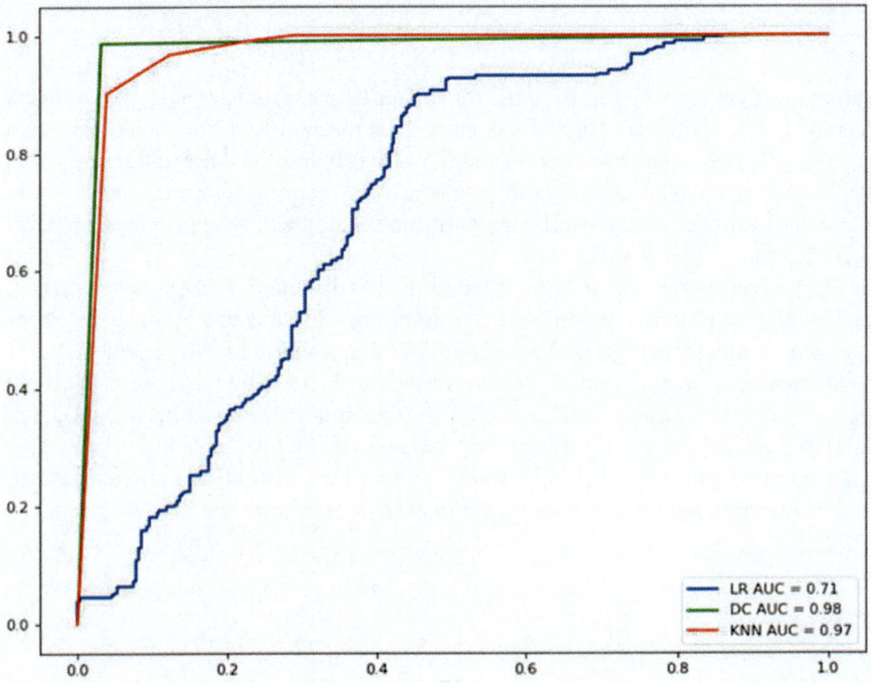

Fig. 4 ROC curve

KNN Classification metrics:

	0	1	Accuracy	Macro avg	Weighted avg
$f1$-score	0.8745	0.903	0.890785	0.888917	0.889212
Precision	1.0000	0.8236	0.890785	0.911846	0.910041
Recall	0.7770	1.0000	0.890785	0.888502	0.890785
Support	287.00	299.00	0.890785	586.00000	586.000000

Decision Tree classification metrics:

	0	1	Accuracy	Macro avg	Weighted avg
$f1$-score	0.9877	0.9883	0.988055	0.988047	0.988053
Precision	0.9929	0.9834	0.988055	0.988201	0.988103
Recall	0.9825	0.9933	0.988055	0.987945	0.988055
Support	287.00	299.00	0.988055	586.00000	586.000000

7 Conclusion

This study lays the preliminary foundation for building a system for detecting students that are at risk of dropping out. We examined the machine learning techniques such as decision trees, logistic regression and KNN to students database and investigated their performance. This research shows that machine learning algorithms are efficient in predicting dropouts, and similar methods can be adopted by every school for early identification of at-risk students.

The approach implemented here can detect signs of student's disengagement from the learning environment for different age groups on different education levels. Proof of concept implementation shows a certain level of viability of this approach.

Moreover, the ranking of contrastive variables obtained by this approach would help to determine which variable affects the learning process the most at a given point of time. By monitoring the critical variables constantly, the learning process is made more adaptive to the student and planning of the institution's curriculum can be considered based on this information in order to minimize the dropout rates.

References

1. Romero C, Ventura S (2010) Educational data mining: a review of the state of the art. IEEE Trans Syst Man Cybern Part C Appl Rev 40(6):601–618. https://doi.org/10.1109/TSMCC.2010.2053532
2. Sateesh M, Sekher TV (2014) Factors leading to school dropouts in India: an analysis of national family health survey-3 data. Int J Res Method Educ 4:75–83.https://doi.org/10.9790/7388-04637583
3. Kominski R (1990) Estimating the national high school dropout rate. Demography 27:303–311. https://doi.org/10.2307/2061455
4. McCaul EJ, Donaldson GA, Coladarci T, Davis WE (1992) J Educ Res 85(4): 198–207
5. Langley P, Simon HA (1995) Applications of machine learning and rule induction. Commun. ACM 38(11):54–64
6. Kotsiantis SB, Pierrakeas CJ, Pintelas PE (2003) Preventing student dropout in distance learning using machine learning techniques. In: Palade V, Howlett RJ, Jain L (eds) Knowledge-based intelligent information and engineering systems. KES 2003. In: Lecture notes in computer science, vol 2774. Springer, Berlin, Heidelberg
7. Yukselturk E, Ozekes S, Türel Y (2014) Predicting dropout student: an application of data mining methods in an online education program. Euro J Open Distance E-Learn 17:118–133. https://doi.org/10.2478/eurodl-2014-0008
8. Aulck LS, Nishant V, Blumenstock JE, West J (2016) Predicting student dropout in higher education. https://ArXiv.org/abs/1606.06364
9. Suh S, Suh J, Houston I (2007) vol 85(2): 131–255. Spring 2007. https://doi.org/10.1002/j.1556-6678.2007.tb00463.x
10. Kleinbaum DG, Klein M (2010) Logistic regression: a self-learning text. Springer, New York, NY. https://doi.org/10.1007/978-1-4419-1742-3
11. Scott M (2001) Applied logistic regression analysis, 2nd edn. SAGE, pp 1–33. https://books.google.co.in/books?id=JbVIDwAAQBAJ
12. Soucy P, Mineau GW (2001) A simple KNN algorithm for text categorization. In: Proceedings 2001 IEEE international conference on data mining. San Jose, CA, USA, pp 647–648

13. Yigit H (2013) A-weighting approach for KNN classifier. In: 2013 international conference on electronics, computer and computation (ICECCO), Ankara, pp 228–231
14. Liaw A, Wiener M (2002) Classification and regression by random forest. R News 2:18–22
15. Bradley AP (1997) The use of the area under the ROC curve in the evaluation of machine learning algorithms. Pattern Recogn 30(7):1145–1159
16. Kloft M, Stiehler F, Zheng Z, Pinkwart N (2014) Proceedings of the 2014 conference on empirical methods in natural language processing (EMNLP), 25–29 Oct 2014. Association for Computational Linguistics, Doha, Qatar, pp 60–65
17. de Santos KJ, Menezes AG, de Carvalho AB, Montesco CAE (2019) Supervised learning in the context of educational data mining to avoid university students dropout. In: 2019 IEEE 19th international conference on advanced learning technologies (ICALT). Maceió, Brazil, pp 207–208
18. Rumberger RW (2001) Why students drop out of school and what can be done. UCLA: the civil rights project/Proyecto Derechos civiles. Retrieved from https://escholarship.org/uc/item/58p2c3wp

Explaining Deep Learning-Based Classification of Textual Tweets

Diksha Malhotra, Poonam Saini, and Awadhesh Kumar Singh

Abstract Social media platforms like Twitter, Facebook, Instagram allow users to share content, express their views, and generate a massive volume of sentiment-rich data. The process of sentiment analysis (SA) classifies a sentiment, opinion, blog, or update into three categories, namely *positive*, *negative*, and *neutral*. Nowadays, deep learning-based classification for such unstructured datasets has become popular due to its high performance. Hence, the proposed model builds a deep learning-based classifier to categorize the sentiments. Nevertheless, the classifier proves to be accurate; the question of why a particular statement is predicted as positive or negative sentiment remains unanswered. In the paper, we propose an explainable sentiment analysis model to explain the outcome of a classifier's instance using a model-agnostic local interpretable model. For experimental purposes, two instances classified as positive and negative sentiments, respectively, have been explained successfully by local interpretable model-agnostic explanations (LIME) explainer along with their analysis.

Keywords Explainable artificial intelligence · Sentiment analysis · Deep learning · LIME

1 Introduction

With the growth and widespread reach of the Internet among users, it has become easy to access the Internet from any part of the world without any spatial or temporal

D. Malhotra (✉) · P. Saini
Department of Computer Science and Engineering, Punjab Engineering College (Deemed to be University), Chandigarh, India
e-mail: Dikshamalhotra.phdcse@pec.edu.in

P. Saini
e-mail: Poonamsaini@pec.edu.in

A. K. Singh
Department of Computer Engineering, National Institute of Technology, Kurukshetra, India
e-mail: aksinreck@gmail.com

A. Khanna et al. (eds.), *Data Analytics and Management*, Lecture Notes on Data Engineering and Communications Technologies 54,
https://doi.org/10.1007/978-981-15-8335-3_18

restrictions. The users can express their views and opinions on any topic on social media platforms such as Twitter and Facebook. In turn, a massive amount of data is generated through blogs, forums, and social media by such opinions and views [3, 11]. This data can be used to understand the user's sentiment (positive, negative, or neutral) about a product from the text of the opinion. Getting insights from such data can be useful for various marketing companies and institutions. Sentiment analysis (SA) is a branch of natural language processing that can be used to predict the underlying sentiments from the opinion data [10]. A sentiment (or an opinion) can be seen as a quintuple, (*e, a, s, h, t*), where *e* is the name of an entity, a is the aspect of *e*, *s* is the sentiment on aspect *a* of entity *e*, *h* is the opinion holder, and *t* is the time when the opinion is expressed by *h* [8]. Generally, the tweets have a limitation of 140 characters, due to which users express their opinions using a lot of abbreviations, slangs, and URLs, along with hashtags, mentions, and markings. Hence, cleaning and preprocessing of such unstructured opinions become essential before sentiment analysis. The success of sentiment analysis depends on the features extracted from the opinions. Deep learning classification models have demonstrated significant results that outperformed conventional models in several domains, such as computer vision and natural language processing; hence, they are currently adopted for sentiment analysis tasks. They can be seen as powerful computational models for extracting sentiments from text without feature engineering [2]. The learned representations can be naturally used as features and applied for classification tasks. Further, due to the inherent feature generation, the deep learning models learn high levels of complexity leading to a lack of clarity on their learned behavior.

The current artificial intelligence (AI)-based systems provide highly accurate results but often lack transparency as they cannot be directly explained to users without strong background information [5, 7]. In other words, high accuracy comes at the cost of low interpretability [6]. Figure 1 shows the explainability versus accuracy graph for various machine learning algorithms. It can be observed that neural networks (deep learning models) achieve high accuracy but lack explainability, whereas simpler models such as classification rules are explainable in nature. In order

Fig. 1 Tradeoff between accuracy and explainability

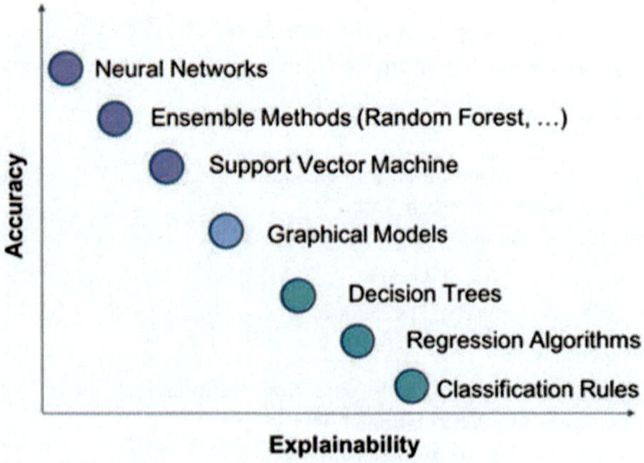

to induce trust in the system, it must be capable of making human-understandable explanations, explain its rationale, and characterize its strengths and weaknesses. Hence, it has become essential to include explainability in existing AI systems. Although deep learning models show remarkable performance, they are still unable to explain the reason behind a particular prediction intrinsically.

The XAI techniques can be classified into two categories, namely *intrinsic* and *post-hoc explainability*. The intrinsic explainability aims to provide explainability within the model, *whereas* post-hoc explainability does not change anything in the inner working of the model. Simpler models are used for interpretable approximation of the complex models [1, 9, 13]. Although a substantial amount of work exists in the XAI domain, as per our knowledge, post-hoc local XAI approaches have not found their way into state-of-the-art literature in the sentiment analysis domain.

Below are the major contributions of this paper:

- The paper proposes a post-hoc explainable deep learning-based sentiment analysis model in order to explain the impact of features on the predicted sentiments for each instance.
- As the proposed model classifies tweets into sentiments using complex deep neural networks, it explains the prediction of instances using a post-hoc model-agnostic explainer, LIME.

The remainder of this paper is structured as follows: Sect. 2 presents an overview of the post-hoc local interpretability technique used. Further, Sect. 3 outlines the system architecture of the proposed model, along with the detailed step-wise process. Section 4 explains the implementation details and results of the proposed model, followed by the conclusion.

2 Local Interpretable Model-Agnostic Explanations (LIME)

Local interpretability techniques aim at answering the following questions:

- Why did the model make a specific decision?
- What was the effect of specific feature value on a prediction?

Local interpretable model-agnostic explanations (LIME) [12] explain individual predictions of any black-box machine learning model. Existing metrics for evaluating the model such as accuracy, and precision may not be suitable. Hence, the user may require the explanations of the predictions. LIME approximates any model's instance locally with an existing interpretable machine learning model such as linear regression, decision trees, rule lists. It generates a new permuted dataset consisting of samples around a particular data instance and their corresponding predictions. LIME can be used on text, tabular data, and images. For textual data, it creates a bogus dataset by perturbing the data points around the class labels. Further, it trains a simpler interpretable model on the bogus dataset so created. The created simpler model and the original model can be compared in order to check the level up to which the simpler model approximates the original model. The explainer is, hence, used to explain a particular instance from the original test dataset.

In the current implementation of LIME, the number of features (K) for deciding the model complexity needs to be provided by the user. LIME can be implemented using lime, *eli5,* and skater libraries in python and lime and *iml* libraries in *R*.

3 System Architecture

The section describes the dataset used for sentiment analysis and the system architecture for predicting sentiments from the tweet text and further explaining the outcome of the sentiment analysis model.

3.1 Dataset

For the experimental purposes, the Stanford sentiment140 dataset [4], a benchmark twitter dataset, is used. The dataset is publically available ground truth data where each message is tagged based on emoticons present inside the tweet. The fields of the dataset include Target, Id, date, User, and text of the tweet. The target field of the data includes two classification classes, *i.e.*, positive (denoted by 4) and negative (denoted by 0). However, only the target and the text fields are required for sentiment analysis. Hence, the extraneous fields are pruned from the dataset for training and testing the model. Figure 2 shows the step-wise illustration of the explanation procedure for the

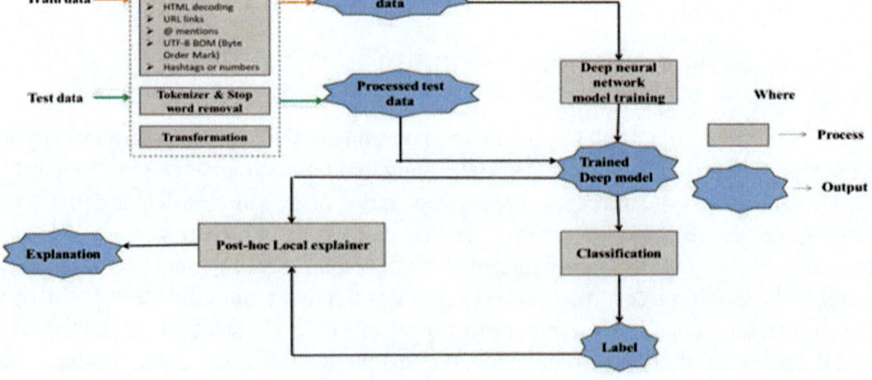

Fig. 2 Explainable sentiment analysis framework

sentiment analysis process.

3.2 Data Preprocessing

The text field of the Sentiment140 dataset consists of the tweets by the users. It contains extraneous content in the text such as URLs, @ mentions, Hashtags, numbers, HTML decoding, '?' marks, extra spaces, etc. The extraneous content is not useful for sentiment analysis; hence, it has been removed in the data cleaning step using natural language processing (NLP) techniques. Next, the text from the data cleaning step is converted into tokens by splitting it into strings and then converting it into integers or feature vectors using word embeddings. Also, commonly used stop function words that have a high frequency of occurrence such as 'a', 'an', 'the', 'is' are removed in the stop word removal process. This process is followed by the transformation process in which the parts of the sentence are annotated; text terms are translated into machine understandable format using standardization techniques such as stemming, lemmatization. The training and testing data are passed through the preprocessing step before training the model to generate processed train and test data.

3.3 Model Training and Explanation

A feed-forward deep neural network is built to predict the sentiments in the sentiment140 dataset. The input layer, hidden layers, and output layer constitute the major parts of a deep neural network, as shown in Fig. 3. Wherein, the hidden layers are added for feature extraction and contribute towards the complexity of the model. The output layer returns the probability of a tweet belonging to a given sentiment ('positive' or 'negative'). Hence, the result of prediction is the target (sentiment) with the highest probability for the corresponding tweet. The processed training data is used

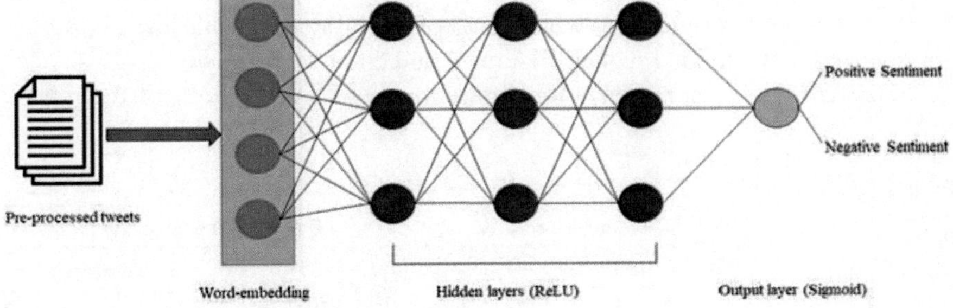

Fig. 3 Feed-forward deep neural network

to train the model. It is split into training and validation by 80% and 20%, respectively. The trained model is then used for classification of test tweets into positive and negative sentiments as labels and further tested for performance using the processed test data. After the testing of the model, the output of the model is explained using a post-hoc local model-agnostic explainer, LIME.

4 Experiments and Evaluation

4.1 Implementation Details

The sentiment140 training dataset consists of 1,045,576 tweets, whereas the testing dataset consists of 360 tweets. The proposed sentiment analysis explanation model is implemented in the Python programming language. The deep learning model has been created using the sequential model from Keras library, which runs on the top of TensorFlow library. The input layer consists of 127,099 nodes,*i.e.*, the vocabulary size of the dataset. There are two hidden layers in the created model with 500 and 250 nodes in the layers, respectively. ReLu activation function is applied to hidden layers, whereas for the output layer, the sigmoid function is used. Hyper-parameters have been tuned accordingly in order to increase the model accuracy and F1 score. The model is trained in 100 epochs with the hyper-parameters. Stochastic gradient descent (SGD) optimizer is used to minimize the objective function while model compilation. The LIME explainer for model explanation is implemented using *Lime* library in python, followed by plotting of the explanation.

4.2 Results

After the sequential model is trained, it is tested for accuracy and F1 score calculations. Since the principal concern of the proposed model is to explain the prediction of a particular instance, the paper does not consider the performance of the model as a major parameter. However, the trained model's performance is comparable with the existing literature and proves out to be 75.6% accurate. Table 1 shows the confusion matrix for the trained model. Figures 4 and 5 show the model's performance (accuracy and loss, respectively) for different epochs. An instance of the model,

Table 1 Confusion matrix

	Predicted: positive	Predicted: negative
Actual: positive	168	37
Actual: negative	53	102

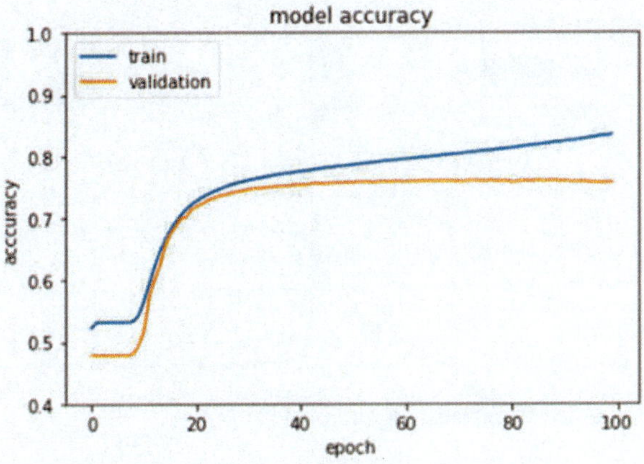

Fig. 4 Model accuracy

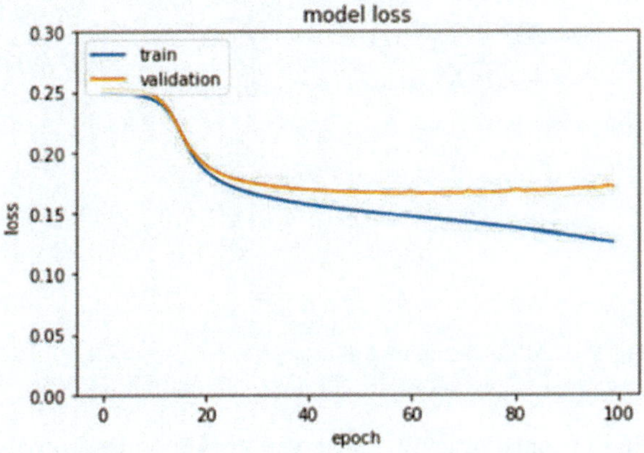

Fig. 5 Model loss

each for positive and negative sentiment, respectively, is then explained using LIME explainer. Table 2 shows the instances to be explained from the Sentiment140 dataset.

The output of the LIME explainer shows the main features which contribute to the prediction of sentiments along with highlighted text for each instance. For the

Table 2 Instances to be explained

S. No.	Tweet	Processed tweet	Sentiment
1	@switchfoot https://twitpic.com/2y1zl—Awww, that's a bummer. You shoulda got David Carr of Third Day to do it.;D	Awww, that's a bummer. You shoulda got David Carr of Third Day to do it	Negative
2	@naughtyhaughty HOW DID I FORGET ABOUT TWO AND A HALF MEN?!?!? I LOVE THAT SHOW!!!	how did i forget about two and a half men i love that show	Positive

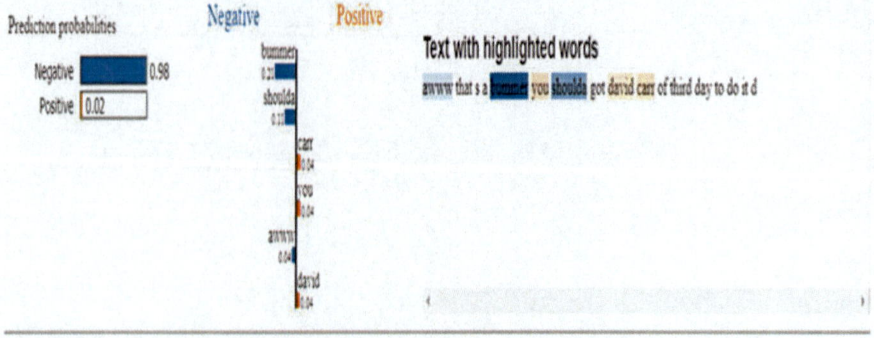

Fig. 6 LIME explanation for instance 1

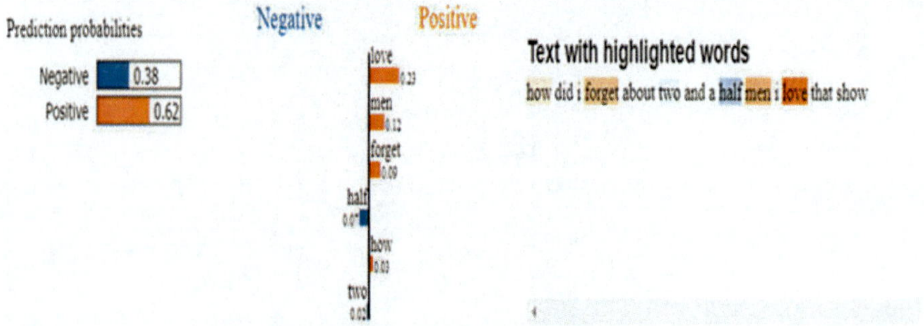

Fig. 7 LIME explanation for instance 2

first instance, it can be observed in Fig. 6 that words such as 'bummer', 'shoulda', and 'aww' contribute towards the prediction of the tweet's sentiment as negative with 'bummer' word contributing majorly (43.75%) towards the negative prediction. Similarly, in the second instance's explanation in Fig. 7, 'love', 'men', 'how', and 'forget' words contribute towards the positive prediction of the tweet where the word 'love' contributes majorly (41.07%) towards prediction.

5 Conclusion

With the growth and widespread use of the Internet, users can easily express their opinions and views on any topic through blogs, discussion forums, micro-blogs, and social media platforms such as Facebook, Twitter, and Instagram. Sentiment analysis has been widely used in literature to predict the sentiments from the opinion text data. The proposed model uses a feed-forward deep learning technique to predict the sentiments. Due to the lack of transparency in the prediction process, the user may not trust the classifier's predictions. Hence, the proposed model explains the prediction made by the classifier for a particular instance using local post-hoc model-agnostic explainer, namely *LIME*. However, it has been observed that the LIME explainer suffers instability when the same instance is explained multiple times. In order to overcome the limitation, as a future work, the researchers aim to explain the model's prediction for an individual instance using Shapley values explainer, which is based on game theory and further, compare the two explainers for the explanation of predictions in textual data.

References

1. Dosilovic FK, Brcic M, Hlupic N (2018) Explainable artificial intelligence: a survey. In: Proceedings of 2018 41st international convention on information and communication technology, electronics and microelectronics, MIPRO 2018
2. Duncan B, Zhang Y (2015) Neural networks for sentiment analysis on Twitter. In: Proceedings of 2015 IEEE 14th international conference on cognitive informatics and cognitive computing, ICCI*CC 2015
3. Giachanou A, Crestani F (2016) Like it or not: a survey of twitter sentiment analysis methods. ACM Comput Surv
4. Go A, Bhayani R, Huang L (2020) Sentiment 140. https://help.sentiment140.com/for-students. Accessed 21 Jan 2020
5. Gunning D, Aha D (2019) DARPA's explainable artificial intelligence (XAI) program. AI Mag.https://doi.org/10.1609/aimag.v40i2.2850
6. Hall P, Gill N (2018) An introduction to machine learning interpretability. o'reilly
7. Lisboa PJG (2013) Interpretability in machine learning - Principles and practice. Int Work Fuzzy Log Appl. 8256: 15–21. https://doi.org/10.1007/978-3-319-03200-9_2
8. Liu B (2012) Sentiment analysis and opinion mining. Synth Lect Hum Lang Technol.https://doi.org/10.2200/S00416ED1V01Y201204HLT016
9. Lundberg SM, Lee S (2017) A unified approach to interpreting model predictions. In: Neural Inf Process Syst: 1–10
10. Pang B, Lee L (2008) Opinion mining and sentiment analysis. Found Trends Inf Retr.https://doi.org/10.1561/1500000001
11. Ravi K, Ravi V (2015) A survey on opinion mining and sentiment analysis: tasks, approaches and applications. Knowl Based Syst.https://doi.org/10.1016/j.knosys.2015.06.015
12. Ribeiro MT, Singh S, Guestrin C (2016) "Why should i trust you?" Explaining the predictions of any classifier. In: Proceedings of the ACM SIGKDD international conference on knowledge discovery and data mining
13. Samek W, Müller KR (2019) Towards explainable artificial intelligence. In: Lecture notes in computer science (including subseries lecture notes in artificial intelligence and lecture notes in bioinformatics)

Effect of Quality of Existing Concrete Structures in Ajdabia Region, Libya

Moustafa Abdulrahim Mohamed and Aslam Amirahmad

Abstract In this research, the authors accomplished the statistical quality control of concrete. Data of 22 projects constructed by different companies have been randomly collected from Almetraka Laboratory for this research. Statistical analysis of the collected data has been completed according to ACI, BS and standards. This study reveals that the required average compressive strength f_{cr}' used in mix design of concrete, of the all projects except one, does not meet the specified compressive strength criteria according to ACI-214 criteria. For example, the standard deviation for Project No. 10 was 1.82, which is less than 2.8 for the excellent construction quality control of the concrete work according to ACI. Nevertheless, it is noted that the calculated value of a is 0.74 much less than 1.73 the approach one of the American standard. Therefore, implemented compressive strength of the concrete had been disapproved. Finally, based on the sustainable performance criteria recommendations had been made to Ajdabia Municipality's decision makers.

Keywords Demolition wastes · Quality control · Concrete strength · Statistical analysis

1 Introduction

It is widely known that the construction industry has significant impact on the environment. Construction and demolition (C and D) waste which are predominating used as illegal burial materials. Wastes are usually produced when new structures are intended to replace existing structures, when structures are innovated, or when existing structures collapse for natural causes such as landslides or abnormal causes

Additional information on the ASTM standards discussed in this article can be found at www.astm.org.

M. A. Mohamed · A. Amirahmad (✉)
Department of Civil Engineering, Faculty of Engineering, Najran University, Najran 66441, Kingdom of Saudi Arabia
e-mail: masalihe@nu.edu.sa; aaahmaad@nu.edu.sa

A. Khanna et al. (eds.), *Data Analytics and Management*, Lecture Notes on Data Engineering and Communications Technologies 54,
https://doi.org/10.1007/978-981-15-8335-3_19

such as explosions as Martin stated [6]. These wastes include materials such as concrete, wood, bricks, metals, parquet, roofing. Disposal of these wastes has become a major concern in recent years, where, often, this waste is disposed of unlawfully and illegally to avoid hauling costs. Disposal sites (landfills) include agricultural land, major residential areas, boreholes and lowlands, which have become a threat to groundwater contamination. In fact, C and D waste does not produce any chemical contamination if the bulk of it is recycled with residual material used after recycling to fill or settle low areas. Some literature review is given as:

- In developed countries, dedicated landfills are established, usually located in abandoned areas. In collaboration with local manufacturers, the Ministry of the Environment has developed a set of policies for best practices that limit the negative impacts of C and D.
- In New Zealand, for example, entities related to the construction industry developed a set of policies that reduce waste by improving the efficiency of resources, taking into account the overall view of the building, from design and construction, to the stage of dismantling/demolishing [7].
- Also in Australia, the National Waste Policy sets a clear direction for Australia for the next 15 years and will update and integrate Australia's policy framework and regulatory framework. The government aims to support the development of best practices across all states and territories through these policies. [5].
- For the above reasons, it is important that decision makers are to be more closely related to this industry to assess the amount of demolition waste produced from construction industry, and then analyze practices and find solutions to reduce and treat waste in order to develop an optimal approach to sustainable construction without any negative impact on the environment.
- This data have been statistically analyzed, verified according to the proposed Libyan concrete strength assessment criteria, introduced by Alazhari and Al Shebani [8]; ACI, and BS standards.

The aim of this paper is to promote better waste management practices in an effort to reduce the trend of increasing the volume of waste generated and disposed of in the Ajdabia area. This can be achieved through tracking and evaluating efficient resources and minimizing waste during project construction phases. Existing reinforced concrete structures are often demolished when elements or some of them fail in their compressive strength, which may result due to poor quality control. So, the authors attempted to assess the statistical quality of concrete in Ajdabia region. To maintain the design compressive strength, mix proportion has to be designed considering basic factors presented in many research articles and text books such as Neville [3].

2 Materials and methods

In this study, the authors assess the statistical quality control of concrete, particularly compressive strength, used in construction field in Ajdabia region. To do so, data of concrete tests for compressive strength of 22 projects constructed by different companies have been randomly gathered from Almetraka Laboratory for concrete and soil testing, taking in the mind that, because no sufficient data available for statistical analysis, results of several concrete compressive strengths that concern other different projects have been discarded. These data have been statistically analyzed, verified according to the proposed Libyan concrete compressive strength assessment criteria, introduced by Alazhari and Al Shebani [8]; ACI, and BS standards.

3 Modeling Approach

3.1 General

For an accurate interpretation of the compressive strength of concrete, statistical quality control of concrete is essential. Minimum of two mold are fabricated from a sample taken from a single batch of concrete is required for each test. In case of standard cube (150 mm), the strength that got may converted to standard cylindrical molds strength multiplying the cube strength by 1.25.

Considering that the curve pattern of concrete tests for strength distributes normally, it is known that there is a good control when the strength test values tend to cluster near to the average value, that is, the histogram of test results is tall and narrow. In contrast, the normal distribution curve for same mean strength and different variability is shown in Fig. 1.

Statistical analysis can be used to propagate the mathematical relation of two statistical parameters as shown below:

$$f'_{\mathrm{cm}} = \overline{X} = 1/n \sum_{i=1}^{n} X_i \tag{3.1}$$

where

X_i Compressive test strength of cylinders of ith sample.
n Number of tests concerns each project in the record. And

$$\sigma = \sqrt{\sum_{i=1}^{n} (X_i - \overline{X})^2 \Big/ n - 1} \tag{3.2}$$

Fig. 1 **a** First picture; **b** second picture

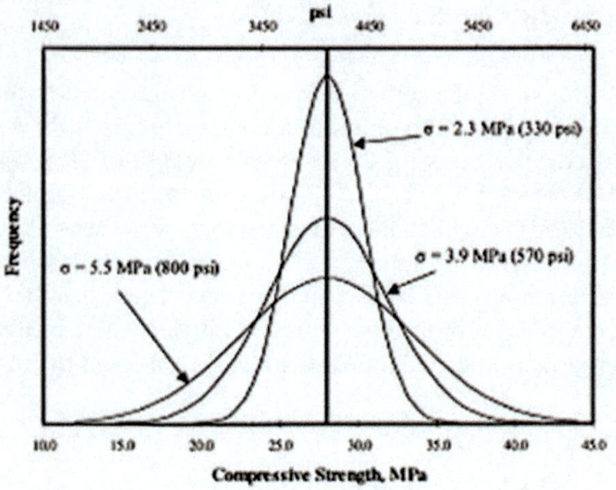

3.2 Standards of Control

The comparison between statistical overall variability and within-test variability is made by the statistical analysis of quality control. Based on statistical analysis of compressive strength of concrete, Tables 1 and 2 show the appropriate standards of control for specified design strength of concrete.

Table 1 Concrete control standards, f'_c < 34.5 MPa (Table 5 in ACI214 [1])

Overall variation					
Class of operation	Standard deviation for different control standards (MPa)				
	Excellent	Very good	Good	Fair	Poor
General construction testing	Below 2.8	2.8–3.4	3.4–4.1	4.1–4.8	Above 4.8
Laboratory trial batches	Below 1.4	1.4–1.7	1.7–2.1	2.1–2.4	Above 2.4
Within-batch variation					
Class of operation	Coefficient of variation for different control standards (%)				
	Excellent	Very good	Good	Fair	Poor
Field control testing	Below 3.0	3.0–4.0	4.0–5.0	5.0–6.0	Above 6.0
Laboratory trial batches	Below 2.0	2.0–3.0	3.0–4.0	4.0–5.0	Above 5.0

Table 2 Concrete control standards, f'_c > 34.5 MPa (Table 6 in ACI214 [1])

Class of operation	Coefficient of variation for different control standards (%)				
	Excellent	Very good	Good	Fair	Poor
Overall variation					
General construction testing	Below 7.0	7.0–9.0	9.0–11.0	11.0–14.0	Above 14.0
Laboratory trial batches	Below 3.5	3.5–4.5	4.5–5.0	5.0–7.0	Above 7.0
Class of operation	Coefficient of variation for different control standards (%)				
	Excellent	Very good	Good	Fair	Poor
Within-batch variation					
Field control testing	Below 3.0	3.0–4.0	4.0–5.0	5.0–6.0	Above 6.0
Laboratory trial batches	Below 2.0	2.0–3.0	3.0–4.0	4.0–5.0	Above 5.0

3.3 *Approved Criteria.*

Defining a is the bias (mean strength to nominal, design strength) factor, $a = {}^{f'_{cm}}/_{f'_c}$ and cov is the coefficient of variation, (cov $= {}^{\sigma}/_{f'_{cm}}$), and according to the approved criteria for the proposed Libyan-code, ACI-code, and BS-code, the following inequalities were adopted.

3.4 *Proposed Libyan-Based Approach*

(a) First approved criteria lead to

$$a \geq 1/(0.87 - 1.304\,\text{cov}) \tag{3.3a}$$

(b) Second approved criteria lead to

$$a \geq {}^{1}/_{(1 - 2.58\,\text{cov})} \tag{3.3b}$$

3.4.1 ACI-Based Approach

(a) First approved criteria lead to

$$a \geq {}^{1}/_{(1 - 1.34\,\text{cov})} \tag{3.4a}$$

(b) Second approved produces

$$a \geq {}^{1}\!/(1 - 2.33\,\text{cov}) \tag{3.4b}$$

3.4.2 BS-Based Approach [4]

The general approved criterion is written as

$$a > 1/\left(1 - \frac{(1.64\,\text{cov})}{\sqrt{n}}\right) \tag{3.5}$$

Actually, it is important to differentiate between the average strength and the required average strength of the concrete. To meet strength-performance requirements, statically, the average strength of the concrete should be in excess of the design compressive strength f_c' . While the required average strength f_{cr}' depends on the expected variability of test results, and on the allowable proportion of tests below the appropriate specified approved criteria.

According to ACI 318-11(clause 5.3), the minimum required average strength f_{cr}' can be computed using Eqs. (3.6a, b) below, depending on whether the coefficient of variation or standard deviation is used. The value of f_{cr}' will be the same for a given set of strength test results regardless of whether the coefficient of variation or standard deviation is used.

$$f_{\text{cr}}' = {}^{f_c'}\!/(1 - z * \text{cov}) \tag{3.6a}$$

$$f_{\text{cr}}' = f_c' + z\sigma \tag{3.6b}$$

Assuming a normal distribution of strength test results, z is selected to provide a sufficiently high probability of meeting the specified strength. For a particular project, the amount required to make the required average strength f_{cr}' exceed the design strength f_c' depends on the degree of confidence followed (i.e., the approved criteria specified).

So, the actual compressive strength $f_{c,\text{act}}'$ of a required degree of confidence for a given strength data of that used in a certain existing project, substituting f_{cr}' with f_{cm}', can be calculated from Eqs. (3.6a, b) as follows:

$$f_{c,\text{act}}' = f_{\text{cm}}'(1 - z * \text{cov}) \tag{3.7a}$$

$$f_{c,\text{act}}' = f_{\text{cm}}' - z * \sigma \tag{3.7b}$$

The main objective of this evaluation is to identify departures from desired target values, and possible, to assist the formulation of an appropriate response.

Therefore, bias factor a will be estimated as following:

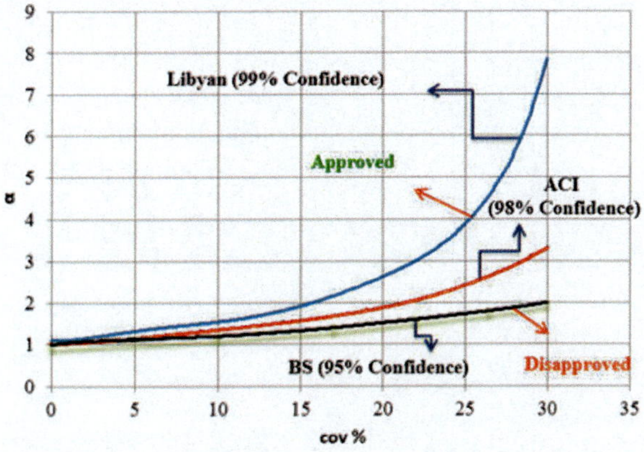

Fig. 2 Acceptance criterion of the adopted standards

$$a = f'_{c,\text{act}} / f'_c \tag{3.8}$$

It should be known that the proposed Libyan-code specifies that 99% of all test data to be approved while ACI and BS codes specify that 98% and 95%, respectively, of all test data to be approved. The comparison between the proposal approved of the three criteria is given in Fig. 2.

4 Results and Discussions

4.1 Standards of Control

Taking into account requirements of ACI 318–11, clause 5.3.1.2 [2], and that adopted in the previous sections, the results of the statistical analysis that carried out on the data of the 22 projects were summarized in Table 3.

In addition, of the symbols defined above, the following ones in Table 4 are defined as follows:

- $f'_{c,\max}$ = the maximum strength of concrete,
- $f'_{c,\min}$ = the minimum strength of concrete, and
- A_b = the ambit for each project, $(A_b = f'_{c,\max} - f'_{c,\min})$.

Note that f'_c for the 22 projects was below 34.5 MPa, so Table 3 was used.

From Table 3, it is cleared that the percentage of (A_b/f'_{cm}) for the 22 projects ranged from (8 to 93.96) %, where the data scatter (given by A_b/σ) ranges between 2.84 and 4.42 with an average equals to 3.611. That means all data included in the range $\pm 3\sigma$. Therefore, the spread of data increases with the lowering and widening of the natural distribution curve.

Table 3 Summary of statistical analysis of the 22 projects strength data

Proj. No.	N	f'_c (MPa)	$f'_{c,\min}$ (MPa)	$f'_{c,\max}$ (MPa)	A_b	f_{cm} (MPa)	σ	A_b/σ	A_b/f_{cm}	cov	Statistic judgment (QC)
1	16	20	16.66	27.33	10.67	21.6	3.24	3.3	49.43	14.9977	V. Good
2	13	20	10.83	26.91	16.08	19.45	4.37	3.68	82.71	22.4611	Fair
3	16	20	17.26	41.26	24	28.54	6.32	3.79	84.11	22.1655	Poor
4	24	20	10.28	31.48	21.25	22.62	4.71	4.42	93.96	21.2708	Fair
5	18	20	19.41	38.78	19.37	27.62	5.28	3.67	70.16	19.1178	Poor
6	13	20	14.25	27.41	13.16	20.12	3.77	3.49	65.42	18.7567	Good
7	15	20	24.39	39.54	15.15	29.78	3.65	4.15	50.85	12.2427	Good
8	15	20	14.01	32.79	18.78	20.86	5.55	3.38	90.02	26.6057	Poor
9	19	20	12.33	29.94	17.61	22.86	5.03	3.5	77.02	21.9872	Poor
10	22	24	16.14	24.05	7.91	21.61	1.82	4.35	36.61	8.41774	Excellent
11	25	24	15.16	16.93	1.77	15.99	0.56	3.16	11.09	3.505	Excellent
12	14	24	14.48	19.16	4.68	17.1	1.39	3.37	27.36	8.11455	Excellent
13	17	24	21.78	25.96	4.18	23.89	1.07	3.9	17.48	4.48308	Excellent
14	16	24	14.66	25.62	10.96	18.7	3.2	3.42	58.59	17.1301	V. Good
15	22	20	15.29	29.38	14.09	20.84	4.5	3.13	67.62	21.5957	Fair
16	14	20	11.28	21.6	10.32	16.54	3.63	2.84	62.38	21.9428	Good

(continued)

Table 3 (continued)

Proj. No.	N	f'_c (MPa)	$f'_{c,\min}$ (MPa)	$f'_{c,\max}$ (MPa)	A_b	f_{cm} (MPa)	σ	A_b/σ	A_b/f_{cm}	cov	Statistic judgment (QC)
17	22	24	16.14	24.05	7.91	21.61	1.82	4.35	36.61	8.41774	Excellent
18	15	24	26.26	26.26	2.2	27.57	0.6	3.68	8.0	2.17459	Excellent
19	20	20	13.19	13.19	5.84	16.36	1.38	4.23	35.71	8.43619	Excellent
20	24	20	20.06	20.06	16.01	26.26	4.64	3.45	60.96	18.0516	Fair
21	23	20	12.89	12.89	8.41	17.83	2.75	3.05	47.12	15.4437	Excellent
22	22	20	20	20	5.27	21.85	1.89	3.13	24.13	7.7131	Excellent

From the results of the study, we can summarize the specific control of the quality of concrete on the sites under study, ranging from poor to excellent according to ACI214 [1] as shown in Table 3.

4.2 Evaluation and Approved of Concrete

Recalling Eqs. (3.3a, 3.4a, 3.5, and 3.8), substituting *z*-values as 2.33, 2.12, and 1.64 for 99%, 98%, and 95% confidences, respectively; ACI 318-11 approved or disapproved of the 22 projects tabulated in Table 4.

This has explained graphically as shown in Figs. 3, 4 and 5 based on Approved Criteria of Libyan Standard, Approved Criteria of ACI Standard. Approved Criteria of BS Standard, respectively.

It is clear that the factor a's, for 21 out of the 22 projects, less than values of criteria's approach that approved for the proposed Libyan-code, ACI-code, and BS-code. This means that the actual compressive strength $f'_{c,\text{act}}$ to the majority of the completed projects is much less than its design strength f'_c , i.e., disapproved.

Here, one can note that, despite the specific control of the quality of concrete at the sites under study varies from poor to excellent according to ACI-214 criteria, the study revealed that the average strength of concrete f'_{cr} of the all projects except one does not meet the design compressive strength approved criteria standards. Indeed, quality control can be achieved under either proper site conditions or improper site conditions. In other words, quality control can be achieved if there is continuation in keeping site conditions as it is; regardless, site conditions are proper or improper.

Monitoring of construction works in many sites in Ajdabia region was done, and the following construction mistakes have been observed:

1. Cements used are delivered from different sources, local or imported from different countries, and no periodic acceptance tests are carried out. Performing of these tests, by independent laboratory, is very important to ensure the viability of the cement used and that it is in limit of the standard used.
2. At sites where concrete being batched and mixed, mixer laborers usually add randomly water without taking care the selection of correct ingredients relative quantities according to the mix design, if designed.
3. Is well known that, and practical experience, that the most important factors affecting the concrete strength lie in the ratio of water to cement, degree of compaction, temperature and age of concrete. Assuming that concrete has a full compaction, normal temperature, and at a certain age, the strength of concrete is inversely proportional to the ratio of water to cement.
4. At sites, placing generally ready-mixed concrete, delayed beyond cement initial setting time, i.e., prolonged mixing. When mixing period time is long, the ready-to-place concrete is exposed to evaporation, which decreases the workability, which leads the supplier to add more water in order to restore workability, lowering the strength and increasing shrinkage.

Table 4 Statistical analysis of the twenty-two projects strength data

Proj. No.	$\acute{f}_c$	BS (95% Confidence)				ACI (98% confidence)				LIBYAN (99% confidence)			
		$\acute{f}_{c,\,actual}$ (MPa)	Cal.a	Approach a	Evaluation of strength	$\acute{f}_{c,\,actual}$ (MPa)	Cal.a	Approach a	Evaluation of strength	$\acute{f}_{c,\,actual}$ (MPa)	Cal.a	Approach a	Evaluation of strength
1	20	16.28	0.81	1.07	Disapproved	14.73	0.74	1.25	Disapproved	14.05	0.70	1.48	Disapproved
2	20	12.29	0.61	1.11	Disapproved	10.19	0.51	1.43	Disapproved	9.27	0.46	1.73	Disapproved
3	20	18.16	0.91	1.10	Disapproved	15.13	0.76	1.42	Disapproved	13.80	0.69	1.72	Disapproved
4	20	14.73	0.74	1.08	Disapproved	12.42	0.62	1.40	Disapproved	11.41	0.57	1.69	Disapproved
5	20	18.96	0.95	1.08	Disapproved	16.42	0.82	1.34	Disapproved	15.31	0.77	1.61	Disapproved
6	20	13.93	0.70	1.09	Disapproved	12.12	0.61	1.34	Disapproved	11.33	0.57	1.60	Disapproved
7	20	23.80	1.19	1.05	Disapproved	22.05	1.10	1.20	Disapproved	21.28	1.06	1.41	Disapproved
8	20	11.76	0.59	1.13	Disapproved	9.10	0.45	1.55	Disapproved	7.93	0.40	1.91	Disapproved
9	20	14.62	0.73	1.09	Disapproved	12.21	0.61	1.42	Disapproved	11.15	0.56	1.71	Disapproved
10	24	18.63	0.78	1.03	Disapproved	17.75	0.74	1.13	Disapproved	17.37	0.72	1.32	Disapproved
11	24	15.07	0.63	1.01	Disapproved	14.80	0.62	1.05	Disapproved	14.68	0.61	1.21	Disapproved
12	24	14.82	0.62	1.04	Disapproved	14.16	0.59	1.12	Disapproved	13.87	0.58	1.31	Disapproved
13	24	22.13	0.92	1.02	Disapproved	21.62	0.90	1.06	Disapproved	21.39	0.89	1.23	Disapproved
14	24	13.45	0.56	1.08	Disapproved	11.91	0.50	1.30	Disapproved	11.24	0.47	1.55	Disapproved
15	20	13.46	0.67	1.08	Disapproved	11.30	0.56	1.41	Disapproved	10.35	0.52	1.70	Disapproved
16	20	10.59	0.53	1.11	Disapproved	8.85	0.44	1.42	Disapproved	8.08	0.40	1.71	Disapproved
17	24	18.63	0.78	1.03	Disapproved	17.75	0.74	1.13	Disapproved	17.37	0.72	1.32	Disapproved
18	24	26.59	1.11	1.01	Approved	26.30	1.10	1.03	Approved	26.17	1.09	1.19	Disapproved
19	20	14.10	0.70	1.03	Disapproved	13.44	0.67	1.13	Disapproved	13.15	0.66	1.32	Disapproved
20	20	18.49	0.92	1.06	Disapproved	16.21	0.81	1.32	Disapproved	15.22	0.76	1.58	Disapproved
21	20	13.32	0.67	1.06	Disapproved	11.99	0.60	1.26	Disapproved	11.42	0.57	1.50	Disapproved

(continued)

Table 4 (continued)

Proj. No.	$\acute{f}_c$	BS (95% Confidence)				ACI (98% confidence)				LIBYAN (99% confidence)			
		$\acute{f}_{c,\text{ actual}}$ (MPa)	Cal.a	Approach a	Evaluation of strength	$\acute{f}_{c,\text{ actual}}$ (MPa)	Cal.a	Approach a	Evaluation of strength	$\acute{f}_{c,\text{ actual}}$ (MPa)	Cal.a	Approach a	Evaluation of strength
22	20	19.08	0.95	1.03	Disapproved	18.28	0.91	1.12	Disapproved	17.92	0.90	1.29968	Disapproved

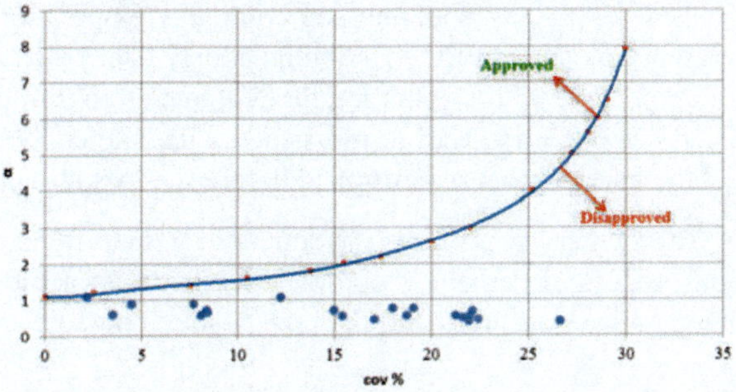

Fig. 3 Approved criteria of LIBYA standards and projects actual strength

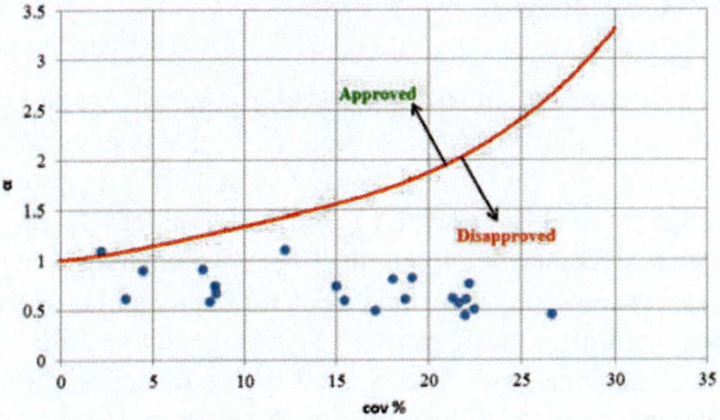

Fig. 4 Approved criteria of ACI standards and projects actual strength

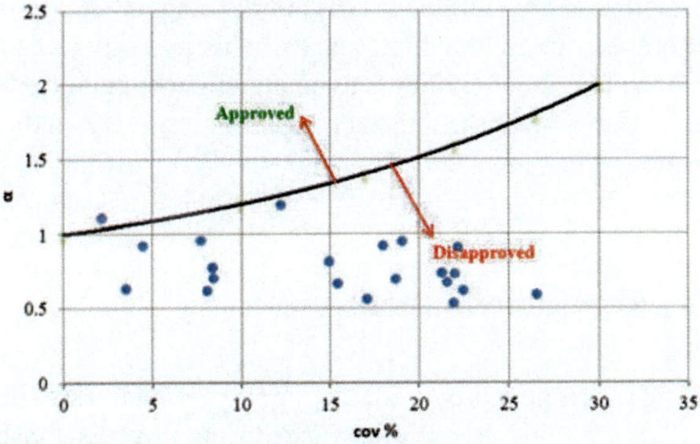

Fig. 5 Approved criteria BS standards and projects actual strength

Without using of admixtures, according to ASTM C 94, the concrete must be placed within 1.5 h of mixing, and according to BS5328, the period of time required for cement to contact with moist aggregate does not exceed 2 h. However, it should remain cohesive and should not segregate.

5. At almost sites, it is observed that concrete compaction, either by hand or by vibration, is ignored.

In fact, placing concrete and compacting operation are simultaneous processes that are accomplished simultaneously. They are necessary to ensure that the requirements for obtaining the required strength of concrete are met.
6. Also, it is observed, in most cases, that site concrete is never cured actively, or the curing process is stopped before the hydration process is completed.

The primary goal of making concrete retain its moisture is to ensure that the hydration process of cement is completed in order to achieve the desired bond between the aggregate particles and then the desired concrete strength.

5 Conclusion

As requirements of sustainable development, an attempted study to curb the trend of growing volumes of demolition waste that is generated and disposed of in Ajdabia region is introduced. The study has focused on waste minimization during projects construction stages which achieved by assessing the statistical quality control of concrete. Data of concrete tests for compressive strength of 22 projects constructed by different companies have been randomly gathered, analyzed statistically and verified according to the proposed Libyan, ACI and BS standards of concrete strength assessment criteria. Due to construction mistakes, the study reveals that the required average strength of the majority of the projects does not meet the specified compressive strength of the approved criteria. Construction mistakes are a clear indication that the professional engineer (structural designer or contractor) does not know enough about the concrete, which results in his inability to calculate the correct proportions of the concrete ingredients to be implemented at the site to achieve the desired workability and strength. This fact impacts negatively on the environment around Ajdabia region.

6 Recommendations

1. Government procurement policies can have a significant impact on overcoming or minimizing demolition waste during projects construction stages. The following guideline suggestions are recommended to be followed:

   - By authority, as technical sector, a body such as Engineering Council should be organized, or activated if existed. Its goals include followings:-
   - Adopting appropriate specifications and criteria standards that conform to the requirements of the local construction needs and conditions.
   - Developing of technical skills of professionals involved in the design, construction, maintenance and demolition cycle of buildings and infrastructure.
   - Executing certificates to practice engineering works.

- Issue of decisions that concern engineering practice fairs.

2. The municipality should have an executive office, such as engineering administration to execute the following tasks:

Checking of compliance of architectural and structural designs, of common and private projects, to the local building code or the adopted criteria standards.

- Preventing any one not awarded certification from Engineering Council to practice engineering works.
- Holding contractors to employee resident engineer for each project under construction.
- Monitoring sites under construction in order to avoid construction mistakes that can be occurred.
- Preventing any structure to be demolished without getting permission from the municipality.

Acknowledgements Thanks are due to the Almetraka Laboratory for Concrete and Soil Testing—Ajdabia, for its co-operation in providing information.

References

1. ACI Committee 214 (2011) Guide for evaluation of strength test results of concrete (ACI 214R-11). American Concrete Institute, Farmington Hills, MI, 16 pp
2. ACI Committee 318-11 (2011) Building code requirements for structural concrete and commentary. American Concrete Institute, Farmington Hills, MI, pp 503
3. Neville M, Brooks JJ (1987) Concrete technology. United States, Wiley Inc, New York
4. BS 5328 (1993) Concrete British Standard Institution. England
5. Department of Sustainability, Environment, Water, Population and Communities (2011) Construction and demolition waste guide—recycling and reuse across the supply chain. Prepared by Edge Environment Pty Ltd. for the Department Australian Government
6. Gauteng Department of Agriculture Conservation and Environment, South Africa (2009) Gauteng provincial building and demolition waste guidelines. Published in the Environmental Impact Assessment Regulations, March 2009
7. Mahara Inglis (2007) Construction and demolition waste—best practice and cost saving. Ministry of the Environment, New Zealand. SB07, Paper Number 057
8. Alazhari MS. Al Shebani MM (2013) Assessment of concrete quality in Libya. Procedia Eng 54:117–126.

Machine Learning and Evolutionary Algorithms for the Diagnosis and Detection of Alzheimer's Disease

Moolchand Sharma, S. P. Pradhyumna, Shubham Goyal, and Karan Singh

Abstract Alzheimer's disease is a chronic neurological disorder that generally progresses gradually and steadily gets worse over time. In 2050, Alzheimer's occurrence will rise to around every 33 s. Hence, early-stage diagnosis is both essential and crucial. Cancer is accountable for 60–70% of cases involving dementia. The early symptom found which is most common is difficulty in remembering recent events. As the disease progresses, signs may include language issues, disorientation (including getting lost easily), mood swings, and lack of energy, not managing self-care and behavioral problems. "Did I forget something this morning?" These types of questions start arising in the patient's mind. As the status of an individual deteriorates, they sometimes withdraw from family and community. Little by little, bodily functions are lost, which eventually leads to death. While the progression rate can vary, the average lifespan after an evaluation is three to nine years. The idea would be to use the evolutionary approach to extract the primary useful information from the Alzheimer dataset and use it in machine learning algorithms to increase the accuracy of the prediction, i.e., declaring whether the patient has Alzheimer's or not. Initially, we applied traditional machine learning algorithms, and we got an accuracy of 72–84%. After feature selection by using different evolutionary algorithms, we got a hike of 5–10% in accuracy, and the best accuracy of 95.71% was provided by particle swarm optimization algorithm and bacterial foraging algorithm with random forest classifier.

M. Sharma (✉) · S. P. Pradhyumna · S. Goyal · K. Singh
Maharaja Agrasen Institute of Technology, New Delhi, Delhi, India
e-mail: moolchand@mait.ac.in

S. P. Pradhyumna
e-mail: pradhyumna.purushottam@gmail.com

S. Goyal
e-mail: shu12.sg@gmail.com

K. Singh
e-mail: karansinghg@ymail.com

A. Khanna et al. (eds.), *Data Analytics and Management*, Lecture Notes on Data Engineering and Communications Technologies 54,
https://doi.org/10.1007/978-981-15-8335-3_20

Keywords Alzheimer's disease · Machine learning · Neurodegenerative disease · Dementia · Evolutionary algorithms · Python · K-nearest neighbor · Random forest classifier · Logistic regression · Decision tree classifier · Support vector machine · Bernoulli naive bayes · Ant colony optimization · Artificial bee colony optimization · Particle swarm optimization · Bacterial foraging · And bio-inspired algorithms

1 Introduction

Dementia is a condition that is defined as affecting various brain functions like memory, thought, comprehension, measuring, learning capacity, language, and judgment. The loss in cognitive function is sometimes followed by a decrease in emotional regulation, social activity, or motivation, and often preceded. Alzheimer's disease is, and perhaps the most common type of dementia contributes to 60 to 70 percent of cases [1]. The most startling characteristic of AD is that the patients lose short-term memory. This starts as a mild forgetfulness and progresses to a condition in which people can't even recognize their loved ones. Patients lose both visual and verbal memory, such as missing telephone calls, names of recent friends, misplacing objects, and losing way in a recognized environment [2].

Dementia can affect a person in various ways, and the disease's progression depends on the effect of the disease itself and the temperament and state of health of the individual. Dementia may be split into three phases:

1. first-stage—one year or two
2. second-stage—two to five years
3. final-stage—five years and later

These intervals are given as an estimated framework, and not all individuals with dementia exhibit the same manifestations. "AD occurs due to abnormal and toxic protein accumulation in neurons. We lose the ability to remove toxic proteins with age, and that's why such disorder manifests in older age," said Rukmini [1, 3].

Usually, people who have Alzheimer's also have a chance of suffering from other mental disorders such as Parkinson's disease in which tremors, loss of involuntary actions such as blinking and smiling, even Huntington's disease in which symptoms which include cognitive disorders such as lack of awareness, movement disorders such as tremors in hands while writing and legs while walking [4, 5].

In this paper, particle swarm optimization (PSO), artificial bee colony optimization (BCO), ant colony optimization (ACO), bacterial foraging algorithm (BFA) are performed. To achieve optimal accuracy, we passed the dataset into all mentioned evolutionary algorithms [6]. After preprocessing, the number of features was reduced, and there is a significant increase in the result. The use of evolutionary algorithms is gainful as it provides a more accurate prediction of whether the patient has Alzheimer's disorder or not. Initially, the dataset was fed to machine learning algorithms, such as k-nearest neighbor (KNN), logistic regression, Bernoulli Naive Bayes, random forest, support vector machine (SVM), and decision tree.

Then, we used bio-inspired optimization evolutionary algorithms for feature selection, and we used particle swarm optimization (PSO), artificial bee colony optimization (BCO), ant colony optimization (ACO), and bacterial foraging algorithm (BFA). The optimum result provided by the algorithm is chosen. After feature selection, the modified dataset is then fed again to the machine learning algorithms.

The key highlights of the paper includes.

- The following paper helps detect and cure Alzheimer's during the initial phases of the disorder.
- Different machine learning techniques and evolutionary algorithms are used on the dataset to provide the best accuracy.
- The best accuracy is provided by particle swarm optimization and bacterial foraging algorithm using random forest classifier.

The paper is lined up as: The study about the disease using several techniques is discussed in Sect. 2. Applying feature selection and pre-processing methods are described in Sect. 3. Discussion about machine learning classifiers and input parameters, and the dataset is explained in Sect. 4. The representation of results obtained by the experiment is in Sect. 5. Finally, Sect. 6 consists of conclusions and future scope, followed by References.

2 Literature Review

Dementia is an overall term related to Alzheimer's which can be considered as a broad aspect of the syndrome, it mainly influences about five percent of the elder community over the age of 65 years and has biased and unexplainable supremacy in women and a low rate in some parts of the world. There are many various types of dementia—Alzheimer's disease, frontotemporal dementia, Lewy bodies, and secondary disease dementia, such as acquired immunodeficiency syndrome (AIDS) dementia. Neurochemical and neurobiological works have contributed to developments in identifying the causes of dementia, and functional scanning, along with the development of hardware solutions, has made it possible to recognize potential biomarkers; a number of new therapeutic strategies have evolved from these. Evolutionary algorithms (EA) have proved to be appropriate for optimization problems because of the feature selection along with a reduction in noise of the dataset. Due to the inherent parallelism, they can capture several solutions concurrently in a single run [3, 7].

The predominance of dementia syndrome increases rapidly in the 70s–80s of the person's lifetime; this ailment affects more than one-fourth population of people who are older than eighty-five years. The onus of this syndrome also affects caretakers and their near and dear ones emotionally and mentally. Among all primary care patients over the age of 65, 1.8–12% have unrecognizable dementia, and 50–66% most of the cases of dementia in key groups of citizens are unchecked.

The Mini-Mental State Examination (MMSE), also known as the Mini-Cog test, is the best-studied, descriptive dementia screening tool. A cut point of 24–26 out of 30 points is normally approved as a definitive screen and will result in retrospective assessment with more history, review, and dementia testing. The scores need to be adjusted to educational success. The MMSE will classify cases of dementia with 70–91% sensitivity and 55–95% precision. Dementia is often curable but seldom healable or reversible. No more than 1.5% of all cases of dementia are fully reversible. About 60% of people with dementia disorder suffer from Alzheimer's disease, and 15% have vascular dementia. Most of the cases of dementia do not have a cure, so the medical perks of primary care detection will be heavily persuaded by the perks of early diagnosis and treatment of Alzheimer's disease, vascular dementia, and another diagnosis, which is incapable of being changed [8]. Most of the real-world challenges require optimization at the same time as many unequal and often encountering objectives. The only cure to the problem is to follow multiple precautions to prevent further damage to the present condition. These strategies are optimal in the wider sense that, provided that all goals of the situation are considered, no other approaches in the computational complexity are superior to these. Such research will be performed in a wide but diverse society of Alzheimer's cases with variations in many factors, including medical, neuropathological nature of the disease, temporal length of the disease, and clinical/neuropsychological profile, to accurately monitor the variability between brains affected by Alzheimer's disease. [3, 9]. There is other evolutionary algorithm as well like whale optimization algorithm [19] and these algorithms can be used to predict pnemonia [21] and neuro disorder [20] efficiently.

In computational science, particulate swarm optimization is a mathematical technique that streamlines the problem by recursively trying to improve the efficiency of response It tackles an issue by having a population of potential solutions, here called particles, and by moving those particles over the particle's location and velocity in the question space as per the basic mathematical formulae. The best-known locale location determines each particle's movement. Even, it is often directed toward the query-space best-known positions, which are remodeled as other particles find better positions. It will push the swarm toward the best possible solutions [10]. In computer science and research, ant colony optimization is an approach relating to the probability that uses graph routing to solve mathematical problems that can be condensed to find efficient paths in graphs. Artificial ants acting as multi-actors copy the behavior of actual ants mathematically. Pheromones are laid by the real ants to direct other ants toward the resources in their environment, which has been analyzed by them. The quality and position of these solutions are recorded by the simulated ants so that ants can locate better solutions in further iterations [11].

Artificial bee colony: the given algorithm is a Swarm-based meta-heuristic algorithm, created in the early 2000s. The model comprises three main components: bees foraging, sources of food for the workers, and the uninsured. The first two parts, working and unemployed foraging bees, are looking for rich food sources, the third component near their hive. The model also describes two leading behavioral modes that are important for self-organization and collective intelligence: recruitment of foragers to rich food sources leading to positive feedback and abandonment by foragers causing negative feedback from poor sources [12].

Bacterial foraging optimization algorithm is influenced by bacteria such as E.coli and M.xanthus foraging activity in units. In particular, the BFOA is affected by the chemotaxis activity of the bacteria, which can sense chemical gradients in the atmosphere (such as nutrients) and move to or away from different signals. Bacteria perceive the food's direction based on biochemical gradients within their atmosphere. Likewise, bacteria naturally produce chemicals that lure and annihilate into the atmosphere and can see each other similarly. Bacteria can move throughout their atmosphere using mechanisms of locomotion (such as flagella), Sometimes it moves chaotically (tumbling and twisting), and other times it moves in a circular manner that can be called a float. Bacterial cells are seen as agents in an environment, using their awareness of food and other cells as motives for motion, and stochastic tumbling and swimming as a displacement movement. Cells can swarm a source of food and, depending on the cell–cell interactions, actively repel, or ignore each other [13].

3 Methodology

This section discusses improvements done in machine learning algorithms and methodology used. The general flowchart for the process of Alzheimer's detection is shown in Fig. 1.

3.1 Initial Setup

Initially, we have made use of machine learning algorithms like k-nearest neighbors (KNN), support vector machine (SVM), logistic regression, decision tree, random forest, and Bernoulli Naive Bayes with an initial accuracy of 72.42, 74.95, 78.16, 76.71, 83.88, and 78.15%. The initial accuracy of the algorithms is stored and represented below in Fig. 2.

But the accuracy obtained by machine learning algorithms is not satisfactory, so we have used further evolutionary algorithms which will extract the important features from the dataset and helps us in increasing the efficiency for better diagnosis and detection of the disease. The evolutionary algorithm extracts the number of features that have been shown below in Fig. 3.

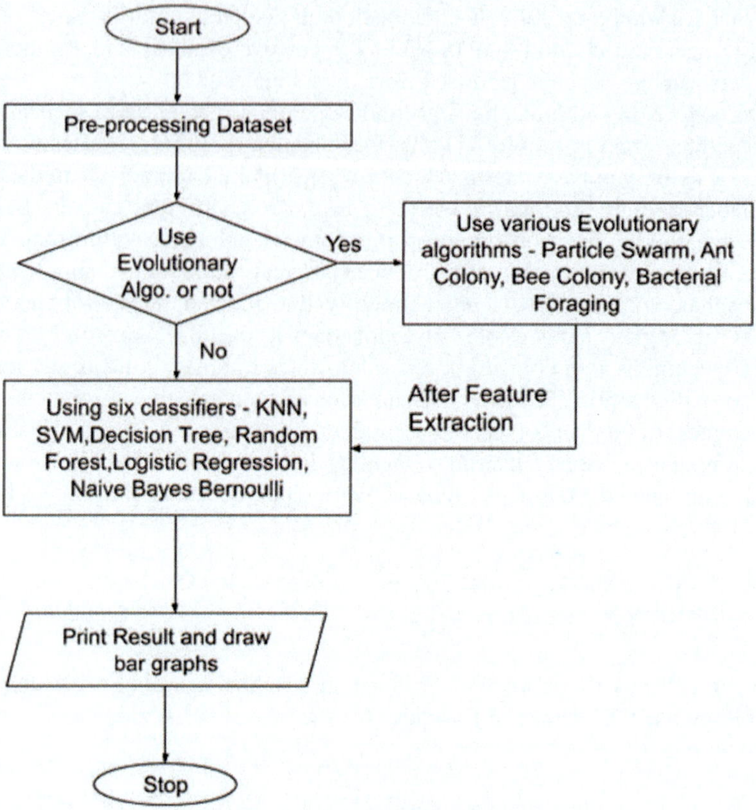

Fig. 1 General flow process of the methodology used

3.2 *Selection of Evolutionary Algorithm*

After applying machine learning without feature selection, as shown in Fig. 1, we used various evolutionary algorithms, such as artificial bee colony optimization (ABCO), ant colony optimization (ACO), particle swarm optimization (PSO), bacterial foraging algorithm (BFA). The following algorithms have reduced the number of features from the real-valued attributes as described in the datasets section, and the same is shown in Fig. 3.

3.3 *Binary Ant Colony Optimization (ACO)*

In computer science and research, ant colony optimization is an approach relating to the probability that uses graph routing to solve mathematical problems that can

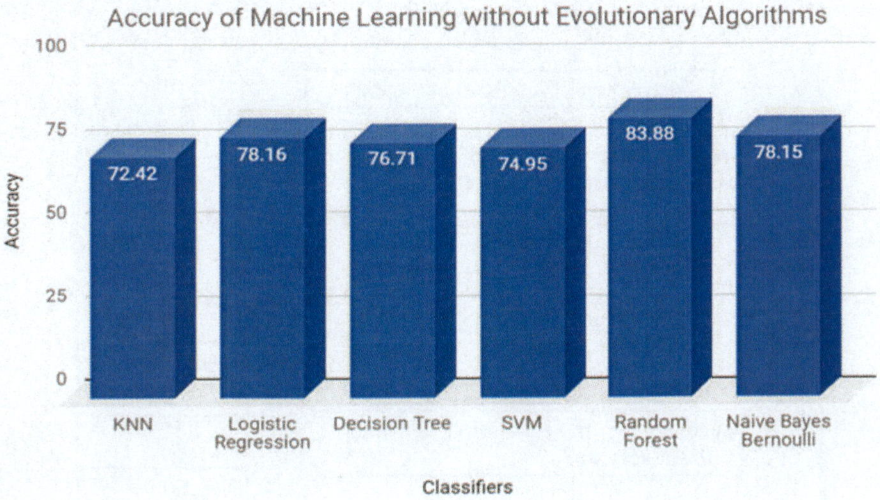

Fig. 2 Accuracy graph without feature selection

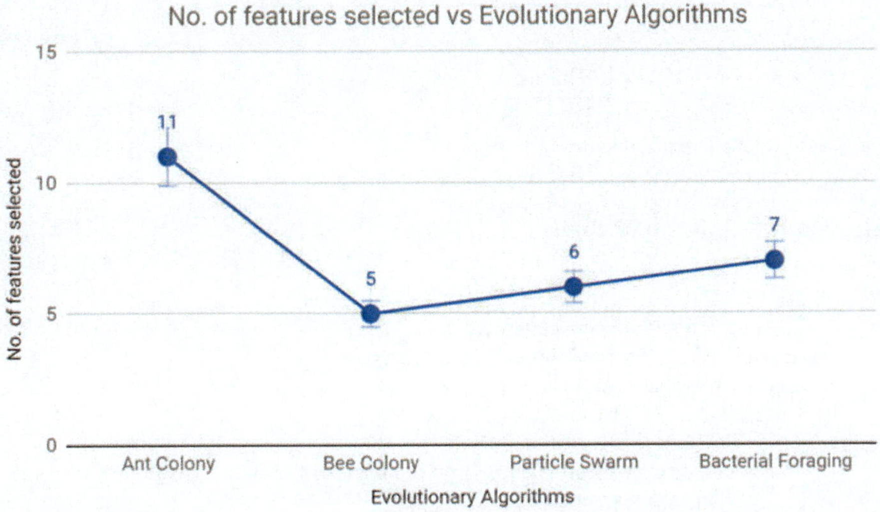

Fig. 3 Number of features selected by each evolutionary algorithm

be condensed to find efficient paths in graphs. Artificial ants acting as multi-ctors copy the behavior of actual ants mathematically. Pheromones are laid by the real ants to direct other ants toward the resources in their environment, which has been analyzed by them. The quality and position of these solutions are recorded by the simulated ants so that ants can locate better solutions in further iterations. A general pseudo-algorithm for the ACO algorithm is shown as Algorithm 1, and the flowchart is shown in Fig. 4.

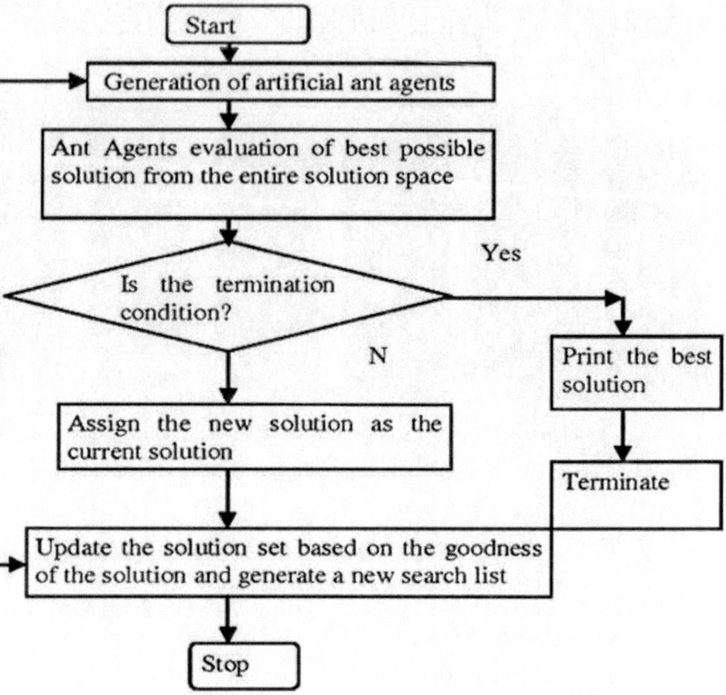

Fig. 4 Ant colony optimization flowchart

Algorithm 1: ACO algorithm

```
Start
  The initial population is generated
 While criteria of termination are not satisfied, do
   Each ant is placed at the starting point
     Repeat
          Perform this for each ant
          The next position is chosen on the basis of the state transition rule
          Pheromone is updated at each step
        End For
      Until a solution is found by each ant
    Best solution is updated
  Pheromone are updated globally
 End while
End
```

3.4 *Particle Swarm Optimization (PSO)*

In computational science, particle swarm optimization is a mathematical technique that streamlines the problem by recursively trying to improve the efficiency of response. It solves a problem by having a population of potential solutions, here called particles, and by moving those particles over the particle's location and velocity in the question space as per the basic mathematical formulae. The best-known locale location determines each particle's movement. Even, it is often directed toward the query-space best-known positions, which are remodeled as other particles find better positions. It will push the swarm toward the best possible solutions. A general pseudo-algorithm for the PSO algorithm is shown as Algorithm 2, and the flowchart is shown in Fig. 5.

Algorithm 2: PSO algorithm

```
Start
    Particles population is initialized
    while the criterion of termination is not satisfied
      for each particle p at particlePosition do
        fitness value f(particlePosition) is evaluated
        if f(particlePosition) is found better than particleBest then
          particleBest = particlePosition
        end if
    end for
    neighbourBestp is defined as the best position found by any of p's neighbors so far
    for each particle p do
         particleVelocity = compute_particle_velocity(particlePosition, particleBest, neighbourBestp)
         particlePosition = update_particle_position(cp, particleVelocity)
     end for
  end while
end
```

3.5 *Artificial Bee Colony Optimization (ABC)*

The given algorithm is a swarm-based meta-heuristic algorithm, created in the early 2000s. The model comprises three main components: bees foraging, sources of food for the workers, and the uninsured. The first two parts, working and unemployed foraging bees, are looking for rich food sources, the third component near their hive.

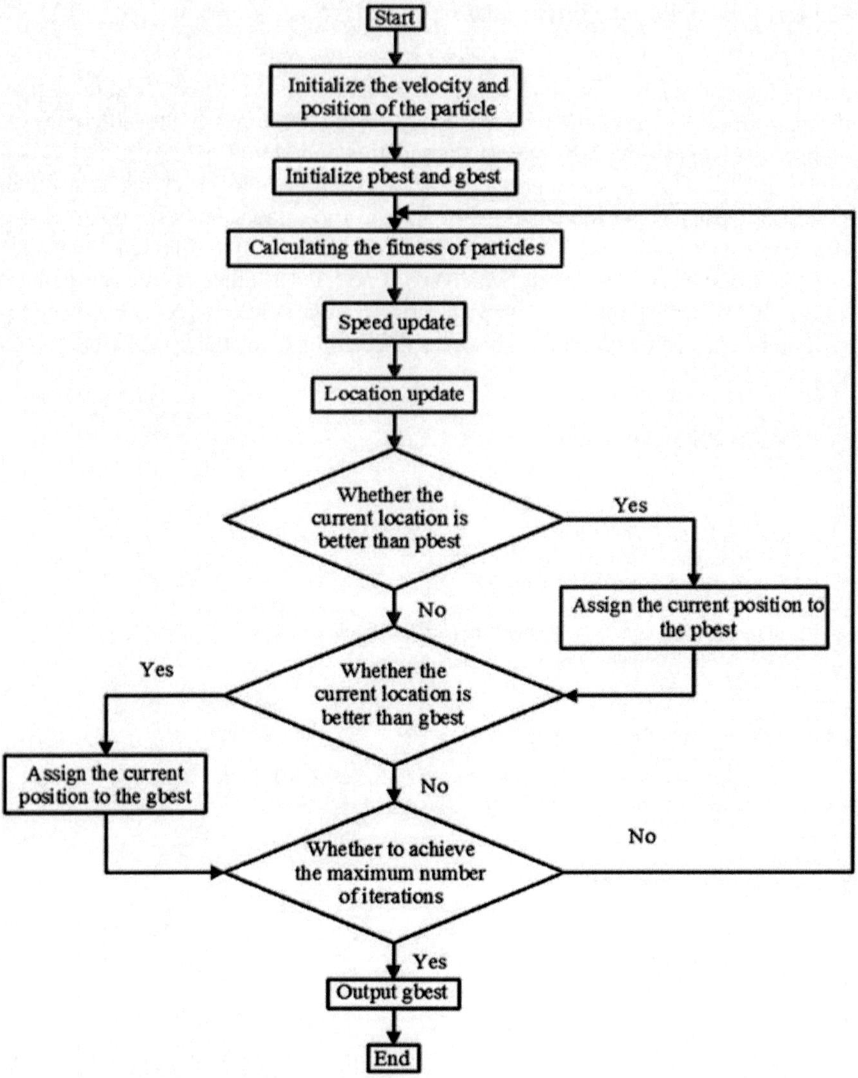

Fig. 5 Particle swarm optimization flowchart

The model also describes two leading behavioral modes that are important for self-organization and collective intelligence: recruitment of foragers to rich food sources leading to positive feedback and abandonment by foragers causing negative feedback from poor sources. A general pseudo-algorithm for the ABC algorithm is shown as Algorithm 3, and the flowchart is shown in Fig. 6.

Algorithm 3: ABC algorithm

Start
    Stage is initialized
    For iter < termination condition, i.e., maximum CPU time
    The phase of Employed bee
    The phase of Onlooker bee
    The phase of the Scout bee
    The best result achieved so far is selected
End For
End

3.6 *Bacterial Foraging Algorithm (BFOA)*

Bacterial foraging optimization algorithm is influenced by bacteria such as E.coli and M.xanthus foraging activity in units. In particular, the BFOA is affected by the chemotaxis activity of the bacteria, which can sense chemical gradients in the atmosphere

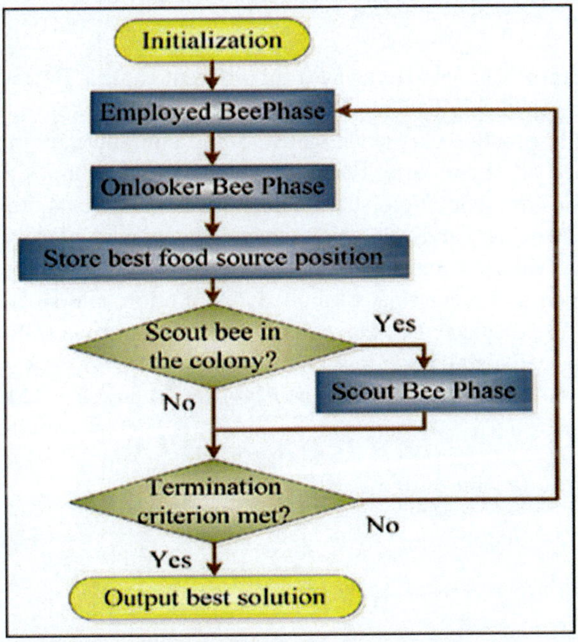

Fig. 6 Artificial bee colony optimization flowchart

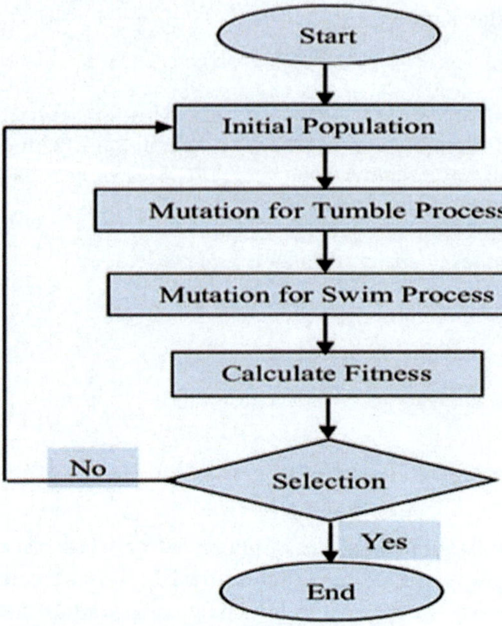

Fig. 7 Bacterial foraging optimization flowchart

(such as nutrients) and move to or away from different signals. Bacteria perceive the food's direction based on biochemical gradients within their atmosphere. Similarly, bacteria naturally produce chemicals that lure and annihilate into the atmosphere and can see each other similarly. Bacteria can move throughout their atmosphere using mechanisms of locomotion (such as flagella), Sometimes it moves chaotically (tumbling and twisting), and other times it moves in a circular manner that can be called a float. Bacterial cells are seen as agents in an environment, using their awareness of food and other cells as motives for motion, and stochastic tumbling and swimming as a displacement movement. Cells can swarm a source of food and, depending on the cell–cell interactions, actively repel or ignore each other. A general pseudo-algorithm for the BFOA shown as Algorithm 4, and the flowchart is shown in Fig. 7.

Algorithm 4: BFO algorithm

```
Bacterial_Foraging(ProblemSize, CellsSize, Ned, Nre, Nc, Ns, StepSize, dattract, wattract, hrepellent,
wrepellent, Ped):
BPopulation = Initialize BPopulation based on ProblemSize, and CellsSize
Loop i in (0:Ned):
    Loop k in (0:Nre):
        Loop j in (0:Nc):
            For each cell ⊂ BPopulation:
              Fitness = Cost(Cell) + Interaction(Cell, BPopulation, dattract, wattract, hrepellent, wrepellent)
                Cell_Health = Fitness
               Cell' = Θ
                For x in range(0:Ns):
                    RandomStepDirection = Create random step of ProblemSize
                    Cell' Take step RandomStepDirection of StepSize
                 Fitness' = Cost(Cell') + Interaction(Cell', BPopulation, dattract, wattract, hrepellent, wrepellent)
                    If (Fitness' > Fitness):
                        x = Ns
                    Else
                        Cell = Cell'
                        Cell_Health = Cell_Health + Fitness'
                    End
                End
            End
            For cell in BPopulation:
                    If Cost(Cell) ≤ Cost(BestCell):
                    BestCell = Cell
                End
            End
        End
        Sort all cells by Health(BPopulation)
        BSelected = BSelected cells based on Health (BPopulation, CellNum/2)
        BPopulation = BSelected
    End
    For cell ⊂ BPopulation:
        If(Rand() < Ped):
            Create cell at random location
        End
    End
End
Output: BestCell
```

4 Implementation

In this section, the proposed algorithms are implemented. The algorithms are run multiple times, with six classifiers, namely k-nearest neighbors (KNN), decision tree, Naive Bayes Bernoulli, random forest, support vector machine (SVM), and logistic regression.

4.1 Input Parameters

For making any prediction during the process of classification, it is necessary to have an efficient model. Classification is a process of supervised learning in which it provides a class or a label to the provided data inputs.

This subsection describes the various classifiers used:

(i) **Random forest classifier (RFC)**: The random forest is a model composed of several trees for decision. Two main principles that differentiate between averaging tree prediction (called as "forest") and RFC are:

   (a) Data points are sampled randomly while trees are constructed.
   (b) While separating the nodes, a subset of features that are accounted for is chosen randomly [14].

(ii) **decision tree classifier (DTC)**: Decision trees is a form of supervised machine learning algorithm where at each iteration, division of data is done on the basis of a particular decision. Two entities may describe the tree, namely the leaves and decision nodes. The leaves are the options, or tests, where the data is divided. [15].

(iii) **K-nearest neighbors (KNN)**: K-nearest neighbors (KNN) algorithm is a form of supervised machine learning algorithm that can be used for problems with predictive regression and classification. However, it is primarily used in the industry for predictive classification problems. The following two properties could well characterize KNN.

   Lazy learning algorithm—KNN is a lazy learning algorithm since it does not have a dedicated training process and uses all the data during classification for training.
   Non-parametric learning algorithm—KNN is also a non-parametric learning algorithm since it assumes nothing about the basic data [16].

(iv) **Naive Bayes Bernoulli (NBB)**: Naive Bayes classifiers are a collection of classification algorithms based on Bayes' theorem. It is not a single algorithm, but a family of algorithms where all of them share a common principle, i.e., every pair of features being classified is independent of each other. The Bernoulli naive Bayes classifier assumes that all our features are binary such that they take only two values (e.g., a nominal categorical feature that has been one-hot encoded) [15].

(v) **Support vector machine (SVM)**: SVM is a supervised machine learning algorithm that can be used for both classification or regression challenges. However, it is mostly used in classification problems. In the SVM algorithm, we plot each data item as a point in *n*-dimensional space (where n is the number of features you have) with the value of each feature being the value of a particular coordinate. Then, we perform classification by finding the hyper-plane that differentiates the two classes very well. Table 1 shows the key parameters used for a specific classifier [15].

Table 1 Input parameters for classifiers

Classifier	Parameter
Decision tree	Maximum depth = 5 Criterion = "Entropy" Minimum sample split = 2
K-nearest neighbors	Neighbors = 5 Leaf size = 30
Random forest	Estimators = 100 Minimum sample split = 2
Bernoulli Naive Bayes	Alpha = 1.0 Binarize = 0.0 Fit_prior = True
Support vector machine	Kernel = 'rbf' (Radial basis function) Max_iteration = 70
Logistic regression	Penalty = 'l2' Max_iterations = 100 Tol = 0.0001

(vi) **Logistic regression (LR)**: Logistic regression is a supervised learning classification algorithm used to predict the probability of a target variable. The nature of the target or dependent variable is dichotomous, which means there would be only two possible classes. In simple words, the dependent variable is binary, having data coded as either 1 (stands for success/yes) or 0 (stands for failure/no). Mathematically, a logistic regression model predicts $P(Y = 1)$ as a function of X. It is one of the simplest ML algorithms that can be used for various classification problems [15].

Table 1 shows the parameters used in this experiment for each classifier.

4.2 Dataset

Magnetic resonance imaging comparison of demented and non-demented adults, and this dataset has been taken from www.kaggle.com; this dataset was updated around 3 years ago. This set comprises a retrospective selection of 150 subjects aged between 60 and 96. On two or more occasions, each subject was examined, for a total of around 400 picture sessions, separated by a minimum of one year. For each topic, three or four individual T1-weighted MRI scans obtained in single scanning sessions are included the topics that refer to males and females. 72 of the subjects were marked as non-demented during research. By the date of their primary visits, 64 of the subjects involved were marked as demented and continued to remain so for subsequent scans, comprising 51 people with Alzheimer's disease mild to moderate. During the time of their initial visit, another 14 subjects were classified as undemented and were then identified as demented at a then encounter.

Table 2 showcases the explanation of various attributes of the dataset that is being

Table 2 Various attributes of the dataset

Real-valued feature	Meaning
Group	This is considered as the output of the project
MR delay	Delay time in MRI in seconds
M/F	Showcasing the gender of the subject
Hand	It tells whether the subject is right-handed or left-handed
Age	The age of the subject
Educ	It tells how much the subject is educated
SES	Socioeconomic status
MMSE	It is a commonly used cognitive function test among elderly people; it requires focus, attention, recollection, language tests
eTIV	Estimated total intracranial volume refers to the estimated volume of the cranial cavity
nWBV	Normalize whole brain volume, refers to the volume inside the cranium, including the brain, meninges, and CSF, can be described
ASF	Atlas scaling factor refers to the scaling factor of one parameter which allows comparison of the estimated total intracranial volume

used after removing some of the attributes which are of no use, such as the ID of subjects.

In Fig. 8, principal component analysis (PCA) of the dataset is displayed as a scatter plot graph showing the distribution of the dataset. In Fig. 8, red dots represent the demented and blue dots represent the non-demented instances of the dataset.

Fig. 8 Principal component analysis of dataset

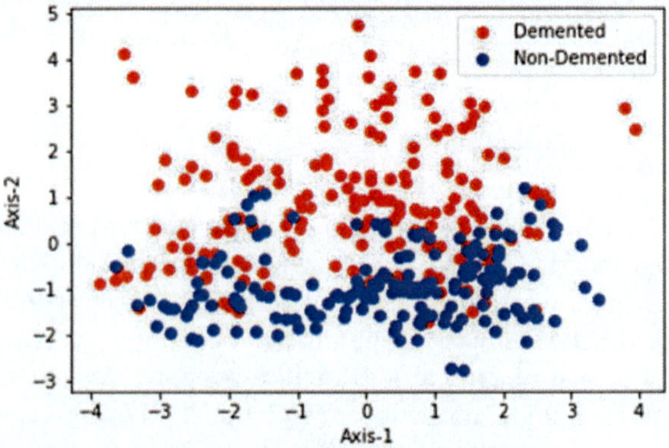

5 Results and Discussion

During the experiment, the termination condition was set by maximum iteration, because it can approximate the utter precision. The finishing state is not set by the required accuracy, because if the algorithm cannot fulfill the accuracy, it can lead to an infinite loop. And the precision limit is hard to estimate because the accuracy varies with the classifier chosen. We have shown Fig. 3 before, which depicts the accuracy of classifiers without evolutionary algorithms. Figure 9 depicts the accuracy of classifiers with particle swarm optimization algorithms.

Figure 10 depicts the accuracy of classifiers with ant colony optimization algorithm.

Figure 11 depicts the accuracy of classifiers with bee colony optimization algorithm.

Figure 12 depicts the accuracy of classifiers with bacterial foraging optimization algorithm.

We used the particle swarm optimization algorithm, and it showed a hike of 10.98%, 16.48%, 15.76%, 19.69%, 11.83%, 16.14% in KNN, logistic regression, decision tree, SVM, random forest, and Naive Bayes Bernoulli, respectively. We used the ant colony optimization algorithm, and it showed a hike of 4.66%, 15.41%, 15.06%, 14.3%, 11.12%, 14.69% in KNN, logistic regression, decision tree, SVM, random forest, and Naive Bayes Bernoulli, respectively. We used the bee colony optimization algorithm, and it showed a hike of 11.15%, 16.48%,15.04%, 19.69%, 11.48%, 16.49% in KNN, logistic regression, decision tree, SVM, random forest, and Naive Bayes Bernoulli, respectively. We used the bacterial foraging optimization algorithm, and it showed a hike of 9.68%, 15.77%, 16.49%, 18.98%, 11.83%, 14.69%

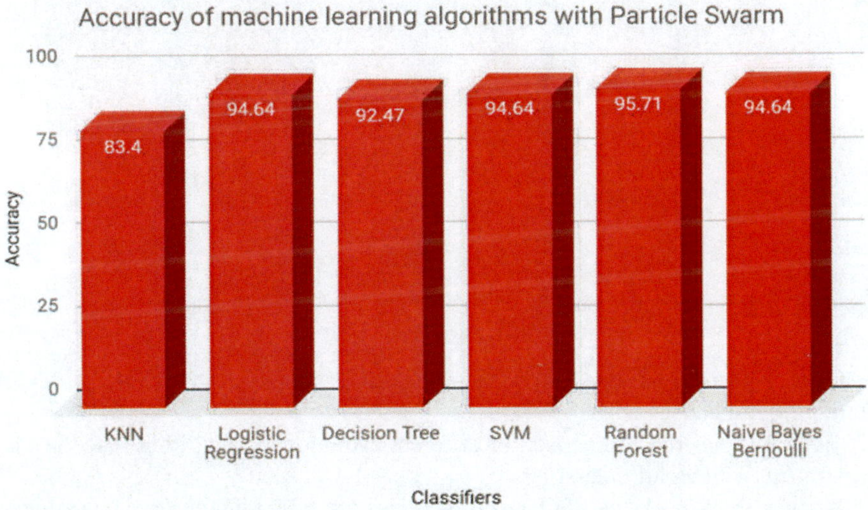

Fig. 9 Accuracy graph of particle swarm optimization algorithm

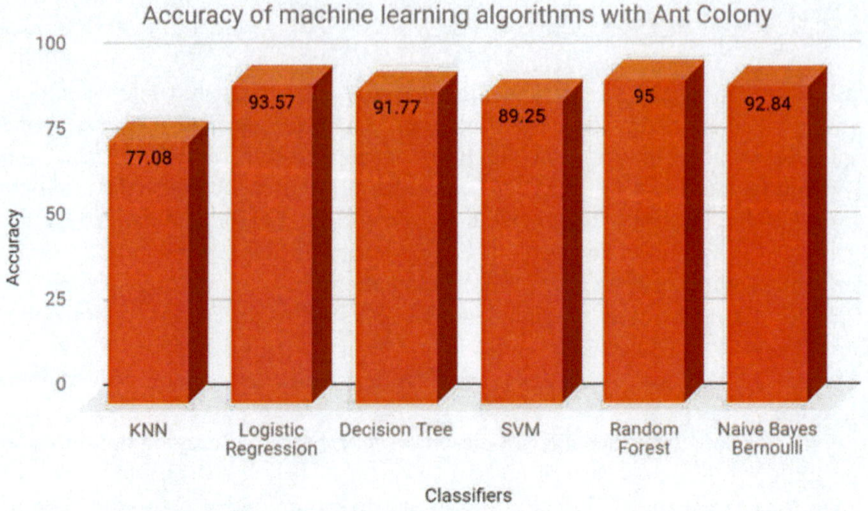

Fig. 10 Accuracy graph of ant colony optimization algorithm

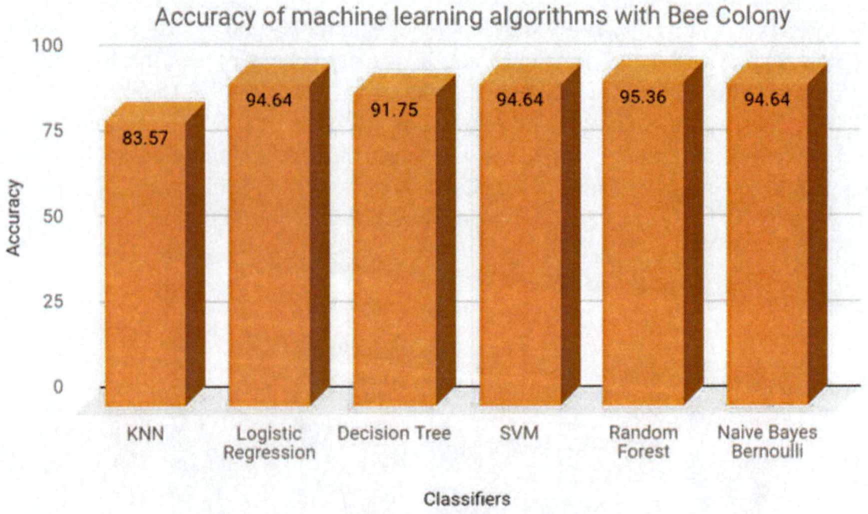

Fig. 11 Accuracy graph of bee colony optimization algorithm

in KNN, logistic regression, decision tree, SVM, random forest, and Naive Bayes Bernoulli, respectively.

Table 3 shows the comparison of different evolutionary algorithms used in the experiment with various classifiers.

As seen in Table 3, the PSO and BFA algorithm with random forest classifiers have shown the best results. Figure 13 depicts a comparison of accuracies of various

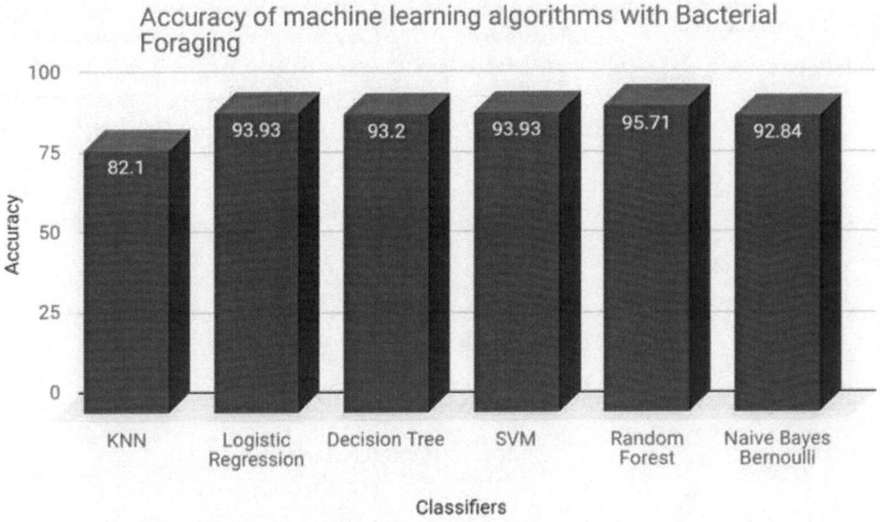

Fig. 12 Accuracy graph of bacterial foraging algorithm

classifiers with and without evolutionary algorithms.

6 Conclusion and Future Scope

The early and accurate detection of Alzheimer's can save the lives of many people across the globe. This paper has proposed a methodology for the accurate detection of Alzheimer's. We have used four evolutionary algorithms, namely bacteria foraging, particle swarm, bee colony, and ant colony. These algorithms are executed on six classifiers, namely k-nearest neighbor, decision tree, support vector machine, random forest, Naive Bayes Bernoulli, and logistic regression. An accuracy of 83.57%, 93.20%, 94.64%, 95.71%, 94.64%, and 94.64% is observed for the six classifiers k-nearest neighbor, decision tree, support vector machine, random forest, Naive Bayes Bernoulli, and logistic regression, respectively. Moreover, the number of features is reduced to 11 features by ant colony optimization, five features by bee colony optimization, six features by particle swarm optimization, and seven features by bacterial foraging.

The study and comparative analysis show that a few features are found in every result, such as MR Delay, Hand, SES, CDR, and e-TIV. These features ought to be considered as they are decisive in the process of detecting Alzheimer's. It is also noticed that aggregation results have increased overall accuracy significantly. In the future, hybridization can be employed for the detection of these type of diseases as they can able to detect and diagnose at early stage with accurate precision in less computational time.

Table 3 Comparison of results

Evolutionary algorithms	Classifier	Accuracy (%)
Ant colony optimization (ACO)	KNN	77.08
	Logistic regression	93.57
	Decision tree	91.77
	SVM	89.25
	Random forest	95.00
	Naive Bayes Bernoulli	92.84
Artificial bee colony algorithm (ABC)	KNN	83.57
	Logistic regression	94.64
	Decision tree	91.75
	SVM	94.64
	Random forest	95.36
	Naive Bayes Bernoulli	94.64
Particle swarm optimization (PSO)	KNN	83.40
	Logistic regression	94.64
	Decision tree	92.47
	SVM	94.64
	Random forest	95.71
	Naive Bayes Bernoulli	94.64
Bacterial foraging algorithm (BFA)	KNN	82.10
	Logistic regression	93.93
	Decision tree	93.2
	SVM	93.93
	Random forest	95.71
	Naive Bayes Bernoulli	92.84

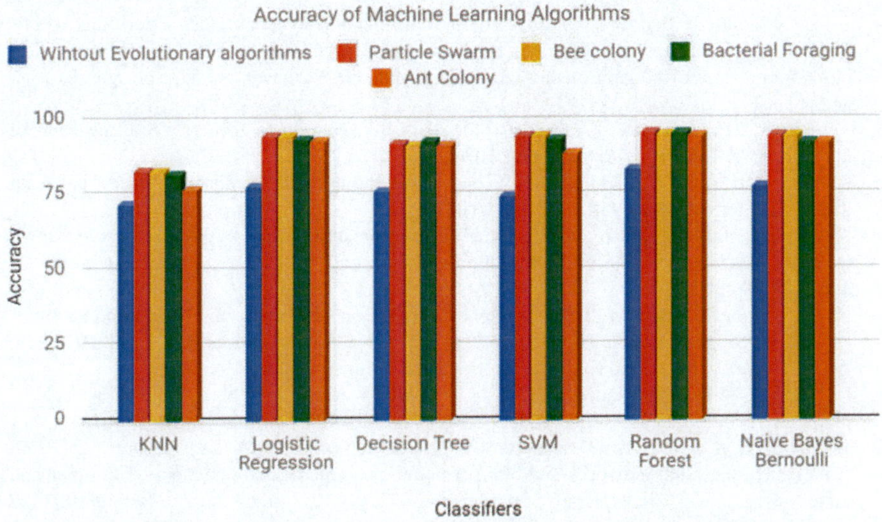

Fig. 13 Comparison of the various classifiers with and without evolutionary algorithms

References

1. Duthey B (2013) Background paper 6.11: Alzheimer disease and other dementias. In: A public health approach to innovation, pp1–74
2. Mathuranath P, Menon R, Ranjith N, Sarma Ps, Verghese J, George A, Justus N, Kumar Ms (2012) Incidence of Alzheimer's disease in India: A 10 years follow-up study. Neurol India 60(6):625. https://doi.org/10.4103/0028-3886.105198
3. Tan K, Lee T, Khor E (2002) Evolutionary algorithms for multi-objective optimization: performance assessments and comparisons. Artif Intell Rev 17:251–290. https://doi.org/10.1023/A:1015516501242
4. Sharma P, Sundaram S, Sharma M, Sharma A, Gupta D (2018) Diagnosis of Parkinson's disease using modified grey wolf optimization. Cogn Syst Res.https://doi.org/10.1016/j.cogsys.2018.12.002
5. Barboza LA, Ghisi NC (2018) Evaluating the current state of the art of Huntington disease research: a scientometric analysis. Br J Med Biol Res 51(3). https://doi.org/10.1590/1414-431x20176299.
6. Sharma M, Gupta S, Deswal S (2020) Modified bio-inspired algorithms for the diagnosis of breast cancer using aggregation. Int J Innov Comput Appl. (in Press)
7. Ritchie K, Lovestone S (2002) The dementias. Lancet 360(9347):1759–1766. https://doi.org/10.1016/s0140-6736(02)11667-9
8. Gulhare KK, Shukla SP, Sharma LK (2017) Overview on segmentation and classification for the Alzheimer's disease detection from brain MRI. Int J Comput Trends Technol IJCTT V43(2):130–132. ISSN:2231-2803
9. De LaCoste M-C, White CL (1993) The role of cortical connectivity in Alzheimer's disease pathogenesis: a review and model system. Neurobiol Aging 14(1):1–16. https://doi.org/10.1016/0197-4580(93)90015-4
10. Zhang Y, Wang S, Ji G (2015) A comprehensive survey on particle swarm optimization algorithm and its applications. Math Probl Eng 2015:1–38. https://doi.org/10.1155/2015/931256

11. Dorigo M, Stützle T (2003) The ant colony optimization metaheuristic: algorithms, applications, and advances. In: Glover F, Kochenberger GA (eds) Handbook of metaheuristics. international series in operations research and management science, vol 57. Springer, Boston, MA
12. Verma BK, Kumar D (2013) A review of the artificial bee colony algorithm. Int J Eng Technol 2(3). https://doi.org/10.14419/ijet.v2i3.1030
13. Chen H, Zhu Y, Hu K (2011) Adaptive bacterial foraging optimization. Abstract Appl Anal 2011:1–27. https://doi.org/10.1155/2011/108269
14. An implementation and explanation of the random forest in Python by Will Koehrsen https://towardsdatascience.com/an-implementation-and-explanation-of-the-random-forest-in-python-77bf308a9b76
15. Dhall D, Kaur R, Juneja M (2020) Machine learning: a review of the algorithms and its applications. In: Singh P, Kar A, Singh Y, Kolekar M, Tanwar S (eds) Proceedings of ICRIC 2019. Lecture Notes in Electrical Engineering, vol 597. Springer, Cham
16. Imandoust SB, Bolandraftar M (2013) Application of K-nearest neighbor (KNN) approach for predicting economic events theoretical background. Int J Eng Res Appl 3(5):605–610
17. He H, Sui J, Du Y, Yu Q, Lin D, Drevets WC, Savitz JB, Yang J, Victor TA, Calhoun VD (2017) Co-altered functional networks and brain structure in unmedicated patients with bipolar and major depressive disorders. Brain Struct Function 222(9):4051–4064. https://doi.org/10.1007/s00429-017-1451-x
18. Liang Y, Wang L (2016) Alzheimer's disease is an important risk factor of fractures: a meta-analysis of cohort studies. Mol Neurobiol 54(5):3230–3235. https://doi.org/10.1007/s12035-016-9841-2
19. Jain R, Gupta D, Khanna A (2018) Usability feature optimization using MWOA. In: Bhattacharyya S, Hassanien A, Gupta D. Khanna A, Pan I (eds) International conference on innovative computing and communications (ICICC2018). Lecture Notes in Networks and Systems, Vol 56. Springer, Singapore
20. Khamparia A, Singh A, Anand D, Gupta D, Khanna A, Arun Kumar N, Tan J (2018) A novel deep learning based multi-model ensemble methods for prediction of neuromuscular disorders. Neural Comput Appl.https://doi.org/10.1007/s00521-018-3896-0
21. Jaiswal AK, Tiwari P, Kumar S, Gupta D, Khanna A, Rodrigues JJPC (2019) Identifying pneumonia in chest X-Rays: a deep learning approach. Measurement 145:511–518

Comparison of Various Word Embeddings for Hate-Speech Detection

Minni Jain, Puneet Goel, Puneet Singla, and Rahul Tehlan

Abstract Word Embedding plays a crucial role in natural language processing, and other related domains. The vast variety of language modelling and feature learning techniques often concludes in a quandary. The motivation behind this work was to produce comparative analysis among these methods and finally use them to flag hate-speech on social media. The progress in these word embedding techniques has led to remarkable results by incorporating various natural language applications. Understanding the different context of polysemous words is one of the features that evolved over time with these word embedding models. A systematic review on varying word embedding methodologies has been performed in this paper. Various experimental metrics have been used and detailed analysis has been done on each word embedding model. It is shown that analysis involves various aspects of the model like dealing with multi-sense words, and rarely occurring words, etc., and finally a coherent analysis report is presented. The various models under analysis are—Word2Vec (Skip-Gram, CBOW), GloVe, Fast-Text and ELMo. These models are then put to a real-life application in the form of Hate Speech detection of twitter data, and their individual capacities and accuracies are compared. Through this paper we show how ELMo uses different word embeddings for polysemous words to capture the context. We show how Hate speech can be better detected by ELMo because such speech requires better understanding of context of words for segregation from normal speech/text.

Keywords Word2Vec · Skip-gram · CBOW · GloVe · Fast-text · Elmo · Word embedding · Hate-speech detection

M. Jain · P. Goel · P. Singla (✉) · R. Tehlan
Department of Computer Engineering, Delhi Technological University, New Delhi 110042, India
e-mail: puniet.singla@gmail.com

M. Jain
e-mail: minnijain@dtu.ac.in

P. Goel
e-mail: goelpuneet25@gmail.com

R. Tehlan
e-mail: rahul.tehlan@gmail.com

A. Khanna et al. (eds.), *Data Analytics and Management*, Lecture Notes on Data Engineering and Communications Technologies 54,
https://doi.org/10.1007/978-981-15-8335-3_21

1 Introduction

Word embedding is a collective name given to a group of techniques to represent words in the form of vectors composed of real values [1]. It is emerging as one of the most important tools in Natural Language Processing (NLP). From a historical standpoint, the growing requirement of assigning numerical values to individual words in a sentence or a document, based on its relation to other words in the form of syntax and semantics, lead to the gradual progress in these techniques. These numerical metrics and vectors were important because it makes it easier to find out correlations and similarities between two words and even help them represent graphically.

Word embedding finds a widespread application in various domains of NLP. Semantic analysis is an important addition as well as an application to the word embedding, where the importance is given to the context of the word [2]. With the ongoing progress in deep learning and NLP, question answering has too evolved in the process with the help of information retrieval procedures. These include using non-negative matrix factorization methods [3] and end-to-end model for question answering [4]. Another important application of word embedding is neural machine translation [3, 5–7] which was a huge improvement over statistical translation which invoked a single neural network. POS tagging [8] and sentiment analysis [9] are also done using word embedding.

The word embedding models have evolved over time from very basic neural network models to the modern ones that can handle better the complexities of the simultaneously evolving natural languages. These evolutions are important because of the shortcomings that each model comes up with. Word2vec is easy to implement but not so efficient in representing extremely rare words. Fast-text is good a projecting these rare words but fail to establish the context in which a word has been used. GloVe can grab the context but is inefficient in dealing with nuanced polysemous words. ELMo excels in managing polysemous words and can bring out minute context difference as well by providing different word vectors, which is still a domain that is evolving with time. In representing words as vectors, sometimes unrelated words appear together and similar ones are found to be at a distance. The focus of this work is:

- To provide understanding of various word embeddings by plotting vectors from each model for polysemous key words.
- To provide comparative analysis of these models on Hate-Speech Detection dataset.

After the introduction which is Sect. 1, the paper is distributed into seven more sections. Section 2 mentions the available literature work related to this paper, which set the base for research and analysis of various word embedding techniques. Section 3 briefly describes each of the model under consideration in this paper, with related formulae and diagrams. Section 4 describes illustration of vector representation using sample data. This section puts the theoretical description of each word embedding model to experiment. It showcases the stepwise process of selecting few keywords and building word representations around these central words. Section 5 provides the graphical representation for each model's word embedding representation around the same keywords. This helps in better understanding the ability of a model to grasp context of a word which is analysed in this section. In Sect. 6, the learning of these models has been put to task. A Hate Speech Detection task has been done using different word embedding models. Section 7 summarise the various comparison parameters of different models in Hate Speech Detection in the form of a table, with model wise insights drawn from these results. Finally, Sect. 8 talks about the conclusion of this work.

2 Related Work

With the evolution of techniques pertaining to word embedding, a constant necessity has been of trying to deal with the sense of a word. In this paper too, we compare how some word embedding models work better in dealing with the context of its use. Camacho-Collados et al. [10] in their paper did a survey on representation of words making use of sense representations. This was done using word sense disambiguation techniques. The feasibility of word2Vec techniques (CBOW, Skip-gram) are demonstrated in the paper.

Another similar work includes the word sense disambiguation using knowledge-based techniques by Agirre et al. [11]. A work on multilingual words and their representation was done by Ammar et al. [12].

Word embedding using Fast-text that take into consideration the multiple sense of a word was demonstrated by Athiwaratkun et al. [13]. A detailed comparison of some word embedding models using parameters like cosine similarity, and making use of methods like outlier detection and concept categorization, has been done in 2019 paper titled "Evaluating word embedding models: Methods and experimental results" [14].

Our paper makes use of the research of earlier word embedding methods (Word2Vec, GloVe, Fast-text) and compare and contrast them with a more recent model, i.e., ELMo and differentiate in their abilities to and shortcomings in dealing with real life applications like Hate Speech Detection.

3 Word Embedding Models

3.1 Word2Vec

Word2Vec is a neural network-based approach which learns the vector representation of a word. There are two vector representation approaches that are proposed under Word2Vec viz. Skip-Gram model and CBOW model.

Skip-gram: Skip-Gram aims at finding the closely related words to a given word in a sentence or a document. It is an unsupervised learning technique. The word around which the related words are extracted is called the target word and the context words are found surrounding the target word. It basically works in reverse manner to the CBOW model (Fig. 1) as here the target word is given as input and context words related to it are found as output [15]. Let $w_1, w_2, w_3 \ldots w_t$. represent the sequence of training words, skip-gram's objective is to make the following average log probability maximum [1].

$$\frac{1}{T}\sum_{t=1}^{T}\sum_{-c\leq j\leq c, j\neq 0} \log p(w_{t+j}|w_t) \tag{1}$$

Here, the size of the training context is represented by c, to which the accuracy is directly proportional. This is because small context word sample size is not given enough for the model to train better. At the same time, the training time is directly proportional to the training sample c. Another important part of the formula iterm $\boldsymbol{log\, p\, (w_{t+j}|w_t)}$, which can be expressed using SoftMax function as following:

$$p(w_O|w_I) = \frac{\exp\left(v_{w_O}^{\prime T}\, v_{w_I}\right)}{\sum_{w=1}^{W} \exp\left(v_w^{\prime T} v_{w_I}\right)} \tag{2}$$

Here, v_w is the representation of input vector and $v_w^{'}$. is output vector representation of the number of words in the vocabulary, i.e., w.

CBOW: Continuous-Bag-of-Words takes a window of size '*n*' and learns by predicting the centre of the window. It predicts the embeddings based on its context.

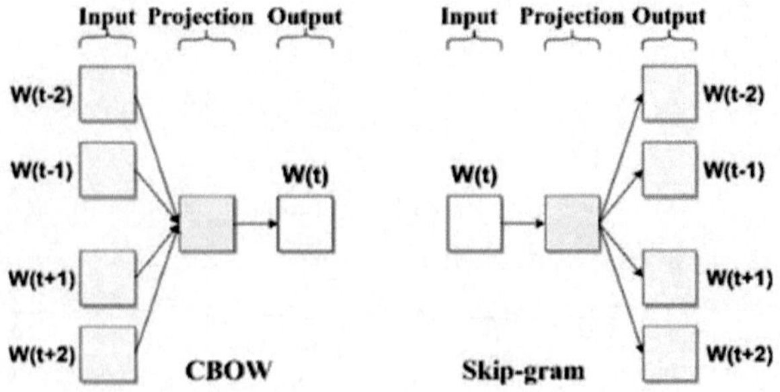

Fig. 1 CBOW and Skip-gram training models

This model uses the probability and maximizes the same for a word in the given context [15]. This is done by minimizing the following objective function:

$$J(u_i) = -u_i^T \hat{v} + \log \sum_{j=1}^{|W|} \exp(u_j^T \hat{v}) \tag{3}$$

Here, $\boldsymbol{u_i}$ is the ***i***th row of matrix $\boldsymbol{U}$, and $\hat{\boldsymbol{v}}$ is the average of embedded input words.

3.2 GloVe

GloVe stands for Global Vectors for word representation. It is a more efficient algorithm as compared to the word2vec method. This approach uses both the global statistics of matrix factorization as used in LSA (Latent Semantic Analysis) and captures the local context of the words as in word2vec [16]. In contrast to using a window of size 'n' to define context, GloVe constructs a statistical model of the word and its co-occurrence with all other words in the corpus. The model is trained to produce better word embeddings.

Initially, it is important to know the frequency at which a word under consideration exists within the sense and context of another word. For this, a word-word co-occurrence matrix is constructed, which in turn helps in maintaining the probability of a word ***j*** in context of another word ***i***, and this probability is denoted by ***P(j\i)***. The model then tries to solve the equation ***fWw:TwYZ= P(j\i)***. Here ***w*** is a word vector. Along with another word vector ***wY***, the function ***f*** relates these vectors to the eventual probability ***P(j\i)***. This can be also be shown as:

$$\hat{J} = \sum_{i=1}^{W} \sum_{j=1}^{W} f(X_{ij})(u_j^T v_i - \log X_{ij})^2 \tag{4}$$

where $\boldsymbol{f}$ $(\boldsymbol{X_{ij}})$ denotes the weighing function, representing the count of word $\boldsymbol{u_j}$'s existence in the context of word $\boldsymbol{v_i}$. Also, $\boldsymbol{u_j^T v_i}$ is the actual co-occurrence probability whereas log $\boldsymbol{X_{ij}}$ is the co-occurrence probability that is predicted by the model. |**v**| is the total size of the vocabulary. The co-occurrence matrix is depicted in Fig. 2.

3.3 Fast-Text

This word embedding model deals with the low frequency words in the corpus. It becomes difficult for primitive word embedding models like word2vec and GloVe to capture context or predict wd vectors of this kind of words [17].

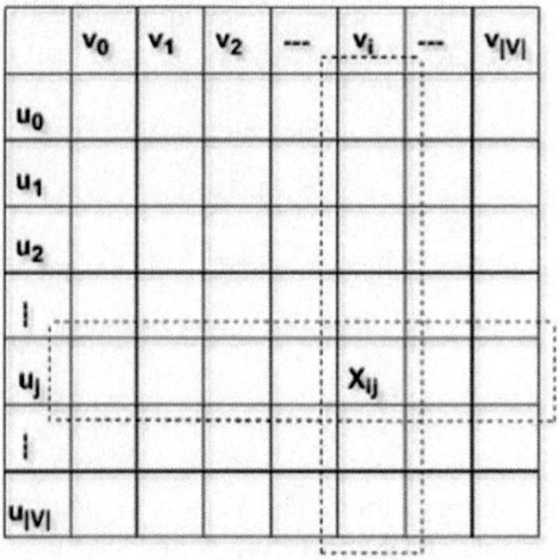

Fig. 2 Co-occurrence matrix using GloVe

Hence, in case there is a word that does not appear in the corpus, primitive models fail to predict the words vector representation. However, this model follows the same approach as word2vec as shown in Fig. 3.

This embedding technique treats words as the minimal entity which is composed of *n*-gram. Simply put, for a word say 'Computer' and the value of *n* be 5 we have ['$comp';'compu';'omput';'mpute';'puter';'uter$'] as its representation. '$' is a special symbol which resembles blank space before and after the word. Also the smallest and largest *n*-gram are assumed to be 5. Finally, the word representation of 'Computer' can be represented as the vector sum of each *n*-gram.

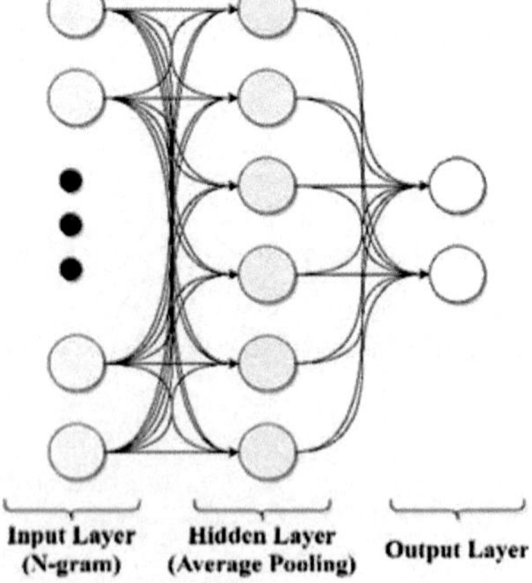

Fig. 3 Fast-text training model

3.4 ELMo (Embeddings from Language Models)

ELMo is based on a recurrent neural network approach using LSTMs. It calculates multiple vector representation of a word depending on the context of that word [18]. It works in a bi-directional manner as depicted in Fig. 4. Rather than computing single word representation it computes a matrix which represents the word in accordance to its co-occurrences with other words. ELMo keeps in mind the context of the word while inferring vector representations on new words. This approach can be used to deal with polysemous words (words with different meanings and same spelling).

The model uses LSTM (Long Short-Term Memory) cells which are made of a feedback neural network connection. The network consists of 'L' layers and they make use of the bidirectional nature in dealing with word sequences. For evolving the prediction process, the model learns by predicting the following word to an already fed and its immediate prior word. Also, character convolution is utilized for a character-based encoding. So, overall $2L + 1$ vectors constitute a word matrix, with L layers of LSTM that are bidirectional, and an extra layer called the token layer.

4 Illustration of Vector Representation of Models on Sample Data

The analysis on the word embedding methods discussed in the last section involved a stepwise procedure. The steps are mentioned below:

(a) **Creation of sample data**
An application of different word embedding models on data helps in bringing out the contrasting results of the models. For this, a sample data from Wikipedia

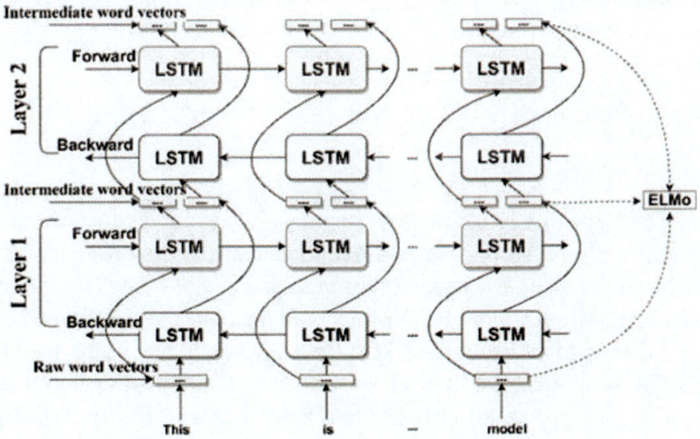

Fig. 4 Bidirectional language model using LSTMs for ELMo

is created using three keywords-**Watch, Book** and **Apple**. The selection of these words is precedented given their polysemous nature.

- **Watch** could mean a device to keep track of time or the act of seeing something.
- **Book** could mean reserving something in advance or it could mean a record consisting of pages.
- **Apple** could refer to a fruit or the tech-company of same name.

(b) Modification of each model
The data is modified according to the input format of each model. For this, data cleaning is performed where stop words are removed and lemmatization is done.

(c) Vector representation of words
After model-wise modification of dataset, it is fed to the models. Each model gives the vector representation of the embedded words. To make visualization more feasible, the obtained vectors were reduced to 2 dimensions by performing dimensionality reduction using Principal Component Analysis (PCA). Hence, only the most prominent features of the vectors are deployed for visualization.

(d) Plotting the graphs
Based on the modified vectors, the words are represented on a 2D graph. This visualization is important because it makes it easier to understand how closely two words are related in our dataset.

(e) Analysis of results
The results are shown in the following section in the form of graphs.

5 Comparison and Graphs

This section contains the graphical representation of each word embedding model, along with a illustrative comparison among them.

5.1 Word2Vec

Skip-gram. The keyword **book** in Fig. 5 is closely surrounded by few words like ‘widely’, ‘thousand’ and so on. An intuitive glance at these words give us a vague meaning of the book as a noun instead of the verb or act of booking (e.g. tickets). This lack of clarity of a skip-gram model to project polysemous nature of words is further confirmed by the words surrounding keywords **apple** (‘hand’) and **watch** (‘tree’, ‘california’). These words give little or no context to the keywords they surround.

CBOW. The plot in Fig. 6 (CBOW) is almost identical to Fig. 5 (Skip-gram). Hence, there’s no drastic improvement in grasping the context of a word from skip-gram to CBOW, and hence in word2vec as a whole.

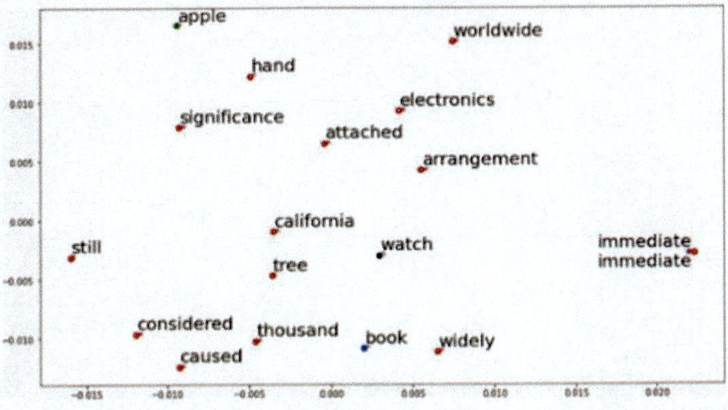

Fig. 5 Representation of vectors obtained from Skip-gram

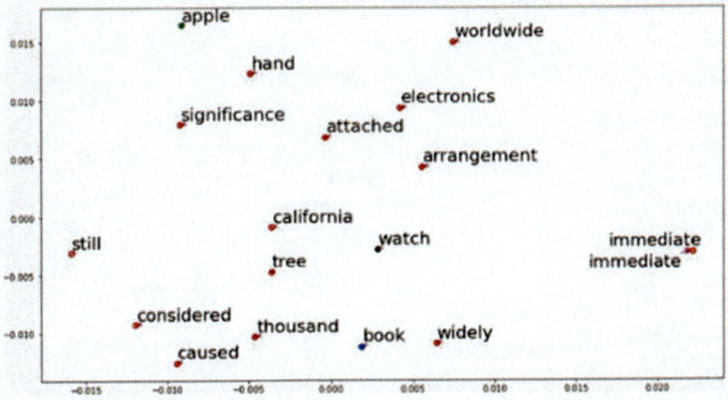

Fig. 6 Representation of vectors obtained from CBOW

5.2 *GloVe*

The keyword **apple** is close to words like 'amazon', 'company' and 'microsoft' in the plot shown in Fig. 7. This shows that the sense of **apple** as a tech-company has completely dominated the other meaning, i.e., a 'fruit', which is at a further distance with the word 'grow'. The keywords **watch** and **book** are too clustered with words that don't bring out their polysemous nature.

5.3 *Fast-Text*

The keyword **apple** is surrounded by words that point to the sense of the keyword as a fruit in Fig. 8. Also, the word **watch** too is surrounded by only one sense words

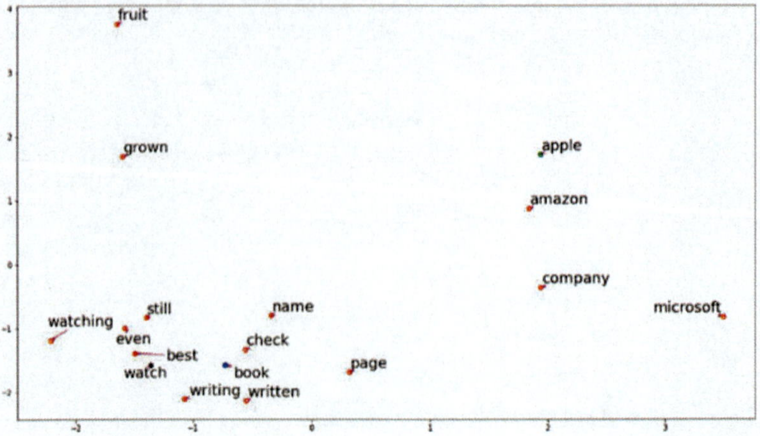

Fig. 7 Representation of vectors obtained from GloVe

Fig. 8 Representation of vectors obtained from Fast-text

instead of bringing out the polysemous nature. At the same time, the keyword **book** is close to words like 'location' (sense: booking a ticket) and category (sense: reading book genre). But the clusters are still not well-defined.

5.4 ELMo

The immense improvement in dealing with polysemous words is shown by ELMo in Fig. 9. There are clusters for each sense. For example, the keyword **watch** is surrounded by words like 'bracelet' and 'pocket' (sense: watch as noun) in one cluster, whereas, it is close to word like 'watching' (verb) in another cluster. Hence, ELMo is better than other word-embedding models in dealing with polysemous words.

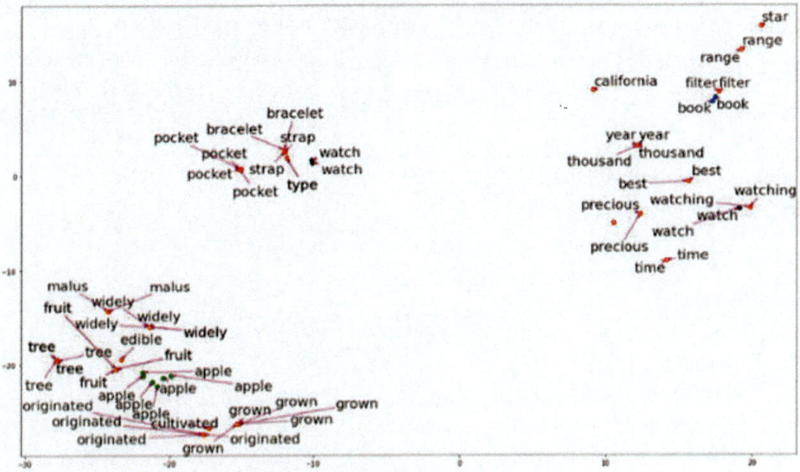

Fig. 9 Representation of vectors obtained from ELMo

6 Hate Speech Detection Using Word Embedding Models

In this section we use hate-speech detection dataset to perform a comparative study among word embedding models. Each model is trained and evaluation results are then compared for better understanding on how these models perform.

The research on the word embedding techniques gave varying results on the sample data. Naturally, it has various real-world applications as well. One such application is Hate Speech Detection. Hate speech [19], is a message or speech delivered in a public domain that promotes hate or violence, usually targeted at a race, religion, sex, etc. It is a very dangerous trend on social media and trying to detect it is even more important for the cyber as well as personal security of the citizens of a nation.

The hate speech detection process using the word embedding models involved following steps:

1. Twitter data was collected from Kaggle[1] consisting of 2000 tweets. A tweet in this data either belonged to the category of hate speech, or not.
2. The data was pre-processed which involved converting the text into lowercase, removing digits, punctuation and removal of stop words. This was followed by tokenization and lemmatization of words in the data. This is important because further tasks in NLP can only be performed after getting rid of the noise in the data.
3. Vector representation of words extracted after pre-processing is obtained from each model.

[1] https://www.kaggle.com/arkhoshghalb/twitter-sentiment-analysis-hatred-speech#train.csv.

Table 1 Comparative analysis of word embedding models on hate speech dataset

Model	Approach	Accuracy	Precision	Recall	F-measure	Produce different vectors for polysemous words?
Skip-gram	Neural network	0.538	0.75	0.14	0.24	No
CBOW	Neural network	0.715	0.75	0.66	0.70	No
Glove	Statistical Co-occurrence matrix	0.793	0.80	0.79	0.80	No
Fast-text	N-grams with neural networks	0.838	0.86	0.82	0.84	No
ELMo	BiLM using LSTMs	0.860	0.83	0.90	0.87	Yes

4. The vectors are of each word in the sentence are then added and normalized to get an average vector of the complete tweet. Normalization rescales the range of the vector to $[-1, 1]$. This is done to rule out unnecessary bias in representations from various models.
5. The data is then split into 80% data as training data and remaining 20% as testing data.
6. The training data is then used to train neural networks with the help of vectors obtained through it.
7. The trained neural network is then deployed on the testing data which flagged the tweets as hate speech or not, each model giving different accuracies.
8. The results are shown in Table 1.

7 Results and Discussion

When it comes to natural language, there are sometimes a lot of ambiguities even in the simplest of sentences. Even in our day to day conversations, we come across words that can have multiple meanings. Normally, because of experience and human intellect, we are able to grasp the correct sense of the words/sentences. However, a machine has no consciousness of its own. Hence, the faults made by a machine can lead to disastrous misinterpretations. While dealing with these syntactic and semantic ambiguities it is important that the system is able to differentiate between the different context in which the word can be used.

As discussed earlier, ELMo was able to differentiate between the context of the words, it can be said that due to this unique property we obtain very high accuracy, precision and recall using this word embedding model.

Fast-text helps us understand the structure of the words by breaking it down to n-grams. This helps us get representation for even those words that occur very few times in the vocabulary.

GloVe instead of using neural networks to train the dataset, optimizes the word embedding directly depending upon the cooccurrences of the word. Rather than storing the co-occurrence probabilities, it stores the ratio of these probabilities. Accordingly, GloVe performed better than neural network approaches like skip-gram and CBOW.

Fast-text also performs considerably well. However, it fails to provide any vector representation of words that do not occur in the model dictionary.

Skip-gram performs exceptionally well with small training data and it helps us represent those words which occur rarely. Comparatively, CBOW trains much faster and has a better accuracy as compared to Skip-gram.

The numerical values of various metrics for word embedding models mentioned in Table 1 are compared graphically in Fig. 10 in the following order: (a) *Accuracy* (b) *Precision* (c) *Recall and* (d) *F-measure.*

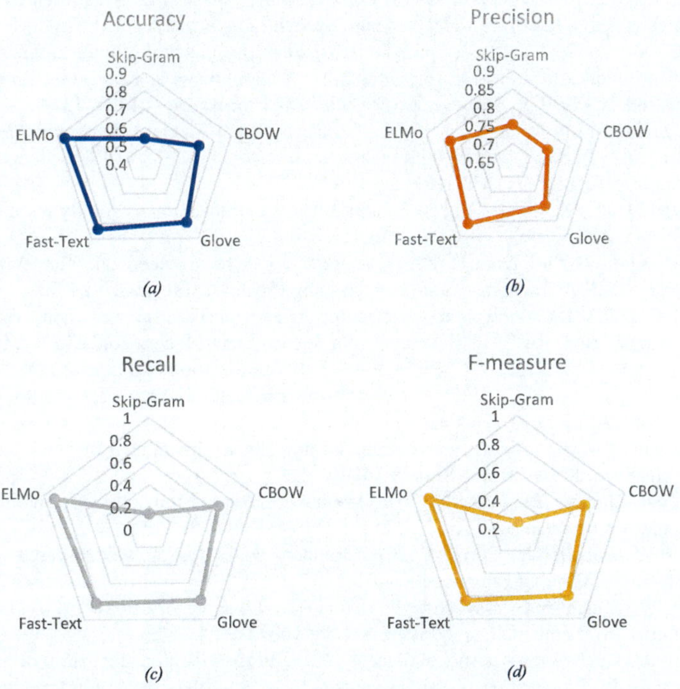

Fig. 10 Graphical representation of: **a** accuracy, **b** precision, **c** recall and **d** *F*-measure; of word embedding models

8 Conclusion

In this work we provided evaluation of various word embeddings models viz. Word2Vec (CBOW and Skip-gram), fast-text, glove and ELMo. Our study offers a valuable analysis on how the models produce vectors with polysemous words. Furthermore, we have seen how these models perform on Hate Speech Detection Dataset. It is worth mentioning here that ELMo was capable context capturing and how it has the highest accuracy because of its unique property of treating polysemous words. A language is filled with ambiguity in syntactic and semantic forms. Clearly, ELMo was able to better tackle these ambiguities.

References

1. Mikolov T, Sutskever I, Chen K, Corrado GS, Dean J (2013) Distributed representations of words and phrases and their compositionality. In NIPS
2. Yu L, Wang J, Lai KR, Zhang X (2018) Refining word embeddings using intensity scores for sentiment analysis. IEEE/ACM Trans Audio Speech Lang Process 26(3):671–681
3. Zhou G, Xie Z, He T, Zhao J, Hu XT (2016) Learning the multilingual translation representations for question retrieval in community question answering via non-negative matrix factorization. IEEE/ACM Trans Audio Speech Lang Process 24(7):1305–1314
4. Hao Y, Zhang Y, Liu K, He S, Liu Z, Wu H, Zhao J (2017) An end-to-end model for question answering over knowledge base with cross-attention combining global knowledge 221–231. https://doi.org/10.18653/v1/P17-1021
5. Bahdanau D, Cho K, Bengio Y (2014) Neural machine translation by jointly learning to align and translate. ArXiv. 1409
6. Zhang B, Xiong D, Su J, Duan H (2017) A context-aware recurrent encoder for neural machine translation. IEEE/ACM Trans Audio Speech Lang Process 25(12):2424–2432
7. Chen K et al (2018) A neural approach to source dependence based context model for statistical machine translation. IEEE/ACM Trans Audio Speech Lang Process 26(2):266–280
8. Li Z, Zhang M, Che W, Liu T, Chen W (2014) Joint optimization for Chinese POS tagging and dependency parsing. Audio Speech Lang Process IEEE/ACM Trans 22:274–286. https://doi.org/10.1109/TASLP.2013.2288081
9. Ravi K, Ravi V (2015) A survey on opinion mining and sentiment analysis: tasks, approaches and applications. Knowl Based Syst 89:14–46
10. Camacho-Collados J, Pilehvar MT (2018) From word to sense embeddings: a survey on vector representations of meaning. J Artif Int Res 63(1):743–788
11. Agirre E, de Lacalle OL, Soroa A (2014) Random walks for knowledge-based word sense disambiguation. Comput Linguist 40(1):57–84
12. Ammar W, Mulcaire G, Tsvetkov Y, Lample G, Dyer C, Smith NA (2016) Massively multilingual word embeddings. https://arXiv.org/1602.01925
13. Athiwaratkun B, Wilson A, Anandkumar A (2018) Probabilistic fasttext for multisense word embeddings. In: Proceedings of the 56th annual meeting of the association for computational linguistics, vol 1 (Long Papers). Association for Computational Linguistics, pp 1–11
14. Wang B, Wang A, Chen F, Wang Y, Kuo C (2019) Evaluating word embedding models: methods and experimental results. APSIPA Trans Signal Inf Process 8:E19. https://doi.org/10.1017/ATSIP.2019.12
15. Mikolov T, Corrado GS, Chen K, Dean J (2013) Efficient estimation of word representations in vector space, pp 1–12

16. Pennington J, Socher R, Manning CD (2014) Glove: Global vectors for word representation. In: EMNLP
17. Joulin A, Grave E, Bojanowski P, Mikolov T (2017) Bag of tricks for efficient text classification, pp 427–431. https://doi.org/10.18653/v1/E17-2068
18. Peters M et al (2018) Deep contextualized word representations. In: Proceedings of the 2018 conference of the North American chapter of the association for computational linguistics: Human Language Technologies, vol 1, pp 2227–2237
19. Roels J (2017) The battle against hate speech and freedom of expression online, 5 Apr 2017. Charles University in Prague Faculty of Law Research Paper No. 2017/I/4

Multimodal Biometric Algorithm Using IRIS, Finger Vein, Finger Print with Hybrid GA, PSO for Authentication

E. Sujatha, J. Sathiya Jeba Sundar, P. Deivendran, and G. Indumathi

Abstract Biometric is emerging and promising technology to identify and authenticate human being. It is more robust, accurate, and accurate. It is hard to imitate, forge, share, distribute and cannot be stolen, forgotten. After September 11, 2001, incident, the biometric technologies are focused more. Integrating more than one biometric trait yields a promising solution to provide more security. It manages the variety of demerits in unimodal biometric systems such as non-universality, noise in sensed data, intra-class variations, distinctiveness, and spoof attacks. The traditional way of authentication a human and their identity is resolved. The proposed method proves with experimental results on multimodal biometric algorithm for authentication using normalized score-level fusion techniques and hybrid Genetic Algorithm and Particle Swarm Optimization for optimization in order to reduce the parameters considered for evaluation as false acceptance rate and false rejection rate and to enhance accuracy. In this proposed research work, it integrates iris, finger vein, and finger print biometric traits chosen for their best biometric characteristics. The experiment is conducted by SDUMLA-HMT database, and the state-of-art algorithm is evaluated by metrics as false acceptance rate, false rejection rate, equal error rate, and accuracy for proving that the claimed identity as genuine or imposter.

Keywords Multimodal biometrics · Genetic Algorithm · Particle Swarm Optimization

---

E. Sujatha (✉) · J. Sathiya Jeba Sundar · P. Deivendran · G. Indumathi
Department of Computer Science and Engineering, Velammal Institute of Technology, Panchetti, India
e-mail: sanjaymohankumar@gmail.com

J. Sathiya Jeba Sundar
e-mail: sundarjose18@gmail.com

P. Deivendran
e-mail: deivendran1973p@gmail.com

G. Indumathi
e-mail: gindumathi84@gmail.com

A. Khanna et al. (eds.), *Data Analytics and Management*, Lecture Notes on Data Engineering and Communications Technologies 54,
https://doi.org/10.1007/978-981-15-8335-3_22

1 Motivation and Literature Review

The recent scenario demands highly secured and robust authentication algorithm to get rid of vulnerabilities and threats. Emerging biometric technology can improve the safety and security. After September 11, 2001, attack, the world realized the importance of public safety and the need for data security. Resetting the forgotten passwords costs makes expensive for the corporate companies in the traditional security system. Biometric-based authentication system is adapted for designing novel authentication algorithm. Multimodal reduces the pitfalls of unimodal biometric systems and other existing traditional authentication systems.

The biometric technology evolution focuses on significance and where the future technology of world travels can be identified by several papers and their authors contributed the research work but only few are quoted.

Introduction, motivation, and scope of multimodal biometrics using score-level fusion techniques and sensors, how more than one biometric trait can be combined to form efficient algorithm, is referred in Veeramachaneni et al. [1], Hong and Jain [2]. Importance of biometrics and its applications, future trends of biometrics, and efficiency of biometric technologies are studied from Frischholz and Deickmann [3], Hong et al. [4]. Iris recognition techniques and its comparison with other techniques are discussed in Vatsa et al. [5]. Chimeric database for evaluation is obtained from SDUMLA-HMT Shandong University, China, with license agreement for use. Score-level fusion techniques importance and effectiveness are obtained from Romaissaa Mazouni et al. [6], Poh et al. [7]. Hybrid Particle Swarm Optimization and Genetic Algorithm is efficient comparatively with other techniques is referred from Dalila et al. [8], Eberhart et al. [9].

2 Objectives and Scope

This research work aims to design a novel authentication algorithm for high security. Iris, finger vein, and finger print biometric traits are chosen for integration with respect to their characteristics. The analysis and importance of combining these multimodal biometrics for authentication algorithm after fusing their respective scores and hybrid optimization technique is adapted for more security. Multimodal biometrics enhances accuracy and improves security. Iris, finger vein, and finger print biometric traits are undergone preprocessing techniques and recognized individually. The matching algorithm produced score. These scores are normalized before applying fusion rules. Normalization of scores brings the compatibility in fusion. Hybrid GA and PSO fusion technology is applied for optimization. It inherits the benefits of GA and PSO in fusion and optimizes the scores, and finally, whether the person claiming as authenticated identity is accepted as genuine or rejected as imposter. The calculated value of false acceptance rate and false rejection rate is reduced. Hence, the efficiency and accuracy of the algorithm are enhanced. If false acceptance rate and false rejection rate reduce, it improves the accuracy and reduces the equal error rate.

3 Proposed System

The proposed architecture diagram depicts that the system has several stages as preprocessing, recognition, normalization, fusion and optimization, and analysis stage. In preprocessing, the image is fetched from the database for eliminating noise. Recognition of each biometric trait is done, and the scores for each trait are obtained. Normalization makes the compatibility between scores of multimodal biometrics. It is mandatory before fusion. Fusion and optimization stage is adapted for inheriting the benefits of GA and PSO in fusion and optimization of scores.

3.1 *Iris Recognition*

It is accurate and reliable over 200 millions of comparisons. It is easy to detect artificial irises. It possesses good biometric characteristics. Preprocessing steps consists of localization and normalization. Localization includes canny edge algorithm as smoothing—to filter out the noise, Gaussian filter is used (Fig. 1).

$$H_{ij} = \frac{1}{2\pi\sigma^2}\exp\left(-\frac{(i-(k+1))^2+(j-(k+1))^2}{2\sigma^2}\right);\ 1 \le i,\ j \le (2k+1)$$

- **Finding gradients**—used to identify in the image about the horizontal, vertical, and diagonal edges.
- **Non-maximum suppression**—also known as edge thinning technique where it assists to hide all gradient values except local maxima at some extent.
- **Double thresholding**—determines weak edge pixel by verifying between highest threshold and lowest threshold values.
- **Edge tracking by hysteris**—weak edge pixel caused by noise/color variation is removed by this method.

Preprocessing of iris image by canny edge detection algorithm using Laplacian of Gaussian filter and normalization by pupil dilation is known as iris localization. During preprocessing, color image is localized by converting from color to gray scale image, and the centre of the pupil is identified by binarization of image. The polar to cartesian coordinates transformation is done as:

$X = r\cos\theta$

$Y = r\sin\theta$

This conversion helps in noise removal. Gray scale matrix values are training data. The radius of the pupil is obtained through this technique. The fact that the distance

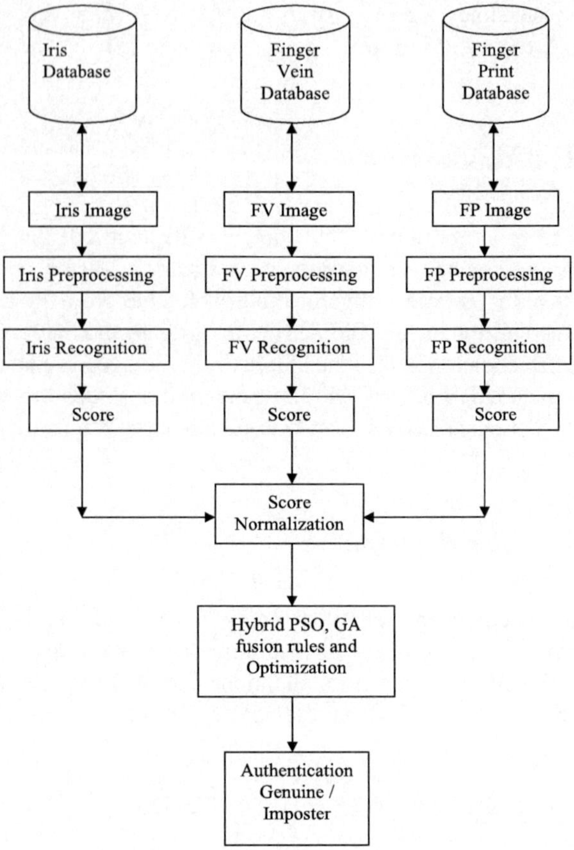

Fig. 1 Architecture diagram for proposed algorithm

between pupil and the boundary of iris is unique in all irises and also proved that inner and outer boundaries are not concentric in nature. Pupil size normally differs from 10 to 80% with respect to its iris diameter. So, iris proves highly robust and reliable recognition algorithm. Iris template size varies from 256 to 512 bytes. Matching algorithm by hamming distance is done to obtain score. The score is normalized using z-score normalization technique. The SDUMLA–HMT database is used for evaluation of algorithm, and it is obtained from five different sensors from 106 individuals which is known as chimeric database. Iris adapts hamming distance for matching algorithm as:

$$\mathrm{HD} = \frac{\sum_{j=1}^{N} x_j (\mathrm{XOR})\, Y_j\, (\mathrm{AND}) X n'_j (\mathrm{AND})\, Y n'_j}{N - \sum_{k=1}^{N} X n_k (\mathrm{OR})\, Y n_k}$$

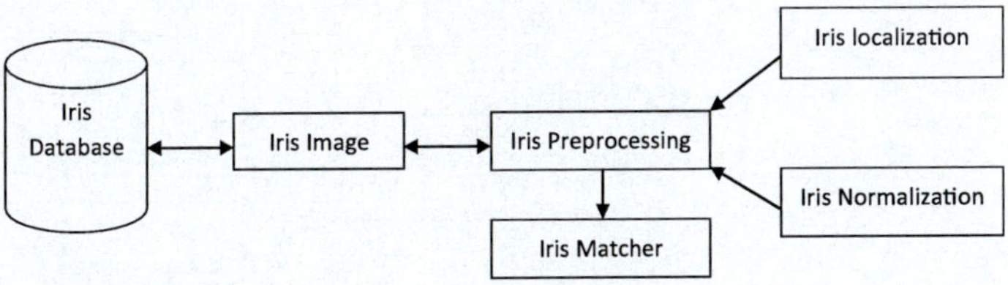

Fig. 2 Iris recognition

The above given formula is for calculating hamming distance HD, and it is clarified that X_j and Y_j are the bit by bit model to compare, and the noise masks are represented by Xn_j and Yn_j, respectively, for X_j and Y_j, and N is used to represent the number of bits in each model. It produces score (Fig. 2).

3.2 Finger Vein Recognition

SDULA–HMT database is the first open finger vein database which consists of 3816 images from six fingers of 106 individuals. At Wuhan University, Joint Lab for Intelligent Computing and Intelligent systems designed the device used to capture finger vein images. All the images are stored in "bmp" format which has size with 320 × 240 pixels. The total finger vein database size is 0.85 G bytes. It is unique for identical twins, triplets, quadruplets, and quintuplets. It can be done only if the person is alive. Skin condition does not affect the accuracy of the reading.

Several preprocessing techniques are used in finger vein authentication. The proposed algorithm consists of:

- **Segmentation**—To increase the accuracy, we first make the background of the image black by setting those background pixels to zero.
- **Aligning the finger horizontally**—which means keeping the finger in a tube-like structure device or using software algorithm can also align the images. The edges in the finger image are identified, and the image is rotated in order to detect the horizontal edges.
- **Image enhancement using contrast limited adaptive histogram equalization**–in order to improve the contrast of the image and the characteristics of brightness, and to reduce the content of noise and/or sharpen the details. Transforming the values in the intensity image helps in increasing the contrast of images. The histogram gray levels in the output are black in majority. If the median value of the output window is 50% gray window around, then each pixel is generated first. The cumulative sum is used to map the input pixel gray level to output of the same, if a pixel has low gray level than all others in the surrounding window (Fig. 3).

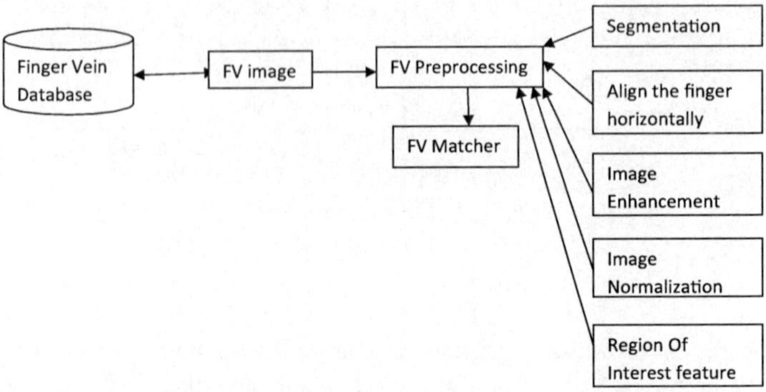

Fig. 3 Finger vein recognition

- **Image normalization**—This resizes the original image into one fourth of its original size which is the optimum value of scaling factor with respect to vein databases and set the optimum scaling factor as 0.6.
- **Region of interest (ROI) extraction**—In order to increase accuracy, unwanted area in the finger vein image is removed in this process. So set the edge points in height which are from 30 to 60% of the image. Crop the image vertically at cropping points. Now crop the image horizontally from 5% of left border 5% to 10% of right border.

Matching algorithm uses Hamming distance. Score is normalized before fusion.

3.3 Finger Print Recognition

Finger prints are unique even in identical twins and others such as triplets, quadruplets, and quintuplets. Ridge structures are not affected by burns, abrasions, and cuts. The preprocessing steps are as:

- **Enhancing FP image**—Gabor filters are utilized to capture both local orientation and frequency information from a fingerprint image for optimization. The Gabor filter is constructed by after determining the ridge orientation and ridge frequency in the image. Gabor filter must be tuned to specific frequency and orientation values for fingerprint enhancement. Gabor filter also used to preserve the ridge structure to enhance the ridges in the direction of local orientation in good fashion (Fig. 4).
- **Binarize the threshold**—It converts the image of gray scale into the binary. The threshold value is preset for binarizing an image. White represents that the pixels lower than this threshold value. Black represents that the pixels values are higher than the threshold value.

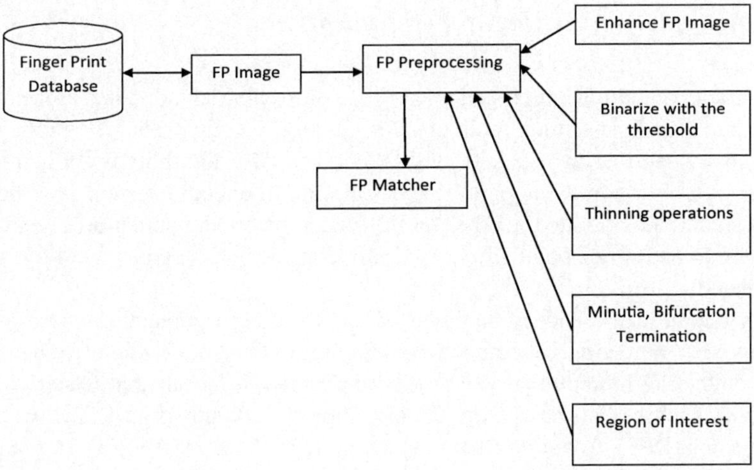

Fig. 4 Finger print recognition

- **Thinning operations**—are used to iteratively preserve the connectivity of the ridges.
- **Minutia, bifurcation termination**—During thinning, if the features of the fingerprints wants to not to get distorted, bifurcation is to be done.
- **ROI**—full image is considered for region of interest.
- **Matching algorithm**—direct comparison of two images pixel wise is done to obtain the score.

3.4 Normalization

Score-level fusion approach is used for its accuracy and optimal performance. It can be easily combined with other biometric trait. It has enhanced response time and system performance. It provides lower communication bandwidth, easy to process and easily cessible. Z-score normalization method is used, and its formula is as follows:

where x'_k: the kth matching score before normalization
x_k: the kth matching score after normalization
μ: mean
σ: standard deviation.

3.5 Score-Level Fusion and Optimization

The hybrid GA and PSO approach manages to pick N initial individuals for random generation. The fitness function is used for sorting N individuals. The probability, *Pk*, set is divided into two subsets $\{\psi G, \psi P\}$. The PSO algorithm is applied to the first subset ψ_G to adjust the particles. GA algorithm operations such as selection, crossover, and mutation to the other set ψ_G. After that, both population's results are integrated to one single population of N individuals which are then sorted for whole set repeatedly.

Both deterministic and probabilistic rules are used in population-based search methods. It is more efficient computationally. It provides social, cognitive behavior. After conducting literature survey on existing works, it reveals that Genetic Algorithm (GA) and the Particle Swarm Optimization (PSO) outperforms than the Brute Force Search (BFS), Adaptive Neuro Fuzzy Inference System (ANFIS), and Support Vector Machine (SVM). According to their research work suggestions, in this proposed work, it is implemented that the hybrid Genetic Algorithm and Particle Swarm Optimization optimize authentication algorithm. The output score weights are assigned to the different biometric modalities fused at the score level in order to increase performance and accuracy. The fitness function is designed with performance metric as equal error rate (EER), and its goal is to minimize accuracy and ERR, which is to be optimum.

. The entire population is split into two subsets out of which one for the Genetic Algorithm operations such as selection, crossover, and mutation processes, and other subset for Particle Swarm Optimization with respect to velocity and position update. It is repeated until the offspring fitness cost is to produce optimum value as minimum EER value. This hybrid GA and PSO algorithm provides the combined advantages of both the algorithms such as faster, less computational time, and robust technique.

Fitness function = Equal error rate

$$Sf_i = \sum_{m=1}^{M} W_m \times S_i^m$$

$$Wm \in [0, 1], \sum_{m=1}^{M} W_m = 1$$

Algorithm: Hybrid GA-PSO for fusion and optimization

```
Initialize the parameters
        Create initial population in weights randomly
          While k < max do
    Iteration starts, sort EER values
    For i=1 to m
       Update particle's p_best and g_best
       Update particle's velocity and position
    Endfor
    For i=m+1 to end
       Select each parents to reproduce
       Create nextgeneration through crossover,mutation
    Endfor
   Merge both sub-populations into single population
   If condition to stop then
      Go to end
   Endif
Endwhile
       Return best EER
```

4 Results and Discussions

The experimental results of the proposed algorithm are discussed, and their results are shown. To evaluate the efficiency of the algorithm, the following database is utilized.

In all biometric systems for security, set the reference threshold value to assess the authentication. Authentication means whether the person is genuine or unauthorized person. The performance of the state-of-the-art algorithm is discussed, and their results are exhibited. Threshold plays vital role in authentication. If the threshold value which set increases, then FAR decreases, FRR increases, and vice versa. Hence, the value of FAR must be small as FRR values. Threshold may be chosen based on the following three reasons:

i. FAR and FRR is equal
ii. FAR is minimum
iii. For fixed value of FAR.

Low FRR leads to high usability, and low FAR provides high security. To evaluate the efficiency of the algorithm, the following database is utilized. Fusion multiple biometric traits is a emerging technology and trend research topic in recent years. In 2010, at Shandong University (SDUMLA-HMT) set up the homologous multimodal

Table 1 Comparisons between the different combinations of FAR and FRR values to fused (iris, finger vein, palm print, face) with unimodal biometric systems

FAR %	FRR %				
	Iris	Finger vein	Palm print	Face	Fused (iris, finger vein, palm print, face)
1.0	0.04	2.8	1.70	14.6	0.91
0.1	0.69	5.9	3.02	38.4	2.1
0.01	1.80	9.6	6.05	59.8	4.5
0.001	2.96	12.6	7.06	61.2	7.3

traits database which is named SDUMLA–HMT database in the machine learning and data mining laboratory. In this, SDUMLA-HMT database comprises of five biometric traits including face, finger vein, gait, fingerprint, and iris. SDUMLA–HMT will provide the SDUMLA–HMT database open to all biometrics recognition researchers in order to promote research. The SDUMLA–HMT database contains of seven different view angles from face images, six different finger vein images, six view angles from gait behavior videos, iris images and fingerprint images acquired with five different sensors. The database is obtained from 106 individuals. It is a database in which all the multimodal biometric samples are collected from the same person. It helped the research work in wide scope. The original database had obtained from the laboratory. I am an authenticated user of this database.

Relating operating curve (ROC) is represented on x-axis when FRR $\neq 0$, FAR $= 1$, and on y-axis when FRR $= 0$, FAR $= 0$, respectively.

Genuine score < threshold = FRR

FRR = No. of false rejections/No. of genuine accesses

FAR = No. of false acceptances/No. of imposter accesses

The performance of unimodal biometric system such as iris, finger vein, palm print, and face is compared with proposed multimodal biometric system. Table 1 proves that the multimodal biometric system with fused iris, finger vein, palm print, and face lowers FAR and FRR values that are compared to unimodal biometric systems. If FAR and FRR values are low, then the accuracy of the system is improved. Hence, this multimodal biometric system can be applied for robust, reliable authentication systems.

When FAR = 1.0, 0.1, 0.01, 0.001, the corresponding FRR values of unimodal biometric systems such as iris, finger vein, palm print, and face is compared with the multimodal biometric system such as fused iris, finger vein, palm print, and face. The values are explicitly proved that the proposed multimodal biometric system is more robust, reliable, and secured for authentication (Fig. 5).

In this proposed method, comparisons of multimodal biometric traits with different combinations and methodologies used for recognition and matching yield different threshold values and their corresponding FAR and FRR values. Hence, accuracy of the system is enhanced.

The SDUMLA–HMT database is used for conducting experiment for the performance of the algorithm. The threshold is set, at point where FAR and FRR are

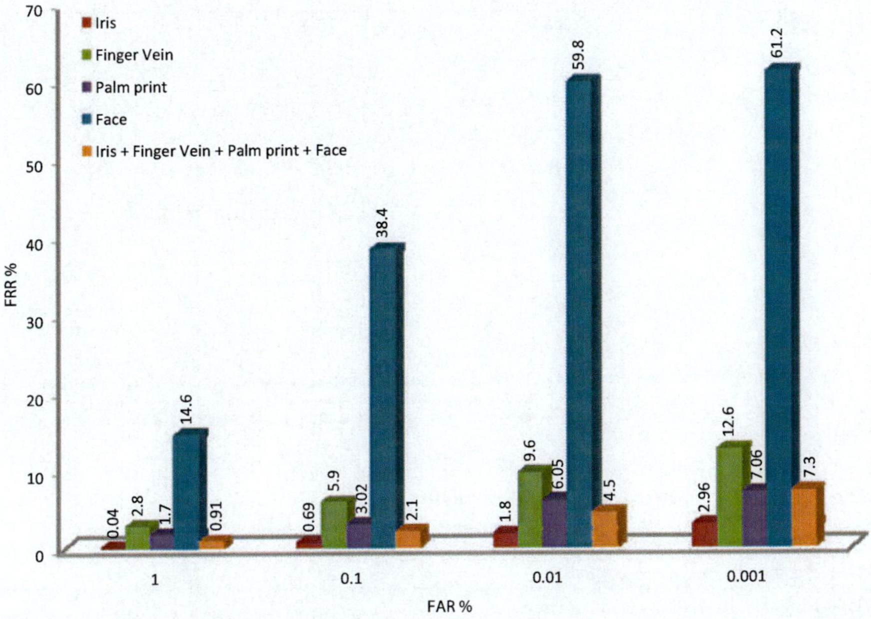

Fig. 5 Comparisons between the different combinations of FAR and FRR values to fused iris, finger vein, palm print, and face with unimodal biometric systems

minimized. Table 2 shows that the result of proposed method's FAR and FRR values of reference threshold 0.03 is highlighted in bold when FAR value is 0.00 and FRR is 0.02. It is too low value when compared to other multimodal biometric systems discussed (Fig. 6).

Table 3 shows that the overall comparisons of this research work. Iris, palm print, face, signature-based multimodal biometric algorithm yields 0.003, 0.250 of FAR, FRR, respectively. Fused iris, finger vein, palm print, and face-based multimodal

Table 2 Proposed method threshold versus FAR and FRR values

Threshold	FAR %	FRR %
0.010	0.00	99.00
0.015	0.00	72.15
0.020	0.00	30.71
0.025	0.00	20.05
0.030	**0.00**	**0.02**
0.035	10.05	0.01
0.040	40.19	0.00
0.045	70.10	0.00
0.050	99.00	0.00

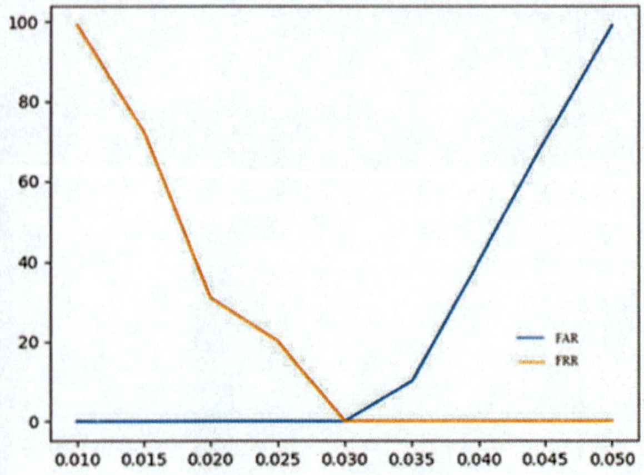

Fig. 6 FAR versus FRR

Table 3 Overall comparisons of multimodal biometric systems

Biometric	FAR (%)	FRR (%)
Iris + Palm print + Face + Signature	0.003	0.250
Iris + Finger vein + Palm print + Face	1.0	0.91
Proposed method (Iris + Finger vein + Finger print)	0.00	0.02

biometric algorithm provides 1.0 and 0.91 values of FAR, FRR, respectively. But iris, finger vein, finger print-based multimodal biometric algorithm produced 0.00 and 0.02 FAR and FRR values, respectively. Since FAR, FRR values directly influences the equal error rate and accuracy, it is mandatory to observe the FAR and FRR values that are low (Fig. 7).

Proposed methods for FAR and FRR values are listed in Table 4.

If threshold increases, then FAR decreases and FRR increases and vice versa. Equal error rate (ERR) is defined as FAR and FRR are equal. Devices are accurate only when ERR is low.

The equal error rate comparison Table 5 shows that the proposed method is more efficient since EER decreases as performance increases.

The equal error rate is shown in Fig. 8.

Accuracy formula is presented as:

$$\text{Accuracy} = 100 - \left(\frac{\text{FRR} + \text{FAR}}{2}\right)$$

The accuracy is compared with the existing biometric technologies as follows (Table 6).

The comparison of accuracy is shown in Fig. 9.

Fig. 7 Overall comparisons of multimodal biometric systems

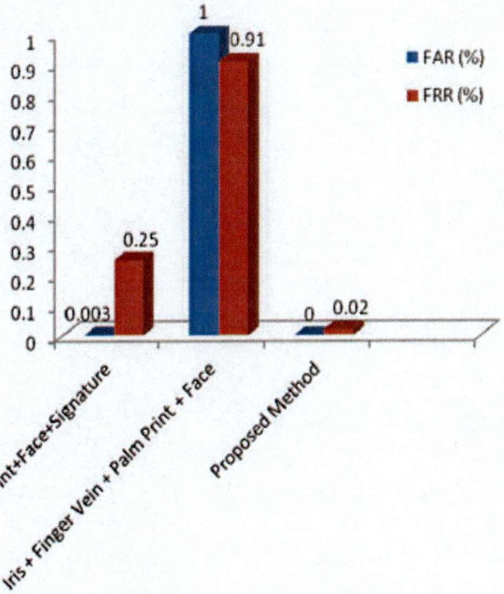

Table 4 FAR and FRR values

Threshold	FAR	FRR
0.010	0.00	99.00
0.015	0.00	72.15
0.020	0.00	30.71
0.025	0.00	20.05
0.030	**0.00**	**0.02**
0.035	10.05	0.01
0.040	40.19	0.00
0.045	70.10	0.00
0.050	99.00	0.00

Table 5 Equal error rate (EER) comparison

Multimodal biometrics	Equal error rate
Finger print	0.5
Finger vein	0.145
Iris + Finger print	0.038
Iris + Finger vein	0.128
Finger print + Finger vein	1.21
Proposed method (Iris + Finger print + Finger vein)	0.03

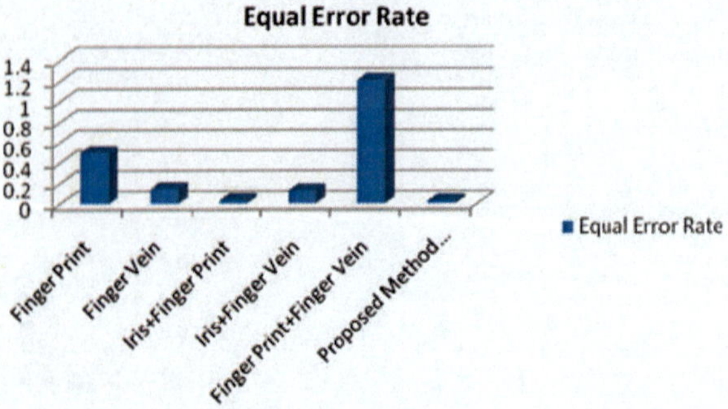

Fig. 8 Equal error rate

Table 6 Comparison of accuracy

Multimodal biometrics	Accuracy
Finger print	85
Iris + Finger print	99.975
Finger print + Finger vein	98.78
Proposed method (Iris + Finger print + Finger vein)	99.99

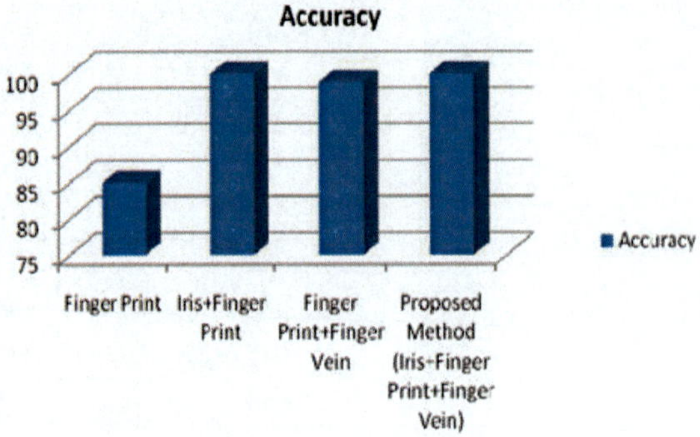

Fig. 9 Comparison of accuracy

5 Conclusion

Multimodal biometric algorithm is designed using score-level fusion techniques with hybrid Genetic Algorithm and Particle Swarm Optimization for developing optimized authentication system with enhanced accuracy and security.

This research work integrates more than two biometric traits for developing robust authentication algorithm. Iris, finger vein, and finger print biometrics are integrated and opted out for their best biometric characteristics. It provides reliable and most promising solution for security systems. It also eliminates the disadvantages and limitations of unimodal biometric systems. In the proposed method, hybrid Genetic Algorithm operations and Particle Swarm Optimization are used to construct fusion rules. It optimizes the accuracy and enhances security.

Preprocessing techniques are adapted in each biometric trait in order to improve the quality of image for recognition to obtain good score as output. While recognition the image, the matching algorithm processes the image and produces score. The heterogeneous biometric traits scores are normalized before fusion. To make it homogeneous, normalization method is adapted. In order to optimize the fusion scores, hybrid Genetic Algorithm and Particle Swarm Optimization is implemented. The decision is taken whether the claimed identity is genuine or imposter.

MATLAB software is used for implementation of preprocessing, matching, normalization, and optimization. The SDUMLA–HMT database is utilized for validating the authentication system with respect to metrics. This database is chimeric database obtained from the same person. This research work makes use of this database because of its chimeric nature.

Finally, the performance metrics used in this proposed system are false acceptance rate (FAR) and false rejection rate (FRR), equal error rate (EER), and accuracy. When the threshold is shoots up, the dalse rejection rate increases. When the threshold is maintained low than the reference, then the false acceptance rate increases. Hence, the threshold is set to remain low FAR, FRR values. The equal error rate (EER) is calculated when FAR and FRR are equal. To improve the accuracy of the system, maintain EER value as low.

World's largest existing biometric database is India's national ID program called Aadhaar to identify citizenship. It is linked with life time activities of a person such as bank transaction and other significant activities in order to make digitial world.

Biometrics makes password-less world in near future. In future, biometrics will be the door way to all the accessible systems.

6 Future Enhancement

Biometric is emerging trend in the research era, enhancement can be done in all dimensions. Online feature extraction could be adapted for each biometric trait using multiple sensors.

Multi-instance: Various instances of the same biometric trait can be considered as input for recognition.

Multi-sensor: A single biometric trait captured using multiple sensors with different information.

Multi-algorithm: Different algorithms are implemented for the same biometric trait.

Multi-sample: More samples by a single sensor of same biometric trait is considered.

Cancellable biometrics: Any biometric trait image can be cancelled like a password.

Cryptographic algorithms: If biometrics concepts can be fused with cryptographic algorithms, the system would be more robust, reliable, and accurate.

Hybrid systems: Any combination of above methods may be implemented for high security systems.

Integrated approach: Combination of biometric scheme with non-biometric scheme such as possession or knowledge-based schemes can be implemented for better performance of the system.

Recent biometric traits such as DNA, brainwave, body odor, finger nail bed, feart patterns, aging facial, vascular pattern recognition, and dorsal hand vein recognition can be used for multimodal biometrics.

Fuzzy logic can be applied at fusion level in order to implement secured system for specific applications.

Various other factors can also be considered such as order of algorithms to improve speed, biometric transaction time, and biometric enrollment rate etc.,

Biometric traits can be applied to medical applications for identifying the symptoms of disease in order to focus prevention.

References

1. Veeramachaneni K, Osadciw LA, Varshney PK (2005) An adaptive multimodal biometric management algorithm. IEEE Trans Syst Man Cybern Part C Appl Rev 35(3):344–356
2. Hong L, Jain A (1998) Integrating faces and fingerprints for personal identification. IEEE Trans Pattern Anal Machine Intell 20(12):1295–1307
3. Frischholz RW, Dieckmann U (2000) BiolD: a multimodal biometric identification system. IEEE Comput 33(2):64–68
4. Hong L, Jain AK, Panikanti S (1999) Can multibiometrics improve perfomance? In: Proceedings of AutoID, Summit, NJ, pp 59–64
5. Vatsa M (2004) Comparison of iris recognition algorithms. Department of CSE ZIT Kanpur, India mayank-richa@ yahoo.com 0-7803-8243-91041$17.000, IEEE, pp 354–358
6. Mazouni R, Rahmoun A (2011) On comparing verification performances of multimodal biometrics fusion techniques. Int J Comput Appl 33(7):24–29
7. Poh N, Bengio S (2005) Database, protocol and tools for evaluating score-level fusion algorithms in biometric authentication. Pattern Recogn 39(2):223–233
8. Dalila C, Imane H, Amine NA, Multimodal score-level fusion using hybrid ga-pso for multibiometric system. Informatica. 39:209–216. International Journal of Engineering Research and Development e-ISSN: 2278-067X, p-ISSN: 2278-800X, www.ijerd.com
9. Eberhart R, Shi Y (1998) Comparison between genetic algorithms and particle swarm optimization. In: Presented at the 7th annual conference evolutionary programming, San Diego, CA
10. Ross A, Jain AK (2003) Information fusion in biometrics. Pattern Recogn Lett 24(13):2115–2125
11. Ross A, Nandakumar K, Jain, AK, Handbook of multibiometrics. Springer, Berlin
12. Woodward JD (2005) Biometrics: facing up to terrorism. In: Biometrics consortium conference
13. Benaliouche H, Touahria M (2014) Comparative study of multimodal biometric recognition by fusion of iris and fingerprint. Sci World J. https://doi.org/10.1155/2014/829369
14. Mahri N, Suandi SA, Rosdi BA (2010) Finger vein recognition algorithm using phase only correlation. 978-1-4244-7065-5/10/$26.00 ©2010 IEEE
15. Srivastava PC, Agrawal A, Mishra KN, Ojha PK, Garg R (2013) Fingerprints, Iris and DNA features based multimodal systems: a review. Int J Inf Technol Computer Sci (IJITCS) 5(2):88–111. I.J. Information Technology and Computer Science, Published Online January 2013 in MECS (http://www.mecs-press.org/) https://doi.org/10.5815/ijitcs.2013.02.10

16. Agarwal M (2007) Design approaches for multimodal biometric system. Master's thesis, IIT Kanpur
17. Yin Y, Liu L, Sun X, SDUMLA-HMT: a multimodal biometric database. In: Sun Z (ed) CCBR 2011, LNCS 7098, © Springer, Berlin Heidelberg, pp 260–268. School of Computer Science and Technology, Shandong University, Jinan, 250101, China
18. Shruthi BM, Pooja M, Ashwin RG (2013) Multimodal biometric authentication combining finger vein and finger print. Int J Eng Res Dev 7(10):43–54
19. Snelick R, Indovina M, Yen J, Mink A (2003) Multimodal biometrics: issues in design and testing. In: Proceedings of the 5th international conference on multimodal interfaces (IMCI)
20. Sujatha E, Chilambuchelvan A (2017) Multimodal biometric authentication algorithm using iris palm print, face, signature with encoded DWT. Wireless Pers Commun 99:23–34. https://doi.org/10.1007/s11277-017-5034-1
21. Sujatha E, Chilambuchelvan A (2016) Neural network based normalized fusion approaches for optimized multimodal biometric authentication algorithm. Sci Res Circ Syst 7:1199–1206
22. Sujatha E, Chilambuchelvan A (2015) Multimodal biometric algorithm using normalized score level fusion rules with support vector machine optimization. Aust J Basic Appl Sci 9(21):47–51
23. Subbarayudu VC, Prasad MVNK (2008) Multimodal biometric system. In: Proceedings of 1st international IEEE conference emerging trends engineering technology, pp 635–640
24. www.biometrics.gov, National Science and Technology Council, Committee on Technology, Committee on Homeland Security, Subcommittee on Biometrics

Employing Real-Time Object Detection for Visually Impaired People

Kashish Naqvi, Bramah Hazela, Sumita Mishra, and Pallavi Asthana

Abstract Visually impaired and blind people face several difficulties in their daily life. This was the primary motivation of this work as to create and assemble an object detector that can assist people with visual impairments using OpenCV and TensorFlow API on Raspberry Pi and provide an audio output for the detected objects using Espeak; Text-to-Speech Synthesizer. Single Shot Detector (SSD) model with MobileNet v2 has been employed to perform the detection with high accuracy and processing speed. The scripts are written in Python which utilizes the model to recognize the objects with boxes and provide class of the objects. The recognized image category is extracted and stored in a text file. The developed system provides aid to a visually impaired person for performing tasks independently using real-time object detection and identification technology. Developed system can successfully provide information about detected object in the form of an audio output to the visually impaired person.

Keywords Raspberry Pi · TensorFlow · Faster-RCNN · OpenCV · Python · Text-To-Speech Synthesizer

---

K. Naqvi
Department of Computer Science & Engineering, Amity University, Lucknow Campus, UP, India
e-mail: kashishnaqvi10@gmail.com

B. Hazela (✉) · S. Mishra · P. Asthana
Department of Computer Science & Engineering and Electronics & Communication Engineering, Amity University, Lucknow Campus, UP, India
e-mail: bramahhazela77@gmail.com

S. Mishra
e-mail: smishra3@lko.amity.edu

P. Asthana
e-mail: pallaviasthana2009@gmail.com

A. Khanna et al. (eds.), *Data Analytics and Management*, Lecture Notes on Data Engineering and Communications Technologies 54,
https://doi.org/10.1007/978-981-15-8335-3_23

1 Introduction

Mankind is experiencing the inception of the futuristic age of deep learning accompanied by embedded devices. Factual and real-time functioning applications in several fields such as recognition of entities in any form of digital visuals, autonomous cars, finer health monitoring systems, etc., are being implemented and continuously in the process of improvement through the deployment of latest technologies [1]. Technology is becoming user friendly, and it can be effectively provide relief to people suffering from disabilities. People who are born with or get after birth complete blindness or any sort of visual impairments suffer from difficulties in their daily lives [2]. It is estimated that presently, there are around 39 million people without eye sight and about 246 million are suffering from low eye sight. Such individual suffers from an extensive variety of issues and struggles in order to perform communication with the entities that are existent in surrounding and encounter many problems in their daily lives. *This* saddening state of the people with no or poor eye sight plays a vital role in building such a project that would reduce the *handicap* of these blind persons. A great number of complications and difficulties in computer vision-based models and applications were there in performing saturation on their exactness or precision about ten years ago. Nevertheless, due to the advent of deep learning abilities and techniques, the exactness of the models extraordinarily improved. Therefore, the primary motivation in this project is to create and assemble an entity identifier which would support detection and identification of objects employing TensorFlow Object Detection API in accordance with Raspberry Pi. The main objective of the paper is to provide ample amount of information for training a detection classifier for objects using the regional convolutional neural network and deep learning [3]. It would recognize the entities, and after performing the identification, it would be able to 'speak' the relevant information about the identified object with a certain extent of accuracy and confidence.

This system would provide aid in real-world applications in a productive and cost-effective fashion which would support visually impaired or blind people.

Configuration of this system is based on following technology:

- It employs Espeak [4] which provides offline Text-to-Speech Synthesis as compared to other systems with a related objective which incorporate gTTs [5] that is online and thus fails to work in network-restricted environments.
- SSS_mobilenet [6] is employed in system instead of Faster RCNN [7] inception, thus providing faster and accurate outcomes.
- It utilizes TensorFlow API [8] of Object Detection which is a powerful tool that makes the system work efficiently and effectively.
- It can provide much faster and real-time performance as compared to previous works.

The paper progresses as follows, section following introduction provides a literature review of past and related projects and models. The next section utilizes the concepts discussed in literature survey to define a better system model. The succeeding section discusses the materials and methods for the development of the system followed by results and discussions, and finally, the last section concludes this paper followed by acknowledgement and references.

2 Literature Survey

Various researches have been carried out to create automated tools to help the people with visual impairments. Still, numerous visually impaired people depend on their own memory or the benevolence of others for their everyday life.

A group of researchers [9] performed initial research on installing a real-time entity identification system that functions at fast rate of frames per second on a resource restricted body like a Raspberry Pi-based embedded system and cell phones. It focused on the detection of traffic-related things such as walkers, vehicles and other such objects. The proposed system provided by the contributors employed the trials numerous models for performing detection, plus the YOLO2 [10]. Main drawback of entity detection methods deployed by authors is that, despite being precise, this method is not efficient in real-world use because of huge model size and slow speed.

There was another somewhat similar research performed by another group of researchers in where they build a visual surveillance system employing a Raspberry Pi as well as Arduino. [11] Permitted persons or users were allowed to gain means of access in the system that could monitor through remote set-up connection over Internet by employing a cell phone so that the user can keep an eye at the site. If a trespasser tries to gain access to the system, a snapshot of the invader's face is taken by using a camera, and this snapshot is then mailed as an attachment through an email to the email address of the authorized person. In addition to this, there is also a function which makes the information to be sent as an SMS message to the authorized person's cell phone number, therefore identifying the intruder having no valid identity. A related project which focuses at entity detection in Raspberry Pi has been taken into consideration [12]. This work relates to a desktop application which allows the client to provide an order to capture a scene. This scene is then set as the target scene. The robot captures the scene at present, and then, the captured scene is put into comparison with the target scene. It performs utilization of region-based correlation for identification.

The navigation assistance in another project provides aid to the blind people. [13] The system requires a Text-to-Speech Synthesizer that requires online support which may not be suitable for network constrained environments. It makes use of Google Text-to-Speech Synthesizer. This project is unique as it is trained for the visually impaired people to identify the objects of daily core in an offline, real-time environment. It uses a simple speech synthesizer to convert the identified image into speech, thus making the process very simple and user friendly.

3 System Design

Raspberry Pi camera module is installed to take the images of the surrounding entities of the person for input in the scene and identify in real-time environment. Architecture of system consists of the following: (i) Raspberry Pi 3, (ii) a Raspberry Pi camera module version 1, (iii) utilizing AI through TensorFlow Object Detection API, (iv) speaker or set of earphones and (v) power bank to fulfil power requirement [14]. Figure 1 demonstrates the hardware set-up of the developed

Block diagram as shown in Fig. 2 shows the sequence of the different tasks like collection of data, object detection, object identification and converting the identified object into speech signal through the tools.

3.1 Hardware Specifications

(i) Raspberry Pi 3: Raspberry Pi 3 is a third-generation Raspberry Pi, and due to useful features, it has been employed in a advanced applications like Internet of things, security monitoring, gaming, as a wireless access point, as a server for cloud, etc. It has a 1.2 GHz ARM Cortex and 64-bit processor. It can be connected to 802.11 b/g/n LAN and Bluetooth 4.1. It has a memory of 1 GB and works on the LINUX operating system of Windows 10 IoT. Its power requirement can be fulfilled through micro-USB socket 5V1, 2.5A. It has a 10/100 Base T Ethernet socket for connectivity. It has a 40-pin GPIO connector. It can serially interface with (CSI-2) with 15-pin camera connector [15].

Fig. 1 Hardware set-up of the provided system

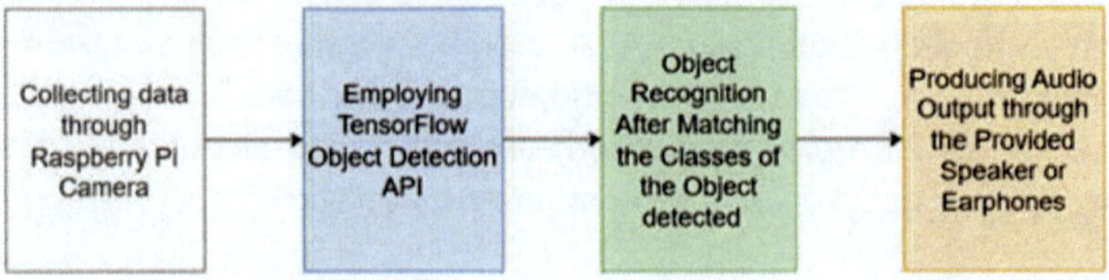

Fig. 2 Depiction of system model

(ii) Raspberry Pi camera: Raspberry Pi camera version 1 has been used in module. This is a small camera with the dimensions 25 × 24 × 9 mm and 3 gm of weight. It has a still resolution of 5 mega pixels, and it can work in video modes of 1080p30, 720p60 and 640 × 480p 60/90. It uses V4L2 driver based on Linux integration. It has Omnivision OV5647 with the sensor image area of 3.76 × 2.74 mm and sensor resolution at 2592 × 1944 pixels [16].

(iii) Speaker or earphones: Raspberry Pi 3 has HDMI interface for transmitting audio signals, so headphone can be configured through this interface. Bluetooth headsets can also be connected as it has internally connected low-energy Bluetooth. In this project, we have used headset that is connected through HDMI.

3.2 *Software Requirement*

(i) OpenCV Python:

It is an all-available set of packages and scripts for computer vision. Therefore, it is a collection of programs and software packages that are made universally accessible that provides programming operations and tasks mostly targeted for real-time activities and applications of computer vision. This collection of programs and software packages that is universally accessible is able to be employed on various types of computer systems or being employed with a wide range of software packages. It is free to be employed in various systems and devices in the all available BSD license OpenCV.

(ii) Espeak TTS:

The Text-to-Speech Synthesizer employed in the system is Espeak. It provides an offline synthesis of speech from written text and can be used with Python efficiently in HDMI.

(iii) TensorFlow API of Object Detection:

TensorFlow is drafted and authored from Google Group Brain. It provides a large gathering of models for detection which are pre-trained on COCO data set, known as Common Object in Context, KITTI dataset, and Open Images Dataset. This API lets us to express the models for detection of object employing the configuration files. The API of TensorFlow for Detection of Objects is a sole package, permitting us to rapidly iterate over various sequences using the TF back-end. It is also accountable for structuring all other required and important features together in a single package.

SSD:

Single Shot Detector consists of:

(i) Single Shot: Object localization and detection are done in a single forward pass of the network.

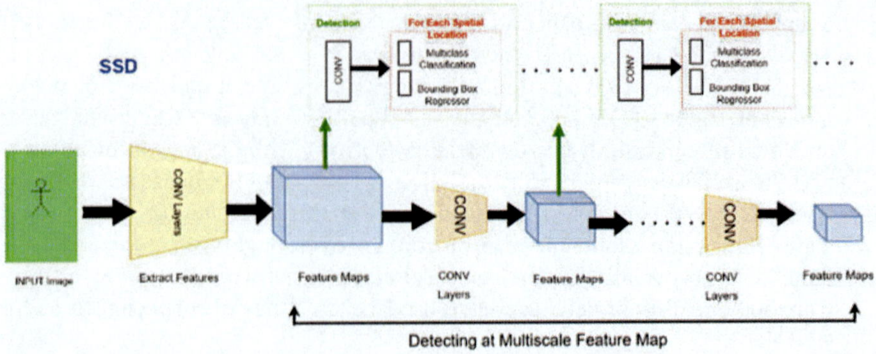

Fig. 3 Single Shot Detector (SSD) architecture

(ii) Multibox: Technique for bounding box regression.
(iii) Detector: It is an object detector and classifier network.

SSD has inclusive features like Risk Priority Number (RPN) and multi-scale CONV highlights for a faster location performance and apex identification quality of the detected entities. Figure 3 depicts the process of a high-level architecture of SSD demonstrating a process of generic entity identification. It shows various auxiliary CONV layers which are continuously diminishing in measurements. SSD utilizes VGG-16 as convolutional layer network. It has a strong performance in image classification. In SSD multibox architecture, these layers are not fully connected, but auxiliary layers are added to enable the extraction of features at multiple scales that would decrease the size of input to each subsequent layer, thus speeding up the process [18] SSD is a one-shot locator. It has no designated region proposition network and predicts the boxes and the classes legitimately from feature maps in a single pass. To compensate the drop-in precision, SSD applies a couple of upgrades including multi-scale highlights and default boxes. Object is detected by employing a single deep neural network and merging it with proposals of regions and attribute drawing.

SSD_MobileNet:

In SSD MobileNet, convolutional layer is split into two subtasks; depthwise convolutional followed by pointwise convolution. This layer combines the filtered value to create new features. These two convolutions form a depthwise separable convolution block that is faster than traditional convolution. Convolution is followed by batch normalization. MobileNet uses ReLU6 as a activation function because it prevents the activations from becoming too large. Global average pooling layer is connected at the end which is followed by fully connected classification layer and a Softmax function for the probability distribution [17, 18]. SSD MobileNet has a faster rate of detection, which is essential in this application which is developed for visually impaired people.

4 Image Classification

Image classification in computer vision is expressed as predicting or foretelling the class in context of the graphic [19, 20] The TensorFlow API for Object Detection needs to employ the particular directory antinomy. There is also a need for a number of several supplementary packages for Python, particular supplements to PYTHONPATH variable and several additional instructions for set-up and to train and execute the detection classifier [21]. TensorFlow has many models for object detection inside its model zoo. Packages of python are also used, such as Pillow, lxml, Cython, Jupyter, Matplotlib, pandas and OpenCV. Last two packages are not required by the TensorFlow; they are needed in scripts of Python to generate TF records. The compilation of the Protobuf files, for configuration of the model and instruction parameters. There is a requirement of a great number of images of an object to perform a suitable detector for objects. For training a classifier, the images having random objects with a variety of environmental, lighting and background conditions must be selected.

(i) Pre-processing of data: Images have been taken from the website sources that provide datasets for research and educational purpose [22]. Overlapped images are also selected to more efficient training of network. Size of the image needs to kept under 250 KB each for faster speed. Figure 4 shows the collection of 'bottle' objects. 20% of them are moved to the test directory, and 80% to the train directory. There must be a good variety in both the directories for training.
(ii) Labelling of object of Interest—Object of interest in the images is labelled through LabelImg tool. As shown in Fig. 5, it will label all the images and have their .xsv versions stored in the test and train directories, respectively. The .xml

Fig. 4 Collection of 'bottle' object

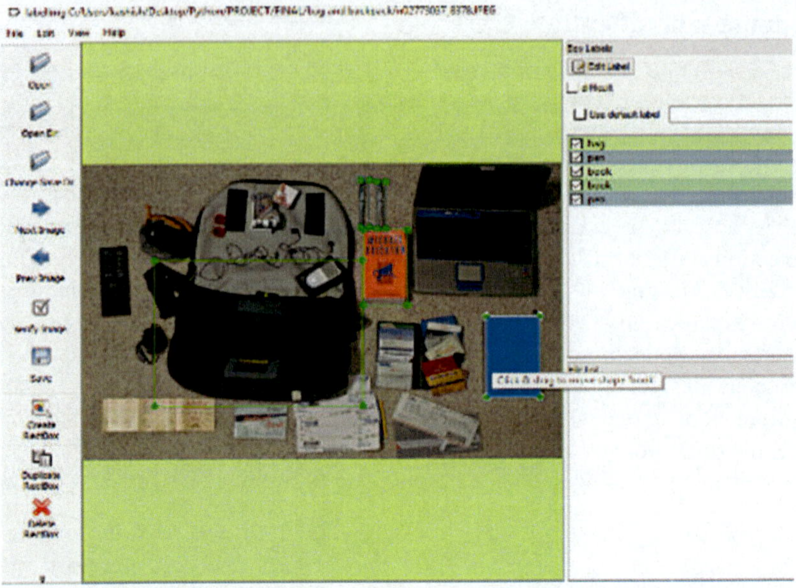

Fig. 5 Labelling of gathered pictures

files are be used for generate TFRecords and act as one of the several inputs to the trainer employing TensorFlow API for detection of objects. Now the images have been labelled, and the TFRecords have to be generated that are provided as input data for the TensorFlow trainer. The TFRecord scripts are generated with instruction in the object detection folder to generate a train.record and a test.record file in the same folder. The hindmost step before starting the training is to produce a label map and the training configuration executable file in Python. A label map will report the trainer about the object by mapping class names to class ID numbers, a new file is created and saved as labelmap.pbtxt in the training folder, and the label map ID numbers must be similar to that of in the tfrecord.py file. [12]

(iii) Model selection for classification—The training pipeline for detection of object is configured to identify the model whose parameters are taken into consideration to train the classifier. Create num classes in the ssdlite_mobilenet_v2.config file and train input and label map paths. This data returns through eval reader section in their respective folder. Training command is provided in the prompt of object_detection directory. Time for initialization is 30 s, before training starts. Figure 7 shows the graph depicting the loss in training; initially it is high and it is gradually decreasing. Training of the dataset would continue till the loss is less than 5%. Checkpoints are saved at every five minutes during training. Checkpoints having maximum steps are provided for the generation of frozen inference graph which is present in object detection directory as a frozen_inference_graph.pb file folder of object detection folder. The .pb file

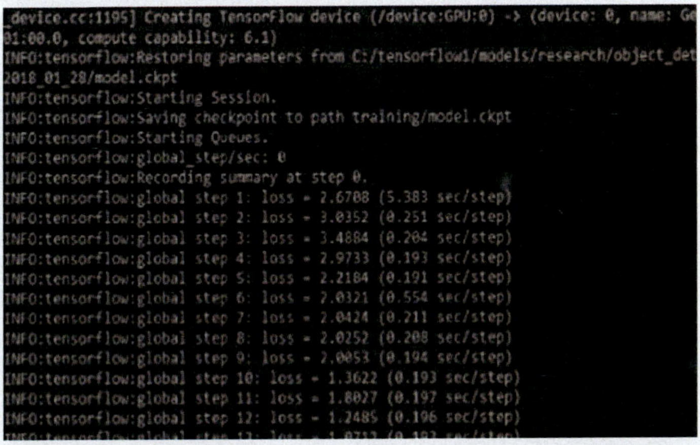

Fig. 6 Training of the object detector

Fig. 7 Loss graph

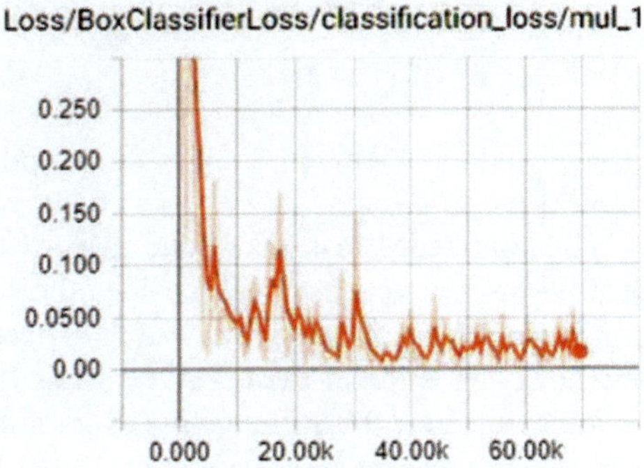

consists of the object detector for real time, image or video feeds. The training is done till the number of steps is more than 56,000, and the inference graph is created.

Loss function ($L(x, c, l, g)$) for SSD is the combination of classification loss and regression loss, where N is the number of positive match and α is the weight for the localization loss. [19] $L_{conf}(x, c)$ is confidence loss, and $L_{loc}(x, l, g)$ is localization loss. [11]

$$L(x, c, l, g) = \frac{1}{N}(L_{conf}(x, c) + \alpha L_{loc}(x, l, g)). \tag{1}$$

The localization loss is the difference in between the ground truth box and the anticipated boundary box. Single Shot Detector just considers predictions from positive matches. We need the forecasts from the positive matches to draw nearer to the ground truth. The non-positive ones can be overlooked. The localization loss used here is Smooth-L1 loss, between predicted l and ground truth g boxes, with cx and cy

as offset of boundary boxes of width and height *w* and *h* [19]. Category is *p*, default box of boundary is *d*, s is default box *i*, and t is ground truth box *j*.

$$L_{\text{loc}}(x, l, g) = \sum_{i \in Pos}^{N} \sum_{m \in (cx, cy, w, h)} z_{ij}^{k} \text{smooth}_{L1}\left(l_i^m - \hat{g}_j^m\right) \tag{2}$$

$$\hat{g}_j^{cx} = \frac{g_j^{cx} - d_i^{cx}}{d_i^w} \tag{3}$$

$$\hat{g}_j^{cy} = \frac{g_j^{cy} - d_i^{cy}}{d_i^h} \tag{4}$$

where $\hat{g}_j^w = \log \frac{g_j^w}{d_i^w}$ $\hat{g}_j^h = \log \frac{g_j^h}{d_i^h}$

$$L_{conf(x,c)} = -\sum_{i \in Pos}^{N} x^p \log\left(\hat{c}_i^p\right) - \sum_{i \in Neg} \log\left(\hat{c}_i^0\right) \tag{5}$$

$$x_{ij}^p = \begin{cases} 1 \text{ if } IoU > 0.5 \text{ between } s \text{and } t \text{on class } p \\ 0 \text{ otherwise} \end{cases}$$

The confidence loss is the one in performing a prediction of a class. For each positive match forecast, we consider the loss as per the confidence score of the comparing class. For negative match expectations, we consider the loss as per the level of confidence of the class ‘0’: class ‘0’ shows that no entity is detected. We have calculated as the Softmax loss over multiple classes confidence c, and the weight term α is set to 1 by cross validation.

$$-L_{conf(x,c)} = -\sum_{i \in Pos}^{N} x^p \log\left(\hat{c}_i^p\right) - \sum_{i \in Neg} \log\left(\hat{c}_i^0\right) \tag{6}$$

where $\hat{c}_i^p = \frac{\exp(c_i^p)}{\sum_p \exp(c_i^p)}$.

By this time, the entity identification will be performed well on the laptop, to use the same on the Raspberry Pi. The procedure adopted is to first and foremost update the Raspberry Pi. At the beginning, the Raspberry Pi requires to be entirely up to date. Have a terminal being opened and script up the commands to update it accordingly. Based on the condition, the amount of time the Raspberry Pi was updated the last time it was kept in use, and the amount of time required to upgrade it can be in between sixty seconds to about sixty minutes. After this, another thing that is to be performed is to deploy the TensorFlow API for Object Detection in the Pi. There is a need to have it downloaded or else it would take up a considerably large amount of time as it is about a hundred megabytes; hence, it would take a sometime. The TensorFlow API for Object Detection also demands for the collection of several packages and programs, precisely the LibAtlas one. Deploy the same by scripting the necessary commands. After performing the aforementioned steps, there was deployment of OpenCV, Protobuf and other packages, then set up the path in an identical manner on the Pi. Following this, it requires to add the model that was trained in the aforementioned steps for the daily objects by putting in the frozen inference graph inside the object-detection directory. Finally perform alterations on the path of the model as being provided in the script [23]. Embed a TTS code into the Raspberry Pi to attain the audio outcomes for the identified items. We have employed Espeak to perform the TTS operation and have embedded it into the Python script [24]. The detector is able to identify the given objects correctly and draw boxes around these in videos, images or in a webcam feed in real time, and there are Python scripts to test it out on any of the three mentioned forms. The object detector takes 10 s for initialization and displays the window and identifies and object that is required to be detected. This object discovery classifier which would let anyone of the specified day-by-day entities is perceived, and after that these can be spoken up by the suggested system to perform object recognition with an audio output with Raspberry Pi and then provide aid in to persons with no or poor eye sight. As it was developed with a few daily objects and their images, it can detect only a few objects for which we performed the training for. In order to increase the objects, we chose to select the pre-trained coco data set for ssd_mobilnetv2 which consists of 90 object classes. These provide a better coverage to a larger arena for the blind people. Therefore, we utilized the pre-trained model and embedded it into the Raspberry Pi and performed the above steps for on the Pi for the pre-trained ssd-mobilenet_v2_coco model. This works well and provides optimum accuracy for a real-time system and wider range of objects.

5 Results and Discussions

The proposed system can be employed on embedded devices for performing entity sighting and identification followed by audio generation. This system has imparted well enough to our comprehension as novice in the arenas of deep learning accompanied by embedded devices. Figures 8, 9, 10, 11, 12, 13, 14, 15 and 16 demonstrate the results of how the objects are detected and recognized after several trials.

The detector is able to identify the given objects correctly and draw boxes around them in real time. The object detector takes 10 s for initialization and displays the name and identifies the object to produce an audio-based output.

With the use of advanced techniques like TensorFlow API of Object Detection, OpenCV collection of packages and software, model was trained by performing and then undergone testing for several rounds. Written scripts of code were embedded in the Raspberry Pi that provided the outcome to be employed by people with poor or no eye sight in the formalism of speech as audio signals in order to inquire any item or article for which network was trained.

Fig. 8 Object detection of bottle

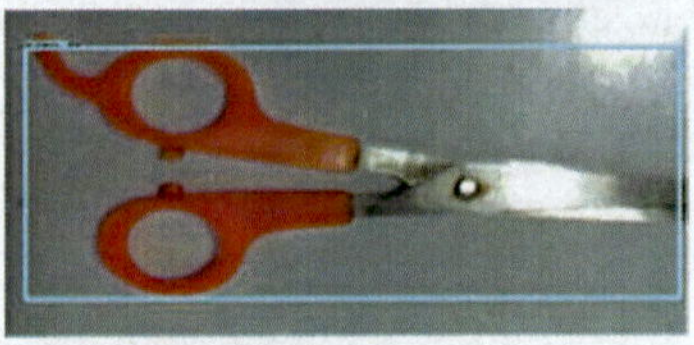

Fig. 9 Object detection of scissors

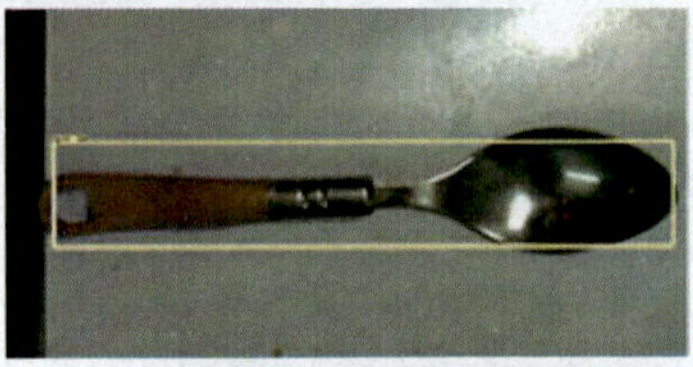

Fig. 10 Object detection of spoon

Fig. 11 Object detection of chair

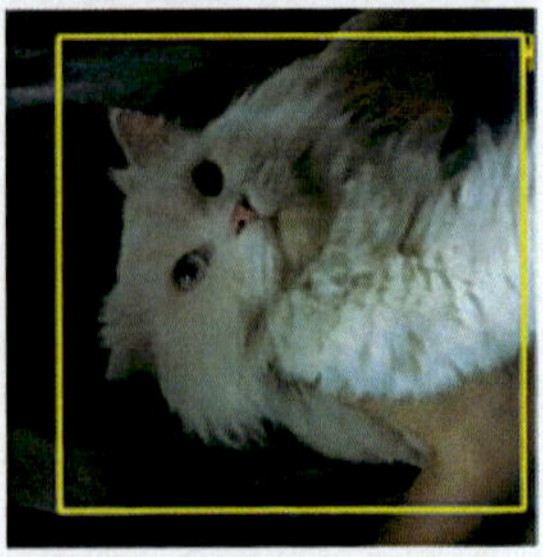

Fig. 12 Object detection of cat

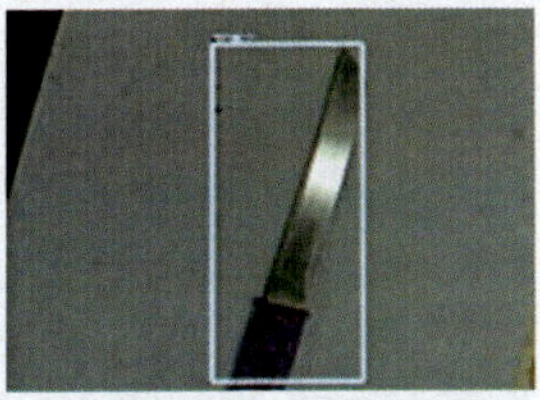

Fig. 13 Object detection of knife

Fig. 14 Object detection of wine glass

Fig. 15 Object detection of phone

Fig. 16 Object detection of mouse and keyboard

In future, speed in frames per second may be improved for desired level of precision and exactness. Table 1 provides details of results and demonstrates the accuracy percentage of each detected entity. If there are two entities in one single frame, it takes one with higher accuracy percentage. [13]

Table 1 Results

	Captured category	Classified category	Evaluation accuracy (%)	True/False
1.	['cell phone']	['cell phone']	95	True
2.	['mouse']	['mouse']	75	True
3.	['person', 'bed']	['bed']	85	True
4.	['table']	['bed']	70	False
5.	['clock', 'clock']	['clock']	88	True
6.	['tie'. 'tie']	['tie']	90	True
7.	['teddy bear']	['teddy bear']	89	True
8.	['suitcase']	['suitcase']	90	True

6 Conclusions

The proposed system will provide assistance to the persons with no or poor eye sight so that they can also live a life without being dependent on other people. This system will provide aid to them through an audio-based assistance in accordance with the article or entity placed in front of them. They capture the images of objects present in the surroundings of the user and then perform the computation to identify the items for which the model has been trained upon. The suggested model is uncomplicated to deploy and is reliable in terms of speed. It made possible for the people with visual impairments to identify the nearby objects. Thus, producing an audio signal is mainly performed by using the person's blind stick or through the use of sensing by hands as touch with no audio assistance. Altogether, the built model can provide a good and acceptable computational way in order to find the solution of a real-time problem with optimum performance. One of the major aims for future would be to produce a more compact and lighter version of the system so as to make it possible for the persons with no or very little eye sight to wear a wristband or put it on their sticks which provide help to them for walking.

7 Acknowledgement

The authors would like to thank CSE and ECE departments, Amity School of Engineering and Technology, Amity University Lucknow, for providing research facilities to carry out the research work.

References

1. Abbas Q, Ibrahim MEA, Arfan Jaffar M (2019) A comprehensive review of recent advances on deep vision systems. Artif Intell Rev 52(1):39–76
2. Brady E et al (2013) Visual challenges in the everyday lives of blind people. In: Proceedings of the SIGCHI conference on human factors in computing systems
3. Huang J, Rathod V, Sun C, Zhu M, Korattikara A, Fathi A, Fischer I, Wojna Z, Song Y, Guadarrama S (2016) Speed/accuracy trade-offs for modern convolutional object detectors. arXiv preprint arXiv:1611.10012
4. Espeak, Available: http://espeak.sourceforge.net/index.html, Accessed:23Jan2020
5. Google Text-to-Speech—Apps on Google Play
6. SSD_mobilenet Available: https://github.com/tensorflow/models/tree/master/research/object_detection/models
7. FasterRCNN_inception. Available:https://github.com/tensorflow/models/tree/master/research/object_detection/models
8. TensorFlow Object Detection. Available: www.tensorflow.org
9. Zhang Y, Peng H, Hu P (2017) A report on towards real-time detection and camera triggering
10. Redmon J, Farhadi A (2016) YOLO9000: better, faster, stronger. arXiv:1612.08242. Available from https://pjreddie.com/darknet/yolov2
11. Vijaya NM, Kiran G (2017) Automatic surveillance using raspberry pi and arduino. IJESRT
12. Kirpan OR, Baviskar PI, Khawase SD, Mankar AS, Ramteke KA (2017) Object detection on raspberry pi. Int J Eng Sci Comput 7(3)
13. Rajalakshmi R, Vishnupriya K, Sathyapriya MS, Vishvaardhini GR (2018) Smart navigation system for the visually impaired using Tensorflow. IJARIIE
14. Nishajith A, Nivedha J, Nair SS, Mohammed Shaffi J (2018) Smart cap—wearable visual guidance system for blind. In: ICIRCA
15. https://www.raspberrypi.org/magpi-issues/Beginners_Guide_v1.pdf
16. https://www.raspberrypi.org/documentation/hardware/camera/
17. Ghoury S, Sungur C, Durdu A (2019) RealTime diseases detection of grape and grape leaves using Faster R-CNN and SSD MobileNet architectures. In: ICATCES
18. Hui J (2018) SSD object detection: single shot multibox detector for real-time processing. Available:https://medium.com/@jonathan_hui/ssd-object-detection-single-shot-multibox-detector-for-real-time-processing-9bd8deac0e06
19. Khamparia A, Singh KM (2019) A systematic survey on deep learning architectures and applications. Exp Syst. https://doi.org/10.1111/exsy.12400
20. Caesar H, Jasper U, Ferrari V (2016) Region-based semantic segmentation with end-to-end training, computer vision–ECCV. Springer International Publishing
21. Juras E EdjeElectronics TensorFlow-Object-Detection-API-Tutorial-Train-Multiple-Objects-Windows-10. Available: https://github.com/EdjeElectronics/TensorFlow-Object-Detection-API-Tutorial-Train-Multiple-Objects-Windows-10
22. Stanford Vision Lab (2016) Stanford University, Princeton University. Available at: www.image-net.org)
23. Juras E (2020) Edje electronics object detection Github with raspberry pi and TensorFlow API. Available at: https://github.com/EdjeElectronics/TensorFlow-Object-Detection-on-the-Raspberry-Pi
24. (2020) Speech recognition in python (text to speech). [Online]. Available at: https://pythonprogramminglanguage.com/text-to-speech/
25. Ren S, He K, Girshick R, Sun J (2016) Faster R-CNN: towards real-time object detection with region proposal networks. arXiv:1506.01497

DDoS Attacks Impact on Data Transfer in IOT-MANET-Based E-Healthcare for Tackling COVID-19

Ashu, Rashima Mahajan, and Sherin Zafar

Abstract The Covid-19, a pandemic situation, effects the economy of the whole world severely and is gaining much huge attention in the field of research currently across the globe. The Internet of things (IoT) technology is playing a great role for taking care of the patients by monitoring and controlling the symptoms and is very much essential for the developing countries, where monitoring of health of huge population has its own challenges. So the IoT and its amalgamation with mobile ad hoc network (MANET) acts as base of networks where devices send information among each other wirelessly thus also named as wireless mesh networks (WMN), in which various nodes are either stationary or allied with static position. Sensors and different other devices involved in e-healthcare sector used in WMN converse wirelessly and hence become the main gate to a numerous susceptibilities. The main aim of this research study is to evaluate the performance of reactive, secured and hybrid routing protocols for throughput as one of the important quality of service (QoS) parameters in absence as well as in presence of distributed denial of service (DDoS). The NS-2(network simulator) is used to simulate AODV (ad hoc on-demand vector, SAODV (secured AODV) and hybrid wireless mesh protocol) in scenario of changing nodes. The comparative analysis concludes the HWMP as most suitable protocol among the other two routing protocols with impact on throughput for handling DDoS attacks. This research study aids in providing implications to enhance existing protocols and alleviate the consequence of DDoS instigated by such attacks.

Keywords AODV · HWMP · Throughput · IoT-MANET ad hoc networks for e-healthcare

---

Ashu (✉)
Manav Rachna International Institute of research and Studies, Faridabad 122017, India
e-mail: ashuone@gmail.com

R. Mahajan · S. Zafar
Department of CSE, SEST, Jamia Hamdard, Delhi, New Delhi 110062, India
e-mail: rashima.mahajan@fet.mriu.edu.in

S. Zafar
e-mail: zafar.sherin@gmail.com

A. Khanna et al. (eds.), *Data Analytics and Management*, Lecture Notes on Data Engineering and Communications Technologies 54,
https://doi.org/10.1007/978-981-15-8335-3_24

1 Introduction

The COVID-19 is the term widely used to explain a new of kind of virus which is a threat to whole human kind and no medicine exist till date for curing this coronavirus as per the record of World Health Organization [1]. However, some symptoms have been shared by physicians for the patients suspected with high fever. Many researches have been done for identification of role of IoT along with AI and ML techniques for handling this pandemic situation created by coronavirus [2]. There has been substantial decrease in the IoT technology implications on consumer devices; however, COVID-19 impacts the usage of IoT technology in health monitoring of patients at their home during the quarantine, in hospital room for watching vital parameters of patients. The other major impact is also on the utilizing the IoT for checking the movement of equipment in hospitals. The usage of drones is other area for IoT which can help us to handle this situation of coronavirus across the globe, mainly in monitoring the areas of outbreaks at public places, as the main step to prevent this is to control the public gatherings as the virus is transferable from one person to other [3]. Also, the mobile application like Arogya Setu recently launched by government for tracing the infected persons. Thus, IoT is having major influence on populated countries like India, where citizens are obeying the quarantine rule as per the directives issued by government, for checking the supply and demands of required things also known as management of supply chain, by tracking and tracing the containments of supply [4]. The modern methods of taking decision via e-governance are also based on mobile based applications connected to cloud environment. The most commonly used devices, for transferring the important data (monitoring related), are wireless based and network formed by them is categorized as wireless mesh network. GAN has been applied to predict Covid-19 [5].

The wireless healthcare system networks are incredibly susceptible to numerous categories of existing attacks because of their intermediate medium, which is universal wireless access and constrained assets. The bandwidth on which mesh networks work is 2.4 GHz, which is an unlicensed; thus, attacker can easily target to avert services from its normal operational efficiency either temporally or indefinitely. One of the most infectious attacks is DDoS type which stops the legitimate users to use network resources [6]. Thirty-three percent of organizations were hit by DDOS assaults in 2017, about twofold the number of assaults in 2016. Eighty-two percent of associations have confronted various DDOS assaults [7]. Since there has been increase in the utilization of mobile applications for monitoring of health conditions of patients, such attacks may harm wireless healthcare application network, which can headed toward damage of patient's life. For successful monitoring, it is very much essential that all the data file reach efficiently. Thus, routing is a very crucial facility for the success of end-to-end announcement. In many research studies carried over years so far, many reactive and hybrid routing protocols have been suggested for networks based on QoS parameters [8]. The QoS parameter helps us to analyze the efficient handling of a particular routing protocol. Therefore, because of this convergence of MANET and WSN, MANET is appropriate to be integrated with IoT for healthcare environment.

1.1 MANET and DDOS Attacks

MANET is one of the broad category of wireless networks in which nodes communicate among themselves and create a network without the presence of any infrastructure or centralized administrative support. Innumerable characteristics of MANET had also raised serious concerns for its security and their related challenges. MANET comprehends a number of imperfections which make its prone to various attacks and thus becomes an easy target for attackers [9]. The irrational activity of nodes in MANET makes changes unpredictable in the topology. In this research work, the SYN flood attack which belongs to the one of the categories of DoS attack focused [10]. SYN flood attack is one class of DoS assaults [11]. The chief concern as regards to flooding attack is that the malicious flooder node floods the entire network. Flooding attacks main goal is to make higher power consumption in context to battery usage and bandwidth resulting in degradation in performance of the network. SYN flood: In this attack, malicious flooder node directs the numeral synchronization packets toward the destination node. Hence, the large amount of memory will be consumed through this attack caused by manipulating the handshake contrivance of a TCP connection depicted in Fig. 1. An intruder or flooder node propels a series of SYN requests to the destination node or device [12]. Therefore, the node or the device beneath attack is incapable to finish the complete three-way protocol of handshake for establishing a reliable server–client connection.

Introduction section of the research paper has focussed on concepts of effects of coronavirus and IoT-based networks and attacks; further in Sect. 2, the layers of e-healthcare have been discussed for transferring the data in healthcare, WMN and MANET. Section 3 depicts broad area that focuses on various protocols used in transmission layer. Section 4 includes the scenario of simulation model used in this research work followed by results based on two quality of service (QoS) parameters. The discussions and conclusion on the research work conducted are listed in Sect. 5.1, followed by future and current work of research and references are listed in the end.

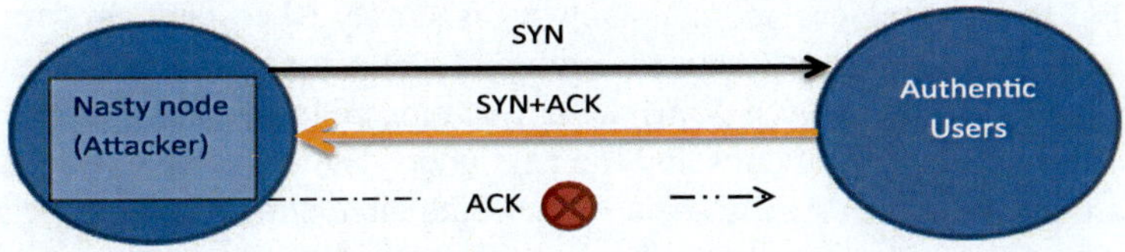

Fig. 1 Schematic depiction of SYN flood

2 E-Healthcare

This section elaborates the different layers used in remote monitoring of services. Healthcare system architecture broadly consists of expert layer which could be the physician of patient or any healthcare professionals, proficient for allocating procedure for treatment and could approve the recommendations from analysis layer. Secondly layer is comprised of application layer and incorporates both local processing analysis and storage of data as well as complex processing. Typically, this layer collects the data and also integrates the same with external data. This cloud layer is very much susceptible to many kinds of attacks, ultimately resulting in DDOS at each layer [13]. Thirdly private clients layer consisting of body sensor networks (BSN) layer, which is responsible for collecting the important vital information of physiological variables and then passing on to cloud gateway via routing decision layer. The routing decision layer comprises of IoT devices and sensor controller unit. The IoT devices have a direct link with the local processing, storage and analysis layer via Wi-Fi,GSM, Ethernet 2G/3G, etc. The smart phone is an IoT device. These sub-layers are capable for processing sensor output and retrieving context, integrating with cloud and providing feedback to and from the patient [14]. The enormous growth of the Internet has led to greater awareness and attention in network.

3 AODV, SAODV and HWMP Routing Protocols

In this section, various routing protocols used in WMN are analyzed. The enormous growth of the Internet has led to greater awareness and attention in network security and routing. Three-level Internet security needs have been recognized: quality of service (QoS), secure network structure and security of the end system [15]. End-to-end security is well suitable to offer confidentiality, authentication and integrity. Protected QoS presents a numeral of novel security challenges: authentication and authorization of operators who require affluent network assets, mutually to prevent resource holdup and to avoid denial of service due to unapproved traffic, etc. An influence on system arrangement can cause denial of service from the user's viewpoint; however, for network engineer, the invader is taking benefit of the absence of authenticity, integrity and perhaps privacy [16].

The most universally used reactive protocol is ad hoc on-demand distance vector (AODV) [17] that produces and maintains paths simply when they are demanded. In comparison to the dynamic source routing, in which a node can cache multiple route for same destination, AODV maintains a routing table and each node will have a information about the previous and the next node. The unique characteristics of AODV are the dynamic establishment of the route table entries, which means that only nodes of the active routes are involved in maintain routing information. It is also emphasized that the routing table entry is expired if not used recently. The timer associated with each entry of the routing table, and this will specify that at which time that entry should be removed from the routing table. Another unique characteristic is the presence of destination sequence number, is often used a time stamp, indicating at what time we have received the path information from the destination node. In case if the information is old, then the destination sequence number can be used to check whether the incoming information is fresh or not.

AODV works in an implicit 'trust your neighbor' mode. There are two main security requirements (a) node authentication and (b) message integrity. Secure AODV (SAODV) adds an extension to different AODV packet formats to incorporate digital signature for protecting the non-mutable information and hash chains to protect the mutable information (hop count). SAODV also has two mechanisms divided as route discovery and secondly maintenance of the route similar to like that of AODV. The major alteration amid them is about the process of route detection process. SAODV increases the process of directly verifying the destination node by using the exchange of random numbers [18].

Hybrid protocols conglomerate features of proactive and reactive protocols for achieving enhanced results. Such routing protocols fix the overhead routing of proactive processes and postpone reactive processes. In this, networks are separated into zones and appropriate for large networks. Hybrid protocols practice reactive routing, for route discovery scenario. However, for table maintenance process, the proactive protocols are widely utilized by hybrid routing protocols. The kind of hybrid routing protocols that are most widely used for discovering routes are: ZRP, hybrid clustering routing (HCR) and ant-based hybrid routing algorithm for mobile and ad hoc networks (ANTHOCNET) [19, 20]. They have the prospective to deliver higher scalability in comparison to pure reactive or proactive protocols. Hybrid procedures can deal with greater scalability in comparison to traditional ones [21].

4 Simulation Results

This research study is implemented on mobility model for evaluating AODV, SAODV, HWMP routing protocols by utilizing the NS-2. The important characteristic of ad hoc networks is the frequent changes in incoming and outgoing nodes for sending the information from source to destination. Simulation environment setup for mobility-based model with varying number of connection on three protocols AODV, SAODV, HWMP MANET is shown above in Table 1.

The simulation results for AODV, SAODV and HWMP are presented by varying no. of nodes for throughput as a QoS parameter. The throughput helps us to know the actual data received to destination node. So time taken by the receiver node to acquire the last information is called throughput. Comparison of throughput for three routing protocols is shown in Table 2. Figure 2 represents the bar graph. Throughput is high both under attack and in absence of attacks in HWMP in comparison to two other routing protocols. Figuress 3 and 4 represent the variation of throughput in absence of attacks and under influence of attack respectively.

Table 1 Simulation environment with changing number of nodes on routing protocols

Parameters	Value used
No. of nodes	30, 60, 90, 120, 150
Area	1800 × 840
Traffic	CBR
Simulation time	60 s
No. of connections	20 4 packets/s
Traffic rate	Traffic rate
Speed	20 m/s
Packet size	1024

Table 2 Assessment of throughput with changing number of nodes for routing protocols

Routing protocols	Throughput for varying no. of connections (30, 60, 90, 120, 150)				
	30	60	90	120	150
AODV without attack	193.92	144.74	183.52	148.03	135.52
AODV with attack	164.83	117.23	152.32	119.9	111.126
SAODV without attack	221.45	160.92	195.6	171.4	159.29
SAODV with attack	179.1	147.414	165.2	151.23	127.54
HWMP without attack	276.36	231.27	255.22	241.73	199.35
HWMP with attack	201.82	183.201	197.15	188.62	173.26

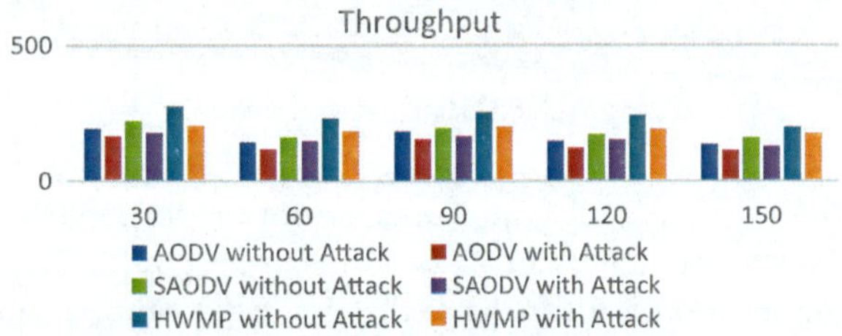

Fig. 2 Comparison of throughput with respect to number of nodes for AODV, SAOD and HWMP routing protocols

5 Discussions and Conclusion

For ensuring privacy and security of patients data during transmission in healthcare sector, MANET could be used as one of the important technologies combined with IoT for handling COVID-19, especially when the data is to be gathered from mobile applications. For dealing with end-to-end data transmission of patients sensitive data, MANET protocol plays a vital role for sending the information securely. The effect of DDoS attacks has been shown and compared. The simulation results of NS-2 for the throughput, as QoS parameter, evaluate that as the number of nodes reaches 150,

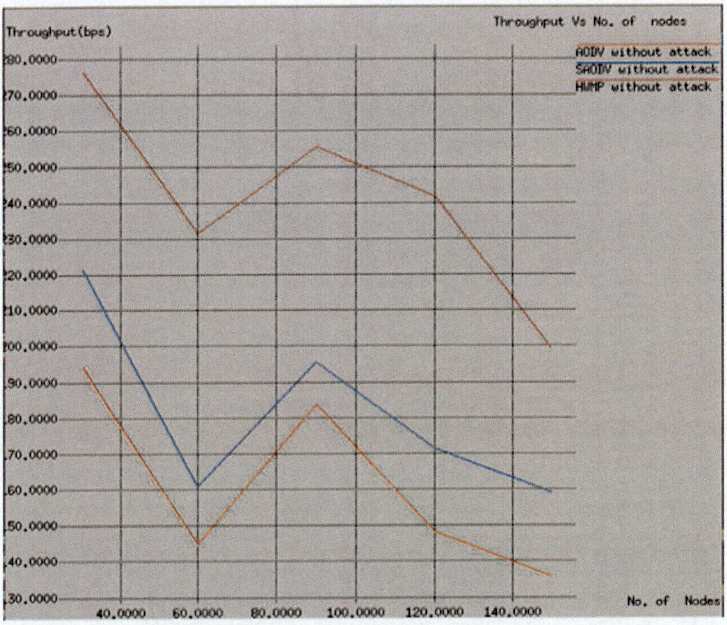

Fig. 3 Throughput in absence of attack for AODV, SAODV, HWMP routing protocols

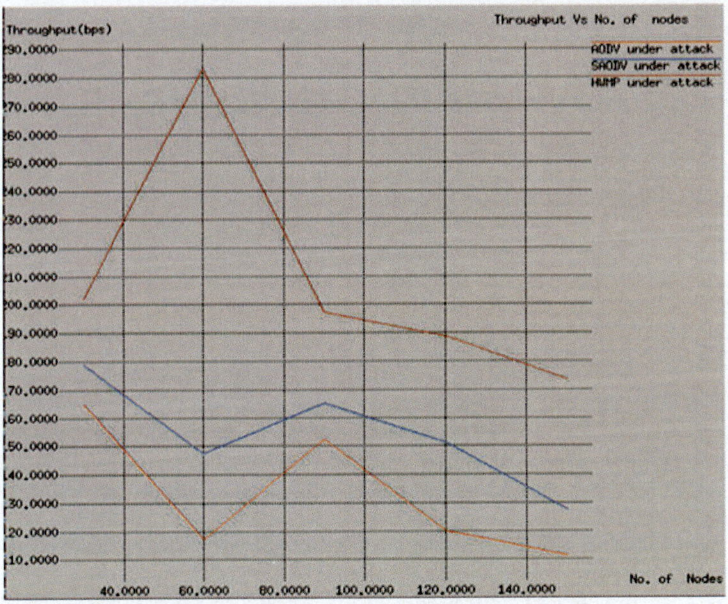

Fig. 4 Throughput under influence of attack for three routing protocols

HWMP outperforms from the other two traditional routing protocols considered in the simulation. In absence of DDoS attack, HWMP is 47.10% and 25.14% better than AODV and SAODV routing protocols, respectively. Under the influence of DDoS attacks, also HWMP is better than 55.91% and 35.84% from AODVA and SAODV, respectively. The result emphasizes that hybrid routing protocol should be preferred in situations where data to be transmitted is very important and any packet loss during transmission is not at all desirable specifically related to healthcare sector. For the successful implementation of IoT-MANET techniques based on mobile applications for tackling coronavirus, it is necessary to make hybrid routing protocol more secure and efficient.

5.1 Current and Future Developments

In future, this research study will be seeing the procedures to the whole thing on enrichment of safety feature of HWMP routing protocols as security being the primary concern in all the healthcare related information exchange of patients. Thus, to enhance security, block chain will be utilized to ensure the legitimated smart devices, for monitoring vital characteristics of patients, which is the key issue in e-healthcare data transfer.

References

1. Peeri NC, Shrestha N, Rahman MS, Zaki R, Tan Z, Bibi S, Haque U (2020) The SARS, MERS and novel coronavirus (COVID-19) epidemics, the newest and biggest global health threats: what lessons have we learned? Int J Epidemiol
2. Mohammed MN, Syamsudin H, Al-Zubaidi S, AKS, Ramli R, Yusuf E (2020) Novel Covid-19 detection and diagnosis system using IoT based smart helmet. Int J Psychosoc Rehabil 24(7)
3. Tun SYY, Madanian S, Mirza F (2020) Internet of things (IoT) applications for elderly care: a reflective review. Aging Clin Exp Res. https://doi.org/10.1007/s40520-020-01545-9
4. Garg S, Bhatnagar N, Gangadharan N (2020) A case for participatory disease surveillance of the COVID-19 pandemic in India. JMIR Public Health Surveill 6(2):e18795
5. Waheed A, Goyal M, Gupta D, Khanna A, Al-Turjman F, Pinheiro PR CovidGAN: data augmentation using auxiliary classifier GAN for improved Covid-19 detection. IEEE Access. https://doi.org/10.1109/access.2020.2994762
6. Musiat P, Yang Y, Maeder A, Bidargaddi N (2020) A digital infrastructure for storing & sharing Internet of Things, wearables and app-based research study data. Stud Health Technol Inf 268:87–96
7. Majumder M, Mandl KD (2020) Early transmissibility assessment of a novel coronavirus in Wuhan, China. China
8. Available from https://www.msspalert.com/cybersecurity-research/kaspersky-lab-study-average-cost-of-enterprise-ddos-attack-totals-2m/
9. Islam SMR, Daehan Kwak MD, Kabir H, Hossain M, Kwak K-S (2015) The internet of things for health care: a comprehensive survey. IEEE Access 3:678–708

10. Alvarez J, Stephane M, Zaïdi F (2017) Monitoring dynamic mobile ad-hoc networks: a fully distributed hybrid architecture. In: 31st international conference on advanced information networking and applications (AINA), 2017 IEEE, pp 407–414
11. Degirmencioglu A, Erdogan HT, Mizani MA, Yılmaz O (2016) A classification approach for adaptive mitigation of SYN flood attacks: preventing performance loss due to SYN flood attacks. In: NOMS 2016-IEEE/IFIP network operations and management symposium, pp 1109–1112
12. Džaferović E, Sokol A, Almisreb AA, Norzeli SM (2019) DoS and DDoS vulnerability of IoT: a review. Sustain Eng Innov 1(1):43–48. ISSN 2712–0562
13. Islam SMR, Kwak D, Kabir MDH, Hossain M, Kwak K-S The internet of things for health care: a comprehensive survey. IEEE Access 3:678–708
14. Oh S, Cha J, Ji M, Kang H, Kim S, Heo E, Han JS et al (2015) Architecture design of healthcare software-as-a-service platform for cloud-based clinical decision support service. Healthcare Inf Res 21(2):102–110
15. Latif R, Abbas H, Assar S (2014) Distributed denial of service (DDoS) attack in cloud-assisted wireless body area networks: a systematic literature review. J Med Syst 38(11):128
16. Govindasamy J, Samundiswary P (2017) A comparative study of reactive, proactive and hybrid routing protocol in wireless sensor network under wormhole attack. J Electr Syst Inf Technol
17. Prabha R, Krishnaveni M, Manjula SH, Venugopal KR, Patnaik LM (2015) QoS aware trust metric based framework for wireless sensor networks. Procedia Comput Sci 1(48):373–380
18. Sharma AK, Trivedi MC (2016) Performance comparison of AODV, ZRP and AODVDR routing protocols in MANET. In: 2016 second international conference on computational intelligence & communication technology (CICT), IEEE, pp 231–236
19. Sengupta A, Sengupta D, Das A (2017) Designing an enhanced ZRP algorithm for MANET and simulation using OPNET. In: 2017 third international conference on research in computational intelligence and communication networks (ICRCICN), IEEE, pp 153–156
20. Walikar GA, Biradar RC (2017) A survey on hybrid routing mechanisms in mobile ad hoc networks. J Netw Comput Appl 1(77):48–63
21. Brill C, Nash T (2017) A comparative analysis of MANET routing protocols through simulation. In: 12th international conference for internet technology and secured transactions (ICITST), IEEE, pp 244–247

BER Performance Analysis of MMSE with ZF and ML Symbol Detection for Hard Decision MU-MIMO LTE on Rayleigh Fading Channel

Jyoti, Vikas Nandal, and Deepak Nandal

Abstract Before moving towards the 5G systems, we have to improve the symbol estimation techniques in the present 4G network to intensify the system performance. There are many symbol estimation techniques for MIMO LTE-A systems; among these techniques and after many researches, it is found that zero-forcing, maximal ratio combining and the minimum mean square error are mostly used. We have focused on the analysis of these techniques. In the MU-MIMO system, we tried to double the data rate and minimize the bit error rate (BER) using MMSE estimation using channel interpolation in the frequency domain. Here, we compared the performance on an AWGN and Rayleigh channel of channel estimation techniques using "ZF", "ML" and "MMSE" on a 2 × 2 MIMO LTE system with BPSK, QPSK and 16 QAM using the VITERBI hard decision method of analysis. Then we modified the system with a 4 × 4 MU-MIMO LTE system and calculated the BER and analysed. Simulation results showed that among these techniques, "ML" is a salient features in characterizing the performance of the data channel and the LS and MSME behaved very similar to each other on a 2 × 2 MIMO system. Further, simulating the results we have showed that the MMSE MIMO detector slightly outperforms as compared to least square (LS) on SINR.

Keywords MIMO-OFDM · MMSE · ML · ZF · BER · AWGN · Rayleigh channel

Jyoti (✉) · V. Nandal
Department of Electronics & Communication Engineering, Maharshi Dayanand University, Rohtak, Haryana, India
e-mail: jyotidalal779@gmail.com

V. Nandal
e-mail: nandalvikas@gmail.com

D. Nandal
Department of Computer Science, Guru Jambheswar University, Hisar, Haryana, India
e-mail: gju.dpknandal@gmail.com

A. Khanna et al. (eds.), *Data Analytics and Management*, Lecture Notes on Data Engineering and Communications Technologies 54,
https://doi.org/10.1007/978-981-15-8335-3_25

1 Introduction

In the world of digitalization, the number of users is increasing day by day and we have limited bandwidth in addition to fading. Different types of fading channels are considered to reduce these fading effects, and among them a Rayleigh and RICIAN fading channel are used in main to broadcast the info over the air in a cellular network. Rayleigh fading [1] is better for the non-line of sight of communication. To reduce these fading effects and to increase received signal power, we use various diversity techniques like receive diversity (where a receiver uses two antennas) and transmit diversity (where a transmitter uses two and more than two antennas) which is further divided into two types:

- Closed-loop transmitter diversity
- Open-loop transmitter diversity.

In a closed-loop, we use the pre-coding matrix indicator (PMI) which determines the phase shift and generate PMI feedback to the transmitter. In open-loop transmit diversity, the transmission energy is enhanced and provides high reliability [2] in which a major consideration is ALAMOUTI codes where the BER is reduced, without reducing the data. In "ALAMOUTI'S technique", two symbols are transmitted in two-successive time-steps, but these diversity techniques have some limitations in transmitting symbols. Hence, to overcome these limitations we use another technique known as multiple-input, multiple-output (MIMO) where the multiple transmitter and receiver is used at both ends to increase data rates. In MIMO-OFDM, pre-coding is done at the transmitter and post-coding is done at the receiver. According to [3], pre-coding technique is used to predict the channel condition at the transmitter side using the beam-forming techniques to achieve transmission diversity. Pre-coding techniques eliminate the fading effects and improve the efficiency in terms of performance of the system. Nonlinear decoding techniques [4] are versatile than a linear decoding technique due to its BER performance for example: successive interference cancellation (SIC).

Linear techniques [5] are preferred in comparison with nonlinear because of the intricacy of nonlinear techniques which are performed using post-coding at the receiver. In linear equalization decoding techniques, all other signal is removed by proceeds as an interference signal while the original signal is contemplated. The wireless power antenna has distinctive frequencies that can clarify the research economical, and analysis is associated based on RF frequencies in terms of MIMO [6]. Two types of MIMO technology are used for diversity processing,

- SU-MIMO (Single User MIMO)
- MU-MIMO (Multi-User MIMO).

- In **SU-MIMO**, as the name indicates a single user transmits bits to the single receiver but using multiple transmitters and multiple receivers for a single user using multiple paths. On these different paths, the data is sent via different routes and all possible routes are known as channel matrix. So, the channel matrix

is used for the estimation of symbols using different channel estimation techniques. Maximal ratio combining (MRC) and the least square error (LS) are the two methods that we are used for the diversity combining. In the maximal ratio combining, the signal from each channel is added in which the gain is symmetrical to the RMS value and oppositely symmetrical to the squaring of average of noise in a particular channel, while the least square concludes the minimal value to the best fit curve.
- In **MU-MIMO** for uplink transmission, we consider multiple users that want to access data to a single user which has the same transmission time and the same carrier frequency without using any pre-coding technique. Here, we use the minimum mean square error (MMSE) detection technique to disparate the transmission. This technique is suitable for free users who are far apart from each other. In the case of downlink transmission we use a beam-forming technique, where a single transmitter transmits the information to various receivers.

In this paper, Sect. 1 is the introduction and literature review in Sect. 2. In Sect. 3, the system model of different techniques and different types of algorithms is used to detect the bit error rate among these techniques using symbol estimation. Section 4 shows the simulation results obtained by these parameters with the theory of techniques; in Sect. 5, the results are obtained using MATLAB; and in Sect. 6, the conclusion is derived.

2 Literature Review

According to Minango et al. [7], as in MMSE detection an inversion of the matrix is used so, the author used the Gauss–Seidel method for massive MIMO in which with the help of this method rank of matrix is reduced to achieve top-notch BER performance.

Feng et al. [8], to reduce the iterations in channel estimation, we have to deal with the problem of constellation distance for which a hard output detector is recommended to increase the reliability in comparison with soft techniques used for the Internet of things (IoT).

Ji et al. [9] used quantum version for massive "MIMO" using "MMSE" detection for uplink where he proposed phase estimation, Hamiltonian simulation, etc., to reduce the "BER" for uplink detection so, that to make the system compatible to all users.

According to Vasavada et al. [10], for "SIMO" diversity using "OFDM" for various channel estimation schemes used to support symbol estimation. Their errors are estimated for the amplitude of signal and root mean square (RMS) error for the channel matrix is performed.

Yunida et al. [11] proposed a joint deign method using half-duplex mode for non-orthogonal “AF-MIMO” relaying system using the “MMSE” detection technique in which upper bound is used for the pre-coding matrix where singular value decomposition (SVD) technique is used.

Datta et al. [12] used bee colony optimization algorithm on mutation-based using maximum likelihood (ML) technique for detection in “GSM-MIMO” where for both, perfect and imperfect channel “MMSE” achieves near ML performance. Bit error rate performance for the GSM-MIMO system is also equal for the CSI error scenarios.

Said et al. [13] suggested using SM-MIMO in comparison with MU-MIMO for the user link transmission using RF chains SM transmission schemes and group subspace pursuit (GSP) is proposed for signal detection.

3 System Model

In SU-MIMO, as shown in the system model, multiple signals are transmitted simultaneously; as a result, the transmission rate is increased. Firstly, using the pre-coding technique with the help of antenna mapping one bit is dispatched for each antenna which is obtained by modulation.

At the receiver end, the first chore is to retrieve the channel element which is done with the help of Eq. (1), and the bits are estimated using different channel estimation techniques.

$$Y(I) = \sum_{I=1}^{n} K(S, I)A(I) + N(S). \tag{1}$$

Here, for a MIMO system in MMSE $K(S, I)$ is the independent and identically distributed (IID) complex random variable where “I” is the number of input variable and “S” is the received variable, “n” is the number of the transmitter and $N(S)$ is I.I.D noise samples (Fig. 1).

Here, a 2×2 MIMO system is considered [14] (number of transmitter and receiver both are two on each side) in which the number of paths for the radio transmission is four, i.e. H_{11}, H_{12}, H_{21} and H_{22}.

So, channel matrix for a 2×2 MIMO system is:

$$H = \begin{bmatrix} H_{11} & H_{12} \\ H_{21} & H_{22} \end{bmatrix} \tag{2}$$

Hence, Eq. (3) using the value of “H” (channel matrix) for a 2×2 MIMO system can be written as:

$$\begin{bmatrix} Y_1 \\ Y_2 \end{bmatrix} = \begin{bmatrix} H_{11} & H_{12} \\ H_{21} & H_{22} \end{bmatrix} \begin{bmatrix} X_1 \\ X_2 \end{bmatrix} + \begin{bmatrix} N_1 \\ N_2 \end{bmatrix} \tag{3}$$

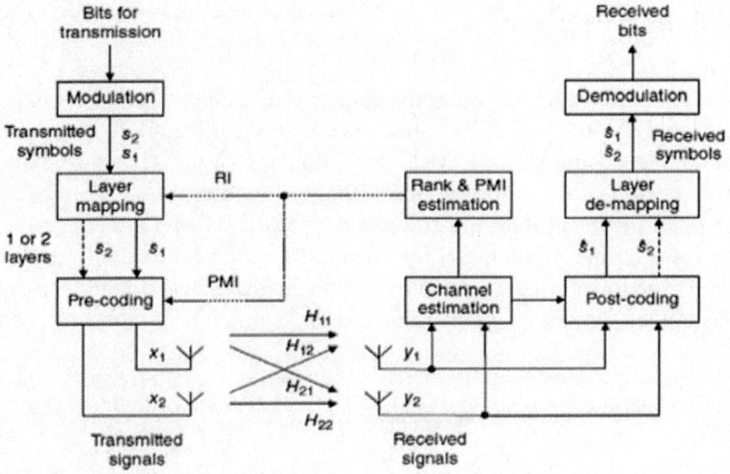

Fig. 1 A "2 × 2" MIMO system model

After the indication of rank according to bits obtained (rank of channel matrix), the bits are transferred for the symbol estimation where post-coding techniques are used to find the bit error rate (BER). By increasing the transmitters and receivers, the channel capacity is also enhanced, but the calculations for channel matrix were getting difficult to calculate the channel estimation.

However, in a 4 × 4 MU-MIMO system as the number of input and output are four on each side the number of elements of the channel matrix increased from four to sixteen. Hence, the complexity of the system is also increased along with the increment in the capacity of the system.

So, we used the VITERBI hard decision method to decode the bitstream where HAMMING distance (where two strings with the different number of symbols of equal length) is used as a matrix, which helps different error detection techniques to reduce the calculations and also increment in the number of iterations for rank indication. After error detection and correction layer, de-mapping is used where the received symbol vector (Z) can be written as:

$$Z = M.H.A.s + M.N \tag{4}$$

where in Eq. (4), "A" is the pre-coding matrix and "M" is the post-coding matrix. "s" is the transmitted symbol at the input and "H" is the channel matrix with noise "N". The received symbol vector is then demodulated and transmitted data is obtained as a result.

4 Simulation Analysis

The simulation parameters that are considered for the different techniques are shown in "Tables 1 and 2". In "Table 1", parameters are used for comparison with the basic theoretical models using a 2 × 2 MIMO system, and in the "Table 2", parameters are for comparison of these techniques with one another to find the most suitable technique to minimize a bit error rate using 4 × 4 MU-MIMO system.

As, we consider the Rayleigh fading channel (a statistical model for the non-line of sight of communication for circulation in the environment) to find the simulation parameters. When we send data over this channel some bits are lost in transmission

Table 1 Basic simulation parameters

Simulation parameter	Type of range
Fading channel	Rayleigh
Modulation techniques	BPSK, QPSK
Noise in the channel	AWGN
Detection algorithm	MMSE, ZF and ML
Number of transmitting and receiving antennas	2
E_b/N_o (dB)	Varied up to 0–25
Base station height	26 m
The distance of inter sites	1100 M
PL exponent	3.7

Table 2 4 × 4 MU-MIMO system simulation parameters

Simulation parameter	Type of range
Fading channel	Rayleigh
Modulation technique's	QAM
Noise in the channel	AWGN
Detection algorithm	MMSE and ZF
Number of transmitting and receiving antennas	4
E_b/N_o (dB)	Varied up to 0–30
Base station height	26 m
The distance of inter sites	1100 M
PL exponent	3.7
Symbol time	10 μs
Cyclic prefix length	1.5 μs
OFDM subcarrier spacing	80 kHz
Channel bandwidth	20 MHz
Length of OFDM symbols	32

due to interference, noise and bit synchronization, which is known as "BER" (bit error rate). So, we use different modulation techniques (BPSK, QPSK, QAM) to remove BER and to estimate the transmitted signals for which we use different error detection techniques, i.e. zero-forcing (ZF), minimum mean square error (MMSE) and maximum likelihood (ML) detection. The zero-forcing (ZF) technique is the basic technique among all these techniques. In this technique, we ignore the noise and interference in the received signal and tried to achieve original data. Hence, the equation for the ZF equalizer becomes $Y = HX$ when noise and interference are neglected. So, at the receiver end the transmitted data is retrieved for a 2×2 MIMO system by using equation, i.e.

$$X_1 = \frac{H_{22}Y_1 - H_{12}Y_2}{H_{11}H_{22} - H_{21}H_{12}} \tag{5}$$

$$X_2 = \frac{H_{11}Y_1 - H_{21}Y_2}{H_{11}H_{22} - H_{21}H_{12}} \tag{6}$$

But there is a limitation along with this technique's; i.e. when we obtain the same transmitted data at both receivers end, the resultant of this technique goes to "zero" or some small values less than the actual data that results dreadfully and makes the device useless. The BER of a 2×2 MIMO LTE is shown in Fig. 2 where the number of trans-receiver both are two and for BPSK modulation in ZF equalizer (where the noise is neglected) techniques using the Rayleigh fading channel. So, to overcome the limitations of these techniques we use another technique named maximum likelihood (ML) detection.

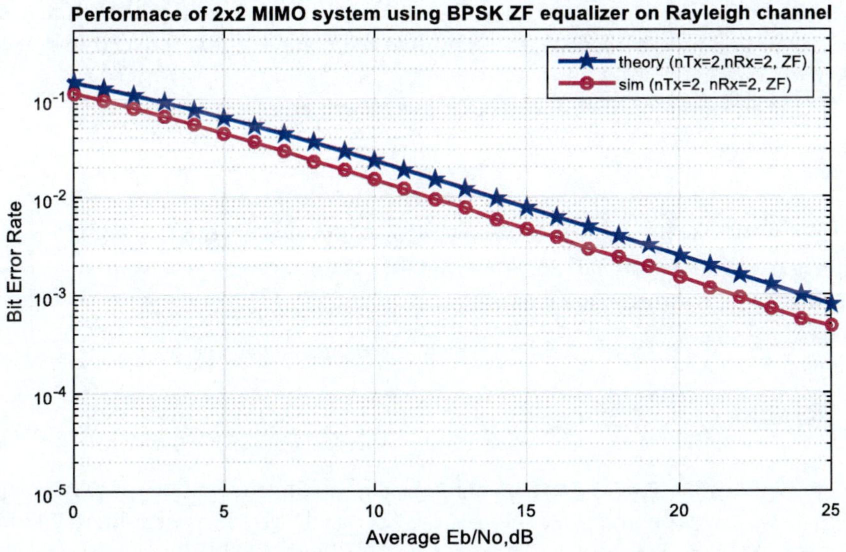

Fig. 2 A comparison of the ZF technique with the theory

In maximum likelihood (ML) detection [12], we estimated the value of the signal nearer to the average value that should be maximum symmetrical about the mean values in a hypothesis and keep shifting the value to find the strongest signal so that the maximum data can be achieved at a particular time.

In maximum likelihood (ML) detection, the number of the active set of the transmitted antenna is denoted by "J" as shown in Eq. (7).

$$J = \left| Y - H\hat{X} \right|^2. \tag{7}$$

Here, the values are the same that are provided by the estimation of different channels. So, Eq. (8) can be written as:

$$J = \left\| \begin{bmatrix} Y_1 \\ Y_2 \end{bmatrix} - \begin{bmatrix} H_{11} & H_{12} \\ H_{21} & H_{22} \end{bmatrix} \begin{bmatrix} \hat{X}_1 \\ \hat{X}_2 \end{bmatrix} \right\|^2 \tag{8}$$

Now, we compared the results of ML with actual theory, and results are obtained in Fig. 2 where we compare BER using BPSK modulation for a Rayleigh fading. But, when the maximum number of observations and the size of samples are below the limit of detection this method abate in retrieving the signals.

So, in this situation to conquer from this we use another technique named "MMSE" (minimum mean square error) in which channel matrix is invert, and then the channel is retrieved with the help of channel estimation. In "MMSE" detection, different types of algorithms/detectors are used to calculate inputs sent by the transmitter some of them are the Richardson method, the Jacobin method, the Gauss–Seidel method any many more. In "MMSE", the conversion/inverse of a matrix has huge complexity to calculate the input bits. In [7], the Gauss–Seidel method reduces the number of iterations yield by the channel matrix so that the complexity is reduced to makes calculation easier.

So, the equation for MMSE detection can be written as:

$$Z = [H^{-1}]Y \tag{9}$$

Here, "H^{-1}" is the oppositely symmetrical of the channel matrix and "Z" is the estimated transmitted signal. So, as a result we get the more precise result in MMSE as shown in Fig. 3 where the ZF technique is unable to find these parameters (Fig. 4).

5 Results

Firstly, as shown in Figs. 5 and 8 a comparison of all three techniques is performed. Here, as we can see both the techniques, i.e. "ZF" and "MMSE" performed superior than the "ML" detection techniques, but in "ZF" and "MMSE" technique results are vague due to overlapping. So, we are unable to calculate the results distinctly.

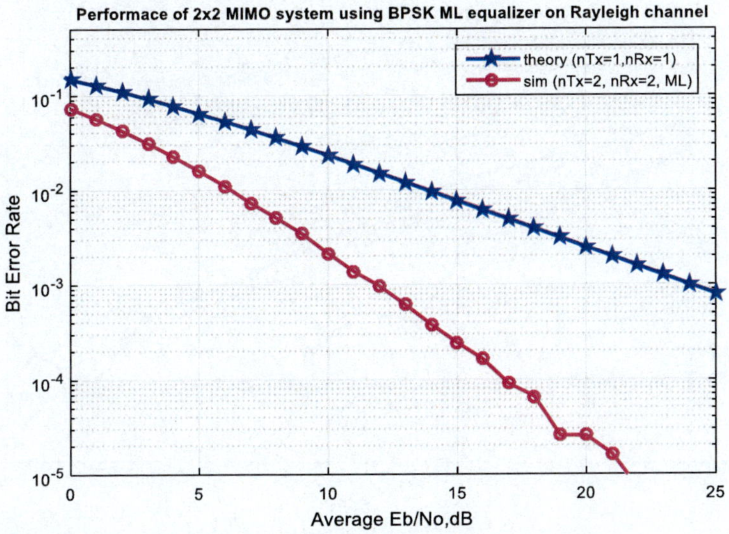

Fig. 3 A comparison of ML detection technique with theoretical parameters

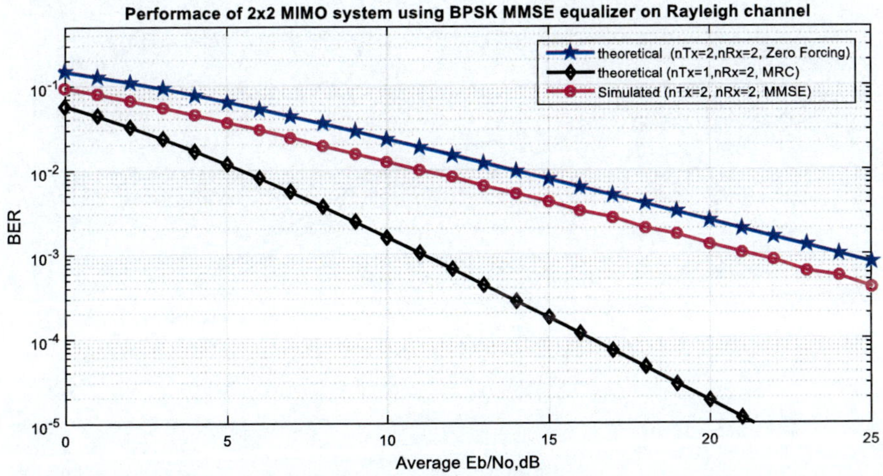

Fig. 4 MMSE equalizer results with MMSE theoretical results

Hence, we compare these two techniques separately using the 16-QAM modulation technique on the advance 4×4 MU-MIMO channel as shown in Figs. 6 and 7, where we obtained results in the commendation of MMSE.

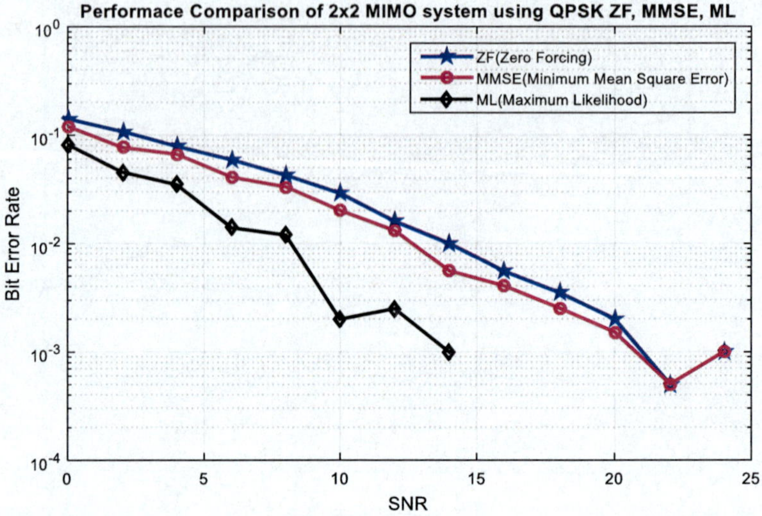

Fig. 5 Comparison of a 2 × 2 MU-MIMO system for "ZF", "MMSE" and "ML" using QPSK modulation

Fig. 6 A comparison between ZF and MMSE using a 16-QAM modulation for 4 × 4 MU-MIMO system

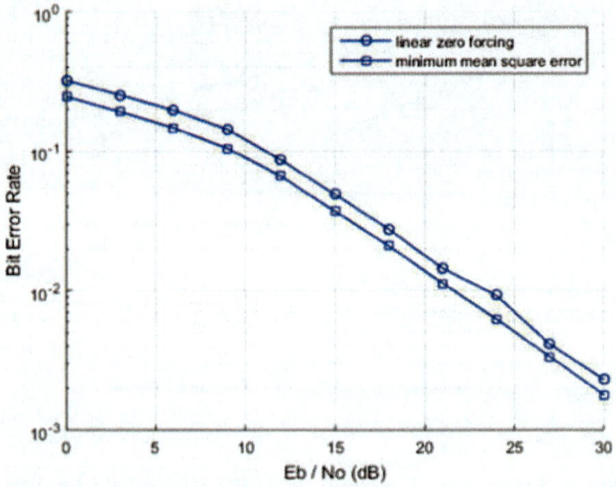

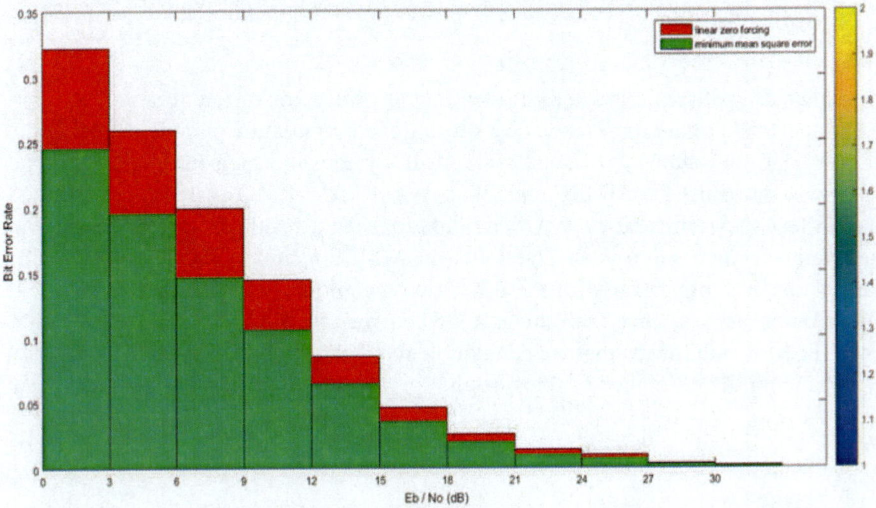

Fig. 7 Performance comparison between ZF and MMSE using a 16-QAM modulation for 4 × 4 MU-MIMO system

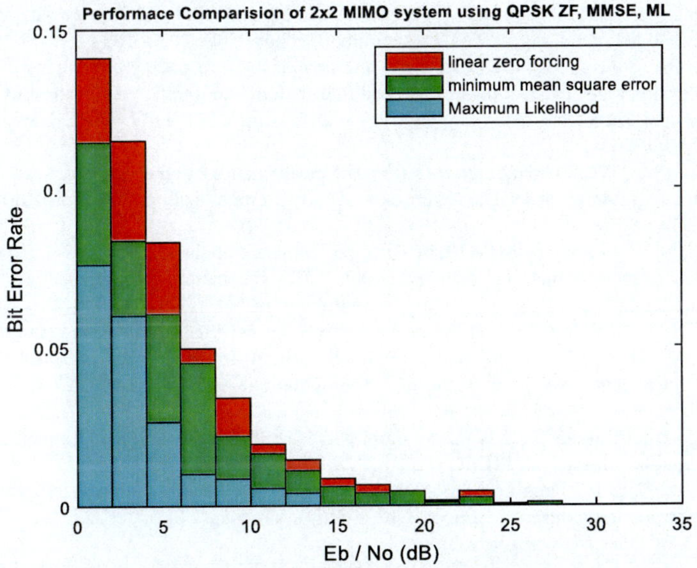

Fig. 8 Performance comparison among ZF, MMSE and ML using QPSK modulation

6 Conclusion

In this work, we have extracted the theoretical results for different error detection techniques and collate their results by varying their simulation parameters using the Rayleigh fading channel for a 2×2 MU-MIMO system, in which due to the imbricate of signals in results for MMSE and ZF, a 4×4 MU-MIMO system is considered to manifest the results using QAM modulation. As a result, from the simulation, theory and comparison it is analysed that the MMSE technique is more optimal for the system in comparison with the other two techniques, i.e. ML and ZF. MMSE shows better performance than linear ZF when we use 16-QAM modulation rather than BPSK, which means that we can double the data rate using the same bandwidth by enhancing the efficiency of the system.

References

1. Eduru S, Rangaswamy N, BER analysis of massive MIMO system under correlated Rayleigh fading channel. In: IEEE-43488, 9th ICCNT 2018 IISC Bangalore, India
2. Rehman Z, Savitha HM (2014) BER performance comparison of MIMO systems using OSTBC with ZF and ML decoding. ICTACT J Commun Technol 5(4), 1030–1034
3. Sharma PK, Vijaysai T (2018) Bit error rate analysis of pre-coding techniques in different MIMO systems. In: Proceedings of the 2nd international conference on electronics, communication and aerospace technology (ICECA 2018), pp 1144–1147. IEEE Explore ISBN: 978-1-5386-0965-1
4. Berceanu MG, Voicu C, Halunga S (2019) The performance of an uplink large scale MIMO–SIC detector. In: International conference on military communications and information systems (ICMCIS), pp 1–4
5. Chaudhary SR, Thombre MP (2014) BER performance analysis of MIMO–OFDM system using different equalization techniques. In: IEEE international conference on advanced communication control and computing technologies (ICACCCT), pp 637–677
6. Grover H, PathaniaR (2018) Performance analysis of MMSE equalizer and time-frequency block code for MIMO-OFDM under single-RF system. In: Proceeding of 2nd international conference on inventive communication and computational technologies (ICICCCT), pp 1277–1280
7. Minango J, de Almeida C (2017) Low complexity MMSE detector based on Refinement Gauss-Seidel method for massive MIMO system. 978-1-5386-2098/17 ©2017 IEEE
8. Feng J, Hu D, Li P, Zhang J (2018) An enhanced MMSE detection with constellation distance correction for the Internet of things with multiple antennas. Springer Science + Business Media, LLC, part of Springer Nature
9. Ji Y, Meng F, Jin J, Yu X, Zhang Z, YouX, Zhang C (2020) Quantum version of MMSE –based massive MIMO uplink detection. © Springer Science + Business Media, LLC, part of Springer Nature 2020
10. Vasavada YM, Beex AA, Reed JH (2018) Iterative channel and symbol estimation for OFDM and SIMO diversity. 978-1-5386-3821-7/18, 2018 IEEE
11. Yunida Y, Nasaruddin N, Muharar R, Away Y (2018) Linear precoder design for non-orthogonal AF MIMO relaying systems based on MMSE criterion. EURASIP J Wirel Commun Netw 2018:285

12. Datta A, Mandloi M, Bhatia V (2019) Mutation based bee colony optimization algorithm for near-ML detection in GSM-MIMO. In: Rawat BS et al (eds) Advances in signal processing and communication, Lecture Notes in Electrical Engineering 526, ©Springer Nature Singapore Pte Ltd. 2019. https://doi.org/10.1007/978-981-13-2553-3_13125
13. Said S, Saad W, Shokiar M, El-Araby S (2020) BER performance of signal detection for massive multi-user spatial modulation systems. Int J Wirel Inf Netw © Springer Science + Business Media, LLC, part of Springer Nature 2020
14. Christopher COX (2014) Multiple antenna techniques. An introduction to LTE: LTE, LTE-advanced SAE, VOLTE and 4G mobile communications, 2nd ed. Willey, pp 87–104. https://doi.org/10.1002/9781118818046.ch.5

Traffic Congestion Analysis and Occupancy Parameter in India

Tsutomu Tsuboi

Abstract This study focuses on the traffic congestion analysis in the developing country, especially in a major city in India and introduces the traffic congestion analysis method by using occupancy which is one of traffic flow parameter. In general, it is hard to make the traffic flow analysis in the developing countries because their road infrastructure is not able to catch up with their demand of rapid growing transportation condition. Therefore, their traffic congestion becomes heavier and causes the serious negative impact such as traffic accidents, air pollution, unnecessary fuel consumption, health problem, and so on. The uniqueness of this study is the following three items. At first, this study is based on one month measured traffic data in the city and shows its traffic flow characteristics comparing with the traffic flow theory. In the second, it introduces combination traffic congestion analysis method by calculating the traffic occupancy from the compared traffic flow characteristics. In the third, it shows the validation of the new traffic congestion analysis method by comparing the occupancy measurement data.

Keywords Traffic flow · Traffic congestion · Traffic fundamental characteristics

1 Introduction

This study has started since October 2014, when the Japanese Government founded program allowed to install the traffic monitoring cameras and the electrical traffic sign board "Variable Message Sign" or VMS, which displays the current traffic condition. The installation location is one of major city in India, which is Ahmedabad city of Gujarat state in India—where its population is about 6 million; and its economics are rapidly growing. It was the first traffic management system in the city for showing its traffic condition through the display based on the traffic monitoring camera data. In this system, there are three-level traffic condition for showing, i.e., smooth traffic,

T. Tsuboi (✉)
Nagoya Electric Works Co. Ltd, Ama, Japan
e-mail: t_tsuboi@nagoya-denki.co.jp

A. Khanna et al. (eds.), *Data Analytics and Management*, Lecture Notes on Data Engineering and Communications Technologies 54,
https://doi.org/10.1007/978-981-15-8335-3_26

Fig. 1 Traffic monitoring camera location in Ahmedabad city

little crowed, and heavy traffic and shows the traffic condition neighboring roads for considering detour route choice.

The traffic management system consists of 10 traffic monitoring cameras and 4 VMS with traffic monitoring cameras, and the total number of camera is 14. Figure 1 shows the camera installation place in Ahmedabad west side of city. Author took 11 camera data through camera number 1–10 and VMS number 1. In Fig. 1, Cam#1 means CCTV number 1, and the VMS#1 means VMS number 1.

2 Traffic Measurement and Theory

2.1 *Traffic Flow Measurement*

In terms of the measurement data, Fig. 2 shows the daily base traffic flow and traffic speed trend from 7:00 am to 6:00 am of next day at Camera#1 and #2, for example. As shown in Fig. 2, each traffic condition is different. In case of Camera#1, there are two peak traffic volumes in the morning and the afternoon. The average vehicle speed of each time frame does not so much. On the other hand, the average vehicle speed of Camera#2 drops, especially in the evening. Among all other location, the average vehicle speed reduction of Camera#2 is most significant.

From Fig. 2, the average speed is one of key parameter for judging the traffic congestion condition. In general, the traffic volume is easy to consider their traffic condition in general. It is important to make a judgment of the traffic congestion to use both parameters the traffic volume and its average vehicle speed. In the next

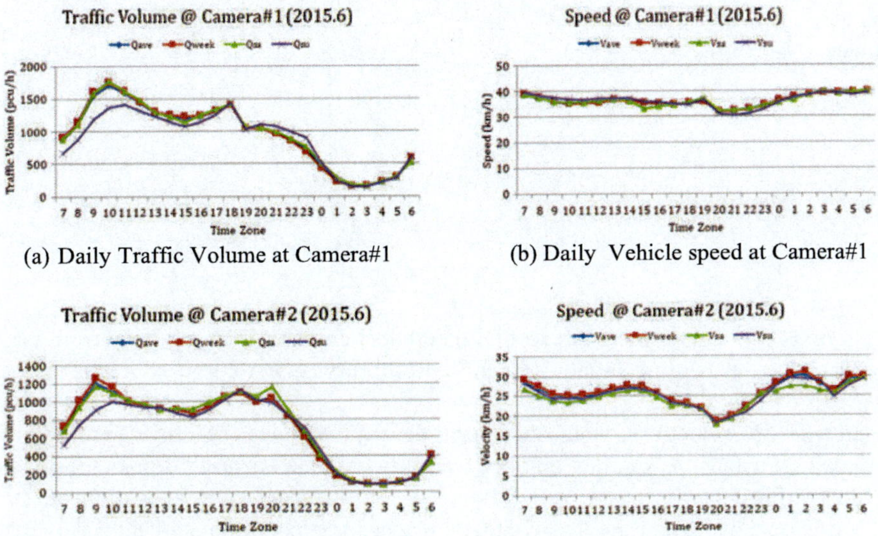

(a) Daily Traffic Volume at Camera#1 (b) Daily Vehicle speed at Camera#1

(b) Daily Traffic Volume at Camera#2 (d) Daily Vehicle speed at Camera#2

Fig. 2 Daily traffic volume and average speed trend

section, it explains how the traffic congestion is defined form the traffic theory and how difficult making the judgment from measurement data.

2.2 *Traffic Congestion from Theory and Measurement*

In the traffic flow theory, there are three fundamental characteristics. The first one is the traffic density (k) to the vehicle speed (v)—k–v curve. And the next one is the traffic density (k) to the traffic volume (q)—k–q curve. The last one is the traffic volume (q) to vehicle speed (v)—q–v curve. Figure 3 illustrates k–v curve and k–

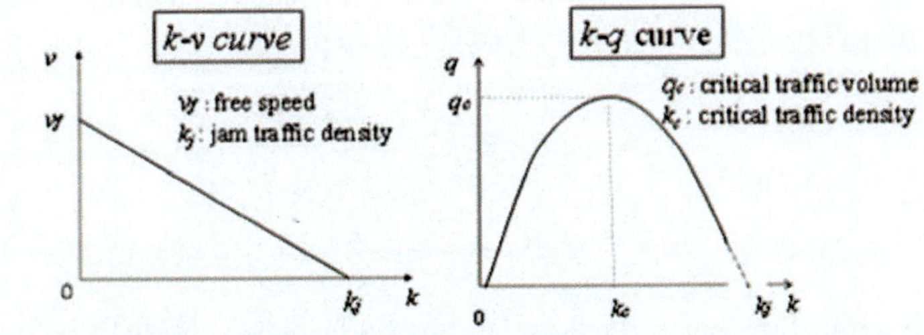

(a) Traffic Density to Vehicle speed curve (b) Traffic density to Traffic Volume curve

Fig. 3 Traffic flow fundamental characteristics (k–v curve and k–q curve)

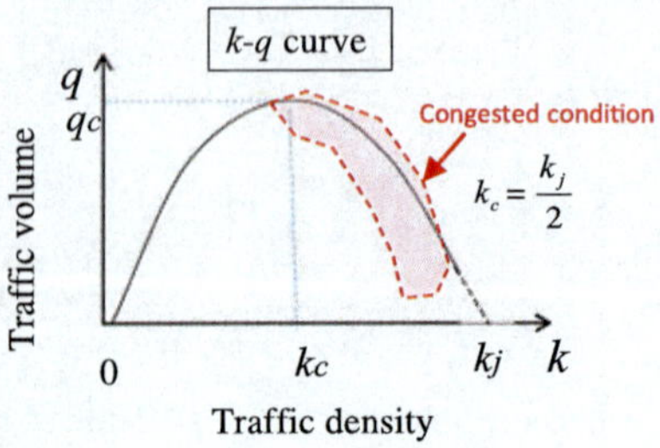

Fig. 4 Traffic congestion in theoretical k–v curve

q curve. In k–v curve, there are several theoretical curves which all come from the experimental law. Figure 3a shows the most useful curve of k–v curve of Greenshields [1].

In Fig. 3b k–q curve, when the traffic density becomes the critical condition (k_c), which means the running point of the traffic volume becomes drops. After the critical traffic density (k_c), the traffic volume becomes lower according to more dense of traffic density. When the traffic density reaches jam condition so-called the jam density (k_j), its traffic volume becomes theoretical zero. Therefore, it is able to say that the traffic congestion condition starts after the critical traffic density (k_c), which is illustrated in Fig. 4.

The traffic flow equation from Greenshields is defined in Eq. (1).

$$v = v_f\left(1 - \frac{k}{k_j}\right) \tag{1}$$

From the traffic flow continuity, the traffic volume (q) is defined in Eq. (2).

$$q = k \times v \tag{2}$$

Equation (3) is obtained by inserting Eq. (2) to Eq. (1).

$$q = -\frac{v_f}{k_j}\left(k - \frac{k_j}{2}\right)^2 + \frac{v_f k_j}{4} \tag{3}$$

From Fig. 4, Eqs. (5) and (6) are obtained.

$$q_c = \frac{v_f k_j}{4} \tag{4}$$

$$k_j = 2k_c \tag{5}$$

In case of collecting real traffic flow data, the traffic density (k) to the vehicle speed (v) and the traffic density (k) to the traffic volume (q) characteristics at Camera#1 are shown in Fig. 5, for example.

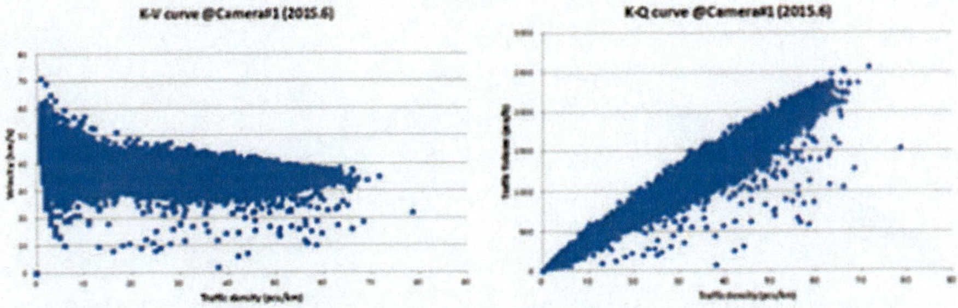

(a) k-v curve from measurement (b) k-q curve from measurement

Fig. 5 Traffic flow characteristics from measurement at Cam#1

As shown in Fig. 5b, it is difficult to define its traffic congestion condition from the measurement data. The number of plot data is 43,200 (= 60 min × 24 h × 30 days) in June 2015. In the next section, it is introduced how to get the traffic flow characteristics from the measurement data.

2.3 *Definition of the Traffic Flow Characteristics from Measurement*

In Fig. 5, we use actual all measurement data which includes many variation of driving vehicles such as regular four wheelers, three wheelers or rickshaw, buses, and tricks. In order to define the traffic flow condition, we use the boundary observation method which focuses on the edge of the boundary line of the traffic flow characteristics. From Fig. 5, it is clear that there is no measurement data outside of the boundary line in k–v curve and k–q curve. In order to define its boundary line, we take the traffic flow fundamental equation which is described in Sect. 2.2. When it fits Eq. (1) to Fig. 5a graph and Eq. (3) to Fig. 5b graph, then we are able to get the traffic flow boundary curve. The boundary curve is shown in Fig. 6 as red color line.

From the boundary observation method, it is important to describe the traffic flow parameter for each roads such as the free speed (v_f), jam traffic density (k_j), critical traffic density (k_c), critical traffic volume (q_c), and so on. The summary of each calculated parameter and traffic flow equation is shown in Table 1 [2].

As shown in Table 1, we have the quantitative traffic flow characteristics by the boundary observation. But it is still difficult to define their traffic congestion. In the next section, it focuses on the traffic congestion investigation.

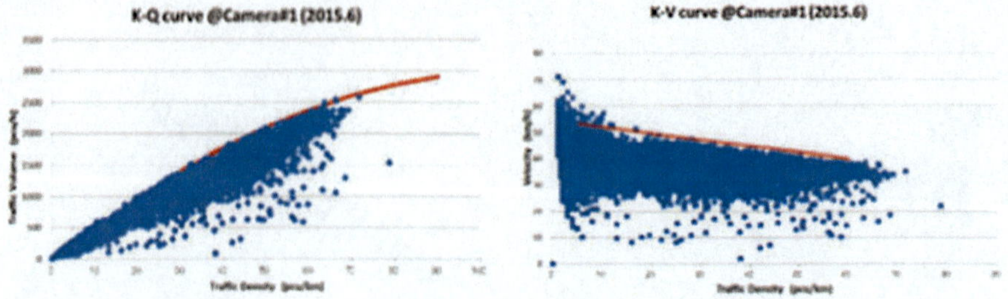

(a) k-v curve with boundary line at Cam#1 (b) k-q curve with boundary line at Cam#1

Fig. 6 Traffic flow characteristics with boundary line

Table 1 Summary of traffic flow equation and parameters

Location	Traffic flow parameter				
	v_f/k_j	k_c	q_c	Traffic volume equation	v_f
Cam#1	0.2479	110	3000	$q = -0.2479(k - 110)^2 + 3000$	54.545
Cam#2	0.1556	150	3500	$q = -0.1556(k - 150)^2 + 3500$	46.667
Cam#3	0.2153	120	3100	$q = -0.2153(k - 120)^2 + 3100$	51.667
Cam#4	0.3200	100	3200	$q = -0.3200(k - 100)^2 + 3200$	64.000
Cam#5	0.3704	90	3000	$q = -0.3704(k - 90)^2 + 3000$	66.667
Cam#6	0.2367	130	4000	$q = -0.2367(k - 130)^2 + 4000$	61.538
Cam#7	0.2361	120	3400	$q = -0.2361(k - 120)^2 + 3400$	56.667
Cam#8	0.3200	100	3200	$q = -0.3200(k - 100)^2 + 3200$	64.000
Cam#9	0.4898	70	2400	$q = -0.4898(k - 70)^2 + 2400$	68.571
Cam#10	0.3438	80	2200	$q = -0.3438(k - 80)^2 + 2200$	55.000
VMS#3	0.2361	120	3400	$q = -0.2361(k - 120)^2 + 3400$	56.667

3 Traffic Congestion Analysis

In the previous session, we able to define the traffic flow parameter with using the boundary observation method. However, as we see the difficulty to define the traffic congestion, even we have the traffic flow parameter. In the actual traffic flow, there are many variations of the vehicles driving on the road. In order to avoid variation from the measurement data, it takes the two-dimension (2D) histogram between the traffic density and average speed for each Camera#1 and #2, and it is shown in Fig. 7. From Fig. 7 2D histogram of Camera#1 and #2, the average vehicle speed of Camera#1 is stable against its traffic density change. On the other hand, the average vehicle speed of Camera#2 has two areas against its traffic density. The lower vehicle speed zone is located at high traffic density zone from Fig. 7b. In order to define the traffic congestion, we have to introduce the different way to show its traffic congestion condition. In Sect. 2.1, the vehicle speed and the traffic volume are key for the traffic

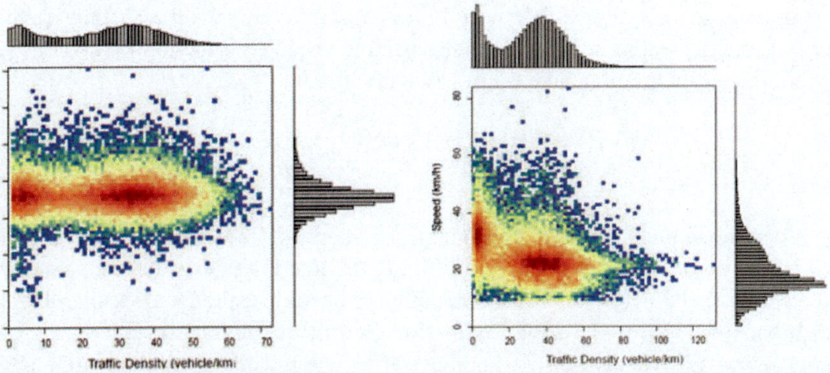

(a) 2D histogram k and v at Cam#1 (b) 2D histogram k and v at Cam#2

Fig. 7 2D histogram between traffic density (k) and vehicle speed (v)

congestion judgment. Therefore, in this section, we focus on these two parameters for the traffic congestion analysis.

3.1 Traffic Congestion Analysis

As the traffic congestion parameter, the vehicle speed (v) is essential from the time travel point of view. When it takes the inverse of vehicle speed as one of the traffic flow parameters, it takes the traffic volume (q) as another traffic flow parameter. In Fig. 8, the traffic volume (q) to the inverse of vehicle speed (v) curve is illustrated [3]. In terms of the relationship between the traffic volume and inverse of vehicle

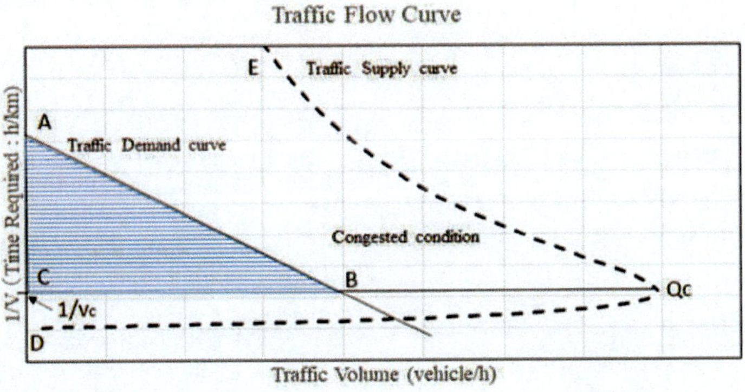

Fig. 8 Traffic flow curve between q and $1/v$

speed, from Eq. (2), the vehicle speed (v) is the inverse of the travel time (t). When $v = 1/t$ and the traffic density (k) is the vehicle speed (v) function ($k(v)$), Eq. (2) is differentiated by time (t). Then, Eq. (6) is obtained.

$$\frac{\mathrm{d}q}{\mathrm{d}t} = -\left(v\frac{\mathrm{d}k}{\mathrm{d}v} + k\right)\frac{1}{t^2} \tag{6}$$

When the vehicle speed is high, the traffic density (k) becomes small and $\mathrm{d}k/\mathrm{d}v < 0$. Then, $\mathrm{d}q/\mathrm{d}t$ becomes positive. In Fig. 8, the line D to Qc is this line. And when the vehicle speed is low, the traffic density (k) becomes small and $\mathrm{d}k/\mathrm{d}v > 0$. Then, $\mathrm{d}q/\mathrm{d}t$ becomes negative. In Fig. 8, the line Qc to E is this line. In Fig. 8, the traffic supply curve is given by Eq. (3), and the traffic demand curve should be given by the actual traffic data, which is the traffic congested condition area from the traffic flow theory. From this scenario, the area ABC shows the traffic congested volume.

From our previous study [4], the traffic congested occurs from 2/3 free speed (v_f). Figure 9 shows the different time zone basis traffic flow curve at Camera#2. The time zone is divided by 4 h from 7:00 am to 6:00 am of the next day as T1 through T6. T1 means from 7:00 am to 10:59 am; then, T2 means from 11:00 am to 14:59, for

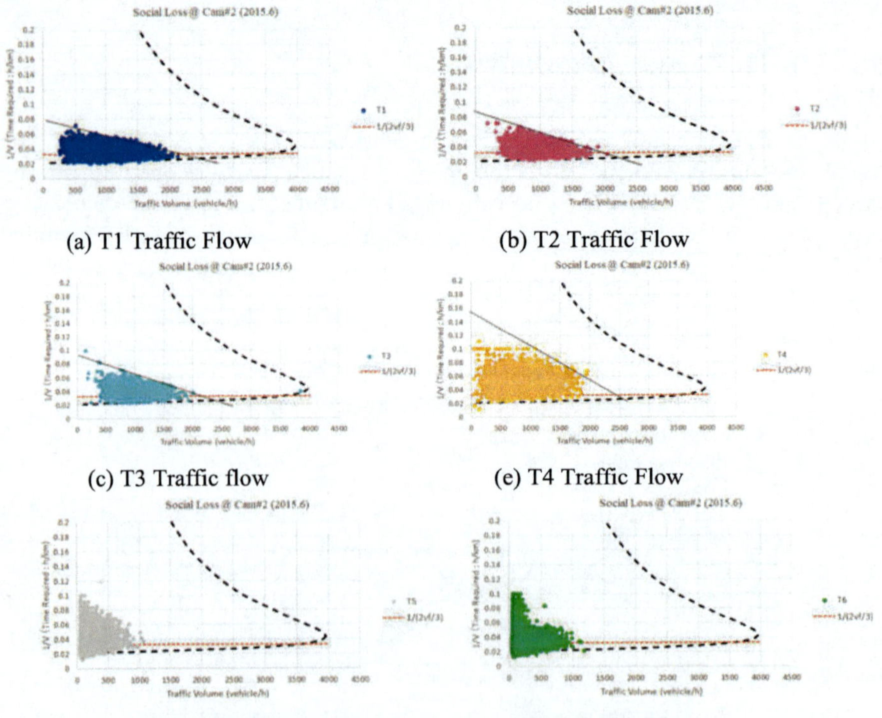

(a) T1 Traffic Flow (b) T2 Traffic Flow

(c) T3 Traffic flow (e) T4 Traffic Flow

(f) T5 Traffic Flow (g) T6 Traffic Flow

Fig. 9 Different time zone basis traffic flow

example. The area ABC in Fig. 8 means the volume caused by the traffic congestion from its definition. The ABC is calculated by the congested traffic volume (q) times to the inverse of vehicle speed (v). When the average vehicle length is defined, the area ABC times to average vehicle length equals to the occupancy as Eq. (7) [5, 6]

$$\mathrm{OC}(\%) = 100 \times q \times \frac{1}{v} \times l_{\text{ave}} \tag{7}$$

3.2 Comparison Between Congestion Loss and Measurement

The summary of the calculated congestion loss and occupancy is shown in Table 2, and the average vehicle length puts 3 m in this calculation. As for evaluation of the calculated congestion occupancy, Fig. 10 shows the comparison with the measurement occupancy at Camera#2. The average vehicle length here is virtual value here. So, the calculated congestion occupancy is not exactly same as the measurement value. But the trend of the congested condition is able to provide the idea about the traffic congestion condition. In Fig. 10, the comparison between the calculated occupancy loss and the measurement occupancy at Camera#2 is shown. If the average driving vehicle length is bigger than 3 m, then the calculated occupancy becomes large volume. As for observation at the location, main vehicles are regular four

Table 2 Summary of traffic congestion loss and occupancy at Camera#2

Time zone	Congestion loss	OC (%)
T1	55.1	16.5
T2	57.6	17.3
T3	61.7	18.5
T4	151.1	45.3
T5	48.5	14.6
T6	48.5	14.5

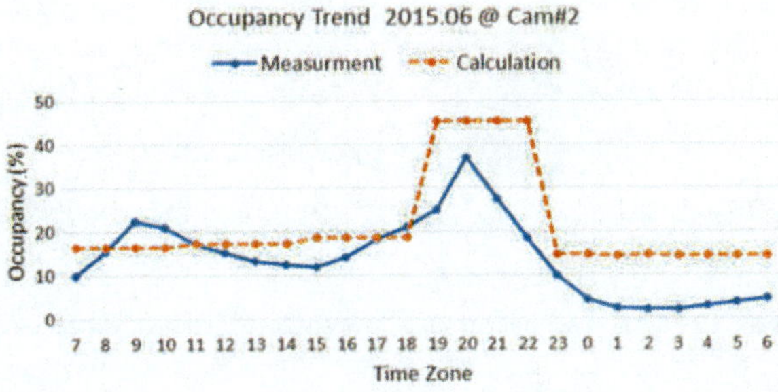

Fig. 10 Comparison between the calculated congestion loss and the measurement

wheelers and three wheelers such as rickshaw, so the length of general small four wheelers is 3.3 m and that of rickshaw is 2.6 m.

4 Conclusion

This study proposes the traffic congestion condition by the traffic congestion loss calculation with the traffic flow theory. This congestion loss calculation method is valid in the developing county as the traffic congestion analysis. As for future work, it is necessary to evaluate more hourly congestion calculation and evaluate at the other locations. And it is worth to continue collect the traffic data in Ahmedabad city through whole year; it will be good reference about the traffic congestion trend and be useful for the future traffic congestion estimation.

Acknowledgements This research is part of SATREPS program 2017 (ID: JPMJSA1606)) between India and Japan.

Appendix

Figure 11 shows that the total time zone base traffic flow in Fig. 9. The time zone T4 is most critical for the traffic congestion condition. After reviewing with local traffic management authority, this congestion is mainly caused by parking problem. In the evening after work, there are many people visits shopping and restaurants. Therefore, there is not enough parking space around such those shops. Then, after parking their cars along the street, it creates the traffic congestion, especially at Camera#2 area.

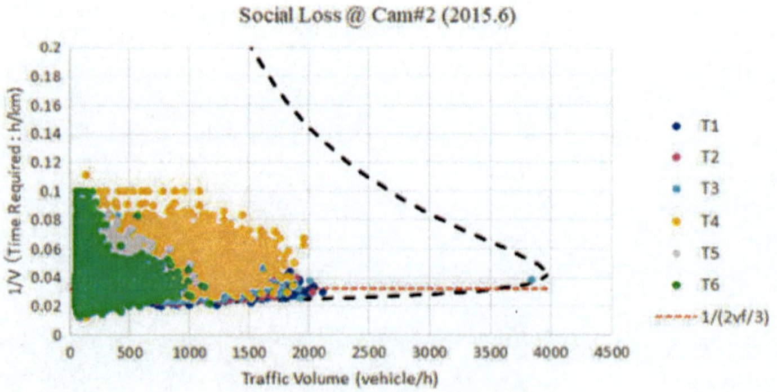

Fig. 11 Time zone base traffic flow characteristics at Camera#2

References

1. Greenshields BD (1935) A study of traffic capacity. Highw Res Board Proc 14:448–477
2. Tsuboi T, Oguri K (2016) J Soc Inf Process Eng 57(12):2819–2826
3. Kuroda T et al (2008) Urban and regional economics. Yuhikaku Book, pp 265–268
4. Tsuboi T, Yoshikawa N (2017) Traffic flow analysis in emerging country (India). CODATU XVII
5. Tsuboi T (2019) Traffic congestion visualization by traffic parameters in India. In: Proceedings of ICICC 2019, vol 2
6. Kawakami S, Matsui H (2007) Transport engineering. Morikita Publishing Co., pp 108–109

Performance Evaluation and Comparison Study of OFDM in AWGN and Colored Noise Environment

Nikita Goel and Amit Kumar Ahuja

Abstract In modern wireless communication system Orthogonal Frequency Division Multiplexing (OFDM) gathers interest by reason of its numerous advantages like high spectrum efficiency, immunity against Additive White Gaussian Noise (AWGN), impulsive noise, multi-path fading and high bit rate capacity etc. The AWGN channel is theoretical, simple, and is generally considered as the starting point to develop the basic system for the performance evaluation purpose. In practical situations, however, noise is not AWGN and can rather be better approximated as colored noise. Thus it becomes important to evaluate the performance of communication systems under the colored noise environment and compare the same with the case when noise is AWGN. This is the objective of this work, and, is carried out in this paper. MATLAB R2014 is used for performance evaluation and comparison study. For lower values of SNRs, SNR $\leq$ 0 dB, $BER_{Colored} < BER_{AWGN}$, (BDF > 0) is observed, however, for SNR > 0 dB, the $BER_{Colored}$ is found to be almost similar to BER_{AWGN} leading to small BDF. This study is important in view of analysing the OFDM system performance in colored noise and comparing it with the case when channel is theoretical, that is, AWGN.

Keywords OFDM · Colored noise · BER · AWGN · Cyclic prefix

1 Introduction

For various professional applications and also in our daily life, high data rate wireless communication has become increasingly important. OFDM is an emerging technique

N. Goel
Department of Electronics and Communication Engineering, K.I.E.T Group of Institutions, Delhi NCR, Ghaziabad 201206, India
e-mail: nikita.goel@kiet.edu

A. K. Ahuja (✉)
Department of Electronics and Communication Engineering, M.I.E.T, Meerut 250005, India
e-mail: amit.ahuja@miet.ac.in

A. Khanna et al. (eds.), *Data Analytics and Management*, Lecture Notes on Data Engineering and Communications Technologies 54,
https://doi.org/10.1007/978-981-15-8335-3_27

now a days due to its high spectral efficiency, reduced Inter Symbol Interference (ISI), allowable Bit Error Rate (BER), maximum delay etc. [1, 2]. OFDM is a multi-carrier modulation technique in which the complete bandwidth is divided among multiple sub-carriers in such a way that the sub-carriers are orthogonal with respect to each other. OFDM is bandwidth-efficient, as due to orthogonality, sub-carrier is closely packed or partially overlapped to each other. Data is carried by a large number of these orthogonal sub-carriers [1, 2]. OFDM usage is found in a number of applications. Efficient implementation of a pipeline Fast Fourier Transform/ Inverse Fast Fourier Transform (FFT/IFFT) processor for OFDM applications has been presented in [3]. In [4, 5] performance of OFDM over Rayleigh distributed AWGN channel have been discussed. Evaluation of Peak to Average Power Ratio (PAPR) reduction performance of BCH coded OFDM has been done in [6]. In [7] authors analysed the effect of PAPR reduction in OFDM using Goppa codes. In [8] power law noises ($\frac{1}{f}$) have been discussed in detail. Authors in [9] presented a novel algorithm and code for power law noise simulation analysis and also review the past simulation techniques. Applicability of the Blind source separation algorithm over OFDM is justified in [10]. Performance analysis of color shift keying over AWGN and colored noise environment have been discussed in [11]. Linear minimum mean square error estimator for MIMI OFDM system in colored channel have been derived in [12]. In [13] analysis of BER performance of OFDM over a channel corrupted by impulsive noise and multi-path effect has been carried out. In [14] authors review the performance of VLC for 5G considering MIMO-OFDM. A receiver has been derived for Multiple Input Multiple Output (MIMO) OFDM considering colored noise affected Co-Channel Interference (CCI) in [15].

The AWGN channel is theoretical, simple, and is widely considered for performance evaluation purpose in many systems. However, in practical situations, noise is not AWGN and can rather be better approximated as colored noise. Thus it becomes important to evaluate the performance of communication systems under the colored noise environment and compare the same with the case when noise is AWGN. This is the objective of this work. The following is carried out in this paper:

- BER performance of OFDM system in a practical, colored noise environment, is evaluated and analyzed,
- BER performance of OFDM system in theoretical AWGN channel is evaluated and analyzed,
- Both the above BERs are plotted and compared,
- Above steps are repeated for different colored noise environments.

Rest of the paper is structured as per the following. The OFDM system model and brief description of AWGN and colored noise is presented in Sect. 2. Different parameters that are considered for this simulation study and analysis are also discussed in this section. In Sect. 3, performances of OFDM, in both colored and AWGN environment, are evaluated and compared. The paper is concluded in Sect. 4.

2 System Model

In OFDM the complete bandwidth (B) is divided into N number of narrowband sub-carriers at equidistant frequencies. Following the principle of orthogonality, the sub-carriers are overlapping each other partially, to efficiently utilize the channel bandwidth [1]. Bandwidth allocated to each sub-carrier is denoted by $\frac{B}{N}$ Hz. A bit sequence with the rate 'R' is divided into 'N' sub-channels (in parallel), each with different frequency [1]. The total bit rate is equally divided into N parts thus making bit rate of each sub-carrier as $\frac{R}{N}$. So, in the OFDM system the single high rate data symbol is subdivided into low rate symbols. Each sub-carrier is modulated and transmitted. This is done simultaneously in a superimposed and parallel form. There are N adjacent and orthogonal sub-carriers that are spaced by the frequency distance $\frac{B}{N}$. Each OFDM symbol is denoted by $S(t)$ and defined as given below [1]

$$S(t) = \text{Real}\left[\sum_{n=-\infty}^{+\infty}\sum_{m=0}^{N-1} Z_{nm} e^{j2pif_m t} P(t - nT)\right] \quad (1)$$

where $T_s = \frac{N}{R}$. In (1), OFDM signal $S(t)$ can be written as the summation of each information symbol denoted by Z_{nm} being carried in the mth sub-carrier within the nth OFDM symbol. In (1), $P(t)$ denotes an ideal square pulse of length T_s, N represents the number of sub-carriers and f_m is the mth sub-carrier frequency as indicated in Table 1 [1].

Table 1 List of Notations

Symbols	Description
B	Bandwidth
N	Number of sub-carriers
$S(t)$	OFDM signal
Z_{nm}	Information symbol used in OFDM
f_m	mth sub-carrier frequency
T_s	Symbol duration
t_{cp}	Cyclic prefix duration
BDF	BER deviation factor
$\text{BER}_{\text{colored}}$	BER in colored noise environment
BER_{AWGN}	BER in AWGN environment

2.1 OFDM Transmitter

The block diagram of OFDM system is represented in Fig. 1. For parallel data transmission serial to parallel converter is used, which collects the serial incoming data symbols and convert it into parallel data stream [1, 2]. The symbol vector can be written as in (2)

$$M(K) = [M_0(k), M_1(k), M_2(k), \ldots, M_{M-1}(k)]. \tag{2}$$

The parallel data streams are then modulated independently by using any digital modulation techniques such MPSK, QAM, QPSK etc. The modulation scheme used for the study and analysis of this paper is Binary Phase Shift Keying (BPSK). After modulation, the resulting complex vector $S(k)$ can be defined in (3).

$$S(k) = [S_0(k), S_1(k), S_2(k), \ldots, S_{M-1}(k)] \tag{3}$$

The superposition of independent modulated sub-carriers is typically performed by IFFT where the input channels are spaced equivalently [1, 2]. As the symbols to be transmitted are defined in frequency domain, so for time domain conversion the modulated symbols vector is applied at the input of N-point IFFT, resulting in a set of N complex time domain samples. During the process of N-point IFFT, it may be possible that the number of modulated symbols 'M' is less than the value 'N', means $M < N$, then it is required to do zero padding to equalize the length of modulated symbol vector to the length of IFFT (N) before IFFT process [1, 2]. The resulting time domain vector after IFFT can be written as in (4)

$$s(k) = [s_0(k), s_1(k), s_2(k), \ldots, s_{M-1}(k)] \tag{4}$$

Next important process in OFDM signal generation is Cyclic Prefix (CP) insertion, which is briefly described as follows. To eliminate the effect of ISI from previous symbol, a guard band is inserted between the adjacent symbols. Here the cyclic prefix is performing as guard interval [1, 2]. The cyclic prefix is generated by duplicating the last H samples (approx 10 % to 25 %) of IFFT output or OFDM symbol $s(k)$ and appending them at the beginning of $s(k)$ [16]. Figure 2 represents the cyclic prefix insertion process. Resulting symbol is written in Eq. (5).

$$s(k) = [s_{N-H}(k), s_{N_H-1}(k), s_{N-H-2}(k), \ldots, s_{N-1}(k)] \tag{5}$$

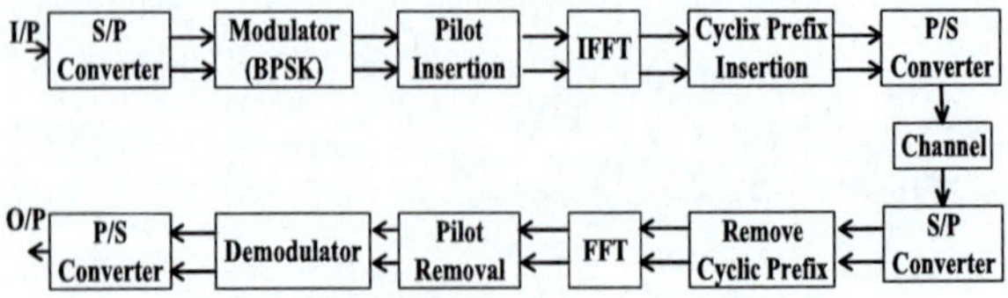

Fig. 1 Block diagram of OFDM system [1]

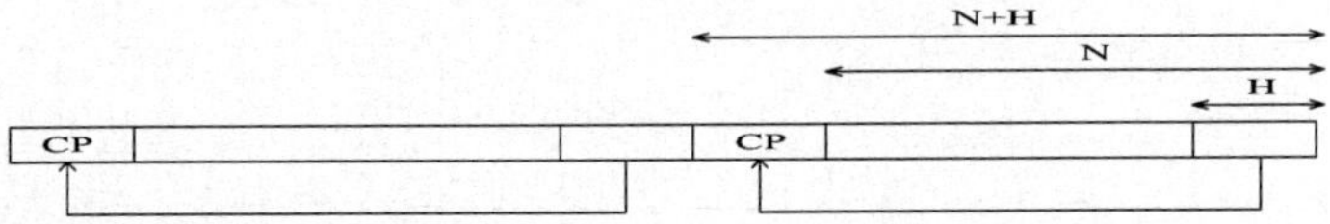

Fig. 2 Cyclic prefix insertion [1]

The advantage of adding cyclic prefix is that the duration of symbol is increased without sending any additional data. After adding cyclic prefix, the OFDM symbol duration become $T_s + t_{cp}$, where t_{cp} is the extension of the symbol period due to the cyclic prefix. This way, the linear convolution of the channel is converted into the circular convolution. This is suitable for Discrete Fourier Transform (DFT) processing. After the cyclic prefix addition, the output is converted into serial data by using parallel to serial converter [1, 2]. The serial data is transmitted to the destination through the transmission media or channel.

2.2 *OFDM Receiver*

The reverse process of transmitter is performed at the receiver side. Let $x(t)$ be the transmitted signal. The received signal after passing through the channel is defined as given below

$$r(t) = x(t) * h(t) + z(t) \tag{6}$$

where $r(t)$ is the received signal, $h(t)$ denotes continuous time impulse response of the channel and $z(t)$ is additive noise. In receiver side firstly the cyclic prefix is removed from the incoming signal, such that only an ISI free block of samples is passed through IFFT [1].

2.3 *AWGN and Colored Noise*

The effect of noise on the behaviour of communication system has been a subject of interest for many theoretical and practical studies. Due to constant Power Spectral Density (PSD) over the complete range of frequency, AWGN is considered to be a theoretical noise for simulation and analysis purpose. In many applications, it is required to analyse the system performance in non white noise environment. Colored noise is the stochastic process whose PSD is not constant but proportional to $\frac{1}{f^{\beta}}$, where f is the cyclic frequency and β is real number usually between -2 and 2. Because of this spectral density function the colored noise is also termed as power law noise. On the basis of analysis of various spectral profiles in reference to the colors of light with approximately similar spectra, color names like pink, red, and

Table 2 Types of colored noise [9]

Values of β	Colored noise
$\beta = 0$	White noise
$\beta = 1$	Pink noise/flicker noise
$\beta = -1$	Blue noise
$\beta = -2$	Voilet noise

blue were given to the various noise signals [9]. For instance, depending on the values of β the colored noise can be characterize as in Table 2.

In this work, the performance of OFDM system is evaluated by considering different colored noise environment and corresponding results are compared with the case when transmission medium is AWGN. For this purpose, pink noise, blue noise, purple noise and brown noise have been considered. OFDM signal is transmitted through these channels and the transmitted information is recovered from its corrupted version at the receiver. The simulation study and analysis is carried out using MATLAB R2014a® and the obtained results are compared appropriately. The deviation in performance for the two cases; (1) when AWGN channel is used and (2) when colored noise channel is used, is evaluated using BER-Deviation-Factor (BDF) which is defined in (7)

$$\mathrm{BDF} = \left[1 - \frac{\mathrm{BER}_{\mathrm{colored}}}{\mathrm{BER}_{\mathrm{AWGN}}}\right] \tag{7}$$

where $\mathrm{BER}_{\mathrm{colored}}$ is the BER at the receiver output for colored noise channel and $\mathrm{BER}_{\mathrm{AWGN}}$ is the BER at the receiver output for AWGN channel. Higher deviation between the two cases is reflected by higher value of BDF and vice versa. Parameters used for the simulation purpose are given in Table 3.

Table 3 Simulation parameters

Simulation parameters	Values
Total data bits	52×10^5
Number of frames (L)	10^5
Number of sub-carriers (S)	52/frame
Frame length	80 bits
Cyclic prefix	16 bits
Length of IFFT	64
Pilot insertion	12
Modulation scheme	BPSK

3 Simulation Results—Comparison and Discussion

As discussed earlier, the performance of OFDM system is evaluated by considering different colored noise environment and corresponding results are compared with the case when transmission medium is AWGN. BER versus SNR curves are plotted and compared. In order to evaluate the deviation in BER, at a given SNR, BER-Deviation-Factor (BDF), as defined earlier, has been evaluated. The corresponding results are shown from Fig. 3 to Fig. 6.

In almost all the considered cases of colored noise channels, the higher deviation in $BER_{Colored}$ and BER_{AWGN} is observed corresponding to lower SNR values. For lower values of SNRs, SNR ≤ 0 dB, corresponding to all considered cases of colored noises, $BER_{Colored}$ at a given SNR is found lower as compared to BER_{AWGN} (refer to Fig. 3 to Fig. 6). This is an expected result since colored noise is less severe than the AWGN noise. Further, at a given SNR with SNR > 0 dB, the $BER_{Colored}$ is found

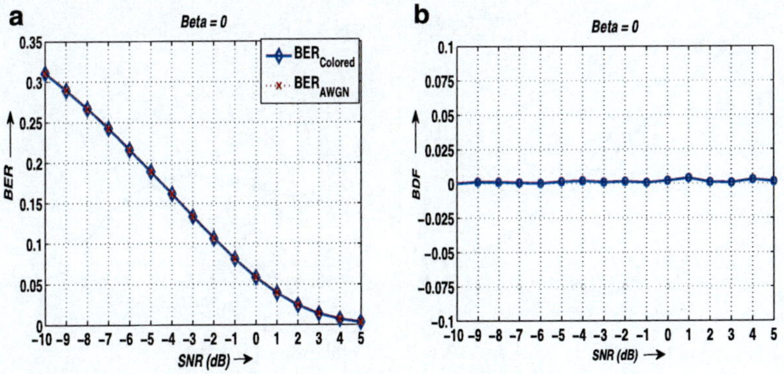

Fig. 3 **a** BER and **b** BDF w.r.t SNR for AWGN and white noise ($\beta = 0$)

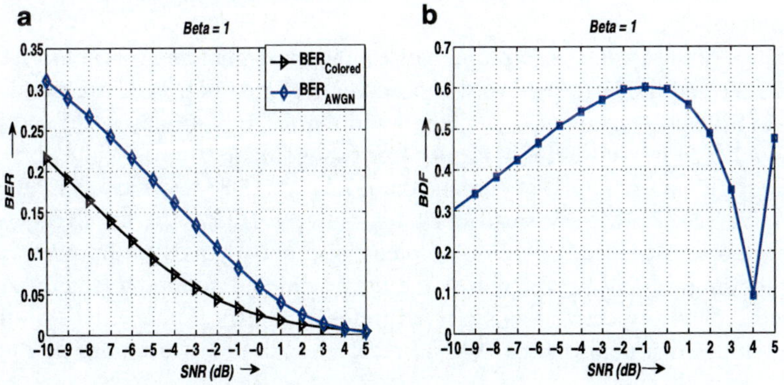

Fig. 4 **a** BER and **b** BDF w.r.t SNR for AWGN and pink noise ($\beta = 1$)

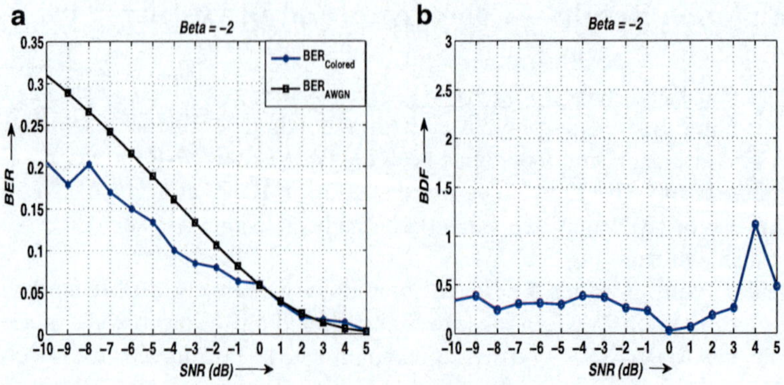

Fig. 5 **a** BER and **b** BDF w.r.t SNR for AWGN and purple noise ($\beta = -2$)

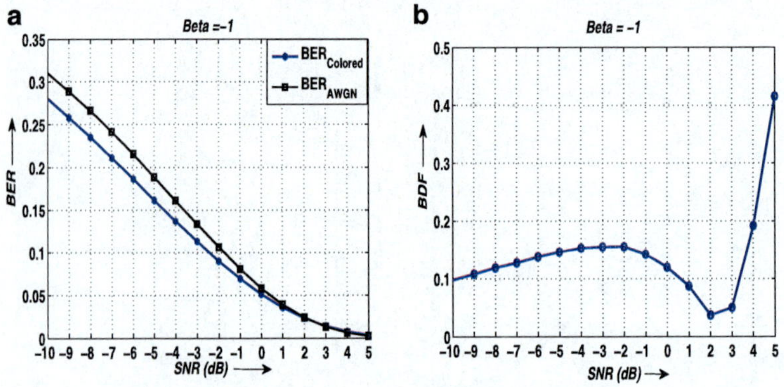

Fig. 6 **a** BER and **b** BDF w.r.t SNR for AWGN and blue noise ($\beta = -1$)

to be almost similar to BER_{AWGN}. This is observed for all the considered cases of colored noise channels.

Corresponding to $\beta = -2$ (purple noise), the maximum and minimum BDF of around 1.11 and 0.023 respectively is observed (refer to Fig. 3). Corresponding to $\beta = -1$ (Blue noise), the maximum and minimum BDF of around 0.415 and 0.036 respectively is observed (refer to Fig. 4). Corresponding to $\beta = 0$ (White noise), the maximum and minimum BDF of around 0.0037 and 0.00083 respectively is observed (refer to Fig. 5). Here, very small deviation between $BER_{Colored}$ and BER_{AWGN} is an expected result since for $\beta = 0$ colored noise corresponds to the white noise. Corresponding to $\beta = 1$ (Pink noise), the maximum and minimum BDF of around 0.60 and 0.0908 respectively is observed (refer to Fig. 6). Future extension to this work includes restructuring the AWGN receiver such that it becomes optimal for colored noise channel.

4 Conclusion

This work is concerning performance evaluation of OFDM system in different colored noise environments and comparison study with the case when channel is AWGN. Purple, Blue, White and Pink noises are considered for this study. Performance comparison study is carried out using BER versus SNR, and, BDF versus SNR curves that are evaluated for different colored noise environments. For lower values of SNRs, SNR $\leq$ 0 dB, $BER_{Colored} < BER_{AWGN}$ is observed. Further, for SNR > 0 dB, the $BER_{Colored}$ is found to be almost similar to BER_{AWGN} leading to small BDF. These are observed corresponding to all the considered cases of colored noise. The work is important as it depicts the simulation of OFDM in more practical environment, that of, colored noise and helps to understand its performance analysis and comparison with that of the theoretical case, that is, AWGN.

References

1. Sesia S, Toufik I, Baker M (2011) LTE-the UMTS long term evolution: from theory to practice. Wiley
2. Holma H, Toskala A (2009) LTE for UMTS: OFDMA and SC-FDMA based radio access. Wiley
3. Yu C, Yen MH, Hsiung PA, Chen SJ (2011) A low-power 64-point pipeline FFT/IFFT processor for OFDM applications. IEEE Trans Consum Electron 57(1):40–40
4. Agrawal M, Raut Y (2011) BER analysis of MIMO OFDM system for AWGN & Rayleigh fading channel. Int J Comput Appl 34:33–37
5. Li Z, Xu H, Chen D, Qu D (2020) Phase rotation to avoid imaginary interference leakage in multi-user MIMO-OFDM/OQAM systems. Wirel Pers Commun 110(4):1963–1984 Citeseer
6. Kumar P, Ahuja AK, Chakka R (2018) BCH/hamming/cyclic coding techniques: comparison of PAPR-reduction performance in OFDM systems. In: International conference on intelligent computing and applications. Springer, pp 557–566
7. Sengupta S, Lande BK (2020) An approach to PAPR reduction in OFDM using Goppa codes. Procedia Comput Sci 167:1268–1280 Elsevier
8. Kasdin NJ (1995) Discrete simulation of colored noise and stochastic processes and 1/f/sup/spl alpha//power law noise generation. Proc IEEE 83:802–827
9. Kasdin NJ, Walter T (1995) Discrete simulation of power law noise (for oscillator stability evaluation). In: Proceedings of the 1992 IEEE frequency control symposium, pp 274–283
10. Sriyananda MGS, Joutsensalo J, Hämäläinen T (2012) Blind source separation for OFDM with filtering colored noise and jamming signal. J Commun Netw 14:410–417 IEEE
11. Durukan F, Güney BM, Özen A (2019) Performance analysis of color shift keying systems in AWGN and color noise environment. In: 27th Signal processing and communications applications conference (SIU), IEEE, pp 1–4
12. Wong TF, Park B (2004) Training sequence optimization in MIMO systems with colored interference. IEEE Trans Commun 52:1939–1947 IEEE
13. Ma YH, So PL, Gunawan E (2005) Performance analysis of OFDM systems for broadband power line communications under impulsive noise and multipath effects. IEEE Trans Pow Deliv 20:674–682 IEEE
14. Vishwaraj, Ali L (2019) Hybrid MIMO-OFDM system for 5G network using VLC-a review. In: 2019 IEEE International conference on electrical, computer and communication technologies (ICECCT), pp 1–5

15. Choi J (2007) An EM-based iterative receiver for MIMO-OFDM under interference-limited environments. IEEE Trans Wirel Commun 6:3994–4003 IEEE
16. Ancora A, Toufik I, Bury A, Slock D (2009) Orthogonal frequency division multiple access (OFDMA). In: LTE–The UMTS long term evolution: from theory to practice, Wiley Online Library, pp 111–134

Performance Improvement Using Spline LS and MMSE DFT Channel Estimation Technique in MIMO OFDM Using Block-Type Pilot Structure

Neha Sharma, Vikas Nandal, and Deepak Nandal

Abstract Multiple input multiple output (MIMO) is a very important and prominent key technology in wireless system which is extensively used in 4G systems. However, accurate channel estimation poses a challenge in reducing error rate in MIMO-LTE system. For choosing a correct estimate for the MIMO-LTE system, there are many aspects for implementation which include time variation, computation, and performance of the channel like Rayleigh or Rican. The least square (LS) and minimum mean square error (MMSE) are the two well-known techniques for estimating the channel. In this paper, we used these techniques along with linear and cubic spline interpolation techniques. In this work, Discrete Fourier Transform (DFT)-based channel estimation technique is presented for improvement in the performance of MMSE and LS estimation techniques using block-type pilot arrangement. Signal-to-noise ratio and BER performance of all channel estimation schemes have been evaluated with modulation techniques, MQAM and MPSK over Rayleigh and AWGN noise channels. This paper presents that DFT channel estimation scheme improves the performance of LS and MMSE channel estimation via noise reduction outside channel delay for 2×2 MIMO system.

Keywords LS · MMSE · DFT · MIMO · OFDM · Interpolation

N. Sharma (✉) · V. Nandal
Department of Electronics & Communication Engineering, Maharshi Dayanand University, Rohtak, Haryana, India
e-mail: neha17008@gmail.com

V. Nandal
e-mail: nandalvikas@gmail.com

D. Nandal
Department of Computer Science, Guru Jambheswar University, Hisar, Haryana, India
e-mail: gju.dpknandal@gmail.com

A. Khanna et al. (eds.), *Data Analytics and Management*, Lecture Notes on Data Engineering and Communications Technologies 54,
https://doi.org/10.1007/978-981-15-8335-3_28

1 Introduction

MIMO-LTE technology has been emerged for satisfying the rising demand of transmission rate. MIMO system has drawn the attention of the researchers in last decades. Capacity of transmitting data and reliability can be increased via using antennas in larger number. Also, OFDM technology has advantage of mitigating ISI and multipath fading via applying cyclic prefix during channel estimation [1, 2]. Normally, in MIMO-LTE system, fading channel classification can be done as:

- Two-dimensional (2D) signal (into frequency and time domain)
- One-dimensional (1D) signal
  - But, practically, it is very complicated to implement such 2D design. Therefore, generally in OFDM system one-dimensional (1D-comb and block type) pilot-based methods are used for estimating channel [3, 4]. Basically, channel estimation (CE) techniques are categorized as:

- Block pilot-based method
- Comb-type pilot-based method
  - In block pilot method, transmission of pilot symbols is done periodically. In comb-type method of channel estimation, pilot symbols are transmitted continuously [5, 6]. For estimating the channel, even spacing is provided on the subcarriers. The comb pilot technique interpolates channel for channel estimation. Comb pilot and block pilot arrangements are displayed in Fig. 2. Here, block-type arrangement for pilot insertion is used.
  - Block-type arrangement is based upon two sorts of CE techniques [5]:

- Least square (LS) channel estimation
- Minimum mean square (MMSE) channel estimation

LS being the simplest technique for estimating the channel can be implemented easily. Channel information is not required in advance for implementation of channel. But, LS technique has higher MSE and can be affected via noise [6]. MMSE has much better performance than LS [7, 8] channel estimation technique but suffers from higher complexity in computation. DFT channel estimation technique is used with MMSE and LS channel estimation techniques for reducing noise [9]. Different one-dimensional interpolation techniques have been used for frequency response estimation at missing tones of pilot. A. Zaib et al. discussed various channel estimation techniques [10]:

- Linear interpolation
- Polar interpolation
- Adaptive linear interpolation

Linear and cubic spline interpolation techniques are simplest. These techniques are one-dimensional and hence, are less complex. Shubhangi R. Chaudhary et al. did BER analysis of various equalization techniques. Authors proposed zero-forcing

(ZF) equalizer for linear equalization. Inversion of channel is applied to received signal via ZF equalizer for restoration of signal, prior channel [11]. In this paper, Sect. 2 describes system model for 2 × 2 MIMO system, Sect. 3 provides various channel estimations techniques for LTE downlink system, Sect. 4 explains proposed methodology, and Sect. 5 contains results and analysis. Finally, the conclusion is specified in Sect. 6.

2 System Model

In MIMO, multiple antennas lead to diversity that leads to increased reliability. In addition to diversity, MIMO uses much high data rates. It is feasible via transmitting multiple information stream in parallel between transmitter and receiver, also known as spatial multiplexing. Various techniques such as adaptive precoding and layer mapping based upon rank and PMI estimation are used for making MIMO system more robust when different kinds of channel impairments are present. In this paper, 2 × 2 MIMO system model with two transmitting antennas and two receiving antennas illustrated in Fig. 1 is considered [12]. Received signal is expressed into matrix notation for N_T number of transmitting antennas and N_R number of receiving antennas as:

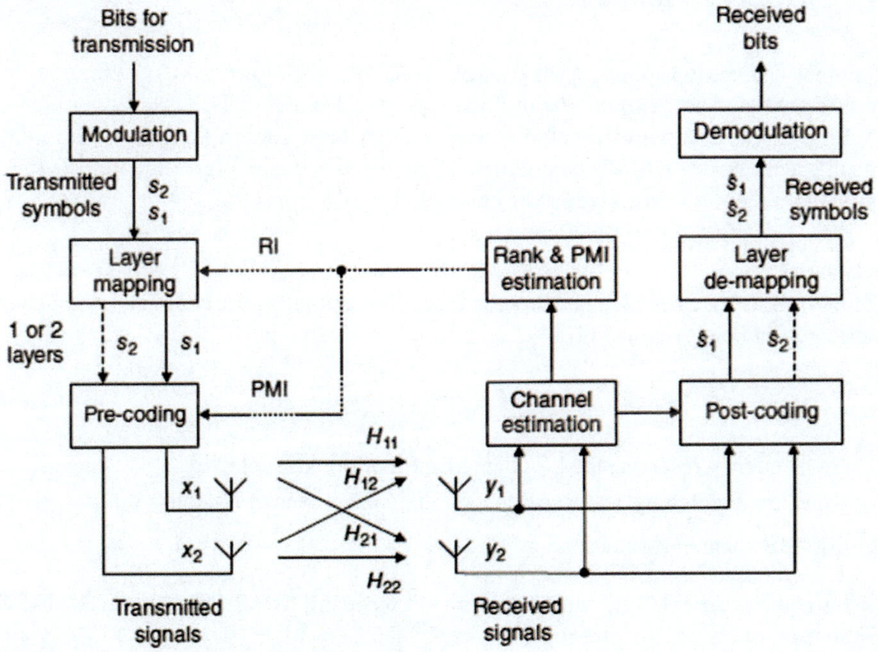

Fig. 1 System model for 2 × 2 MIMO-LTE system

$$Y_R = H \cdot X + N \tag{1}$$

Here, signals which are sent by N_T transmit antennas are contained by X column vector. At N_R receiving antennas, resulting signals are contained by Y column vector, and noise is contained by N column vector. Phase shifts and amplitude changes which are introduced by the air interface are represented by channel matrix H. Here, we are considering system model for 2 × 2 MIMO-LTE system, so we can rewrite the equations. Further, equations can be expressed as:

$$\begin{bmatrix} Y_1 \\ Y_2 \end{bmatrix} = \begin{bmatrix} H_{11} & H_{12} \\ H_{21} & H_{22} \end{bmatrix} \begin{bmatrix} X_1 \\ X_2 \end{bmatrix} + \begin{bmatrix} N_1 \\ N_2 \end{bmatrix} \tag{2}$$

After applying singular value decomposition (SVD), we can rewrite H (channel matrix) as:

$$H = R \cdot \Sigma \cdot U^{\dagger} \tag{3}$$

Here, $N_T \times N_T$ matrix is denoted via U. $N_R \times N_R$ matrix is denoted via R. U and R both are unitary matrices.

3 Channel Estimation

Channel estimation poses a challenge in LTE [13] downlink system due to interference and noise added via multipath propagation (Fig. 2).

Block-type channel estimation is based upon least square (LS) and minimum mean square error (MMSE) techniques. Here, we will discuss least square, MMSE, and DFT schemes of estimating the channel:

1. LS channel estimation:

Pilot symbols have been located on subcarriers for estimating the channel. LS channel estimates can be given as [14],

$$h_p^{\mathrm{LS}} = x_p^H y_p \tag{4}$$

where frequency response for LS channel estimation is denoted by h_p^{LS}. Transmitted symbols are denoted by x_p^H. Received symbols have been denoted by y_p.

2. MMSE channel estimation:

Mean square error (MSE) can be minimized by using MMSE technique. MSE for estimation of channel is given as [15],

$$Q\{|r|^2\} = Q\left\{|h - \hat{h}|^2\right\} \tag{5}$$

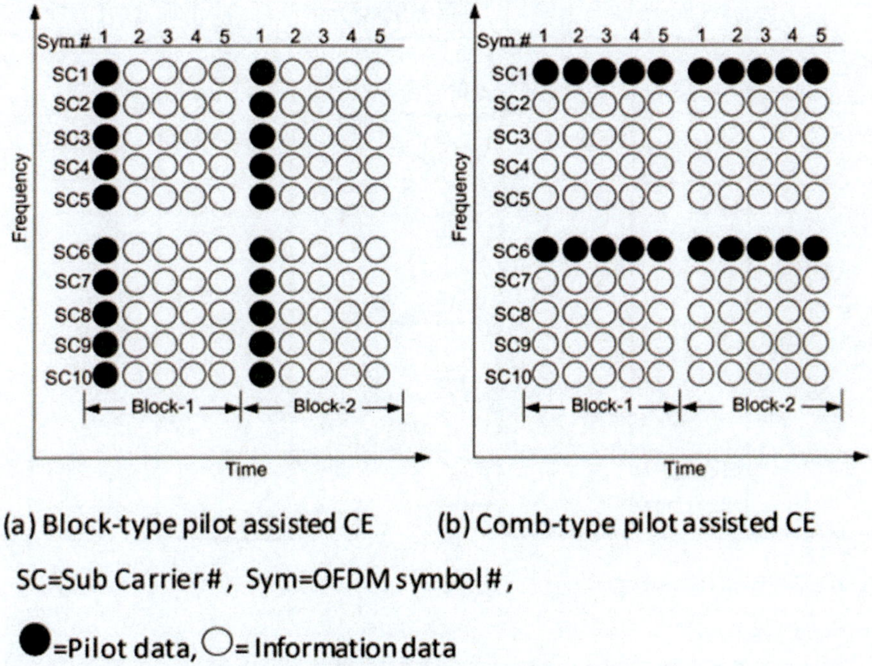

Fig. 2 Block-type and comb-type pilot-assisted CE

where expectation is defined by $Q\{\}$. Actual or original channel estimation parameter is denoted by h. Raw parameter for channel estimation is denoted by $\hat{h}$. Error for estimating the channel is denoted by r.

3. DFT channel estimation:

Using DFT and spline interpolation, we can achieve higher accuracy in the estimation techniques like LS and MMSE as there is noise reduction outside the channel delay (L). Block diagram for DFT technique for estimating the channel has been displayed in Fig. 3.

There are many steps for computing DFT technique:

- Compute $h_{\mathrm{LS1}}/h_{\mathrm{MMSE1}}$ for MMSE and LS CE techniques.
- IDFT should be taken for MMSE and LS CE techniques and $h_{\mathrm{LS}}/h_{\mathrm{MMSE}}$ should be converted into time domain by using IDFT. Equation can be given as [8],

$$\mathrm{IDFT}\left\{h_{\mathrm{LS1/MMSE1}}[S]\right\} = h[a] + O[a] \triangleq \hat{h}[a] \tag{6}$$

where noise component is denoted by $O[a]$.

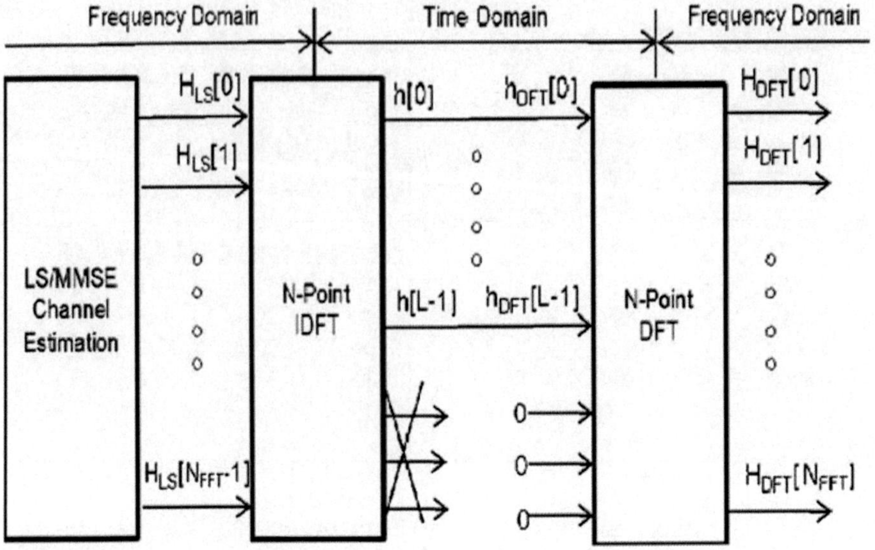

Fig. 3 DFT-based channel estimation

- Ignoring the coefficients $\hat{h}[a]$ we can now define the maximum channel delay (L) as

$$\hat{h}_{\mathrm{DFT}}[a] = \begin{cases} h[a] + O[a], \; a = 0, 1, 2, 3, \ldots, L-1 \\ 0, \qquad\qquad \text{otherwise} \end{cases} \tag{7}$$

- Using DFT, $\hat{h}_{\mathrm{DFT}}[a]$ is converted in frequency domain

$$H_{\mathrm{DFT}}[s] = \mathrm{DFT}\{\hat{h}_{\mathrm{DFT}}[a]\} \tag{8}$$

4 Methodology

A. Interpolation Techniques

Using the block pilot CE method (Fig. 5), interpolation is done into time domain for estimating the channel and we obtain N_{FFT} points which are used in two types of interpolation methods used in this paper.

(a) Linear interpolation has lesser complexity and we use two consecutive pilot symbols for the channel estimation, such as:

$$\overset{\frown}{h}(s) = \overset{\frown}{h}(MN_F + 1) \tag{9}$$

$$\overset{\frown}{h}(s) = (\overset{\frown}{h}(M + 1) - \overset{\frown}{h}(M))\frac{1}{N_F} + \overset{\frown}{h}(M) \tag{10}$$

(b) Spline interpolation type has better and smooth performance than above technique, but it still needs to be proved. It has higher-order interpolation. Comparison is shown in Fig. 4.

The equation for it can be described as

$$\overset{\frown}{h}(s) = \overset{\frown}{h}(aN_F + l) \tag{11}$$

$$\overset{\frown}{h}(s) = \propto_1 (\overset{\frown}{h_p}(M + 1)) - \propto_0 (\overset{\frown}{h_p}(M) + N_i \propto_1 \overset{\frown}{h}(M + 1) - N_i \propto_0 \overset{\frown}{h}(M)) \tag{12}$$

In this paper, comparison of LS and MMSE CE schemes is done with actual channel. DFT-based channel estimation techniques are used along with LS and MMSE CE techniques. Linear and spline interpolation methods are used with LS CE technique. LS and spline interpolation methods are applied for obtaining channel transfer function for N_{FFT} points. Here, we are using block-type pilot insertion arrangement which has been presented in Fig. 5. The effect of estimation and

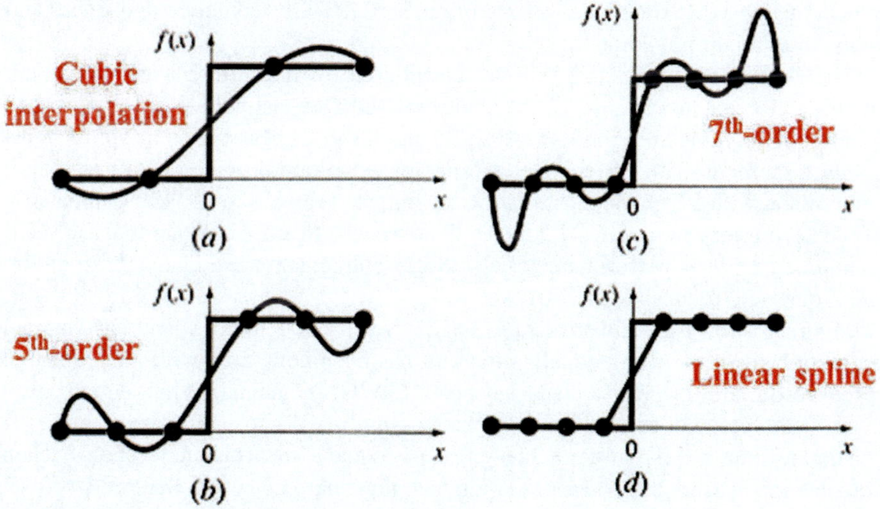

Fig. 4 Spline versus linear

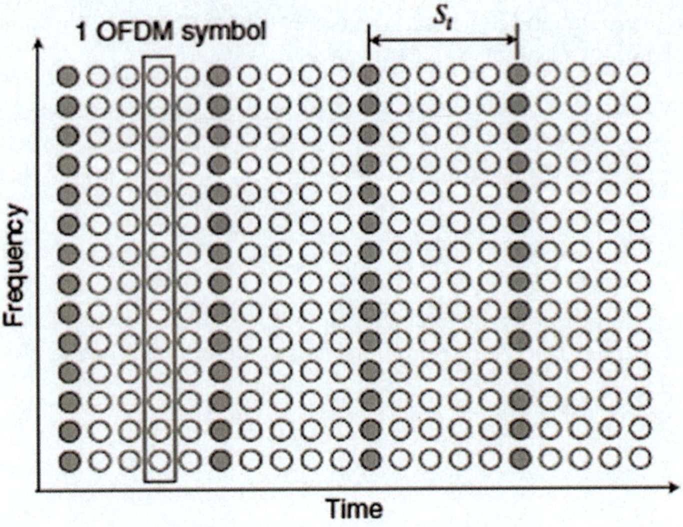

Fig. 5 Block-type pilot arrangement

compensation has been displayed in Fig. 12 for 16 QAM OFDM system.

5 Results and Analysis

Here, simulation results have been shown using MATLAB® for evaluating performance of LS, MMSE channel estimators using DFT CE technique and linear and spline interpolation methods.

The simulation parameters that are considered for the different techniques are shown in Table 1. BPSK, 16 QAM along with OFDM technology have been used for establishing transmission link. According to results, linear and cubic spline interpolation methods with LS channel estimation technique provided better results in comparison with LS channel estimation technique. Table 2 shows MSE comparison of CE techniques with and without DFT approach. Results indicate that DFT CE technique improves performance of all LS, LS-spline, and MSE channel estimation techniques and reduces MSE.

Figure 6 shows the bit error rate (BER) versus SNR performance of channel estimation response using equalization and has been compared with ideal channel response on additive white Gaussian noise (AWGN) channel. When equalization is not used, channel estimation response has high BER. It is observed that better channel estimation response is achieved, nearly equal to ideal channel response when equalization is used. 8 QAM modulation technique has been used. BER versus SNR comparison of CE techniques with and without equalization is presented in Table 3.

Figures 7 and 8 show comparison of LS channel estimation scheme using linear

Table 1 Basic simulation parameters

Simulation parameter	Type of range
Number of OFDM symbols, m	2560
Length of OFDM symbols, N	100
Constellation order, M	16
Pilot arrangement	Block type
CE techniques	LS, MMSE, and DFT
Interpolation	Linear and spline cubic
Pilot frequency	20
Pilot energy	10 dB
Cyclic prefix length	512
Channel length (number of taps), L	20
Modulation technique	16 QAM, BPSK
Number of FFT points	2048
Equalization type	Zero forcing (ZF)
System model	2 × 2 MIMO system

Table 2 MSE comparison of CE techniques with and without DFT approach

Channel estimation techniques	MSE without DFT	MSE with DFT
LS-linear	0.0064	0.0014
LS-spline	0.0054	0.0011
MMSE	1.4236	1.1694

and cubic spline interpolation, respectively, with ideal channel response. Number of symbol errors is 1746. LS scheme has high MSE. Therefore, by using interpolation scheme, better results are obtained. It is observed that LS-linear and LS-spline provide good response near to ideal channel response till 28 subcarrier index. After that further performance of LS channel estimation is improved by using Discrete Fourier Transform (DFT) approach along with spline interpolation method. Figure 9 shows the comparison of LS-spline using DFT-based approach with ideal channel response. Results indicate that LS-spline using DFT-based approach is superior to LS and LS-spline channel estimation scheme. It provides good performance over entire 30 subcarrier index.

Figures 10 and 11 show the comparison of MMSE channel estimation and MMSE using DFT-based approach with ideal channel response, respectively. Results show that MMSE with DFT approach is superior to MMSE channel estimation. MMSE along with DFT-based approach attains much better results than MMSE channel estimation. Complexity of MMSE channel estimation is reduced by using DFT approach with it.

Figure 13 shows SNIR response of LS-linear CE technique and of MMSE CE technique. Linear interpolation is used for exploiting LS estimates and SNIR is

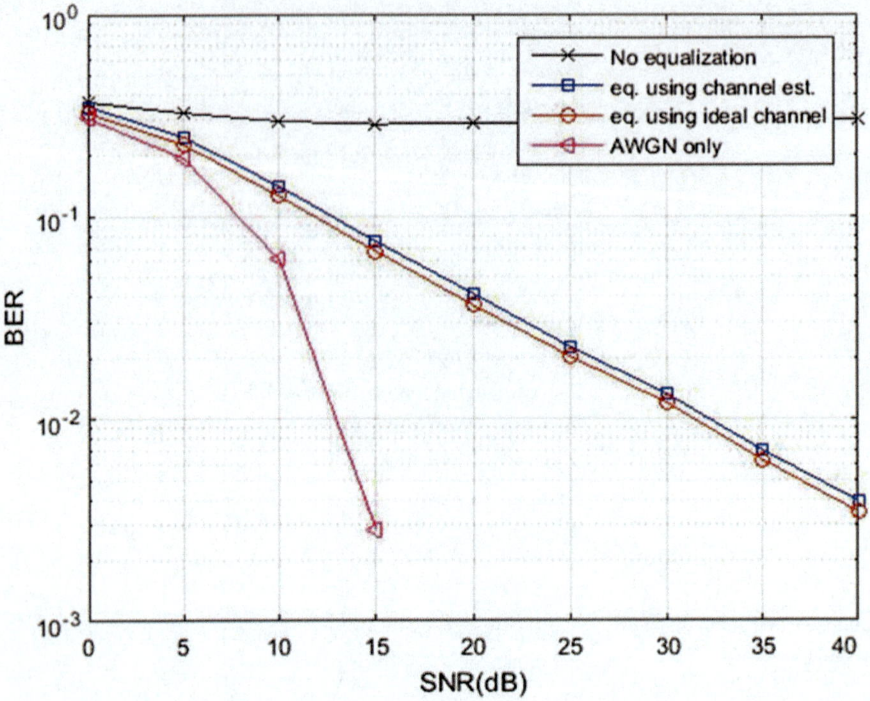

Fig. 6 Comparison of channel estimation using equalization with ideal channel estimation in AWGN (in terms of BER vs. SNR)

Table 3 BER versus SNR comparison of CE techniques with and without equalization

BER	SNR 0 dB	SNR 5 dB	SNR 10 dB	SNR 15 dB	SNR 20 dB	SNR 25 dB	SNR 30 dB	SNR 35 dB	SNR 40 dB
CE without equalization (BER)	0.3692	0.3265	0.2961	0.2864	0.2898	0.2958	0.3007	0.3013	0.3019
CE with equalization	0.3502	0.2477	0.1414	0.0754	0.0413	0.0226	0.0133	0.0070	0.0039
Ideal channel with equalization	0.3241	0.2265	0.1261	0.0672	0.0362	0.0203	0.0120	0.0062	0.0034

compared with MMSE. LS-linear has very high SNIR than MMSE scheme. The above figures show us that the DFT-based estimation of symbols is much more efficient than the linear LS and MMSE techniques of estimation. Also, by comparing Figs. 10, 11, 12, and 13, it is observed that the MMSE CE performance is better than that of LS CE in terms of MSE. Figure 14 shows illustration of insertion of pilot symbols. Figure 15 shows compared analysis of LS, MMSE CE techniques with ideal

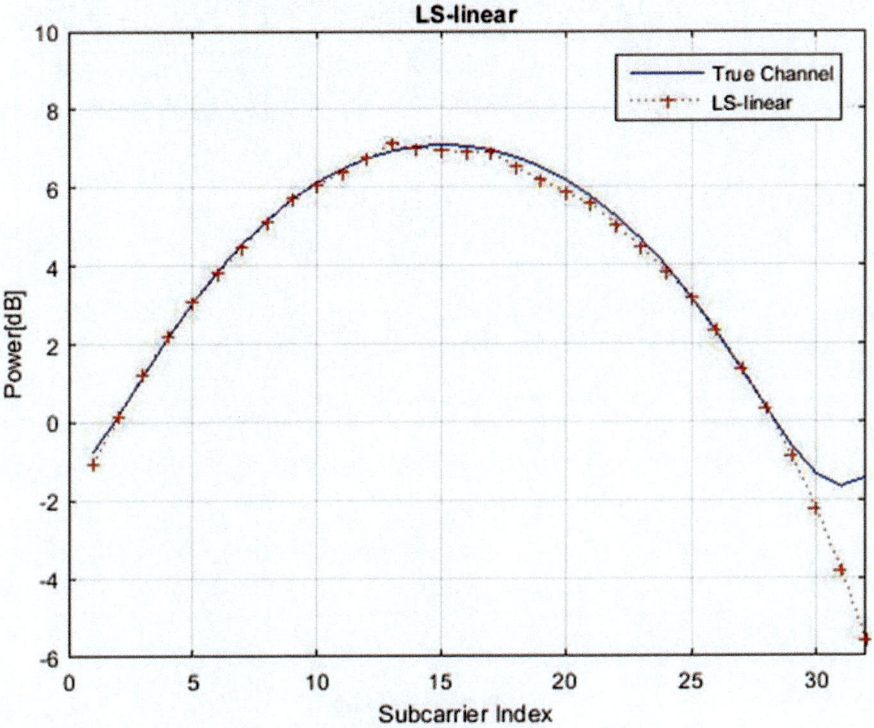

Fig. 7 Comparison of LS-linear with ideal channel estimation

channel estimation technique. Figure 13 indicates SNIR comparison of LS, MMSE CE techniques with ideal or actual channel response. LS and MMSE CE schemes exhibit similar SNIR.

6 Conclusion

In this work, SNIR and BER performances of LS and MMSE CE techniques are evaluated. According to results, it is observed that MMSE CE technique has lower MSE than LS CE technique. DFT CE technique with LS-spline interpolation provides better results in comparison with LS and LS-linear and LS-spline CE techniques. DFT CE scheme improves the BER and SNIR performance of LS and MMSE with $L = 20$ using 16 QAM over Rayleigh fading channel for 2×2 MIMO system and provides better results as compared to BPSK and 8 QAM modulation techniques. Also, better MSE performance is obtained via using zero-forcing equalization technique. In future, two-dimensional interpolation techniques can be used for improving the performance.

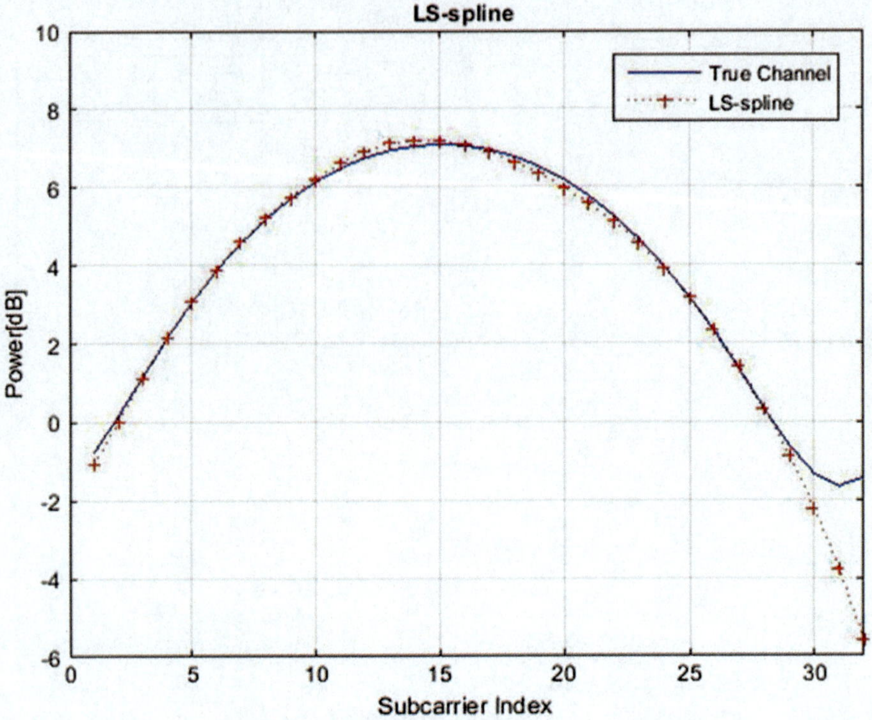

Fig. 8 Comparison of LS-spline channel estimation technique with ideal channel estimation

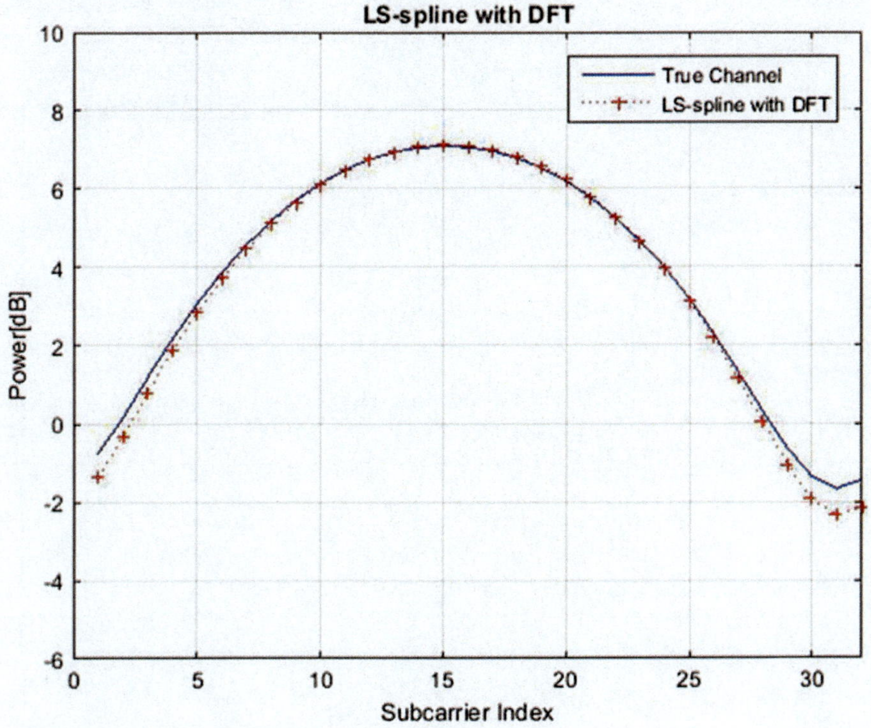

Fig. 9 Comparison of LS-spline using DFT approach with ideal channel

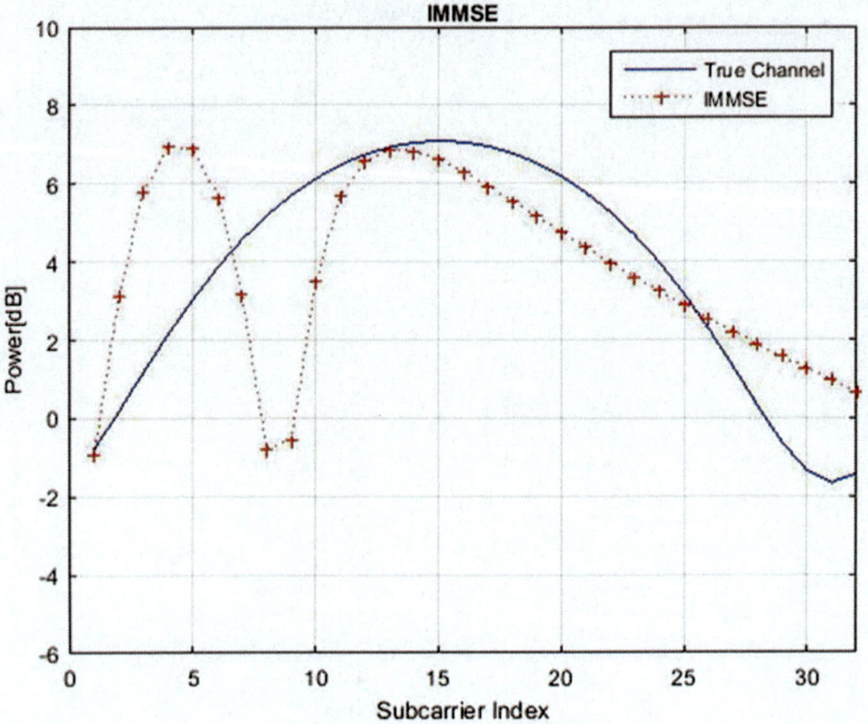

Fig. 10 Comparison of MMSE channel estimation with ideal channel

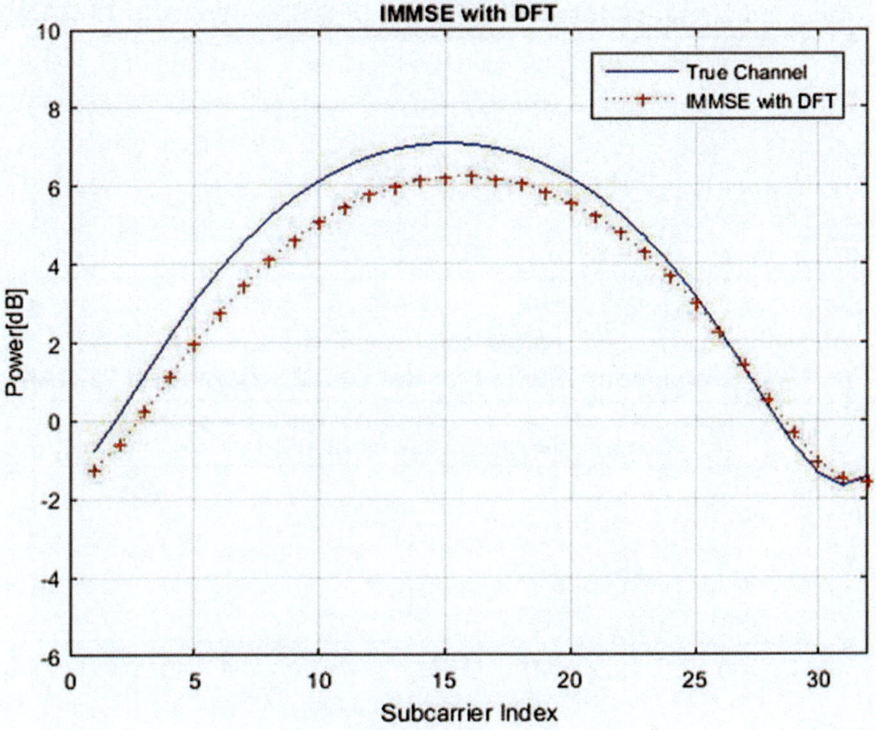

Fig. 11 Comparison of MMSE-DFT CE with ideal channel

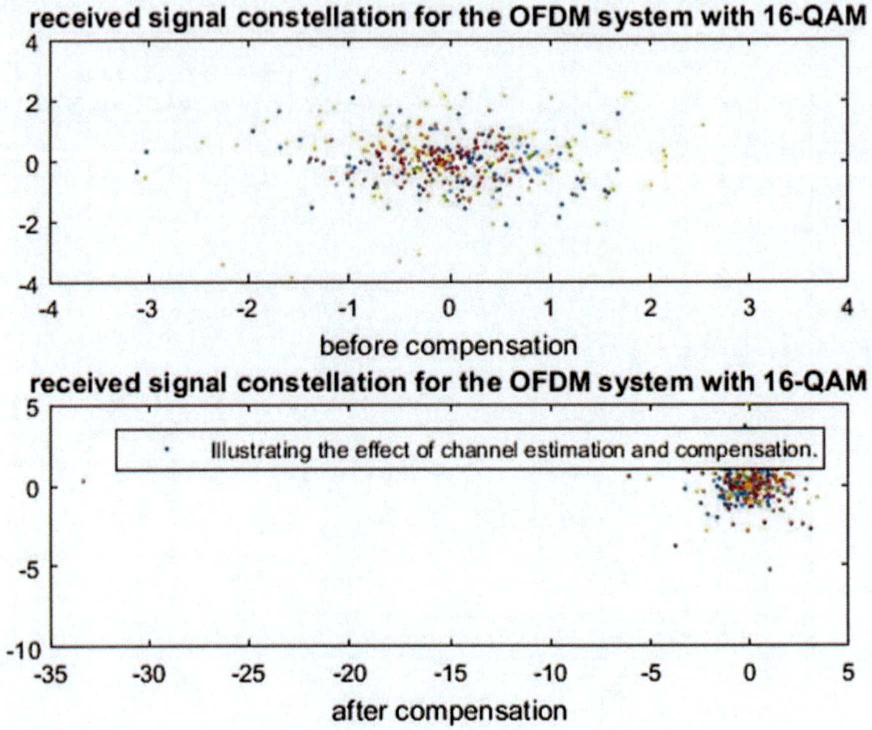

Fig. 12 Illustration of effects of channel estimation and compensation

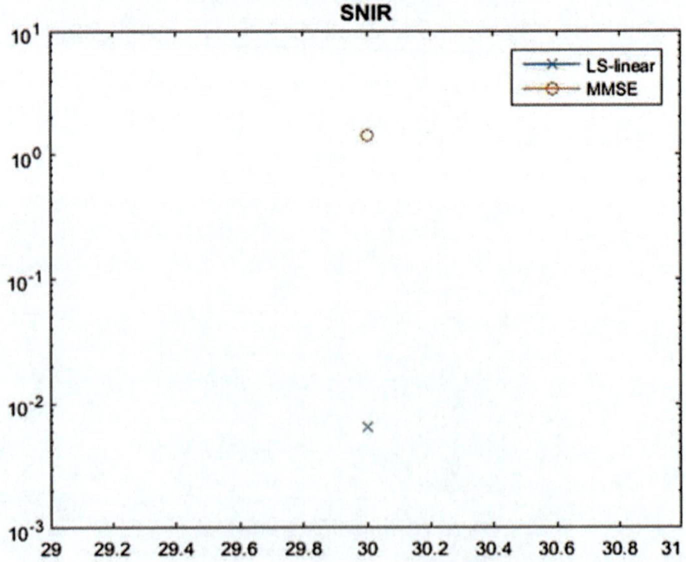

Fig. 13 SNIR comparison of LS-linear with MMSE

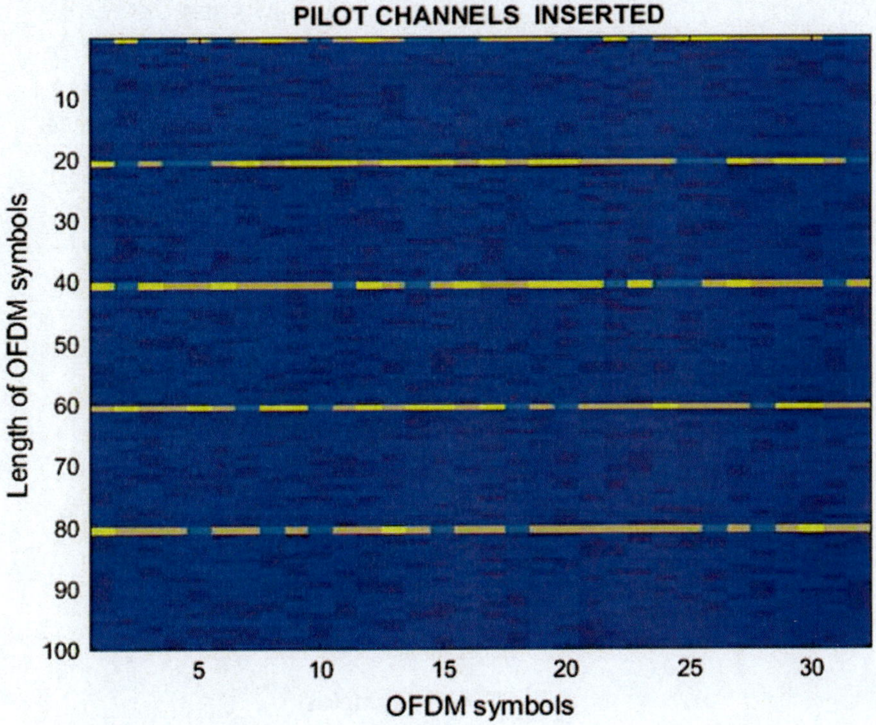

Fig. 14 Illustration of insertion of pilot channel

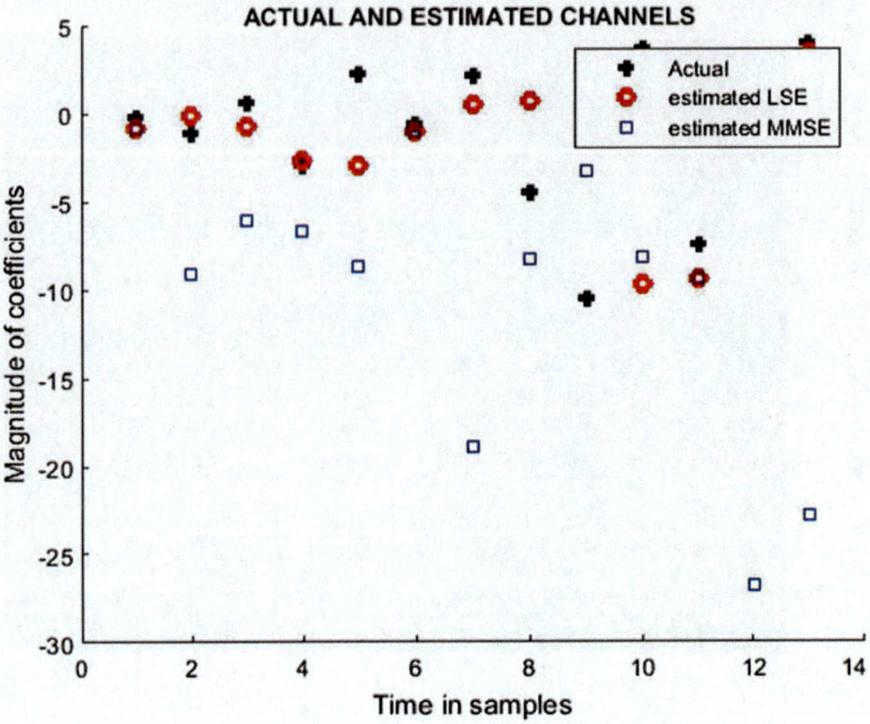

Fig. 15 Comparison of LS, MMSE, and actual channel estimation techniques

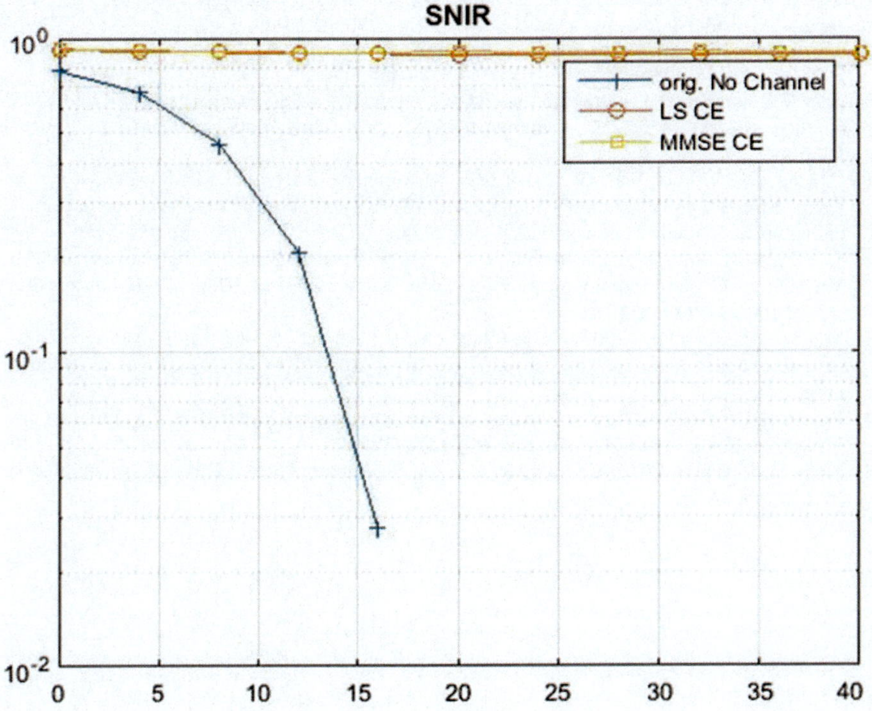

Fig. 16 SNIR comparison of LS, MMSE, and original channel estimation techniques

References

1. Drieberg M, Min YK, Jeoti V (2005) A new practical MIMO channel estimation technique for IEEE802.16a: performance analysis. In: TENCON 2005—2005 IEEE Region 10 conference, Melbourne, Qld, 2005, pp 1–6. https://doi.org/10.1109/tencon.2005.300941
2. Usha M, Mahesh B (2020) Least square channel estimation for image transmission with OFDM over fading channel
3. Sutar M, Patil V (2019) LS and MMSE Estimation With Different Fading Channels For OFDM system
4. Shrivas AK et al (2015) A comparative analysis of LS and MMSE channel estimation techniques for MIMO-OFDM system. Int J Innov Res Sci Technol 1(8):44–48
5. Ozdemir MK, Arlsan H (2007) Channel estimation for wireless OFDM systems. Commun Surv Tutor IEEE 9:18–48
6. Ijiga O, Ogundile O, Akpakwu G, Oloyede M (2019). Analysis of block type channel estimation techniques for NC-OFDM transceiver over frequency selective Rayleigh fading channel
7. Ye L (2000) Pilot-symbol-aided channel estimation for OFDM in wireless systems. IEEE Trans Veh Technol 49:1207–1215
8. Khan AM, Jeoti V, Zakariya MA (2013) Improved pilot-based LS and MMSE channel estimation using DFT for DVB-T OFDM systems. In: 2013 IEEE symposium on wireless technology & applications (ISWTA), Kuching, 2013, pp 120–124. https://doi.org/10.1109/ISWTA.2013.6688752

9. Al-Salihi H, Nakhai MR, Le TA (2018) DFT-based channel estimation techniques for massive MIMO systems. In: 2018 25th international conference on telecommunications (ICT), St. Malo, 2018, pp 383–387. https://doi.org/10.1109/ict.2018.8464829
10. Zaib A, Khattak S (2017) Structure-based low complexity MMSE channel estimator for OFDM wireless systems. Wirel Pers Commun 97:5657–5674. https://doi.org/10.1007/s11277-017-4800-4
11. Chaudhary SR, Thombre MP (2014) BER performance analysis of MIMO–OFDM system using different equalization techniques. In: IEEE International conference on advanced communication control and computing technologies (ICACCCT), 2014, pp 637–677
12. Cox C (2014) Multiple antenna techniques. An introduction to LTE: LTE, LTE-advanced, SAE, VoLTE and 4G mobile communications, 2nd edn. Wiley, Hoboken, NJ, pp 87–104. https://doi.org/10.1002/9781118818046.ch5
13. You-Seok L, Kim HN (2007) Noise-robust channel estimation for DVB-T fixed receptions. In: International conference on consumer electronics, 2007. ICCE 2007. Digest of technical papers, pp 1–2
14. Parmar KA et al (2016) A survey on channel estimation algorithms for LTE downlink systems. Int J Innov Res Comput Commun Eng 4(2):1251–1258
15. Singh H, Bansal S (2016) Comparison of channel estimators for OFDM channel estimation. Int J Comput Appl 975:8887

Deep Learning Approach for Speech Emotion Recognition

M. Kalpana Chowdary and D. Jude Hemanth

Abstract Emotion recognition plays a dominant role in man-machine interaction. This research paper proposes the detection of human emotions through the study of speech signals. The main motivation of recognition of emotions from speech signals is to increase the communication between man and machines that is advantageous in numerous fields like call centers, lie detection machines, clinical studies, computer games, etc. In this work, to extract the features, the well-known Mel-frequency Cepstral Coefficient (MFCC) feature extraction technique is used, and convolutional neural networks are used to classify the emotions. According to the results, the model achieves the accuracy of 92% for eight emotions on the RAVDEES dataset.

Keywords Emotions · Man-machine interaction · Feature extraction · MFCC · Convolution neural networks

1 Introduction

The most natural and fastest communication between human beings is speech signals. Nowadays, speech emotion identification systems are playing major roles in the fields of human-machine communication [1], psychiatric diagnosis, customer voice review [2], etc. The main elements of any SER are the extraction of features from speech signals and classifying into different emotions. Some of the most used speech features are pitch, formants, energy, spectral features, and MFCC features. After the feature extraction, different classifiers like Support Vector Machines (SVM) [3], Gaussian Mixture Models (GMM), decision trees, Bayesian Network models, Nearest Neighbor [4], and Ensemble methods are used for a classification task [5].

M. Kalpana Chowdary · D. Jude Hemanth (✉)
Department of ECE, Karunya Institute of Technology and Sciences, Coimbatore, India
e-mail: judehemanth@karunya.edu

M. Kalpana Chowdary
e-mail: mkalpana306@gmail.com

A. Khanna et al. (eds.), *Data Analytics and Management*, Lecture Notes on Data Engineering and Communications Technologies 54,
https://doi.org/10.1007/978-981-15-8335-3_29

Nowadays, researchers are paying more attention to deep learning techniques. Some of the deep learning applications are [6] chatbots, image and voice recognition, text sentiment analysis, automated driving, and data mining. The authors of [7] give a detailed review of deep learning techniques for SER and also mentioned the performance and limitations of various deep learning methods. In [8], the authors used MFCC, pitch, energy, and Log-Mel spectrogram features for SER. And then, the features are given to four different networks of LSTM, DNNs, CNNs, and HMM for classification and comparison. The authors of [9] proposed a ConvNet model with a joint random forest model. In this CNNs are used to extract the features, and the random forest model is used for classifying the emotions. Damodar et al. [10] proposed voice-based emotion recognition by using MFCC features with CNN and decision tree classifiers. In this work, decision tree achieved a very low accuracy of 38% and CNNs achieved better accuracy up to 70%. Tzirakis et al. [11] have presented the combination of both recurrent and convolution networks for speech expression identification. In this work, the features are extracted by using CNNs, and the extracted features are fed to long short-term memory for prediction.

- The major contribution of this paper is to overcome the problems in the aforementioned literature and to increase the speech emotion recognition rate by using deep learning techniques.

The contents of the paper are further arranged into the subsequent sections. Section 2 outlines the speech emotion recognition system along with the database and MFCC feature extraction. Section 3 deals with convolution neural networks. Section 4 presents the experimental results of the proposed model, and lastly, Sect. 5 is the conclusion.

2 Speech Emotion Recognition Structure

The speech emotion recognition process mainly depends on three modules: preprocessing, extraction of features, and classifying the emotions. The schematic representation of the speech emotion recognition process is displayed in Fig. 1.

The input audio files are taken from the RAVDEES database. In the proposed technique, Mel-frequency Cepstral Coefficients are taken into consideration as speech features. The MFCC features are given to convolution neural network for classification.

Fig. 1 Schematic diagram of speech emotion identification process

2.1 Database

The RAVDEES speech and song emotional database is used in this work [12]. In total, the database contains 7356 files in three modalities of video only, audio only, and audio-visual. The information was recorded from twenty-four actors of twelve males and twelve females with eight expressions. The total number of files used in this work is 2452 in that 1440 are audio speech files and 1012 are audio song files.

2.2 MFCC Features

MFCC is the most frequent technique used to separate speech components [13]. The advantages of MFCC are less complex, more effective, and suitable in various conditions [14]. The main blocks involved in MFCC are shown in Fig. 2.

The following are the steps to extract the MFCC features [15, 16]:

Step 1: Pre-emphasis

The process to raise the strength of the speech signal at higher frequencies is called pre-emphasis. The equation is shown below

$$Y[t] = X[t] - 0.95X[t-1] \tag{1}$$

Step 2: Framing

The segmentation of speech signal into short time intervals of 20–30 ms is called framing. Generally, the speech signal is broken up into N samples and adjoining frames are separated through M. The same old values for $M =$ one hundred and $N = 256$.

Step 3: Windowing

To limit the disturbances within the starting and quit of the frame windowing approach is used, and the input body is multiplied with the windowing function. The windowing signal result is shown below.

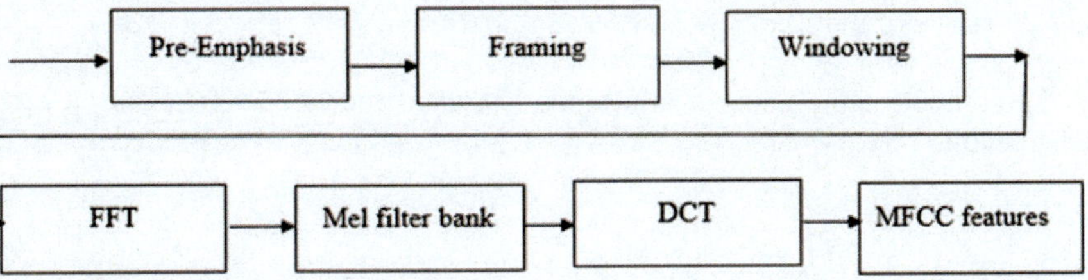

Fig. 2 Flow diagram of MFCC

$$Y(t) = X(t)W(t) \tag{2}$$

The most popularly used windowing technique is a Hamming window. The mathematical representation of the window is

$$W(t) = 0.54 - 0.46\cos\left[\frac{2\pi t}{N-1}\right] \quad 0 \le t \le N-1 \tag{3}$$

where $Y(t)$ = output signal
$X(t)$ = input signal
$W(t)$ = windowing function
N = number of samples in every frame.

Step 4: Fast Fourier Transform

The conversion of the time area to frequency area is carried out by the use of Fast Fourier Transforms. FFT is particularly carried out to get the significant frequency response of each frame.

$$X_k = \sum_{n=0}^{N-1} x(n)e^{\frac{-j2\pi kn}{N}} \tag{4}$$

Step 5: Mel filter bank

To get a smooth magnitude spectrum, we need to multiply the magnitude frequency reaction by a fixed of 20–30 triangular band skip filters.

$$\text{mel}(f) = 2595\log_{10}\left(\frac{f}{700} + 1\right) \tag{5}$$

Step 6: Discrete Cosine Transform

The DCT is used to transform the Mel spectrum again to the time domain. The mathematical equation to represent the DCT is shown below.

$$C_m = \sum_{k=1}^{N} \cos\left[m \times (k-0.5) \times \frac{\pi}{N}\right] \times E_k \tag{6}$$

The output of the above conversions is considered as Mel-frequency Cepstral Coefficient.

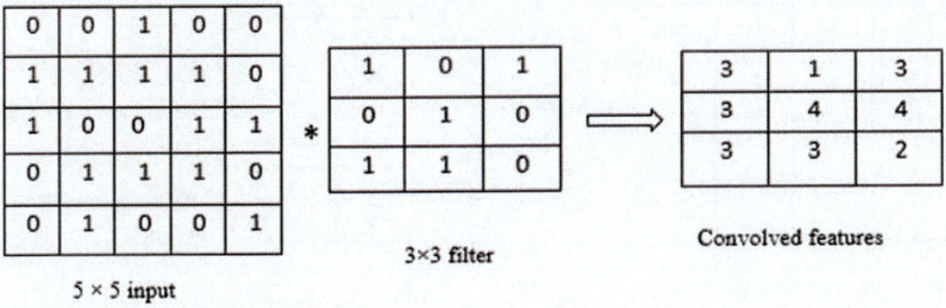

Fig. 3 Convolution operation of 5 × 5 matrix with 3 × 3 filter

3 Convolution Neural Networks

3.1 Overview of Convolution Neural Networks

CNNs are one of the most important deep learning models. CNNs are succeeding in many areas like image processing [17], analyzing documents, computer vision, object recognition, and text classification. The main constructing layers utilized in convolution neural networks are the convolution stage, pooling stage, and fully connected layer [18].

3.1.1 Convolutional Layer

To excerpt, the features from the inputs convolution layer are used. These feature maps are produced by the multiplication of input data with so many filters [19]. The illustration of the convolution procedure is shown in Fig. 3.

3.1.2 Pooling Layer

The motive of the pooling layer is to lessen the scale of convolved features. By considering only the dominant information, the computation power required to progress the data is likewise reduced. Generally used pooling strategies are Max pooling and Average pooling [20]. From the selected portions, the maximum value is obtained by using the Max pooling process, and the average values are obtained by the Average pooling process. The instance of Max and Average pooling is shown in Fig. 4.

3.1.3 Fully Connected Layer

It takes the input from the preceding layers and produces the 1D array output. The dimensions of the output rely upon the number of classes [21].

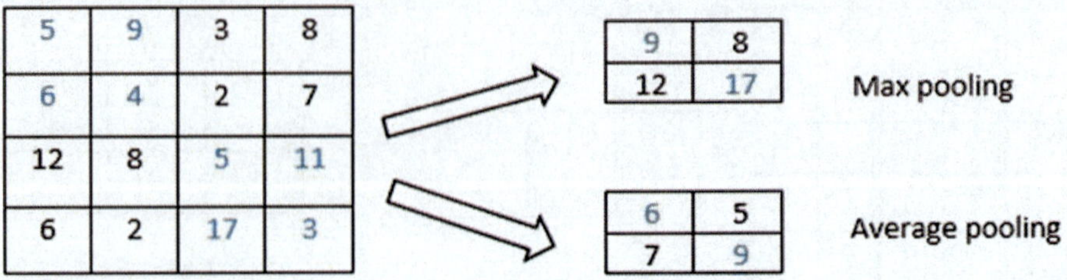

Fig. 4 Pooling with stride 2

Table 1 Design of the suggested CNN

Proposed convolutional neural network
CONV1D-256
RELU
DROPOUT
MAXPOOL 1D
CONV1D-256
RELU
DROPOUT
FLATTEN
FULLY CONNECTED
SOFTMAX

3.2 *Proposed Convolution Neural Network*

In the proposed convolution neural network model, the primary convolution layer contains 256 filters of size 5×5 and ReLu activation function accompanied by way of Max pooling layer of size 8. And then next the second convolution layer also contains the equal quantity of filters and the ReLu activation function. After that, we used the flatten layer and end with a dense layer of 8 classes with a softmax activation function. Table 1 suggests the structure of the proposed convolution neural network.

The Keras implementation of the proposed ConvNet with the entire range of trainable and non-trainable parameters is displayed in Fig. 5.

4 Experimental Results

In this work total, 2452 files are taken from the RAVDEES database in that 1440 are emotional speech files and 1012 are emotional song files. All the files are in .wav format. In total 2452 files, 1642 files are used for training and 810 files are used for

```
Layer (type)                 Output Shape              Param #
=================================================================
conv1d_14 (Conv1D)           (None, 40, 256)           1536
_________________________________________________________________
activation_21 (Activation)   (None, 40, 256)           0
_________________________________________________________________
dropout_14 (Dropout)         (None, 40, 256)           0
_________________________________________________________________
max_pooling1d_9 (MaxPooling1 (None, 5, 256)            0
_________________________________________________________________
conv1d_15 (Conv1D)           (None, 5, 256)            327936
_________________________________________________________________
activation_22 (Activation)   (None, 5, 256)            0
_________________________________________________________________
dropout_15 (Dropout)         (None, 5, 256)            0
_________________________________________________________________
flatten_7 (Flatten)          (None, 1280)              0
_________________________________________________________________
dense_7 (Dense)              (None, 8)                 10248
_________________________________________________________________
activation_23 (Activation)   (None, 8)                 0
=================================================================
Total params: 339,720
Trainable params: 339,720
Non-trainable params: 0
```

Fig. 5 Summary of the proposed model

testing. It contains eight emotions: sad, happy, fear, neutral, surprise, calm, angry, and disgust. Table 2 shows the confusion matrix for the test data of 810 samples.

By using the confusion matrix, the performance metrics of sensitivity, specificity, and accuracy are calculated in this work. Exact recognition of true values is measured by sensitivity, and the exact recognition of negative values is measured by specificity. Finally, accuracy is the proportion of all true values to all values. All this sensitivity, specificity and accuracy are measured in percentages (%). Below are the formulas to calculate the performance metrics.

Table 2 Confusion matrix for test data

	Neutral	Calm	Happy	Sad	Angry	Fearful	Disgust	Surprised
Neutral	38	6	0	7	2	0	1	2
Calm	7	105	4	4	1	3	4	2
Happy	2	7	92	1	11	5	0	4
Sad	9	6	5	77	4	20	2	2
Angry	1	1	6	2	106	5	7	4
Fearful	3	1	8	14	5	82	2	2
Disgust	3	2	2	4	5	1	40	8
Surprised	3	0	6	4	5	2	5	38

$$\text{Accuracy} = \frac{\text{TN} + \text{Tp}}{\text{FP} + \text{FN} + \text{TN} + \text{Tp}} \tag{7}$$

$$\text{Sensitivity} = \frac{\text{TP}}{\text{TP} + \text{FN}} \tag{8}$$

$$\text{Specificity} = \frac{\text{TN}}{\text{TN} + \text{FP}} \tag{9}$$

where TP = true positive, FP = false positive, TN = true negative, and FN = false negative.

From the above calculations, the accuracy of the proposed convolutional neural network for all the eight emotions is 92%.

Compared to some of the existing methods, our method achieved the best accuracy of 92% using convolution neural networks.

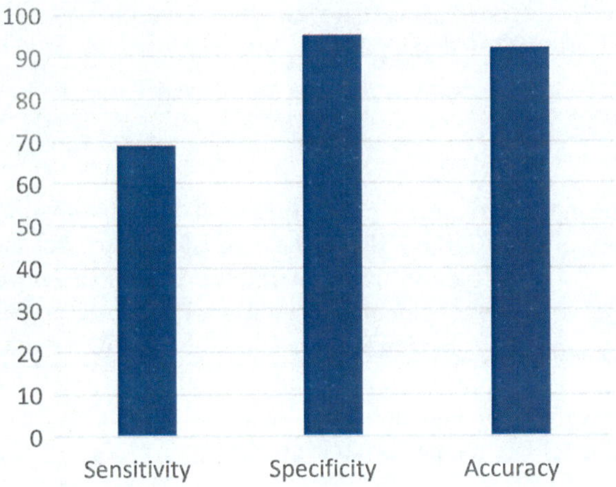

Fig. 6 Graphical representation of performance metrics

Table 3 Performance measures calculation

	TP	TN	FP	FN	Sensitivity	Specificity	Accuracy
Neutral	38	726	28	18	0.67	0.96	0.94
Calm	105	657	23	25	0.80	0.96	0.94
Happy	92	657	31	30	0.75	0.95	0.92
Sad	77	649	36	48	0.61	0.94	0.89
Angry	106	645	33	26	0.80	0.95	0.92
Fearful	82	657	36	35	0.70	0.94	0.91
Disgust	40	724	21	25	0.61	0.97	0.94
Surprised	38	723	24	25	0.60	0.96	0.93
Average results					0.69	0.95	0.92

Table 4 Comparison with existing models

S. No.	Work	Accuracy (%)
1	Lim et al. [22]	86.6
2	Likitha et al. [15]	80
3	Basu et al. [23]	80
4	Damodar et al. [10]	72

5 Conclusions

In this work, we have given a new ConvNet architecture for speech emotion recognition. MFCC features are used as speech signal features that are fed to convolution neural networks for classifying the emotions. The experiment was conducted on the RAVDEES database and achieved an accuracy of 92%. The proposed approach can be helpful in so many applications such as social robots, call center conversations, lie detection. In the future, we would like to study deeper ConvNets and recurrent models for a better identification rate.

References

1. Cowie R, Douglas-Cowie E, Tsapatsoulis N, Votsis G, Kollias S, Fellenz W, Taylor JG (2001) Emotion recognition in human-computer interaction. IEEE Signal Process Mag 18:32–80
2. Lugović S, Dunđer I, Horvat M (2016) Techniques and applications of emotion recognition in speech. In: 39th international convention on information and communication technology, electronics and microelectronics. IEEE, pp 1278–1283
3. Pan Y, Shen P, Shen L (2012) Speech emotion recognition using support vector machine. Int J Smart Home 6:101–108
4. Praksah C, Gaikwad VB (2015) Analysis of emotion recognition system through speech signal using KNN, GMM & SVM classifier. IOSR J Electron Commun Eng (IOSR-JECE) 10(2):55–67
5. Chen L, Mao X, Xue Y, Cheng LL (2012) Speech emotion recognition: features and classification models. Digit Signal Process 22(6):1154–1160
6. Dargan S, Kumar M, Ayyagari MR, Kumar G (2019) A survey of deep learning and it's applications: a new paradigm to machine learning. Arch Comput Methods Eng 1–22
7. Khalil RA, Jones E, Babar MI, Jan T, Zafar MH, Alhussain T (2019) Speech emotion recognition using deep learning techniques. A review. IEEE Access 7:117327–117345
8. Venkataramanan K, Rajamohan HR (2019) Emotion recognition from speech. arXiv preprint arXiv:1912.10458
9. Zheng L, Li Q, Ban H, Liu S (2018) Speech emotion recognition based on convolution neural network combined with random forest. In: Chinese control and decision conference (CCDC). IEEE, pp 4143–4147
10. Damodar N, Vani HY, Anusuya MA (2019) Voice emotion recognition using CNN and decision tree. Int J Innov Technol Expl Eng 8(12):4245–4249
11. Tzirakis P, Zhang J, Schuller BW (2018) End-to-end speech emotion recognition using deep neural networks. In: IEEE international conference on acoustics, speech and signal processing (ICASSP). IEEE, pp 5089–5093

12. Livingstone SR, Russo FA (2018) The Ryerson audio-visual database of emotional speech and song (RAVDESS). A dynamic, multimodal set of facial and vocal expressions in North American English. PLoS ONE 13(5)
13. Ghadge SA, Janvale GB, Deshmukh RR (2010) Speech feature extraction using mel–frequency cepstral coefficient (MFCC). In: International conference on emerging trends in computer science communication and information
14. Majeed SA, Husain H, Samad SA, Idbeaa TF (2015) Mel frequency cepstral coefficients (MFCC) feature extraction enhancement in the application of speech recognition: a comparison study. J Theor Appl Inf Technol 79:38–56
15. Likitha MS, Gupta SRR, Hasitha K, Upendra Raju A (2017) Speech based human emotion recognition using MFCC. In: 2017 international conference on wireless communications, signal processing and networking (WiSPNET). IEEE, pp 2257–2260
16. Saste ST, Jagdale SM (2017) Emotion recognition from speech using MFCC and DWT for security system. In: International conference of electronics, communication and aerospace technology (ICECA), vol 1. IEEE, pp 701–704
17. Bhandare A, Bhide M, Gokhale P, Chandavarkar R (2016) Applications of convolutional neural networks. Int J Comput Sci Inf Technol 7(5):2206–2215
18. Wu J (2017) Introduction to convolutional neural networks. National Key Lab for Novel Software Technology, Nanjing University, China, p 5
19. O'Shea K, Nash R (2015) An introduction to convolutional neural networks. arXiv preprint arXiv:1511.08458
20. Indolia S, Goswami AK, Mishra SP, Asopa P (2018) Conceptual understanding of convolutional neural network—a deep learning approach. Procedia Comput Sci 132:679–688
21. Albawi S, Mohammed TA, Al-Zawi S (2017) Understanding of a convolutional neural network. In: 2017 international conference on engineering and technology (ICET). IEEE, pp 1–6
22. Lim W, Jang D, Lee T (2016) Speech emotion recognition using convolutional and recurrent neural networks. In: 2016 Asia-Pacific signal and information processing association annual summit and conference (APSIPA). IEEE, pp 1–4
23. Basu S, Chakraborty J, Aftabuddin M (2017) Emotion recognition from speech using convolutional neural network with recurrent neural network architecture. In: 2017 2nd international conference on communication and electronics systems (ICCES). IEEE, pp 333–336

Detection of Cache Pollution Attacks in a Secure Information-Centric Network

Akanksha Gupta and Priyank Nahar

Abstract The future of Internet architectures is information-centric networking structure, to solve the problems of content spoofing attacks in the current Internet structure, making it more useful for IoT-based applications. The ICN is structured with the Internet forwarding state technology which is an advanced technology with a comparative structure. In this paper, we are concentrating on the Internet forwarding strategy which uses data forwarding in NDN-based networking. It understands content priority and prefixes the content parameter and passes through the named data network to deliver the packet based on the demands. Also, future Internet router cache could face the problem of overflowing with non-popular content due to cache pollution attack (CPA); i.e., the router keeps receiving requests for vulnerable content. The detection and defense against such spoofing attacks are especially difficult due to cache pollution attack's similarities with every other consumer request. Based on the hobby content priority, named records networking accelerates the process and decreases the traffic to reach the request with low latency site visitors. We thereby address the present-day measures, arrangements and endeavors of the apparatus's applied clustering approach to discover and defend against CPAs. Finally, we recommend the improved decision tree method where once any attack is detected an assault table will be updated to report any abnormal requests. While such requests are nevertheless forwarded, the corresponding content chunks are not cached. We carry out the above technique simulations with the aid of ndnSIM.

Keywords Information-centric networking (ICN) · Content pollution attack · Clustering · False locality attack · Location distribution attack · Internet of Things · Smart agriculture · Attack detection

A. Gupta (✉) · P. Nahar
Shri Venkateshwara University, Gajraula, Uttar Pradesh, India
e-mail: agupta.du@yahoo.com

P. Nahar
e-mail: priyank.nahar@gmail.com

A. Khanna et al. (eds.), *Data Analytics and Management*, Lecture Notes on Data Engineering and Communications Technologies 54,
https://doi.org/10.1007/978-981-15-8335-3_30

1 Introduction

In recent years, several projects have been designed in promising architectures for investigating new network architectures based on named data networking. These futuristic approaches of architecture design integrate with information-centric networking (ICN). In general, these structures are designed to overcome the traditional IP-based networking issues which cause poor mobility, scalability and low-efficient content distribution. Additionally, the proposed structure has been the ideal solution for replacing the traditional TCP/IP-based Internet model. Moreover, it does support content-based Internet services. Content-centric networking (CCN) is one of the best examples for named data network architecture which carries out the communication via content rather than TCP/IP-based hosting [1]. Instead of host-centric architecture, the Internet future architecture has more features such as high dissemination power and traffic management as well. The future Internet architecture shall fulfill the lack of new content-based IoT agriculture applications. The main feature of the NDN is network caching in multipath communication with secure data retrieval in content-oriented IoT agriculture applications. Named data networking routers distribute large scale of content based on incoming packet. In order to fulfill NDN consumer request, the router forwards the subsequent content against flow of retrieval. These are the processes which improve the NDN in various sequences; efficient multipath communication, quick content retrieval, network traffic control and fault tolerance, though dense caching causes security problems against various spoofing attacks and cache pollution. Finally, network caches register malicious activities in terms of violating the content locality. Further, it will affect cache misses and latency of content response throughout the consumers. One of the main security features of NDN content-centric network is interested about what kind of content is stored in the network structure, where is the IP layer structured, and where to store the content. The process of current Internet structure faces security-related issues with respect to content storing. There are two types of cache pollution attacks: false locality attack and location distribution attack. Location distribution attack starts the request of distinct unpopular content through which the transmission path occupies unpopular content and pushes out the popular content. Finally, the user does not fulfill the requirement. With continuation of LDA attack, the network follows up an FLA attack where the malicious user frequently sends requests for unpopular content due to which the router keeps unpopular content in the cache and it does affect the original users.

To design an advanced detection system for cache pollution attack, understanding the impacts of attacks throughout the network is more important. The current detection symptoms are not efficient to detect the pollution attack with respect to indistinguishable unpopular content from router cache. The router content is difficult to distinguish, and it also increases the computation time to respond to legitimate requests. On the other hand, the network does receive frequent requests from attackers without interval which may cause large memory usage and computation time. Cache pollution attack scales routing topologies without knowing network servers which will restrict large number of NDN application deployments. [2–4] The proactive countermeasures will protect the NDN network from adversarial attack moments. The existing system does not maintain protective circumstances detection approach and mitigates pollution attack.

Our proposed solution shall consider mentioned aspects and contribute the better computation and response to users. The clustering mechanism is one of the prominent methods to overcome the difficulties. Named data network security mechanism would have been the solution for discussed challenges than in current host-centric method. The host-centric authentication security mechanism should not provide security for content, but it does secure the hosting part. The host-centric model basically describes how to open the communication channel to start the host basic Interest approach and through that way it makes security for hosts, channel security and end-to-end device security possible. The proposed solution should consider the quality of the content since it provided well enough security mechanism. Due to the security aspect the network channel restricts some quality features in terms of memory and interest traffic. Unexpected arrivals of data packets in NDN router queue which does not have capacity to handle the interest traffic will affect the response quality. Also, attack detection system should analyze the interest queue because when the legitimate user sends the interest request to the NDN data structure, the system immediately responds back via router content store. Since CS is utilized for content, cache sends it through via NDN router. The existing DoS/DDOs and cache pollution attacks describe the unpopular cache content requests which might reduce the intensity of content quality.

Further, in the advanced NDN infrastructure, the proposal needs to be flexible, application adaptable and auto learning system, which can play a prominent role in future real time IoT applications. The system takes care and provides sufficient security against various content-based attacks. Achieving high response of content, private cache availability and identifying unknown spoofing attacks with mitigation, are the areas that need enhancement, to provide NDN quality measurements.

In this paper, we investigate through various research references on NDN and present the algorithm we rely on in our proposal design. We present short overview of unknown threat detection and mitigation design and evaluate content quality measurements with NDN simulation results.

2 NDN Overview

Named data networking is a future Internet architecture which has been designed with pull model and inserts the content into the router based on the consumer's request. Even then, the consumer request has been fulfilled through interests and content objects. The content object basically constructs binary data together combined with its name, bits of confirmation and data signature [5, 6]. In hierarchical structure, NDN content has more than one component. High volume content is spited with frame fragments, and it is easily identifiable via its name. The optional control information provides the filter facility to send desired content by the same name which is stored in the router. The NDN router forwards consumer requested interest name toward the network content producer since the content is stored in the producer with its name using prefix where as in host-centric IP prefix utilizes for requesting the content [7, 8].

Every step of data process will hold your forwarding information base (FIB). It is called lookup table, and it supports multipath delivery towards frequent forwarding incoming entries in the same prefix name. The NDN router maintains the pending interest table (PIT), and it stacks outstanding consumer interest. The final step of data process is to retrieve the content from content producer and traverse the same path to reach the consumer. The above architecture clearly describes about content cache in router content store. The refined structure maintains a cache of a copy of content in each router content store. Every consumer request does not need to be responded to from content producer and further then request will get the response from nearest content availability.

3 Literature Review

Li proposed architecture-based encryption concept for ICN control [9], where the subject and content will be shared to all clients who are authorized in the system and all the authorized clients will maintain a symmetric key. The symmetric key will encrypt the content name and subject. The publisher creates the AC policy which can be third-party accessible. Whenever the publisher sends the content and the subject to the client, the third party will encrypt the data and send to the client and the symmetric key encrypts the details and shares it to the client. AC policy does the encryption on the process and indicates the details to the clients. The process of the ICN architecture will maintain the system. The information-centric networking would be working under Internet scenarios of users, but normally the host and IP-based architecture would be in place. Here usage scenario will accommodate the data which is information-centric networking. Consider the case when a popular Internet page has been saved at different locations and is accessed by a bunch of users, would reduce the traffic and load significantly and the users will be able to access the data regularly. Based on users request, the routers shall reduce the traffic overall in the network. NDN content-centric network based on information and data to UDP/TCP to implement ideal features of caching and naming reduces the congestion through this method and will get to optimize the resources [10]. Reza Tourani, Travis Mick proposed that security protocols integrated with client and significant mitigation might be investigated [7]. Every investigation's suspicious interface is identified and prefixed with the attacks. By following the content cache, the privacy-analyzed schemes affect the clients significantly [11]. The propositional way network load will be increased, and there is increased download latency in traffic. The network will maintain privacy mechanism which is routed to wide network approach and not fulfilled in cache scheme. Zhang proposed a content encryption method, the process of the encryption is notified with public key and random key, every client communication will pass through the random key k1, and request has reached to random key k2, which is located at another communication end of the process [8]. Now each communication will end at an edge router which encrypts the random key content name and publisher identity. The publisher key has decrypted the content and identity and finally encrypts the name and identity of the communication. These processes will take 10 s time for small content. The large content of the request will flow through the edge routers for multimedia IoT agriculture applications. Marica Amadeo, Claudia Campolo proposed priority-based data delivery in the interest vehicles via named data networking. The content of named data networking communication performs caching and accurate movements on the time based on the request from the other vehicles or RSUs. The selection and forwarding will decide the interest's low-priority and high-priority contents with the prefix names of Low and High. The consumers will send the request at the same time it will deliver the packets and simultaneously forwards both the requests. This will analyze the latency and reliability of the data and secure the sensitive data. Tan proposed the protected content into two portions; one is a large cacheable portion, and the second is small cacheable. Every client retrieves the smaller portion from publisher which is used to reconstruct the content and check the verified clients. The malicious user passes the smaller content to illegal user who wants to check the authorized clients, meaning the system will not allow passing of the details. Also, the process needs online connectivity until it completes the entire process.

4 Novel Based K-Means Clustering Technique

We present the attack detection system usually target to content-based locality in named data network caches. Clustering is the most easily available option to detect the attack through hierarchical or partitioning methodology. We would like to exploit the novel based k-means clustering technique and then present our scheme in details:

Overview

- Compute interest metrics.
- If interest is received, the router will update the popularity table for this interest.
- All the interests are classified into different clusters by using novel-based k-means clustering technique.
- Based on the clustering result, we can determine whether the router is under LDA or FLA by using an 'improved decision tree'.
- Once location-disruption attack (LDA) or false-locality attack (FLA) is detected, an attack table will be generated to record those suspicious interests and broadcasted to the neighbors hop by hop. All the routers receiving the attack table will not cache the content chunks in the corresponding data packets.

Detection Contribution

- Interests clustering is performed by the proposed novel k-means clustering technique.
- To detect whether it is LDA or FLA, use proposed improved decision tree method.

Proposed Flow

Compute Interest metrics:

- First, we compute total number of all interests, total count of the interests for the same content and time interval of two consecutive interests for the same content.
- Each and every router computes interest probability and time interval for same content.
- The router will update the popularity table for corresponding interest.

Clustering:

- All interests are classified into different clusters by using NBAC.
- In existing method like k-means, a structure that is more informative than the unstructured set of flat clusters. But, our method is basically hierarchical, and it provides hierarchical output.
- However, hierarchical clustering method is very sensitive to initial value.
- In our method, the initial cluster and number of clusters are computed by mean-shift clustering.

Detection:

- After clustering based on the clustering result, we can determine whether the router is under LDA or FLA by using IDT method.
- Normally decision tree method, without proper pruning or limiting tree growth, tends to over fit the training data, making them somewhat poor predictors.
- To avoid this issue, in our proposed method, the tree growth is computed by butterfly optimization algorithm (BOA).

The proposed flow chart describes the flow of interest with content store in pending interest table. Based on the table, the interest gets discarded or moved forward to forward interest table. The core step of the proposed flow in the clustering technique further then makes the detection part involved in decision tree process. The detection part will check the attack table with respect to the response of the cache content or forward will carry over the process. The simulation process executes the attack detection with dropped packets. In order to execute the process, the total number of packets, total received packets, duration of simulation and transmitted bits are examined. The pseudocode steps mention the clear execution step of entire interest forward details, pending interest table, forwarding information base and further then will send interest and discard the collapsed interest (Fig. 1).

Algorithm 1 Interest Process

1. start

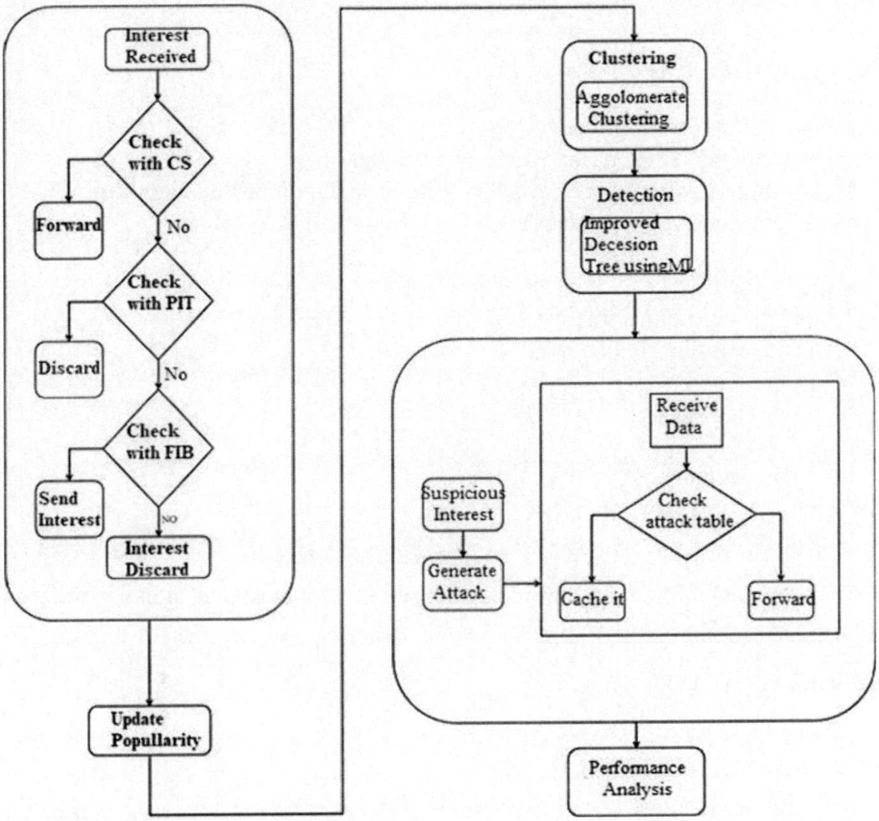

Fig. 1 Smart agriculture attack detection flow chart

```
Interest Received//I
If (I ==CS)
{
    Forward data to interface
}
else if(I ==PIT)
{
    Interest discarded
}
else if(I ==FIB)
{
        Send Interest
}
else
{
        Interest discarded
```

```
18. }
```

Algorithm 2 Popularity Table Process

```
1. Update Popularity Table
2. clustering by Novel based K-Means technique
3. Detection of decision tree growth by Butterfly Optimization algorithm (BOA)
4. if (Data Received ==Attack)
5. {
6. Cache the Data
7. Send the suspicious interest
8. Generate Attack Table
9.  }
10. else
11. {
12.     forward the data
13.  }
14. End
```

The above algorithms follow described interest process and popularity table updates.

Algorithm 3 Attack Detection

```
1. Start
2. for(i = 0;i < nm;i ++)
3. {
4.  for(j = 0;j < nm;i ++)
5.  {
6.     if(nm[i].packets.equals(drop))
7.     {
8.       Blockedlist[i] = nm[i];
9.       message("Attack Detected !! Packetdroped")
10.       }
11.       else
12.       {
13..        message("Packet Sent")
14.       }
15.  }}
16.  End
```

Table 1 Simulation parameters

Parameters	Values
Network size	600 * 600
Maximum speed of nodes	16 m/s
Transmission smooth factor	0.5
Content store	500 packets
Data rate	6 Mbps
Node size	2.00
Speed	1.000
Zoom	2.000
Nodes	11
Simulation time (s)	600 s
String value	1024
X axis proposal	50
Y axis proposal	20

When implementing the proposed flow in ndnSIM simulator [12] the transmission smoothness is 0.5, node size is 2.00 and string value 1024. The total number of nodes utilized for this scenario is 11, and total simulation time is 600 s. The data rate is 6 Mbps, and content store capability is 500 packets. The maximum speed of the node is 16 m/s (Table 1).

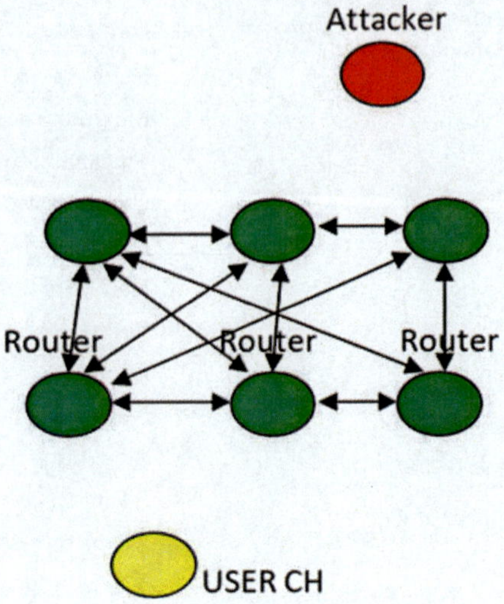

Fig. 2 ndnSIM python viewer window

Figure 2 shows the simulation of the attack in python viewer.

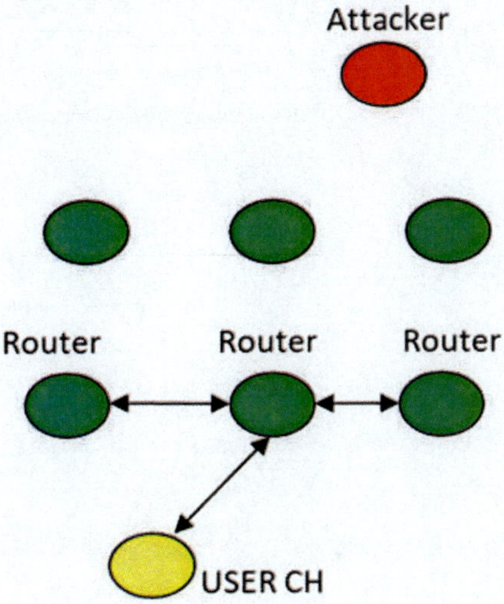

Fig. 3 Packet transmission to cluster head

Figure 3 depicts the simulation in the net animator. The position of the network nodes and characters is classified with different colors. Red-colored node is an attacker node. The cluster head forms and the nodes sent packet to the CH.

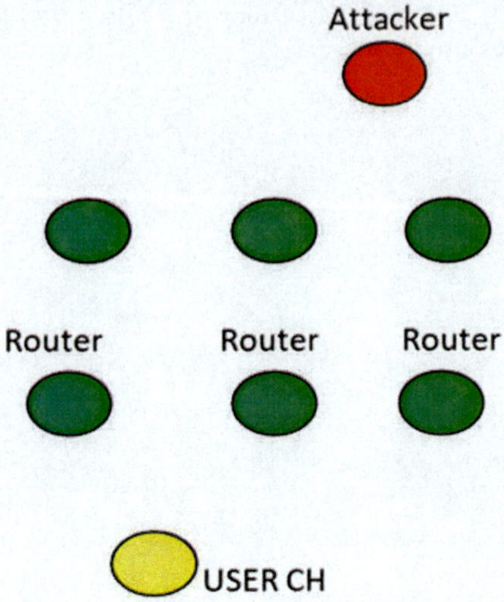

Fig. 4 Node location analyzed

Figure 4 depicts the scenario with router, user and the attacker. The figure clearly shows that there is one attacker, nine routers and one user. We have given the node numbers and simulation time for this ndnSIM to get the parameters.

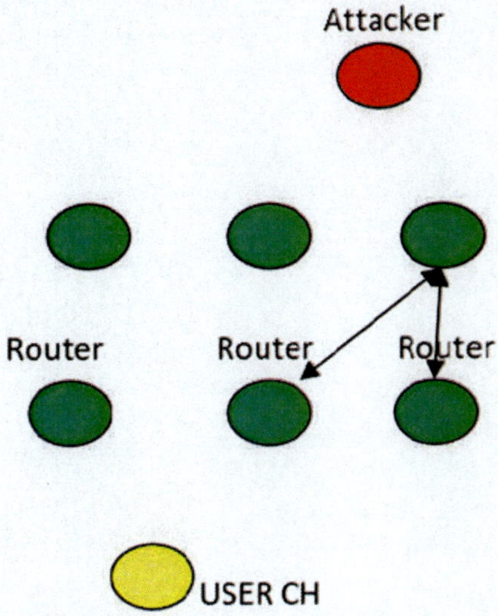

Fig. 5 Router-to -router transmission

Figure 5 depicts the simulation in the net animator. The cluster head forms and the nodes send packet to the CH.

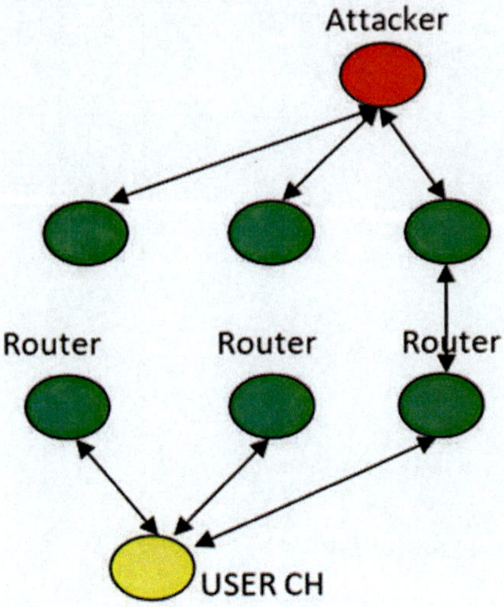

Fig. 6 Packet dropped position

Figure 6 depicts the simulation in the net animator. The cluster head forms and the nodes send packet to the CH. The attacker node sends the packet and the packet drops. The above scenario describes the packet dropped.

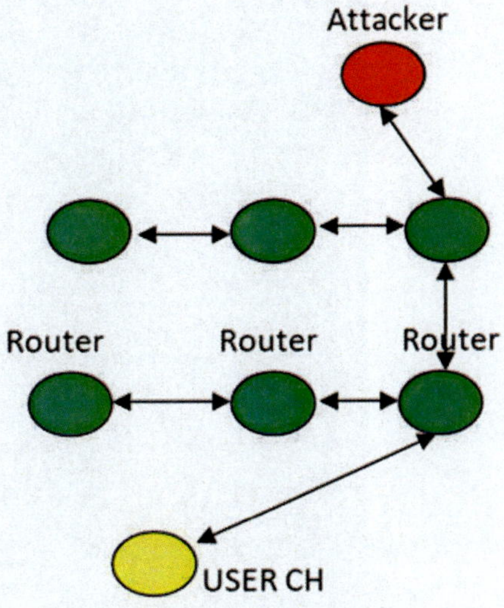

Fig. 7 Packet dropped position

Figure 7 depicts the simulation in the net animator. The cluster head forms and the nodes send packet to the CH. The overall process is clearly seen in above image, and it connects the user to the router and attacker.

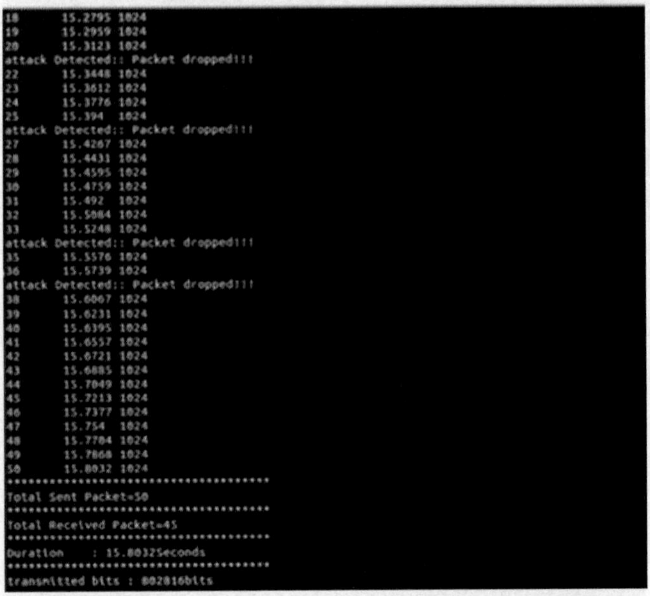

Fig. 8 Packet dropped position

Figure 8 shows the output of the simulation. The output describes total packet sent and received and the duration for the packets. The total sent packets are shown against total duration. The entire simulation time is mentioned, and attack detection is visualized with packet dropped. Finally, the visualization ends with transmitted bits.

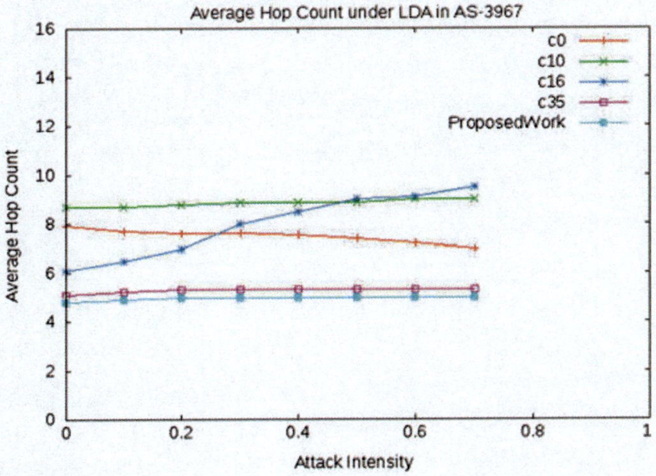

Fig. 9 Attack intensity

Figure 9 depicts the comparison of hop count of the node to reach the user and proposed system in existing system. The proposed work impact is clearly visible in above figure.

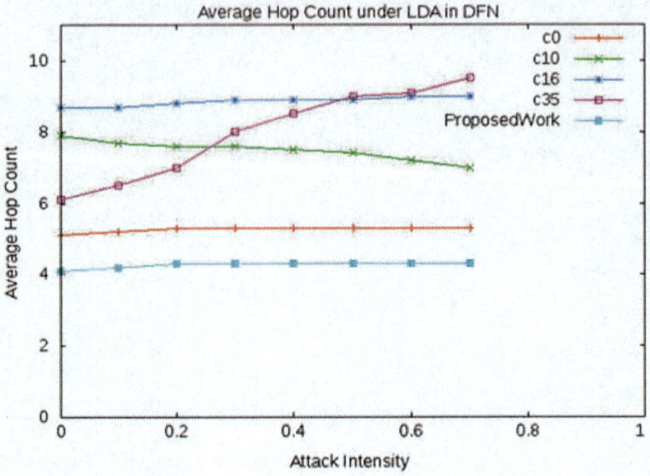

Fig. 10 Attack intensity in LDA

The average hop count is decreased in the proposed work which means the attack intensity downgraded in proposed method. The image has taken average hop count against attack intensity.

Figure 10 depicts the comparison of hop count of the node to reach the user and proposed system in existing system (by comparing in LDA DFN). The location distribution attack is taken, and the values of average hop count against attack intensity.

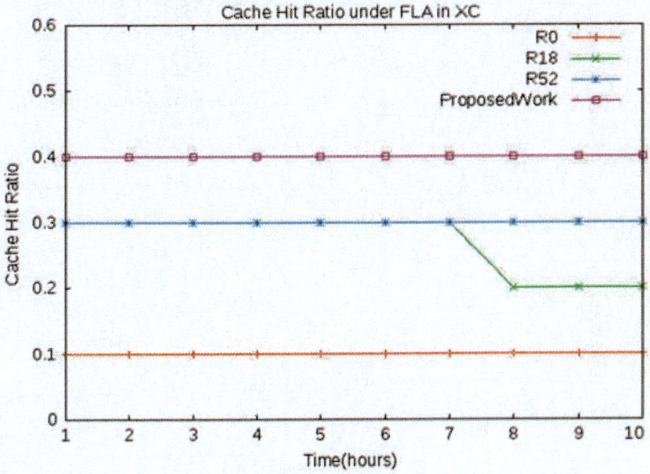

Fig. 11 Cache hit in FLA

Figure 11 depicts the comparison of cache hit ratio of and proposed system and in existing system. The false locality attack has taken cache hit ratio against time of the process.

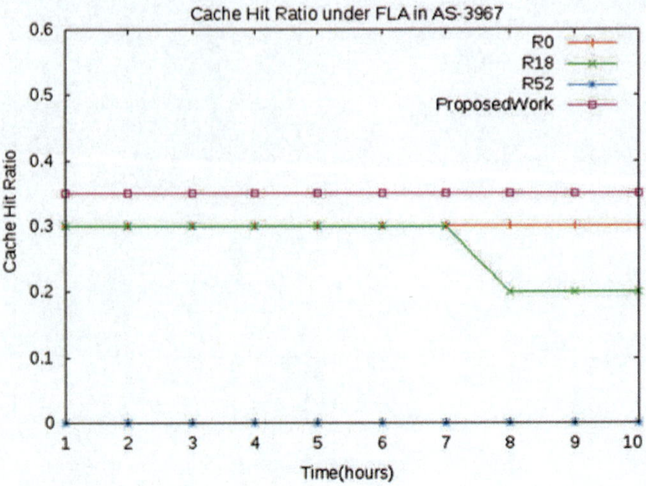

Fig. 12 Cache hit ratio in AS

Figure 12 depicts the comparison of cache hit ratio of the node to reach the user in the proposed system and in existing system.

5 Conclusion

Data retrieving is important task of Internet of Things in agriculture using NDN, and this helps in networking of the various popular contents. So, cache memory is more vulnerable from the different kinds of spoofing attacks. This attack is called cache pollution attack, and this attack is overcome in our proposed system. In this work, we have proposed an algorithm which is more efficient in defense and deduction of attacks in the system. In this system, we have differentiated various requests which help us to differentiate malicious and regular requests. We have proposed k-means clustering technique a cluster-based technique to find the abnormal requests. We have used named data network (NDN) for simulating the entire scenario. In our system, the cache hit ratio is increased up to 60%, detection of attack ratio increased up to 25% when compared to previous algorithms. Also, we have increased the memory level, CPU utilization when compared to the previous system.

References

1. Xie M, Widjaja I, Wang H (2012) Enhancing cache robustness for content-centric networking. In: Proceedings of the international conference on computer communications (INFOCOM'12), vol 131, no 5, pp 2426–2434

2. Park H, Widjaja I, Lee H (2012) Detection of cache pollution attacks using randomness checks. In: IEEE international conference on communications, pp 1096–1100
3. Xu Z, Chen B, Wang N, Zhang Y, Li Z (2016) ELDA: towards efficient and lightweight detection of cache pollution attacks in NDN. In: Local Comput Networks, 2016, pp 82–90
4. Gao Y, Deng L Kuzmanovic A, Chen Y (2006) Internet cache pollution attacks and countermeasures. In: IEEE international conference on network protocols, ICNP 2006, Santa Barbara, CA, USA, 12–15 Nov 2006, pp 54–64
5. Gao Y, Deng L et al (2006) Internet cache pollution attacks and countermeasures. In: ICNP. IEEE
6. Xie M, Widjaja I et al (2012) Enhancing cache robustness for content-centric networking. In: INFOCOM. IEEE
7. Salah H, Alfatafta M, SayedAhmed S, Strufe T (2017) CoMon++: preventing cache pollution in NDN efficiently and effectively. In: Proceedings of the 42nd IEEE conference on local computer networks (LCN'17), pp 1–9
8. Zhang G, Liu J, Chang X, Chen Z (2017) Combining popularity and locality to enhance in-network caching performance and mitigate pollution attacks in content-centric networking. IEEE Access 5:19012–19022
9. Roberts J (2009) The clean-slate approach to future internet design: a survey of research initiatives. Ann Telecommun 64(5):271–276
10. Conti M, Gasti P et al (2013) A lightweight mechanism for detection of cache pollution attacks in named data networking. Comput Netw 57(16):3178–3191
11. Gupta R, Chi C-H (1990) Improving instruction cache behavior by reducing cache pollution. In: Supercomputing. IEEE
12. Afanasyev A, Moiseenko I et al (2012) ndnSIM: NDN simulator for NS-3. Technical report, UCLA

Effects of Social Distancing on Spread of a Pandemic: Simulating Trends of COVID-19 in India

Minni Jain, Aman Jaswani, Ankita Mehra, and Laqshay Mudgal

Abstract Most of the diseases spread from human-to-human which makes it dangerous for anyone who comes in contact with an infected being. The key point to note here is that such a spread gives rise to a network where nodes represent humans and edges show if two humans came in contact with each other or not. Studying and analysing pandemic networks help in managing the spread of disease efficiently. The recent coronavirus (COVID-19) outbreak that was first identified in December 2019 in Wuhan, Hubei province, China according to WHO reports, is an apt example of a deadly contagious disease. India has been fighting the virus since February 2020 relentlessly, with the scientists experimenting to find a cure for the disease, and healthcare personnel along with other essential workers ensuring that all necessary preventive and protective measures are being taken to reduce health risks. In our proposed study, we aim at forecasting the spread of COVID-19 in India with the help of SEIR-DH and linear regression model, by simulating the dynamics of disease spreading in a large population. We also aim to mathematically depict how increasing the severity of social distancing can affect the spread of the disease. The results of our study indicate that increasing the strictness of social distancing measures can help reduce the overall number of infected patients and also help flatten the epidemic curve of COVID-19 spread. The curve depicts the number of infected patients requiring healthcare for combating the disease over time.

Keywords Coronavirus (COVID-19) · Pandemic · India · Social distancing · Contagious diseases · SEIR-DH model

M. Jain · A. Jaswani · A. Mehra (✉) · L. Mudgal
Delhi Technological University, Shahbad Daulatpur, Main Bawana Road, Delhi 110042, India
e-mail: mehra.ankita99@gmail.com

M. Jain
e-mail: minnijain@dtu.ac.in

A. Jaswani
e-mail: ajaswani11@gmail.com

L. Mudgal
e-mail: laqshay.mudgal@gmail.com

A. Khanna et al. (eds.), *Data Analytics and Management*, Lecture Notes on Data Engineering and Communications Technologies 54,
https://doi.org/10.1007/978-981-15-8335-3_31

1 Introduction

Coronavirus disease is an illness that is caused by severe acute respiratory syndrome coronavirus-2 (SARS-CoV-2). Common symptoms include fever, cough, fatigue, etc., while more viral symptoms include pneumonia, multi-organ failure, etc. The pandemic that was first reported in China [1] has now severely affected the entire world. The world still awaits the production of a vaccine that can treat the virus. Thus, the countries have only one option left that is to obstruct the disease spread. In a large country like India with a population of more than 130 crore people, it becomes more necessary to take all the precautions possible to restrict the spread of virus. The main motive of various social distancing guidelines is to flatten the curve [2] that essentially means to hinder the growth in the number of infected cases. This is done so that more number of people can get better access to timely healthcare. Since the Indian healthcare system is not very advanced to deal with colossal number of patients simultaneously, it is necessary to reduce the burden on hospitals. Government has been trying its best to save its citizens by imposing lockdowns [3] and restricting movement of people. It, however, boils down to whether people are willing or determined enough to follow all the regulations without any violation. GAN has already been applied to predict COVID-19 [4].

To study the effects of variations in the implementation of social distancing on spread of the virus, we have tried to simulate the trends of COVID-19 in India. The major contributions of our research work are listed as follows:

(i) We have used an SEIR-DH model which is an advancement of the simple SIR (susceptible-infected-recovered) model [5]. This model more accurately depicts the dynamics of the disease spreading by taking into account the number of deaths and hospitalized cases. The working of the model is explained in Sect. 3.2.

(ii) We have predicted the estimated number of infected, exposed, death and hospitalization cases using this model at different severities of social distancing measures from May ’20 to May ’21. These values have been predicted using real COVID-19 data in India gathered between January ’20 and April ’20.

The remainder of the paper is organized as follows. Section 2 gives an elaborate study on the various works done in the related field. Section 3 explains the datasets and the model used. Section 4 presents the results of the study along with important findings. Section 5 concludes the research work.

2 Literature Review

Many researchers have previously studied the effects of social distancing on spread of a virus. In most of the cases, data of transmission and infection was transformed into a network structure because of the ease of working with graphs. Newman [6] observed that the data distributions show different patterns in different scenarios.

The distribution structure of the degrees of vertices in a typical disease propagation network has been discussed in [7].

Population density also plays a pivotal role in understanding spread of the disease. In [8], P. M. Tarwater and C. F. Martin have explained how a decrease in the susceptible rate of contact adversely affected the scale of outbreak and, thus, the disease spread. The nature of the population is also a key factor, as explained in [9]. A homogenous network assumes that each node or person has the same probability of catching the infection, whereas a heterogeneous mixture not only takes into account the probabilities of contacts made but also studies the influence of the entire network on any pair of nodes.

An improvement in disease prevention mechanisms has been discussed in [10] where geographic information system (GIS) systems have been employed that display early warnings to tackle infectious disease and also help report the events of infection to the organization that uses GIS systems.

3 Experimental Design

In this section, we discuss the different aspects of the proposed methodology. Section 3.1 elaborates about the various datasets used to gather information about real-time COVID-19 statistics of India. Section 3.2 explains the working of the SEIR-DH model that has been employed to run simulation of various levels of social distancing. Section 3.3 elucidates the functioning of linear regression model whose performance has been compared with that of SEIR-DH model.

3.1 Dataset

The data containing the number of confirmed cases, recovered cases and deaths in world is obtained from Johns Hopkins University's repository [11]. These three files contain the time-series data for the real-time number of COVID-19 confirmed, recovered and death cases for all countries starting from 22 January 2020 to 27 April 2020. This repository is updated every day with the new cases found. Only the data for India was selected from these files. The first COVID-19 case in India was detected on 30 January 2020. So, for our study we have used the time frame between 22 January 2020 and 27 April 2020.

The number of active cases for each day is calculated as Active = Confirmed − (Recovered + Deaths). Mortality rate dataset [12] and Indian age distribution dataset [13] are also collected in order to approximate the number of hospitalized cases (per age cohort) for each day and the normal mortality rate of infected patients in the hospital.

3.2 SEIR-DH Model

This model is an extension of the susceptible-infected-recovered (SIR) model which has been extensively used in past studies to model the spreading dynamics of diseases like dengue fever, HIV, Bubonic plague, etc. [14–16]. According to the simple SIR model, a 'susceptible' person is one who has not yet been exposed to the disease but is at the risk of catching it sometime in the near future. An 'infected' person is one that has caught the disease and can now spread it to other susceptible people in the population. Finally, a 'recovered' person is that individual that has now recovered from the disease and is considered to have developed immunity against that disease. This basic SIR model is governed by two principal parameters $\boldsymbol{\beta}$ and $\boldsymbol{\gamma}$, where $\boldsymbol{\beta}$ is defined as the average contact rate in population (i.e. $\boldsymbol{\beta}$ = probability of transmission × number of contacts) and $\boldsymbol{\gamma}$ is the mean recovery rate (calculated as the inverse of the mean infectious period).

The new terms E, D and H refer to the number of exposed, dead and hospitalized cases, respectively. An 'exposed' person is one who has been exposed to the disease but has not yet shown any symptoms. A third parameter $\boldsymbol{\delta}$ also comes into the picture where $\boldsymbol{\delta}$ is calculated as the inverse of the incubation period. Incubation period is the time elapsed between exposure and when symptoms and signs are first apparent in a person. The number of 'hospitalized' cases is the number of people seeking medical treatment for the disease from a hospital. Finally, a 'death' case occurs when the infected person's life could not be saved despite healthcare or in the case when Indian healthcare system reached its maximum serving capacity and people could not avail these services. Maximum serving capacity reaches when the number of infected hospitalized patients (in a day) becomes greater than 713,986 because that is the number of hospital beds available in India according to [17]. The basic SEIR-DH model is depicted with the help of a flowchart in Fig. 1.

The ratio $\mathbf{R_0} = \boldsymbol{\beta}/\boldsymbol{\gamma}$ is known as the reproduction rate, and it is necessary for a disease to have $\mathbf{R_0} > 1$ such that it can spread in a large population otherwise it will die out. We have taken $\mathbf{R_0}$ to be 2.5 with reference to the work published by the

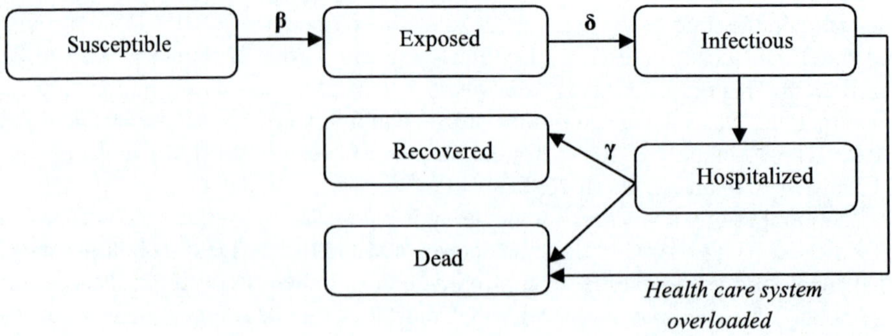

Fig. 1 SEIR-DH model

Centres for Disease Prevention and Control (CDC) [18]. The mean infectious period is taken to be 21 days [19] and therefore = 0.0476. Finally, $\boldsymbol{\beta} = \mathbf{R}_0 \times \boldsymbol{\gamma} = 0.119$. The mean incubation period in our study is taken to be 6 days [19] and therefore = 0.166.

The number of hospitalized cases is calculated as follows—Hospitalized cases (on day t) = Hospitalization rate (per age cohort) * Active infected cases (on day t).

The number of deaths is predicted as follows—if healthcare facility is overloaded, deaths (on day t) = (CDR * non hospitalized infected cases) + (MR * hospitalized infected cases), else, deaths (on day t) = (MR * hospitalized infected cases). Here, CDR is the critical death rate which is estimated to be 0.122 in reference to [20], and MR is the normal mortality rate.

The SEIR-DH model can be modelled with the following equations

$$S = -\boldsymbol{\beta} SI \tag{1}$$

$$E = \boldsymbol{\beta} SI - \boldsymbol{\delta} E \tag{2}$$

$$I = \boldsymbol{\delta} E - \boldsymbol{\gamma} I - D \tag{3}$$

$$R = \boldsymbol{\gamma} I \tag{4}$$

$$N(\text{total population}) = S + E + I + R + D \tag{5}$$

If social distancing is implemented and large gatherings of people are controlled (even to a limited extent), then spread of the disease can be reduced. This will impact the contact rate $\boldsymbol{\beta}$. To imitate this situation, we introduce a new variable that captures our social distancing effect. $\boldsymbol{\rho}$ is a uniform value between 0 and 1, where 0 indicates everyone is locked down and quarantined, while 1 is equivalent to no social distancing followed.

According to this new model, Eqs. (1) and (2) are modified as follows—

$$S = -\boldsymbol{\rho\beta} SI \tag{1'}$$

$$E = \boldsymbol{\rho\beta} SI - \boldsymbol{\delta} E \tag{2'}$$

We run six different simulations of social distancing in our study to determine how much the disease can spread in the country for 1 year after the days for which data is available, with different percentages of social distancing being followed. The six simulations consider the social distancing percentages as—0%, 10%, 25%, 50%, 75% and 90%, i.e. $\boldsymbol{\rho}$ value = 1, 0.9, 0.75, 0.5, 0.25 and 0.1, respectively.

3.3 Regression Model

Regression modelling is a form of predictive modelling technique which is used to estimate or predict the target or dependent variable on the basis of independent variables. Our study uses linear regression for prediction of COVID-19 cases in India. Linear regression is a simple model which is used to find the relation between a dependent and an independent variable using a best fit straight line (also known as the regression line), hence, using the value of intercept and slope to predict the output variable, number of confirmed cases of COVID19, in our study. Equation (6) shows the relationship between a dependent and independent variable in a linear regression model. In Eq. (6), β_0 and β_1 are two independent variables which represent intercept and slope, respectively, and ϵ is the error.

$$Y = \beta_0 + \beta_1 + \epsilon \tag{6}$$

4 Results and Discussions

Historical trend shows that in India, the first case of COVID-19 was reported on 30 January 2020, but the number of cases remained consistently low below 5 during the month of February. However, the significant increase in the spread of the disease started in March '20. Figure 2 shows the historical trend of data of number of

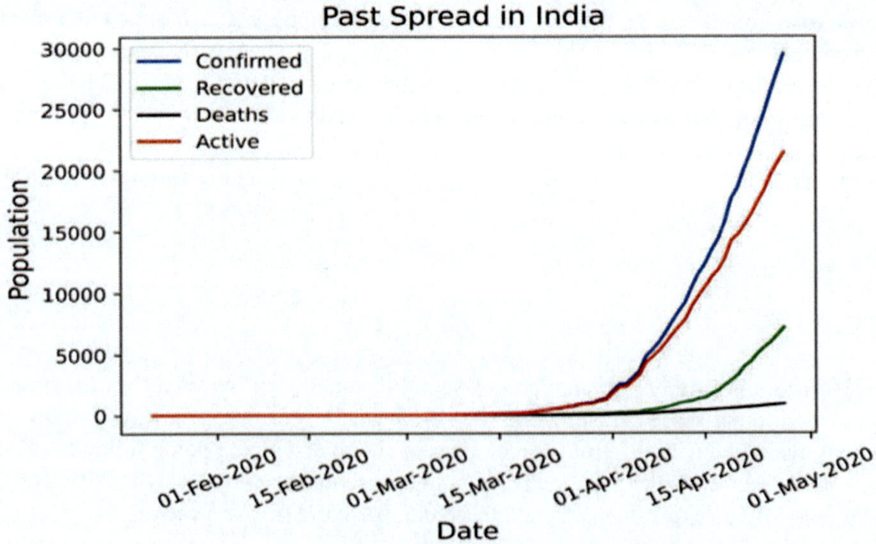

Fig. 2 Historical trend of data of number of confirmed COVID-19 cases in India

confirmed cases on COVID-19 in India.

In our current analysis, we have used data till 20 April 2020 as our training data and data from 21 April 2020 to 27 April 2020 as the test/evaluation data. We provide a solution to two pressing questions—the first question that we address is how different levels of nationwide restrictions affect spread of the disease. Here, we aim to mathematically and objectively analyse the effect of nationwide restriction on internal movement (also referred to as 'lockdowns'). The second question is how the future trend of the spread of disease will look like.

To understand how different levels of nationwide restrictions affect the spread of the disease, we analysed six different scenarios, ranging from the trend when no restrictions are implemented to when extreme restrictions (leading to 90% social distancing) are implemented. We show data for the next twelve months. In the graph plots, red line denotes the number of recorded infected cases, blue line denotes the number of recorded exposed cases, black line denotes the number of recorded death cases, and green line denotes the number of recorded hospitalized cases.

No Social Distancing In this case, our model predicts that India will reach the peak of infected rate on 8 November 2020 when there will exist around ~29 crore active cases. As a result, we see that the 'Total Deaths' curve, shown in Fig. 3, is on a significant rise from 24 November 2020.

10% Social Distancing This represents the case of relaxed social distancing, where lockdown and quarantine measures were first adopted, starting 25 March 2020. In this case, we see a slightly longer timeline. Although there is only slight change in the number of deaths, Fig. 4 shows that the peak value of number of infected persons is now ~24.9 crore and is reached on 1 December 2020.

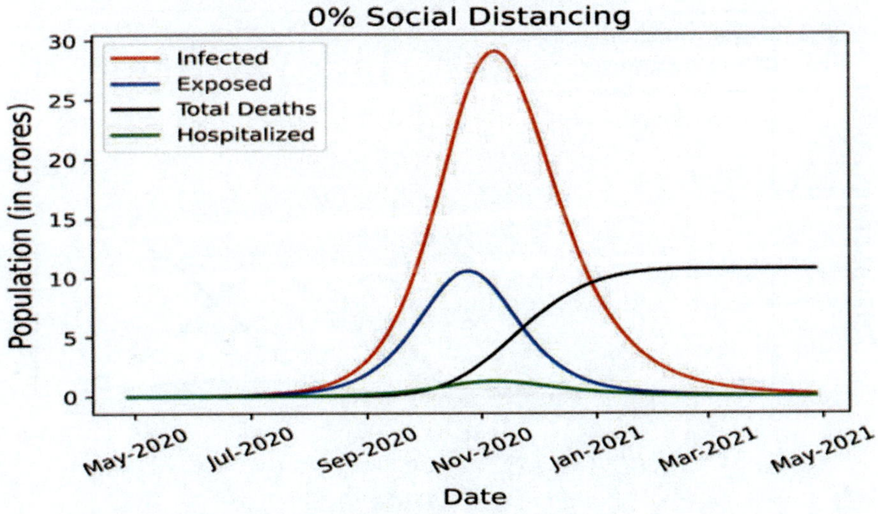

Fig. 3 SEIR-DH simulation for no social distancing

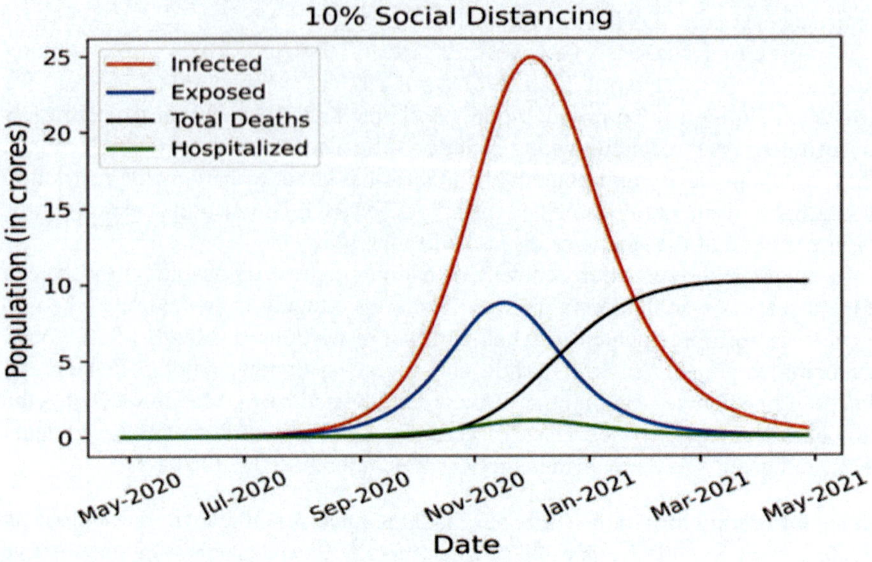

Fig. 4 SEIR-DH simulation for 10% social distancing

25% Social Distancing The increase in social distancing from 10% to 25% drastically reduces the number of deaths from 10.1 to 8.6 crore, as shown in Fig. 5. This causes a further decrease in the peak value of number of infected persons to ~17.9 crore, reached on 25 December 2021.

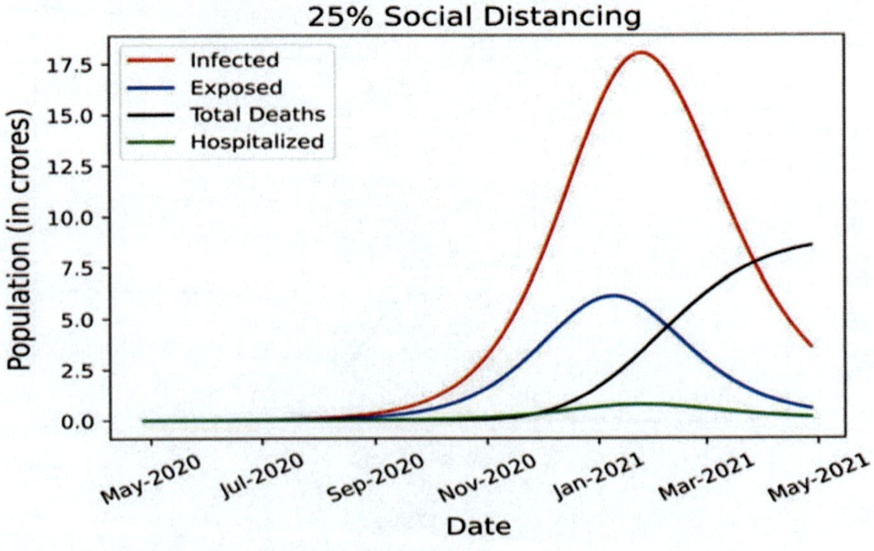

Fig. 5 SEIR-DH simulation for 25% social distancing

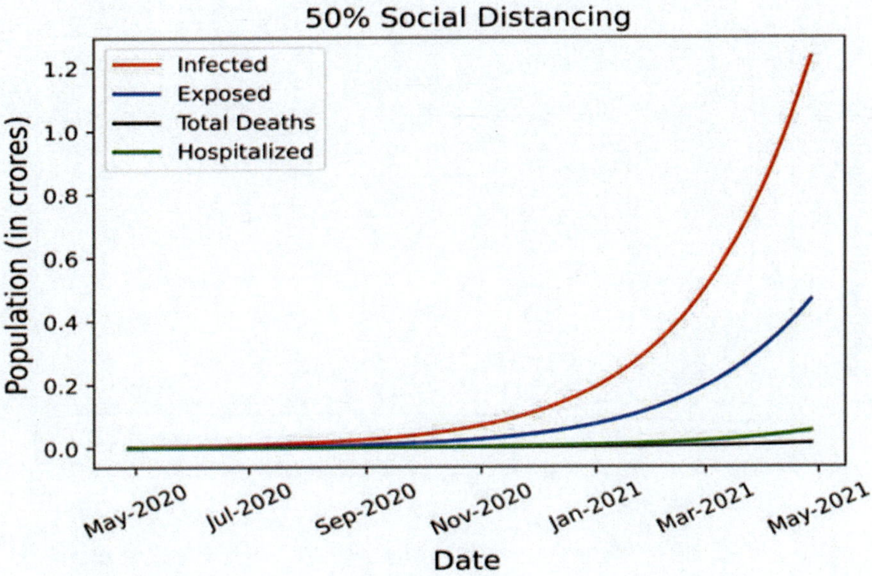

Fig. 6 SEIR-DH simulation for 50% social distancing

50% Social Distancing The increase in social distancing from 25 to 50% dramatically decreases the number of infected persons, from 17.9 to 1.23 crore. Figure 6 shows that in this case, the maximum value is recorded at an even later stage—27 April 2021.

75% Social Distancing The increase in social distancing from 50 to 75% then further decreases the peak value of number of infected persons, from 1.23 to 29,315. Figure 7 shows that in case, the peak is reached on 17 May 2020.

90% Social Distancing This perfect level of social distancing of 90% reduces the number of deaths to 2068 and creates a significant decrease in the peak value of number of infected persons to 27,502—reached on 4 May 2021. It is illustrated in Fig. 8.

Table 1 displays a tabular comparison among the results of predictions of infected, exposed, hospitalized and death cases for different levels of social distancing implemented. SEIR-DH model was used to obtain the results by performing simulations of the effects of variations in the extent of social distancing on spread of COVID-19.

We addressed the second issue of understanding future trends, by forecasting of COVID using SEIR-DH and regression model. Figure 9 graphically depicts the comparison of the predicted data from 21 April 2020 to 27 April 2020. It is observed that the two models that have been employed compare well with actual data. We used root mean squared log error (RMSLE) to check the performance of our model—the value of RMSLE for the SEIR-DH model was calculated to be 0.215 and it was 0.122 for the regression model.

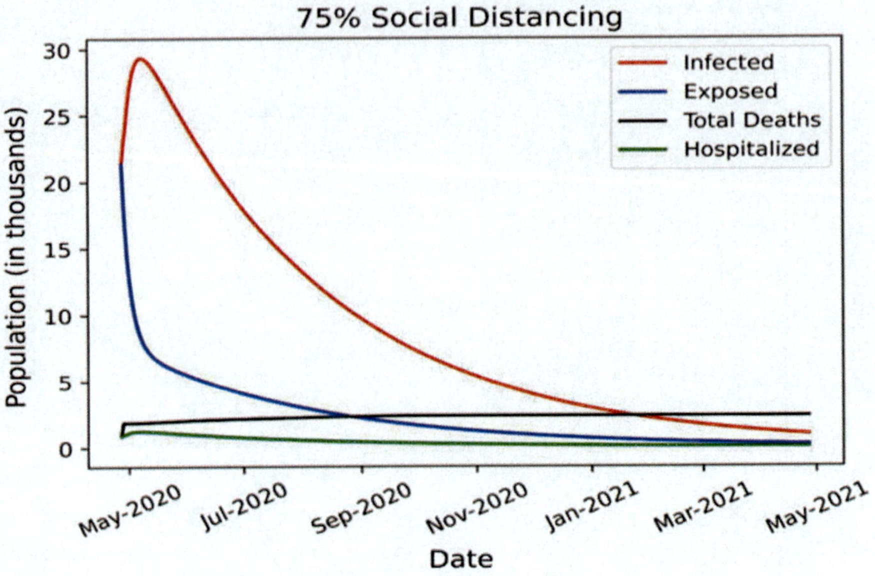

Fig. 7 SEIR-DH simulation for 75% social distancing

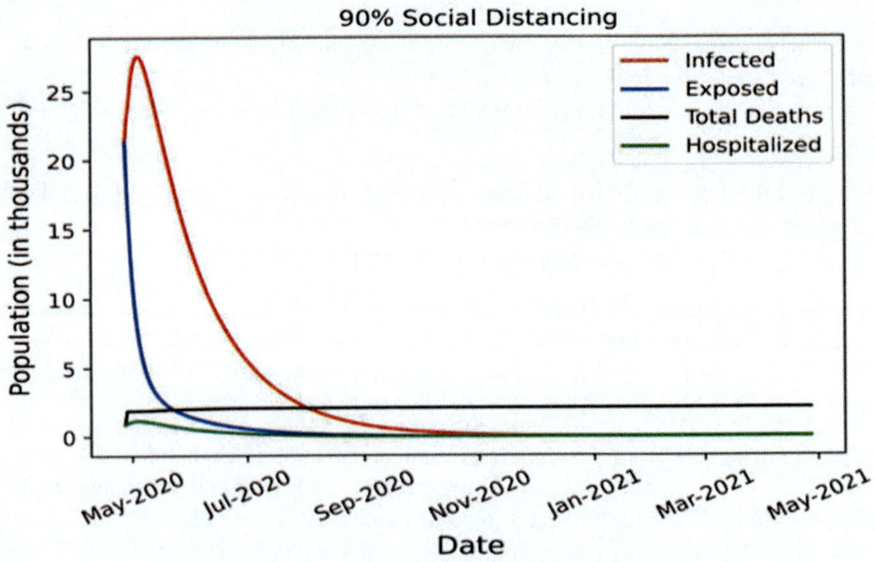

Fig. 8 SEIR-DH simulation for 90% social distancing

Table 1 Comparison of COVID-19 spread on the basis of different levels of social distancing

Level of social distancing (%)	SEIR-DH predictions			
	Infected	Exposed	Hospitalized	Deaths
0	29.0125 crores	10.5159 crores	1.23669 crores	10.713 crores
10	24.9134 crores	8.79039 crores	1.06196 crores	10.143 crores
25	17.984 crores	6.08468 crores	76.658 lakhs	8.6906 crores
50	1.2315 crores	46.584 lakhs	524972	130,054
75	29,315	21,375	1249	2365
90	27,502	21,375	1173	2068

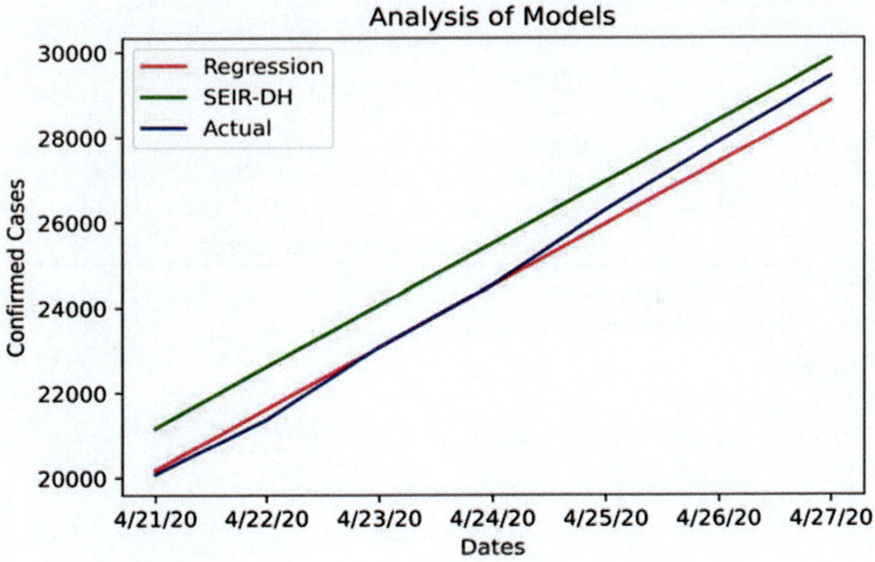

Fig. 9 Analysis of the performance of the models employed

Table 2 and Fig. 10 show the prediction of future trend of number of confirmed COVID-19 cases for the first two weeks of May '20 using the two models—SEIR-DH model and linear regression model.

Our study also presents an analysis of the upcoming healthcare crisis, as shown in Fig. 11, by taking into account the number of hospital beds available in the country. According to [17], 713,986 hospital beds are available in India. In Fig. 11, we show the comparison of the number of infected cases at different levels of social distancing.

We observed that with further increase in social distancing, we were able to successfully move towards 'flattening the curve'. India also reaches the peak of number of infected cases at a much later date, which in turn gives enough time for development and implementation of strategies and healthcare facilities to prevent deaths in the country.

Table 2 Prediction of confirmed COVID-19 cases for the first two weeks of May '20 in India

Date	SEIR-DH predictions	Regression model predictions
2020-05-01	29,905.0	34,624.0
2020-05-02	31,528.0	36,070.0
2020-05-03	33,101.0	37,515.0
2020-05-04	34,648.0	38,961.0
2020-05-05	36,193.0	40,406.0
2020-05-06	37,749.0	41,852.0
2020-05-07	39,329.0	43,297.0
2020-05-08	40,944.0	44,742.0
2020-05-09	42,602.0	46,188.0
2020-05-10	44,309.0	47,633.0
2020-05-11	46,071.0	49,078.0
2020-05-12	47,894.0	50,524.0
2020-05-13	49,783.0	51,969.0
2020-05-14	51,741.0	53,415.0

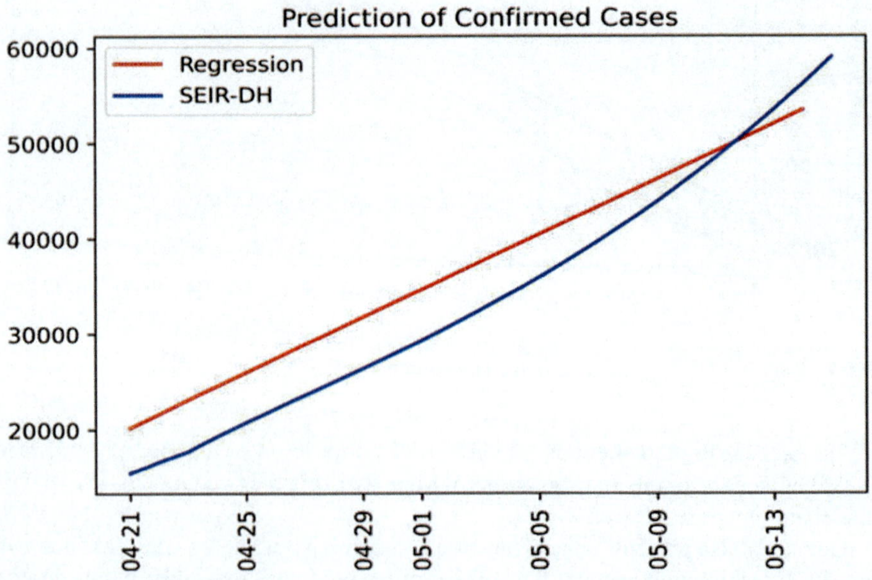

Fig. 10 Prediction of confirmed COVID-19 cases for the first two weeks of May '20 in India

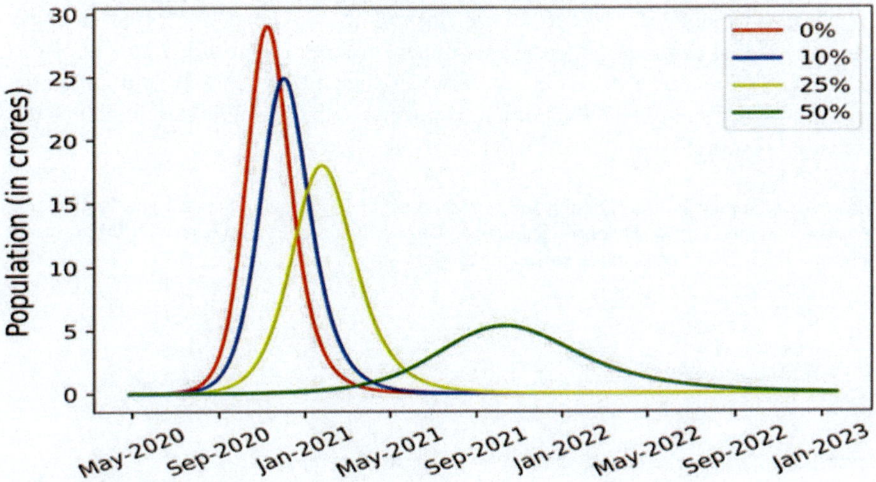

Fig. 11 Flattening the curve with social distancing

Also, Table 1 shows that at levels of 0 and 10% social distancing, there is immense burden on the healthcare system, evidenced by the drastic difference in number of hospitalized and infected cases. Thus, an increase in social distancing positively impacts multiple aspects of COVID-19 prevention and control.

It is to be noted that implementing 75 and 90% social distancing measures also leads to successful flattening of the curve, but due to the difference in scale, they are not shown in Fig. 11.

5 Conclusion

In our study, we were able to demonstrate that the COVID-19 growth rate declined with the increasing severity of nationwide social distancing measures. Not only that, we also noticed a substantial increase in the duration at which the peak for infected cases is reached. This allows the government and health workers to have more time to research and develop vaccinations and reduces the amount of load on the hospitals of the country, which in turn reduces the number of deaths. We used two models—SEIR-DH model and linear regression model—to forecast the spread of COVID-19 in India and estimated the number of infected/death cases to be expected with different percentages of social distancing measures supposed to be followed over the next one year. We were also able to show the impact of social distancing on the number of hospitalized cases in India.

We further showed that the trend of prediction remains almost linear; however, India must stay prepared even for an unexpected exponential increase and employ suitable and timely measures to control loss of lives. The policy of strict monitoring and social distancing measures is indispensable in controlling spread of this pandemic. The research work clearly shows that such policies will help flatten the curve efficiently and reduce the burden on hospitals, thereby allowing judicious use of existing resources.

Acknowledgements We are grateful to the Department of Computer Science and Engineering (Software Engineering) at Delhi Technological University for presenting us with this research opportunity which was essential in enhancing our learning.

References

1. WHO (2020) Novel coronavirus (2019—nCoV) situation report, 1
2. India locks down 1.3 billion people in biggest isolation effort, Bloomberg. https://www.bloomberg.com/news/articles/2020-03-24/india-to-impose-nationwide-lockdown-from-midnight-pm-modi-says. Last accessed 25 Apr 2020
3. Coronavirus: what is 'flattening the curve' and will it work? Livescience. https://www.livescience.com/coronavirus-flatten-the-curve.html. Last accessed 25 Apr 2020
4. Waheed A, Goyal M, Gupta D, Khanna A, Al-Turjman F, Pinheiro PR (2020) CovidGAN: data augmentation using auxiliary classifier GAN for improved Covid-19 detection. IEEE Access. https://doi.org/10.1109/access.2020.2994762
5. Zhao L, Cui H, Qiu X, Wang X, Wang J (2013) SIR rumour spreading model in the new media age. Phys A Stat Mech Appl 392:995–1003
6. Newman MEJ (2002) Spread of epidemic diseases on networks. Phys Rev E Stat Non-linear Soft Matter Phys 66(1 pt 2):016128
7. Pastor-Satorras R, Vespignani A (2001) Epidemic spreading in scale-free networks. Phys Rev Lett 86:3200
8. Tarwater PM, Martin CF (2001) Effects of population density on the spread of disease. Complexity 6:29–36
9. Tsui K-L, Wong ZS-Y, Goldsman D, Edesess M (2013) Tracking infectious disease spread for global pandemic containment. IEEE Intell Syst 28:60–64
10. Pramudyo R, Susilo R, Aprilia D, Albarda (2015) Improving infectious diseases prevention system: the case study of Department of Health Sragen, pp 1–6
11. CSSEGISandData/COVID-19. https://github.com/CSSEGISandData/COVID19/tree/master/csse_covid_19_data/csse_covid_19_time_series. Last accessed 27 Apr 2020
12. Report 9—Impact of non-pharmaceutical interventions (NPIs) to reduce COVID-19 mortality and healthcare demand, Imperial College London. https://www.imperial.ac.uk/mrc-global-infectious-disease-analysis/covid-19/report-9-impact-of-npis-on-covid-19. Last accessed 27 Apr 2020
13. Population pyramids of the world from 1950 to 2100. PopulationPyramid.net, https://www.populationpyramid.net/india/2019. Last accessed 27 Apr 2020
14. Side S, Noorani M (2015) A SIR model for spread of dengue fever disease (simulation for south Sulawesi, Indonesia and Selangor, Malaysia) (2015)
15. Zakary O, Rachik M, Elmouki I (2016) On the impact of awareness programs in HIV/AIDS prevention: an SIR model with optimal control. Int J Comput Appl 133 (2016)
16. Keeling MJ, Gilligan CA (2000) Bubonic plague: a metapopulation model of a zoonosis. Proc R Soc B Biol Sci 267(1458):2219–2230
17. COVID-19: is India's health infrastructure equipped to handle an epidemic? Brookings. https://www.brookings.edu/blog/up-front/2020/03/24/is-indias-health-infrastructure-equipped-to-handle-an-epidemic. Last accessed 27 Apr 2020

18. Sanche S, Lin YT, Xu C, Romero-Severson E, Hengartner N, Ke R (2020) High contagiousness and rapid spread of severe acute respiratory syndrome coronavirus 2. Emerg Infect Dis 26
19. WHO (2020) Report of the WHO-China joint mission on coronavirus disease 2019 (COVID-19)
20. Mizumoto K, Chowell G (2020) Estimating risk for death from 2019 novel coronavirus disease, China. Emerg Infect Dis 26

Implementation of EAODV-Based SON for Balanced Energy-Efficient Routing Using Tree for WSN

U. Hariharan, K. Rajkumar, and Nilotpal Pathak

Abstract Wireless Sensor Networks (WSN) the primary investigation hotspot for WSN is power conservation. As electrical power drains more quickly, the system lifetime additionally reduces. Self-organizing networks (SON) are simply the option just for the above-discussed issue. SON may right away configure themselves, come across an optimal remedy, detect as well as self-heal to some degree. This effort has been created to utilize a self-organization network to balance the power and decrease power drowning. This particular process utilizes various neighbouring enduring energy and the nodes since the requirements for significant group awareness elections develop a tree-based network. The threshold for recurring vitality as well as distance is identified to determine the route of the information transmission that is energy efficient. The enhancement manufactured in selecting strong details for group awareness election and cost-efficient details transmission leads to reduced power use. The setup on the suggested process is carried through in NS2 atmosphere. A different number of nodes does the test as 20, 40, and 60 nodes and 2 pause situations 5 and 10 ms. The evaluation of the result suggests that the proposed SON has 17.6% much less power compared to current techniques.

Keywords EAODV · SON · NS2 · Balanced energy routing · Threshold · WSN

U. Hariharan (✉) · K. Rajkumar · N. Pathak
Department of Information Technology, Galgotias College of Engineering and Technology, Greater Noida, Uttar Pradesh, India
e-mail: udhayakumarhariharan@gmail.com

K. Rajkumar
e-mail: rajonline7@gmail.com

N. Pathak
e-mail: nilotpal.pathakbit15@gmail.com

A. Khanna et al. (eds.), *Data Analytics and Management*, Lecture Notes on Data Engineering and Communications Technologies 54,
https://doi.org/10.1007/978-981-15-8335-3_32

1 Introduction

Wireless sensor networks (WSNs) are a variety of receptors deployed within a specific location for capturing as well as overseeing the actual physical details of the ecosystem. The information captured of WSNs is organized as well as saved systematically from a main place. The benefits of these networks are they're not costly, simple to put into action, supports various kinds of topologies, and most importantly, they consume less electricity [1]. Nevertheless, receptors within the WSNs have natural restrictions with regard to energy, for example mind, computational speed, energy, and then correspondence bandwidth due to the small size of its as well as capability.

The best obstacle inside a WSN could be the electricity of nodes. To get a system continually running as well as be effective, the power of the node needs to be held to the best possible fitness level. We have seen massive amount tests in connection with power usage found WSNs, and many scientists have suggested tips to bring down electricity usage as well as boost community lifetime. SONs are found to become a great strategy to boost power usage [2].

Below, SON is utilized to bring down power usage. SON is the network which could instantly configure themselves, discover optimum fixes and also which could identify as well as mend to some extent. There are three sub-functions of SON.

1.1 Self-configuring

SON must be in a position to configure brand-new community components [2]. For instance, when a brand-new foundation station is put in, it must instantly configure, set up connectivity and also incorporate itself into the current community. This is often completed by the "plug as well as play" paradigm. If the brand-new foundation station is driven on, it is instantly realized by other facilities and so they create a mention of specialized details.

1.2 Self-optimization

SON must be in a position to configure brand-new community components [1]. For instance, when a brand-new foundation station is put in, it must instantly configure, set up connectivity and also incorporate itself into the current community. This is often completed by the "plug as well as play" paradigm. If the brand-new foundation station is driven on, it is instantly realized by other facilities and so they create a mention of specialized details.

1.3 Self-healing

SON must be in a position of identifying some node disaster or may be hyperlink breakages within the networking and get remedial steps. The SON could additionally be programmed to notify the end user when absolutely no remedial steps could be used. For instance, another cause could be decided to attain the spot in the event of website link disaster.

In this paper, we focus on the self-optimization sub-function for decreasing electricity use to balance power to extend the lifetime. The power balancing steps are brought on within the system when activities that call for greater power are recognized immediately. AODV is needed as it is a regular routing process that is doable with WSN.

To enhance the system lifetime and the functionality of the WSN, "Enhanced AODV dependent Self Organized tree for Energy balanced Routing (EASER)" is suggested that it's a typical design to balance the power usage among nodes in suggested WSN.

The goals of the suggested method are:

- To establish a self-organized tree-based bunch by electing a strong bunch mind.
- In order to carry out a cost-effective detail's transmission version.
- In order to carry out a cost-effective detail's transmission version.
- To confirm the electricity usage on the suggested device is much better when compared with the current technique while using the NS2 simulator.

2 Literature Survey

Routing approaches with clustering methods is an effective method to fix the power effectiveness problem in deep WSN. The job contained [3] details the concepts of clustering designs as well as briefly classifies the clustering routing strategies above WSNs. Further, it systematically analyses the usual clustering routing strategies and also compares a variety of cluster-based routing techniques making use of varied routing metrics. The job contained [4] offers a comprehensive survey of clustering routing strategies above WSNs.

When it comes to [5], many energy-efficient routing algorithms for hierarchical routing process in wireless sensor networks are already talking about depending on the clustering techniques. These methods of clustering algorithms are distributed, centralized or even hybrid.

The job contained [6] proposes a dynamic clustering and distance aware routing (DDAR) routing way of WSN which thinks distance metric found CH choice. This particular effort elects the nodes which are nearest to sink node as CH, and also the powerful number of super cluster head (SCH) prolongs the WSN lifetime substantially.

The job contained [7] presents a routing mechanism referred to as chain cluster-dependent mixed routing (CCM) over WSN. The CCM exploits the benefits of essential LEACH process to improve WSN efficiency. It separates the WSN nodes into several chains, called as clusters, as well as routes the information packets within two quantities. At first, sensor nodes inside a bunch send out information packets to the CH node of theirs in parallel as outlined by the routing measures. Next, the CCM forms clusters among CH nodes inside a self-organized manner. Additionally, CH fuses information given through the cluster, and even it transmitted the fused information to the sink node to find the routing track.

The job contained [8] proposes a smart framework in order to bunch the WSN nodes and also to enhance the places of sensor nodes within WSN for extending realizing coverage as well as decreasing the power depletion of the whole community.

A multi-weight-based clustering algorithm (MWBCA) in [9] resolves the imbalanced power spending of the problem of WSN. In [10], an adaptable cluster-based routing protocol (ACRP) enables sensor nodes to create clusters within a self-organized manner. Inside ACRP, the CH nodes set aside some time openings for each and every sensor node for information forwarding. Additionally, CH fuses have been given info through the cluster of its and even transmit the fused info to sink node around to find routing track. The sink node occasionally changes the clusters within a network, and the sensor node selects a new CH node with energy highly based upon being given information.

A many-dimensional tree-based routing technique was offered around [11] over multi-sink WSNs. This effort exploits ant colony SEO (ACO) to improve the routing procedure. A delay-aware energy balanced dynamic routing protocol (DA-EBDRP) in [12] mostly is designed to balance the power depletion of WSN. The DA-EBDRP optimizes the routing progression and also evaluates the WSN efficiency utilizing a number of routing metrics which are throughput, end-to-end hold off, portion of living node (PLN), as well as community lifetime.

The author [13] designs an energy-balanced routing protocol (EBRP) that creates various practical opportunities in the industry by considering the level, node density, so the rest of the node power consumption should be ideal. Largely, the nodes are forced by the EBRP to switch towards an impressive, lively place to speak with sink as well as improve the WSN lifetime by saving the nodes with least recurring power. Additionally, it proposes enhanced strategies to rectify routing loop problems above WSN.

A tree-based routing protocol (TBRP) in [14] constructs a tree structure within a particular region, along with every sensor node estimates the distance between the sink. Additionally, it assigns a quality depending on the distance. The sensor nodes make use of a join REQ to migrate a single degree on the subsequent fitness level. The information forwarding procedure is set up as a result of the low-level nodes inside a tree building. The interaction requires directly into numerous amounts, and also, it is halted once the information gets to the location. The selected root node allocates some time schedules to kid nodes called low-level nodes for information transmission.

For proposing technique, the election of node top by taking into consideration ideal details lead to an effective tree-based bunch building. Furthermore, the threshold electricity as well as distance determines the road for information interaction in between bunch participants as well as bunch mind that results to the enhancement of community lifetime leading to a lesser amount of power used. With this research, the current WSN programme and also the suggested method is when compared, plus it indicates the proposed device is much better when compared with the current programme of terminology of power ingestion [15].

3 Proposed Technique

The primary goal is reducing usage of power of the nodes and also to boost the system lifetime. The suggested product could be considered two parts: the very first component is definitely the self-organizing mechanism for tree-based bunch development; another component will be the information transmission stage. Each of the phases must be energy efficient to get the goal.

3.1 Sequential Process Explanation

During self-organized tree-based bunch development, the E-ASER divides the sensor nodes into a number of clusters as well as elects a CH for every bunch by picking out the node with top recurring power amount and also the very last level of neighbour nodes. The CH is recognized as being a parent node that records and aggregates the information sensed from all the cluster members of it [15]. Moreover, the E-ASER selects a root node for energy-efficient details transmission. In effective details transmission stage, the E-ASER estimates a threshold printer based upon distance and energy for transmitting information effectively. The choice if the node should transmit the information through the CH or even straight to the root node is dependent on this particular threshold printer. Figure 1 shows the flow chart of E-ASER, about how the CH is elected, and then forwards it to the starting station. Figure 1 shows the flow chart of E-ASER.

3.2 Source Code of E-ASER

Figure 2 gives the pseudocode of the E-ASER process. The original values are established by the electricity type prior to the information transmission begins. The power effectiveness is attained by pairing the degree and residual energy of neighbouring nodes as conditions for that choice on the mom or dad node.

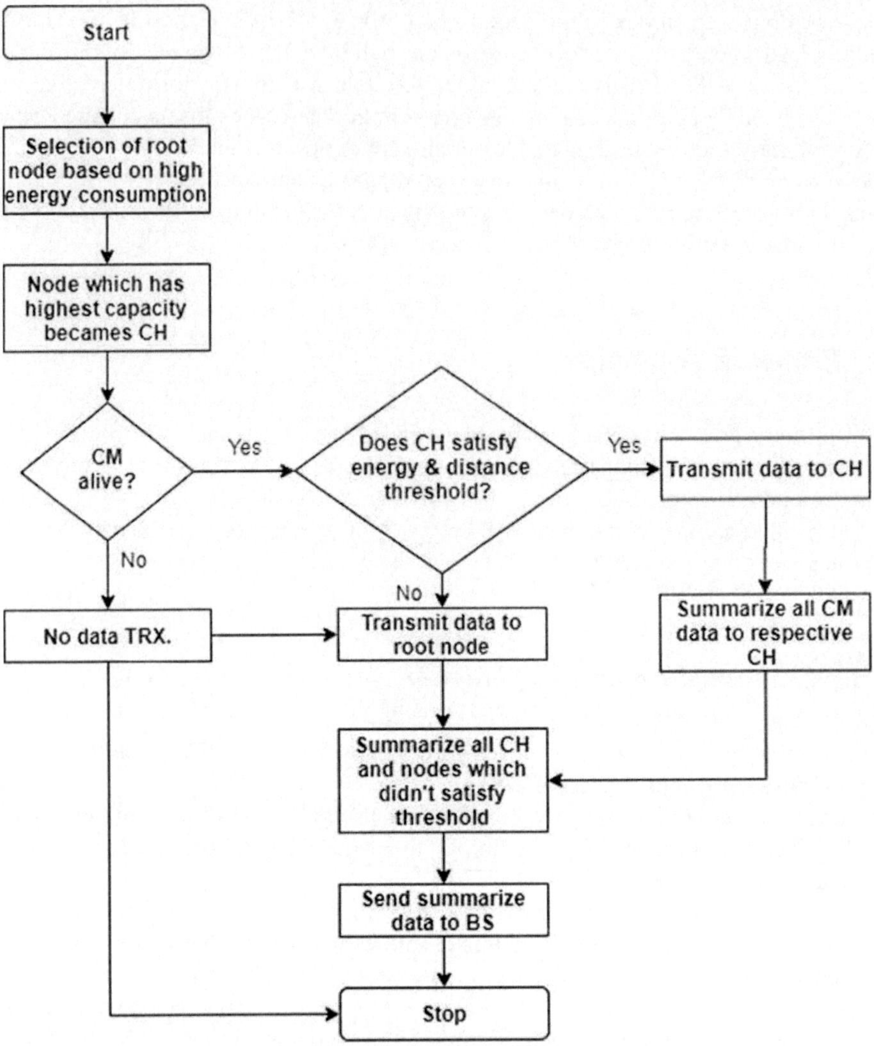

Fig. 1 Sequential process of E-ASER

Additionally, the pseudocode proceeds with all the information transmission stage. The threshold is utilized to determine if the information ought to be routed towards the bunch mind or even towards the root node.

```
Initialize energy parameter
while (request for data transmission arrives)
for (each node i) //select the root node
determine its neighbour and their residual energy
end for
root node <- node with high residual energy
for (each node i within each cluster) //to select cluster head
COMPUTE capacity (residual energy and degree of neighbours taken together)
end for
select node with highest value as cluster head for each cluster
//data transmission conditions
if (parent node energy > threshold_energy and distance form node to parent > threshold_distance from node to parent > threshold_distance)
transmit the data to the parent node for aggregation
else
transmit the data directly to the root node
```

Fig. 2 Source code for E-ASER

3.3 *Tree-Based Cluster Formation*

The BS at first broadcasts a package to each node in the system to have the routing procedure. The greatest recurring power that contains node is initially picked when the root node. Within bunch development, each and every node switches a hello note, such as the own ID of its, recurring power amount as well as neighbours matter. The nodes that lie in the proximity on the correspondence assortment are supposed to be on the respective clusters. Each and every node evaluates the capacity of it for CH election utilizing Eq. (1).

$$C_i = E_i + N_i \tag{1}$$

where

C_i = Capacity of the node,
E_i = Highest residual energy,
N_i = Number of neighbouring nodes.

Within situation (one), C_i could be the capability of the node I. The recurring vitality as well as neighbours matter of node I am represented as N_i and E_i, respectively. Each node broadcast its calculated capability to other neighbouring nodes, and every node compares the capacity of it with its neighbour's capability. The node together with the greatest electrical capacity is going to be picked when the bunch top for every bunch. In case two bunch heads have exactly the same electrical capacity, subsequently the tie is broken off by deciding one arbitrarily. Each one of the bunch heads articles on the root node which communicates towards the starting station. The potential for electrical power gap issues is eradicated by introducing the root node. Inside scenarios in which the starting station is situated miles away coming from the bunch heads or maybe the goal location due to that the power becomes exhausted rapidly. The reason behind the fast power empty is the fact that the bunch top has got the process of aggregating the gathered-up information then transmitting this particular with the starting station that is situated a long way away. The aggregated information might be forfeited within trying to transport the aggregated information during a lengthy distance to attain the starting station top towards the big energy gap issue. Root helps you to deal with the issue. The root node is in between the bunch top as well as the starting station and that cuts down on the big energy usage on the cluster top within attaining the base station. The root requires the duty of mailing the aggregated details in the bunch top of the starting station.

3.4 *Efficient Data Transfer*

The root, the cluster and cluster head participants are picked along with a tree-dependent bunch is formed. Each node calculates the distance in between itself as well as bunch head and also itself plus the root node. It establishes the recurring power on the bunch top with the purpose of information transmission. A threshold great is made, the decision thinking about the application programme necessity. When it comes to the experiment objective, we have regarded as 40% of the first power to function as the threshold power on the bunch mind. The threshold distance is 60% of the correspondence radius on the bunch. Each and every node compares inspections the power on the bunch mind. When the power of the bunch mind is in excess of 40% of original distance and energy in between the bunch fellow member as well as the cluster head is under 60% on the bunch correspondence assortment, the cluster part directs the information on the cluster mind. The power threshold conditions create certain the bunch top has sufficient power to admit the information, aggregate as well as advanced it with the root node.

In the event that possibly the threshold power or maybe the threshold power is not pleased, the bunch part specifically sends the information on the root node. This process removes the re-election of bunch hence cutting back on a considerable level of power. When the recurring power is much less compared to the threshold electrical power or perhaps generally if the distance is in excess of the threshold distance, the information is transmitted straight to the root node with all the intention which the power of the bunch mind is able to extend more.

The bunch top records the information at all of the bunch participants that gratify the threshold requirements. As soon as it has the information obtained as a result of almost all the cluster members of it, it aggregates the information as well as forwards it to the root node. However, the information delivered straight to the root node is going to be aggregated as soon as root node has been given the information from just about all the cluster head of it. The root aggregates the bunch part information with all the aggregated information delivered by all of the bunch heads and also directs them to the starting station.

The topology of the system is disrupted when any kind of node expires within the system. Thus, any kind of node that is intending to deplete the power must inform any other nodes within the system prior to it is able to give out. When every round has ended, the nodes are going to broadcast package on the whole community. Many of the additional nodes that happen to be continually overseeing the channel on getting the package will conduct an ID test and even alter the tables of theirs appropriately. In the event that absolutely no this kind of package is gotten, it implies that every node has enough power, and also the system will begin the subsequent round. The intention of eliminating node ahead is to preventing some loss within the network, becomes zero is preventing some loss.

The current technique DSTEB is missing the election of a great bunch top with the tree system that obtains treatment within the suggested method.

The information transmission from the current product is restricted towards the bunch mind, while it is given to point transmission to the root node underneath the conditions within the suggested method. These enhancements enhance the system lifetime by decreasing the power usage.

4 Results and Discussion

The system simulation device is utilized for simulating the system scenarios. NS2, a readily accessible simulation resource, is utilized for simulating equally wired as well as wireless networks. It may be utilized for computing different functionality variances as throughput, hold off, package shipping and delivery ratio, community lifetime, etc.

The system is modelled by arbitrarily deploying wireless sensor nodes near to the starting station inside a rectangular region. The starting station will be the sole fixed component with infinite power. The sensor nodes have movement and therefore are energy constrained. The sensor nodes take part in a tree-based bunch building as well as information transmission just so long as they have electricity. The favourite AODV routing process is definitely the foundation with the method version. The down below Table 1 has the device design employed for the simulation.

The device design is diverse through 20, 40, 60 nodes as well as two pause occasions 5, 10 ms to learn the actions of sensor nodes beneath various scenarios.

4.1 Energy Utilization

The deployed sensor nodes within the device are governed by power restrictions aside from the starting station. The original power is given prior to the information

Table 1 Simulation parameters

Simulator	NS2
Topology	Random
Interface type	Phy/wirelessphy/802.15.4
MAC type	IEEE 802.15.4
Queue type	Drop tail/priority queue
Length of queue	50 packets
Antenna type	Omni
Propagation type	Two-ray ground
Routing protocol	AODV
Transport agent	UDP
Application agent	CBR

transmission begins. The main our difference in power, we are subtracting the actual power which every node is consuming with balanced energy, i.e. power consumed for data transfer is knowing as energy utilization.

$$\text{Energy Utilization} = \text{Initial energy} - \text{Balance Energy}$$

By equally Fig. 3, and also Fig. 4, the power usage of E-ASER is drastically smaller compared to DSTEB due to good parameter choice with the bunch mind election. If the mom or dad node has exhausted beneath the threshold, the energy drains quicker since it has to aggregate the information. Information aggregation and transmission of bunch mind is eradicated once the recurring power is under the

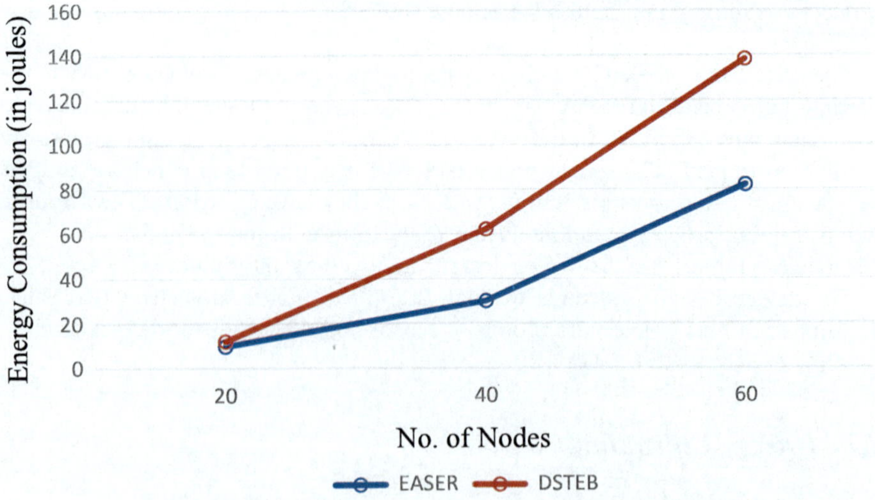

Fig. 3 Number of nodes versus energy consumption with pause time = 5 ms

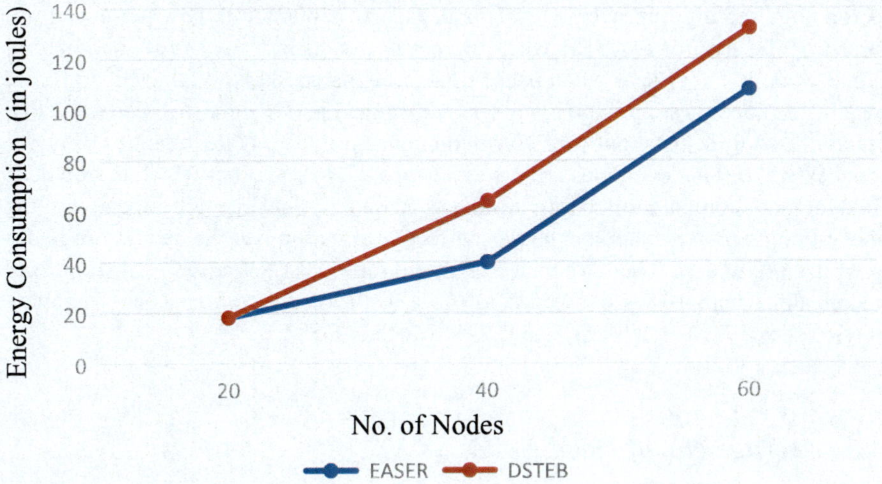

Fig. 4 Number of nodes versus energy consumption with pause time = 10 ms

threshold or even if the distance in between bunch part as well as bunch head is above threshold distance. By moving the information on the root node each time, a threshold key element is not happy outcomes to come down with much less overhead along the mom or dad node. Bunch mind is able to extend the system lifetime for this explanation.

When the pause period is 5 ms, nodes shift gradually as well as there is a time that is enough for interaction, while during 10 ms, the nodes shift quicker as well

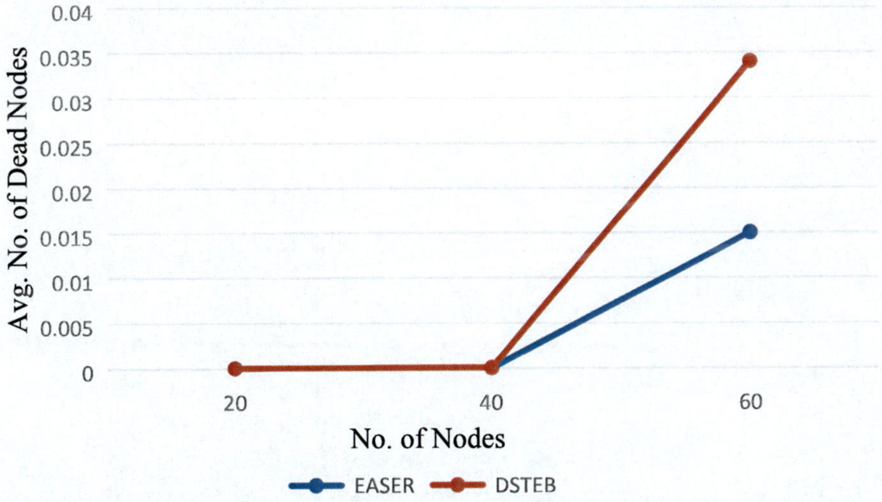

Fig. 5 Number of nodes versus dead nodes at pause time = 5 ms

as topology rapidly improvements. This particular explanation, how the power use improves linearly for E-ASER when the pause period is 5 ms as revealed around Fig. 3 while it's not linear when pause time is 10 ms, as shown within Fig. 4. If the quantity of nodes is simply too much less, (20 nodes) nodes are distributed away that make it hard to achieve out there for neighbouring nodes. If the quantity of nodes is excessive, (60 nodes) nodes are extremely around one another which in turn calls for a lot more computation within turning up at bunch mind election. Because of the above-described explanations, the most effective situation is if the quantity of nodes is typical (40 nodes). Thus, we visit a next, dip for E-ASER found Fig. 4 from forty nodes, along with 10 ms pause period. As a result of the common benefit gotten, power use is 17.6% much less found E-ASER than DSTEB.

4.2 *Average No. of Dead Nodes*

When the sensor begins taking part on tasks as initialization, setup, aggregation, transmission of packets, reception of packets, they drop the power as outlined to the power version. The express whenever the power of the node turns into 0 is known as an old node. The old node does not participate in the network because the node dead. The primary goal is balancing the power within the nodes like they do not deplete power fast to get old nodes. Much more the variety of old nodes, cheaper the system lifetime will be. Furthermore, the package fall fee is going to be significant as the variety of old nodes grows.

Via Figs. 6 and 5, the variety of old nodes is practically zero up to forty nodes

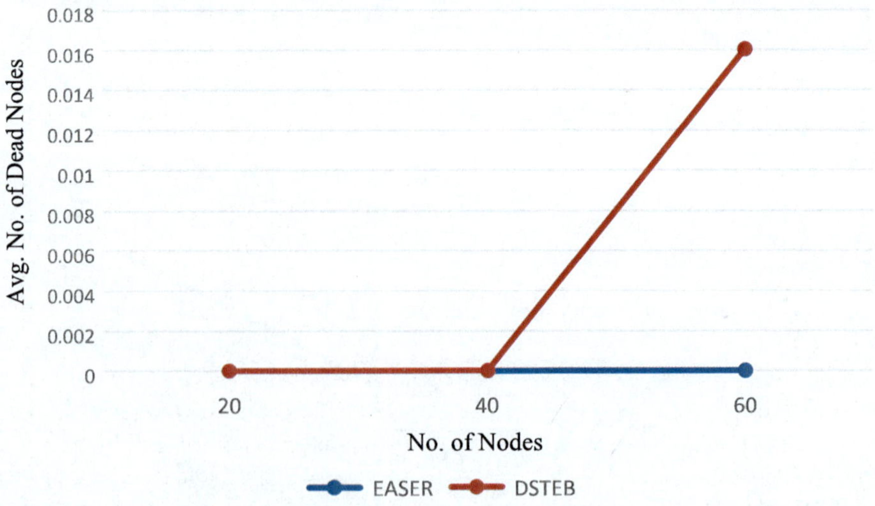

Fig. 6 Number of nodes versus dead nodes at pause time = 10 ms

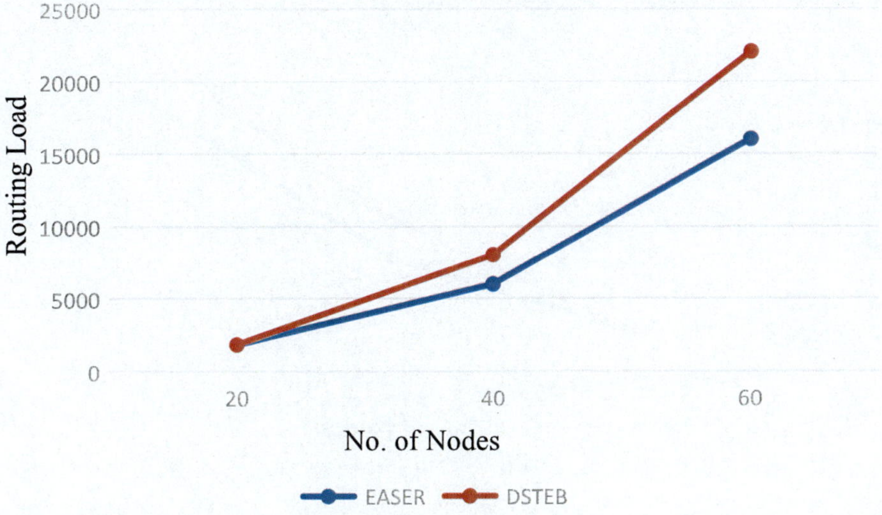

Fig. 7 Number of nodes versus routing load at pause time = 5 ms

for DSTEB and E-ASER with pause time = 5 ms as well as 10 ms. Nevertheless, DSTEB has typical old node of 0.016 but is almost zero from 10 ms for sixty nodes in E-ASER. This is since the assortment on the bunch mind is grounded on the recurring as well as level of neighbouring nodes. Bunch top as a much better lifetime when each metrics are believed to be. Additionally, information transmission whenever the quantity of nodes is sixty is sleek because the nodes close by can easily be bought. Merely recurring power is recognized as in DSTEB and that results to bad bunch mind election and also drains the power of nodes fast since regular re-election of bunch mind happens. Nodes found close to power empty inform all of the nodes prior to beginning the other cycle and also do not take part in the following round. This removes information damage within the E-ASER. Due to the above-mentioned factors, the common old node in E-ASER is under DSTEB.

4.3 Path Load

For each package being transmitted of source of energy to the desired destination, there ought to be sufficient routing info. Routing, ton will be the variety of routing packets needed per information package to effectively get to the location. Every forwarded package is regarded as a single transmission. The routing ton could thus be described as the ratio of the amount of routing packets delivered to the number of packets obtained in the location. In the event that you discover a number of routing packets for a single data package, it brings about an overburden on the system thus minimizing power faster. The routing ton must be much less to be able to bring down

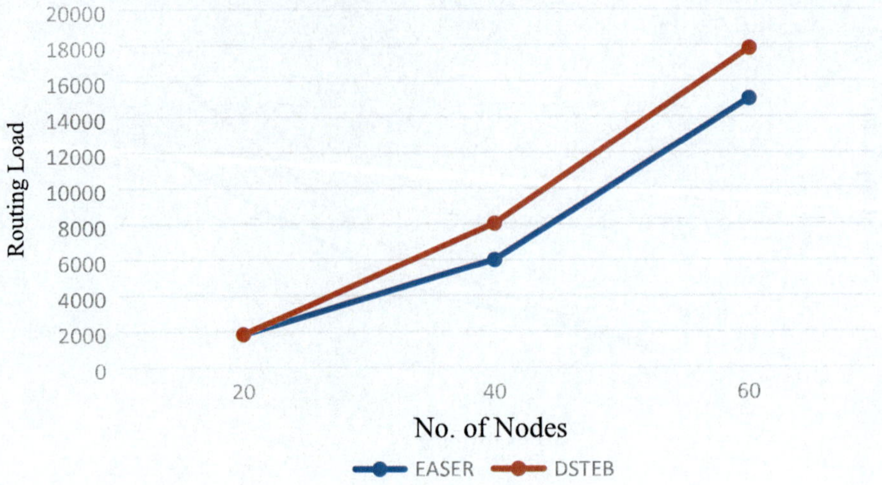

Fig. 8 Number of nodes versus routing load at pause time = 10 ms

power usage.

Path Load = No. of routing packets/Packets gotten in the destination

Via Figs. 8 and 7, the routing ton is considerably less found E-ASER as when compared with DSTEB. Every node sends information with neighboring nodes, regardless of the recurring power of all of the clusters. Inside E-ASER, power is well balanced by mailing the information on the root node whenever the cluster does not adequate electricity. The root node requires theduty to aggregate the information on the nodes (cluster member) not to gratify the power threshold or maybe the distance threshold. The routing ton is practically similar for 10 ms and 5 ms pause occasions, indicating a sound community. The routing ton is 23% in E-ASER as when compared with DSTEB.

4.4 *Lifetime*

The system lifetime could be described as the time period in that the nodes within the system are practical. Cheaper the power usage, greater the system lifetime.

Lifetime = Number of nodes ∗ (initial energy/energy consumption)

There is a little enhancement within the system lifetime between DSTEB and E-ASER as revealed in Fig. 9 as well as Fig. 10. Community lifetime depends upon various community scenarios. Generally, there might be nodes within the device that

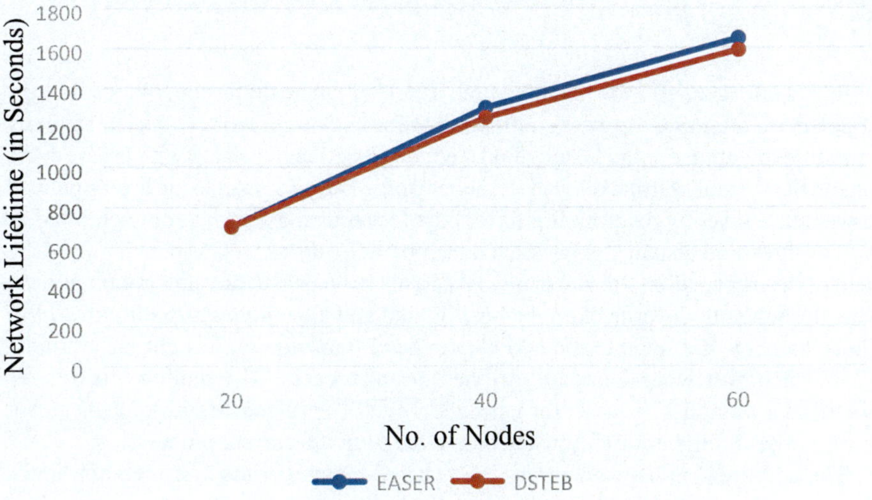

Fig. 9 Number of nodes versus network lifetime at pause time = 5 ms

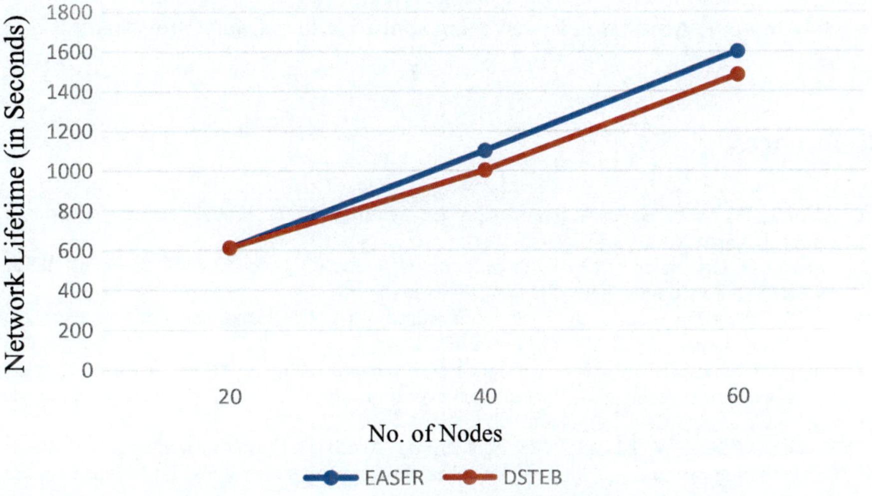

Fig. 10 Number of nodes versus network lifetime at pause time = 10 ms

are not taking part in the information transmission while you will find not many nodes that are thoroughly moving information.

An additional aspect to think about may be the variety of nodes. Having a greater number of nodes, the intermediate nodes can easily be bought for information transmission, while, by way of a reduced number of nodes [16], the quick node for information transfer may be at a farther distance. Thus, the system lifetime gets better with a rise in the quantity of nodes.

5 Conclusion

With this effort, an effort is manufactured to self-organize the system of terminology of power for attaining a greater community lifetime. Power use is cut back by picking a mixture of details for the bunch mind election. Electing bunch top with great details extend the system lifetime of all of the nodes. Information transmission is pronounced power successful by determining thresholds for recurring power for bunch mind as well as threshold distance in between bunch participants as well as bunch mind. The road of the information transmission makes a decision on the threshold requirements. This decreases the overhead on the bunch mind and also prolongs the lifetime of it. Great number of efficient data and cluster head transmission has cut back around 17.6% of power when compared to the current process. The routing ton or even overhead is around 32% under the current process. The system lifetime enhancement is somewhat of higher quality when compared with the current process.

The additional enhancement may be reached by taking into consideration much better metrics for bunch mind election. The tree-dependent cluster framework is usually analysts talking about a lot of amounts of the tree to take a look scalability. The E-ASER is simply a process for testing a brand-new concept. This particular process could be utilized in uses in which the substantial powerful topology is engaged. Heading more, E-ASER is designed being applied with real-time programmes.

References

1. Akyildiz IF, Su W, Sankarasubramaniam Y, Cayirci E (2002) Wireless sensor networks: a survey. Comput Netw 38(4):393–422
2. Osterbo O, Grondalen O (2012) Benefits of self-organizing networks (SON) for MOBILE OPERATOrs. J Comput Netw Commun 2012:1–16
3. Liu X, Shi J (2012) Clustering routing algorithms in wireless sensor networks: an overview. KSII Trans Internet Inf Syst 6(7):1735–1755
4. Singh SP, Sharma SC (2015) A survey on cluster based routing protocols in wireless sensor networks. In: Proceedings of international conference on advanced computing technologies and applications (ICACTA), vol 45
5. Hassan A, Shah W, Iskandar MF, Mohammed A (2017) Clustering methods for cluster based routing protocols in wireless sensor networks: comparative study. Int J Appl Eng Res 12(21):11350–11360
6. Yan A, Wang B (2017) An adaptive WSN clustering scheme based on neighborhood energy level. In: 2017 IEEE 3rd information technology and mechatronics engineering conference (ITOEC), Chongqing, pp 1170–1173
7. Tang F, You I, Guo S, Guo M, Ma Y (2012) A chain-cluster based routing algorithm for wireless sensor networks. J Intell Manuf 23:1305–1313
8. Awad F (2012) Energy-efficient and coverage-aware clustering in wireless sensor networks. Wirel Eng Technol 3:142–151
9. Fan Z, Jin Z (2012) A multi-weight based clustering algorithm for wireless sensor networks. College of Computer Science & Educational Software Guangzhou University
10. Rahmadhani MA, Yovita LV, Mayasari R (2018) Energy consumption and packet loss analysis of LEACH routing protocol on WSN over DTN. In: 2018 4th international conference on wireless and telematics (ICWT), Nusa Dua, pp 1–5

11. Alnaham MYM, Abdelrahaman IAM (2018) Improving QoS in WSN based on an optimal path from source to sink node routing algorithm. In: 2018 international conference on computer, control, electrical, and electronics engineering (ICCCEEE), Khartoum, pp 1–6
12. Osamy W, Khedr AM, Aziz A, El-Sawy AA (2018) Cluster tree routing based entropy scheme for data gathering in wireless sensor networks. IEEE Access 6:77372–77387
13. Alaparthi N, Parvataneni SR, Kanderao NVS, Jagtap SV, Donthu A, Korivi H (2019) Performance comparison of CSMA, MACA, Generic MAC and Sensor MAC channel access protocols for ZigBee WSN with RIPv2 as Routing protocol. In: 2019 innovations in power and advanced computing technologies (i-PACT), Vellore, India, pp 1–5
14. Singh M, Sethi M, Lal N, Poonia S (2010) A tree based routing protocol for mobile sensor networks (MSNs). Int J Comput Sci Eng 2(1):55–60
15. Suresh VA, Gopinath PG (2015) Distributed self-organized tree based energy-balanced routing protocol for wireless sensor networks. Int J Innov Res Sci Eng Technol 4(6)
16. Choudhary S, Sharma L, Kaushik AK, Mishra A (2019) Novel approach to reduce the replication of information and to increase the reliability of end to end data transmission in WSN. In: 2019 2nd international conference on intelligent communication and computational techniques (ICCT), Jaipur, India, pp 83–86. https://doi.org/10.1109/icct46177.2019.8968785

An Efficient Technique for Traffic Estimation Using Virtual Trip Lines in Probe Vehicles

Teena Goud, Ajay Dureja, and Aman Dureja

Abstract Estimating traffic continuously is an important principle serviceability of urban communities. In order to decrease the expenses of arrangement and activity, traffic estimation with vehicles has been broadly studied. The main aim of this paper is to propose a traffic observing framework utilizing vehicles. The proposed calculation has the benefit of having an extremely low data processing expenses, permitting the majority of the pre-preparation to be done in the vehicles and consequently making conceivable the incorporated assortment of countless estimations. The proposed framework is made out of two calculations; a virtual trip line that evaluates the traffic and a traffic information authority that totals the data from numerous vehicles and consolidates with data authorities. The framework has been tried in a genuine situation, contrasting its exactness and a customary congestion demodulator, demonstrating its precision.

Keywords GPS · Guide coordinating · Traffic observing framework · VTLD

1 Introduction

These days, street systems, and particularly urban areas, experience the ill effects of security, biological and comfort issues because of the tremendous utilization of private vehicle in our everyday schedule. Traffic blockage, which regularly follows known examples, has become an ordinary issue in urban conditions, bringing about observable transient and gigantic burdens. This has driven numerous urban areas to implement certain estimates. Urban metropolis manages the above issues in a

T. Goud (✉) · A. Dureja · A. Dureja
Department of Computer Science & Engineering, PDM University, Bahadurgarh, Haryana, India
e-mail: teenagoud16@gmail.com

A. Dureja
e-mail: ajaydureja@gmail.com

A. Dureja
e-mail: amandureja@gmail.com

A. Khanna et al. (eds.), *Data Analytics and Management*, Lecture Notes on Data Engineering and Communications Technologies 54,
https://doi.org/10.1007/978-981-15-8335-3_33

progressively complex and less invasive manner, however, to have the option to direct the traffic, and an essential necessity is to have continuous observing of the traffic conditions.

The conventional techniques utilized by open organizations are fixed estimation gadgets that gather information about traffic, for example, inductive circle locators, pneumatic street tubes, camcorders, laser and other vehicle identifiers. Such gadgets are fit for gathering data like vehicle type, vehicle measurements or speed, contingent upon the sort of gadget utilized. Be that as it may, these gadgets can just gather information on the particular segment of street where they are introduced. Moreover, traffic observing frameworks worked with these gadgets need an overwhelming speculation to create, send and look after them. In most recent years, we have seen a sensational increment within the sight of gadgets is worked with GPS ability. In this work, the emphasis is on advanced probe vehicles, which for the most part has two area suppliers (GPS and the cell/WiFi arrange).

These gadgets have two fundamental points of interest contrasted with fixed gadgets—dynamic inclusion zone and zero extension cost. Vehicle area information authorities are regarded legitimate for traffic detecting purposes, since as long as there is an adequate infiltration rate, they will give precise estimations of the traffic flow, that is, the example of clients with traffic sensors over the vehicles which entered the objective street ought to be in any event 3–5% [1, 2]. With the expanding development in the quantity of vehicles, the information got from the traffic observing has got more accurate. Simultaneously, this expanded dependability includes some significant downfalls in cost power. Traffic checking frameworks must total and procedure data to high units of demodulator at present. Further, it is viewed as a major information [3–6] issue, with respect to size and speed, and the information gives rise to exceptional calculation systems.

For this situation, the gigantic size of information is because of the enormous units of gadgets providing traffic data. Hence, so as to build the limit, the handling enforced per report is limited. Information revealed by sensors is typically an area and a timestamp [7–10]. Hence, the crude information cannot be utilized to gauge the traffic information without pre-handling. The area information should initially be related to a particular street or road. This procedure is called guide coordinating and has elevated data processing expense. Various guide coordinating strategies have been distributed [11], in spite of the fact that relatively few [12] have demonstrated to run continuously. Those that accomplish work progressively [13–15] are frequently utilized widely in business turn-by-turn route gadgets (both devoted and advanced cell based). A few techniques are not in any case unveiled (or just halfway uncovered) as it is licensed [16, 17].

The next estimation is given for perusing the timestamps. Usually utilized to calculate the number of users passing through a street per time interval. Once more, the above data needs few pre-handlings for building the data processing expense per estimation. A few productions have proposed information combination as a method for upgrading the precision of traffic evaluation [18–21]. In this study, the proposed size to effectively total information from numerous sources is the traffic inhabitance on a virtual trip circle (VTC). It is essential to call attention to that traffic inhabitance, which quantifies the time that a particular street fragment is involved by vehicles, ought not be mistaken for vehicle inhabitance, which gauges the quantity of travelers per vehicle and is out of the extent of this paper. The important principle is that the size can be processed in the individual portable estimating gadgets and afterward utilized in a unified area to appraise the general traffic and to combine it with other information sources [22] with a insignificant pre-preparing cost. The primary commitment is to propose an estimated traffic utilizing gadgets. The strategies used in this paper are to reduce the data processing expense of every area in the incorporated assortment point. This in turn provides high server prerequisites and expansion of the monetary expense in the organization and diminishes its versatility. If there is any decrease in the expenses on each test, then permit either the utilization of commoditized gear (permitting littler open associations to utilize the framework) or the adaptability of the framework into enormous information domain (permitting huge topographical territories to be observed without a moment's delay).

In this paper, the expense per test is diminished by depending on the vehicle for the vast majority of the calculations rather bringing together assortment point, discharging assets for extra estimations. In particular, the guide coordinating and traffic inhabitance computations are implemented in the portable ends. Thus, it includes the prerequisite of few calculations that limits vitality utilization. This paper includes the proposed guide coordinating calculation that run progressively and simultaneously by using vehicles without expending numerous assets. For the traffic estimation, the VTL [23] calculation is utilized.

This calculation emulates conventional estimating gadgets per second spending at some endpoint, that is, the traffic inhabitance. Trial of each step shows the legitimacy of the technique and capacity to work with flat inspecting rates and thus lessens the power utilization in vehicle. These experiments are done in a genuine arrangement, henceforth approving the calculation to use in real-time world. The rest of this task is organized as given. In Sect. 2.1, the issues of checking gadget area are introduced. In Sect. 2.2, the proposed arrangement is depicted. In Sect. 2.3, the way toward recovering traffic information from a solitary vehicle is introduced. In Sect. 2.4, the traffic information calculation is introduced. When the difficulties and proposed arrangements are appeared, these arrangements will be tried, and the outcome will be appeared in Sect. 3. At last, a few decisions about the current work are outlined in Sect. 4.

2 Materials and Methods

2.1 Gadget Observing Difficulties

In this, two difficulties in traffic observing frameworks by means of area gadgets are portrayed: off base area and track calculation.

2.1.1 Off Base Area

The area required in traffic observing is given by the GPS and the system supplier. The two components include mistake in the assessed area. From one perspective, in GPS frameworks, the primary blunder source is because of wrong estimation by clock of the recipient gadget. The recipient and satellite clock segments are increased by the velocity of light(c).

Consequently, as a result, a little clock mistake can affect an extremely huge program and stage blunder. For instance, a clock blunder of 1.1 ns converts into 0.2 m in coverage inaccuracy, though 1 μs infers a mistake of 320 m. Expanding the collector clock exactness by atomic clocks is costly so it isn't a attainable choice for commercial gadgets like PDA's. Then again, the system area supplier utilizes GPS and cell towers to known vehicles to inexact the area of a client. At the point when the area supplier is surveyed, the IDs of the GPS in the region are sent through Web to the virtual area server, a database with area data on GPS. The virtual area server restores the estimated area of the client. GPS permits an exactness of 10–50 m, while vehicle arranges just permit a precision more prominent than 50 m.

Thus, system supplier will be mistaken in zones without GPS. So, whatever the location provider is, the mistake must be taken into consideration in arrange to assess the real area of the client.

2.1.2 Track Evaluation

The vehicles are customized to intermittently allow area reports. In any case, a base or consistent recurrence for the area report is not ensured, because of various issues like inclusion gaps (on arrange or on the spot supplier), jamming in the network or client information imposition. In particular, the challenge now is accomplishing an adequate exactness in map-coordinating with a low recurrence inspecting. The lower the area testing information the higher the trouble of the guide coordinating issue. Then again, a low recurrence testing limits battery utilization and system use.

The guide coordinating issue can be characterized as finding the way that interfaces an arranged rundown of areas on a street organize and that is the most likely followed direction. With huge separations between back to back areas, the trouble for map-coordinating is higher in light of the fact that there are progressively potential directions that interface these sequential areas, and in this manner, there are increasingly potential directions among two areas.

2.2 Proposed Work

An incorporated framework to gather traffic updated information is proposed in the given Fig. 1. The fundamental block totals information from a few information sources in the observing traffic framework. Section 2.4 gives typical configuration

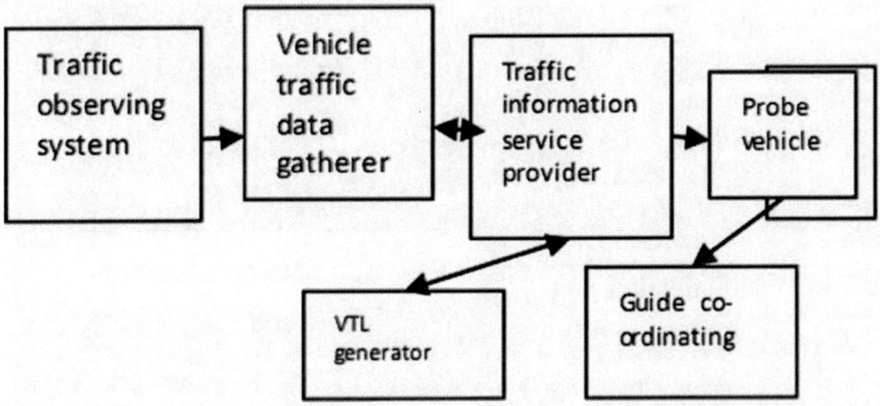

Fig. 1 Framework for traffic monitoring by VTL

of information sent to the checking traffic framework thus characterized. Hence, regular configuration will extraordinarily lessen the handling required per test, in this manner expanding the limit of the principle block. Prerequisites subtleties are to adjust the information sources by providing inside structure of the checking traffic framework.

2.3 *Single Vehicle Location*

Revealed area and VTLs will once in a while be at a similar point, because of various factors, for example, the location mistake or the recurrence of area estimations. Along these lines, the course succeeding a traveler in the vehicle is determined with guide coordinating. When the course is known, it is conceivable to infer when the client experienced an estimation point (VTLs). If many gadgets in the vehicle conclude its area, copied VTLs crossing the trip line will send the records, respectively. Since the gadgets have no chance to know about one another, the discovery of such identical is stored in VTL server.

2.3.1 Guide Coordinating

Guide-Coordinating is the method of associating a sorted list of areas to the street network on a digital outline. An advanced outline is represented by a numerical graph G. A chart is a portrayal of a lot of items or vertices where a few sets of these articles are associated by joins. These items are ordinarily considered hubs, and the connections that associate them are called bends. Figure 2 shows the guide coordinating result.

Fig. 2 Guide coordinating result

$$G = \{V, E\}$$

Here, V is the number of hubs, and E is the number of bends.

The point of a guide coordinating calculation is to remake from an area grouping the way determined by a vehicle out and about system. The primary challenges in this errand get from the mistakes of area suppliers situating estimations and from the vulnerability presented by the inspecting of the information. Figure 2 represents the guide matching problem.

Latest works identified with map-coordinating are centered around settling, with perpetually precision and dependability, a similar issue of keeping track continuously of the right situation of the client on a map. Thus, the situating information utilized for this sort of undertaking for the most part has a high inspecting recurrence. A wide range of systems have been created to settle this sort of guide coordinating issue, running from basic geometrical contemplations to further developed surmising techniques yet they are normally sorted for effortlessness into following gatherings—geometric, topological, probabilistic and progressed [24], and hence, geometric-based methodology is followed in this paper.

The lower nature of this sort of guide coordinating calculation permits it to run easily. At the point when the client is going via vehicle, the followed telephone is situated in the vehicle as well, so area supplier inclusion is regularly low and, in this manner, the area blunder is ordinarily high. Moreover, it must be considered that the inspecting recurrence may not be high [25, 26]. High examining recurrence depletes the battery of the vehicles and may be exorbitant relying upon the information tax. Subsequently, it is important to discover a tradeoff between precision, computational expense and battery utilization.

2.3.2 Virtual Trip Line Detector

So as to gauge the traffic status, when the path from a client and invested time are known, a virtual trip line detector (VTLD) procedure is proposed. The VTLD is a fragment of a curve of a computerized map, which recreates the activity or a genuine trip line detector. At the point when a VTLD is crossed by the client gadget, the application in the gadget creates a traffic information bundle to send to the vehicle traffic data collector. This abstains from sending pointless information to the server, advancing the handling limit of the framework. The traffic information parcel incorporates the estimation time stamp, inhabitance, rundown of voyaged curves, time spent and voyaged separation, albeit just the inhabitance is used to assess traffic in the monitoring traffic system. The remainder of the information is put something aside for conceivable future employments.

During the execution, it has been seen that the area supplier mistake infers terrible areas causing wrong estimations in the accompanying cases:

$$\text{speed} = \text{distance}/\text{time}$$

2.4 Traffic Information Gatherer

Detailed information of solitary client is not strong enough to appraise metropolis traffic. A solitary monitor could stop for various reasons. Be that as it may, if the revealed information from all clients in each circular segment at a given time interim is consolidated, a superior way to deal with the traffic conditions can be gotten. On the other hand, customer application sends the traffic information determined to the framework through a Web API. Besides, this information is given to traffic information window where they are joined with information from different clients in a similar circular segment and schedule vacancy. The necessary current circular segment data for this procedure is mentioned from (OSM [24]). OSM is an open peruse and compose get to stage about road maps information.

Likewise, copy of the documentation (report of the same VTLD with fundamentally the same as timestamps) ought to be strained.

This should be possible with two sifting rules relying upon the quantity of paths in the VTLD. The copied reports are done on a solitary path VTLD: Both will have the equivalent or fundamentally the same as timestamp; henceforth, just the primary report to arrive at the traffic information authority will be thought of

$$
\begin{aligned}
di &= B - A \\
dif &= I - A \\
\text{proj} &= di \cdot *dif/|di| \\
d(I, AB) &= d(I, A)
\end{aligned}
$$

where

A: one endpoint
B: another endpoint
I: provides information of an area
$d()$: distance across endpoints.

The copied reports are done on a various path VTLD: This circumstance is increasingly mind boggling, so once two reports have the equivalent timestamp, a confirmation procedure must be propelled to check in the event that they are situated in various vehicles. This confirmation procedure will check if the last NR reports of the gadgets are likewise copies.

2.4.1 Traffic Information Evaluation

There is no normalized definition for traffic jamming [23]. In this manner, there are diverse similarly substantial estimations so as to consider the traffic blockage [26]. In this paper, the accompanying standardized traffic pointers are picked. Traffic gauge is given by Table 1,

$$F(v/h) = \text{vehicle}_{ij}$$

$$I(\%) = \sum T v_{ij}/\text{time-slot\_ length} * 100$$

where

F = force of the traffic in a particular portion.
vehicle_{ij} = quantity of vehicles in VTLD.
$I = I$ is the traffic inhabitance in a particular fragment of street (VTLD).

Table 1 Traffic information evaluation

Time	Power	Occupancy (%)
50	$\frac{32}{3} = 50$	30
120	$\frac{3.60}{3} = 60$	36.1
180	$\frac{2.60}{3} = 40$	22.2

Tv_{ij} is the time for the vehicle in VTLD at schedule opening.

A VTLD gauges traffic in all paths of street (with the exception of paths for extraordinary use, e.g., a transport path). The quantity of paths in every street is extraordinary, so it is critical to standardize these estimations to the quantity of paths.

At last, force and traffic inhabitance are accounted for estimated triggered module to the outer traffic observing server. These pointers specifically are valuable since they permit to think about, even total the traffic estimation revealed by various estimating gadgets [23]. In any case, consider that the power is only a fractional estimation, since it just mirrors the force of vehicles that have an estimation gadget, which is only an example of the aggregate. This infers the outright estimation of the power will not really be helpful, yet its general conduct (development or decrease) might be of enthusiasm for certain applications. Also, it is conceivable to get these markers from different places with straightforward changes.

3 Results and Discussion

By testing whole arrangement appeared, a lot of tests has been characterized, reporting various modules of the proposed framework.

In this area, first the presentation of the guide coordinating strategy introduced in Sect. 2.3 is assessed with various area inspecting periodicity. At that point, the traffic estimation exactness is broken down by contrasting the returned information and the information announced by the pneumatic street counter.

3.1 Guide Coordinating Area Frequency

The point of the accompanying investigation is to show the exhibition of guide coordinating utilizing the course estimator calculation. The investigation relates the exactness in the estimation and the recurrence utilized in area demands. Guide coordinating is centered around the amendment of the inherent area blunder and fitting the area to the course. The course estimation is centered around limiting the mistake gave by the absence of information area. In spite of the fact that it appears to be clear that high area inspecting frequencies can accomplish more exactness in the course estimation, this improvement might be less significant in the guide coordinating execution because of the inherent blunder by area suppliers. Figure 3 represents the VTLD arcs across the location.

This execution examines the area supplier with GPS and scope of area recurrence refreshes a set up between 1.2 and 0.018 Hz. This tests results, when gadget went more than 80 km in various kinds of locales and speed—close to high structures, regions without structures and congested roads. The consequences of this investigation show

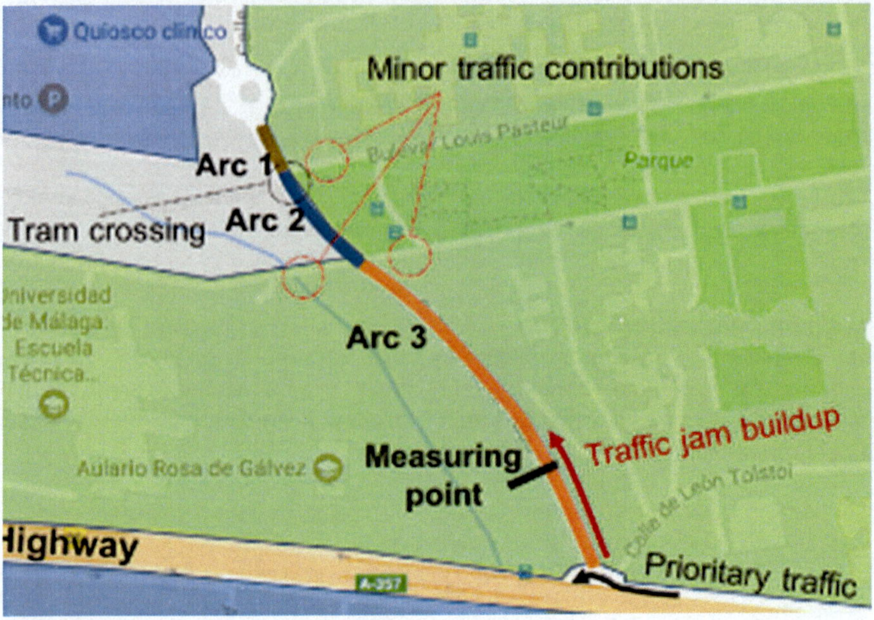

Fig. 3 Representation of VTL associated with arcs

that the normal precision is in the scope of 60–85% as appeared. The exactness of evaluated course keeps an eye on a lineal condition relative to the area examining frequency.

$$\text{Accuracy}(\%) = \text{number of curves}/\text{total curves in path} * 100.$$

As we see the outcome, the exactness is more awful when compared with time frame between tests. This is basically because of the expansion of the potential courses when the first and last point are a long way from one another. At the point when the period between tests is low, the separation between the estimations is likewise low, and the quantity of conceivable elective courses among them is typically just one.

Then again, when the period is high, the separations between the estimations might be longer (contingent upon the velocity), and subsequently, the quantity of potential courses might be multiple, accordingly expanding the quantity of mistakes. Figure 4 demonstrates the average accuracy rate at a given period.

Fig. 4 Guide coordinating to measure accuracy

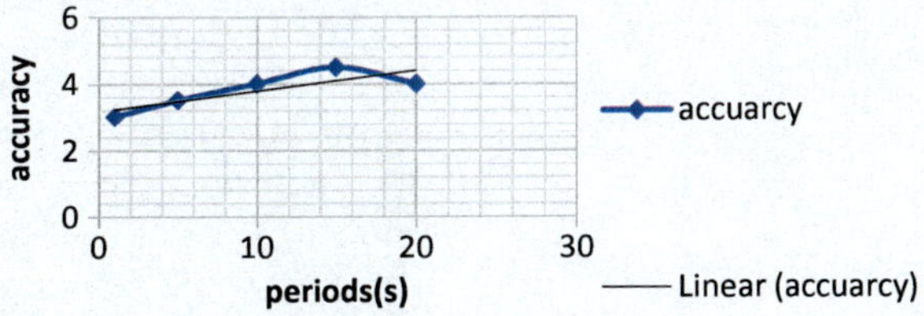

3.2 *Traditional Technique VS VTLD*

In this test, the assessed traffic inhabitance gave by the VTLD technique is considered. However, the investigation of the total estimation of the force is pointless on the grounds that it is evaluated by tallying the gadgets associated with the framework in each bend, not the all-out number of vehicles. For example, if 30 vehicles travel along some street in one hour however just 15 of them are associated with the framework, at that point, the VILD reports a force of 15 vehicles for each hour rather than 30 vehicles for every hour. In any case, the pattern of this halfway power may mirror the general conduct of the genuine force, particularly if the pace of the vehicles that have an estimating gadget is known. Figure 5 represents the traffic inhabitance scaled for 30 samples.

Figure 6 represents the MSC of traffic inhabitance for 30 devices, and hence, the MSC is proportional to the devices that means the error with least samples is not executed.

Fig. 5 Traffic inhabitance (30 samples/h)

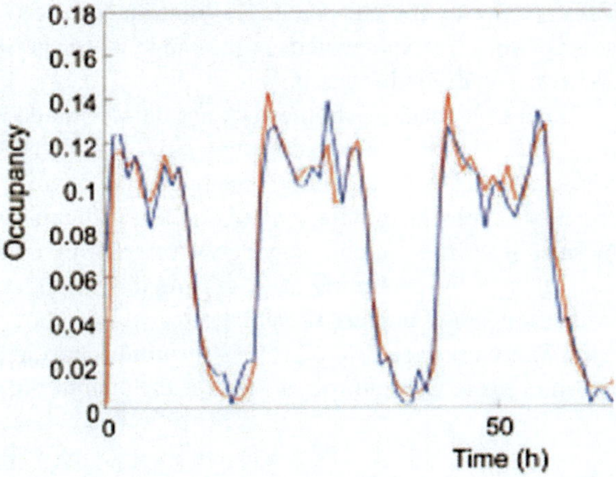

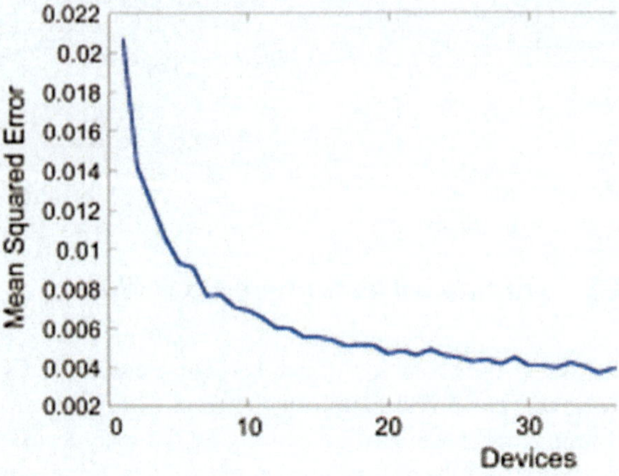

Fig. 6 MSC of traffic inhabitance

4 Conclusion

A total framework for observing traffic status utilizing vehicles has been established in this paper. The performed tests give the precision and adequacy of the proposed approach in evaluating path and traffic inhabitance. It also reveals assessed path is contrasted and the genuine followed path and the inhabitance in a street with the estimations of the approved traffic observing gadget are performed by simulation.

At this point, the guide coordinating calculation has been demonstrated by providing a high exactness with various inspecting frequencies, considering the blunder gave by the GPS area. The consequences of guide coordinating show right coordinating rate somewhere in the range of 66 and 86% with tests each moment and consistently individually.

The traffic checking framework has demonstrated by giving finite exactness and inhabitance in clock evaluation. Velocity of immense traffic has never tried in light of the fact that the estimation is relative to the quantity of gadgets detailing in every street as referenced previously. This angle is difficult for upgrading the future strategy because it utilizes numerous gadgets for emitting data. An answer, further work is consolidated by estimating the data from different sources. Thus, traffic inhabitance evaluation shows at least 11 vehicles reveal in an hour as an outcome.

Future work includes improving the guide coordinating outcomes and producing calculations to gauge through vehicles the quantity of vehicles voyaging.

References

1. Herrera JC, Work DB, Herring R, Ban XJ, Jacobson Q, Bayen AM (2010) Evaluation of traffic data obtained via GPS-enabled mobile phones: the mobile century field experiment. Transp Res C Emerg Technol 18:568–583

2. Zan B, Hao P, Gruteser M, Ban X (2011) VTL zone-aware path cloaking algorithm. In: Proceedings of the 2011, 14th international IEEE conference on intelligent transportation systems (ITSC), Washington, DC, USA, 5–7 Oct 2011, pp 1525–1530
3. Abolfazli S, Sanaei Z, Bojanova I (2015) Internet of cores. IT Prof 17(3):5–9
4. Cordeschi N, Shojafar M, Baccarelli E (2013) Energy-saving self-configuring networked data centers. Comput Netw 57(17):3479–3491
5. Urgaonkar R, Kozat UC, Igarashi K, Neely MJ (2010) Dynamic resource allocation and power management in virtualized data centers. In: Proceedings of IEEE NOMS, pp 479–486
6. Balakrishnan P, Tham C-K (2013) Energy-efficient mapping and scheduling of task interaction graphs for code offloading in mobile cloud computing. In: Proceedings of 6th IEEE/ACM international conference on utility and cloud computing, pp 34–41
7. Baccarelli E, Cordeschi N, Mei A, Panella M, Shojafar M, Stefa J (2016) Energy-efficient dynamic traffic offloading and reconfiguration of networked data centers for big data stream mobile computing: review, challenges, and a case study. IEEE Netw 30(2), 54–61
8. Mouftah HT, Kantarci B (2014) Communication infrastructures for cloud computing. IGI Global
9. Hajibaba M, Gorgin S (2014) A review on modern distributed computing paradigms: cloud computing, jungle computing and fog computing. J Comput Inf Technol 22(2):69–84
10. Suryawanshi R, Mandlik G (2015) Focusing on mobile users at the edge of internet of things using fog computing. Int J Sci Eng Technol Res 4(17):3225–3231
11. Song J, Cui Y, Li M, Qiu J, Buyya R (2014) Energy-traffic tradeoff cooperative offloading for mobile cloud computing. In: Proceedings of 22nd IEEE international symposium of quality of service, pp 284–289
12. Miettinen AP, Nurminen JK (2010) Energy efficiency of mobile clients in cloud computing. In: Proceedings of 2nd USENIX conference on Hot topics in cloud computing, pp 4–11
13. Balasubramanian N, Balasubramanian A, Venkataramani A (2009) Energy consumption in mobile phones: a measurement study and implications for network applications. In: Proceedings of ACM IMC'09, pp 280–293
14. Huang J, Qian F, Gerber A, Mao ZM, Sen S, Spatscheck O (2012) A close examination of performance and power characteristics of 4G LTE networks. In: Proceedings of ACM MobiSys'12, pp 225–238
15. Jagadeesh GR, Srikanthan T (2016) Heuristic optimizations for high-speed low-latency online map matching with probabilistic sequence models. In: Proceedings of the 2016 IEEE 19th international conference on intelligent transportation systems (ITSC), Rio de Janeiro, Brazil, 1–4 Nov 2016, pp 2565–2570
16. Cordeschi N, Patriarca T, Baccarelli E (2012) Stochastic traffic engineering for real-time applications over wireless networks. J Netw Comput Appl 35(2):681–694
17. Wu C-M, Chang R-S, Chan H-Y (2014) A green energy-efficient scheduling algorithm using the DVFS technique for cloud datacenters. Fut Gener Comput Syst 37:141–147
18. Vaughn D (1996) Vehicle speed control based on GPS/MAP matching of posted speeds. U.S. Patent 5485161, 16 Jan 1996
19. Yuan Y, Van Lint H, Van Wageningen-Kessels F, Hoogendoorn S (2014) Network-wide traffic state estimation using loop detector and floating car data. Int J Intell Transp Syst Res 18:41–50
20. Treiber M, Kesting A, Wilson RE (2011) Reconstructing the traffic state by fusion of heterogeneous data. Comput Aided Civ Infrastruct Eng 26:408–419
21. Van Lint J, Hoogendoorn SP (2010) A robust and efficient method for fusing heterogeneous data from traffic sensors on freeways. Comput Aided Civ Infrastruct Eng 25:596–612
22. Engel JI, Martin J, Barco R (2016) A low-complexity vision-based system for real-time traffic monitoring. IEEE Trans Intell Transp Syst 18:1–10
23. Melnikov VR, Krzhizhanovskaya VV, Boukhanovsky AV, Sloot PM (2015) Data-driven modeling of transportation systems and traffic data analysis during a major power outage in the Netherlands. Procedia Comput Sci 66:336–345
24. Haklay M, Weber P (2008) OpenStreetMap: user-generated street maps. IEEE Pervasive Comput 7:12–18

25. Hashemi M, Karimi HA (2014) A critical review of real-time map-matching algorithms: current issues and future directions. Comput Environ Urban Syst 48:153–165
26. Hashemi M, Karimi HA (2016) A weight-based map-matching algorithm for vehicle navigation in complex urban networks. Int J Intell Transp Syst Res 20:573–590

Automated Attendance Management Using Hybrid Approach in Image Processing

Arun Saharan, Munish Mehta, Piyush Makwana, and Sanju Gautam

Abstract Image Processing, which is a significant field of Machine Learning and falls under the adage of Artificial Intelligence, is used worldwide to achieve face detection and recognition. Image Processing focuses on two tasks, namely the enhancement of image for our interpretation and reading image data to store, transmit, and represent image perception via machines. The project focuses on the implementation of Machine Learning and Image Processing to automate attendance management in a corporate/personal space. In the current scenario, most corporate spaces make use of biometric systems to mark attendance, which is both time-consuming and costly. Setting up a working space with the biometric system also requires an individual with professional knowledge of the same. Our primary focus is to build a robust and automated security system capable of monitoring attendance without much human interference.

Keywords AdaBoost · Eigenfaces · Haar cascade · LDA · PCA · Viola-Jones algorithm

---

A. Saharan · M. Mehta · P. Makwana · S. Gautam (✉)
Department of Computer Applications, NIT Kurukshetra, Kurukshetra, India
e-mail: sanjugautam615@gmail.com

A. Saharan
e-mail: arun1316sharan@gmail.com

M. Mehta
e-mail: munish.mehta80@gmail.com

P. Makwana
e-mail: piyushmakwana74@gmail.com

A. Khanna et al. (eds.), *Data Analytics and Management*, Lecture Notes on Data Engineering and Communications Technologies 54,
https://doi.org/10.1007/978-981-15-8335-3_34

1 Introduction

Digital Image Processing, by and large, means applying algorithms and analyses to an image for quantifying the image. It involves implementing "3 D parametric maps" to calculate the values of the image. The recent developments in Image Processing have been done to automate the process to as much extent as possible. Take, for example, neuroimaging. It is used in place of visual analysis (which may not be able to detect abnormalities as minutely as the former) to detect abnormal tissues in the brain and indirectly help improve the wellbeing of an individual. The most well-known real-life implementation of Image Processing can be seen in neuroimaging to improve the detection rate of abnormal brain tissues, including irregularities that may not be recognizable through bare visual analysis.

Image Processing is being used by a wide range of individuals and organizations in the current scenario. With a minimal requirement of investment and access to digital cameras and computers, one can amplify contrast, diagnose edges, assess the intensity, and implement Image Processing by applying a variety of mathematical and logical algorithms.

Talking about the detailed process of Image Processing, it is a well-defined collection of computational techniques for analyzing, compressing, augmenting, and redesigning images. The main elements of Image Processing involve importing, in which an image is captured through a scanner or a digital camera; manipulation and analysis of the image which is accomplished using various cutting-edge software packages; and output. Image Processing has widespread applications, including monitoring systems, security protocols, medical treatments, industrial robotics, astronomy, and remote sensing by satellites.

2 Existing System

2.1 Face Recognition-Based Attendance System [1]

Several face recognition algorithms and techniques have been developed in recent times. Authors of this paper [1] have implemented deep learning techniques for face recognition. Deep learning technique aids in the transformation of the video into still frames consisting of student's faces. The working model uses Convolution Neural Network (CNN). This system consists of an input, an output, and a hidden layer having multiple convolution layers, fully connected layers, pooling layers, and normalization layers. The number of convolution layers is directly proportional to the accuracy of the system; however, the detection time increases with the addition of convolution layers. While using suitable libraries and under ideal conditions, CNN can achieve an accuracy of around 98% (Fig. 1).

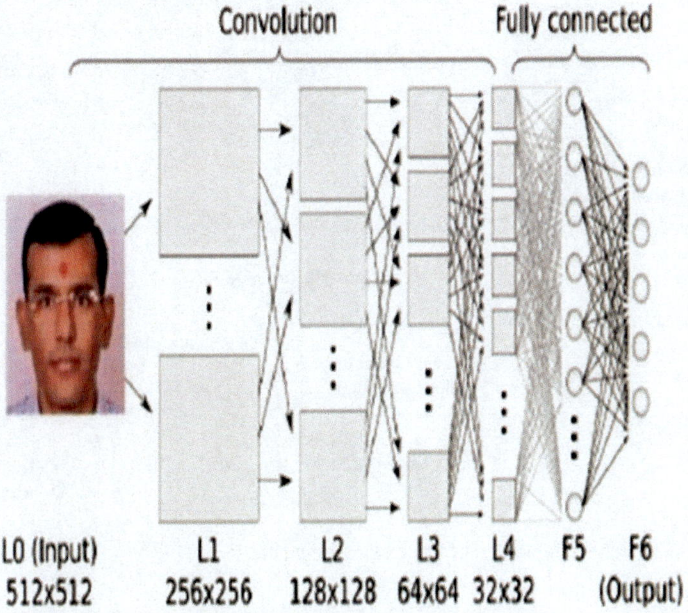

Fig. 1 Network diagram to explain the working of the CNN model

2.2 *Face Detection System for Attendance of Class' Students [2]*

Face detection and recognition technologies have attracted market potential and application value on a vast scale. Authors of [2] have proposed a refreshing new algorithm to provide a solution for attendance management. They have introduced a real-time face detection algorithm combined with Learning Management System (LMS) that detects a student attending the lecture and updates his/her attendance automatically on the database. The system designed using the mentioned approach is a child of Libri (an in-house built learning management suite) constructed using 3 modules: image capturing module, face detector module, and face recognizer module (Fig. 2).

2.3 *Ear-Based Attendance Monitoring System [3]*

Two geniuses decided to take an unorthodox approach for face recognition. The authors of this paper [3] proposed a technique using an ear for attendance management. The system captures a photo of the student's ear, which is then supplied to the computer for processing. In Image Processing, edge detection is carried out on the

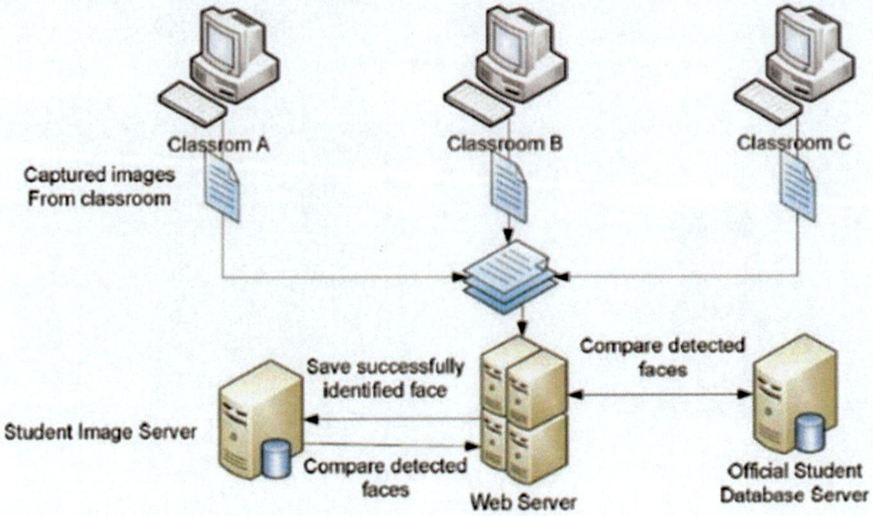

Fig. 2 Physical architecture of the proposed system in [2]

ear. These extracted edges are then used to derive a reference line with respect to other identified features. Afterward, these derived features are stored in the form of a vector, each representing a unique record in the database (Fig. 3).

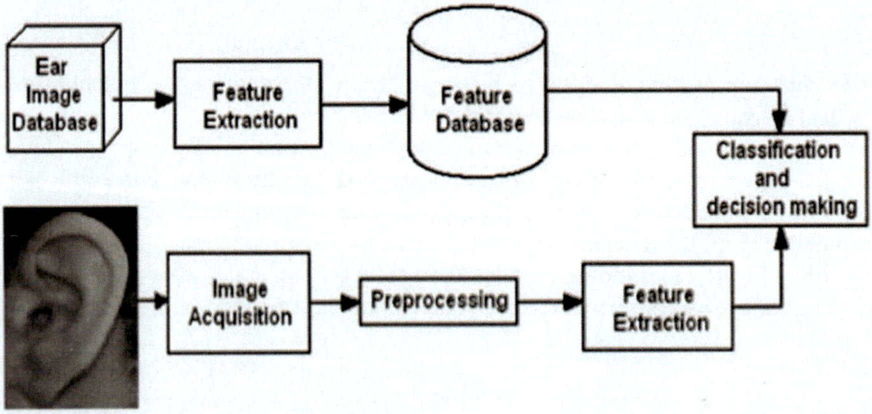

Fig. 3 Flow chart of the proposed framework in [3]

2.4 Face Recognition Using Histogram of Oriented Gradients

Face recognition has been a challenging task in the field of computer vision. In recent times, Histogram of Oriented Gradients (HOGs) have emerged out to be an effective solution for object recognition in general and facial recognition in particular. Authors [4] have invented a robust approach of Histogram of Oriented Gradients for face recognition. The proposed methodology solves three major problems: (1) To increase the accuracy of the recognition system under poor illumination and pose variations by extracting HOG descriptive features from a regular grid. (2) A combination of HOG descriptors at different facial scales allows us to efficiently capture important features for face recognition. (3) Recognize the need for performing dimensionality reduction to reduce noise and make the classification procedure less inclined to overfitting.

2.5 Face Recognition Using a Fuzzy Fisherfaces Classifier

In reference [5], authors have proposed a hybrid approach for facial recognition using the Fuzzy Fisherfaces Classifier and its dedicated set-based augmentation. It is a combination of the popular Fisherfaces method and Principal Component Analysis (PCA). The proposed method has been tested on ORL, CNU (Chungbuk National University), and Yale face database. The experimental results have shown significant improvement over PCA and Fisherfaces technique.

2.6 An Automatic Attendance System Using Image Processing

This paper [6] by Aziza Ahmedi and Dr. Suvarna Nandyal proposes an automatic attendance management system in a classroom environment. For detection of faces, the authors have used the AdaBoost algorithm and Histogram of Oriented Gradients (HOGs) and Local Binary Pattern (LBP) to further extract features once a face is successfully detected. After successfully detecting a face and extracting mentioned features, these features are compared with faces already existing in the student database using the Support Vector Machine (SVM) classifier. The student gets attendance if a potential match is found in the database. As mentioned in the paper's conclusion, the proposed system relies heavily on the quality of the image capture from the camera, and that makes it heavily dependent on the camera used in the model.

3 Challenges Faced by Existing Systems/Limitations of Existing Work

The study and analysis of facial recognition pose a large number of challenges that can directly affect the computational and analysis process.

1. The subject's facial components like beard and mustache can lead to abnormalities in the deductions due to the absence of any structuring element. The image can be tainted due to facial alterations, accessories, and disguises.
2. Since head rotates egocentrically, its movement can lead to significant fluctuations in the results since varying face appearance can make facial recognition slightly tricky.
3. The variations in an individual's facial expressions are correlated with their emotional states. The most important application of automated facial recognition is recognizing these expressions, something that machines are not well versed with.
4. The changes in illumination conditions can also affect AFR's performance. The illuminations, such as low lighting in the background or the foreground, are known to impact how an image is processed. The varying levels of lighting can make it harder for facial recognition to perform since shadows can make a face unnoticeable. Even if your face is overexposed, facial detection can meet out wrong results.

4 Working Model/Architecture of the System

In the general setting, most institutions/organizations prefer manual or biometric fingerprint-based attendance systems. With this process, it makes the whole attendance procedure lengthy as well as involves a lot of human interference. To automate the attendance monitoring system, the proposed method uses a facial recognition-based surveillance system. We have utilized the concepts of Haar cascades in stage 1 of face detection over already existing methods due to its exceptionally well accuracy. Stage 2 consists of facial features extraction, which is then fed to stage 3 for final facial recognition.

4.1 Face Detection

Detection using the Viola-Jones Algorithm For the course, we are only considering the detection segment and skipping the training one. The first thing to note here is: Viola-Jones produces the best results for frontal faces. Since we are designing this system to operate in a classroom environment, attentive faces will have frontal orientation and will be easy to detect. Before the detection takes place, the image/still

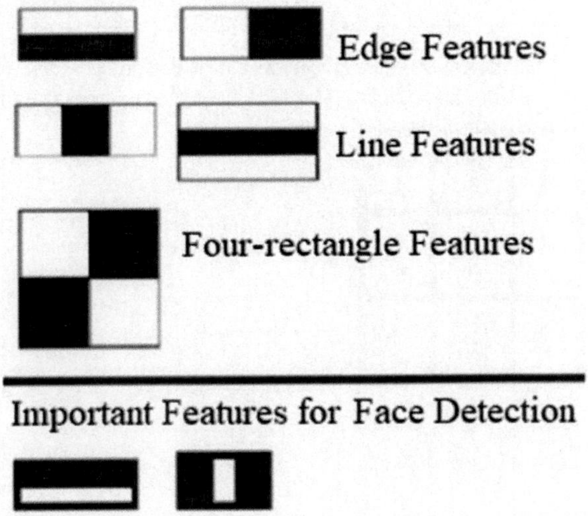

Fig. 4 Haar-like features

frame from a video is converted into grayscale since they are easier to work with and have less amount of data to process. The Viola-Jones framework then detects the face on the grayscale image and finds its location on the original image afterward.

The algorithm highlights a box and slides it throughout the image to scan for faces. In this process, the box is primarily searching for Haar-like features (explained later) to fall into the box. One can manipulate the box size for achieving higher accuracy, depending on the situation.

Haar-like features Named after famous Hungarian mathematician Alfred Haar, the features mentioned below highlight two sides, a dark side and a light side. Using these features, the framework identifies what kind of feature it is. For instance, a darker side than the other can be considered as an eyebrow (Edge feature). Similarly, a shinier box surrounded by lighter ones can be determined as a nose (Line feature) (Fig. 4).

There are 3 types of Haar-like features:

1. Four-sided features
2. Edge features
3. Line features.

Each Haar feature result corresponds to a single value and is calculated via the following formula:

$$\begin{aligned}\text{Feature Result} = (&\text{Sum of pixels under black rectangle}\\ &- \text{Sum of pixels under white rectangle})\end{aligned}$$

Haar classifiers define over 160,000 features for a sample image. Computing Haar result for so many features can be computationally inefficient. Therefore, Viola-Jones

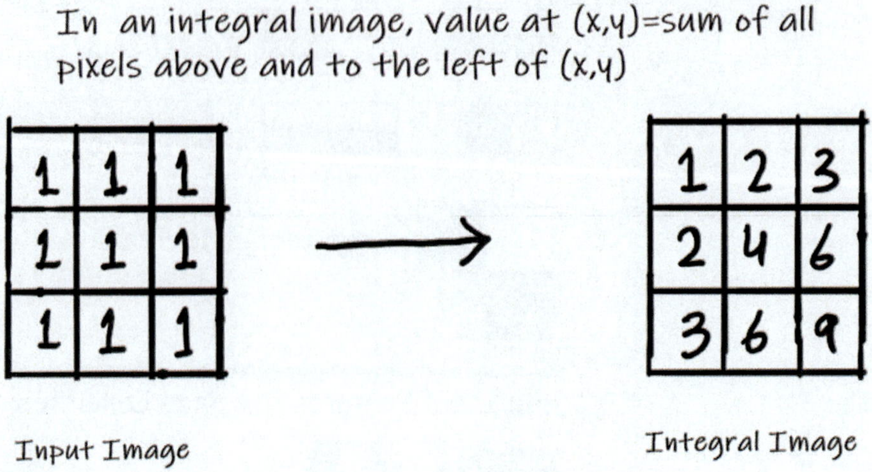

Fig. 5 Input image to integral image

proposes the concept of integral image, where the result can be calculated just by using 4 corner values of the patch instead of all individual pixels. Here is how it is done:

Further in the process, we use the AdaBoost algorithm for shortlisting the most relevant features out of all 160,000 features. AdaBoost selects the best features and reduces the number to 2500. Once these features have been decided, a weighted linear combination of all these features is used to decide whether a window contains a face or not. These features are called weak classifiers, and a linear combination of these features is used to form a strong classifier. To delineate this step (Fig. 5):

$$F(x) = \alpha_1 f_1(x) + \alpha_2 f_2(x) + 3_1 f_3(x) + \alpha_4 f_4(x) + \cdots$$

where $F(x)$ is the strong classifier and $f_i(x)$ is the weak classifier.

Finally, cascading is done to boost the algorithm further. Cascade classifiers are defined, consisting of stages where each stage contains a strong classifier. All the features are divided into different stages containing a certain number of features. Each stage checks the possibility of a sub-window for a face, and it is immediately discarded if it does not contain a face, otherwise passed on to the next stage in case of a possible face.

Here are the results of the detection performed on a sample image from our college with acceptable accuracy (Figs. 6, 7, and 8).

The above results show how scaling factor variations can improve the accuracy of the system (Fig. 9).

Fig. 6 Result of face detection when scaling factor value is 1.1. Scaling factor = 1.1. Total number of faces found = 72. False positives = 4. True negatives = 1

Fig. 7 Result of face detection when scaling factor value is 1.15. Scaling factor = 1.15. Total number of faces found = 70. False positives = 3. True negatives = 1

Fig. 8 Result of face detection when scaling factor value is 1.18. Scaling factor = 1.18. Total number of faces found = 68. False positives = 1. True negatives = 1

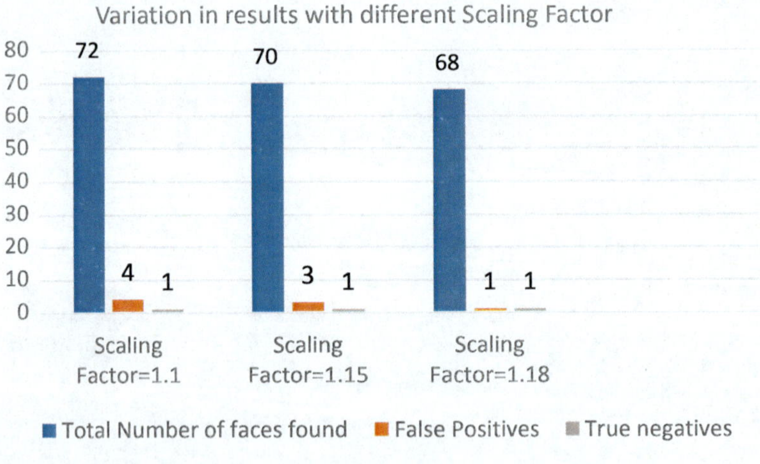

Fig. 9 Compiled graphical representation of the above results

4.2 *Face Recognition*

For feature selection and face recognition, we are using PCA and Fast PCA (where PCA stands for Principal Component Analysis) combined with Linear Discriminant

Analysis (LDA). The face recognition module evaluates Eigenfaces value with the help of PCA and then applies the LDA technique on the outputs generated by PCA.

PCA (Principal Component Analysis) PCA is considered as one of the most effective and easy approaches for face recognition. In the primary step, PCA transforms each face into Eigenfaces (descriptive characteristics), which leads to the initial set of learning images. During recognition, a new image is superimposed on the learning image (called Eigenfaces subspace), post which the face is distinguished by comparing its position in the Eigenfaces space with the relative positions of known characteristics. The primary advantage of PCA is its ability to ignore minute changes in the face (effective in cases where the subject has grown facial hair or is wearing some accessories). Here is a diagrammatic representation of the PCA algorithm (Fig. 10).

Linear Discriminant Analysis (LDA) LDA is an enhancement to the PCA algorithm and introduces the concept of classes. It overcomes the limitations faced by PCA by implementing linear discriminant criterion. Due to the high dimensionality of image space, several LDA-oriented applications first use the PCA algorithm to project an image to lower-dimensional face space and then using the LDA algorithm to maximize the discriminatory efficiency.

Step 1: Image representation in matrix form

Step 2: Vector expansion ϕ $R\ m \times n$ for each image and sub-image, beginning with 2D $m \times n$ array of I(x, y) intensity values.

Step 3: Computation of within-class matrix S_w and between-class scatter matrix $\mathbf{S_b}$:

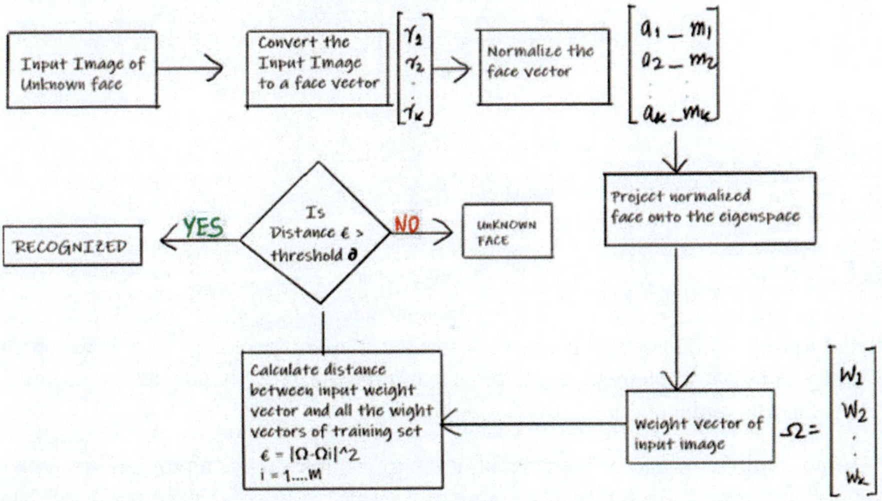

Fig. 10 Diagrammatic representation of the PCA algorithm

$$S = \sum_{j=1\,i=1}^{N} = \left(r_i^j - \mu_j\right)\left(r_i^j - \mu_j\right)^T$$

$$S = \sum_{j=1}^{C} = (\mu_j - \mu)(\mu_j - \mu)^T$$

where c is the number of classes, μ represents the mean of all classes, μ_j is the mean of class j, N is the number of samples in class j, and r_i^j represents the ith sample of class j

Step 4:

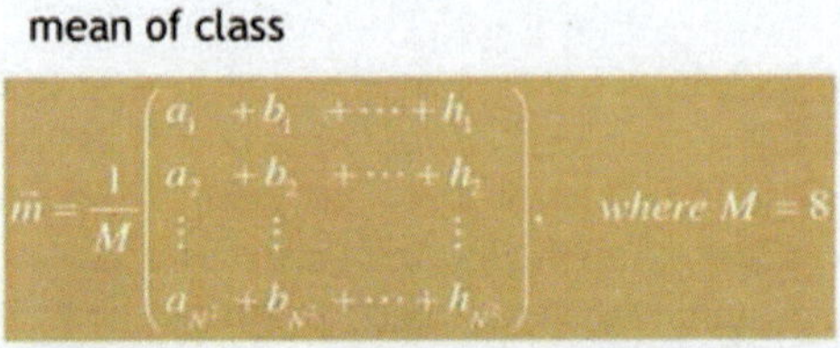

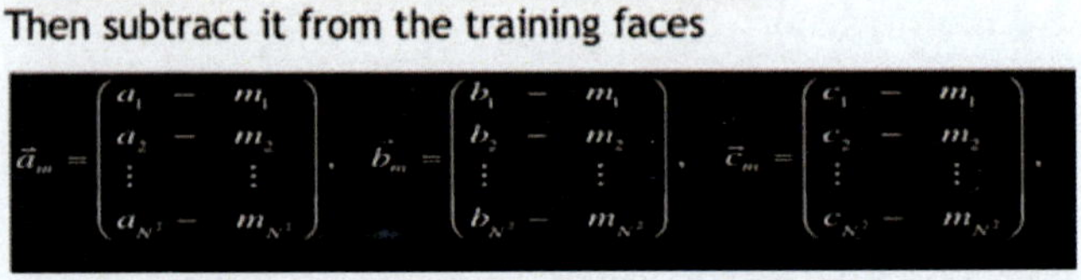

Step 5: LDA subspace spanning by a set of vectors w, where w is given by

$$W = \arg\max = \operatorname{mod}\left[\frac{w^T s_b w}{wT s_w w}\right]$$

5 Tools and Software Used

5.1 Tools

Scikit-image Scikit-image is an open-source Python package that works with NumPy arrays. It implements algorithms and utilities for use in education, research, and industrial applications.

Numpy NumPy is one of the core libraries in Python programming and provides support for arrays. An image is a standard NumPy array containing pixels of data points. By using NumPy operations, we can modify the pixel values of an image.

SciPy Another core library in Python programming, SciPy can be used for some entry-level image manipulation.

Pillow Expanded as Python Imaging Library, Pillow is an open-source Python library that provides support for saving, opening, and manipulating different image file formats.

Programming Language Python.

5.2 *Software*

IDE PyCharm is a dedicated IDE mainly used for development using Python language.

6 Conclusion

The concepts of Haar cascades and Viola-Jones algorithm [7] have been used for face detection in one of the modules. Further, the dominant concepts of a fusion algorithm consisting of PCA and LDA have been utilized for facial recognition. The proposed face detection may encounter slow training but is blazing fast in detecting faces from a still image. The combination of PCA and LDA makes the system one of the most consistent systems for face recognition in case of faces with less variation. Since the system is meant to be deployed in an educational institution or a professional workplace, the variation in faces remains minimal, and the proposed system achieves high accuracy.

References

1. Nandhini R, Duraimurugan N, Chokkailngam SP (2019) Face recognition bases attendance system. Int J Eng Adv Technol (IJEAT) 2(3s). ISSN: 2249-8958
2. Fuzail M, Nouman HMF, Mushtaq MO, Raza B, Tayyab A, Talib MW (2014) Face detection system for attendance of class' students. Int J Multidisc Sci Eng 5(4)
3. Jawale JB (2011) Ear based attendance monitoring. In: System proceedings of ICETECT 2011. IEEE. 978-1-4244-7926-9/11/$26.00
4. Deniz O, Bueno G, Salido J, De la Torre F (2005) Face recognition using histograms of oriented gradients. Pattern Recogn 38(10):1717–1732
5. Kwak K-C, Pedrycz W (2005) Face recognition using a fuzzy fisherface classifier. Pattern Recogn 38(10):1717–1732

6. Aziza A, Suvarna N (2015) An automatic attendance system using image processing. Int J Eng Sci (IJES) 4:1–8
7. Yuvaraj CB, Srikanth M, Santhosh Kumar V, Srinivasa Murthy YV, Koolagudi SG (2017) An approach to maintain attendance using image processing techniques. In: Proceedings of tenth international conference on contemporary computing(IC3), Aug 2017

Real-Time, YOLO-Based Intelligent Surveillance and Monitoring System Using Jetson TX2

Prashant Kumar, S. Narasimha Swamy, Pramod Kumar, Gaurav Purohit, and Kota Solomon Raju

Abstract In recent years, real-time surveillance and monitoring system is getting importance because of security reasons. Government organizations, residential areas, commercial complexes, schools and colleges, industries, borders, and others require a dedicated surveillance system. The traditional surveillance systems do not provide real-time object identification and alerts. This paper aims to design and develop a real-time object detection and alert system (ODAS). The proposed system could be used as a surveillance and monitoring system. The object detection system uses the YOLO framework to detect objects within images and live videos. Object identity and time stamp, along with bounding box image, are marked by the detection system and stored locally. Also, the collected information is transmitted to the server in real time. At the server, the graphical user interface (GUI) application continuously gathers information from different nodes, analyzes it, and fires an alert message if the anomalies/targeted object is found. GUI also facilitates us to analyze the obtained information of an individual node or in the combination that infers the direction of movement made by a detected object and its current position. ODAS also takes the services of the real-time database to provide real-time updates of anomalies on the android phone/tablet.

Keywords YOLO · Object detection · Surveillance system · Jetson TX2 · IoT

P. Kumar (✉) · S. Narasimha Swamy · P. Kumar · G. Purohit · K. S. Raju
CSIR-Central Electronics Engineering Research Institute, Pilani, India
e-mail: prashant.mnnit10@gmail.com

S. Narasimha Swamy
e-mail: narasimha.rmgm@gmail.com

P. Kumar
e-mail: pramod.tanwar@gmail.com

G. Purohit
e-mail: gp.bits@gmail.com

K. S. Raju
e-mail: kotasolomonraju@gmail.com

Academy of Scientific and Innovative Research (AcSIR), Ghaziabad, India

A. Khanna et al. (eds.), *Data Analytics and Management*, Lecture Notes on Data Engineering and Communications Technologies 54,
https://doi.org/10.1007/978-981-15-8335-3_35

1 Introduction

Nowadays, the rate of crimes has been increased. Consequently, the public has started to think over the significance of the surveillance system [1]. According to MarketsandMarkets, the video surveillance industry is expected to grow 74.6 billion by 2025 [2]. Electronic devices like closed-circuit televisions (CCTVs) and digital video recorders (DVRs) are commonly used in surveillance to capture live videos. High-end panoramic 360° cameras, storage devices, network routers, and display monitors are the key elements of the traditional surveillance systems. The organization of these resources makes the system more expensive and sophisticated, and consumes more power, which results in high maintenance costs and occupies more area [3]. Also, most of the surveillance systems existing today require human intervention to keep track of the activities [4]. However, few real-time surveillance systems use major technologies like IoT, image processing, and computer vision techniques to reduce human intervention in the surveillance systems [5].

IoT [6, 7] plays a significant role in transforming the traditional surveillance systems to real-time intelligent surveillance systems by providing anywhere and anytime services using the Internet. IoT is responsible for capturing and sharing videos across the globe using the sensor and communication technologies, respectively. Hence, the adoption of characteristics of IoT reduces the crime rates in smart homes, cities, and offices [8]. Furthermore, blockchain-based networks are also gaining importance to transfer the sensor data securely over the globe [9, 10].

Human's perception efficiency is not good as cameras and loses the observations quickly even when they are vigilant [11]. Human perception reduces due to lousy weather and visual changes. Recent advancements in image processing and computer vision (artificial intelligence and machine learning) algorithms solve these problems. Computer vision algorithms play a crucial role in processing, analyzing, and distinguishing the objects in real-time videos.

1.1 Contribution of the Paper

- An overall picture of the traditional surveillance system is provided.
- We proposed the decentralized architecture for intelligent surveillance and monitoring system.
- Use of decentralized architecture (edge computing architecture) reduces the bandwidth consumption and latency and increases the object detection rate.
- The proposed system takes advantage of both IoT and the YOLO framework to deliver low-cost, small size, and high-speed real-time surveillance systems.
- We implemented the graphical user interface (GUI) for the proposed system, which provides benefits like on-demand analysis, real-time alerts, and visualization.

The rest of the paper is organized as follows: Sect. 2 describes the related work in the field of traditional surveillance systems. Section 3 deals with the proposed system. Section 4 presents the implementation of the proposed real-time surveillance system. Section 5 emphasizes the results. Finally, Sect. 6 highlights the future scope of the proposed system.

2 Related Work

Remote surveillance has been on the rise in recent times. Starting from small homes to the skyscrapers, shopping complexes, borders, etc., are inclined toward a surveillance system. In this section, we discuss various existing solutions adopted in the surveillance systems. Tanya et al. [12] proposed a home security system using IoT. The authors used a simple Raspberry Pi board and PC to monitor unusual events using image processing. Jain et al. [13] proposed a security surveillance system, in which the authors analyzed suspicious activity with encryption and decryption, which would add additional computational complexity. Sruthy and George [14] have explained node MCU and Raspberry Pi-based surveillance systems, in which they transmit the live video and captured image to the client. In paper [15], the authors have explained the Raspberry Pi-based surveillance system, in which camera image is displayed using a Web-based application. In articles [6, 16], the authors have explained the same Raspberry Pi-based surveillance system except images are being uploaded on the cloud server. Gulve et al. [17] have proposed Raspberry Pi- and Arduino-based video surveillance system, in which they are sending video via email. The authors have also used the GSM module for SMS alert. Kumar and Solanki [18] have used TeamViewer to access the home system, on which live feed is available, by the remote system. Virendra et al. [19] proposed a Raspberry Pi- and OpenCV-based home surveillance system, in which the authors transmit 100-frame video clips via email as an alert. Juhana and Anggraini [20] explained the PC-based home surveillance system, in which the authors have used PC as an image processing system. They also transmit the whole image to the client dropbox account as the intruder is detected. In paper [21], the authors proposed an intelligent surveillance system; they use Raspberry Pi along with PIR sensor. They are also sending the whole image and videos to the cloud for further analysis. Jyothi and Vardhan [22] explained Raspberry Pi along with a GSM modem-based surveillance system. They have also given camera rotation functionality. As the motion detected, the camera captures the image or video and sends it to the cloud. Virginia et al. [23] developed one OpenCV-based system, in which the motion is detected, light is on, and the picture is taken by the camera. Later, the whole image is uploaded to the Web site.

In outline, the majority of the systems are transmitting/uploading images and recordings on the server, Web-based application, and cloud storage. As we probably are aware, multimedia data devours more network bandwidth [18]. Likewise, it takes more time to transfer/download images and videos. It makes an extra burden on the IoT network. Additionally, few authors are executing the object detection algorithm on Raspberry Pi utilizing OpenCV. It is challenging to run a sophisticated object detection algorithm on Raspberry Pi because of the constrained assets. Besides, practically all systems have demonstrated surveillance using a single node. Surveillance and monitoring using the network of multiple edge nodes were not discussed. Likewise, none of the articles has dealt with data analysis.

Such kinds of gaps encouraged us to build up a system that could send just the required information to the remote server for additional investigation. This prompts a decrease in bandwidth utilization and latency. Also, the proposed system sends only the bounding box image (Fig. 5b) of the identified object as opposed to sending the entire frame. Additionally, the powerful edge node runs a sophisticated object detection algorithm in our system. The fastest object detection algorithm is executed at the edge node to get the real-time object detection and analysis experience. Eventually, our system provides a GUI to manage the edge nodes and other information.

3 Proposed Work

The architecture of the proposed system is shown in Fig. 1. It incorporates the object detection and alert system, which is utilized as the surveillance and monitoring system. In this work, a Jetson TX2 development board is used as an edge node. The Jetson TX2 is exceptionally intended to run machine learning algorithms. Moreover, to collect real-time videos or images, USB cameras and a Jetson TX2 onboard camera are used.

As appeared in Fig. 1, the YOLO object detection algorithm executes on each edge nodes, and it extracts and saves the features locally from the live video if an object is detected. The standard features are bounding box image of an object, type of object, time, day, date, and so on. Afterward, the locally saved data is transmitted to the server. At the server, continuously running GUI records the data with the

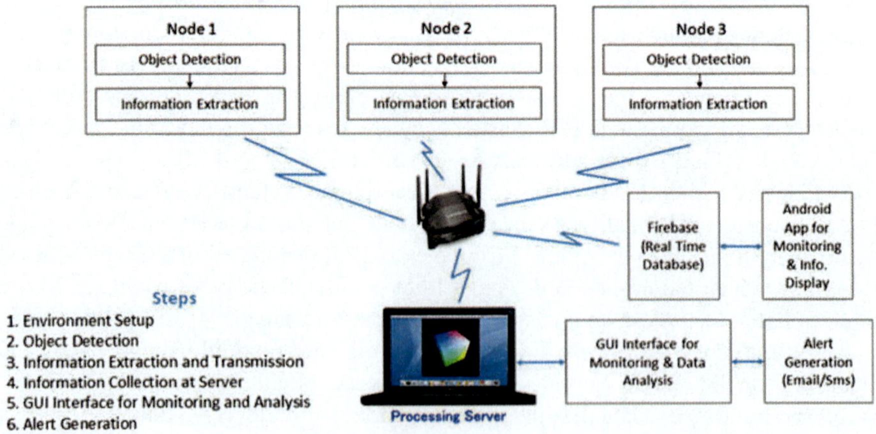

Fig. 1 Proposed architecture of object detection and alert system

node's identity. If anomalies/targeted objects are identified, it consequently fires an email/message with the detected object information like the kind of object, an object image, date, time, and so forth. GUI additionally encourages examination of the data on the whole (all node data) or just as exclusively (specific node). Likewise, it plots different diagrams for information gathered from the edge nodes.

Eventually, the proposed system sends continuous anomaly information to smartphones by exploiting the services of the real-time database, i.e., Firebase.

4 System Implementation

The proposed system comprises five steps, and the following section covers each step in detail.

4.1 Environmental Setup

The Jetson TX2 development board is used as an edge node in the development of real-time surveillance and monitoring system. Technical specification of the Jetson TX2 board is given in Table 1, and the setup will take around 1.30 h. To setting up Jetson TX2, we need a dedicated host, and it is loaded with the Ubuntu 16.04 operating system. The Jetson TX2 is connected to the router that must have Internet connectivity. During installation, we need to join the target machine (Jetson TX2) to the host machine using the mini HDMI. The entire environmental setup of the YOLO framework is described by Catherine Ordun [24]. These nodes are deployed wherever surveillance and monitoring are needed, and latitude and longitude information of each node is recorded.

Table 1 Technical specification of Jetson TX2

CPU	HMP Dual-Core Denver 2/2 MB L2 + Quad ARM® A57/2 MB L2
Display	2x DSI, HDMI 2.0, 2x DP 1.2
Memory	8 GB 128-bit LPDDR4, 59.7 GB/s
Storage	32 GB eMMC, SATA, SDIO
GPU	256 CUDA Cores, NVIDIA Pascal
USB	USB-3, USB-2
Connectivity	802.11ac WLAN, Gigabit Ethernet, Bluetooth

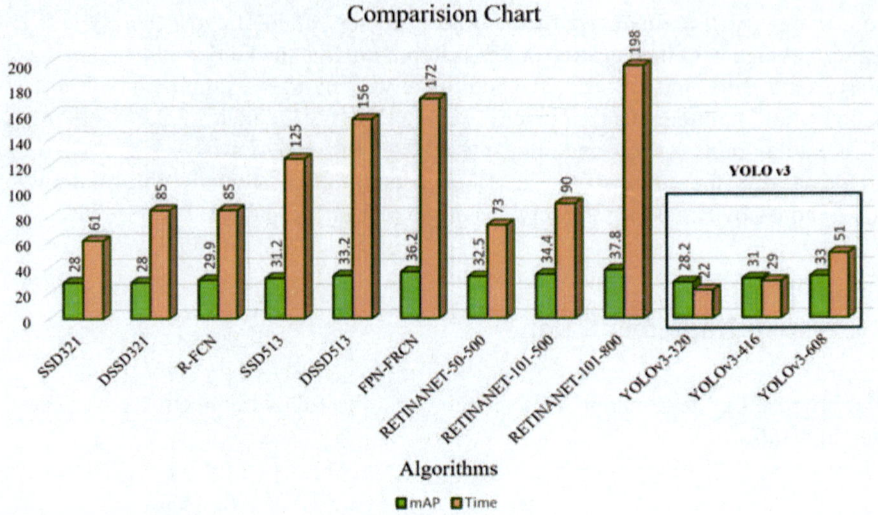

Fig. 2 YOLO v3 runs faster than other object detection algorithms [26]

4.2 *Object Detection with YOLO V3*

Once the Jetson TX2 is ready, a pre-trained You Only Look Once (YOLO) real-time object detection framework is deployed on it. Likewise, all edge nodes are prepared. It is also possible to train the YOLO deep learning algorithm on our dataset. But, already the huge and tested datasets of different kinds of objects are available like Pascal VOC dataset, ImageNet, and COCO dataset. COCO dataset (2014) has 120,000 images (Training + Validation) of 80 types of distinct objects [25]. These datasets also satisfy the requirements of this project. That is why the pre-trained YOLO framework is used. YOLO is capable enough to process the images at 30 frames per second (FPS) on a Pascal Titan X. YOLO has an mAP of 57.9% on COCO test-dev. YOLO v3 is extremely fast and accurate [26], and the comparative performance is shown in Fig. 2. You can easily trade off between accuracy and speed, only by changing the size of the model, and no retraining is required. YOLO setup details are described by Joseph Chet Redmon [25].

4.3 *Information Extraction and Transmission*

First of all, for information/feature extraction of a detected object, a code is appended in the "image.c" file located inside the darknet folder of YOLO framework. During object detection, the following information is extracted and saved at each node: the bounding box coordinates, the type of object, the confidence of object detected, date and time of object detection, etc. A bounding box image of detected objects is also extracted. Secondly, all the extracted textual information along with the bounding box image of the objects is transmitted by each edge node to the remote server using 802.11(Wi-Fi) standard. Consequently, minimal network bandwidth is used by our proposed system.

4.4 *Information Uploading on Real-Time Database for Android App*

To get an alert on the mobile app in real time, services of a real-time database, i.e., Firebase is exploited. The code snippet for the interfacing edge node with the Firebase real-time database is given. As the real-time database is updated, the same information will be reflected on the android app.

```
config = {
            'apiKey': "AIzaSyBLY77lmfRiyB5qJMW-s9zwZGT1q_KsgAk",
            'authDomain': "pks123-6637 f.firebaseapp.com",
            'databaseURL':"https://pks136637f.firebaseio.com",
            'projectId': "pks1233-6637f",
            'storageBucket': "pks123-667 f.appspot.com",
    }
    firebase = pyrebase.initialize_app(config)
    db = firebase.database()
    while True:
        data = db.child("info/ob1").set(input(num))
```

4.5 *Information Collection and Analysis at Server*

Once you start getting the data at the server from multiple nodes continuously and simultaneously, that raised several questions like how data would be analyzed? How can we monitor the activities of objects which is detected at a remote location? How would a decision be made by the data collected at the server? When should an alert be sent? How can we automate the actions? By keeping all these questions in mind, a GUI tool is developed that tackles all the above-raised issues. Figure 3 depicts the GUI of the proposed system. The GUI tool can display the data coming from nodes individually or in a combination. It can also generate various plots by considering the data of an individual node or by combining data of all nodes. When the combined data is analyzed, it assists in object track and current position determination.

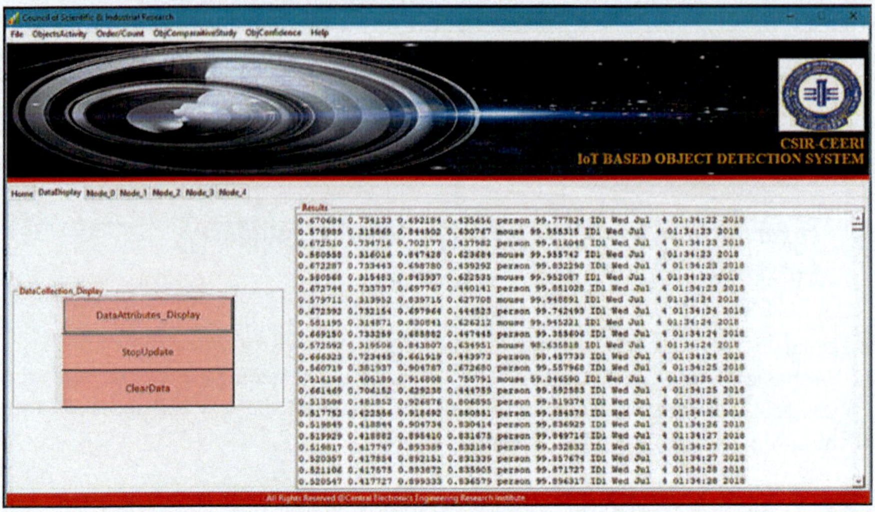

Fig. 3 Cumulative data collection from different nodes

5 Results

Here, the output of the system with original images obtained by running the proposed system is explained. Figure 4a exhibits the various objects detected at a single edge node versus time plot; likewise, individual object activity at numerous nodes could be determined; it appears in Fig. 4b. Figure 5a depicts the confidence of object recognition. Additionally, the system in working condition along with bounding boxes appeared around the objects is shown in Fig. 5b. Moreover, by watching Fig. 4b,

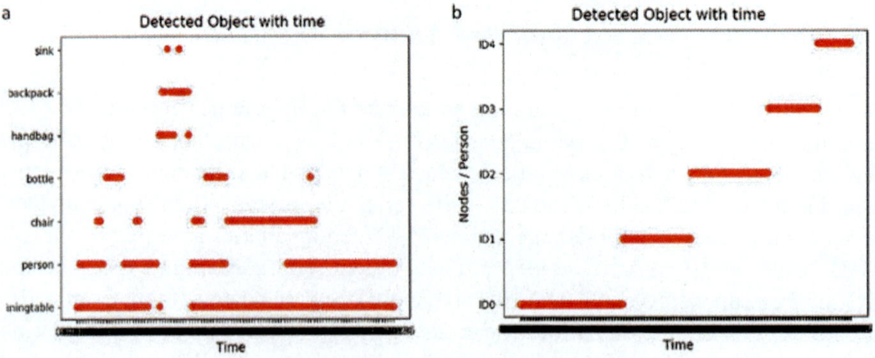

Fig. 4 **a** Objects detected on single node versus time; **b** comparative study: single object versus multiple nodes

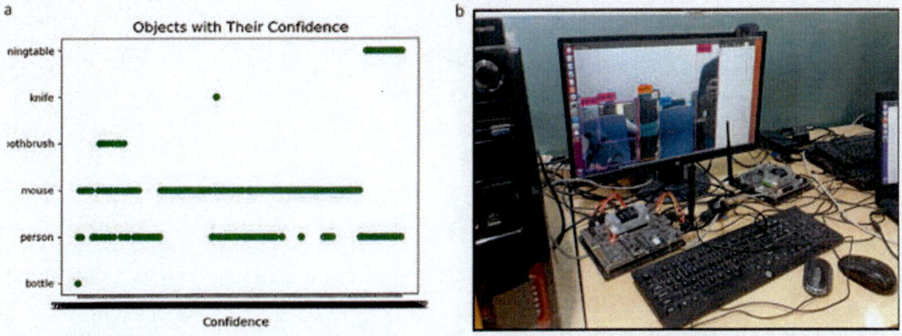

Fig. 5 **a** Objects detected at Node_0 versus confidence; **b** system setup and bounding box appeared around the objects

the individual object (person) track can easily be determined along with its current position because while deploying the edge nodes, latitude and longitude information is already recorded.

6 Future Scope

In this article, we center around object detection in the image and live video streaming by utilizing the YOLO framework. The surveillance systems likewise require a suspicious activity recognition feature to identify anomalous human activities. Later on, we will attempt to stretch out this work to activity recognition like strolling, sitting, fighting, running, and suspicious activity recognition. Moreover, in the current work, all bounding box images of detected objects could not be saved locally; in the future, this issue would also be considered.

7 Conclusion

In this paper, object detection and alert system (ODAS) is proposed, and it could be used as a surveillance and monitoring system. It has been implemented using YOLO framework. In this system, distributed edge computing is used to do real-time analysis and decision making. Once the analysis has been done, the useful information like bounding box details, bounding box images, and date and time information is transmitted to the server. This reduces the additional traffic on the network and processing load on the server. Also, it can detect multiple distinct objects simultaneously. For visualization, GUI has been developed, and it is used to access and analyze the data from different edge nodes distributed over a large geographical area. GUI also triggers the alert when a targeted object found in the bounding box image.

References

1. Surantha N, Wicaksono WR (2018) Design of smart home security system using object recognition and PIR sensor. Procedia Comput Sci 135:465–472
2. Video Surveillance Market. Marketsandmarkets, 2020. [Online]. Available: https://www.marketsandmarkets.com/PressReleases/global-video-surveillance-market.asp
3. Lo Presti L, La Cascia M (2008) Real-time object detection in embedded video surveillance systems. In: 2008 ninth international workshop on image analysis for multimedia interactive services, pp 151–154
4. Santamaria A, Raimondo P, Tropea M, De Rango F, Aiello C (2019) An IoT surveillance system based on a decentralized architecture. Sensors 19(6):1469
5. Abdulraheem Fadhel M, Al-Shamaa O, Husain Taher B (2019) Real-time detection and tracking moving vehicles for video surveillance systems using FPGA. Int J Eng Technol 7(2.31):117
6. Shrivastava G, Peng S-L, Bansal H, Sharma K, Sharma M (2020) New age analytics: transforming the internet through machine learning, IoT, and trust modeling. Apple Academic Press, New York
7. Sinha A, Shrivastava G, Kumar P (2019) Architecting user-centric internet of things for smart agriculture. Sustain Comput Inform Syst 23:88–102
8. Sultana T, Wahid KA (2019) IoT-guard: event-driven Fog-based video surveillance system for real-time security management. IEEE Access 7:134881–134894
9. Kumar P, Shrivastava G, Tanwar P (2019) Demistifying ethereum technology: application and benefits of decentralization. In: Forensic investigations and risk management in mobile and wireless communications. IGI Global, pp 242–256
10. Ahmad F, Kumar P, Shrivastava G, Bouhlel MS (2018) Bitcoin: digital decentralized cryptocurrency. In: Handbook of research on network forensics and analysis techniques. IGI Global, pp 395–415
11. He N, Zhao J, Archbold CA (2002) Gender and police stress. Polic An Int J Police Strateg Manag 25(4):687–708
12. Tanaya, Vadivukarasi K\, Krithiga S (2018) Home security system using IoT. Int J Pure Appl Math 119(15):1863–1868
13. Jain A, Basantwani S, Kazi O, Bang Y (2017) Smart surveillance monitoring system. In: 2017 international conference on data management, analytics and innovation (ICDMAI 2017), pp 269–273
14. Sruthy S, George SN (2017) Wi-Fi enabled home security surveillance system using raspberry pi and IoT module. In: 2017 IEEE international conference on signal processing, informatics, communication and energy systems (SPICES 2017), pp 1–6
15. Ali S, Quadri I, Engineering C (2017) IoT based home automation and surveillance system. Int Res J Eng Technol 5(5):861–866
16. Patil N, Ambatkar S, Kakde S (2017) IoT based smart surveillance security system using raspberry pi. In: Proceedngs of 2017 IEEE international conference on communication and signal processing (ICCSP), pp 344–348
17. Khoje S, Gulve SP, Pardeshi P (2016) Implementation of IoT-based smart video surveillance system. In: International conference on CIDM, pp 771–780
18. Kumar S, Solanki SS (2016) Remote home surveillance system. In: Proceedings-2016 international conference on advances in computing, communication and automation, ICACCA 2016, pp 1–4

19. Virendra, Shete V, Ukunde N (2016) Intelligent embedded video monitoring system for home surveillance. In: Proceedings of international conference on inventive computation technologies, ICICT 2016, pp 1–4
20. Juhana T, Anggraini VG (2016) Design and implementation of smart home surveillance system. In: Proceeding 2016 10th international conference on telecommunication systems, services, and applications (TSSA), pp 1–5. Spec. Issue Radar Technol
21. Haribaabu V, Joseph James S (2016) Intelligent surveillance system using internet of things. Int Sci Press 9(37):313–317
22. Jyothi SN, Vardhan KV (2016) Design and implementation of real time security surveillance system using IoT. In: 2016 international conference on communication and electronics systems (ICCES)
23. Menezes V, Patchava V, Gupta MSD (2015) Surveillance and monitoring system using raspberry pi and simpleCV. In: 2015 international conference on green computing and internet of things (ICGCIoT), pp 1276–1278
24. Ordun C Setting up the NVIDIA Jetson TX2. https://tm3.ghost.io/2018/07/06/setting-up-the-nvidia-jetson-tx2/
25. Redmon J, Farhadi A (2018) YOLO: real-time object detection. [Online]. Available https://pjreddie.com/darknet/yolo/. Accessed 25 May 2020
26. Redmon J, Farhadi A (2018) YOLOv3: an incremental improvement

Optimized Resource Allocation Technique Using Self-balancing Fast MinMin Algorithm

Deepak Kumar Sharma, Kartik Kwatra, Manan Manwani, Nimit Arora, and Aarti Goel

Abstract With rising IT industry demands, there is a huge need of computational power and services for purposes such as research, software development, content hosting, and infrastructure management among others. Such tasks require expensive hardware on a large scale that has huge maintenance and other costs associated with it. Cloud computing is the delivery of computational services like servers, databases, networking tools, storage, software, analytic tools, and intelligence over the Internet which offers economies of scale, flexible and efficient resources, and technological innovation. Cloud computing systems consist of a number of servers, each hosting a number of instances of virtual machines (VMs). When a job arrives, it needs to be allocated to one of the running VMs. This allocation of jobs to VMs is a NP-complete problem. This paper presents an efficient self-balancing fast MinMin (SBFMM) algorithm devised to improve upon the shortcomings of the MinMin algorithm by keeping the best machine available for relatively heavier tasks. SBFMM algorithm reduces makespan and balances load better among the VMs as compared to traditional MinMin. SBFMM achieved an approximate 12–15% reduction in makespan as compared to MinMin algorithm for the data tested.

D. K. Sharma (✉) · A. Goel
Department of Information Technology, Netaji Subhas University of Technology, New Delhi, India
e-mail: dk.sharma1982@yahoo.com

A. Goel
e-mail: aartigoelcse@gmail.com

K. Kwatra · M. Manwani · N. Arora
Division of Information Technology, Netaji Institute of Technology, New Delhi, India
e-mail: kartikkwatra2016@gmail.com

M. Manwani
e-mail: manan.manwani@yahoo.in

N. Arora
e-mail: nimit17arora@gmail.com

A. Khanna et al. (eds.), *Data Analytics and Management*, Lecture Notes on Data Engineering and Communications Technologies 54,
https://doi.org/10.1007/978-981-15-8335-3_36

Keywords Cloud computing · Resource allocation · MinMin algorithm · Makespan · Load balancing

1 Introduction

Cloud computing is a shared pool of technical resources, which are rented by a company to complete their technological requirements. Elston et al. [1] described three service models, namely infrastructure as a service (IaaS), platform as a service (PaaS), and software as a service (SaaS). IaaS at the lowest end of the managed cloud services spectrum, IaaS providers deploy pre-configured hardware and are also responsible for managing it, thereby enabling users to create virtual machines and varied computing resources without incurring any labor costs or hardware management costs. PaaS solutions provide another layer of services which include app development, dashboards, coding platforms, etc., above IaaS for users who want to code, test, and deploy applications on their own in the cloud platform. This also includes managing security patches and operating system updates. SaaS solutions are ready to use software built on the cloud platform which provide users a host of different services that can run directly on the cloud and uses cloud's computation to work. The following section briefs about cloud computing characteristics.

1.1 Characteristics of Cloud Computing

1. *Cost:* Cost factor is one of the most important aspects of cloud computing. Cloud services eliminate the requirement of IT hardware and servers being set up on premise and hence reduce maintenance costs significantly.
2. *Speed:* Cloud services are given in self-service and on-demand modes, so large amounts of computational resources can be provisioned in a very short span on time, typically within minutes, giving companies large amounts of flexibility and reducing off capacity planning pressures.
3. *Global Scale:* Cloud computing is an extremely scalable solution by providing adequate amount of IT resources such as variable computing power and bandwidth storage whenever needed by the company and from the best geographic location.
4. *Reliability:* Cloud provides services like disaster recovery, data backup, and business continuity in a relatively easier and cheaper form which becomes helpful for the companies and helps them in managing data properly.
5. *Security:* Security is an extremely important requirement in any IT infrastructure. Protecting data, encryption, and safeguarding sensitive information are necessary for any company. Cloud computing provides a secure method to store and use data which being compliant with the government rules and regulation.

1.2 Advantages of Cloud Computing

The following are some of the advantages of cloud computing

1. *Easy Maintenance*—Cloud computing provides an alternative to physical servers which require a huge amount of space and maintenance within company premises.
2. *Machine Learning and AI Tools*—Apart from storage, cloud computing provides a number of other services like machine learning and AI tools which help the company derive analytical results from data and use the results to improve company efficiency.
3. *Data Safety*—Cloud computing also provides backup services for data. Physical servers are prone to getting damaged, and hardware needs to be replaced every few years. With cloud, data is stored in data centers which provide better safety and security.

1.3 Limitations

Endo [2] listed challenges and limitations in cloud computing as follows

- *Downtime:* Service outage is one of the biggest disadvantages as it can lead to system failures if the system faces any downtime.
- *Security and Privacy*: Security is one of the most important aspects, and if security is compromised in a cloud system, personal and confidential information would be leaked, and this can cause serious issues.
- *Limited Control and Flexibility:* Customer control over the cloud service remains minimal as it is the cloud service provider who owns the hardware servers and the corresponding software.
- *Support Services*: Support is required by companies to implement cloud services on their computers, and due to high complexity, it becomes difficult to train employees without good support from the cloud service provider.

Resource scheduling is a NP-complete optimization problem. This paper proposes a new algorithm for resource scheduling in cloud computing environment and compares it against existing traditional resource scheduling algorithms.

2 Literature Survey

Cloud computing is a growing technology with advancements taking place daily. Due to its developing nature, there are not many standard resource allocation algorithms present till date. Constant research on effectiveness of the standard algorithms is leading to improved versions of these algorithms which are being applied in specific cases to optimize cost, efficiency, and time.

Yu and Yu [3] introduced a cloud computing MinMin algorithm which was based on grid computing. Its main target was to achieve the minimum makespan while allocating resources. This algorithm focused on selecting cloudlets having minimum completion time on a virtual machine and allocating it to the respective virtual machine.

Minimum execution time (MET) algorithm focuses on execution time of the resources. Elzeki et al. [4] proposed the algorithm in which resources are allocated to VMs which lead to their minimum execution time. This algorithm though focusing on minimizing makespan leads to heavy imbalance of load on each virtual machine.

Bala and Chana [5] worked on a readiness-based task scheduling algorithm called opportunistic load balancing (OLB) algorithm. This algorithm allocates resource which is going to be ready next to the cloudlet. In case two or more VMs get ready at the same time, resource is allocated arbitrarily.

Deepa and Deepa [6] presented a resource allocation technique for getting improved energy efficiency in the cloud system. Zhuang and Huang [7] have presented an overview of resource allocation strategies in cloud computing and have analyzed the benefits and disadvantages of different methods.

Alworafi et al. [8] proposed an improved version of the shortest job first (SJF) algorithm known as modified shortest job first (MSJF) which includes the load balancing aspect and assigns longest task to the better machine. The performance of MSJF came out to be better than the traditional SJF but significantly increases the overall makespan time of the system.

Gawali and Shinde [9] introduced a new heuristic approach which combined other algorithms like longest expected time preemption and divide and conquer techniques to perform effective task scheduling in cloud systems.

There have also been attempts at utilizing the shorter makespan quality of MinMin to generate balanced results. Amalarethinam and Kavitha [10], and Anousha and Ahmadi [11] proposed enhanced algorithms based on MinMin algorithm which combines the properties of MinMin and other resource-aware algorithms to produce an efficient model.

3 Existing Algorithm (MinMin Algorithm)

MinMin algorithm is a simple resource allocation algorithm which assigns tasks to VMs that completes it within the minimum time. In MinMin algorithm, the job with minimum completion time is allocated to the VM giving the respective minimum completion time [12]. Then, the completion time of all other cloudlets on the selected VM is updated using Eq. 3.1. The process is repeated until all cloudlets have been allocated.

Completion time, CT, of a cloudlet i on VM_k is given by Eq. 3.1 [12]

$$\mathrm{CT}(i) = \mathrm{ET}(i)_k + r(k) \tag{3.1}$$

where $ET(i)_k$ is the execution time of cloudlet i on VM_k. It is calculated by Eq. 3.2 [12]

$$ET(i)_k = \text{Cloudlet Length}(i)/(\text{Pes}(k) * \text{Mips}(k)) \tag{3.2}$$

where PEs(k) are the number of processing elements in VM_k and Mips(k) is the processing power of each core in million instructions per second.

Makespan is defined as the sum of execution time of each cloudlet [13].

3.1 Advantages of MinMin Algorithm

1. MinMin algorithm is extremely fast and effective for small-sized tasks which are lesser in number.
2. The delay in processing of tasks is minimum in MinMin algorithm.

3.2 Disadvantages of MinMin Algorithm

As noted by Goudarzi et al. [14] as the size of tasks increases, MinMin suffers from poor load balancing and higher waiting time for the largest tasks.

4 Proposed Algorithm: Self-balancing Fast MinMin Algorithm (SBFMM)

If the size of tasks is large, the traditional MinMin algorithm fails to balance the load and will result in higher waiting time for larger tasks. The SBFMM algorithm works on the principle that the best VM should be reserved for heavier jobs. For assigning the tasks to VMs, this algorithm follows the same principle as MinMin algorithm; i.e., the job having the minimum completion time is allocated to the VM giving the respective minimum completion time. Best VM is the one having maximum computing power (CP).

Computing power (CP) of VM_k is given by Eq. 4.1

$$CP_k = (\text{Pes}(k) * \text{Mips}(k)) \tag{4.1}$$

where PEs(k) are the number of processing elements in VM_k and Mips(k) is the processing power of each core in million instructions per second.

In this algorithm, a coefficient N is used which is known as the reducing factor. This reducing factor determines the number of jobs for which the best VM is inactive; i.e., we assign the jobs to the second-best VM. The value of N ranges from 0 to size of task set (S). For each value of N, the MinMin algorithm is applied by making the best VM inactive for N number of jobs and after that best VM gets activated, tasks are assigned using MinMin, and makespan is calculated. The value of N giving the minimum makespan is selected as the reducing factor.

4.1 Algorithm for SBFMM

1. For all tasks in task set (S).
2. For all resources R_k.
3. for $N = 0$ to Size (S):

   i. Find the best VM (VM_{best}) using Eq. 4.1 and remove it from the resource list R_k.
   ii. For $j = 0$ to N:
      a. Assign the job (T_j) having the minimum completion time to the VM giving the respective minimum completion time.
   iii. Add VM_{best} to the resource list R_k.
   iv. For $j = N + 1$ to Size(S):
      a. Assign the job (T_j) having the minimum completion time to the VM giving the respective minimum completion time until all the jobs are assigned.
   v. Calculate the makespan and update the minimum makespan.

4. Select the N giving the minimum makespan.
5. Terminate.

The flowchart for the SBFMM algorithm is shown in Fig. 1.

5 Results

To measure the effectiveness of the SBFMM algorithm, the following experiments were performed in order to draw a comparison between MinMin algorithm and SBFMM. These experiments were conducted on CloudSim simulator.

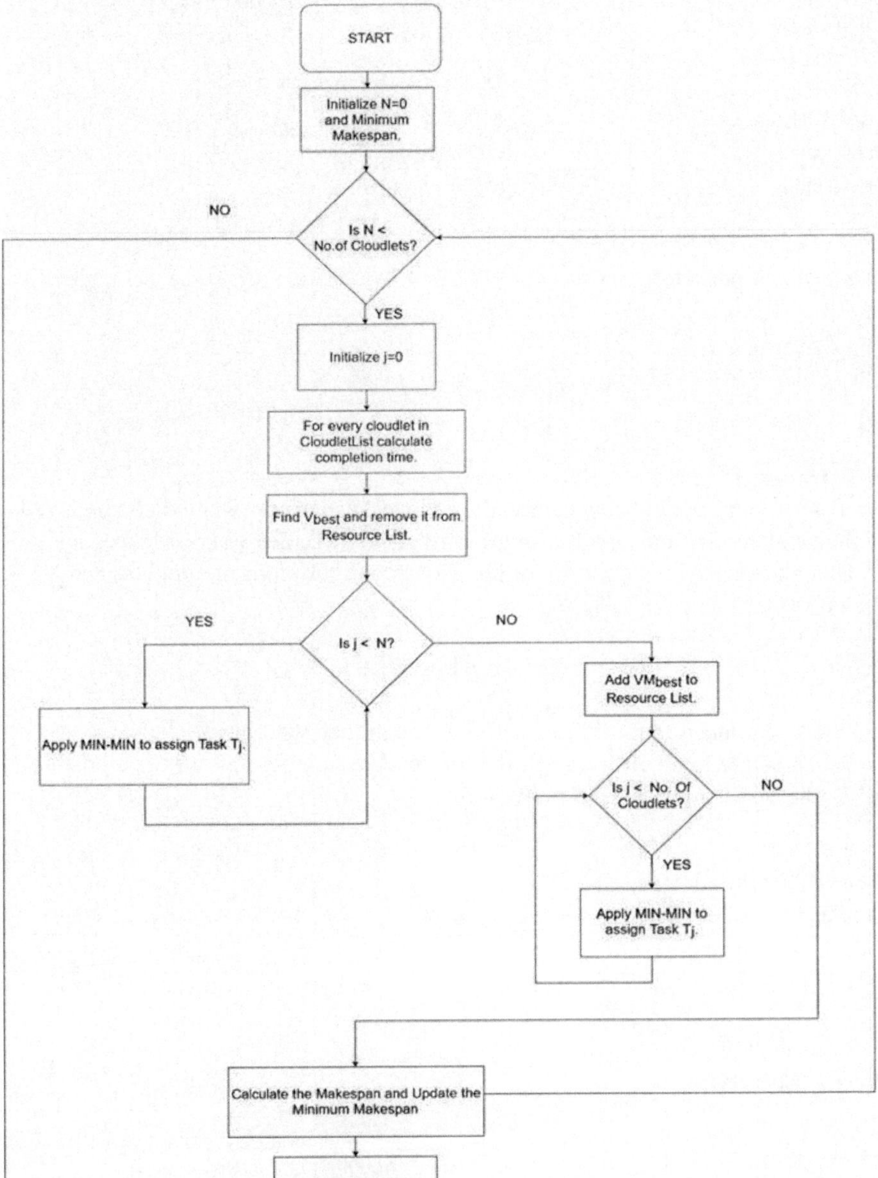

Fig. 1 Flowchart of SBFMM algorithm

VM Characteristics

Mips	4X No of VMs provisioned
RAM	2048 MB
Storage	1,000,000 megabytes
Bandwidth	1000

Datacenter Characteristics

Architecture	X86
OS	Linux

Experiment 1

First experiment was conducted to analyze the makespan with 8 cloudlets and 2 VMs. VM characteristics and cloudlet lengths are given in Tables 1 and 2, respectively.

Using the MinMin algorithm for the above configurations of cloudlets and VMs, the makespan came out to be 61.9 s.

$$\text{Makespan} = 61.9\,\text{s}$$

The VM allocation for this algorithm is mentioned in Table 3.

When SBFMM algorithm was used for the above configurations of cloudlets and VMs, the makespan came out to be 52.5 s.

Table 1 VM characteristics

VM-ID	Mips	PEs (processing elements)	RAM (MB)	Bandwidth (Mbps)
0	500	1	512	1000
1	100	1	512	1000

Table 2 Cloudlet lengths

Cloudlet	Lengths (million instructions)
0	2500
1	6000
2	5000
3	4500
4	1200
5	3700
6	1000
7	7000

Table 3 VM allocation for MinMin algorithm

VM-ID	Number of cloudlets allocated
0	8
1	0

Table 4 VM allocation for SBFMM algorithm

VM-ID	Number of cloudlets allocated
0	5
1	3

$$\text{Makespan} = 52.5\ \text{s}$$

The VM allocation in this algorithm is mentioned in Table 4.

It is evident from Tables 3 and 4 that load is balanced better when SBFMM algorithm is used. The traditional MinMin algorithm overloads VM 1 by allocating all of the 8 cloudlets to it, whereas the SBFMM algorithm allocates 5 cloudlets to VM 1 and 3 cloudlets to VM 2.

When the value of N (reducing factor) changes, then the value of makespan also changes. Table 5 shows the effect on makespan as the value of *N* changes.

Comparative analysis between MinMin algorithm and SBFMM in terms of load balancing and makespan is shown in Figs. 2 and 3, respectively. Figure 4 shows effect of changing the value of *N* on makespan.

Experiment 2

Second experiment was conducted to analyze the makespan with 12 cloudlets and 4 VMs. VM characteristics and cloudlet lengths are given in Tables 6 and 7, respectively.

Using the MinMin algorithm for the above configurations of cloudlets and VMs, the makespan came out to be 72.57 s.

$$\text{Makespan} = 72.57\ \text{s}$$

Table 5 Effect on makespan on changing *N*

Value of *N*	Makespan (s)
0	61.9
1	59.5
2	57.5
3	52.5
4	84.1
5	129.1
6	179.1
7	239.1

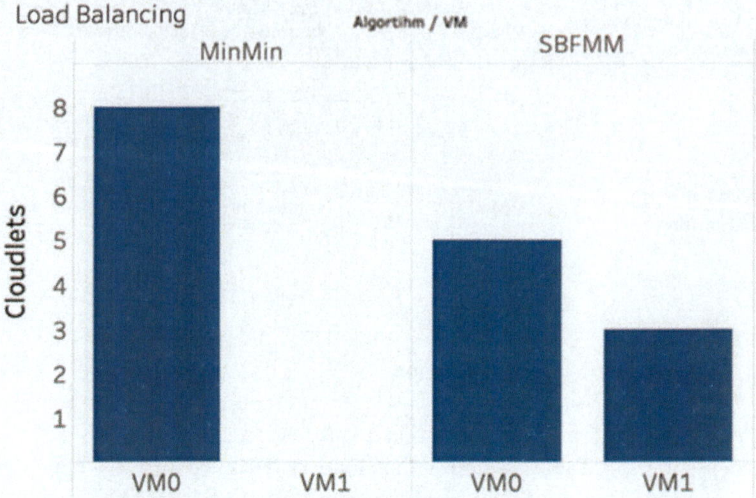

Fig. 2 Load balancing in MinMin versus SBFMM

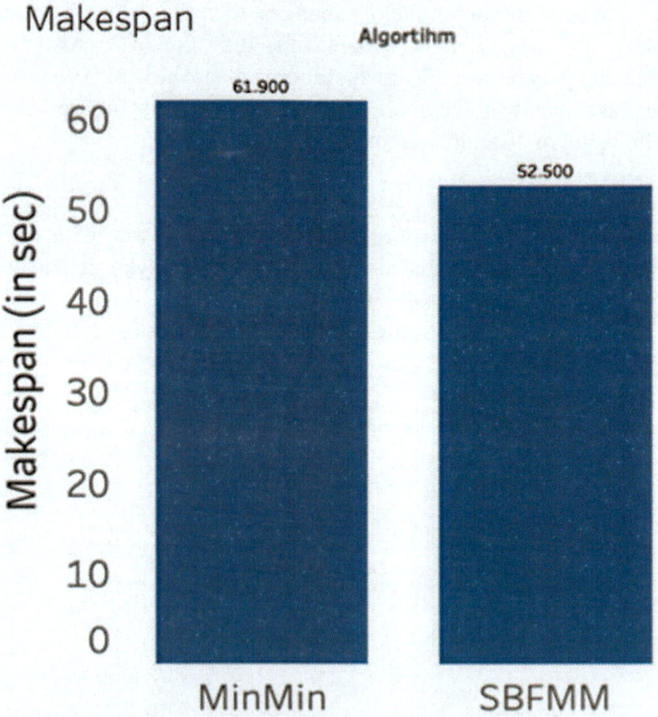

Fig. 3 Makespan of MinMin versus SBFMM

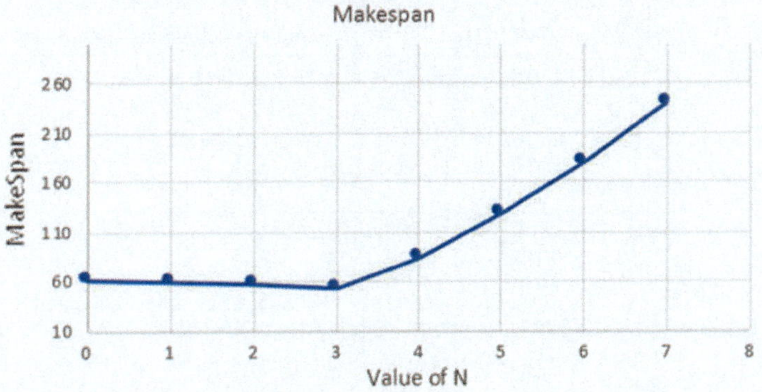

Fig. 4 Makespan versus value of *N*

Table 6 VM characteristics

VM-Id	Mips	PEs	RAM (MB)	Bandwidth (Mbps)
0	80	1	512	1000
1	320	1	512	1000
2	130	1	512	1000
3	270	1	512	1000

Table 7 Cloudlet lengths

Cloudlet	Lengths (million instructions)
0	2500
1	6000
2	5000
3	4500
4	1200
5	3700
6	1000
7	7000
8	2600
9	1080
10	5490
11	2300

The VM allocation in this algorithm is mentioned in Table 8.

When SBFMM algorithm was used for the above configurations of cloudlets and VMs, the makespan came out to be 62.58 s.

$$\text{Makespan} = 62.58\text{ s}$$

Table 8 VM allocation for MinMin algorithm

VM-ID	Number of cloudlets allocated
0	0
1	6
2	1
3	5

Table 9 VM allocation for SBFMM algorithm

VM-ID	Number of cloudlets allocated
0	1
1	3
2	2
3	6

The VM allocation in this algorithm is mentioned in Table 9.

When the value of N (reducing factor) changes, the value of makespan also changes. Table 10 shows the effect on makespan while changing the value of N.

Comparative analysis between MinMin algorithm and SBFMM in terms of load balancing and makespan is shown in Figs. 5 and 6, respectively. These results show that SBFMM algorithm improves load balancing without compromising on the makespan as compared to MinMin algorithm. The makespan is even lower in some cases when using the SBFMM algorithm.

Lastly, both the algorithms were run on 20 cloudlets and 4 VMs. Figure 7 shows the results obtained.

Table 10 Effect on makespan on changing N

Value of N	Makespan (s)
0	72.57
1	70.46
2	72.31
3	64.15
4	65.09
5	70.01
6	71.83
7	68.81
8	73.49
9	62.58
10	71.54
11	78.16

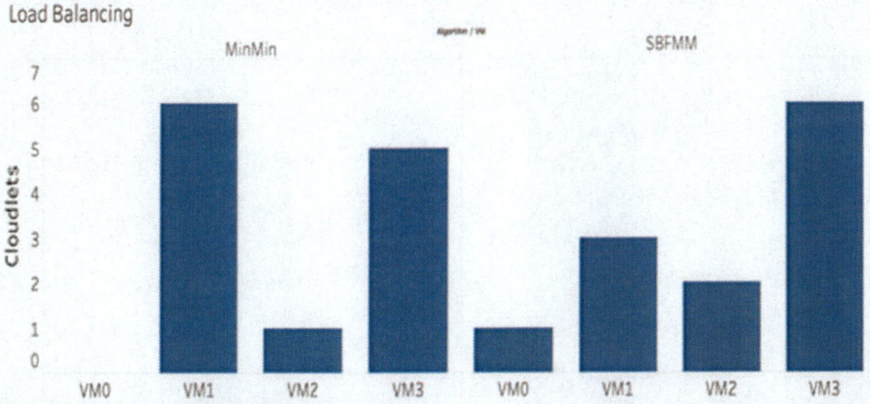

Fig. 5 Comparison between MinMin algorithm and SBFMM algorithm on the basis of load balancing

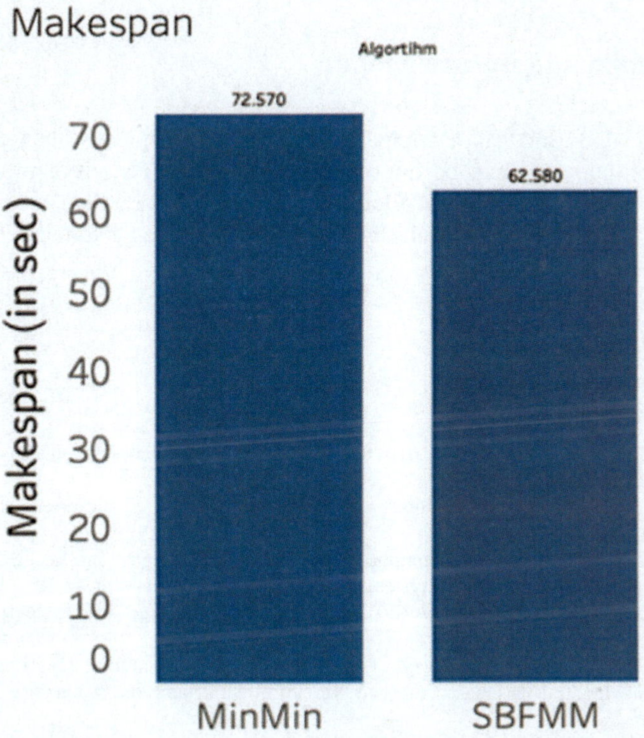

Fig. 6 Comparison between MinMin algorithm and SBFMM algorithm on the basis of makespan

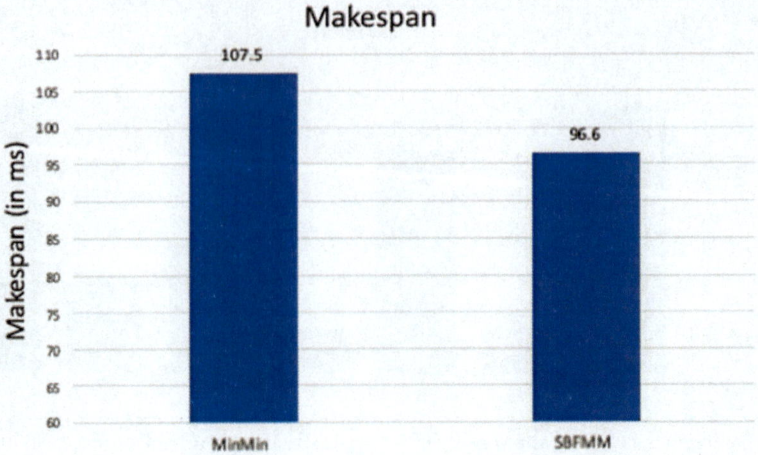

Fig. 7 Performance comparison with 20 cloudlets and 4 VMs

6 Conclusion and Future Scope

The SBFMM algorithm has a comparatively higher complexity than traditional MinMin algorithm. This can be improved in the future. The problem of finding the number of tasks to be saved for allocating later can be researched upon to find an optimized solution. Heuristic search algorithms like genetic algorithm can be applied for the same.

References

1. Tsai W-T, Shao O, Sun X, Elston J (2010) Service-oriented cloud computing. In: World congress on services, pp 473–478
2. Endo PT (2011) Resource allocation for distributed cloud: concept and research challenges. IEEE, pp 42–46
3. Yu X, Yu X (2009) A new grid computation-based Min-Min algorithm. In: FSKD'09 sixth international conference on fuzzy systems and knowledge discovery, pp 43–45
4. Elzeki OM, Reshad MZ, Elsoud MA (2012) Improved max-min algorithm in cloud computing. Int J Comput Appl 50(12):22–27
5. Bala A, Chana I (2011) A survey of various workflow scheduling algorithm in cloud environment. In: 2nd national conference on information and communication technology (NCICT)
6. Deepa R, Dheeba J (2019) Resource allocation in cloud computing for energy efficiency. Int J Eng Adv Technol (IJEAT) 8(6). ISSN 2249-8958
7. Zhuang W, Huang L (2019) Overview of cloud computing resource allocation and management technology. In: 2019 6th international conference on systems and informatics (ICSAI), Shanghai, China, pp 713–718. https://doi.org/10.1109/icsai48974.2019.9010101
8. Alworafi M, Dhari A, Al-Hashmi A, Darem AB, Suresha (2016) An improved SJF scheduling algorithm in cloud computing environment

9. Gawali MB, Shinde SK (2018) Task scheduling and resource allocation in cloud computing using a heuristic approach. J Cloud Comput 7
10. Amalarethinam DI G, Kavitha S (2016) Enhanced Min-Min algorithm for meta-task scheduling in cloud computing. IJCTA 85–91
11. Anousha S, Ahmadi M (2013) An improved Min-Min task scheduling algorithm in grid computing. Springer, Berlin Heidelberg, pp 103–113
12. Kumar S, Mishra A (2015) Application of Min-Min and max-min algorithm for task scheduling in cloud environment under time shared and space shared VM models. Int J Comput Acad Res (IJCAR) 4(6):182–190
13. Sharma N, Tyagi S, Atri S (2017) A comparative analysis of Min-Min and max-min algorithms based on the Makespan parameter. Int J Adv Res Comput Sci
14. Goudarzi H, Pedram M (2011) Maximizing profit in cloud computing system via resource allocation. In: IEEE 31st international conference on distributed computing systems workshops, pp 16

Routing Protocol Based on NSGA-II for Social Opportunistic Networks

Ritu Nigam, Deepak Kumar Sharma, and Satbir Jain

Abstract In the Social opportunistic networks (SONs), nodes inspect the transmission area concurrently for data forwarding and discover relevant relay neighbors to accomplish effective message transmission. The context information of the nodes like distance, contact duration, encounter, centralities, tie strength, etc. are analyzed to find out right relay nodes that support in designing an efficient routing protocol. Developing such kind of routing protocols that optimize multiple conflicting objectives is computationally strenuous. This paper proposes a Non-dominated sorting Genetic algorithm-II based routing protocol for social opportunistic Networks (NSRP). NSRP uses NSGA-II for optimizing the node's average forwarding delay degree, and the collaboration probability objective functions to get non-dominated solutions for message passing. This proposed protocol is robust and expected to produce high delivery ratio by optimizing multiple parameters of the SONs.

Keywords NSGA-II · Average forwarding delay degree · Collaboration probability · Relay nodes · Social opportunistic networks

---

R. Nigam
Division of Computer Engineering, University of Delhi (Netaji Subhas Institute of Technology), New Delhi, India
e-mail: ritu.nigam2106@gmail.com

D. K. Sharma (✉)
Department of Information Technology, Netaji Subhas University of Technology (Formerly Netaji Subhas Institute of Technology), New Delhi, India
e-mail: dk.sharma1982@yahoo.com

S. Jain
Department of Computer Engineering, Netaji Subhas University of Technology (Formerly Netaji Subhas Institute of Technology), New Delhi, India
e-mail: jain_satbir@yahoo.com

A. Khanna et al. (eds.), *Data Analytics and Management*, Lecture Notes on Data Engineering and Communications Technologies 54,
https://doi.org/10.1007/978-981-15-8335-3_37

1 Introduction

The opportunistic networks, also called OppNets, are a kind of multi-hop ad-hoc network which has turned up in recent years [1]. The origin of OppNets comes from delay-tolerant networks (DTN) [2]. The leading feature to differentiate the opportunistic networks from ad-hoc networks is that no end to end path exists ever. A new network, famous as the Social opportunistic networks [3, 4], has been emerged based on the application environment of opportunistic networks, where people with mobile gadgets can communicate sporadically via Bluetooth or WiFi in there available transmission range [5]. OppNets count on the store-and-carry forward pattern to transmit a message from source to destination successfully. Sender node waits for appropriate relay nodes to pass the message to next hop. Additionally, social networks application follow the same pattern with their mobile nodes. For example, big data media communications prefer good relationship people to transmit much data, which shows nodes cooperation approach for data transmission. However, many features of social nodes, such as node's attributes, centralities, encounter history, distance, etc. are utilized to determine the useful relay nodes for message transmission in SONs. Many routing protocols designed for social OppNets so far have examined several types of combinations of these parameters for the process of next-hop selection; still, no work has performed for their optimization. In this view, this paper exploits a multi-objective optimization-based approach for optimized forwarding. The proposed work utilizes the Non-dominated sorting genetic algorithm-II [6] to choose the optimize relay nodes based on two objectives, i.e., minimizing the Average forwarding delay degree and maximizing the Collaboration Probability [7, 8] of the network nodes. NSGA-II helps in achieving a non-dominated set of relay nodes for routing the message towards their destination. More precisely, the following are the main points of this paper:

1. Analyze the critical parameters, the Average forwarding delay degree, and the Collaboration Probability in SONs.
2. Design an efficient routing protocol that provides a set of near-optimal relays by satisfying specific parameters in SONs by employing elitist non-dominated sorting genetic algorithm (NSGA-II).
3. Theoretically analyze the proposed NSRP protocol.

The rest of the article is systematized as follows. Section 2 takes the overview of typical routing protocols; the system model is explained in Sect. 3. In Sect. 4, the proposed work and NSRP algorithm are discussed, and Sect. 5 concludes the NSRP scheme.

2 Related Work

Currently, the main focus in the field of Social opportunistic networks is designing influential routing protocols. Many existing methods of different areas can be applied appropriately to improve transmission. Here, few currently developed algorithms are discussed in the opportunistic networks as follows.

Shubham et al. [9] introduced an intelligent routing mechanism based on Intelligent Water Drop (IWD) Algorithm, which is clubbed with the artificial neural networks to formulate a next-hop selection strategy. In this approach, the IWD algorithm serves adaptiveness and neural network supports in taking smart routing decisions. The training data for the Neural Network model consists of the node's features like the number of successful deliveries, buffer space, the delivery probability with transitivity and energy. The method used for training has two phases; one is the feed-forward phase, and another is the weight update phase. IWDNN outperforms as compared to current protocols in terms of overhead, delivery probability, drop ratio, and latency.

Sharma et al. [10] brought in machine learning techniques in OppNets for message forwarding. MLPROPH takes advantage of decision trees and a machine learning model based on neural networks to resolve message routing issues. The model uses many features like the hop count, buffer space, PROPHET probability, and energy of the node to train the proposed model. The delivery probability predicted by the model is compared with the normalized PROPHET probability to select a better next hop to forward the message. Here, PROPHET probability is taken from the PROPHET protocol.

Sharma et al. came up with KNNR routing [11], which utilizes the K-Nearest Neighbor (KNN) classifier approach to anticipate the best next-hop relay node for message transmission. This scheme works on two phases named as training and application phase. A dataset is created based on the node's encounter history to train the KNN classifier in the training phase. The application phase exploits this trained classifier to take the intelligent routing decision for message transmission.

A new epidemic protocol EpSoc [12] for Opportunistic Mobile Social Network proposed by Lenando et al. recently. The EpSoc is a hybrid form of standard Epidemic routing strategy which employs the social attribute degree centrality. In this work, Messages' TTL is used, which is adjusted as per the degree centrality value of nodes. For controlling the replication, a message blocking procedure is used. As per the simulation, the EpSoc shows better results in terms of delivery, overhead, average latency, and hop counts in contrast to Epidemic and Bubble Rap.

Chen et al. [13] suggested an Adaptive Routing Optimization Algorithm in Community-Oriented Opportunistic Networks, which first analyses the node's social relationship and then introduces a new metric based on average message forwarding delay for fussy message delivery. This new metric also helps in designing a local community detection scheme and identifying ordinary nodes and ferry nodes in the networks. This paper forms two guidelines for the transmission procedure of the message: (1) ordinary nodes forward the message inside the community while the ferry nodes transmit the message between the communities. (2) an encountered node is only chosen if it is more active than the source node or one of the ferry nodes. Some of the recent works that use Machine learning in Internet of Things (IoT) Opportunistic Networks are presented in [14, 15].

3 System Model

Consider that the social opportunistic networks environment has framed the N mobile social nodes that have expected to be mutually cooperative and carry adequate energy to take part in the forwarding of the messages. It has also considered that nodes are not malicious, and to store nodes context information, the nodes carry enough buffer space. In the proposed work NSRT node's activity degree, inclination degree and average forwarding delay degree is used as parameters to decide on the next-hop nodes for message forwarding.

3.1 Collaboration Probability

Collaboration Probability includes two parameters, which are defined below.
(i) Node's activity degree: The node's activity degree demonstrates the count of links between nodes a and b in time T is the relative amount of the total count of contacts in the social OppNets corresponding to the other nodes.

$$\mathrm{NA}_{(a,b)} = \frac{c_{(a,b)}(T)}{c_b(T)} \tag{1}$$

(ii) Nodes Inclination degree: The node inclination degree contrasts from the encounter frequency of the node, which specifies the time possession ratio between nodes a and b building contact link and forwarding a message in a period. The formula of this particular measurement is as follows:

$$\mathrm{Incl}_{(a,b)} = \frac{\sum_{e=1}^{n} t_{a,b}^{e}}{\sum_{e=1}^{n} t_{\mathrm{others},b}^{e}} \tag{2}$$

By using the degree of node inclination and activity, cooperation probability values between two nodes can quantify as follows:

$$\mathrm{Co}_{(a,b)} = \alpha \mathrm{NA}_{(a,b)} + \beta \mathrm{Incl}_{(a,b)} \tag{3}$$

where, α and β are the coefficients of node inclination and activity where $\alpha + \beta = 1$.

3.2 Average Forwarding Delay Degree

This new metric utilizes past encounters that measure the reciprocal of average forwarding delay of the messages in the time T.

$$\mathrm{fd}_{(a,b)} = \frac{\frac{2T}{\sum_{c=1}^{n} t_{c:(a,b)}^{2}}}{\frac{2T}{\sum_{c=1}^{n} t_{c:(\text{others},b)}^{2}}} \tag{4}$$

3.3 Multi-objective Evolutionary Algorithms

This paper assumes that in the proposed work, the optimization functions are depicted as a maximization problem. In the multi-objective optimization (MOO) of p functions $f(X) = (f_1(X), f_2(X), \ldots, f_p(X))$, a candidate solution X_1, can perform better than another solution, X_2 which belong to same objective f_a (i.e., $f_a\,(X_1) > f_a\,(X_2)$, but can perform worse which belong to f_b objective (i.e., $f_b\,(X_1) < f_b(X_2)$). Accordingly, the candidate solutions are inspected by the following notion of dominance in MOO: a candidate solution X_1 dominates X_2, expressed by $X_1 > X_2$, if and only if the successive two conditions hold:

1. $\forall$ a $\in \{1, 2, \ldots, p\} | \mathrm{f}_a(X_1) \geq f_a(X_2)$; and
2. $\exists$ b $\in \{1, 2, \ldots, p\} : \mathrm{f}_b(X_1) > f_b(X_2)$.

The multi-objective evolutionary algorithms (MOEAs) have various variants that have a common goal of containing the non-dominated solutions produces during the search. In this work, famous NSGA-II has adopted, which can keep up an improved spread of solutions as well as better converge in the acquired non-dominated front in comparison with other MOEA approaches [16, 17]. NSGA-II does not use any external archive to accumulate the provided non-dominated solutions. Alternatively, by using Pareto ranking and niching approaches, the algorithm partition the population into a few non-dominated fronts [18]. For the next-generation population, mutually non-dominating solutions are compared based on their contribution to the diversity of the population in the selection process. First front solutions have the maximum chance to be a part of the next generation as parent population.

4 Proposed Approach

This paper aims to combine the two contradictory social features of SONs, i.e., to maximize the collaboration degree (Co) and minimize the average forwarding delay degree (fd) for efficient message transmission. This approach considers as a multi-objective optimization problem and applies NSGA-II for efficient message transmission. There is a tradeoff between both the social objectives, which is solved

as multiple non-dominated solutions. This proposed technique decides the next-hop selection of nodes based on a combination of collaborations and average forwarding delay degree objectives. A non-dominated set of solutions is derived using the NSGA-II algorithm for forwarding the data packets towards their destination.
The objective function which, contains a percentage of collaboration degree F_1 and average forwarding delay degree F_2 to each solution $P_a \in P$ population. Here,

$$F_1(P_a) = \sum_{a=1}^{n} \mathrm{Co}_{(a,b)} \tag{5}$$

$$F_2(P_a) = \sum_{a=1}^{n} \mathrm{fd}_{(a,b)} \tag{6}$$

4.1 NSRP Algorithm and Its Details

The proposed algorithm NSRP uses the $NSGA - II$ algorithm to discover a set of Pareto optimal solutions as a relay for the sender node a, which helps to transmit the message to destination *d*. In the beginning, initialize the population by generating it randomly. The size of the chromosome is *k*, which is fixed. After that, each individual calculates its objective functions based on two fitness functions, $Co_(a, b)$ collaboration degree F_1 and $\mathrm{fd}_(a, b)$average forwarding delay degree F_2. Later, apply non-dominated sorting on initial parent population P_g to generate offspring Q_g, here g depicts the number of generation. Combine both the parent and child population to make the population $R_g = P_g \cup Q_g$ of size $2N_p$.
Subsequently, the R_g population exploits non-dominated sorting and crowding distance approaches to get the fronts. To form the New parent population P_g, we take the solutions from the first front and add them till size reaches to N_p. Crowding distance is also used to sort and accept the last front. Afterward, this $P_g + 1$ population further perform selection, crossover, and mutation to produce a new population $Q_g + 1$ of size N_p. This paper uses a one-point crossover operator. After performing crossover, perform the mutation operator, which randomly alters few relay nodes in genes of the chromosomes of the offspring. A casually played mutation on any gene may cause an invalid path, so to select a proper gene to apply mutation operator require a standard method. This work considers the rate of crossover at 0.8 and mutation at 0.2 for maintaining excellent and adequate chromosomes. This whole procedure repeatedly continues until the variation of fitness values between the present Pareto optimal set, and the former one is less than a chosen threshold (θ). The most significant set of solutions is selected to forward the message towards the destination node. The procedure of algorithm is described in Algorithm 1.

Algorithm 1 NSRP routing protocol

Input Source Node a, message m, destination node d, number of generation g
Begin
initialize g = 0, N_p = Populationsize;
Generate initial population P_g;
Compute objective functions F_1 and F_2 for each solution in P_g ;
F_r = Do Elitist_non_dominated_sorting_procedure; /* Produces non-dominated fronts (F_{r1}, F_{r2},...) */
Q_g = Apply crossover and mutation operator to produce offspring from P_g;
/* choosing the optimal relay node generated by the NSGA-II */
$R_g = P_g \cup Q_g$;
F_r = Do Elitist_non_dominated_sorting_procedure (R_g);
initialize $P_g + 1 = \phi$; i=1;
while Fronts_Size($P_g + 1 + F_{ri}$) $<= N_p$ **do**
    Crowding_distance_method(F_{ri});
    $P_g + 1 = P_g + 1 \cup F_{ri}$;
    i = i+1 ;
end while
Compute objective functions F_1 and F_2 for each solution in $P_g + 1$;
F_r = Do Elitist_non_dominated_sorting_procedure;
if $fitnessvalue\_changes < \theta$ **then**
    Get Pareto Optimal relay nodes;
    Break;
end if
g = g + 1;
Until g < MaxGeneration;

5 Conclusion and Future Work

This paper works on a multi-objective optimization to develop an efficient routing protocol in social OppNets by exploiting NSGA-II [18]. Optimizing a specific objective function might let go of the optimization of the secondary conflicting objective. The proposed protocol efficiently optimizes the parameters, node's average forwarding delay degree, and the collaboration probability nodes, which will support to increase the message delivery. For keeping a functional diversity of solutions in the parent population, the NSRP algorithm uses crowding distance as a measure in decision space. As the NSRP algorithm used the famous Elitist NSGA-II technique, which has a fast convergence rate and accumulates the most deserving solutions by using elitism for choosing the best relay nodes, so this will surely increase the message delivery probability. Additionally, This will control message replication as message copies are only replicate to non-dominated solutions; thus, the overhead will also decrease.

In the future, the simulation will be performed on the NSRP algorithm by using ONE simulator. Moreover, comparison analysis will also perform with the present protocols of social OppNets to check the performance of the NSRP protocol.

References

1. Hu Y, Liu A (2015) An efficient heuristic subtraction deployment strategy to guarantee quality of event detection for wsns. Comput J 58(8):1747–1762
2. Guo H, Wang X-W, Huang M, Jiang D-D (2012) Adaptive epidemic routing algorithm based on multi queue in dtn. J Chin Comput Syst 33(4):829–832
3. De Rango F, Socievole A, Marano S (2015) Exploiting online and offline activity-based metrics for opportunistic forwarding. Wirel Netw 21(4):1163–1179
4. Zhu Y, Jiang R, Zhao J, Ni LM (2015) Correlating mobility with social encounters: Distributed localization in sparse mobile networks. Wirel Netw 21(1):201–215
5. Tuncer D, Sourlas V, Charalambides M, Claeys M, Famaey J, Pavlou G, De Turck F (2016) Scalable cache management for isp-operated content delivery services. IEEE J Sel Areas Commun 34(8):2063–2076
6. Deb K (2001) Multi-objective optimization using evolutionary algorithms, vol 16. Wiley
7. Wu J, Chen Z, Zhao M (2019) Information cache management and data transmission algorithm in opportunistic social networks. Wirel Netw 25(6):2977–2988
8. Yu G, Chen Z, Wu J, Wu J (2018) A transmission prediction neighbor mechanism based on a mixed probability model in an opportunistic complex social network. Symmetry 10(11):600
9. Kumaram S, Srivastava S, Sharma DK (2020) Neural network-based routing protocol for opportunistic networks with intelligent water drop optimization. Int J Commun Syst 33:e4368
10. Sharma DK, Dhurandher SK, Woungang I, Srivastava RK, Mohananey A, Rodrigues JJ (2016) A machine learning-based protocol for efficient routing in opportunistic networks. IEEE Syst J 12(3):2207–2213
11. Sharma DK, Sharma A, Kumar J et al (2017) Knnr: K-nearest neighbour classification based routing protocol for opportunistic networks. In: 2017 Tenth international conference on contemporary computing (IC3). IEEE, pp 1–6
12. Lenando H, Alrfaay M (2018) Epsoc: Social-based epidemic-based routing protocol in opportunistic mobile social network. Mob Inf Syst
13. Chen W, Chen Z, Cui F (2019) Adaptive routing optimization algorithm in community-oriented opportunistic networks for mobile health. Sensors 19(8):1876
14. Vashishth V, Chhabra A, Sharma DK (2019) Gmmr: a gaussian mixture model based unsupervised machine learning approach for optimal routing in opportunistic iot networks. Comput Commun 134:138–148
15. Vashishth V, Chhabra A, Sharma DK (2019) A machine learning approach using classifier cascades for optimal routing in opportunistic internet of things networks. In: 2019 16th Annual IEEE international conference on sensing, communication, and networking (SECON). IEEE, pp 1–9
16. Srinivas N, Deb K (1994) Muiltiobjective optimization using nondominated sorting in genetic algorithms. Evol Comput 2(3):221–248
17. Deb K, Pratap A, Agarwal S, Meyarivan T (2002) A fast and elitist multiobjective genetic algorithm: Nsga-ii. IEEE Trans Evol Comput 6(2):182–197
18. Coello CAC, Lamont GB, Van Veldhuizen DA et al (2007) Evolutionary algorithms for solving multi-objective problems, vol 5. Springer

Connected Public Transportation System for Smart City

P. Pujith Sai, O. Goutham Sai, and Suresh Chavhan

Abstract Raise in the number of mishaps, packed busses, trains roads, turned parking lots and squandering cash on the wrong method of transport is turning into a significant cerebral pain in the vast majority of the metropolitan urban communities in India. At times regardless of whether we pick the correct mode of transport, we may not be on time at our desired destination. The reason for this study is to build up an associated open connected vehicle framework that gives the quickest efficient and the protected method of transport to arrive at our destination in a given circumstance. Utilizing the accessible data about various methods of transport from the official sites, a prototype has been developed which examines various approaches to our destination at the most elevated level of solace at our spending plan. The prototype gets the fundamental information from the client and gives the most ideal approach to arrive at the destination. The model encourages us to diminish the number of mishaps and accidents caused because of wrong decision of transport and lessens the car influx and time taken to travel. It likewise builds the effectiveness of various methods of transport. Further improvements are expected to build up further developed and savvy model which can additionally break down and track the developments of the client in their journey to give greater security during their travel.

P. P. Sai · O. G. Sai
School of Electronics and Communication Engineering, Vellore Institute of Technology, Vellore, India
e-mail: pujithsaip@gmail.com

O. G. Sai
e-mail: obbugouthamsai@gmail.com

S. Chavhan (✉)
Automotive Research Center, Vellore Institute of Technology, Vellore, India
e-mail: suresh.chavhan@vit.ac.in

A. Khanna et al. (eds.), *Data Analytics and Management*, Lecture Notes on Data Engineering and Communications Technologies 54,
https://doi.org/10.1007/978-981-15-8335-3_38

1 Introduction

The number of inhabitants on the planet is around 7 billion (Sumit Malik, 2014), and it is expanding at a quick rate simultaneously; the world economy is additionally developing at an incredible pace. So individuals are accustomed to speeding portability, and when we talk about transportation, individuals need to arrive at their goal at a less expensive rate at most extreme speed with full well-being. Higher the individuals utilizing various methods of transportation, more is blockage and mishaps.

It turns out to be difficult to arrive at a goal in less time. Subsequently, here comes the interest for the best possible sheltered and methodical interest for a transportation framework which can deal with a lot of individuals on wheels securely and ensured it is eco-accommodating.

Overall, different social orders and affiliations have been set up for the advancement of intelligent transportation system; the first was set up in 1991 by US Department of Transportation: alongside, these few models have been proposed in setting for the equivalent, just a couple of executed.

The intelligent transport system is a method to reduce all the difficulties in the field of transportation. The intelligent transportation system has a lot of classification based on applications, such as

(1) Traffic Management Sector;
(2) Traveller Information Sector;
(3) Vehicle Control Sector;
(4) Public Transportation Sector;
(5) Rural Transportation Sector.

1.1 Contributions

The main contributions of this paper are as follows:

1. Developed a connected intelligent transport system prototype;
2. Proposed quickest and efficient protected transport modes to the commuters at various conditions;
3. Discussed how to implement an ITS in India;
4. Implemented the developed prototype and exhaustively test using various performance parameters.

2 Current Situation

At present, just a couple of urban communities are attempting to (Takaaki Hasegawa) actualize the connected transportation and smart traffic framework to their urban communities; however, they are neglecting to execute as a result of no legitimate usage and designers. In the remainder of the urban communities and large towns, there is no smart framework by any means. In those towns where there is no connected transportation framework, individuals are burning through a great deal of cash and time by inclining towards various methods of transports that are not excessively productive to the given circumstance.

2.1 Implementation of ITS

The intelligent transport system has been executed in different nations somewhat, and they have been achievement full somewhat. The nations are Japan, America and Europe.

a. America:

The activities in the USA are Telephonic Data Dissemination, IntelliDriveSM, Next Generation 9-1-1, (Mohammad Bawangaonwala, 2018) Cooperative Intersection Collision Avoidance Systems, Congestion Initiative, Integrated Corridor Management Systems, Clarus Initiative, Emergency Transportation Operations, Mobility Services for All Americans and Electronic Freight Management.

Telephonic information conspires assists with scattering the data about existing voyaging conditions. IntelliDrive permits correspondence between the vehicles. Cutting-edge 9-1-1 takes a shot at the crisis calls, and wellbeing measures along these lines with these plans are been executed in America.

b. Japan:

To actualize the ITS, Japan has created four stages, and each stage will do various employments to keep the ITS working in Japan.

(Mohammad Bawangaonwala1, 2018)

Stage 1: IT is cost assortments and vehicle route framework;
Stage 2: It deals with the crisis and gives the data required to the clients during their movement or about specific transportation.
Stage 3: It deals with the framework and vehicle gear and legitimate frameworks and security of the individuals voyaging.
Stage 4: This deals with the telecommunication between the vehicles and the association. It additionally deals with traffic executives in Japan and observation.

3 Connected Intelligent Transportation System

To accomplish the intelligent transportation system, we need to build up certain phases which do various attempts to accomplish the objective. They are:

Stage 1: This section will have all the data in regards to the current vehicle framework subtleties like transports, trains, taxis, bicycles and each transport medium their timings accessibility and course of the heading this furnishes us with the various choices of how to go to the goal. We can pick the transportation framework where we need to head out to our goal from our boarding point.
(https://mobility.here.com/how-assembleshrewdcity-transport-framework3-stages)
Stage 2: We will give the data in regards to the courses, traffic and climate in that specific course; it causes us to choose a superior course, and this stage additionally controls all the reconnaissance cameras to identify the mishaps or any violations occurring in any of the courses and gives an alarm and encourages them to get the hoodlums, and the emergency vehicle will be called upon if fundamental and sent to the mishap spot on time; this will assist us with taking the individual met with a mishap to the clinic on schedule
Stage 3: This stage furnishes the driver with cautions about the states of the streets, the sum haze or day of the present and shows them if any clumsy territories are available before them, so they can play it safe to dodge mishaps like limited scaffolds and security a specific way to forestall mishaps. It will refresh the client where they are available; their area is constantly refreshed to them utilizing GPS. On the off chance that there are any instances of infringement of traffic rules, at that point, their vehicle number plate is caught and refreshed online for fine in this section.

In the event, we can build up every one of these stages; at that point, we can build up a framework which interfaces all the vehicle frameworks and gives the best and safe approach to arrive at our goal.

3.1 Why India Is not Able to Implement ITS

(1) Financial Crisis:

India needs more cash to purchase the important gear, and programming from various organizations executes ITS for a huge scope; we have to have a ton of sensors and cameras, and we have to employ many individuals to care for it to do all that the administration needs more cash now of time.

(2) Inefficient Road Networks:

The road networks which we have are acceptable in certain areas and urban communities, and in rest of the spots, it is poor; so when we execute the ITS, it needs to propose the client the better method for transport, and if it gives the client roadways and the streets are bad, it will require some investment for the individual to arrive at the goal, and it will prompt disappointment of its regarding time proficiency.

(3) Densely Populated Cities (Over Populated):

Here, the urban areas are overpopulated, and this will expand a great deal of weight on the site which assists with accomplishing the given objectives of its and separated from that the overpopulation will require more transports and prepares which will prompt a ton of traffic and deferral as far as time.

(4) Lack of Resources:

Much subsequent to executing the ITS in certain urban communities of India, the urban areas cannot run the particular ITS since they do not have the best possible system to run it. Like appropriate traffic, the board framework and police network, crisis arranges and so forth.

(5) Breaking of rules by people:

The vast majority of the individuals in India will not adhere to the principles, and it is prompting a ton of bedlam, and it will be a major test for ITS regarding security, timing and trust issues.

3.2 *How to Implement ITS in India*

(1) Controlling traffic and encouraging individuals to take open vehicle framework rather utilizing the private vehicle framework. This will diminish one of the serious issues on roadways and will build the productivity of the ITS as the client gets the chance to arrive at the goal on time securely. (https://www.usenix.org/framework/records/gathering/nsdr12/nsdr12-final2.pdf).
(2) All the streets and railroad systems must be checked altogether and if essential new systems must be set up to meet the developing populace voyaging needs. This will lessen the weight on the current systems and builds productivity.
(3) The government needs to make a legitimate move on the individuals who defies the norms while voyaging whether it is roadways or railroads or aviation routes to build the degree of solace will voyaging. Here and there not having appropriate solace levels in trains prompts individuals utilizing the private vehicle to arrive at their goal regardless of whether it is hazardous and costly.
(4) The government needs to go through some cash to create shipping strategies, and they have to put forth attempts to pull in individuals to utilize the open vehicle framework; at exactly that point, there will be an appropriate significance for actualizing ITS. They additionally need to enlist accomplished individuals to work appropriately to get the greatest yield.

3.3 *Advantages of Connected Transport System*

(1) It reduces pollution as we choose the best way to reach our destination instead of private vehicles (Chavhan, Suresh, and Pallapa Venkataram).
(2) It improves the safety and security of public transportation.
(3) It improves the quality of life.
(4) All the data regarding all transport is available to us, and we are open to a lot of information.

4 Proposed Approach

A prototype has been created for an intelligent transport system. This model is given all the information about the most secure voyaging strategies at given climate conditions at a specific expense. So it will take all the subtleties of the client like where he needs to travel, what is the financial limit for his movement, and what rate the conditions and it will break down the best and the most secure course for the client and recommends him the course. If the client needs to go with a separation with just a restricted sum, then it will check if it is conceivable to travel that separation with the sum he has and lets him know whether unrealistic. If he is utilizing it for clinical reason, at that point, it will propose him the quickest course. The data given to the model is to get the client to the spot at the ideal time with no harm (Fig. 1).

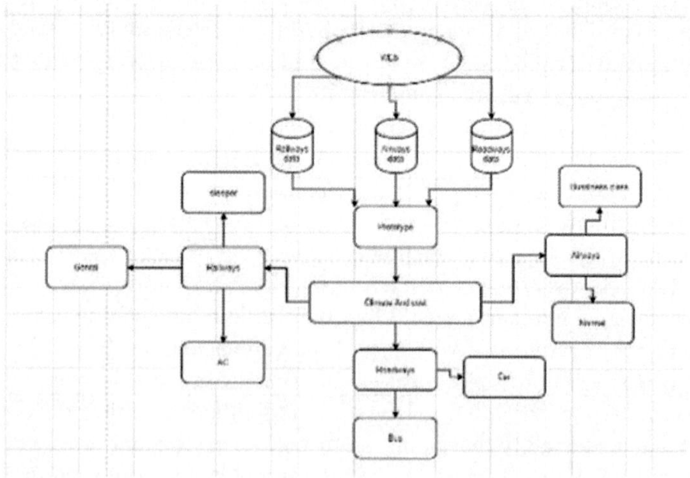

Fig. 1 Architecture of the prototype

4.1 Architecture

The data about the distinctive voyaging courses is taken from the official sites and afterwards given to the model. So when the client utilizes the model, it solicits what is the reason from voyaging on the off chance that it is a clinical reason the model advises him to take the rescue vehicle to get to the emergency clinic.

On the off chance that it is broadly useful, it takes the sum at which he needs to travel it will class the accessible vehicles and afterwards asks the separation and afterwards, it will check what are the accessible courses to venture to every part of the given separation with the given cash. At that point requests, the atmospheric conditions and afterwards picks the most secure approach to travel and proposes the client.

4.2 Flow Diagram

The flow diagram (Fig. 2) shows the step-by-step process that the prototype takes in order to suggest the mode of transport. When the prototype is activated, it takes the details of the customer and his type of travel and his budget to reach the destination, distance and the weather conditions; depending on the input given by the user, it will go through the logic and suggests him the safe, the fastest, and less costly route to travel.

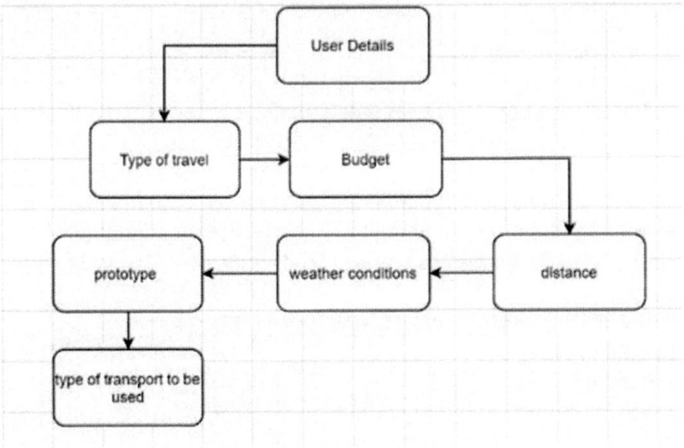

Fig. 2 Flow diagram

5 Results

In this section, we are presenting the implemented prototype model of the proposed system results.

(1) Sample 1 of the prototype:

The user has given his details in Fig. 3 and wants to travel a distance of 1300 km with a budget of Rs 1000.

The prototype suggests travelling a distance 1300 km in a train when it is a foggy day for safe travel.

The graph (Fig. 4) shows the accidents that took place travelling a distance on the road and the accidents took place while travelling on a train during the foggy days

(2) Sample 2 of the prototype:

The user has given his details in Fig. 5 and wants to travel a distance of 302 km with a budget of Rs 1000.

During a sunny day, we can cover a distance of 300 km quickly on-road rather through train at a budget of Rs 1000.

The graph (Fig. 6) shows the time taken to travel the distance on roadways and railways, and the prototype suggests the roadways as it takes less time to reach the destination.

The graph (Fig. 7) suggests that the amount required to get to the destination in a sleeper with full comfort by road is accosting around 1200, whereas an AC sleeper

```
Enter name: Pujith
Enter your age:20
Enter your phone number:9987998721
If it is a medical purpose press 1 or else press 2
2
ENter the distance you want to travel:
1300
What is your budget at what cost you want to travel:
1000
Enter the climate conditions press the number
1:Sunny
2:Rainy
3:Fog
3
Prefer the railways
--------------------------------
Process exited after 35.08 seconds with return value 0
Press any key to continue . . .
```

Fig. 3 Prototype sample

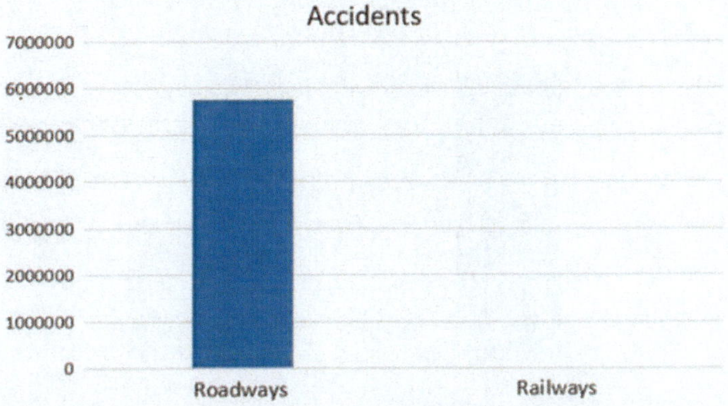

Fig. 4 A graphical representation of accidents in railways and roadways

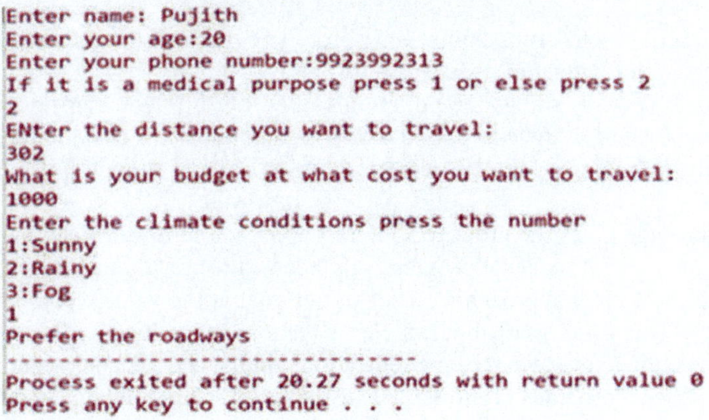

```
Enter name: Pujith
Enter your age:20
Enter your phone number:9923992313
If it is a medical purpose press 1 or else press 2
2
ENter the distance you want to travel:
302
What is your budget at what cost you want to travel:
1000
Enter the climate conditions press the number
1:Sunny
2:Rainy
3:Fog
1
Prefer the roadways
--------------------------------
Process exited after 20.27 seconds with return value 0
Press any key to continue . . .
```

Fig. 5 Prototype sample

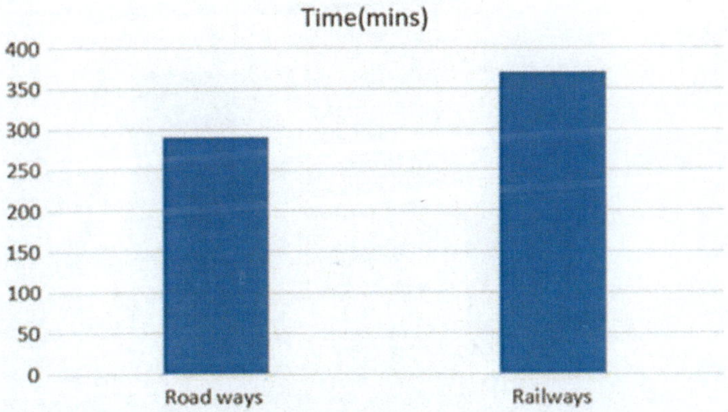

Fig. 6 Graphical data to compare the time taken to travel 300 km in roadways and railways

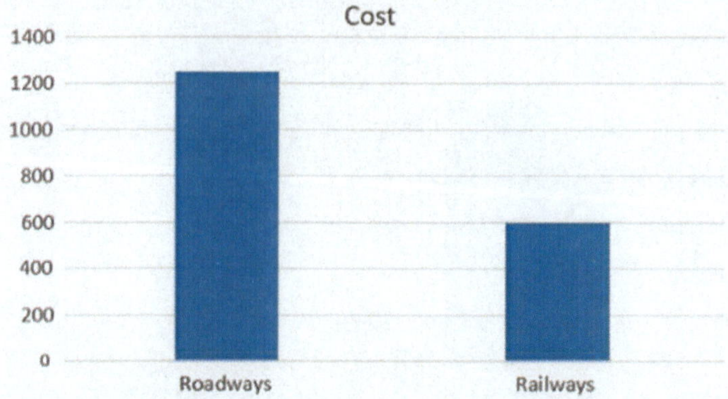

Fig. 7 Graphical data of the cost to travel a distance of 400 km in roadways and railways

in railways will cost us around 600, so the prototype suggests the comfortable mode of transport in our budget as the user wants to travel at 600.

The graph (Fig. 8) shows the pollution caused by different transport methods. The prototype always suggests public transport, and this helps to not only solve the transportation problems but also reduce pollution and saves fuel and reduces the traffic.

The prototype gives the protected and fast approach to travel breaking down all the conditions, and it makes it progressively keen and shrewd thusly; it is useful to choose the most ideal approach to arrive at our goal at the value we need securely and rapidly. This diminishes the traffic on streets as individuals will have another choice to venture to every part of the separation rapidly, and the open vehicle will be energized along these lines; it additionally assists with sparing the fuel and decreases the contamination.

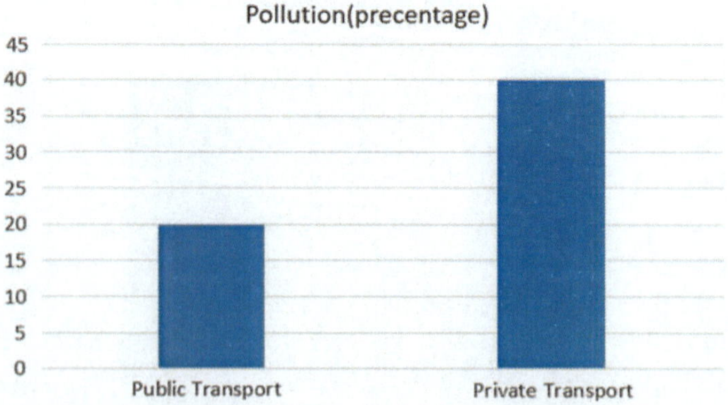

Fig. 8 Graphical data of the amount of pollution in public and private transport

The intelligent transport framework lessens traditional voyaging issues. There are a few positives and negatives in executing the keen vehicle framework. At the point, when ITS is utilized proficiently is utilizes the all the accessible making a trip systems to the greatest degree conceivable and gives the most secure path to the explorers to arrive at the goal lessening the record of mishaps and harm to the administration property. (Chavhan, Suresh, Deepak Gupta, B. N. Chandana, Ashish Khanna, and Joel JPC Rodrigues)

Alongside all the positives, a few negatives are that if there is any disappointment in the framework, it prompts stoppage of the vehicle framework and individuals will be returning to a private vehicle which will prompt a similar situation. So it is imperative to ensure that the framework does not bomb regardless of whether it does we need to ensure that it returns to operational condition as quickly as time permits

So while proposing a technique and creating, we need to deal with all the things and need to create it. May created nations like the USA, Japan, and Europe have grown such things, so we can follow their framework yet need to make a few alterations to make it appropriate for the Indian condition.

6 Conclusion

The created prototype can be useful to decide the most ideal approaches to arrive at the goal in time, in any event, cost and securely. There are a lot of deterrents for executing ITS in India, yet these obstructions are being dealt with by the legislature, and there are a few prospects that very soon, we may see ITS being actualized in India.

References

1. Chavhan S, Gupta D, Chandana BN, Khanna A, Rodrigues JJPC (2019) IoT-based Context-aware intelligent public transport system in a metropolitan area. IEEE Internet Things J
2. Sumit M (2014) Int J Civil Eng Res 5(4):367–372. ISSN 2278-3652
3. Chavhan S, Venkataram P (2019) Emergent intelligence: a novel computational intelligence technique to solve problems. In: Proceedings of the 11th international conference on agents and artificial intelligence, vol 1, pp 93–102
4. Hasegawa T. iatss40_theory-05, https://www.iatss.or.jp/common/pdf/en/publication/commemorative-publication/iatss40_theory_05.pdf
5. Bawangaonwala M, Wadhwa D, Nandeshwar UV, Dhurate S, Ramteke SP, Bante PN, Ansari S (2018) IJCSMC 7(4):12–21
6. Chavhan S, Venkataram P (2019) Transport management for evacuation of victims. IEEE Trans Emerg Top Comput Intell
7. How to build smart city transport system. https://mobility.here.com/how-build-smart-city-transport-system-3-phasesiatss40_theory-05. Accessed on 10/01/2020
8. Chavhan S, Venkataram P (2018) Commuters' traffic pattern and prediction analysis in a metropolitan area. J Veh Routing Algorithms 1(1):33–46

9. Sen R, Raman B (2012) Intelligent transport systems for Indian cities. In: Presented as part of the 6th USENIX/ACM workshop on networked systems for developing regions
10. Balaji PG, Srinivasan D (2011) Type-2 fuzzy logic based urban traffic management. Eng Appl Artif Intell 24:12–22
11. Chavhan S, Venkataram P (2019) Prediction based traffic management in a metropolitan area. J Traffic Transp Eng (English Edition)
12. Kejun L, Yong L, Xiangwu L (2008) Emergency accident rescue system in freeway based on GIS. In: Proceedings of international conference on intelligent computation technology and automation
13. Chavhan S, Venkataram P (2019) Emergent intelligence technique-based transport depot resource management in a metropolitan area. J Veh Routing Algorithms 2(1–4):23–40

Fibroid Segmentation in Ultrasound Uterus Images Using Wavelet Filter and Active Contour Model

K. T. Dilna and D. Jude Hemanth

Abstract The peculiar growth existing in the uterus wall is uterus fibroids. These fibroids can lead to infertility concerns and is a major issue among women nowadays. CT scan, ultrasound scan, and MRI scan are some techniques to detect uterus fibroids. Among these techniques, ultrasound images are precise tool to detect the uterus disorders. Less detectable boundaries, size, and positions are perplexing task in detection of fibroid from ultrasound images. On ultra sound image, it appears as round with areas with a discrete border. A foremost issue is the presence of speckle in ultrasound fibroid image. Speckle noise be subject to the composition of image tissue and parameters of image. It reduces the effectiveness of many image processing steps and decreases human perception of fine details form ultrasound images. In this paper, wavelet filter is applied to shrink this speckle form fibroid scanned image. Active contour method is used to segment fibroid from wavelet-filtered image. It is perceived that fibroids from ultrasound images are segmented accurately, and the accuracy of detection is above 95%. This method segmented the fibroid and extracts some shape-based features also which helps doctors to decide the method of treatment.

Keywords Fibroid · Uterus · Ultrasonic imaging · Active contour · Wavelet

1 Introduction

The female reproductive system of human being is uterus and with 7.5 cm (3 in.) long, 5 cm (2 in.) wide, and 2.5 cm (1 in.) deep uterus is considered as normal uterus. Inside, it is hollow with thick muscular walls. There are many abnormalities present in uterus, and one among them is fibroid which is seen as tumors which are infectious

K. T. Dilna
Department of ECE, College of Engineering and Technology, Payyanur, India
e-mail: dilnageethanand@gmail.com

D. Jude Hemanth (✉)
Department of ECE, Karunya University, Coimbatore, India
e-mail: judehemanth@karunya.edu

A. Khanna et al. (eds.), *Data Analytics and Management*, Lecture Notes on Data Engineering and Communications Technologies 54,
https://doi.org/10.1007/978-981-15-8335-3_39

in nature. Generally speaking, 20–50% of women of reproductive age likely to have fibroids, although not all of them are often diagnosed. These are also known as uterine myomas, leiomyomas, or fibromas. There are three type of fibroids based on position in uterus and are subserosal, submucosal, and intramural. Fibroid which appears on the outside of uterus is known as subserosal and results in compression on the surrounding tissues. Fibroid appears in the wall of the uterus that is known as intramural which causes pressure on the bladder or uterus and can cause infertility or miscarriage. Third type of fibroid submucosal appears in the middle of the womb.

Ultrasound imaging is a most popular technique in detection of uterus fibroid. Ultrasound image system uses high frequency of sound wave to get an image of uterus fibroid. Ultrasound waves travels through soft tissues but it produce echoes on solid objects. Each ultrasound pulse covers a spatial volume which defines the smallest of detectable structures, known as the resolution cells. Fibroid appears as well-defined solid mass, and it appears as hypoechoic. These hypoechoic objects become visible as dark images, since it has low echogenicity, and bright image become visible since it has high echogenicity and is called hyperechoic. The noise occurs in medical image would be during acquisition and transmission usually. Speckle is a multiplicative noise that appears on B-Scan, which reduces quality of image. The occurrence of signal-dependent noise such as speckle degrades utility of ultrasound images. Speckle destroys the target detectability in scanned images and reduces resolution and contrast which help human to differentiate normal and pathological tissue. It also cut down the speed and accuracy of segmentation of ultrasound image. Many speckle reduction filters are proposed but while combining different technique some speckle diagnostic information should be preserved.

Finding of abnormalities by utilizing image processing method are having many steps as

- Image acquisition for processing
- Removal of noise by preprocessing
- ROI detection from preprocessed image
- Segment the desired area of image for analysis
- Detect the feature for classification.

Section 2 discusses with algorithm done in detection of uterus fibroids. Section 3 is about result and discussion. Section 4 gives conclusion.

2 Related Works

N. Sriraam et al. described an automated detection of uterine fibroid by using wavelet features and a neural network classifier [1], and the detection accuracy was 95.1%. Feed forward backpropagation neural network (BPNN) classifier is used for segmentation. Statistical features and frequency domain features are not described in this paper. Yixuan Yuan et al. proposed a novel weighted locality-constrained linear coding (LLC) method for uterus image analysis [2], and the detection accuracy was

96.33%. Leonardo Rundo et al. developed a semi-automatic approach which depends on region growing segmentation technique [3]. Bo Nia et al. used dynamic statistical shape model (SSM)-based segmentation method [4] but it takes more running time. Efficiency and stability are the focal areas of this method. Alireza Fallahi et al. used FCM on MRI image [5] to segment uterine fibroid. Author adopted two step process. First, MRI image is segmented using FCM, and morphological operations are carried out, and fuzzy algorithm is used for refining the output. Fibroid properties are not examined like infarct regions and calcified regions. T. Ratha Jeyalakshmi et al. provided mathematical morphology-based methods for automated segmentation [6]. Here, this method detects fibroid in the inner wall of uterus. Shivakumar K. et al. have used GVF snake method for the segmentation of fibroids in uterus images [7]. Similar ultrasound uterus image analysis methods are available in [8–10]. This work implement a most flexible segmentation technique using active contour to segment uterus fibroid from ultrasound scanned images efficiently and to detect the features of fibroid to finalize treatment method. This method is described by Chan-Vese model, using the classical methods such as thresholding and utilizing gradients [11–14].

3 Materials and Methods

3.1 Image Acquisition

Ideal imaging technique to detect fibroids is ultrasound scanning. Ultrasound adverts to sound waves which cannot be sensed by human ear. Proportional to the acoustic impedance of every soft tissue, sound waves are transmitted. This acoustic impedance is associated to sound transmission velocity and density of tissue. Sound waves are echoed when there is a change in acoustic impedance. This occurs when two dissimilar density tissues placed next to each other. Therefore, ultrasound imaging is a noble tool for soft tissues. Reflected sound waves are absorbed by transducer located directly on the skin, and it converts the echoes to electrical signal. These signals are then demodulated and compressed to outfit for display devices. Fibroid in ultrasound image is commonly appears as distinct, solid hypo echoic masses in ultrasound images. In this work, 256 * 256 gray level images are used with intensity ranges between 0 and 255. Figure 1 shows some sample dataset with uterus fibroids.

3.2 Preprocessing

Medical ultrasound images suffer from speckle noise which is multiplicative in nature. Declination of image resolution and image contrast is the consequence of speckle noise. Boundary edges of image are usually imperfect, misplaced, or transparent at some places. This makes it difficult for the ultrasound images to get

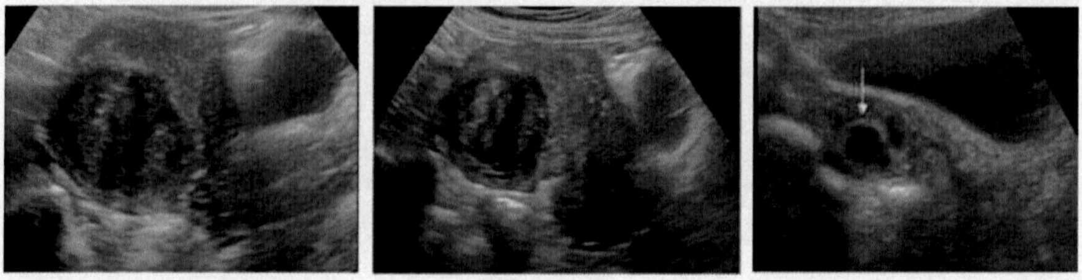

Fig. 1 Sample dataset

segmented. Speckle noise reduction of is very essential to increase ultrasound image quality. Wavelet filtering [15, 16] is adopted in this paper. Wavelet filtering is carried out in three steps

1. Signal decomposition using DWT
2. Image filtering based on thresholding
3. Reconstruction of filtered image using inverse DWT.

DWT decomposes the image into details and approximation parts, and some fragments contains typically contains trivial noise. This can be detached by using thresholding without effecting the image features. Wavelet type and threshold type are the two-filter design parameters used to implement DWT. Wavelet coefficients of a function are, generally, large in uneven regions and small in even regions. If the function is degraded by a noise, then this noise will overlook the wavelet coefficients, and only few large coefficients will be associated to the strong singularity of the fundamental function. One elimination method of most of the noise by conserving large coefficients is thresholding the noisy wavelet. Soft thresholding and hard thresholding are two standard thresholding technique. The threshold level T is in proportion to the noise standard deviation S for both technique. In hard thresholding, all wavelet coefficients of amplitude smaller than T are shifting to zero. Soft thresholding furthermore decreases the amplitude of the other coefficients by the quantity T. In this method, thresholding parameter is used for filtering.

3.3 Segmentation Using Active Contour

Chan-Vese model for active contours is implemented to segment fibroid region, and the algorithm is described as follows. This model identifies boundary areas which are not interpret by gradient.

The basic idea behind CV model is partitioning the image into two regions as object to be detected and background. Let Image $I(x, y)$ is molded by two regions with constant intensity. Let the intensity values be c_1 and c_2. Likewise, assume that C_0 is the boundary of the region which the object to be detected. Then inside (C_0), the intensity of $I(x, y)$ is approximately c_1, whereas outside (C_0) the intensity of $I(x,$

y) is approximately c_2. Then consider the fitting term:

$$F(c_1, c_2, C) = \mu \cdot \text{Length}(C) + v \cdot \text{Area}(\text{inside}(C)) + \lambda 1 \int_{\text{inside}(c)\text{inside}(c)} |(I(x, y) - c_1|^2 \text{d}x\text{d}y + \lambda 2 \int_{\text{outside}(c)} |(I(x, y) - c_2|^2 \text{d}x\text{d}y \quad (1)$$

where C is a curve, and the constants c_1, c_2 are the averages of $I(x, y)$ inside and outside of C, respectively. Length (C) is the length of the closed border, and area (inside (C) is the area of the interior of C. The preferred settings in the paper by Chan-Vese are $v = 0$, $\lambda_1 = \lambda_2 = 1$.

Generally, the length term length (C) could be revised as (Length $(C))^p$ for $p \geq 1$, but usually $p = 1$ or $P = N/N - 1$.

$$\text{Area}(\text{inside}(C)) \leq c \cdot (\text{Length}(c))^{N/N-1} \quad (2)$$

$F(C)$ cannot take the minimum value if the closed border C does not lie on the edge of the two homogeneous regions and take minimum value at the border. The fitting energy is minimized if $C = C_0$.

3.3.1 Level Set Formulation

An alternative in examining for the solution in terms of C, reconsider the problem in the level set. Let Ω be a bounded open set of R^2, with its boundary $\partial\ \Omega$ and $\omega€\ \Omega$ are open, and $c = \partial\ \omega$.

Level sets are developing contours or surfaces that can enlarge, contract, and even split or merge. The fundamental idea behind the level set method is to implant the moving interfaces as the zero level set of a higher dimensional.

Function $\Phi(x, t)$, defined as

$$\Phi(x, t) = \pm d \quad (3)$$

$\pm d$ is the signed distance to the interface from point x. That is, x is outside the interface when $\Phi\ (x, t) > 0$, inside the interface when $\Phi\ (x, t) < 0$, and on the interface when $\Phi\ (x, t) = 0$. The evolution of $\Phi\ (x, t)$ can be represented by a partial differential equation:

$$\Phi' + \Delta\Phi'(t) = 0 \quad (4)$$

We can rewrite $F\ (c_1, c_2, C)$ using Heaviside function H

$$\text{Length}(C) = \int_{\Omega} |\nabla H(\Phi(x, t))| \mathrm{d}x\mathrm{d}y \tag{5}$$

where

$$H(z) = \begin{cases} 1 & \text{for} \quad z \geq 0 \\ 0 & \text{for} \quad z < 0 \end{cases}$$

$$\text{Area}(C) = \int H(x, y) \mathrm{d}x\mathrm{d}y \tag{6}$$

This model cannot be implemented for multi region and can accomplish only for the image which has two types of homogeneous regions. One level set can only denote one background and one object. The disadvantages of the CV model are that multiple objects cannot be detected using one level set. To overcome this problem, it is added by Euler–Lagrange equation. The energy functional F in terms of Φ is F (c_1, c_2, Φ) is

$$\begin{aligned} F(c_1, c_2, \Phi) = \mu \int \delta_0 \Phi(x, y) |\nabla \Phi(x, y)| \mathrm{d}x\mathrm{d}y + \int H\Phi(x, y) \mathrm{d}x\mathrm{d}y \\ + \lambda 1 \int \left| I(x, y) - C_1^2 \right| H\Phi(x, y) \mathrm{d}x\mathrm{d}y \\ + \lambda 2 \int \left| I(x, y) - C_2^2 \right| H\Phi(x, y) \mathrm{d}x\mathrm{d}y \end{aligned} \tag{7}$$

Keeping c_1 and c_2 fixed, and minimizing F with respect to Φ, deduce the associated Euler–Lagrange equation for Φ.

4 Results and Discussion

The proposed algorithm is applied on ultrasound uterus images to detect fibroid. Figure 2 shows a sampled image and the output of preprocessing step using wavelet filter. Figure 3 shows the segmented fibroid region from ultrasound scanned input image. In this paper, Chan-Vese algorithm was implemented on wavelet-filtered image. The features used in this paper are based on shapes, namely area, diameter, accuracy, perimeter, eccentricity, major axis, and minor axis from ultrasound fibroid images. The extracted features help in identifying the size of the fibroids. For a typical case—the smaller ones can be cured by medicines whereas larger sizes would require a surgery.

Table 1 displays the result of feature extraction. It can be noted that the sizes of the fibroids are different for different patients. The treatment planning is based on the size of the fibroids.

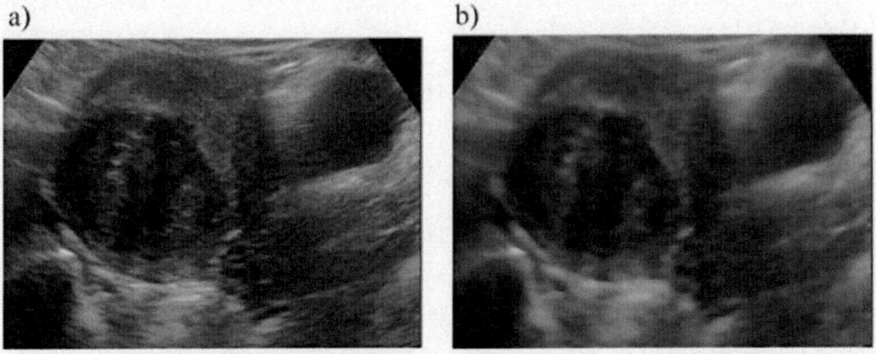

Fig. 2 **a** Ultrasound fibroid image. **b** Filtered image

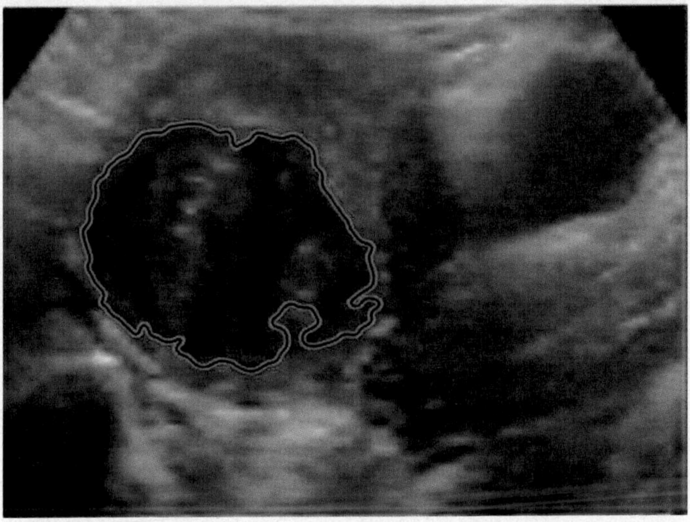

Fig. 3 Segmented fibroid region

Table 1 Shape-based features of segmented fibroid region

Shape-based features	Image 1	Image 2
Area	9754	6027
Perimeter	467.35	357.81
Diameter	111.44	87.6003
Centroid	[98.031, 106.04]	[100.26, 98.537]
Minor axis length	101.2	83.22
Major axis length	126.01	96.44

Table 2 Comparative analysis

	TP	TN	FP	FN	Sensitivity	Specificity	Accuracy (%)
Image 1	5943	61,486	240	3	0.9995	0.9961	99.641
Image 2	9747	57,746	36	2	0.9998	0.9994	99.944
Image 3	7047	60,439	77	2	0.9997	0.9987	99.883
Image 4	6800	60,614	412	2	0.9997	0.9932	99.390
Image 5	8744	58,756	672	146	0.8562	0.8973	89.176

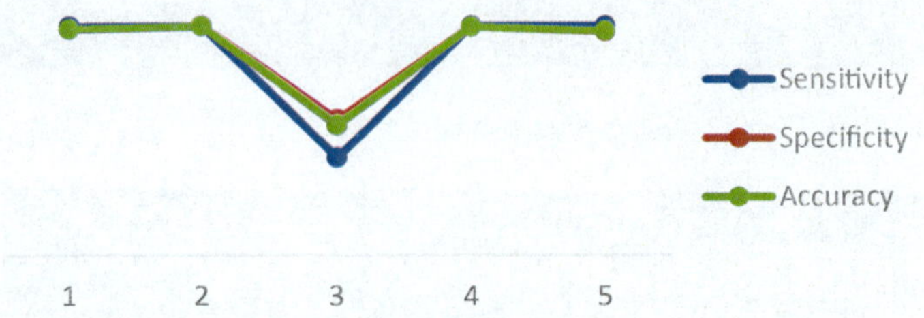

Fig. 3 Graphical representation of comparative analysis

Algorithm performance is compared using evaluation criteria such as TP, TN, FP, FN, sensitivity, specificity, and accuracy Table 2, and Fig. 4 shows the comparative analysis of five images. From the table, it shows that performance of this algorithm is good since it has accuracy, specificity, and sensitivity is above 95%.

5 Conclusion

In this work, active contour algorithm with wavelet filter is implemented for segmentation of fibroid area from ultrasound image. Several features are extracted from these images and used for the segmentation process. Segmented area and features can give treatment plan and state of the fibroid monitoring. The fibroids are extracted with high efficiency which is evident from the experimental results. As a future scope, more different features can be extracted, and multiple fibroids can be extracted, and machine learning techniques can be used for the segmentation process. The target area is segmented, and the evaluation of this tool from the doctor and this helps the doctors in diagnosis, the treatment plan making and state of the fibroid monitoring.

References

1. Sriraam N, Nithyashri D, Vinodashri L, Manoj Niranjan P (2010) Detection of uterine fibroids using wavelet packet features with BPNN classifier. In: IEEE EMBS conference on biomedical engineering & sciences (2010)
2. Yuan Y, Hoogi A, Beaulieu CF, Meng MQH, Rubin DL (2015) Weighted locality—constrained linear Coding for lesson classification in CT images. In: Proceedings of 37th annual international conference of the IEEE engineering in medicine and biology society
3. Rundo L, Militello C, Vitabile S, Casarino C, Russo G, Midiri M, Gilardi MC (2016) Combining split-and-merge and multi-seed region growing algorithms for uterine fibroid segmentation in MRgFUS treatments. Med Biol Eng Comput 54(7):1071–84 (2016)
4. Nia B, Hea F, Yuana ZY (2015) Segmentation of uterine fibroid ultrasound images using a dynamic statistical shape model in HIFU therapy. Comput Med Imaging Graph 46:302–314 (2015)
5. Fallahi A, Pooyan M, Khotanlou H, Hashemi H, Firouznia K, Oghabian MA (2010) Uterine Fibroid segmentation on multiplan MRI using FCM, MPFCM and morphological operations. IEEE International Conference on Computer Engineering and Technology
6. Ratha Jeyalakshmi T, Ramar Kadarkarai K (2010) Segmentation and feature extraction of fluid-filled uterine fibroid-a knowledge-based approach. Int J Sci Technol 4:405–416
7. Harlapur SK, Hegadi RS (2015) Segmentation and analysis of fibroid from ultrasound images. International Journal of Computer Application
8. Yao J, Chen D, Lu W, Premkumar A (2006) Uterine fibroid segmentation and volume measurement on MRI. Proc SPIE 6143 (2006)
9. Alush A, Greenspan H, Goldberger J (2010) Automated and interactive lesion detection and segmentation in uterine cervix images. IEEE Trans Med Imaging 29(2)
10. Padghamod MJ, Gawande JP (2014) Classification of ultrasonic uterine images. Adv Res Electr Electron Eng 1(3):89–92
11. Chan TF, Vese LA, Roberts EF (2001) Active contours without edges. IEEE Trans Image Proces 10(2)
12. Mumford D, Shah J (1989) Optimal approximation by piecewise smooth functions and associated variational problems. Comm Pure Appl Math 42:577–685 (1989)
13. Rousseau O, Bourgault Y (2009) Heart segmentation with an iterative Chan-Vese algorithm. University of Ottawa, Ontario
14. Pan Y, Birdwell JD, Djouadi SM (2006) Efficient implementation of the Chan-Vese Models without solving PDE's. In: International workshop on multimedia signal processing, pp 350–353
15. Nobi MN, Yousur MA (2011) A new method to remove noise in MRI and ultrasound images In.Journal of Science Research
16. Kim YS, Ra JB (2005) Improvement of ultrasound image based on wavelet transform: speckle reduction and edge enhancement. Proc SPIE Med Imaging 5747:1085–1092

YOLOv3 Remote Sensing SAR Ship Image Detection

Yash Chaudhary, Manan Mehta, Nikki Goel, Parul Bhardwaj, Deepak Gupta, and Ashish Khanna

Abstract Deep learning has proved to be useful for ship detection in SAR satellite images. SAR is a satellite that can be used to capture images from Earth's surface even in unfavorable weather conditions. Although many object detection models have been applied to this problem previously, we wanted to use a model that was fast and accurate. In this paper, we use You Only Look Once version-3 (YOLOv3) and compare its results with You Only Look Once version-2 (YOLOv2). The motivation was to improve upon the results of YOLOv2. The results show that YOLOv3 achieved 90.25 average precision (AP) compared to 90.05 AP of YOLOv2. Furthermore, YOLOv3 gave an inference time of 22 ms against 25 ms of YOLOv2. The dataset used is A SAR Dataset of Ship Detection for Deep Learning under Complex Backgrounds; the dataset consists of 43,819 images of 256 × 256 pixels. The dataset contains Gaofen-3 and Sentinel-1 satellite images.

Keywords YOLOv3 · Remote sensing · Synthetic-aperture radar (SAR) images · Object detection · Ship detection

---

Y. Chaudhary (✉) · M. Mehta · N. Goel · P. Bhardwaj · D. Gupta · A. Khanna
Computer Science and Technology, Maharaja Agrasen Institute of Technology, Delhi, India
e-mail: yash612@gmail.com

M. Mehta
e-mail: manan.mehta2@gmail.com

N. Goel
e-mail: nikkigoel20@gmail.com

P. Bhardwaj
e-mail: parulb20@gmail.com

D. Gupta
e-mail: deepakgupta@mait.ac.in

A. Khanna
e-mail: ashishkhanna@mait.ac.in

A. Khanna et al. (eds.), *Data Analytics and Management*, Lecture Notes on Data Engineering and Communications Technologies 54,
https://doi.org/10.1007/978-981-15-8335-3_40

1 Introduction

Ship detection is vital for ocean monitoring for problems like oil spill detection, illegal fishing, maritime piracy and marine traffic management. Presently, automatic identification system (AIS) is used as an automatic tracking system for ships. It requires to have transponders and navigation systems on ships, land stations and satellites. The AIS uses VHF radio frequencies to broadcast ship locations and detect ships. There are many laws and regulations that enforce the installation and use of VHF transponders on ships. It is very effective for monitoring ships. However, ships can turn off their AIS transponders to be undetectable. This can have massive consequences, like trafficking of people or drugs, illegal fishing and piracy. This is where satellite imagery can help. It can be used to detect invisible ships that have turned off their transponder or have no transponder installed. Nowadays, for ship detection, mainly two types of satellite images are used: optical satellite images and radar. Radar captures images by sending continuous microwaves; thus, images can be captured regardless of the weather or lighting conditions. Synthetic-aperture radar (SAR) is a form of radar. We have used SAR images to train our model. SAR is best suited for ship detection, and the continuous nature of observation regardless of the day-night and weather conditions makes SAR satellites perfect for monitoring change like marine traffic. Another major advantage of SAR satellite is that they can cover large areas in short periods of time. Optical imagery might take days to capture an image spanning over hundreds of kilometers, but a SAR satellite can capture such large distances in minutes.

Once we have satellite images, we can use deep learning-based object detection methods to detect ships. There are mainly two techniques for object detection: sliding window algorithm, in which a window slides through an image, and then, each window is fed into a ConvNet for classification. This method is computationally very expensive and not accurate in predicting bounding boxes. The second technique is based on region proposals, in which algorithms like selective search are used to predict regions in image that might contain objects. These regions are then fed into a ConvNet for object classification. Examples of region proposal networks are R-CNN, Fast R-CNN and Faster R-CNN [1] proposed by Girshick et al. In region proposal networks, there are multiple steps within training stages which are time consuming. Thus, the training process and inference become slow. Faster R-CNN, a top detection method, cannot see or understand larger context, hence mistaking background patches as objects. The YOLO architecture is better than Faster R-CNN in this context, and it also has much faster inference time than Faster R-CNN. Unlike the above two approaches, YOLO is a single-pass detector, i.e., the input image requires a single pass through the convolutional neural network to predict bounding boxes and classes. It was proposed by Redmon et al. [2–4], in order to get the high-speed performance and high detection accuracy of real-time operation. On the NVIDIA 2060 GPU, the YOLO model runs in real time at 45 fps (22 ms).

In this paper, we use You Only Look Once version-3 (YOLOv3) on task of ship detection in SAR satellite images. We show that YOLOv3 is better than many existing object detection models. It is fast and accurate.

The main contributions of this paper are:

- The main objective of this paper was to improve accuracy and speed in object detection, which we achieved by using YOLOv3 technique.
- YOLOv3 was compared with other techniques available (Faster R-CNN, YOLOv2) for object detection on various grounds, and the conclusion was drawn that YOLOv3 is indeed better, giving us the accuracy of 90.25 and inference time of 22 ms.

The rest of the paper is organized as follows: In Sect. 2, we introduce some information and related work on YOLOv3 remote sensing SAR ship image detection; in Sect. 3, we describe our methodology for training YOLOv3 model for ship detection and compare it with other various already existing techniques. This section also includes information about the algorithm used as well as the dataset; in Sect. 4, we described our evaluation methods which were used to experiment with to get results for ship detection. In Sect. 5, we present our research and discuss the outcomes of using YOLOv3 for detection over other methods. In Sect. 6, we conclude the paper.

2 Related Work

Many efforts have been made to detect ships in SAR satellite images. The previous works have utilized different object detection models on SAR satellite images. [5] used Faster R-CNN to detect ships. [6] proposed an object detection method based on SSD. [7] used RetinaNet on Gaofen-3 satellite images.

Our work is inspired by the recent work [8] in which authors use YOLOv2 to detect ships in SAR satellite images. YOLOv2 is not the most accurate model for ship detection, but it gives a good combination of accuracy and speed. In this paper, we aim to outperform the previous results by using the latest version of YOLO method, i.e., YOLOv3.

Further, to train our model, we utilize the dataset provided by [9], and it is a collection of 43,819 images. Such a large collection of images has not been used in the previous works. We hope to benefit from using this dataset for training our YOLOv3 model.

The authors of [9] applied various models on this dataset like RetinaNet, SSD, Faster R-CNN. We wish to apply YOLOv3 on this dataset.

3 Methodology

(A) YOLOv3

YOLOv3 [4] is the third version of the popular object detection algorithm "You Only Look Once"; it uses convolutional neural networks for object detection. Being a one-shot learning algorithm, it is one of the fastest object detection algorithms out there. It just needs one pass through the convolutional network to predict all the bounding boxes. Compared to many new object detection algorithms like RetinaNet, SNIPER or Cascade mask R-CNN, YOLOv3 lags behind in accuracy. Nevertheless, it provides decent accuracy with fast results making it an appealing choice.

A couple of years ago, YOLOv2 [3] was the algorithm of choice and is one of the most accurate and quickest algorithms at that time. However, with the development of new algorithms like RetinaNet and Mask R-CNN, YOLOv2 could not match the accuracy of these algorithms. To improve upon the accuracy of YOLOv2, YOLOv3 was introduced. In YOLOv3, the speed was traded off for boosts inaccuracy. The changes were done by improving the underlying CNN architecture called darknet.

According to Kathuria [10], "YOLOv2 used a deep architecture called darknet-19, a 19-layer network augmented with 11 more layers for object detection. With a 30-layer architecture, YOLOv2 struggled with small object detections. Furthermore, YOLOv2's architecture was still lacking some of the most important elements that are now stapled in most of the state-of-the-art algorithms. It had no skip connections, no residual blocks, and no upsampling. YOLOv3 incorporates all of these."

Figure 1 represents the Darknet-53 architecture. "YOLOv3 uses Darknet-53, which originally has a 53-layer network trained on ImageNet; on top of that, it has 53 more layers for detecting objects, giving us a 106-layer architecture" [10].

1. *Detection at three scales*:

With the Darknet-53, YOLOv3 now made detections at three scales. Detections at different layers allow us to detect objects of different sizes. This helps to alleviate a common problem of YOLOv2, i.e., detection of small objects. Our dataset contained images of various sizes.

Some images contained a single large ship covering almost the entire image, while other images had very small ships in the size of dots. Considering the wide variety of object sizes, YOLOv3 seemed a suitable fit.

YOLOv3 makes detections at three places within the network. "The three feature maps downsample the input image by a factor of 32, 16 and 8, respectively. So, for an image of size 416 × 416, detections are made on feature maps of size 13 × 13, 26 × 26 and 52 × 52" [10].

For example, see Fig. 2, the first feature map (stride 32) is responsible for detecting large objects, whereas the second feature map (stride 16) detects the medium objects and the third feature map (stride 8) detects small objects.

	Type	Filters	Size	Output
	Convolutional	32	3 × 3	256 × 256
	Convolutional	64	3 × 3 / 2	128 × 128
1×	Convolutional	32	1 × 1	
	Convolutional	64	3 × 3	
	Residual			128 × 128
	Convolutional	128	3 × 3 / 2	64 × 64
2×	Convolutional	64	1 × 1	
	Convolutional	128	3 × 3	
	Residual			64 × 64
	Convolutional	256	3 × 3 / 2	32 × 32
8×	Convolutional	128	1 × 1	
	Convolutional	256	3 × 3	
	Residual			32 × 32
	Convolutional	512	3 × 3 / 2	16 × 16
8×	Convolutional	256	1 × 1	
	Convolutional	512	3 × 3	
	Residual			16 × 16
	Convolutional	1024	3 × 3 / 2	8 × 8
4×	Convolutional	512	1 × 1	
	Convolutional	1024	3 × 3	
	Residual			8 × 8
	Avgpool		Global	
	Connected		1000	
	Softmax			

Fig. 1 Darknet-53 architecture as illustrated in [4]

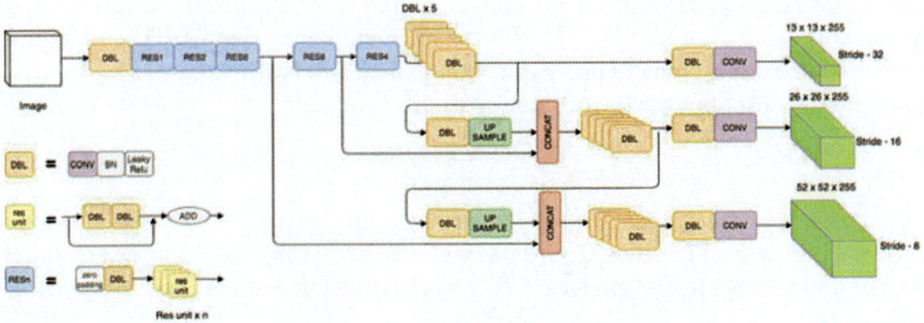

Fig. 2 Detection in different layers

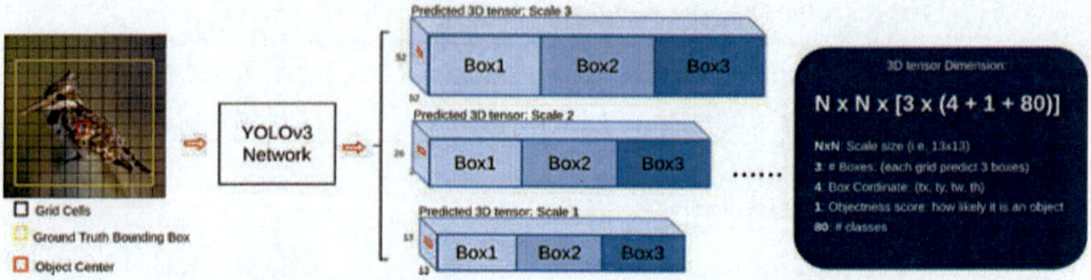

Fig. 3 YOLOv3 feature map [16]

As we can see in Fig. 2, the last layer 52 × 52 has the largest number of grid cells making it capable of predicting small objects that can be missed by the other two layers.

2. *Anchor boxes*:

YOLOv3 uses nine anchor boxes, three for each scale. On the other hand, YOLOv2 used five anchor boxes. A larger number of anchor boxes leads to greater bounding box predictions, an input image of 416 × 416, outputs ([13 × 13] + [26 × 26] + [52 × 52]) × 3 = 10,647 bounding boxes, which is 10 times greater than YOLOv2 (845 boxes). This helps in achieving higher AP.

(B) *The algorithm*

First, the YOLOv3 model is given input images of shape (416, 416, 3) to predict feature maps corresponding to three scales. We expect each grid cell in the feature map to output a bounding box if an object's center falls on that grid cell. In Fig. 3, the red grid cell is responsible for predicting the bounding box of the bird.

Three scales help in detecting objects of various sizes. Let us take the example of a 13 × 13 scale; for this, the input image is divided into 13 × 13 grid cells. After passing through the network, each of the 13 × 13 grid cells produces $B \times (4 + 1 + C)$ depth in the output feature map. "B" represents the number of bounding boxes, "C" represents the number of classes and 4 corresponds to coordinates (x, y, w, h). For our dataset $C = 1$, we have one object (ship). Figure 3 illustrates a feature map with 80 classes. In our network, the feature map will have a depth of 3 × (4 + 1 + 1). Hence, the dimensions of the output feature map will be 13 × 13 x [3 x (4 + 1 + 1)]. Figure 4 describes the overall flowchart of the model.

(C) ***DATASET***

The dataset we used is the "SAR Dataset of Ship Detection for Deep Learning under Complex Backgrounds. This dataset was created using 102 Chinese Gaofen-3 images and 108 Sentinel-1 images" [9]. Figure 5 illustrates some sample images of the dataset.

The Gaofen-3 satellite was launched in mid-2016. It has a resolution of 1 m. "Compared with optical imaging satellites, Gaofen-3 performs better at disaster monitoring

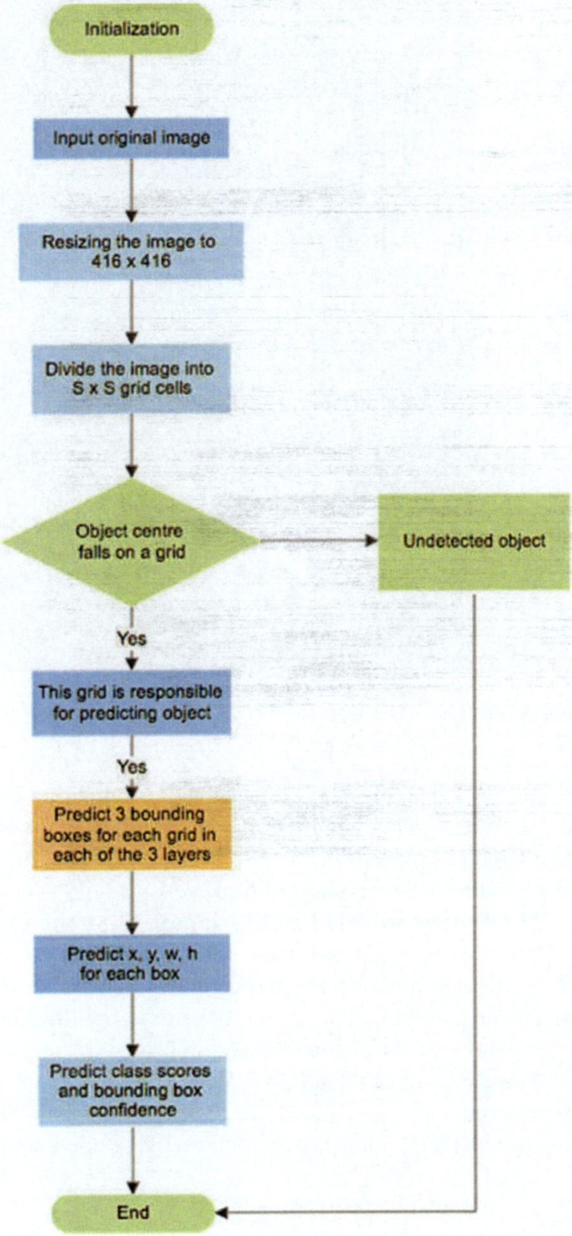

Fig. 4 Flowchart of model

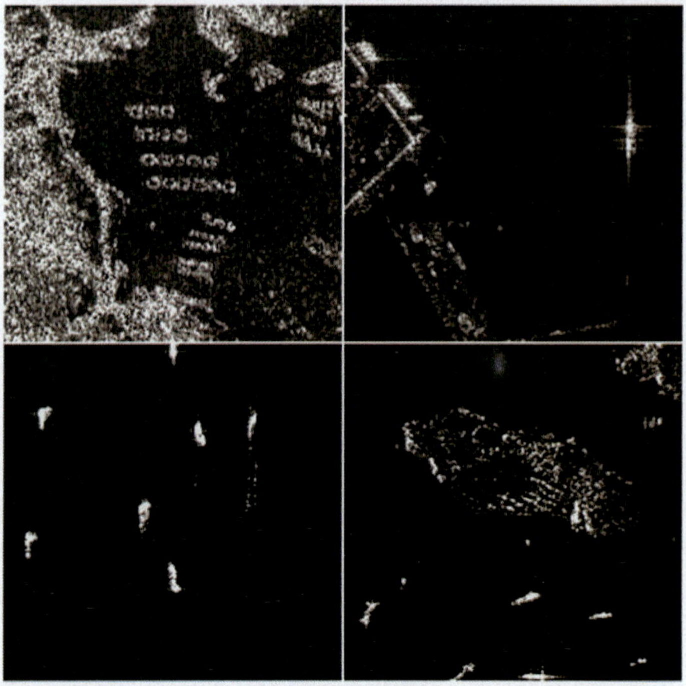

Fig. 5 Sample images from the dataset

as the SAR imaging satellite is capable of imaging in severe weather conditions as it uses microwave transmission. The Sentinel −1 is composed of a constellation of two satellites Sentinel −1A and Sentinel −B" [11]. Both the satellites share the same orbit and have a resolution of 5 m.

The dataset consists of 59,535 ships in 43,819 images. Each image has a dimension of 256 × 256. This is a very complex dataset; ships are located in a variety of distinct environments. The backgrounds consist of adjacent docks, harbors, islands and isolated oceans. This diversity helps in building a robust object detector, capable of generalizing well on new images.

The dataset has a PASCAL VOC format, [12]. PASCAL VOC provides object annotations in XML format. We required a TXT (.txt) file for training purposes. We converted annotations into text format using open-source software from Github [13].

4 Implementation Details

For evaluation, methods used for the YOLOv3 model are well-known techniques like precision, recall, mAP and IOU.

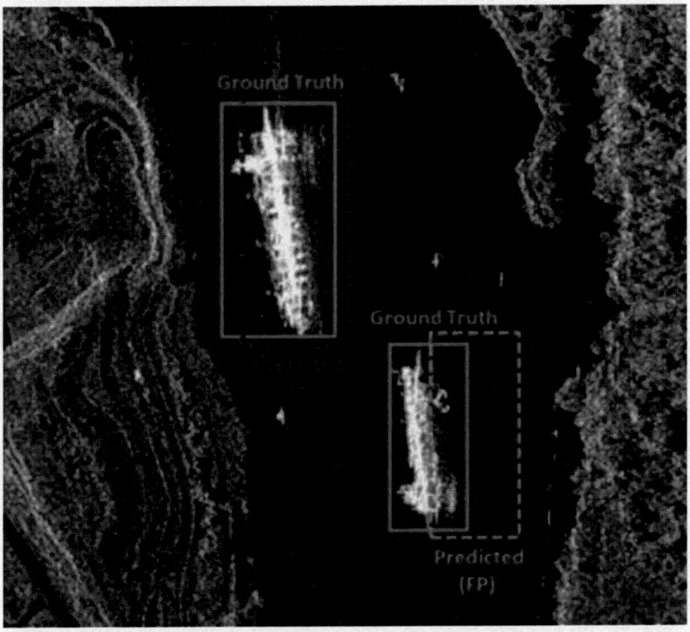

Fig. 6 Example of TP and FP [8]

The mAP is the comparison metric used for object detection. "IoU measures the overlap between two boundaries. We use that to measure how much our predicted boundary overlaps with the ground truth (the real object boundary)" [14]. It measures the related correlation between ground truth and prediction. To calculate IoU, we require the following predicted box and bounding box which is the ground truth. The ratio between the area of overlap and the union area of these two boxes is simply known as IoU (Fig. 6).

IoU is then used to find:

True positive: If IoU exceeds and equal to threshold (0.5), object is classified as "true positive" (TP).

True negative: It is the leftover part of the picture, where no object is predicted by the model.

False positive: If IoU remains under the threshold (0.5), then object is classified as "false positive" (FP).

False negative: When there is a ground truth in the image and the model has failed to predict it, it is classified as "false negative" (FN).

The percentage of accurate predictions is known as precision.

$$\text{Precision} = \frac{\text{TP}}{\text{TP} + \text{FP}} \qquad (1\ [1,\ \text{Eq. } (2)])$$

Recall related to what percentage of actual positives was identified correctly.

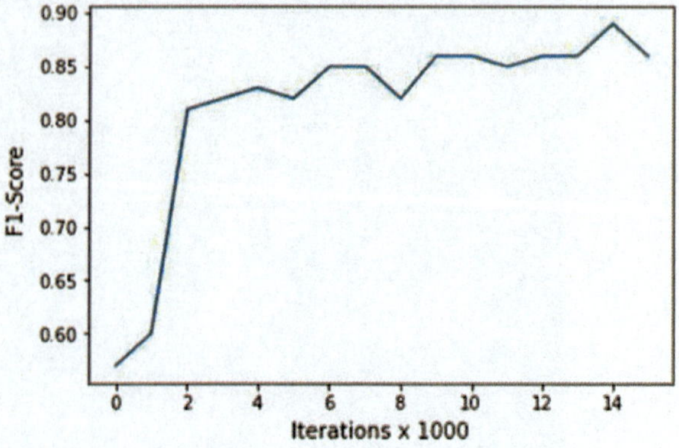

Fig. 7 F1-score graph

$$\text{Recall} = \frac{\text{TP}}{\text{TP} + \text{FN}} \quad (2\ [1, \text{Eq. }(3)])$$

Below figure depicts, F1-score which is the harmonic mean between precision and recall (Fig. 7).

Finally, average precision (AP) leads to mAP that calculated precision and recall. The area under the precision–recall graph is known as AP. AP is determined above a certain threshold (in our case threshold = 0.5, i.e., AP50) and for each class separately. The mean overall APs for each class is called mAP [14].

5 Results and Discussion

In this paper, we wanted to prove the efficacy of YOLOv3 for detecting ships in satellite images. This is a complex task; the model needs to be robust enough to detect ships in a real-world environment. The dataset was complicated; it contained ships of various sizes in a diverse set of environments and backgrounds.

YOLOv3 was chosen as it is good at detecting small objects, and it is fast and accurate. It is better than YOLOv2 which has been very successful on this task previously [8]. Compared to R-CNN, Fast R-CNN and Faster R-CNN, it is faster and achieves higher mAP. It is much better than SSD, and compared to RetinaNet, it four times faster while achieving comparable accuracy. Our results indicate that YOLOv3 is indeed better, as you will see in this result section below.

We performed our research on a PC with the specification of NVIDIA RTX 2060 6 GB, Intel i7-9th gen using CUDA cores. The training was done on the original darknet framework [15] where pretrained Darknet-53 was used. Here, we resized the images to 608 × 608 and 416 × 416 where 608 × 608 dominated in mAP by +4% but was turning out to be computationally heavy. So, we focused on 416 × 416 image size. Prior to training, we generated anchor boxes corresponding to our dataset using the k-means algorithm, and we used data augmentation techniques like

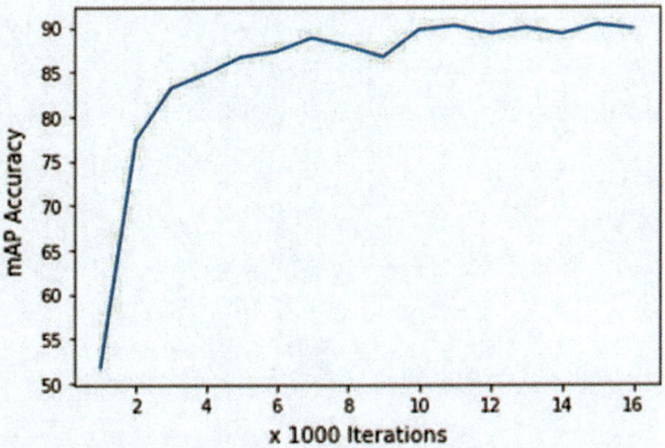

Fig. 8 mAP graph

Table 1 Comparison of YOLOv2 and YOLOv3 on different image sizes [8]

Networks	Image size	AP	Avg. time (ms)
YOLOv2	416 x 416	89.56	15
	544 x 544	90.05	25
YOLOv3	416 x 416	90.25	22

Table 2 Ship detection AP and speed comparison [5, 8]

Networks	AP	Avg. time (ms)
Faster *R*-CNN	70.63	206
YOLOv2	90.05	25
YOLOv3	90.25	22

the flipping of images and image mix-up to get higher mAP. Finally, we trained for 16,000 iterations. We got a max mAP of 90.25 at 15,000 iterations as shown in Fig. 8.

The YOLOv3 outperformed YOLOv2 inaccuracy on 416 × 416 and 544 × 544 image sizes. It was faster than YOLOv2 trained on 544 × 544 image size (Tables 1 and 2).

These results can be further improved by training with higher image resolutions like 544 × 544 or 608 × 608. However, training is computationally heavy, and inference time also increases. Figure 9 illustrates predictions of our model.

6 Conclusion

The main motive of our experiment is to prove the capability of YOLOv3. We evaluate and compare our model YOLOv3 with other existing models on various grounds. We come to this conclusion that YOLOv3 is a very good choice for the task of detecting ships in SAR satellite images. It is better than YOLOv2 and region proposal

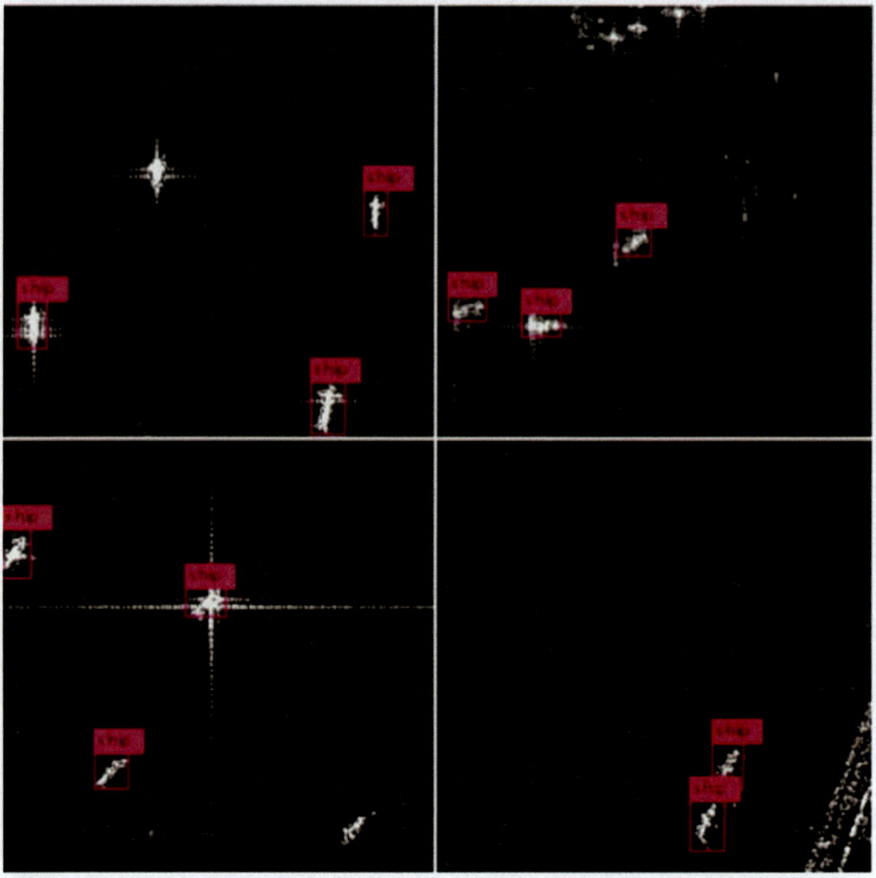

Fig. 9 Inference

networks. And although we did not train RetinaNet and SSD on our dataset, it is known that YOLOv3 is faster and accurate than SSD, and it has comparable accuracy to RetinaNet while being approximately four times faster.

YOLOv3 delivers one of the best ratios of accuracy to speed. We believe the reason behind this is that YOLOv3 uses three layers to detect an object (one output layer and two hidden layers), and this makes it capable of detecting small objects like ships.

According to this study, YOLOv3 outperforms YOLOv2 and provides one of the best trade-offs between accuracy and inference speed for ship detection in SAR images.

References

1. Ren S, He K, Girshick R, Sun J (2015) Faster r-cnn: towards real-time object detection with region proposal networks. arXiv preprint arXiv:1506.01497
2. Redmon J, Divvala S, Girshick R, Farhad A (2016) You only look once: unified, real-time object detection. In: Proceedings of the IEEE conference on computer vision and pattern recognition, pp 779–788
3. Redmon J, Farhadi A (2017) YOLO9000: better, faster, stronger. In 2017 IEEE conference on computer vision and pattern recognition (CVPR), Honolulu, HI, pp 6517–6525
4. Redmon J, Farhadi A (2018) YOLOv3: an incremental improvement
5. Li J, Qu C, Shao J (2017) Ship detection in SAR images based on an improved faster R-CNN. In: 2017 SAR in big data era: models, methods and applications (BIGSARDATA)
6. Wang J, Lu C, Jiang W (2018) Simultaneous ship detection and orientation estimation in SAR images based on attention module and angle regression. Sensors 18:2851. https://doi.org/10.3390/s18092851
7. Wang Y, Wang C, Zhang H, Dong Y, Wei S (2019) Automatic ship detection based on RetinaNet using multi-resolution Gaofen-3 imagery. Remote Sens 11:531. https://doi.org/10.3390/rs11050531
8. Chang Y, Anagaw A, Chang L, Wang Y, Hsiao C, Lee W (2019) Ship detection based on YOLOv2 for SAR imagery. Remote Sensing 11:786. https://doi.org/10.3390/rs11070786
9. Wang Y, Wang C, Zhang H, Dong Y, Wei S (2019) A SAR dataset of ship detection for deep learning under complex backgrounds. Remote Sens 11:765. https://doi.org/10.3390/rs11070765
10. Kathuria A (2018) What'S new. In YOLO V3? [online] Medium. Available at https://towardsdatascience.com/yolo-v3-object-detection-53fb7d3bfe6b. Accessed 14 Mar 2020
11. Ma M, Chen J, Liu W, Yang W (2018) Ship classification and detection based on CNN using GF-3 SAR images. Remote Sens 10:2043. https://doi.org/10.3390/rs10122043
12. Everingham M, Van Gool L, Williams CK, Winn J, Zisserman A (2010) The pascal visual object classes (voc) challenge. Int J Comput Vision 88(2):303–338
13. yifanli8086, yifanli8086/convert2Bo. GitHub, 27 Feb 2018. [Online]. Available https://github.com/yifanli8086/convert2Bo. Accessed 29 Feb 2020
14. Hui J (2019) mAP (mean average precision) for object detection. Medium, 03 Apr 2019. [Online]. Available https://medium.com/@jonathan_hui/map-mean-average-precision-for-object-detection-45c121a31173. Accessed 12 Mar 2020
15. Redmon J, Darknet: open source neural networks in C. [Online]. Available http://pjreddie.com/darknet/. Accessed 13 Mar 2020
16. n.d. [online] Available at https://www.cyberailab.com/post/a-closer-look-at-yolov3. Accessed 13 Mar 2020

Parametric Optimization of Improved Sensing Scheme in Multi-antenna Cognitive Radio Network over Erroneous Channel

P. Bachan, Samit Kumar Ghosh, and Sachin Ravikant Trankatwar

Abstract In this paper, we discussed the parametric optimization of improved sensing schemes in multi-antenna cognitive radio (CR) networks over an erroneous channel. The CR utilizes an improved energy detector (IED) by making the binary decision statistics of the absence or presence of a primary user. From primary users' statistics, the improved energy detector measures power with respect to the threshold so as to make correct spectrum decisions. This decision is further transferred to the fusion center (FC) by the erroneous channel, which decides the complete decision of the existence of PU. For the detection of the spectrum hole and to obtain an optimized number of CRs, we minimize the total error rate (TER) in cooperative spectrum sensing. An optimized value of sensing threshold and arbitrary positive power (p) of each CR is also accessed by considering the TER. The numerical-based simulation results show the validation of correct sensing performance.

1 Introduction

Due to the rapid growth in the traffic and density of communication networks and demand for competent services in 5G networks, the uses of the radio spectrum have become congested. Cognitive radio (CR) has been seen as a proficient technique to meet these demands, which deal with the spectrum scarcity [1, 2]. It also provides high data rate access. CR, i.e., an intelligent software-defined radio, where the secondary user (SU), which is unlicensed, can opportunistically use certain portions

P. Bachan (✉)
Department of Electronics and Communication Engineering, GLA University, Mathura, India
e-mail: p.bachan@gmail.com

S. K. Ghosh · S. R. Trankatwar
Department of Electrical and Electronics Engineering, Birla Institute of Technology and Science, Pilani, Hyderabad Campus, Hyderabad, India
e-mail: samitnitrkl@gmail.com

S. R. Trankatwar
e-mail: strankatwar@gmail.com

A. Khanna et al. (eds.), *Data Analytics and Management*, Lecture Notes on Data Engineering and Communications Technologies 54,
https://doi.org/10.1007/978-981-15-8335-3_41

of the licensed spectrum to a primary user (PU) [3, 4]. Henceforth, optional clients (SUs) called cognitive users (CUs) are permitted to get to the authorized recurrence groups without meddling essential (primary) clients' work [5]. In the cognitive radio environment, to decide the vacant band sensing of the spectrum is a useful task that helps in choosing the empty groups of preoccupied bands [6, 7]. Therefore, spectrum sensing is a substantial feature of CR technology because it is possible to detect the existence of PUs accurately and quickly when the PU signal is unidentified [8, 9]. To moderate these negative impacts and to improve the discovery exactness by consolidating the detecting data from numerous SUs, cooperative spectrum sensing (CSS) has been used in a large number of literature to assist the cognitive radio environment [10–12]. However, in CSS, more SUs partaking range detecting will bring all the more detecting, announcing, and transmission vitality utilization, which can be hindering to the CRNs as the SUs are, for the most part, controlled by a battery with restricted vitality [9, 13]. Still, CSS is preferred in many works of literature as it has more excellent protection to fading and shadowing the wireless environment. Detection threshold assisted optimization for CSS with energy finding system in the CR network is used in [14] to streamline the system parameters to accomplish the better execution with the shrinking multifaceted nature of the network. Xiao et al. [11] were first to introduce the collaboration of perceptive users for spectrum sensing applications with accurate reporting channels. Sharma and Sharma [15] examined and proposed to transmit the diversity-based CSS method to mitigate the effect of a deficient reporting channel on false alarm probability. Gahane and Sharma [16] considered M-antenna CR structure and evaluated the performance of spectrum sensing in a Rayleigh channel environment considering the impact of secondary user mobility. Banavathu and Khan [17] have proposed a fusion-based optimization algorithm so as to optimize CR density for the Bayesian test. Also, they presented a strategy for getting the most modest number of cognitive clients in CSS while accomplishing an objective error in the fusion center. In [18, 19], authors have presented a multi-antenna-based energy detector for CSS assisted cognitive radio network in AWGN and Rayleigh environment. In this paper, we investigate parameter optimization of the CSS scheme with an improved energy detector (IED) using censored CR users to minimize the sum of the cooperative probability of false alarm and missed detection. By considering the decision statistics obtained by IEDs, each CU makes a local observation regarding the absence or presence of the PU. We also derive a mathematical expression for the optimal values of CUs, power, and normalized threshold at each CU to minimize the total error rate. The paper also deals with the graphical results between optimal number of CUs versus normalized threshold by considering a fixed number of SNR. The paper is structured as follows. The analytical framework is described in Sect. 2. In Sect. 3, the false alarm and miss detection probability in the CUs is explained. Parametric optimization for improved sensing performance of CSS schemes, mathematical expressions, and graphical simulation results are described in Sect. 4. Finally, we explain the conclusion obtained from the results in Sect. 5.

2 Analytical Framework

We consider a system framework consisting of a PU, a fusion center (FC), and N number of CUs as shown in Fig. 1 where every PU independently detects the CUs activities. The FC received the decisions from CUs which are cooperating with each other between all CUs. We consider that each CU executes its sensing performance independently after that their decisions are transferred to the reference receiver which can combine all decisions which are available and summarize the absence or presence of PU and based on a binary decision is taken between the subsequent two hypotheses ($\mathbf{H_1}$: signal present, $\mathbf{H_0}$: signal absent) is as follows

$$\mathbf{H_0}\colon y_j(t) = n_j(t) \tag{1}$$

$$\mathbf{H_1}\colon y_j(t) = x(t) + n_j(t) \tag{2}$$

where $x(t)$ is the PU signal, $y_j(t)$ and $n_j(t)$ represent the received signal and additive white Gaussian noise (AWGN), respectively, at jth CU. $j = 1, 2, 3, \ldots, N$ which represents by the antenna index. $x(t) \sim \mathcal{N}(0, \sigma_x^2)$ in which the notation σ_x^2 is used for average transmitted power (PU) of Gaussian signal with zero mean, whereas $n_j(t) \sim \mathcal{N}(0, \sigma_n^2)$ is AWGN with zero mean and variance σ_n^2, respectively. $\sigma_x^2 + \sigma_n^2$ denotes the variance of the received signal at each secondary user under $\mathbf{H_1}$ (PU of the channel is present). It is assumed that each CU contains an improved energy detector [10]. The statistics for deciding the presence of the PU which is utilized by the jth CU is followed.

$$W = |y_j|^p, \quad p > 0 \tag{3}$$

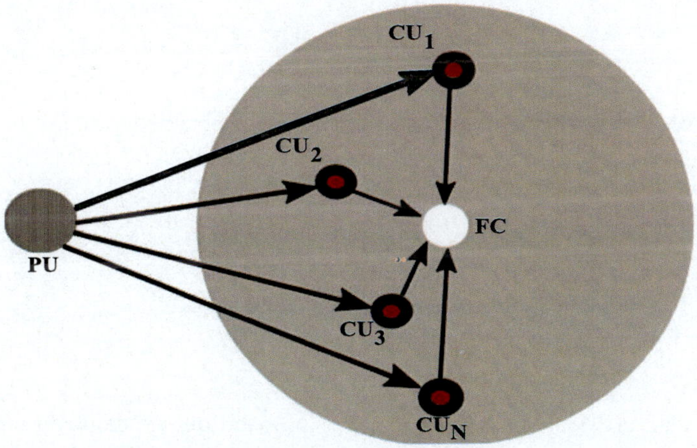

Fig. 1 System framework for CSS model

In (3), W corresponding to the conventional energy detector (CED) if $p = 2$ [20]. Multiple CUs occur in a cognitive radio network in a cooperative sensing system, such that each CU makes independent decisions about the existence or absence of PU. Individual CU makes the binary decision either 1 or 0 indicating the presence or absence of PU which is transferred to the FC over an inadequate reporting channel. Finally, FC takes the final decision (present/absent) based on the status of PU by applying the 'OR' rule [21].

3 False Alarm and Missed Detection Probability in the CUs

For an improved energy detector, the cumulative distribution function (CDF) can be defined as

$$\mathbf{F}_W(y) = \mathbf{P_r}\left(|y_j|^p \le y\right) = \mathbf{P_r}\left(|y_j|^2 \le y^{\frac{2}{p}}\right) \tag{4}$$

where the probability represents by $\mathbf{P_r}(\cdot)$. Differentiating (4) with respect to y gives the probability density function (PDF) of the decision statistics W and is as follows

$$\mathbf{f}_W(y) = \frac{2}{p} y^{\frac{2-p}{p}} \mathbf{f}_{|y_i|^2}\left(y^{\frac{2}{p}}\right) \tag{5}$$

where $\mathbf{f}_{y_i}(\cdot)$ is PDF of received signal under the binary hypothesis $\mathbf{H_0}$ or $\mathbf{H_1}$ and represents by $\mathbf{f}_{|y_i|^2|\mathbf{H_0}}(\cdot)$ and $\mathbf{f}_{|y_i|^2|\mathbf{H_1}}(\cdot)$, respectively, is as follows

$$\mathbf{f}_{|y_i|^2|\mathbf{H_0}}(y) = \frac{1}{\sigma_n^2} e^{-\left(\frac{y}{\sigma_n^2}\right)}, \quad y \ge 0 \tag{6}$$

$$\mathbf{f}_{|y_i|^2|\mathbf{H_1}}(y) = \frac{1}{\sigma_x^2 + \sigma_n^2} e^{-\left(\frac{y}{\sigma_x^2 + \sigma_n^2}\right)}, \quad y \ge 0 \tag{7}$$

The PDF of W under binary hypothesis $\mathbf{H_0}$ and $\mathbf{H_1}$ is obtained from (5), (6) and (5), (7) as follows

$$\mathbf{f}_{W|\mathbf{H_0}}(y) = \frac{2y^{\frac{2-p}{p}}}{p\sigma_n^2} e^{-\left(\frac{y^{\frac{2}{p}}}{\sigma_n^2}\right)}, \quad y \ge 0 \tag{8}$$

$$\mathbf{f}_{W|\mathbf{H_1}}(y) = \frac{2y^{\frac{2-p}{p}}}{p\left(\sigma_x^2 + \sigma_n^2\right)} e^{-\left(\frac{y^{\frac{2}{p}}}{\sigma_x^2 + \sigma_n^2}\right)}, \quad y \ge 0 \tag{9}$$

The false alarm ($\mathbf{P_f}$) and miss detection ($\mathbf{P_m}$) probabilities are defined as follows

$$\mathbf{P_f} = \int_{\lambda}^{\infty} \mathbf{f}_{W|\mathbf{H_0}}(y)\, dy = e^{-\lambda_n^{\frac{2}{p}}} \tag{10}$$

$$\mathbf{P_m} = \int_0^{\lambda} \mathbf{f}_{W|\mathbf{H_1}}(y)\,\mathrm{d}y = 1 - e^{-\left(\frac{\lambda_n^{\frac{2}{P}}}{1+\gamma}\right)} \tag{11}$$

In (10) and (11), λ_n and γ denote the normalized threshold and signal-to-noise ratio (SNR), respectively, i.e., $\lambda_n = \frac{\lambda}{\sigma_n^P}$ and $\gamma = \frac{\sigma_x^2}{\sigma_n^2}$.

4 Parametric Optimization for Improved Sensing Performance

We consider a binary symmetric channel (BSC) with an error probability of 'q' in between each CU and the FC. In this CSS, probability of false alarm ($\mathbf{Q_F}$) and miss detection probability ($\mathbf{Q_M}$) are expressed as

$$\mathbf{Q_F} = 1 - \left[(1 - \mathbf{P_f})(1 - q) + q\mathbf{P_f}\right]^{\mathrm{N}} \tag{12}$$

$$\mathbf{Q_M} = \left[\mathbf{P_m}(1 - q) + q(1 - \mathbf{P_m})\right]^{\mathrm{N}} \tag{13}$$

In CSS scheme, the total error rate (TER) is the summation of the two probabilities ($\mathbf{Q_F}$, $\mathbf{Q_M}$) and is given by

$$\mathbf{E}(\mathrm{N}) \triangleq \mathbf{Q_F} + \mathbf{Q_M} \tag{14}$$

4.1 Cooperative User Density Optimization

Large delays in determining the existence of the spectrum hole are caused in a cognitive network of very large numbers of cooperative CUs. The cooperative cognitive network performance does not depend upon the more numbers of cooperative CUs. Therefore, to decide the presence of the PU, an optimal number of CU is needed, and to obtain the optimal number (N_{opt}) of CUs, the result $\mathbf{E}(N + 1) - \mathbf{E}(N) \triangleq 0$, for a particular value of the signal-to-noise ratio (SNR), positive power (p), and sensing threshold (λ) between the PU-CU link. The optimal value of N is calculated by the following expression

$$N_{\mathrm{opt}} = \left\lceil \frac{\ln \mathbf{f}_2\,(q, \mathbf{P_f}, \mathbf{P_m})}{\ln \mathbf{f}_1\,(q, \mathbf{P_f}, \mathbf{P_m})} \right\rceil \tag{15}$$

where $\mathbf{f}_1\,(q, \mathbf{P_f}, \mathbf{P_m}) = \frac{\mathbf{P_m}(1-q)+q(1-\mathbf{P_m})}{(1-\mathbf{P_f})(1-q)+q\mathbf{P_f}}$, $\mathbf{f}_2\,(q, \mathbf{P_f}, \mathbf{P_m}) = \frac{2q\mathbf{P_f}-q-\mathbf{P_f}}{\mathbf{P_m}-2q\mathbf{P_m}+q-1}$, and the ceiling function denotes $\lceil\cdot\rceil$. The following remarks can be concluded from (15):

Case 1: In case of perfect reporting channel, i.e., for error free $q = 0$, (15) reduces to

$$N_{\text{opt}} = \left\lceil \frac{\ln \mathbf{f}_2\left(q, \mathbf{P_f}, \mathbf{P_m}\right)}{\ln \mathbf{f}_1\left(q, \mathbf{P_f}, \mathbf{P_m}\right)} \right\rceil \simeq \left\lceil \frac{\ln \frac{\mathbf{P_f}}{1-\mathbf{P_m}}}{\ln \frac{\mathbf{P_m}}{1-\mathbf{P_f}}} \right\rceil \tag{16}$$

In (16), if $\mathbf{P_f} = \mathbf{P_m}$, then $N_{\text{opt}} = 1$, i.e., one single CR is required to optimize the TER.

Case 2: In case of deterministic error $q = 1$, (15) reduces to (16). In this case also if $\mathbf{P_f} = \mathbf{P_m}$, then $N_{\text{opt}} = 1$, i.e., one single CR is required to optimize the TER.

Case 3: In case of maximum uncertainty of reporting channel $q = 0.5$, (15) reduces to

$$N_{\text{opt}} = \lim_{q \to 0.5} \frac{\ln \mathbf{f}_2\left(q, \mathbf{P_f}, \mathbf{P_m}\right)}{\ln \mathbf{f}_1\left(q, \mathbf{P_f}, \mathbf{P_m}\right)} \simeq 1 - \left(\frac{\mathbf{P_m}}{1 - \mathbf{P_f}}\right) \tag{17}$$

From (17), it is cleared that if $\mathbf{P_f} \gg \mathbf{P_m}$, then $N_{\text{opt}} \to 1$, and if $\mathbf{P_f} \ll \mathbf{P_m}$, then (17) can be approximated as

$$N_{\text{opt}} = \lim_{q \to 0.5} \frac{\ln \mathbf{f}_2\left(q, \mathbf{P_f}, \mathbf{P_m}\right)}{\ln \mathbf{f}_1\left(q, \mathbf{P_f}, \mathbf{P_m}\right)} \simeq \frac{1 - \mathbf{P_m}}{1 - \mathbf{P_f}} \tag{18}$$

which also implicated that $N_{\text{opt}} \to 1$ if $\mathbf{P_f} \ll \mathbf{P_m}$. Figure 2a shows that erroneous channel reporting error probability versus optimal CUs for SNR= 5 dB. From the above figure, it can be observed that the requirement of CUs is less for decision making in the FC with higher values of q. It is also evident from Fig. 2a that N_{opt} decreases as the value of p increases for the fixed values of SNR. From Fig. 2b, it can be observed that N_{opt} increases with normalized threshold λ_n for a fixed value of SNR, and it is also noted that for a particular value of nomalized threshold and SNR if p is increased, N_{opt} reduces.

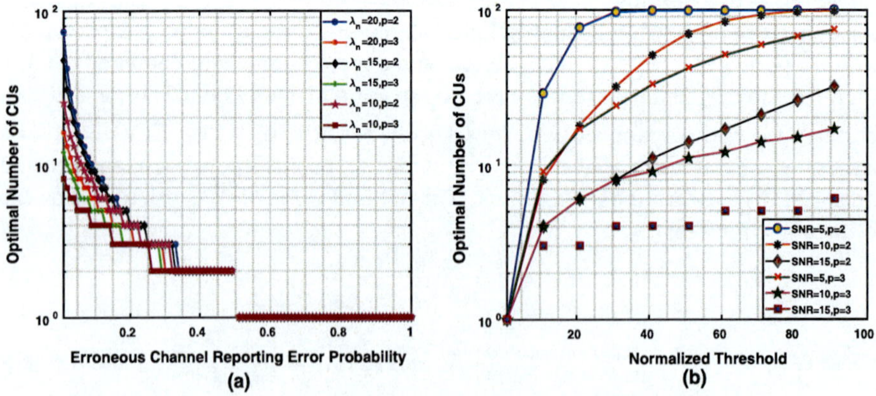

Fig. 2 **a** Erroneous channel reporting error probability versus optimal CUs, **b** normalized threshold versus optimal number of CUs

4.2 Optimization of Normalized Threshold and IED

To get an optimum value of λ_n, we need to differentiate (14) with respect to λ_n with keeping the values of N, q, and γ fixed, and the following expression is obtained.

$$\begin{aligned}\frac{\partial E(N)}{\partial \lambda_n} &= -N(2q-1)\left[(1-\mathbf{P_f})(1-q)+q\mathbf{P_f}\right]^{N-1}\frac{\partial \mathbf{P_f}}{\partial \lambda_n} \\ &+N(1-2q)\left[\mathbf{P_m}(1-q)+q(1-\mathbf{P_m})\right]^{N-1}\frac{\partial \mathbf{P_f}}{\partial \lambda_n}\end{aligned} \tag{19}$$

where $\frac{\partial \mathbf{P_f}}{\partial \lambda_n} = -\frac{2\lambda_n^{\frac{2-p}{p}}\mathbf{P_f}}{p}$, $\frac{\partial \mathbf{P_m}}{\partial \lambda_n} = \frac{2\lambda_n^{\frac{2-p}{p}}(1-\mathbf{P_m})}{(1+\gamma)p}$. Similarly, by keeping N, q, and γ fixed, we obtained the optimized value of p by differentiating (14) with respect to p.

$$\begin{aligned}\frac{\partial E(N)}{\partial p} &= -N(2q-1)\left[(1-\mathbf{P_f})(1-q)+q\mathbf{P_f}\right]^{N-1}\frac{\partial \mathbf{P_f}}{\partial p} \\ &+N(1-2q)\left[\mathbf{P_m}(1-q)+q(1-\mathbf{P_m})\right]^{N-1}\frac{\partial \mathbf{P_f}}{\partial p}\end{aligned} \tag{20}$$

where $\frac{\partial \mathbf{P_f}}{\partial p} = -\frac{2\ln(\lambda_n)\lambda_n^{\frac{2}{p}}\mathbf{P_f}}{p^2}$, $\frac{\partial \mathbf{P_m}}{\partial p} = \frac{2\ln(\lambda_n)\lambda_n^{\frac{2}{p}}(1-\mathbf{P_m})}{(1+\gamma)p^2}$. Figure 3a shows that power p of the received signal versus TER for $N = 1, 2, 3, \ldots, 8$, $\lambda_n = 13$, $q = 0.01$, and $\gamma = 10$ dB and Fig. 3b represents the normalized threshold versus TER for $N = 1, 2, 3, \ldots, 8$, $q = 0.01$, $p = 2.8$, and $\gamma = 10$ dB. From Fig. 3a and b, it can be found that TER can be minimized with respect to λ_n, and p for given values of N, q, and γ. Putting the values of $\frac{\partial \mathbf{P_f}}{\partial \lambda_n}$, $\frac{\partial \mathbf{P_m}}{\partial \lambda_n}$ in (19) and then make $\frac{\partial E(N)}{\partial \lambda_n} = 0$, we obtain $\lambda_{n,\text{opt}}$, i.e., $\lambda_{n,\text{opt}} = \left(\frac{(1+\gamma)}{\gamma}\ln(1+\gamma)\right)^{\frac{2}{p}}$, and similarly, by replacing the values of $\frac{\partial \mathbf{P_f}}{\partial p}$,

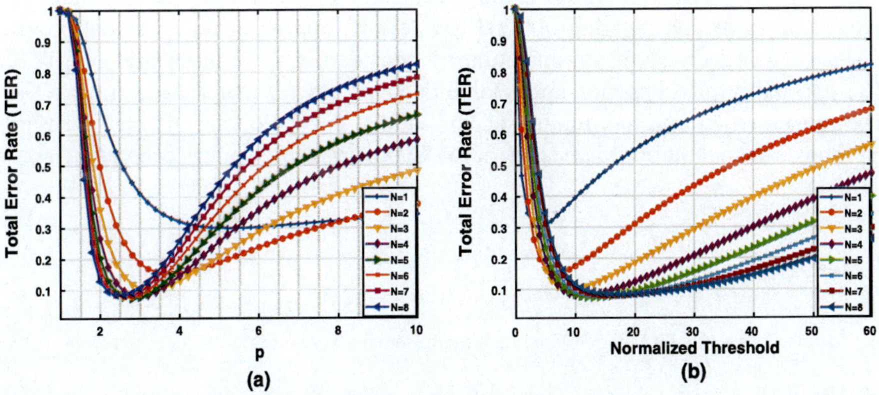

Fig. 3 **a** Power versus TER, **b** normalized threshold versus TER

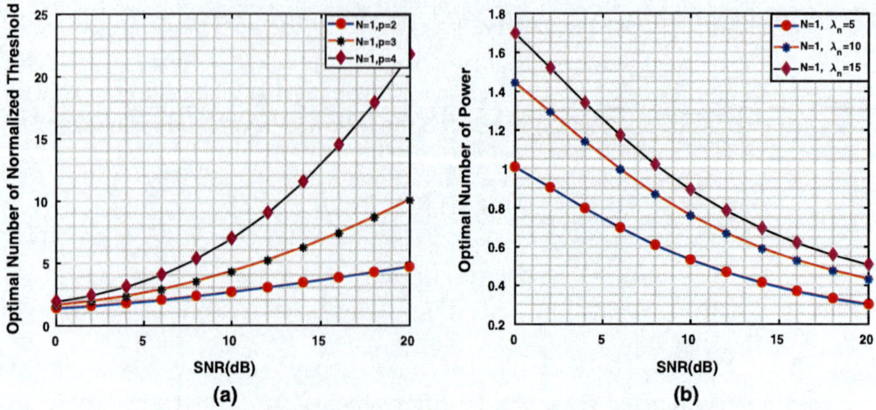

Fig. 4 **a** SNR versus optimal number of normalized threshold, **b** SNR versus optimal number of power

$\frac{\partial \mathbf{P_m}}{\partial p}$ in (20) we obtain p_{opt}, i.e., $p_{\text{opt}} = \frac{2\gamma log\lambda}{(1+\gamma)log(1+\gamma)}$. Figure 4a and b shows the SNR versus optimal number of λ_n and p_{opt}, respectively. It can be seen from Fig. 4a that the value of λ_n drastically increases as higher values of SNR for $p = 2$, $p = 3$, and $p = 4$, respectively, by keeping the constant values of N and q. Figure 4b depicted that as the values of SNR increase p_{opt} decreases for a constant values of q and λ_n, and it also seen that p_{opt} is less for small values of λ_n.

5 Conclusion

In this paper, we have studied the parametric optimization of improved sensing-based CRN by using IED over an erroneous channel. We also derive the mathematical expression from investigating optimal CR density, optimal received signal power, and normalized threshold optimization by considering the minimization of TER. The optimality also depends upon some other parameters, such as average SNR. The impact of an optimal number of normalized threshold values, consistency of reporting channel on optimal values of the IED parameter also has been assessed.

References

1. Mitola J, Maguire GQ (1999) Cognitive radio: making software radios more personal. IEEE Pers Commun 6(4):13–18
2. Ghosh SK, Mehedi J, Samal UC (2019) Sensing performance of energy detector in cognitive radio networks. Int J Inf Technol 11(4):773–778

3. Haykin S (2005) Cognitive radio: brain-empowered wireless communications. IEEE J Sel Areas Commun 23(2):201–220
4. Ghosh SK, Bachan P (2017) Performance evaluation of spectrum sensing techniques in cognitive radio network. IOSR J Electron Commun Eng (IOSR-JECE). e-ISSN 2278-2834
5. Abdullah HN, Abed HS (2018) Implementation of selected spectrum sensing systems for cognitive radio networks using FPGA platform. J Telecommun Inf Technol
6. Yucek T, Arslan H (2009) A survey of spectrum sensing algorithms for cognitive radio applications. IEEE Commun Surv Tutor 11(1):116–130
7. Bachan P, Ghosh SK, Saraswat SK (2015) Comparative error rate analysis of cooperative spectrum sensing in non-fading and fading environments. In: 2015 communication, control and intelligent systems (CCIS). IEEE, pp 124–127
8. Kumar S, Verma PK, Kaur M, Jain P, Soni SK (2018) On the spectrum sensing of gamma shadowed Hoyt fading channel with MRC reception. J Electromagn Waves Appl 32(16):2157–2166
9. Ganesan G, Li Y (2007) Cooperative spectrum sensing in cognitive radio. Part II: Multiuser networks. IEEE Trans Wirel Commun 6(6):2214–2222
10. Chen Y (2010) Improved energy detector for random signals in Gaussian noise. IEEE Trans Wirel Commun 9(2):558–563
11. Xiao S, Zhou X, Li GY, Guo W (2016) Robust resource allocation in full-duplex cognitive radio networks. In: 2016 IEEE global communications conference (GLOBECOM). IEEE, pp 1–7
12. Wang C, Song T, Wu J, Yu Y, Hu J (2018) Energy-efficient cooperative spectrum sensing with reporting errors in hybrid spectrum sharing CRNs. IEEE Access 6:48391–48402
13. Quan Z, Cui S, Sayed AH (2008) Optimal linear cooperation for spectrum sensing in cognitive radio networks. IEEE J Sel Top Signal Process 2(1):28–40
14. Zhang W, Mallik RK, Letaief KB (2009) Optimization of cooperative spectrum sensing with energy detection in cognitive radio networks. IEEE Trans Wirel Commun 8(12):5761–5766
15. Sharma G, Sharma R (2019) Performance improvement of CSS over imperfect reporting using diversity reception in cognitive radio networks. World J Eng
16. Gahane L, Sharma PK (2017) Performance of improved energy detector with cognitive radio mobility and imperfect channel state information. IET Commun 11(12):1857–1863
17. Banavathu NR, Khan MZA (2019) Optimization of *n*-out-of-*k* rule for heterogeneous cognitive radio networks. IEEE Signal Process Lett 26(3):445–449
18. Ranjeeth M, Behera S, Nallagonda S, Anuradha S (2016) Optimization of cooperative spectrum sensing based on improved energy detector with selection diversity in AWGN and Rayleigh fading. In: 2016 international conference on electrical, electronics, and optimization techniques (ICEEOT). IEEE, pp 2402–2406
19. Ranjeeth M, Anuradha S, Nallagonda S (2018) Correction to: Performance analysis of cooperative spectrum sensing network using optimization technique in different fading channels. Wirel Pers Commun 99(1):595–595
20. Urkowitz H (1967) Energy detection of unknown deterministic signals. Proc IEEE 55(4):523–531
21. Singh A, Bhatnagar MR, Mallik RK (2011) Cooperative spectrum sensing with an improved energy detector in cognitive radio network. In: 2011 national conference on communications (NCC). IEEE, pp 1–5

Identification of Diabetic Retinopathy for Retinal Images Using Feed Forward Neural Network

H. Asha Gnana Priya and J. Anitha

Abstract Diabetic retinopathy is an eye disease that mainly affects the people having diabetes, and early identification is required to avoid vision loss. It can be done by trained ophthalmologist, but they are less in numbers; hence, computer-aided diagnosis system is used for automatic screening. In this paper, input image is taken from Indian Diabetic Retinopathy Image Dataset and histogram equalization; top-hat filters are used for enhancing retinal fundus images. Haralick features are extracted from the grayscaled image and filtered image and given to the feed forward neural network classifier. The accuracy obtained from the grayscaled image is 90%, and the accuracy obtained from filtered image is 95%.

Keywords Retinal image · Computer-aided diagnosis · Image enhancement · Retinal disease · Histogram equalization · Top-hat filter

1 Introduction

Human eye is the body's most highly developed sensory organs and 80% of things that we observe comes over the sense of our sight. Healthy brain function requires healthy eye sight and by taking care of eye can reduce the vision loss and blindness. Retina in the eye receives the light focused by the lens, converts into neuro signals, and sends to brain for visual recognition. Few pathologies affecting retina are diabetic retinopathy (DR), central retinal vein occlusion (CRVO), glaucoma, central serous retinopathy

Please note that the CCIS Editorial assumes that all authors have used the Western naming convention, with given names preceding surnames. This determines the structure of the names in the running heads and the author index.

H. Asha Gnana Priya · J. Anitha (✉)
Department of Electronics and Communication Engineering, Karunya Institute of Technology and Sciences, Coimbatore, India
e-mail: anithaj@karunya.edu

H. Asha Gnana Priya
e-mail: ashapriya.h@gmail.com

A. Khanna et al. (eds.), *Data Analytics and Management*, Lecture Notes on Data Engineering and Communications Technologies 54,
https://doi.org/10.1007/978-981-15-8335-3_42

(CSR), choroidal neo vascularization membrane (CNVM), diabetic macular edema (DME), age-related macular degeneration (AMD), and cataract. The most commonly affected eye disease among these is DR which mostly affects the diabetes patients and leads to vision loss. The global prevalence of DR between 2015 and 2019 is 27%. Europe and South East Asia has the lowest prevalence of 20.6% and 12.5%. Prevalence in Africa, Middle East, North Africa is 33.8%, and the highest prevalence is for western pacific region at 36.2%. Retinal disease statistics in India by Analysis Ready Data (ARD) is 2.1 million in 2010, and in 2050, it will be increased to 5.4 million. Between 2010 and 2050, the estimated number of people affected by DR will be 7.7 million to 14.6 million.

Two pilot initiatives concentrating on health system approach to DR and retinopathy of prematurity (ROP) was successfully implemented between 2014 and 2019 around 14 states by The Ministry of Health, Government of India, Public Health Foundation of India (PHFI), The London School of Hygiene and Tropical Medicine (LSHTM), The Queen Elizabeth Diamond Jubilee Trust and partners for avoiding vision loss. 69,970 patients are screened for DR, 4072 are received treatment, and 6475 came for annual second screening. 445 physicians received certificate for DR screening training until 2018, and many clinics and hospitals has been strengthened by this project. Tool for Assessment of Diabetes and DR (TADDS) is developed by World Health Organization (WHO) in 2015 for ensuring timely diagnosis and treating DR. This tool was first used in Nepal, India, for taking actions to reduce the DR level, and many screening camps were held across the world for preventing the vision loss. Research support for DR was initiated by The National Eye Institute (NEI) and the Juvenile Diabetes Foundation International (JDFI). Since 1995, SightFirst has provided US$2.5 million to support diabetic retinopathy screening and treatment projects in eight different countries including India, Brazil, Pakistan, Fiji, Bahrain, Chile, Spain, and Venezuela.

DR is the major cause of blindness and mostly affects the working age people. DR affects the retinal blood vessels and certain feature such as length, width, and branching pattern which provides the techniques to diagnosis the retinal diseases [1, 2]. Fundus images has great significance in the diagnosis of retinal diseases. Automatic parameter selection method via optimization based on imaging control (IC) in gabor filter to obtain the effective image by threshold selection [3]. Vessel outline can be extracted using matched filters and is convolved with 2D kernel. Matched filter can also combine with Contrast Limited Adaptive Histogram Equalization (CLAHE) for enhancing blood vessels [4, 5]. The powerful method of diagnosing vascular diseases is segmentation of retinal blood vessels from the background tissues. Optical Coherence Tomography Angiography (OCTA) provides comprehensive blood vessels maps at different depths of blood flow information [6]. Optimally adjusted morphological operator set is used in DR patients for detecting exudates [7]. Deep convolutional neural networks are the world-shattering method in the field of computer vision for image classification and object detection. Deep Retinal Image Understanding (DRIU) is used for segmentation of blood vessels, and optic disk by trained sets of two specialized layers is highly accurate and efficient [8].

Multiscale convolutional mixture of expert (MCME) is to diagnose retinal pathologies like DME, AMD, and normal retinal image. Applying convolutional neural network (CNN) on MCME will increase the learning training features speed [9]. Multi-layer perception neural network (MLPNN) with features extracted from discrete cosine transform (DCT) can classify the normal and abnormal DR with improved accuracy [10]. Global contrast normalization, zero phase whitening, augmentation using gamma corrections and gamma correction preprocessing steps in deep learning used for the segmentation of blood vessels [11]. Deep convolutional neural network segments the retinal blood thick, thin vessels, and background areas by major modification in the layer. The three-layer architecture 2-4-4 CNN, 4-4-4 CNN, 2-2-2 CNN, is used to detect vessels and achieved segmentation performance accuracy of 94.56% [12]. Dynamic decision thresholding method is used for exudates segmentation combined with heterogeneity, bright, and faint edges. This can be used for real-time applications proved by 9.36 s computation time taken [13]. To overcome the problem of huge variability and less contrast in retinal images, CNN is used equivalent to ensemble learning. Compared to bagged filters, better results have been obtained and is proved that more discriminative filters are produced by boosting [14]. Convolutional neural networks with patch-based pixel-wise segmentation is implemented, and each segment produces a feature map of a specific size, and it combines all the feature map into a single layer [15].

Extraction of blood vessels are crucial and the diagnosis of several diseases becomes complicated related to stroke, hypertension, and cardiovascular diseases. Manual segmentation is difficult compared to automatic segmentation of blood vessels. Automated CNN method is used for blood vessel segmentation, and the screening becomes easier [16]. Tyler Coye algorithm improved with hough line transformation enhanced method can improve the vessel segmentation [17]. Pixel classification utilization and 7-D vector evaluation are composed of gray-level and moment invariants-based features for pixel characterization using CNN. The robustness and effectiveness with simplicity and fast implementation provide this new technique relevant to automated segmentation of blood vessels [18]. Postprocessing morphological method after support vector machine (SVM) training connects the discontinuities in the vessel segmentation and increases the accuracy [19, 20]. Shape-based features like Zernike based on maximum differentiability and statistical features can classify the images with proper segmented blood vessels [21]. Disease features are extracted using CNN for automatic detection of DR [22]. Deep learning with smaller dataset gives better results compared larger dataset [23].

In this paper, Sect. 2 gives the block diagram, Sect. 3 gives the algorithm, Sect. 4 gives the results and discussion, and Sect. 5 gives the conclusion. The main contribution of this paper is

1. The database used is Indian Diabetic Retinopathy Image Datasets (IDRID), and it is the first database with Indian samples and given by ISBI.
2. Automatic detection of diabetic retinopathy.

2 Proposed System

The block diagram for the proposed system is shown in Fig. 1. The proposed system takes the retinal fundus images from the dataset and resized to speed up the process. Then, RGB channels are splitted into red, green, and blue channels. Green channel has more information and is converted into grayscale, and histogram equalization followed by top-hat filter is applied in the grayscaled image. Now, features are extracted from the grayscale and the filtered image and given to the feed forward neural network for classification of DR. Accuracy, sensitivity, and specificity are used for performance analysis, and the system in more detail is explained in the below section.

2.1 Retinal Images

Input retinal fundus images are taken from Indian Diabetic Retinopathy Image Datasets (IDRID). This dataset is taken from the challenge given by ISBI, and this dataset contains 516 images taken from the eye clinic located in Nanded, Maharashtra, India. 81 images are color fundus image with signs of diabetic retinopathy, and 164 images are color fundus image without signs of diabetic retinopathy.

2.2 Image Enhancement

Histogram equalization and top-hat filter is applied in the image for contrast enhancement. To enhance the contrast in an image, intensity distribution of the histogram is modified.

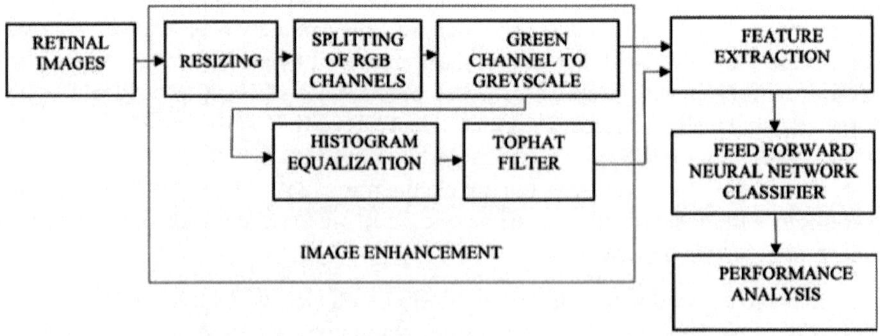

Fig. 1 Block diagram for the proposed system

Histogram equalization modifies image intensity and enhances the contrast in an image. The expression for histogram equalization is given as Eq. 1.

$$P(r_k) = \frac{nk}{n} \tag{1}$$

where

n_k Number of occurrence of gray level
n Total number of pixels in the image.

The top-hat filter is used to enhance bright objects of interest in a dark background which uses the top-hat filter to alien uneven brightness. This filter is an opening operation. The expression for top-hat filter is given as Eq. 2.

$$T_W(f) = f \cdot b \tag{2}$$

where

f Grayscale image
b Structuring element.

2.3 Feature Extraction

The statistical features extracted in the proposed system are contrast, correlation, energy, entropy, homogeneity, inversed differential mode, kurtosis, mean, RMS, skewness, smoothness, standard deviation, and variance. These are the haralick features and the formulas are given in Table 1.

2.4 Feed Forward Neural Network

The feed forward neural network is also known as back propagation neural network. In this network, weights are set initially, and the number of features obtained is given as the input neurons, and hidden neurons can be any numbers. Target is set in the output layer. To improve the network, iterative, recursive, and efficient methods are used for updating the weights until the network is trained for performing the task. The feed forward neural network is shown in Fig. 2.

3 Results and Discussion

The input images are taken from IDRID datasets, and the proposed system takes 20 normal and 20 abnormal fundus images. The image enhancement process includes

Table 1 Feature extraction formulas

Feature extraction	Formulas
Contrast	$S_c = \sum_i \sum_j (i-j)^2 p(i,j)$
Correlation	$S_o = \frac{\sum_i \sum_j (ij)p(i,j) - \mu_x \mu_y}{\sigma_x \sigma_y}$
Energy	$S_N = \sum_{b=0}^{L-1} [p(b)]^2$
Entropy	$S_E = -\sum_{b=0}^{L-1} p(b) \log_2\{p(b)\}$
Homogeneity	$S_H = \sum_{i=0}^{N-1} \sum_{j=0}^{N-1} \left(\frac{P_{ij}}{(1+\|i-j\|)}\right)$
Inversed differential mode	$S_1 = \sum_i \sum_j \frac{1}{1+(i-j)^2} p(i,j)$
Kurtosis	$S_K = \frac{1}{\sigma_b^4} \sum_{b=0}^{L-1} (b-\overline{b})^4 p(b) - 3$
Mean	$S_M = \overline{b} = \sum_{b=0}^{L-1} bp(b)$
Standard deviation	$S_D = \sigma_b = \left[\sum_{b=0}^{L-1} (b-\overline{b})^2 p(b)\right]^{1/2}$
Skewness	$S_D = \sigma_b = \left[\sum_{b=0}^{L-1} (b-\overline{b})^2 p(b)\right]^{1/2}$
Variance	$S_V = \sum_{i,j=1}^{N} (i-j)^2 p(i,j)$

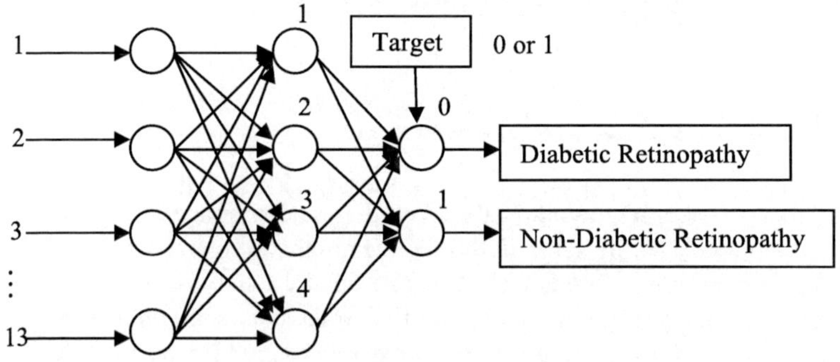

Fig. 2 Feed forward neural network diagram

resizing, splitting of RGB channels, gray scale conversion, histogram equalization, and top-hat filter. The fundus image is resized by 128 * 128 pixel. RGB channels are splitted into R, G, B channels. Green channel has more information and is converted into gray scale. Histogram equalization is applied in the grayscaled image followed by top-hat filter, and the results are shown in Fig. 3.

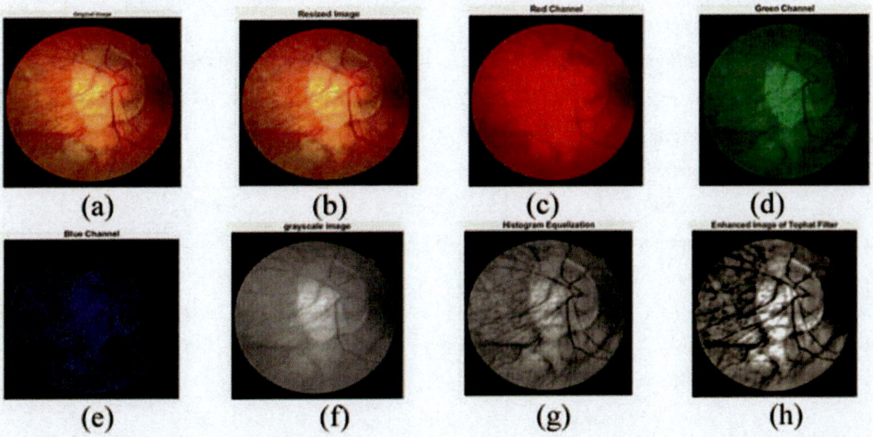

Fig. 3 **a** Input image, **b** resized image, **c** red channel, **d** green channel, **e** blue channel, **f** grayscaled image, **g** histogram equalization image, **h** top-hat filtered image

3.1 Tabulation

The features extracted from the grayscaled image and filtered images are tabulated in Tables 2 and 3. The proposed system used 13 neurons in the input layer, 4 neurons in the hidden layer, and 2 neurons in the output layer. Target is set as 0 and 1. If the output obtained is o, then it is diabetic retinopathy; and if the output is 1, then it is non-diabetic retinopathy. The performance analysis comparison is done by sensitivity, specificity, and accuracy as shown in Fig. 4

Table 2 Feature extracted from grayscaled image

S. No.	Features	Image 1	Image 2	Image 3	Image 4
1	Contrast	0.0535	0.0454	0.0487	0.0449
2	Correlation	0.8858	0.8967	0.9213	0.9153
3	Energy	0.5485	0.5818	0.4710	0.4682
4	Entropy	4.9180	5.0909	5.1769	5.1824
5	Homogeneity	0.9732	0.9773	0.9779	0.9775
6	IDM	255	255	255	255
7	Kurtosis	4.3417	3.5938	2.2539	1.9882
8	Mean	20.9140	21.3295	32.5201	29.9874
9	RMS	11.8083	11.8729	12.0770	12.0282
10	Standard deviation	16.2027	16.7987	22.7226	20.9026
11	Smoothness	1.0000	1.0000	1.0000	1.0000
12	Skewness	0.5054	0.5340	−0.1519	−0.2538
13	Variance	85.0956	78.6702	121.8649	124.2400

Table 3 Feature extracted from filtered image

S. No.	Features	Image 1	Image 2	Image 3	Image 4
1	Contrast	0.7264	0.7584	0.8043	0.8813
2	Correlation	0.8474	0.8673	0.8277	0.8595
3	Energy	0.1819	0.1700	0.1730	0.1518
4	Entropy	5.2783	5.5295	5.5990	5.5721
5	Homogeneity	0.8313	0.8330	0.8259	0.8108
6	IDM	255	255	255	255
7	Kurtosis	4.8334	3.7852	5.2441	3.2145
8	Mean	52.1829	58.3103	53.6780	63.6606
9	RMS	12.5151	12.7975	13.5918	13.0242
10	Standard deviation	53.0649	57.4570	51.2982	59.8397
11	Smoothness	1.0000	1.0000	1.0000	1.0000
12	Skewness	1.2973	1.0673	1.3526	0.8815
13	Variance	1644.87	1451.78	1604.63	1916.88

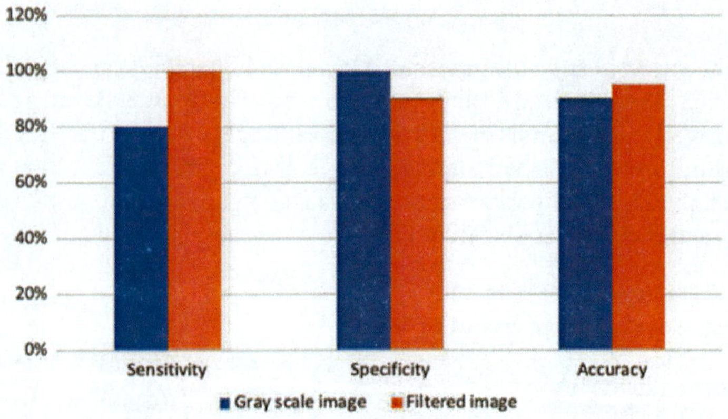

Fig. 4 Comparison of performance analysis like sensitivity, specificity, and accuracy of grayscaled image and filtered image

4 Conclusion

In this paper, IDRID datasets is used, and the features extracted from grayscaled image and filtered image are given to the neural network classifier. The accuracy obtained from the filtered image features given to the classifier is much greater than the grayscaled image features given to the classifier. The accuracy can be increased by different image enhanced techniques, neural network classifiers, and by deep neural networks.

References

1. Franklin SW, Rajan SE (2014) Computerized screening of diabetic retinopathy employing blood vessel segmentation in retinal images. Bio Cybern Biomed Eng 34(2):117–124 (2014)
2. Kumaran Y, Patil CM (2018) A brief review of the detection of diabetic retinopathy in human eyes using pre-processing and segmentation techniques. Int J Recent Technol Eng 7(2):310–320 (2018)
3. Farokhian F (2017) Automatic parameters selection of Gabor filters with the imperialism competitive algorithm with application to retinal vessel segmentation. Biocybern Biomed Eng 37(1):246–254
4. Sofka M, Stewart CV (2006) Retinal vessel centreline extraction using multiscale matched filters, confidence and edge measures. IEEE Trans Med Imaging 25(12):1531–1546
5. Kumar D, Pramanik A, Kar SS, Maity SP (2016) Retinal blood vessel segmentation using matched filter and Laplacian of gaussian. In: International conference on signal processing and communications, pp 1–5. IEEE
6. El Badawi N (2017) Automatic blood vessels segmentation based on different retinal maps from OCTA scans. Comput Biol Medi 89:150–161
7. Sopharak A (2008) Automatic detection of diabetic retinopathy exudates from non-dilated retinal images using mathematical morphology methods. Comput Med Imaging Graphics 32(8):720–727
8. Maninis K-K (2016) Deep retinal image understanding. In: International conference on medical image computing and computer-assisted intervention. Springer, Cham, pp 140–148
9. Rasti R (2018) Macular OCT classification using a multi-scale convolutional neural network ensemble. IEEE Trans Med Imaging 37(4):1024–1034
10. Bhatkar AP, Kharat GU (2015) Detection of diabetic retinopathy in retinal images using MLP classifier. In: International symposium on nanoelectronic and information systems. IEEE, pp 331–335
11. Liskowski P, Krawiec K (2016) Segmenting retinal blood vessels with deep neural networks. IEEE Trans Med Imaging 35(11):2369–2380
12. Khalaf AF, Yassine IA, Fahmy AS (2016) Convolutional neural networks for deep feature learning in retinal vessel segmentation. In: International conference on image processing. IEEE, pp. 385–388
13. Kaur J, Mittal D (2018) A generalized method for the segmentation of exudates from pathological retinal fundus images. Biocybern Biomed Eng 38(1):27– 53
14. Zilly J, Buhmann JM, Mahapatra D (2017) Glaucoma detection using entropy sampling and ensemble learning for an automatic optic cup and disc segmentation. Comput Med Imaging Graphics 55:28–41
15. Song J, Lee B (2017) Development of automatic retinal vessel segmentation method in fundus images via convolutional neural networks. In: 39th annual international conference on engineering in medicine and biology society. IEEE, pp 681–684
16. Thangaraj S, Periyasamy V, Balaji R Retinal vessel segmentation using a neural network. IET Image Process 12(5):669–678
17. Bandara G (2017) A retinal image enhancement technique for a blood vessel segmentation algorithm. In: International conference on industrial and information systems. IEEE, pp 1–5
18. Marín D (2011) A new supervised method for blood vessel segmentation in retinal images by using gray-level and moment invariants-based features. IEEE Trans Med Imaging 30(1)
19. Li M, Ma Z, Liu C, Zhang G, Han Z (2017) Robust retinal blood vessel segmentation based on reinforcement local descriptions. BioMed Res Int 1–8 (2017)
20. Imran A, Li J, Pei Y, Yang J-J, Wang Q (2019) Comparative analysis of vessel segmentation techniques in retinal images. IEEE Access 7:114862–114887
21. Adapa D, Joseph Raj AN, Alisetti SN, Zhuang Z, Ganesan K, Ganesh N (2020) A supervised blood vessel segmentation technique for digital fundus images using Zernike moment based features. Plos One 15(3):1–23

22. Lam C, Yi D, Guo M, Lindsey T (2018) Automated detection of diabetic retinopathy using deep learning. AMIA Summits Transl Sci Proc 147–155 (2018)
23. Sahlsten J, Jaskari J, Kivinen J, Turunen L, Jaanio E, Hietala K, Kaski K (2019) Deep learning fundus image analysis for diabetic retinopathy and macular edema grading. Sci Rep 9(1):1–11

Ameliorating Accuracy Using Dual Dimensionality Reduction on a Classification Data set

Mudita Arya, Shailender Kumar, Monika Vaid, and Shoaib Akhtar

Abstract In this era of globally interconnected networks, we have huge amounts of data. The data is increasing not in gigabytes or terabytes but in terms of trillions of zettabytes and yottabytes. This data can be used in the field of machine learning and artificial intelligence so as to get intelligent machines that can have virtual brains to take and execute decisions. In practice, machine learning is used to train our machines with a specific training set of data in order to develop an efficient algorithm. Once an efficient algorithm is developed, our machine is smart enough to predict the results further. Hence, data is the source food that needs to be supplied to our algorithm and the efficiency of our algorithm is directly proportional to the amount of data that we provide to it. But not all the data is important for the study, and we need to eliminate the unnecessary and redundant information out of that data. In order to do that the concept of dimensionality reduction comes into play. By using dimensionality reduction, we extract only the essential features of our data and use it for our study. We get the motivation behind this study from the fact that the dimensionality reduction techniques used commonly do not increase the accuracy significantly and the problem of overfitting was not rectified too. Hence, in this paper we have used the concept of dual dimensionality reduction to increase the efficiency even more and combat the problem of overfitting. In this paper, we have improved the accuracy of the dimensionality reduction by using the concept of dual dimensionality reduction in which we have used a combination of various other dimensionality reduction algorithms

M. Arya (✉) · S. Kumar · M. Vaid · S. Akhtar
Department of Computer Science Engineering, Delhi Technological University, Delhi, New Delhi 110042, India
e-mail: mudita.arya98@gmail.com

S. Kumar
e-mail: shailenderkumar@dce.ac.in

M. Vaid
e-mail: monika6198@gmail.com

S. Akhtar
e-mail: shoaib1999ak@gmail.com

A. Khanna et al. (eds.), *Data Analytics and Management*, Lecture Notes on Data Engineering and Communications Technologies 54,
https://doi.org/10.1007/978-981-15-8335-3_43

Keywords Machine learning · Artificial intelligence · Training set · Dimensionality reduction · Factor analysis · ICA · SVD · PCA · UMAP · ISOMAP

1 Introduction

Data is the source food that needs to be supplied to our algorithm, and the efficiency of our algorithm is directly proportional to the amount of data that we provide to it. As we know, the more amount of data we provide to our algorithm, the better its performance and efficiency will be [1]. But providing the algorithm with more data means providing more features to it which can make our algorithm very slow and complex. Hence, we need to remove the irrelevant and redundant features from our data set to improve the efficiency of our algorithm [2]. Our goal is to reduce the features while preserving the maximum information. This process of reducing the number of random variables and obtaining certain principal variables is called Dimensionality Reduction [3]. Dimensionality can also be reduced using MWOA [4].

Dimensionality Reduction of data is nothing more than the minimum number of parameters that can give an account for the observed property of data [5]. The dimensionality reduction has its influence in various domain, it reverts the curse of dimensionality and various unnecessary properties of high-dimensional spaces [6]. Overfitting is a problem which happens when the model works only for a specific set of data and does not work for generalized data [1]. Hence, it will work excellently on the training data but fails to show its reliability in real-world data. Speech signal, digital photograph, f-MRI scans all those real-world data which have extremely high dimensionality. As we have to handle this real-world data adequately, we have to reduce its dimensionality. In this paper, we will be using the random forest algorithm for our purpose. The rest of the paper is organized as follows: Sect. 2 describes various algorithms that are being used in this research paper. Section 3 discusses the experimental study which elaborates the whole experimental setup of the research paper. In Sect. 4, the experimental results of the study are included. Query execution time or run time, training time, and testing time are used to evaluate and compare the performance of the proposed model with the query execution time or run time, training time, and testing time with the model containing various combinations. Finally in Sect. 5, we conclude by briefing the contributions of this paper by explaining its scope and conclusions.

2 Background Study

We have seen in the various research papers and conferences that dimensionality reduction was used to improve the accuracy of the model and remove the redundant features. The dimensionality reduction paper given by the Tilburg center for creating computing has proved in their paper that the nonlinear techniques are far better than the linear techniques as they have the capability to deal with the complex nonlinear real-world data [5]. According to a paper "Dimensionality reduction in higher-order signal processing and rank-($R1$, $R2$, …, RN) reduction in multilinear algebra" published in Elsevier, working with the high-dimensional data may be very tedious or even computationally infeasible. Moreover, it is known that low-dimensional estimators often have a smaller variance than high-dimensional estimators, which leads to more accurate results. Important area of application is biomedical engineering, data analysis, image processing, etc. [2]. Motivated by that, here, in this paper, we will prove that if we use the concept of dual dimensionality by using a combination of several algorithms together we can improve the efficiency to a greater extent and eliminate the problem of overfitting in our model.

3 Profuse Methodologies Used to Accomplish the Research

3.1 Random Forests

Random forest algorithm can be used both for the classification and the regression [7]. The forest is made up of the trees, and more number of trees imply the more robust our forest is [8]. We randomly select samples and features and make different–different decision trees on the basis of different–different samples and features of data set. We select the prediction from each decision trees in the forest and use the average of all results produced from different decision trees [9]. An approach to random forests is given in Fig. 1.

Given classifiers $p_1(x)$, $p_2(x)$, …, $p_K(x)$, and with the training set from the distribution of the random vector Y, X, define the margin function as [9]

$$\text{mg} = av_k I(p_k(X) = Y) - \max_{jf=y} v_k I(p_k(X) = j)$$

If mg(X, Y) > 0, then the classification is correct. If mg(X, Y) < 0, then the classification is incorrect

The various algorithms used for dimensionality reduction are:

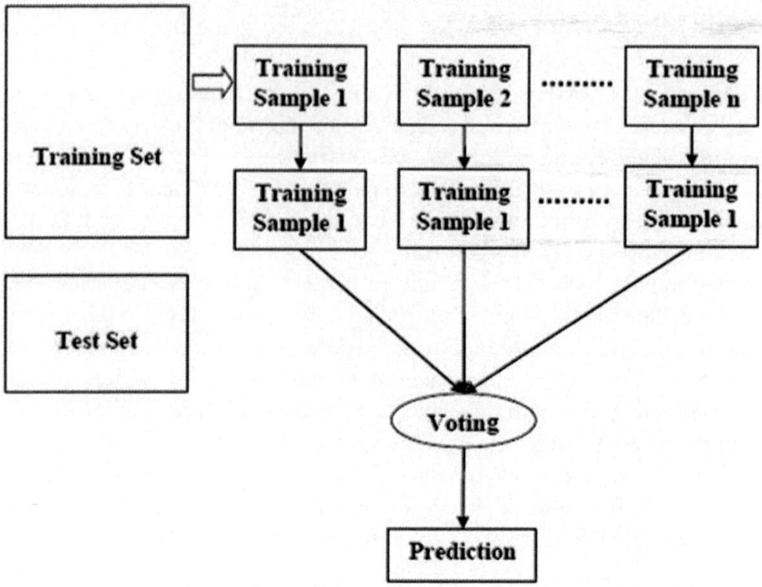

Fig. 1 Random forests approach

3.2 *Factor Analyses*

Factor analysis is data analysis method which analysis the set of variables to form the influential factors out of observed variables. It makes the data interpretations a less tedious task as it reduces the number of variables under consideration. Factor analysis defined the covariance between the set of variables into consideration and divide them into group which is known as factors. Variables are grouped on the basis of the correlation between the variables [10]. In each group, the set of variables will have high correlation between each other. Variables are divided into the groups on the basis of co-relation, and the variable in a group will have high co-relation with each other variable as compared to variables of other group [11]. It is an extension of PCA. It helps in data interpretations by reducing the number of variables under consideration. It reduces the number of variables and makes the analyzation and interpretation easy.

$$Y_i = \beta_{i_0} + \beta_{i_1} F_1 + \beta_{i_2} F_2 + (1) e_i$$

An approach to FA is given in Fig. 2.

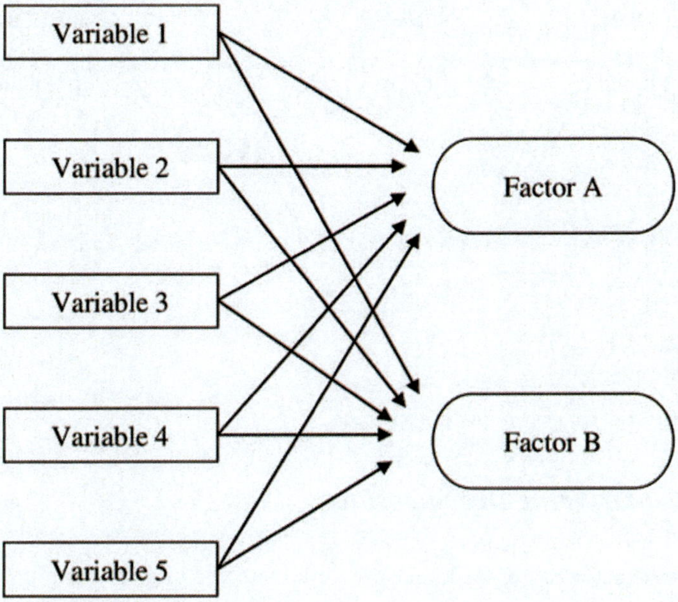

Fig. 2 Factor analysis approach

3.3 Independent Component Analysis

Independent component analysis (ICA) is method in which a multivariate signal consisting of different–different components is separated from each other in underlying components [12]. Then desired component is extracted from the pool of components. It is more powerful in data representation as compared to principal component analysis (PCA) [13]. It is used when we have multivariate signal and we have to extract the desired component out of it efficiently and allows us to better understand data noisy measurement environments [14].

An diagram representing ICA is given in Fig. 3. Assume the random observed vector is given by

P [15]

$$P = [P_1, P_2, \ldots, P_n]^T$$

where $j = 1, 2, 3, \ldots, n$.

n = mixture of n independent elements of random vector Q.

$$Q = [Q_1, Q_2, \ldots Q_n]^T$$

Here, $P = AQ$

$A = n \times n$ matrix mixing matrix.

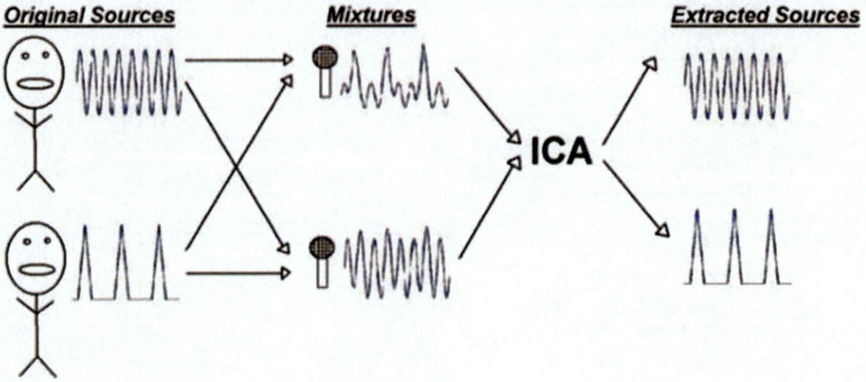

Fig. 3 Independent component analysis approach

3.4 Singular Value Decomposition

It is a matrix factorization technique where a matrix is broken into a product of a square matrix, a diagonal matrix, and another square matrix [16]. It is a very stable and effective method. It can be extremely useful as it can dramatically reduce the amount of data that we are dealing with.

- The high-order matrix can be decomposed using SVD transformation to form the product of three matrices [17].

$$A_{\mathrm{MN}} = U_{\mathrm{MM}} S_{\mathrm{MN}} V_{\mathrm{NN}}^{T}$$

- The columns of the matrix U_{MM} are orthonormal eigenvectors of $A_{\mathrm{MN}} A_{\mathrm{MN}} T$.
- The columns of the matrix V_{NN} are orthonormal eigenvectors of $A_{\mathrm{MN}} T A_{\mathrm{MN}}$.

3.5 Principal Component Analysis

It is the technique used for the dimensionality reduction of the huge data set by increasing the interpretability and decreasing the information loss. It does so by creating new uncorrelated variables that successively maximize variance [18]. It has application in different areas such as face recognition, image compression, and neuroscience. Principal component analysis (PCA) is a mathematical procedure which smaller number of uncorrelated variables called principal components [19].

Eigenanalysis is mathematical technique used for PCA: Eigenvalues and eigenvectors are calculated with sums of cross-products and squares of square symmetric matrix. The direction of the first principal component is same as the eigenvector which has the greatest eigenvalue. Similarly, the direction of second principal component is same as the eigenvector which has the second-highest eigenvalue [20]. The trace of square matrix is equal to sum of all eigenvalues of eigenvectors and number of columns (or rows) in the matrix is given by maximum number of eigenvectors.

The PCA model can be represented by [21]:

$$U_{m\times l} = W_{m\times d}\, X_{d\times l}$$

u, an m-dimensional vector, is a projection of x-the original d-dimensional data vector ($m \ll d$). m projection vectors are given by the eigenvectors e_1, e_2, em.

Covariance matrix is given by S and m is the largest nonzero eigenvalues λ_1, λ_2, ... λ_m.

3.6 t-Distributed Stochastic Neighbor Embedding (t-SNE)

It is a nonlinear dimensionality reduction technique which uses unsupervised learning [22]. It is used for data exploration and visualizing high-dimensional data. t-SNE is different from PCA because it focuses on preserving only small pairwise distances, whereas PCA is concerned with preserving large pairwise distances to get the maximum variance [23].

- Using Gaussian theorem, the similarities between points in high-dimensional space are measured. It is shown in Fig. 4.
- Cauchy distribution with 1 degree of freedom is used. It is shown in Fig. 5.

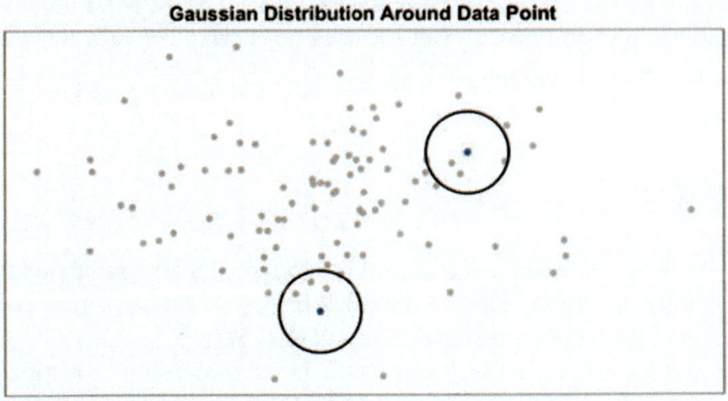

Fig. 4 Gaussian distribution around data point

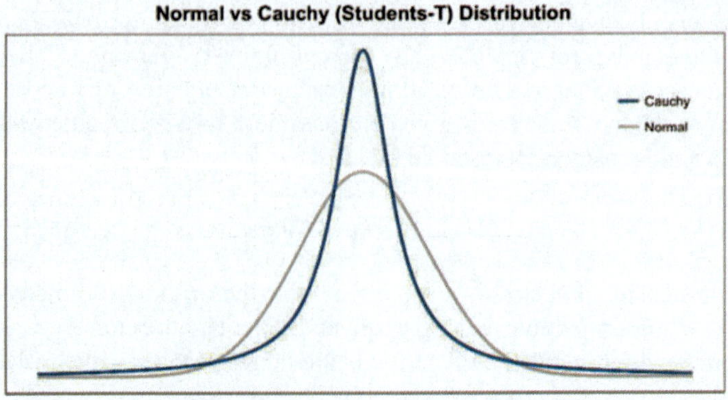

Fig. 5 Normal versus Cauchy distribution

3.7 UMAP

Uniform Manifold Approximation and Projection (UMAP) is a dimensionality reduction technique used for nonlinear structure of data. It is novel manifold learning technique. UMAP and *t*-SNE are competitive to one another but UMAP is preferred nowadays over *t*-SNE. The *t*-SNE technique has some limitations like

1. It cannot be scaled up.
2. Global data structure is not preserved.
3. It is not memory efficient.

UMAP preserves the global structure more as compared to *t*-SNE with even better run-time performance. UMAP is faster than *t*-SNE. UMAP does not apply normalization to probabilities, as UMAP does not normalize so it reduces time of finding high-dimensional graph [24].

3.8 ISOMAP

It is known as isometric mapping. It is nonlinear dimensionality reduction technique. Linear techniques like LDA and PCA are suitable for only linear structure of the data. Figure 6 represents the nonlinear data using PCA [25].

Euclidean distance is used in linear methods for dimension reduction, whereas Geodesic distance approach is used in ISOMAP [26]. Tradition metrics metric multi-dimensional scaling (MDS) has the limitations which are solved by ISOMAP by replacing the Euclidean distance by Geodesic distance. Figure 7 shows the Euclidean distance versus Geodesic distance. ISOMAP works on the following steps [24]:

1. On the basis of manifold distance, neighboring points are connected within a fixed radius.

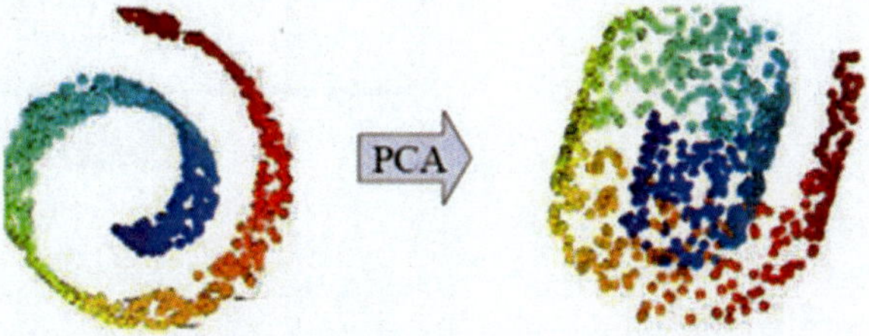

Fig. 6 Nonlinear data using PCA

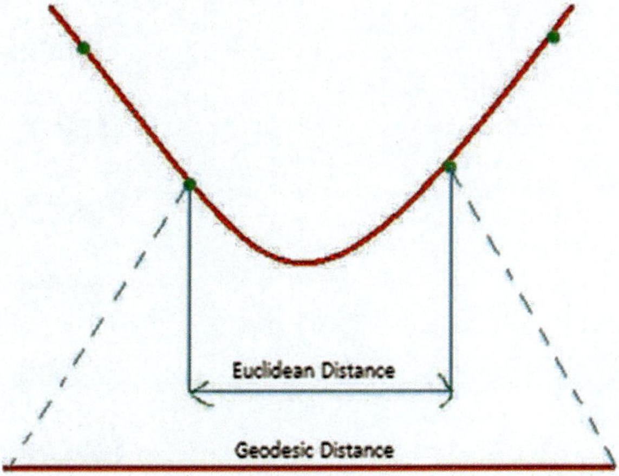

Fig. 7 Euclidean distance versus Geodesic distance

2. Geodesic distance is calculated between the points in this step.
3. On the finally generated distance graph MDS is applied, the points are then optimized with preserving the geometry of point.

4 Experimental Study

Now in this experimental study, we have used various dimensionality reduction algorithms on our data set and various combinations of dimensionality reduction algorithm as well on our data set and find out which dimensionality reduction algorithm alone or with other algorithm works best.

In this, we have used algorithm like factor analysis [10], ICA [12], SVD [16], PCA [18], *t*-SNE [22], ISOMAP [25], and UMAP [24]. On coming to the data set, we have used titanic data set which consists of various attributes. On talking about data set, we have features like name, class, passenger ID, embarked, and various other features. We started the research removing all the NaN value. We find out what percentage of data is NaN, and after that remove all the Indian values after all this removal of extra data set, we have features which are numeric the output of this is whether the person will survive or not.

The data is shown in Fig. 8. The image depicting the red lines showing the NaN

Fig. 8 Original data

	Total	Percent
Cabin	1014	77.46
Age	263	20.09
Embarked	2	0.15
Fare	1	0.08
Ticket	0	0.00
Parch	0	0.00
SibSp	0	0.00
Sex	0	0.00
Name	0	0.00
Pclass	0	0.00
Survived	0	0.00
PassengerId	0	0.00

values is given in Fig. 9. Final data is shown in Fig. 11.

After removing the NaN values, the heat map is.

After that, the variables we have or the features we have in our data set are not in a particular range so with definitely need to normalize our data which helps in the removal of any biasing. Moreover, it also helps in speeding up the calculation of the algorithm. We did standardization by using feature scaling. As we need to classify our data into survived or not survived, we have used random forest with each of the dimensionality reduction algorithm. A random forest is a machine learning inbuilt algorithm which prepares several decision trees, and the result we get from the maximum number of decision tree is considered as the final output of the algorithm. It provides the maximum accuracy when used with classification data set.

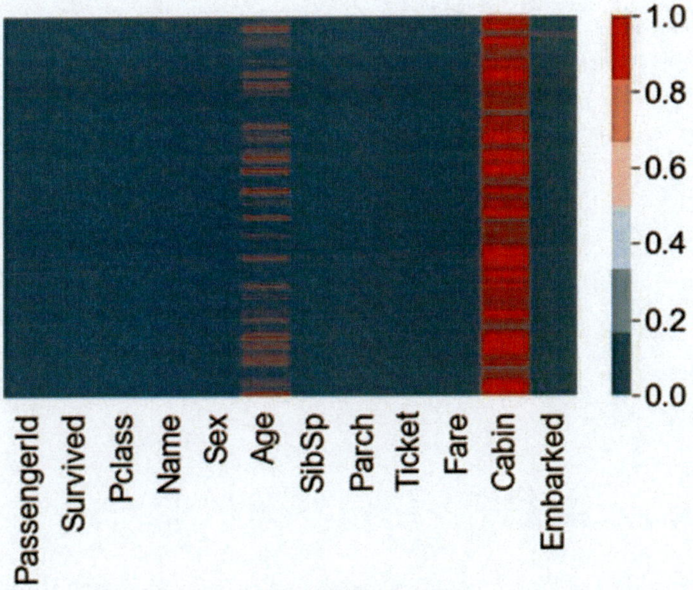

Fig. 9 Red lines showing the NaN values shown in Fig. 10

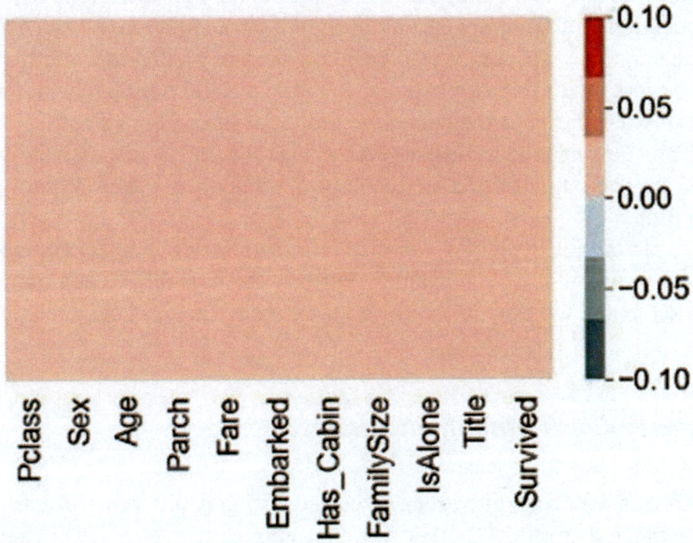

Fig. 10 After removing the NaN values

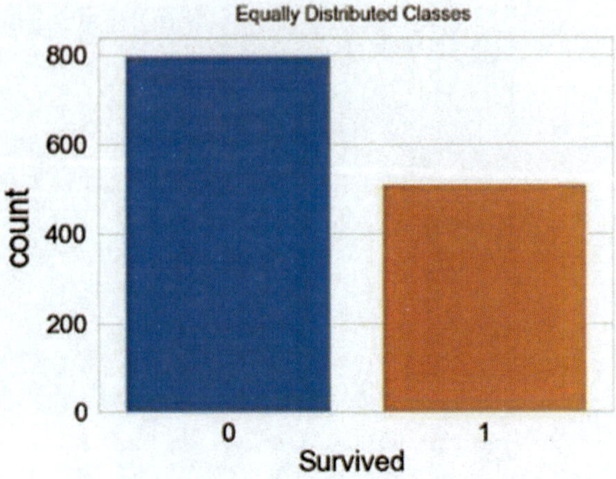

Fig. 11 Final data

4.1 Factor Analysis

Factor analysis is data analysis method which analysis the set of variables to form the influential factors out of observed variables. It makes the data interpretations a less tedious task as it reduces the number of variables under consideration [11]. First, we find out the number of components and the depth of random forest tree such that we could get the maximum accuracy as possible. Then, we plotted how our training and testing data works with two components so that we could have a real pictorial view of our data. After that with the same, we did it for three components as well. The training and testing visualizations are given in Figs. 12 and 13, respectively. Figure 14 shows the variance ratio of fitted factor analyzer component vector. Figure 15 shows the screen plot.

Later we plotted the explained variance ratio with factor analyzer component and later made a screen plot of eigenvalues. Further, the model has been trained to get maximum score and the time, training and the testing score has been recorded for it.

4.2 Independent Component Analyses

ICA is method in which a multivariate signal consisting of different–different components are separated from each other in underlying components [11]. Then desired component is extracted from the pool of components. First you find out the number of components and the depth of random forest which provides the best accuracy. After that we plotted our training and testing data for two components, and then for three components so that we could have a clear view of our data. The training and testing visualizations are given in Figs. 16 and 17, respectively.

After that, we run it by using the number of components and the depth of random forest which provides the best accuracy, so that we could get the best result as possible. It focuses on mutual independence of the components. After plotting all

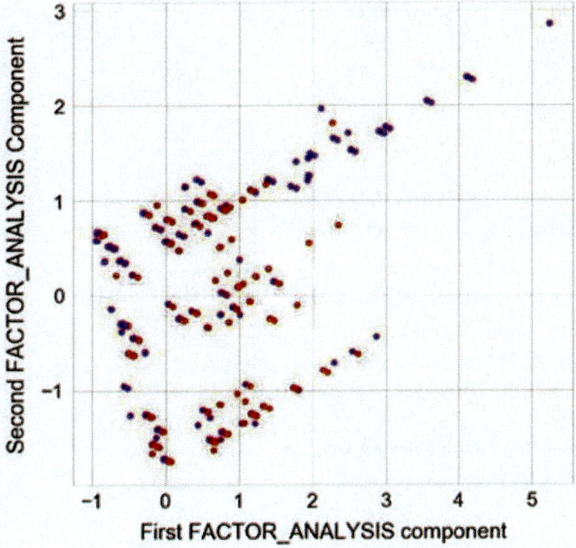

Fig. 12 Training data with 2 component

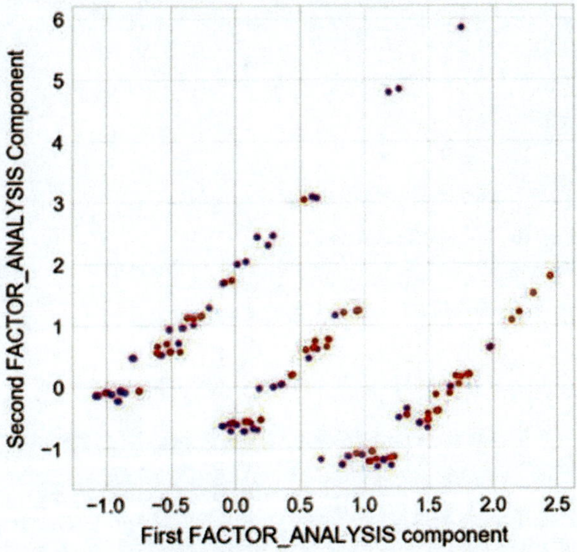

Fig. 13 Testing data with 2 component

the data, we finally train a model by using the components and the depth we received which basically provides the maximum accuracy, and we found the time training score testing score confusion matrix and classification report for training and testing data and saved it for later.

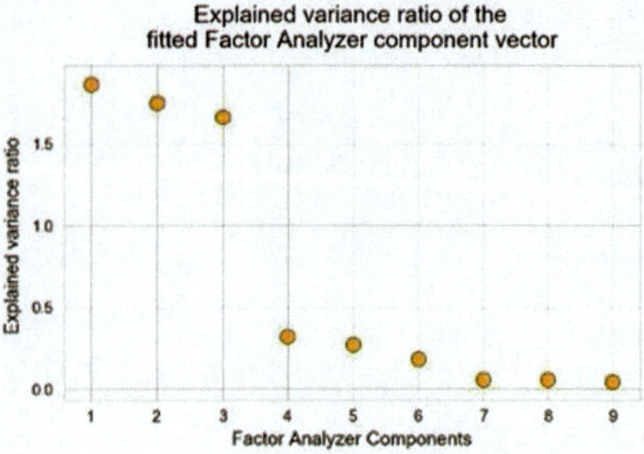

Fig. 14 Explained variance ratio of fitted factor analyzer component vector

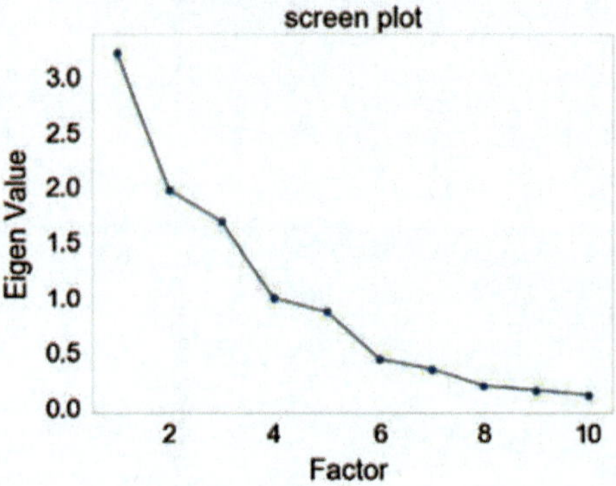

Fig. 15 Graph showing the screen plot

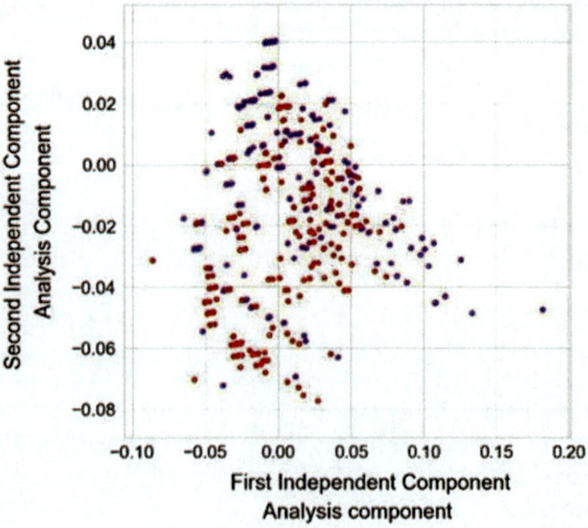

Fig. 16 Training data with 2 component

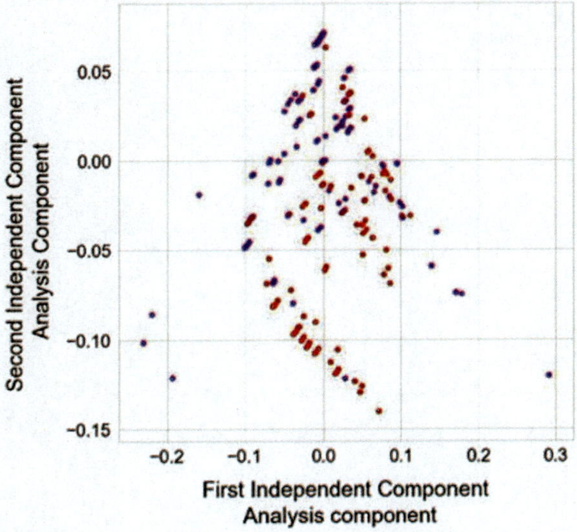

Fig. 17 Testing data with 2 component

4.3 *Singular Value Decomposition (SVD)*

It is a matrix factorization technique where a matrix is broken into a product of a square matrix, a diagonal matrix, and another square matrix. It is a very stable and effective method. It can be extremely useful as it can dramatically reduce the amount of data that we are dealing with [16]. First you find out the number of components and the depth of random forest which provides the best accuracy. After that we plotted our training and testing data for two components and then for three components so that we could have a clear view of our data. The training and testing visualizations are given in Figs. 18 and 19, respectively. The variance is shown in Fig. 20.

After that, we run it by using the number of components and the depth of random forest which provides best accuracy, so that we could get the best result as possible. It focus on mutual independence of the components. After plotting all the data, we finally trains a model by using the components and the depth we received which basically provides the maximum accuracy, and we found the time training score testing score confusion matrix and classification report for training and testing data and saved it for later.

4.4 *Principal Component Analyses (PCA)*

PCA is a mathematical procedure which converts a number of correlated variables into a smaller number of uncorrelated variables [26]. First you find out the number of components and the depth of random forest which provides the best accuracy. After that we plotted our training and testing data for two components, and then for three

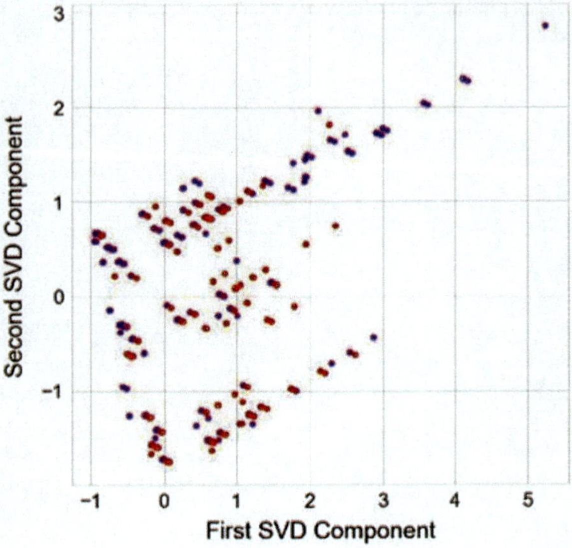

Fig. 18 Training data with 2 component

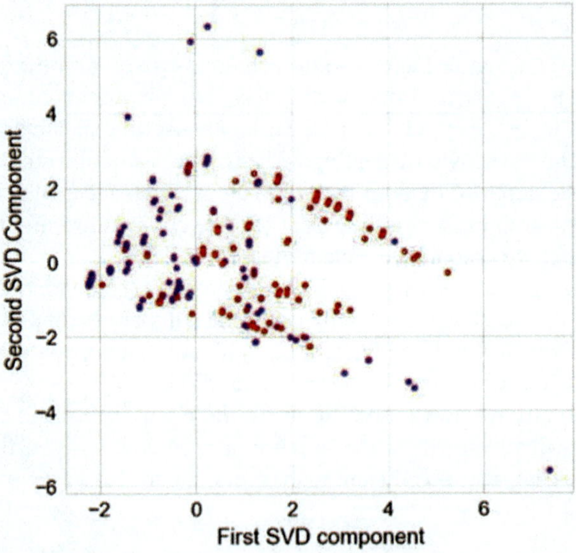

Fig. 19 Testing data with 2 component

components so that we could have a clear view of our data. The training and testing visualizations are given in Figs. 21 and 22, respectively.

After that, we run it by using the component and the depth of random forest which provides best accuracy, so that we could get the best result as possible. It focuses on mutual independence of the components. After plotting all the data, we finally train a model by using the components and the depth we received which basically

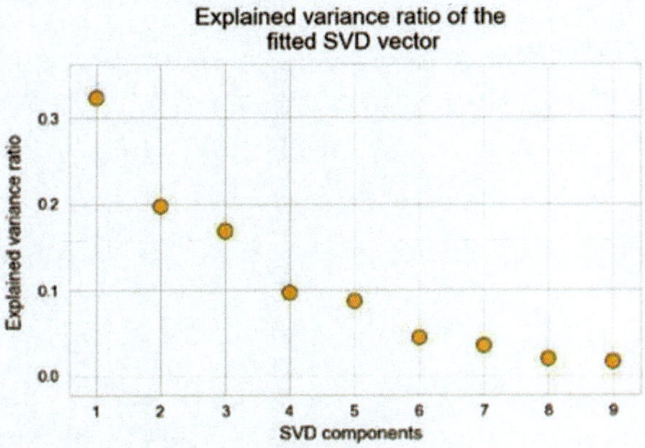

Fig. 20 Variance

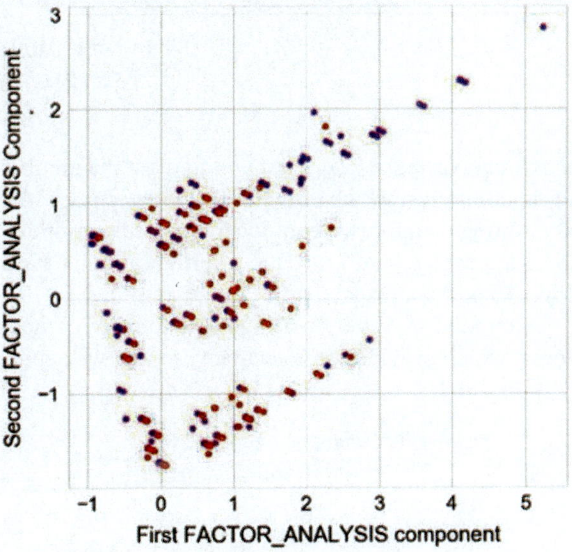

Fig. 21 Training data with 2 component

provides the maximum accuracy, and we found the time training score testing score confusion matrix and classification report for training and testing data and saved it for later.

4.5 t-*Distributed Stochastic Neighbor Embedding* (t-*SNE)*

It is used for data exploration and visualizing high-dimensional data. *t*-SNE is different from PCA because it focuses on preserving only small pairwise distances, whereas PCA is concerned with preserving large pairwise distances to get the

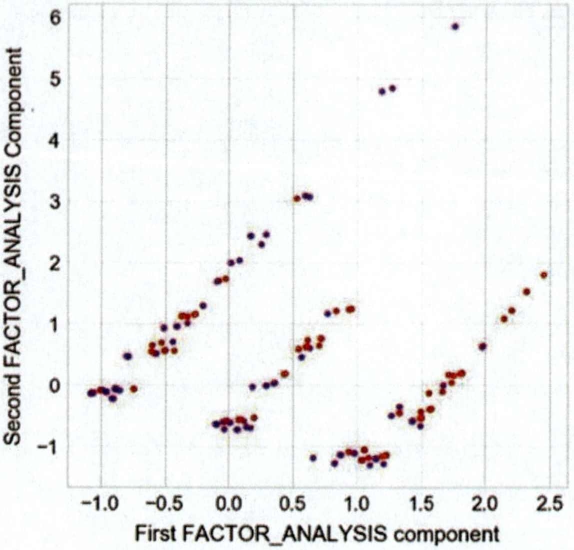

Fig. 22 Testing data with 2 component

maximum variance [23]. First you find out the number of components and the depth of random forest which provides the best accuracy. After that we plotted our training and testing data for two components and then for three components so that we could have a clear view of our data. The training and testing visualizations are given in Figs. 23 and 24, respectively.

After that, we run it by using the number of components and the depth of random forest which provides best accuracy, so that we could get the best result as possible.

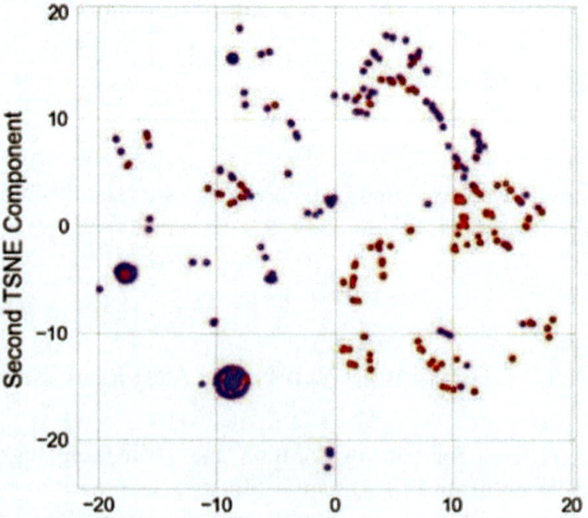

Fig. 23 Training data with 2 component

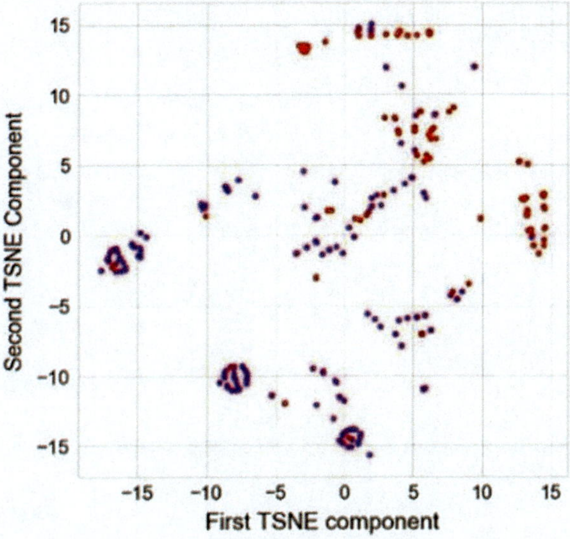

Fig. 24 Testing data with 2 component

It focuses on mutual independence of the components. After plotting all the data, we finally train a model by using the components and the depth we received which basically provides the maximum accuracy, and we found the time training score testing score confusion matrix and classification report for training and testing data and saved it for later.

4.6 UMAP

Uniform Manifold Approximation and Projection (UMAP) is a dimensionality reduction technique used for nonlinear structure of data [24]. It is novel manifold learning technique. UMAP and *t*-SNE are competitive to one another but UMAP is preferred nowadays over *t*-SNE. First you find out the number of components and the depth of random forest which provides the best accuracy. After that we plotted our training and testing data for two components and then for three components so that we could have a clear view of our data. After that, we run it by using the number of components and the depth of random forest which provides best accuracy, so that we could get the best result as possible. The training and testing visualisations are given in Figs. 25 and 26, respectively.

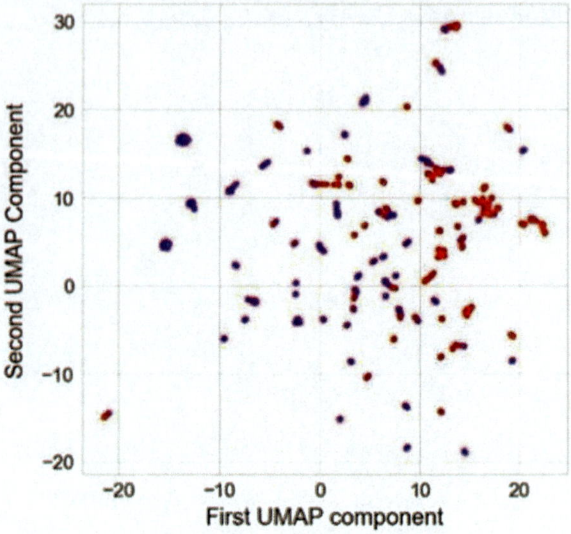

Fig. 25 Training data with 2 components

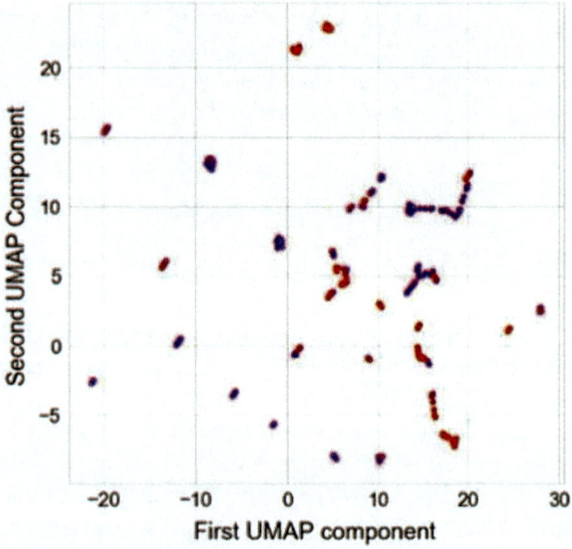

Fig. 26 Testing data with 2 components

4.7 ISOMAP

It is known as isometric mapping. It is nonlinear dimensionality reduction technique. Linear techniques like LDA and PCA are suitable for only linear structure of the data. First you find out the number of components and the depth of random forest which provides best accuracy [26]. After that we plotted our training and testing data for two components and then for three components so that we could have a clear view of our data. After that, we run it by using the number of components and the depth of random forest which provides best accuracy, so that we could get the best result as possible. The training and testing visualization are given in Figs. 27 and 28, respectively.

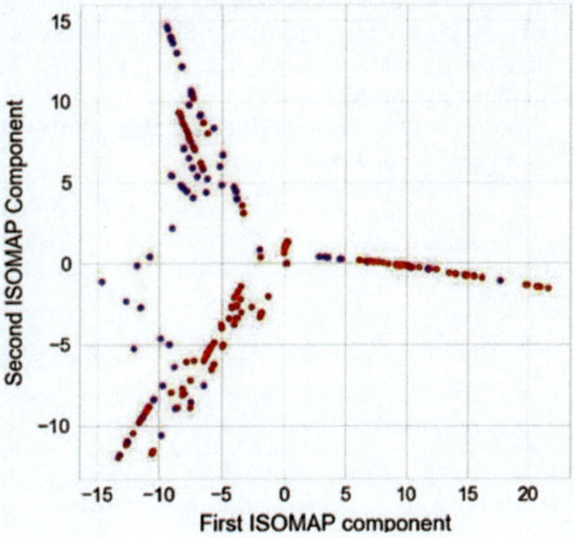

Fig. 27 Training data with 2 components

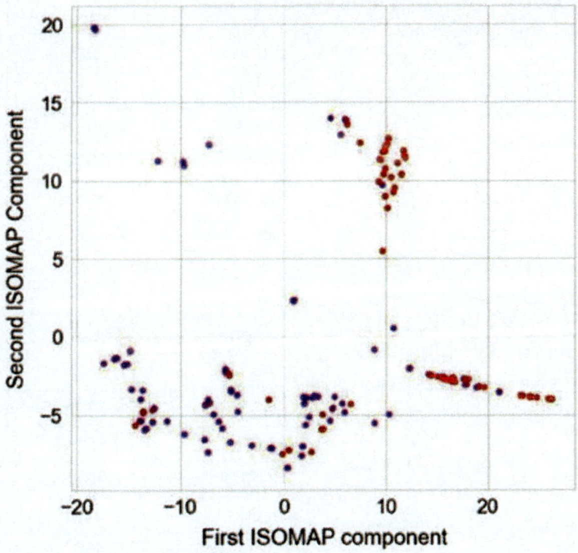

Fig. 28 Testing data with 2 components

5 Results

Now, after using single dimensionality reduction on the data, the result obtained is (Table 1).

As after this, we have found all our FA training data, FA testing data, LCA training data, LCA testing data, SVD training data, SVD testing data, PCA training data, PCA testing data, *t*-SNE training data, *t*-SNE testing data, UMAP training data, UMAP testing data, ISOMAP training data, and ISOMAP testing data with various other algorithms and we get result as

Now, we will be using each of the algorithms with all other algorithms:

See Tables 2, 3, 4, 5, 6, 7, and 8.

Table 1 Dimensionality reduction using different techniques

Algo	Time	Train score	Test score
FA	0.404	0.897	0.856
ICA	0.209	0.724	0.737
SVD	0.413	0.839	0.829
PCA	0.165	0.839	0.829
t-SNE	2.553	0.839	0.723
UMAP	6.957	0.853	0.75
ISOMAP	0.645	0.892	0.338

Table 2 After using FA with all other algorithms

Algo	Time	Train score	Test score
ICA	0.2095	0.8674	0.8414
SVD	0.1719	0.8414	0.8506
PCA	0.2016	0.8297	0.8506
t-SNE	1.3974	0.8664	0.7225
UMAP	5.4688	0.7889	0.75
ISOMAP	0.4782	0.7838	0.5823

Table 3 After using ICA with all other algorithms

Algo	Time	Train score	Test score
FA	0.2155	0.6157	0.5945
SVD	0.3728	0.7176	0.5304
PCA	0.5654	0.8634	0.8262
t-SNE	1.9215	0.8481	0.8018
UMAP	8.0181	0.8501	0.4329
ISOMAP	8.557	0.8114	0.8262

Table 4 After using SVD with all other algorithms

Algo	Time	Train score	Test score
FA	0.2270	0.8522	0.8293
ICA	0.2244	0.8716	0.8323
PCA	0.2278	0.8869	0.8171
t-SNE	1.6945	0.7819	0.4299
UMAP	6.3744	0.7339	0.4752
ISOMAP	0.4720	0.6575	0.5915

Table 5 After using PCA with all other algorithms

Algo	Time	Train score	Test score
FA	0.291	0.8705	0.8231
ICA	0.209	0.8562	0.8292
SVD	0.2312	0.8868	0.8170
t-SNE	1.4119	0.7471	0.7134
UMAP	6.7346	0.9164	0.6158
ISOMAP	0.5956	0.8562	0.7073

Table 6 After using *t*-SNE with all other algorithms

Algo	Time	Train score	Test score
FA	0.2164	0.8083	0.75
ICA	0.1615	0.8287	0.7652
SVD	0.1885	0.8287	0.2957
PCA	0.1562	0.8287	0.3048
UMAP	4.8470	0.6585	0.5792
ISOMAP	0.4539	0.6156	0.5945

Table 7 After using UMAP with all other algorithms

Algo	Time	Train score	Test score
FA	0.2274	0.8491	0.6128
ICA	0.2016	0.7451	0.6585
SVD	0.2126	0.7471	0.6829
PCA	0.2242	0.7451	0.6615
t-SNE	1.6457	0.8674	0.7073

Table 8 After using ISOMAP with all other algorithms

Algo	Time	Train score	Test score
FA	0.2237	0.8165	0.5426
ICA	0.1718	0.7971	0.6280
SVD	0.1746	0.8165	0.4969
PCA	0.1758	0.8165	0.4969
t-SNE	2.1081	0.7206	0.6463
UMAP	3.6003	0.6575	0.5396

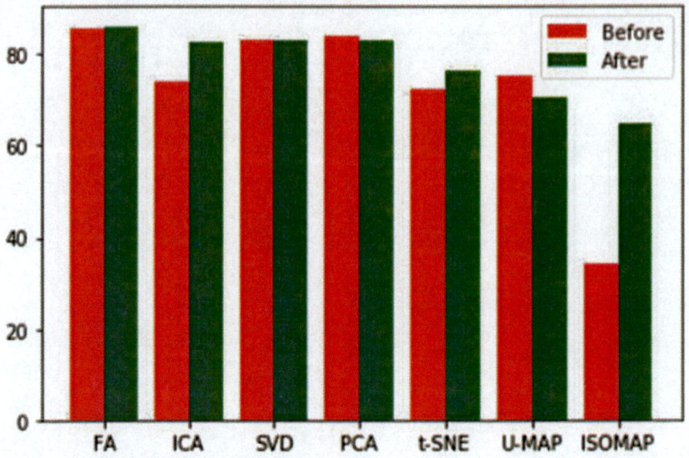

Fig. 29 Graph showing change in accuracy

6 Comparison

In this experiment, the accuracy has been increased which is shown in following figures and tables. This first table shows the previous training time compared with the improved score along with the algorithm which is used in a combination (Fig. 29).

Algorithm	Previous accuracy	Improved accuracy	Combined algorithm
FA	0.856	0.8526	ICA
ICA	0.737	0.8262	PCA
SVD	0.829	0.8323	ICA
PCA	0.829	0.8292	ICA
t-SNE	0.723	0.7632	ICA
UMAP	0.75	0.7073	*t*-SNE
ISOMAP	0.338	64.32	*t*-SNE

In this graph, red color represents the accuracy; we are having while using single dimentionality reduction and the green color shows the improved accuracy.

7 Conclusion and Future Work

In this experimental study, we have combined various dimensionality reduction techniques with each other, and as a result, we can conclude that on using this procedure; we can improve the training and testing score of various algorithms. In this, we have improved the accuracy of factor analysis to 0.856, ICA to 0.8262 from 0.737 by using it in a combination with PCA, the accuracy of SVD is increased to 0.8323 from 0.829 by using it in a combination with ICA, the accuracy of PCA is 0.839, the accuracy of t-SNE is improved to 0.7652 from 0.723, UMAP accuracy is 0.75, and the accuracy of ISOMAP is increased to 0.6463 from 0.338 by using it in combination with t-SNE.

We have seen that dimensionality reduction has tremendous crucial applications, and it can be used in various fields where the amount of dimensions is very large. It extensively reduces the redundancy in data and helps in achieving only the potential features that are of importance to us. Some of the very important features of this technique can be the space reduction and improved accuracy. It also effectively removes the multicollinearity which in turn enhances the interpretation of the underlying parameters of our respective machine learning model.

Also, as mentioned earlier it averts us from the "Curse of Dimensionality." It can be actively used in neuroscience and technology when high-dimensional data is expected to be used. Speech signal digital photograph f-MRI scans all those real-world data which have extremely high dimensionality. As we have to handle this real-world data adequately, we have to reduce it dimensionality, and for doing so, we need this concept of dimensionality reduction. With the wide use of the technique, it is imperative to apply it with careful consideration and to cite pragmatic motivation for its use.

References

1. Jimenez LO, Landgrebe DA (1997) Supervised classification in high-dimensional space: geometrical, statistical, and asymptotical properties of multivariate data. IEEE Trans Syst Man Cybern, Part C (App Rev) 28(1):39–54
2. Lathauwer L, Vandewalle J (2004) Dimensionality reduction in higher-order signal processing and rank-(R1, R2, RN) re- duction in multilinear algebra. Linear Algebra Appl 391:31–55. https://doi.org/10.1016/j.laa.2004.01.016
3. Silva VD, Tenenbaum JB (2003) Global versus local methods in nonlinear dimensionality reduction. Adv Neural Inf Process Syst 15
4. Jain R, Gupta D, Khanna A (2018) Usability feature optimization using MWOA. In: Bhattacharyya S, Hassanien A, Gupta D, Khanna A, Pan I (eds) International conference on innovative computing and communications (ICICC2018). Lecture notes in networks and systems, vol 56. Springer, Singapore
5. Lee J, Verleysen M (2008) Quality assessment of nonlinear dimensionality reduction based on K-ary neighborhoods. J Mach Learn Res Proc Track 21–35
6. Fukunaga K (1990) Introduction to statistical pattern recognition. Academic Press Professional Inc., San Diego, CA, USA
7. Breiman L (2001) Random Forests. Mach Learn 45:5–32
8. Gao X, Wen J, Zhang C (2019) An improved random forest algorithm for predicting employee turnover. Math Probl Eng 2019:1–12. https://doi.org/10.1155/2019/4140707
9. Denil M, Matheson D, de Freitas N (2013) Narrowing the gap: random forests in theory and in practice. In: 31st international conference on machine learning, ICML 2014, vol 2
10. Bryant FB, Yarnold PR (2001) Principal-component analysis and exploratory and confirmatory factor analysis
11. Bouzalmat A, Kharroubi J, Zarghili A (2014) Comparative study of PCA, ICA, LDA using SVM classifier. J Emerg Technol Web Intell 6. 10.4304/jetwi.6.1.64–68

12. Ikhlas AQ, Krause V, Abu-Amara F, Abudayyeh O (2014) Comparative study of deconvolution algorithms for GPR bridge deck imaging. WSEAS Trans Signal Proc 10:9–20
13. Saidi S, Holland C, Kreil D et al (2004) Independent component analysis of microarray data in the study of endometrial cancer. Oncogene 23:6677–6683
14. Langlois D, Chartier S, Gosselin D (2010) An introduction to independent component analysis: in-foMax and FastICA algorithms. Tutorials Quan Methods Psychol 6. https://doi.org/10.20982/tqmp.06.1.p031
15. Makeig S, Bell AJ, Jung T, Sejnowski TJ (1995) Independent component analysis of electroencephalographic data. NIPS
16. Sadek R (2012) SVD based image processing applications: state of the art, contributions and research challenges. Int J Adv Comput Sci Appl IJACSA 3. https://doi.org/10.14569/ijacsa.2012.030703
17. Li Y, Wei M, Zhang F, Zhao J (2017) Comparison of two SVD-based color image compression schemes. PLoS ONE 12(3):e0172746
18. Labib K, Vemuri VR (2006) An application of principal component analysis to the detection and visualization of computer network attacks. Ann. Télécommun. 61:218–234
19. Shlens J (2014) A tutorial on principal component analysis. Educational 51
20. Tao P, Wen X, Wen C (2018) Image recognition based on two-dimensional principal component analysis combining with wavelet theory and frame theory. J Control Sci Eng 2018:1–7. https://doi.org/10.1155/2018/9061796
21. Jolliffe IT, Cadima J (2016) Principal component analysis: a review and re- cent developments. Philosophical Trans R Soc A Math Phys Eng Sci 374:20150202. https://doi.org/10.1098/rsta.2015.0202
22. van der Maaten L, Hinton G (2008) Visualizing data using t-SNE. J Mach Learn Res 9:2579–2605
23. Oliveira FHM, Andrade AA (2018) On the use of t-distributed stochastic neighbor embed-ding for data visualization and classification of individuals with Parkinson's disease. https://doi.org/10.13140/rg.2.2.13290.77764
24. Becht E, McInnes L, Healy J et al (2019) Dimensionality reduction for visualizing single-cell data using UMAP. Nat Biotechnol 37:38–44
25. Lee JA, Lendasse A, Verleysen M (2002) Curvilinear distance analysis versus Isomap. ESANN
26. Bengio Y (2004) Out-of-sample extensions for LLE, Isomap, MDS, Eigenmaps, and spectral clustering. Proc Adv Neural Inf Process Syst 16:177–184

Data Imputation in Wireless Sensor Network Using Deep Learning Techniques

Shweta Rani and Arun Solanki

Abstract Missing information in the sequence of data provided by the wireless sensor network is a prevalent issue. There can be many reasons which may be responsible for this like network loss, sensor maintenance, and sensor failure. Retrieving the missing data from the time series data obtained by the wireless sensor network proves to be a difficult task. There have been many methods which try to recover this data, but limitation still exists. The proposed work discusses the use of deep learning algorithms for the time series prediction of WSN data generated from real-time sensor devices. This paper examines various techniques available in deep learning for time series forecasting and analyzes the results obtained from various methods and at the end, gives the best hybrid combination of which contains the bidirectional LSTM layer for a deep examination of the pattern in the data. The main motivation behind the work is to improve the working environment of the WSN fields, automate various processess of maintenance, and provide an effective method of data imputation in the wireless sensor network in a real-time environment when there is a scarcity of data and also, develop a method for effective forecasting. Another approach that is examined is a CNN layer to find the positional pattern in the data, attention mechanism to put more focus on the relevant part of the sequence, an LSTM layer which makes it an encoder–decoder model, and a dense layer at the end to produce the output in the desired shape. The model is trained and tested on Beijing Air Pollution PM2.5 Dataset. To overcome the problem of lower availability of data, VLSW algorithm is used, which helped in generating the large sample of training data from the limited available datasets. After analyzing and studying the various models and their results with various hyperparameters tried, it is concluded that the model with the LSTM attention mechanism with the encoder–decoder model along with VLSW works best for long missing time series data imputation. Other studied models and their results are summarized too. This work concludes based on the practical and real-time application of data imputation that the CNN model with online training

S. Rani · A. Solanki (✉)
Gautam Buddha University, Greater Noida, India
e-mail: ymca.arun@gmail.com

S. Rani
e-mail: shwetanimesh700@gmail.com

A. Khanna et al. (eds.), *Data Analytics and Management*, Lecture Notes on Data Engineering and Communications Technologies 54,
https://doi.org/10.1007/978-981-15-8335-3_44

works best as it takes less time and resources to train. The results obtained are better from the existing solution available. The SSIM model with VLSW performed 31% more efficient than other methods, while CNN without VLSW performed 76% more efficient than other methods.

1 Introduction

Wireless sensor network comprises the spatially scattered and committed sensors for observing and recording the states of being of the surroundings and arranging the gathered information at a focal area [1]. Sensors provide data for various purposes using which various decisions are taken by our systems. Usually, sensors provide a stream of data related to the time. Data generated by the WSN devices is time series data that finds significance in many critical applications [2, 3]. Due to various reasons like communication outage, sensor maintenance or failure, data sequence may be missing, which leads to various consequences like unavailability of data to research or to make a decision upon [4]. Various approaches are available to deal with missing sequence data, which are discussed further. Linear regression, KNN [5, 5], ARIMA, SARIMA, etc., are the machine learning methods that are used to deal with missing time series data [6–8]. ANN, RNN [9], LSTM, CNN, etc., are the deep learning methods that can be used for the imputation of missing time series data [10–12]. There is not much research about using the deep learning models in time series data imputation in the WSN field because there are some limitations in doing that. WSN is a resource-constrained device whose capabilities are limited, so they are not able to produce large amounts of data enough to be used for deep learning. And without much data, deep learning performance cannot be considered to be acceptable for real application. Also, as the resources are limited for WSN devices, there is a minimal possibility of applying deep learning on end-node devices. The computation and time problem can be easily solved by using cloud computing for processing of data, and to deal with the limited data, VLSW protocol is used. Models proven in the NLP task are applied to solve the imputation problem [13]. In this study, the dataset is taken from an open-source UCI machine learning repository. Dataset consists of values of multiple parameters like pm2.5, TEMP, PRES, etc., that are necessary to detect the level of air pollution in Beijing city [14]. Time series imputation and forecasting find its importance in many real-world applications of WSN especially where they need some scheduling of maintenance work depending on the condition of the environment monitored by the WSN devices, and there are very fewer possibilities of human interventions like thermal power plants where the temperature is much high for a human to tolerate. Agricultural fields are vast enough for humans to not intervene.

2 Time Series Forecasting and Imputation

Times series data is a sequence of data present in a known time gap. Time series forecasting is a part of ML in which future values are predicted using the previously observed values [15, 16]. There are two ways to deal with missing data; the first is to omit the entire record that contains missing values. Another one is to impute the missing values. The main problems in using deep learning in time series forecasting are that due to the low computing power of the devices, it is not possible to run online training on them and there is significantly little data provided by the WSN devices [17–20].

3 Organization of Research Paper

The research paper starts with an introduction about the WSN domain, which covers its applicability with the use of machine learning techniques. In the second section, data imputation and time series forecasting are discussed. The third section consists of the organization of the research paper, which is followed by the fifth section, i.e., literature review. The fifth section discussed the architecture of the proposed system is discussed. The sixth section is having a process flow diagram of the proposed system. The seventh section is about the implementation of the proposed system, and the results are discussed in the eighth section. The ninth section has a conclusion.

4 Literature Review

Kim and Kim [21] proposed a hybrid model, LSTM-CNN model, to predict the prices of stock by combining features from various representations of data. This model outperforms here in comparison with single models because here LSTM+CNN both used to extract temporal features and image features [22]. Verma and Kumar [23] worked on LSTM-based deep learning imputation model for exact missing time series forecasting for health care. The author uses the MIT-BIH database for training the model. The LSTM model proposed by the author is for both five-step and ten-step forecasting. The performance achieved by LSTM mode is far better than in comparison with both the models, Gaussian process regression and linear regression model. The currently known methods of ARIMA, SARIMA, and other machine learning techniques try to fit the trend in the data using nonlinearity. Further combination of ARIMA and Kalman filter [24, 25] is used by the space model and provides better results. Multivariate imputation by chained equations [26] initializes the values which are missing in time series data randomly, and then, elimination of each missing variable is done using the chain equations. Yi et al. [27] proposed a method in which multiple views of learning are used for predicting the missing time series values of air quality data, and the score of 17% mean relative error is achieved. Wang et al. [28] applied collaborative filtering method to the recommendation system in order to fill the missing time series values. Yuan et al. [29] applied the LSTM model on air pollution time series data and prove that the LSTM model outperforms the statistical method like mean, moving average method, etc., and achieve better accuracy in predicting PM2.5 values of time series data. Che et al. [30], in his paper, used imputation technique called GRU-D to be applied for healthcare data which performed very well. This method is based upon GRU for missing time series data imputation. Sutskever et al. [31], in this paper, introduced a new unique approach in which vector-to-vector translation was done using sequence-to-sequence learning in which deep multi-layer LSTM is used to decode the final result. The authors used a sentence dataset which contains the English sentences to convert it to French. BLEU score of 34.8 is obtained on the test dataset. The authors also analyzed that reversing the source sentences can lead to an increase in the performance of LSTM by making optimization easier [24]. Cho et al. [32], the creator proposed the recurrent neural network-based encoder–decoder model. This model is comprised of two recurrent neural networks. One of the RNN would be utilized to encode input sentences into a fixed-length vector, and the other would be utilized to disentangle the fixed-length vector. Lachtermacher et al. [18] focus on a major problem of training an artificial neural network, a deep learning model, which is lack of data due to which forecasting in real world with accuracy has become very difficult. The author uses an idea of identification of lag components in time series, compacting the structure of the network and doing validation with the help of the Box–Jenkins method. This approach significantly reduces the requirement for data. Normally, data is stored on cloud in some database like SQL, which can be populated using some WSN sensor [33].

In the above discussion, various literatures in the field of missing time series data imputation have been discussed. With the study of these works, it is analyzed that machine learning or deep learning techniques [34, 35] can play an important role in the imputation of missing time series data [19, 36]. It is also analyzed and understood that this work would be done by implementing encoder–decoder architecture of sequence-to-sequence learning [12, 37, 38].

5 Model Architecture

Two architectures of deep learning models are used with various combinations with VLSW, and hyperparameters are introduced. Also, few hybrid combinations of the two are also implemented [39].

Figure 1 depicts that model used in architecture is encoder–decoder model using LSTM module. In contrast, Fig. 2 depicts that the model used in architecture is the CNN model and the rest of the modules are same in both figures; only the model module is different in the architecture of both figures [39, 40].

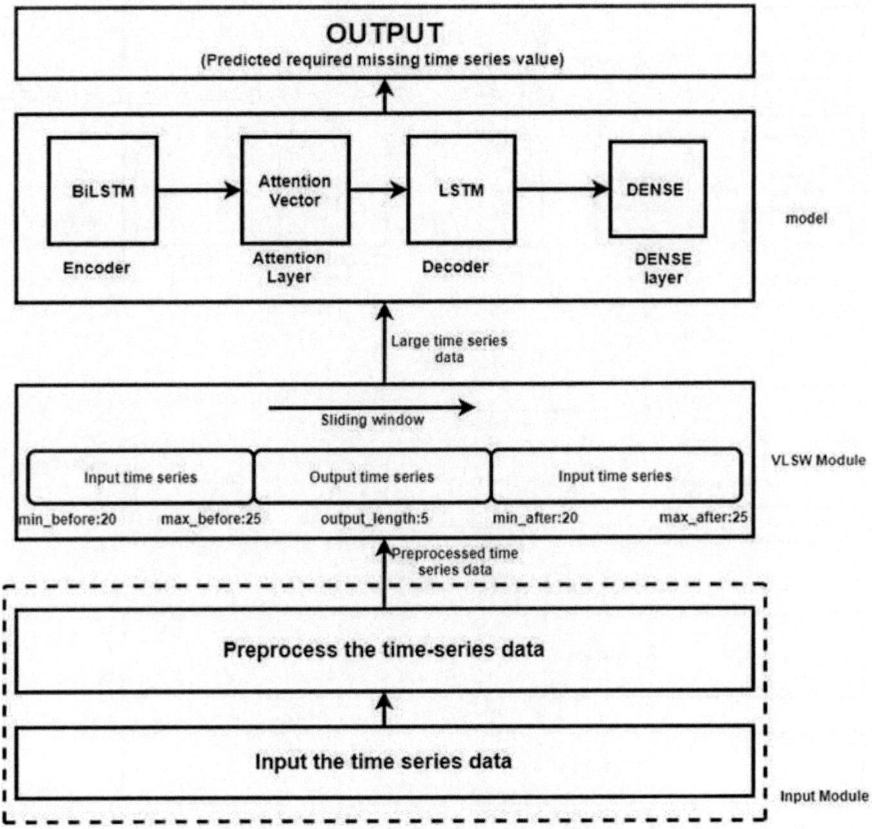

Fig. 1 Architecture of the sequence-to-sequence imputation model)

5.1 Input Module

This input module consists of two submodules:

1. Input time series data
2. Preprocess the time series data

In the first submodule, time series data is fed. This submodule passes the time series data into the next submodule named as preprocess time series data. This submodule preprocesses the time series data to make it able to input into the next modules of the system.

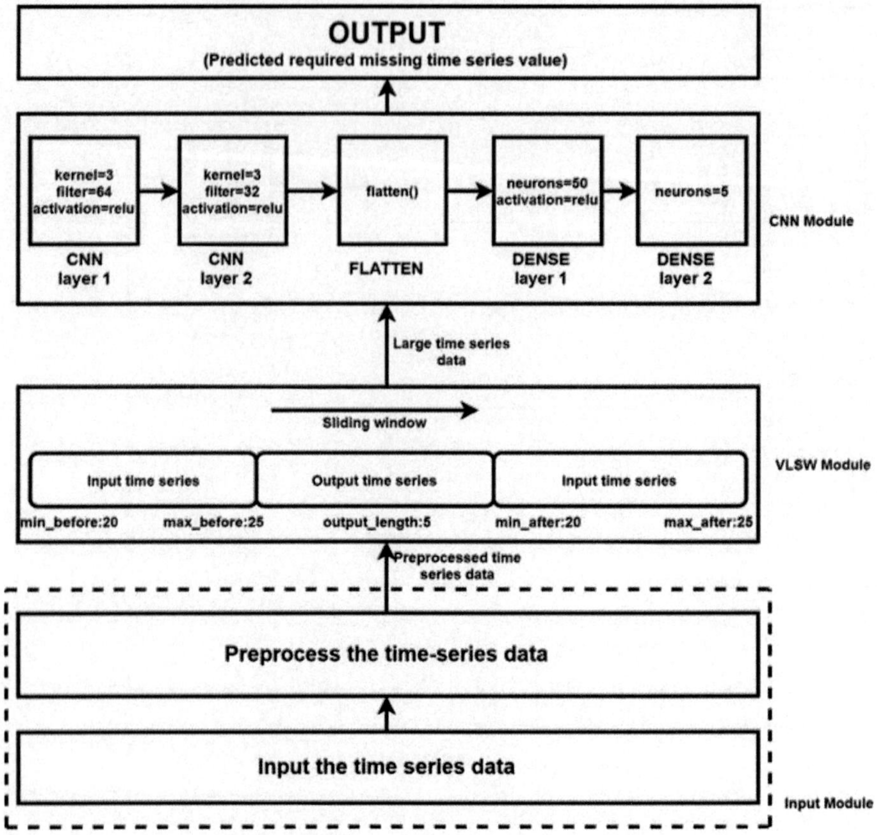

Fig. 2 Architecture of the CNN model with VLSW protocol

5.2 *Variable Length Sliding Window*

This is a method which is used to extend the time series data; it uses both the previous and afterward data [41].

5.3 *Model Module*

A large number of data obtained from VLSW module is fed into a deep learning model module, i.e., CNN OR SSIM model module to train the proposed model.

CNN model module consists of five processes which are:

1. CNN Layer 1
2. CNN Layer 2

3. FLATTEN
4. DENSE Layer 1
5. DENSE Layer 2

Both CNN layer 1 and CNN layer 2 have a kernel size of 3×3 and "relu" activation function. In these layers, a convolution operation is done in which elements of a 2D array of weights called kernel of size 3×3 are multiplied with the corresponding elements of the input time series data. Then, the results of the multiplication are added together. Then, the result created is a feature map which is then passed through a nonlinearity, i.e., relu. After that, a flatten layer is used before a dense layer to reduce the feature maps into 1D array. Then, a fully connected neural network (dense layer) is applied with the number of perceptron being 50 that interprets the features extracted by the convolution part. After that, another dense layer of five neurons which is equal to the length of output time series in VLSW module is applied to get the prediction as an output of the model with the given input.

And the SSIM model module consists of four processes which are:

1. Encoder
2. Attention layer
3. Decoder
4. Dense layer

BiLSTM is used as the encoder which processes the data sequence in both directions, i.e., in the forward direction and in the backward direction. It checks for the pattern and learns its attention mechanism that is used to focus on a particular part of the data sequence, which is necessary to generate the output sequence.

5.4 Output Module

The output module provides an array of values (same as the number of output specified by the model) as the final prediction.

This model is largely inspired by the models used for text processing, machine translation and sentiment analysis problems [42–45].

6 Flowchart

See Fig. 3.

Step 1: Data gathering Data is collected from wireless sensor network devices. The collected data is loaded in our system. And various libraries like NumPy, pandas, Scikit learn, Matplotlib, etc., are imported.

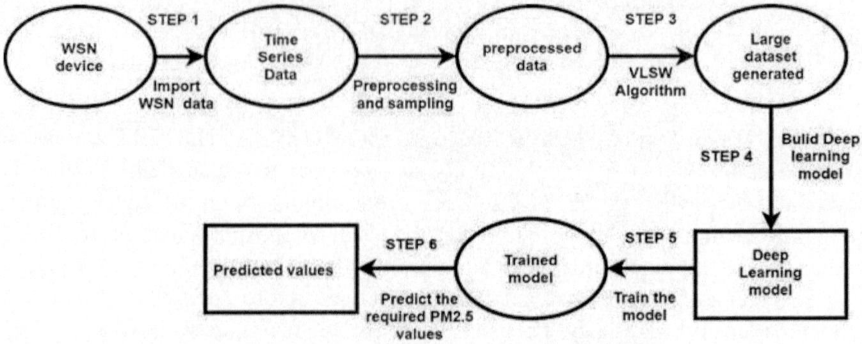

Fig. 3 Process flow diagram of the proposed system in the step following steps

```
import numpy, matplotlib, pandas
data = pandas.reasd_csv("datset_location")
```

Step 2: Data preprocessing and sampling Data is processed to make it able for training. Also, data is divided into training and testing samples.

```
One Hot Encoding
Eliminating missing data from the training set
StandardScaler(Data)
train_test_split(X, y)
```

Step 3: VLSW algorithm Apply variable length sliding window algorithm to increase data sample for training

```
VLSW(X,y)
```

Step 4: Building the model Using the TensorFlow and Keras library of Python, the above-mentioned models are built using the specified parameters and hyperparameters. Two model architectures explained in the diagram are built and trained.
Step 5: Train the model Model created in the last step is trained with data processed in step 4. Training is done for the 20 epochs.

```
model.fit(X, y, epochs=20)
```

Step 6: Predict the required PM2.5 value The trained model is used to predict the pm2.5 values from the testing set, and the actual and predicted values are compared for the accuracy of the model.

```
y_hat = model.predict(X_test)
plt.plot(y_hat, y_test)
```

7 Implementation of Proposed Work

This work used an open-source time series dataset from the UCI machine learning repository. The dataset was collected by placing the PM2.5 sensors at the US Embassy in Beijing. Also, the data from Beijing Capital International Airport is included in this. This dataset contains information about the amount of PM2.5 (ug/m^3) concentration of particles from 36 monitoring stations in the air of the US Embassy in Beijing. These are particles with a diameter of fewer than 2.5 μm that are floating in the air. The measurements are hourly collected from 2/01/2010 to 31/12/2014. This study used 70% of the data as the test data and 30% of the data as the training data.

Table 1 summarizes the dataset characteristics which consist of average max and standard deviations.

The model is trained for 20 epochs with 25 GB RAM on Intel(R) Xeon(R) CPU @ 2.20 GHz processor.

The real implementation of such time series data-based deep learning models can be seen in various fields, including industries and agriculture. These models are used to recover the missing time series data when wireless sensor networks unable to capture the data may be due to any reason like maintenance of sensor, communication problem or any human error. And by using these models, analysis on obtained time series data from WSN can be done, and future values can be predicted. So that, decision on various issues like future failure of machines can be taken.

In industries, various parameters of thermal power plants like temperature can be controlled using the present data values and future values of the condition of thermal plants easily by sitting far away from the thermal plant.

In this way, in agriculture, a greenhouse can be maintained by analyzing the conditions of water level, sunlight and temperature.

Table 1 Dataset characteristics

Parameters	Average	Max	Min	Standard deviation
PM 2.5	97.784	129	0	91.398
Dew point	1.828	28	−40	14.43
Temperature	12.46	42	−19	12.2
Pressure	1016.45	1046	991	10.27
Cumulative wind speed (LWS)	23.89	585.6	0.45	50.02
Cumulative hours of snow (LS)	0.053	27	0	0.76
Cumulative hours of rain (LR)	0.2	36	0	1.42
Wind direction 1	0.323	1	0	0.467
Wind direction 2	0.349	1	0	0.477
Wind direction 3	0.214	1	0	0.41

8 Result Analysis

This section contains the analysis of the results obtained.
Libraries and Framework Used:

- Numpy—Numerical Python is a library for numerical and mathematical computations in Python language.
- Matplotlib—Library to plot graphs of various kinds. This library is used to plot the graphs of various results.
- Pandas—This library of Python is used for data loading and preprocessing. This library is used for preprocessing the data.
- Scikit-Learn—Library used for scaling the data and for regularization.
- Tensorflow—Library to build deep learning models.
- Keras—This is a deep learning library built over TensorFlow.

Implementation in a real-world scenario: After training the models for 20 epochs, the following is the training versus validation loss curve and actual versus predicted graph of pm2.5 values.

Figure 4 shows the loss per epoch (20) while training the CNN model on 5000 samples using VLSW. It signifies that the loss decreases exponentially over the epochs. Graph 2 shows a plot of actual vs predicted PM2.5 values. Predicted values are generated by the CNN model on 5000 samples using VLSW. As the two plots overlap each other mostly, this signifies that the CNN model mostly predicted the correct PM2.5 values.
Effect of applying VLSW for extending the dataset. This study found that the resulting graph that is using VLSW, results get better in comparison to without VLSW as it can be seen in graph 3 and graph 4. The following table summarizes the results of all the models built and trained on the same dataset. For the comparison, the loss values are shaded according to their value. Darker color means higher value, and lighter background color means lower values which means better performance.

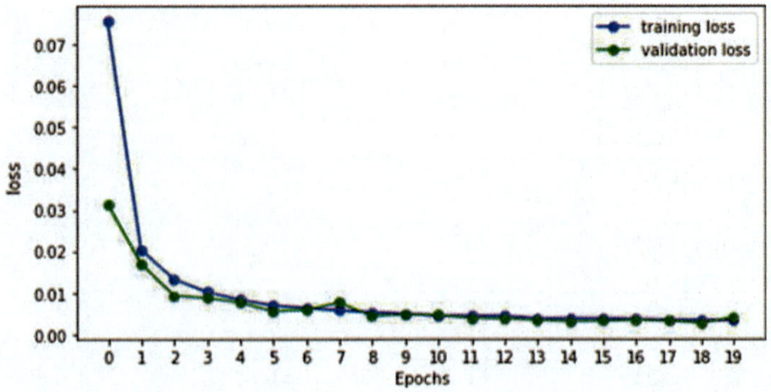

Fig. 4 Training versus validation loss curve over the training epochs

Table 2 Summary of all the models trained and tested for loss

	With VLSW		Without VLSW	
	Training loss	Validation loss	Training loss	Validation loss
LSTM (full dataset)	[a]	[a]	0.0514	0.0821
LSTM (5000)	0.0023	0.0017	0.0717	0.0944
CNN (full dataset)	[a]	[a]	0.0306	0.0764
CNN (5000)	0.0033	0.0025	0.0129	0.0862
SSIM (full dataset)	[a]	[a]	0.0527	0.0775
SSIM (5000)	0.0015	0.0014	0.0956	0.1337
LSTM+CNN (full dataset)	[a]	[a]	0.0482	0.0759
LSTM+CNN (5000)	0.0027	0.0024	0.0516	0.1119
BiLSTM+CNN (full dataset)	[a]	[a]	0.0375	0.0729
BiLSTM+CNN (5000)	0.0022	0.0016	0.0405	0.1059

[a] Not feasible to apply VLSW with a full dataset as it will generate an extremely large dataset

The following is the inference which can be drawn from Table 2.

1. Overall performance has increased when VLSW protocol is used.
2. SSIM model performs best with VLSW.
3. But when a model is trained with fewer data without VLSW, the SSIM model performs the worst.
4. Without VLSW for training with fewer data, CNN model performs best.
5. Some hybrid models with LSTM and CNN are also trained for better performance, but they perform in between not worst but not best also.

Table 3 is the table summarizing the model configuration of all the models trained (Figs. 5, 6 and 7).

9 Conclusion

By analyzing and studying various models of deep learning for time series forecasting and imputation, it concludes that when there is a scarcity of data, then the SSIM model with VLSW protocol is best because 0.0014 (mse loss for SSIM with VLSW) is less

Table 3 Summary of configuration all the models trained and tested for loss

S. No.	Model	Configuration
1	LSTM (full dataset)	LSTM(128), LSTM(64), LSTM(5), optimizer=adam, loss=mse
2	LSTM (5000)	LSTM(128), LSTM(64), LSTM(5), optimizer=adam, loss=mse
3	CNN (full dataset)	Conv2D(filters=64, kernel_size=3, activation='relu'), Conv2D(filters=32, kernel_size=3, activation='relu'), Flatten(), Dense(50, activation='relu'), Dense(5)
4	CNN (5000)	Conv2D(filters=64, kernel_size=3, activation='relu'), Conv2D(filters=32, kernel_size=3, activation='relu'), Flatten(), Dense(50, activation='relu'), Dense (5)
5	SSIM (full dataset)	Bidirectional(LSTM(128, return_sequences=True)), SeqSelfAttention(attention_activation='sigmoid'), LSTM(64), Dense(5), optimizer=adam, loss=mse
6	SSIM (5000)	Bidirectional(LSTM(128, return_sequences=True)), SeqSelfAttention(attention_activation='sigmoid'), LSTM(64), Dense(5), optimizer=adam, loss=mse
7	LSTM+CNN (full dataset)	LSTM(128, return_sequences=True),(Conv1D(filters=32, kernel_size=3, activation='relu'), Flatten(), Dense(O), optimizer=adam, loss=mse
8	LSTM+CNN (5000)	LSTM(128, return_sequences=True), Conv1D(filters=32, kernel_size=2, activation="relu')), Flatten(), Dense(5), optimizer=adam, loss=mse
9	BiLSTM+CNN (full dataset)	Bidirectional(LSTM(128, return_sequences=True)), Conv1D(filters=32,kernel_size=2,activation='relu'), Flatten(), Dense(5), optimizer=adam, loss=mse
10	BiLSTM+CNN (5000)	Bidirectional(LSTM(128, return_sequences=True)), Conv1D(filters=32,kernel_size=2,activation='relu'), Flatten(), Dense(5), optimizer=adam, loss=mse

than (0.0017, 0.0025, 0.0024, 0.0016) compared to other models. When there is sufficient data to train a deep learning model, and it is not feasible to apply VLSW, then, CNN model performs best and should be used because 0.0306 (mse loss for CNN without VLSW) is less than (0.0514, 0.0527, 0.0482, 0.0375) compared to other models. Also, it is more feasible to train online the CNN model in comparison with SSIM so that it should also be considered when choosing the solution. When there is a chance of batch training and no need for online training, SSIM performs better. By all the study, this can be concluded that the encoder–decoder model generally used for NLP problems can be successfully applied to the time series forecasting problem, which proves to be an advantage. VLSW proves to be useful for training

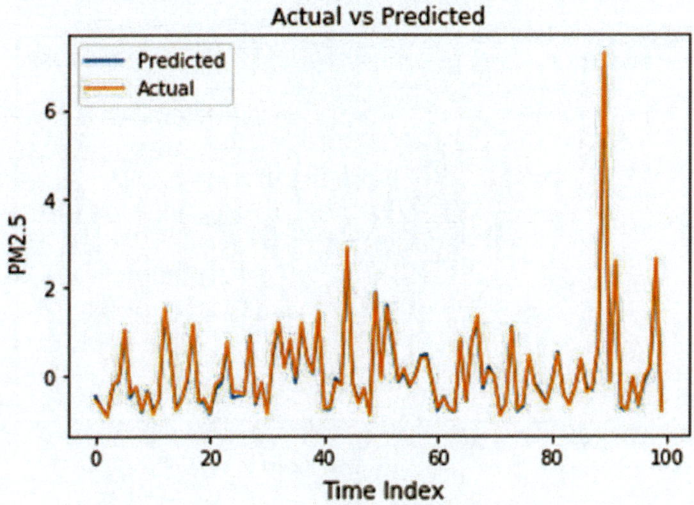

Fig. 5 Actual versus predicted graph of PM2.5 values

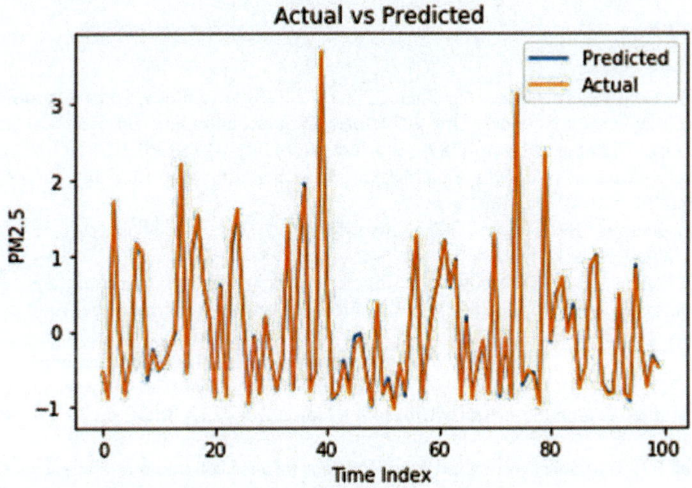

Fig. 6 SSIM with VLSW

deep learning model when data is scarce. SSIM model proves to be best only when used with small data with VLSW; otherwise, CNN performs best when sufficient data is available.

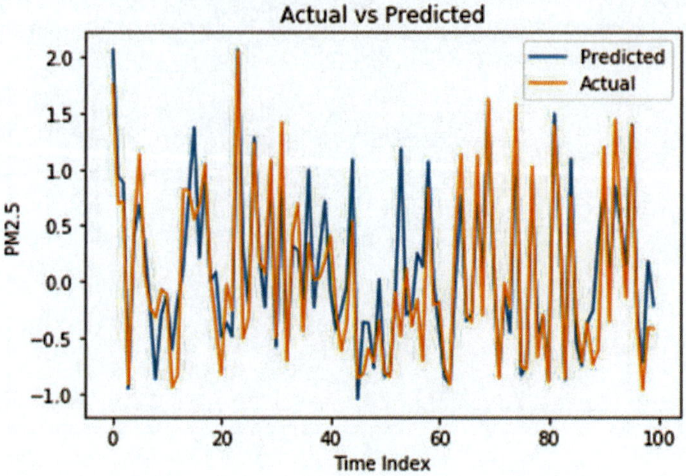

Fig. 7 SSIM without VLSW

References

1. Wang H, Yang G, Xu J, Chen Z, Chen L, Yang Z (2011) A novel data collection approach for Wirelsee Sensor Networks. In: 2011 international conference on electrical and control engineering, Yichang, 2011, pp 4287-4290. https://doi.org/10.1109/ICECENG.2011.6057687.
2. Agarwal A, Solanki A (2016) An improved data clustering algorithm for outlier detection. Self-organology 3(4):121–139
3. Box GE, Jenkins GM, Reinsel GC, Ljung GM (2015) Time series analysis: forecasting and control. Wiley, Hoboken, NJ
4. Zhou J, Huang Z (2018) Recover missing sensor data with iterative imputing network. In: Proceedings of workshops 32nd AAAI Conference on Artificial Intelligence, New Orleans, LA, USA, Feb 2018, pp 209–215
5. Ahuja R, Solanki A (2019) Movie recommender system using K-Means clustering and K-Nearest Neighbor. In: Confluence-2019: 9th international conference on cloud computing, data science & engineering, Amity University, Noida, vol 1231, no 21, pp 25–38 (accepted for publication)
6. Lv P, Yue L (2011) Short-term wind speed forecasting based on non-stationary time series analysis and ARCH model. In: 2011 international conference on multimedia technology, Hangzhou, 2011, pp 2549–2553. https://doi.org/10.1109/ICMT.2011.6002447
7. Thissen U, Brakel RV, Weijer APD et al (2003) Using support vector machines for time series prediction. Chemometr Intell Lab Syst 69(1–2):35–49
8. Natarajan VA, Karatampati P (2019) Survey on renewable energy forecasting using different techniques. In: 2019 2nd international conference on power and embedded drive control (ICPEDC), Chennai, India, 2019, pp 349–354. https://doi.org/10.1109/ICPEDC47771.2019.9036569
9. Singh T, Nayyar A, Solanki A (2020) Multilingual opinion mining movie recommendation system using RNN. In: Singh P, Pawłowski W, Tanwar S, Kumar N, Rodrigues J, Obaidat M (eds) Proceedings of first international conference on computing, communications, and cyber-security (IC4S (2019). Lecture notes in networks and systems, vol 121. Springer, Singapore
10. Che Z, Purushotham S, Cho K et al (2018) Recurrent neural networks for multivariate time series with missing values. Sci Rep 8:6085. https://doi.org/10.1038/s41598-018-24271-9

11. Xue N, Triguero I, Figueredo GP, Landa-Silva D (2019) Evolving deep CNN-LSTMs for inventory time series prediction. In: 2019 IEEE congress on evolutionary computation (CEC). https://doi.org/10.1109/cec.2019.8789957
12. Zhang GP (2003) Time series forecasting using a hybrid ARIMA and neural network model. Neurocomputing 50:159–175
13. Wang B, Sun Y, Xue B, Zhang M (2018) Evolving deep convolutional neural networks by variable-length particle swarm optimization for image classification. arXiv preprint arXiv:1803.06492
14. Du S et al (2018) Deep air quality forecasting using hybrid deep learning framework. arXiv preprint arXiv:1812.04783
15. Du S, Li T, Gong X, Yang Y, Horng SJ (2017) Traffic flow forecasting based on hybrid deep learning framework. In: 2017 12th international conference on intelligent systems and knowledge engineering (ISKE), Nov 2017, pp 1–6
16. Siami-Namini S, Tavakoli N, Siami Namin A (2018) A comparison of ARIMA and LSTM in forecasting time series, pp 1394–1401. https://doi.org/10.1109/ICMLA.2018.00227
17. Graham JW (2009) Missing data analysis: making it work in the real world. Annu Rev Psychol 60:549–576 Jan
18. Lachtermacher G, Fuller JD (1995) Back propagation in time-series forecasting. J Forecast 14(4):381–393
19. Hyndman RJ, Koehler AB (2006) Another look at measures of forecast accuracy. Int J Forecast 22(4):679–688
20. Rana S, John AH, Midi H (2012) Robust regression imputation for analyzing missing data. In: 2012 international conference on statistics in science, business and engineering (ICSSBE), Langkawi, 2012, pp 1–4. https://doi.org/10.1109/ICSSBE.2012.6396621
21. Kim T, Kim HY (2019) Forecasting stock prices with a feature fusion LSTM-CNN model using different representations of the same data. PLoS ONE 14(2):e0212320
22. Cho K, van Merriënboer B, Gulcehre C, Bahdanau D, Bougares F, Schwenk H, Bengio Y (2014) Learning phrase representations using RNN encoder–decoder for statistical machine translation
23. Verma H, Kumar S (2019) An accurate missing data prediction method using LSTM based deep learning for health care. In: Proceedings of 20th international conference on distributed computing and networking, pp 371–376
24. Harvey AC (1990) Forecasting, structural time series models and the Kalman filter. Cambridge University Press, Cambridge
25. Fung DS (2006) Methods for the estimation of missing values in time series
26. Azur MJ, Stuart EA, Frangakis C, Leaf PJ (2011) Multiple imputation by chained equations: what is it and how does it work? Int J Methods Psychiatr Res 20(1):40–49
27. Yi X, Zheng Y, Zhang J, Li T (2016) ST-MVL: filling missing values in geo-sensory time series data
28. Wang J, De Vries AP, Reinders MJ (2006) Unifying user-based and item-based collaborative filtering approaches by similarity fusion. In: Proceedings of the 29th annual international ACM SIGIR conference on research and development in information retrieval. ACM, pp 501–508
29. Yuan H, Xu G, Yao Z, Jia J, Zhang Y (2018) Imputation of missing data in time series for air pollutants using long short-term memory recurrent neural networks. In: Proceedings of the ACM International Joint Conference on Pervasive and Ubiquitous Computing and Symposium on Wearable Computers, 2018, pp 1293–1300
30. Che Z, Purushotham S, Cho K, Sontag D, Liu Y (2018) Recurrent neural networks for multivariate time series with missing values. Sci Rep 8(1):6085
31. Ilya S, Oriol V, Le QV (2014) Sequence to sequence learning with neural networks. In: Advances in neural information processing systems 27: annual conference on neural information processing systems Montreal. Quebec, Canada, 8–13 Dec 2014, pp 3104–3112
32. Cho K, Van Merriënboer B, Gulcehre C, Bahdanau D, Bougares F, Schwenk H, Bengio Y (2014) Learning phrase representations using RNN encoder-decoder for statistical machine translation. arXiv preprint arXiv:1406.1078

33. Singh G, Solanki A (2016) An algorithm to transform natural language into SQL queries for relational databases. Selforganizology 3(3):100–116
34. Leke C, Twala B, Marwala T (2014) Missing data prediction and classification: the use of auto-associative neural networks and optimization algorithms. arXiv preprint arXiv:1403.5488
35. Tayal A, Köse U, Solanki A, Nayyar A, ve Marmolejo Saucedo JA (2019) Efficiency analysis for stochastic dynamic facility layout problem using meta-heuristic, data envelopment analysis and machine learning. Comput Intell (Basımda). https://doi.org/10.1111/coin.12251
36. Moghar A, Hamiche M (2020) Stock market prediction using LSTM recurrent neural network. Procedia Comput Sci 170:1168–1173. https://doi.org/10.1016/j.procs.2020.03.049
37. Yang J, Nguyen MN, San PP, Li X, Krishnaswamy S (2015) Deep convolutional neural networks on multichannel time series for human activity recognition. In Ijcai, 2015, vol 15, pp 3995–4001
38. Yang Y, Guizhong L (2001) Multivariate time series prediction based on neural networks applied to stock market. In: 2001 IEEE international conference on systems, man and cybernetics. e-systems and e-man for cybernetics in cyberspace (Cat. No. 01CH37236), vol 4. IEEE
39. Lin T, Guo T, Aberer K (2017) Hybrid neural networks for learning the trend in time series. In: Proceedings of the twenty-sixth international joint conference on artificial intelligence, IJCAI-17, pp 2273–2279
40. Pandey S, Solanki A (2019) Music instrument recognition using deep convolutional neural networks. Int J Inf Technol 13(3):129–149
41. Zhang Y-F, Thorburn P, Xiang W, Fitch P (2019) SSIM—a deep learning approach for recovering missing time series sensor data. IEEE Internet Things J 1. https://doi.org/10.1109/jiot.2019.2909038
42. Kaur N, Solanki A (2018) Sentiment knowledge discovery in twitter using CoreNLP library. In: 8th international conference on cloud computing, data science and engineering (confluence), vol 345, no 32, pp 2342–2358
43. Rajput R, Solanki A (2016) Real-time analysis of tweets using machine learning and semantic analysis. In: International conference on communication and computing systems (ICCCS-2016). Taylor and Francis, at Dronacharya College of Engineering, Gurgaon, 9–11 Sept, vol 138, issue 25, pp 687–692
44. Rajput R, Solanki A (2016) Review of sentimental analysis methods using lexicon based approach. Int J Comput Sci Mob Comput 5(2):159–166
45. Priyadarshni V, Nayyar A, Solanki A, Anuragi A (2019) Human age classification system using K-NN classifier. In: Luhach A, Jat D, Hawari K, Gao XZ, Lingras P (eds) Advanced informatics for computing research. ICAICR 2019. Communications in computer and information science, vol 1075. Springer, Singapore

Analysis of COVID-19 Data Using Machine Learning Techniques

Rashmi Agrawal and Neha Gupta

Abstract Coronovirus (COVID-19) has impacted the entire world and researchers across the globe are working day and night to identify and predict the patterns related to it. Hundreds of clinical trials are underway to generate the possible cure of the disease. The devastating and uncontrolled worldwide spread of COVID-19 triggered unprecedented global lock-downs and a massive burden on healthcare systems. WHO has recommended an immediate research study of the existing data to understand the care and measures required for COVID-19. In this paper we have listed various machine learning approaches that have been used in the past for the formulation of pandemics e.g., Ebola, H1N1 influenza, Zika, norovirus. Paper also discusses the analysis of COVID-19 patients' data to classify and predict people based on their vulnerability or resistance to potential COVID-19 infection. Recommendation of various machine learning models to predict the pattern of the COVID-19 related parameters has also been presented. We have also analyzed real-time COVID-19 dataset having data from countries across the globe to understand the pattern of the outbreak of coronavirus.

Keywords COVID-19 · Machine learning · Artificial intelligence · Pattern identification · Clustering · Classification

1 Introduction

Coronavirus (COVID-19) is the new form of viral disease that is detected in human beings in 2019. Because it has never before been reported earlier that is why it is known as the novel coronavirus. Coronavirus includes a huge family of viruses that cause diseases in patients varying from common Cold to advanced breathing

R. Agrawal · N. Gupta (✉)
Manav Rachna International Institute of Research and Studies, Faridabad, India
e-mail: neha.fca@mriu.edu.in

R. Agrawal
e-mail: rashmi.fca@mriu.edu.in

A. Khanna et al. (eds.), *Data Analytics and Management*, Lecture Notes on Data Engineering and Communications Technologies 54,
https://doi.org/10.1007/978-981-15-8335-3_45

syndromes, for example Middle East Respiratory Syndrome (MERS-COV). On 12 January 2020, the World Health Organization (WHO) confirmed the presence of coronavirus in a cluster of people in Wuhan city of China that was the cause of a respiratory illness in those bunches of people. The COVID-19 virus spreads primarily through droplets of saliva or discharge from the nose when an infected person coughs or sneezes. No particular vaccine or therapy is currently available for COVID-19. However, several clinical trials are under ways that are testing alternative therapies to develop the medicine or the vaccine for the disease [1]. Thousands of research teams across the globe are integrating their data collection activities and working hard to get the possible solutions. The aim of this paper is to study the effect of several attributes on the spreading of coronavirus CoVID-19 across the globe on real-time collected data and published reports. The paper will discuss the use of various machine learning approaches that can be used to build forecasting models (Yang et al. 2020). Section 1 of the paper is introductory in nature having an emphasis on the basics and background of coronovirus. Section 2 will discuss various AI and machine learning approaches to be used to predict the patterns of Covid-19 based on associated parameters. The last section of the chapter presents data analysis on a real-time dataset to show the correlation between the variables.

2 Pattern Identification for COVID-19

Seeing the current scenario, it is very important to identify the patterns in patient infection. These patterns can be identified on the basis of below mentioned parameters:

1. Patient Gender
2. Patient Age
3. Pre-Existing Conditions of a patient
4. Demographic profile of a patient
5. Nationality of patient
6. Response to a drug
7. Death rate and its relation with age and gender
8. Socio Economic status of a patient
9. Quarantine measures.

We have tried to list some of the general parameters. Advanced parameters can also be taken into consideration such as:

1. Virus-host-protein interaction
2. Protein folding
3. Drug combination and testing
4. Virus seasonality.

Most of the governments are collecting the clinical data to identify and understand the pattern of the virus. For example:

As published by Indian website (Covid19india.org), from a sample of 5207 patients across the country, 66% of the infected patients are male while only 33% are female which largely shows the inclination of males to get infection easily.

Similarly the data is analyzed to check the effectiveness of quarantine measures. A MIT team has developed a machine learning model using coronavirus data and a neural network to assess the efficiency of quarantine steps and to predict the progression of the disease.

3 Use of Machine Learning and AI in Medical Field to Predict the Patterns

Artificial intelligence (AI) detects trends from big data and can play an important role in detecting and analyzing the patterns of COVID-19. COVID-19 is a global pandemic and will expose some of the key shortcomings of AI. Machine learning works by identifying patterns in historical training data [2]. Under current situations, gathering the new training data is a tedious task. However AI has the potential to exceed humans not only through speed but also by detecting patterns in that training data that humans have overlooked.

Machine learning analysis of genetic variants from asymptomatic, mild or severe COVID-19 patients can be performed to classify and predict people based on their vulnerability or resistance to potential COVID-19 infection, by which the machine learning model can also return those prioritized genetic variants, in their decision-making the process as important features for functional and mechanistic studies [3] (Fig. 1).

ML approaches have been used in the past as well for the formulation of pandemics e.g., Ebola, cholera, pig fever, H1N1 influenza, dengue fever, Zika, norovirus, but in the literature there is a divergence from the COVID-19 peer-review articles [4]. Table 1 demonstrates impressive ML methods used in the prediction of past outbreaks. For example Random Forest technique has been used to predict the outbreak of swine flu [5].

4 Suggested Machine Learning and AI Approaches to Predict the Patterns of COVID-19

In this section we have tried to integrate machine learning approaches with the COVID-19 patterns identified in Sect. 2 of the paper. Table 2 will suggest various AI and machine learning approaches that can be used to predict and analyze the patterns related to COVID-19.

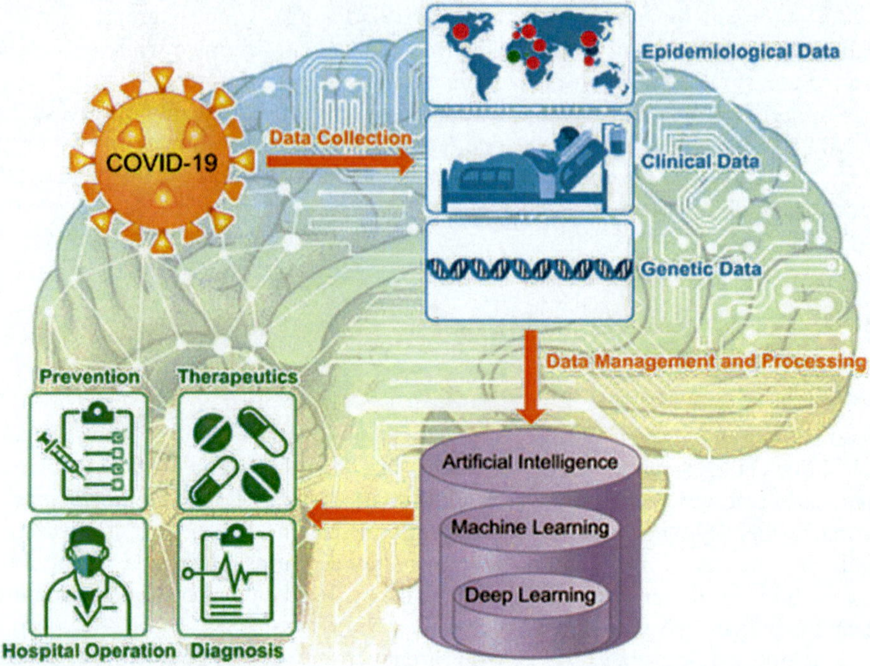

Fig. 1 Application of artificial intelligence and machine learning in the fight against COVID-19

5 Analysis of COVID-19 Data Using Real Datasets

In this section we have analyzed the data related to Covid-19 using real time data set which was downloaded during the corona outbreak. Data set contains the corona virus instances of the countries across the world.

Dataset Source: https://www.ecdc.europa.eu/en/publications-data/download-todays-data-geographic-distribution-covid-19-cases-worldwide

Step by step demonstration of the analysis is given below:

Step 1: Data Downloaded on April 23, 2020 based on the geographic distribution of COVID-19 cases worldwide.

Snapshot of data (First few lines of data is shown in Fig. 2).

Figure 3 shows the last few instances of the data.

So we can see that data of all countries are arranged as per the name of the country. It is also clear that we have 12,595 instances in the dataset.

Step 2: Counting number of instances:

By executing the command df.count (), we can have the details of instances available in each column which is also shown in Fig. 4.

Step 3: Displaying statistical summary of the dataset

Table 1 ML methods used in the prediction of outbreaks

S. No.	Disease	Machine learning approach used outbreak prediction
1	Swine fever (published in Transboundary and Emerging Diseases Journal) [5]	Random forest
2	Dengue fever (published in Geospatial Health journal) [6]	Neural network
3	Influenza (published in BMC Research Notes journal) [7]	Random forest
4	Dengue/Aedes (Journal of Public Health Medicine) [8]	Bayesian network
5	Dengue (published in Informatica journal) [9]	LogitBoost
6	H1N1 flu (published in Global Ecology and Biogeography journal) [10]	Neural network
7	Oyster norovirus (published in Environment International journal) [11]	Neural Network
8	Oyster norovirus (published in Water Research journal) [12]	Genetic programming
9	Dengue (published in Current Science journal) (Agrawal et al. 2019)	Adopted multi-regression and Naïve Bayes
10	Dengue (published in Infectious Disease Modelling Journal) [13]	Classification and regression tree (CART)

We analyzed the statistical summary of the dataset using python. Statistical summary of any dataset is a very powerful feature to know the details of the dataset. Here we can find out the count, mean value, standard deviation, min and max value and the quartiles of the dataset. The details are shown in Fig. 5.

Step 4: Finding Correlation between variables:

The Table 3 displays the correlation of number of cases and number of deaths with respect to number of days, months and year

Step 5: Generating Heat map of the Correlation:

Heat map of correlation is generated to show a 2D correlation matrix (table) between two discrete dimensions or event types. Here the heat map is generated to show the correlation between number of cases w.r.t days, months and year.

Df2 = df[df['cases'] >=100]

df2.count()

It gives the value as 1751 means that there are 1751 instances available on which the case count is more than 10. By performing the filter

df2[df2['countriesAndTerritories'] =='India']

we find that there are 27 days on which the number of covid cases have been reported more than 100 (Fig. 6).

Table 2 Various AI and machine learning techniques and their effectiveness in predicting parameters of COVID-19

S. No.	AI or machine learning technique	Parameter to be predicted
1	Time series analysis	Effect of quarantine measures Linear and non linear patterns of the virus Virus seasonality Death rate Virus-host-protein interaction Bouncing of economy
2	Classification algorithms	To monitor the number of active cases Number of cases treated at home Realtionship of death with age and gender Pre-existing Conditions of a patient Demographic profile of a patient Nationality of patient Areas with community transmissions
3	Clustering algorithms	To monitor the geographic distribution of cases Socio Economic status of a patient Areas with community clusters
4	Logical regression	To measure the death rate Disease prevention steps
5	Linear regression	Virus seasonality
6	Binary classification algorithm	Treatment using a particular drug Virus-host-protein interaction
7	Decision trees	Risk mitigation
8	Correlation	Effect of regular sanitization of places

	https://www.ecdc.europa.eu/en/novel-coronavirus-china/sources-updated	day	month	year	cases	deaths	countriesAndTerritories	geoId	countryterritoryCode	popData2018	continentExp
0	23-04-2020	23	4	2020	84	4	Afghanistan	AF	AFG	37172386.0	Asia
1	22-04-2020	22	4	2020	61	1	Afghanistan	AF	AFG	37172386.0	Asia
2	21-04-2020	21	4	2020	35	2	Afghanistan	AF	AFG	37172386.0	Asia
3	20-04-2020	20	4	2020	88	3	Afghanistan	AF	AFG	37172386.0	Asia
4	19-04-2020	19	4	2020	63	0	Afghanistan	AF	AFG	37172386.0	Asia

Fig. 2 Snapshot of data showing first few lines of data

	https://www.ecdc.europa.eu/en/novel-coronavirus-china/sources-updated	day	month	year	cases	deaths	countriesAndTerritories	geoId	countryterritoryCode	popData2018	continentExp
12591	25-03-2020	25	3	2020	0	0	Zimbabwe	ZW	ZWE	14439018.0	Africa
12592	24-03-2020	24	3	2020	0	1	Zimbabwe	ZW	ZWE	14439018.0	Africa
12593	23-03-2020	23	3	2020	0	0	Zimbabwe	ZW	ZWE	14439018.0	Africa
12594	22-03-2020	22	3	2020	1	0	Zimbabwe	ZW	ZWE	14439018.0	Africa
12595	21-03-2020	21	3	2020	1	0	Zimbabwe	ZW	ZWE	14439018.0	Africa

Fig. 3 Snapshot of data showing last few lines of data

```
https://www.ecdc.europa.eu/en/novel-coronavirus-china/sources-updated    12596
day                                                                      12596
month                                                                    12596
year                                                                     12596
cases                                                                    12596
deaths                                                                   12596
countriesAndTerritories                                                  12596
geoId                                                                    12556
countryterritoryCode                                                     12462
popData2018                                                              12493
continentExp                                                             12596
dtype: int64
```

Fig. 4 Instances in dataset

	day	month	year	cases	deaths	popData2018
count	12596.000000	12596.000000	12596.000000	12596.000000	12596.000000	1.249300e+04
mean	15.653938	2.935773	2019.994681	205.486662	14.513179	5.747734e+07
std	8.471115	1.269893	0.072741	1479.377532	115.632723	1.882847e+08
min	1.000000	1.000000	2019.000000	-9.000000	0.000000	1.000000e+03
25%	9.000000	2.000000	2020.000000	0.000000	0.000000	2.886376e+06
50%	16.000000	3.000000	2020.000000	1.000000	0.000000	1.018318e+07
75%	22.000000	4.000000	2020.000000	24.000000	0.000000	3.797855e+07
max	31.000000	12.000000	2020.000000	37289.000000	4928.000000	1.392730e+09

Fig. 5 Statistical summary of dataset

Table 3 Correlation table

	Day	Month	Year	Cases	Deaths	popData2018
Day	1.000000	−0.057451	−0.132481	−−0.005611	−0.011824	−0.012526
Month	−0.057451	1.000000	−0.521988	0.079564	0.083343	−0.081604
Year	−0.132481	−0.521988	1.000000	0.010138	0.009179	−0.010857
Cases	0.005611	0.079564	0.010138	1.000000	0.830943	0.107075
Deaths	−0.011824	0.083343	0.009179	0.830943	1.000000	0.073174
popData2018	−0.012526	−0.081604	−0.010857	0.107075	0.073174	1.000000

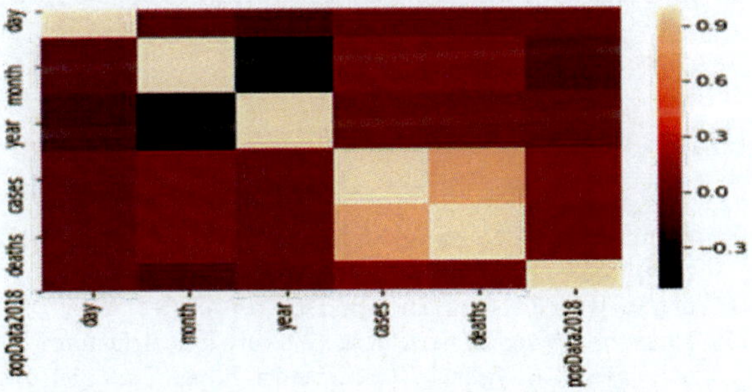

Fig. 6 Heat map

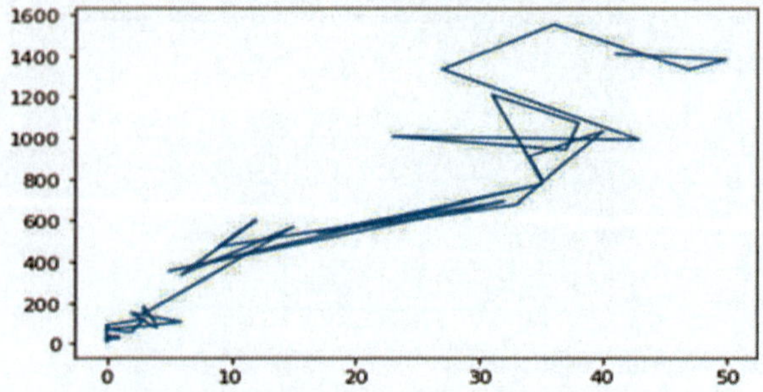

Fig. 7 Growth rate in cases in India (day wise)

Step 6: Graph Representing growth rate in cases in India (day wise)

Figure 7 shows the growth rate of corona virus per day in India. It can be easily concluded that at present the virus is spreading at a comparatively lower rate in India but by the end of May 2020 it will also spread exceedingly with this consistent growth. Sufficient measures need to taken care for this. Much more in-depth analysis of data will also be required in future.

6 Conclusion

The worldwide pandemic of Coronavirus 2 (SARS-CoVID-19) the respiratory syndrome has become many nations' primary national security issue. Providing insights into the spread and effects of the infection is important in creating accurate prediction models for the outbreak. The present paper recommends the use of various AI and machine learning models to predict the pattern of the disease based on various parameters. Analysis of data and finding correlation between the variables based on the dataset have also been provided. The present work proposes ML as an efficient tool to model the time series of outbreaks in relation to the observations reported in this paper. The present work is providing the initial review to demonstrate the potential of machine learning in future research of COVID-19.

References

1. Zumla A, Hui DS, Azhar EI, Memish ZA, Maeurer M (2020) Reducing mortality from 2019-nCoV: host-directed therapies should be an option. Lancet 395(10224):e35–e36. https://doi.org/10.1016/S0140-6736(20)30305-6.9-4393-y
2. Huang C, Wang Y, Li X et al (2020) Clinical features of patients infected with 2019 novel coronavirus in Wuhan, China. Lancet. https://doi.org/10.1016/S0140-6736(20)30183-5
3. Guan W, Ni Z, Hu Y et al (2020) Clinical characteristics of coronavirus disease 2019 in China. New Engl J Med. https://doi.org/10.1056/NEJMoa2002032
4. Shao Y, Wu J (2020) IDM editorial statement on the 2019-nCoV. Infect Dis Model 5:233–234

5. Anno S, Hara T, Kai H, Lee MA, Chang Y, Oyoshi K, Mizukami Y, Tadono T (2019) Spatiotemporal dengue fever hotspots associated with climatic factors in taiwan including outbreak predictions based on machine-learning. Geospatial Health 14:183–194. https://doi.org/10.4081/gh.2019.771
6. Liang R, Lu Y, Qu X, Su Q, Li C, Xia S, Liu Y, Zhang Q, Cao X, Chen Q et al (2020) Prediction for global African swine fever outbreaks based on a combination of random forest algorithms and meteorological data. Transboundary Emerg Dis 67:935–946. https://doi.org/10.1111/tbed.13424
7. Tapak L, Hamidi O, Fathian M, Karami M (2019) Comparative evaluation of time series models for predicting influenza outbreaks: Application of influenza-like illness data from sentinel sites of healthcare centers in Iran. BMC Res Notes 12. https://doi.org/10.1186/s13104-019-4393-y
8. Raja DB, Mallol R, Ting CY, Kamaludin F, Ahmad R, Ismail S, Jayaraj VJ, Sundram BM (2019) Artificial intelligence model as predictor for dengue outbreaks. Malays J Public Health Med 19:103–108
9. Iqbal N, Islam M (2019) Machine learning for dengue outbreak prediction: a performance evaluation of different prominent classifiers. Informatica 43:363–371. https://doi.org/10.31449/inf.v43i1.1548
10. Koike F, Morimoto N (2018) Supervised forecasting of the range expansion of novel non-indigenous organisms: Alien pest organisms and the 2009 H1N1 flu pandemic. Global Ecol Biogeogr 27:991–1000. https://doi.org/10.1111/geb.12754
11. Chenar SS, Deng Z (2018) Development of artificial intelligence approach to forecasting oyster norovirus outbreaks along Gulf of Mexico coast. Environ Int 111:212–223. https://doi.org/10.1016/j.envint.2017.11.032
12. Chenar SS, Deng Z (2018) Development of genetic programming-based model for predicting oyster norovirus outbreak risks. Water Res 128:20–37. https://doi.org/10.1016/j.watres.2017.10.032
13. Riou J, Althaus CL (2020) Pattern of early human-to-human transmission of Wuhan 2019 novel coronavirus (2019-nCoV), December 2019 to January 2020. Euro surveillance 25(4)

CapGen: A Neural Image Caption Generator with Speech Synthesis

Akshi Kumar and Shikhar Verma

Abstract Computer vision and natural language processing play a crucial role in automatic depiction of contents in an image. Image captioning is a fundamental problem in the field of artificial intelligence. Machines cannot imitate the human intelligence or the ways humans interact with one another, therefore it stays a continuous undertaking. Automizing the process of understanding and processing the large quantity of multimedia inclusive of images and videos generated everyday have become a major concern in today's market. In this paper, we anticipate present a generative model based on a deep recurrent neural network structure that joins the latest advances in computer vision and natural language translation so as to generate coherent sentences describing an image. In this endeavor, a hybrid framework is proposed utilizing the multilayer convolutional neural network model to produce human apprehensible descriptions of the pictures using a long short-term memory to precisely structure coherent sentences utilizing the partial captions produced from the considered dataset. The convolutional neural network model looks at the objective picture to an enormous dataset of preparing pictures, to produce a precise portrayal utilizing the prepared inscriptions of data. The model is prepared to amplify the probability of the target words based on maximum likelihood estimation using the given training set. We feature the effectiveness of our proposed model utilizing the Flickr8K dataset. Apart from human evaluation, we were able achieve a cumulative score of 0.693 in the BLEU metric using 200 images at random from the validation and test dataset.

Keywords Computer vision · Artificial intelligence · Deep learning · Image captioning

A. Kumar (✉) · S. Verma
Department of Computer Science and Engineering, Delhi Technological University, Delhi, New Delhi 110042, India
e-mail: akshikumar@dce.ac.in

S. Verma
e-mail: shikhar.verma03@gmail.com

A. Khanna et al. (eds.), *Data Analytics and Management*, Lecture Notes on Data Engineering and Communications Technologies 54,
https://doi.org/10.1007/978-981-15-8335-3_46

1 Introduction

Being content creators, we come across a large number of images on a daily basis. However, that is just not limited to us. People from all walks of life come across a large number of images and content from miscellaneous sources. These sources may just be in the forms of newspaper articles or billboards across the street to content on the Internet such as posts and advertisements. All the images which these sources contain would have to be interpreted by the audience themselves. Without having a description linked to them, humans can largely understand them without their detailed captions. One way to generate a brief description, to make the content more comprehensible, is by translating them into a language which most audience is familiar with, in this case, English. This is where image captioning comes into the question [1, 2].

Caption generation is an intriguing computer vision issue where a description in a defined language is created for a given picture. In order to achieve this, the process as defined by Kumar et al. includes the functions from the field of computer vision to comprehend the features in the picture and a language model from the field of NLP to translate the features in the picture into the most probable words and arrange those words semantically based on the rules of the language, in this case, English, to form a coherent sentence [2]. Figure 1 explains the task of image captioning.

Generated caption: Man and woman sitting on a rock near the lake.

True caption: Man and woman sitting together on the large rocks by the lake.

In the recent times, deep learning strategies have accomplished state-of-the-art craftsmanship results on problems based on the problems of image captioning. Based on Karpathy's work [3], on increasing the memory accessible to the GPU's used to

Fig. 1 Input image

prepare the model as well as increasing the training time, we can overcome most of the constraints encountered when dealing with a deep learning architecture model.

In this work, we take step toward enhancing generation of captions by integrating the following into our proposed model:

- GloVe Algorithm:

Most of the encoder–decoder models used for the study of image captioning use word2vec, or own word vectors on a custom corpus, and we decided to go with GloVe which stands for Global Vectors. GloVe is a pretrained model mainly used for vector mapping of words in a given corpus.

- ResNet Convolution Neural Networks:

On comparing different convolutional neural networks, we decided on using ResNet50 for our study not only because it won for being computationally the most efficient in the ImageNet Competition, held in 2015 but also because it has much less parameters in comparison with VGGNet resulting in sparing of computational assets.

- On Flickr8k dataset:

Since the amount of information increases as we use comparatively larger dataset, this led to more refined results. However, working with a baseline benchmark dataset, i.e., Flickr8k, enabled us to work upon the features of the images within our computation throughput and also motivated us to get enhanced results compared to the previous studies on the same.

2 Related Work

The caption generation from images is a very old problem in computer vision and several solutions to this problem have been proposed. Researchers and data scientists from different parts of the world have come up with different algorithms and procedures to conquer this problem to be able to find a unified solution.

This problem was more focused on generating natural language descriptions, or captions, in short from visual data but mainly for video, for which complex systems consisting of visual primate recognizers were studied [2]. These models, however were trained in very specific domains, such as traffic control, sports, and biomedicine.

In the recent times, still image descriptions are slowly becoming the topic of interest. Krizhevsky et al. [4] built up a neural network utilizing non-immersing neurons and a dreadfully productive novel technique GPU execution of the convolution function. They prevailing with regard to decreasing overfitting by making use of a regularization technique called as Dropout. Their neural network model comprised of maxpooling layers with a dense layer of 1000-way softmax for classification. In order to find out intrinsic details of the internal correspondence's visual information and language, Karpathy et al. [1] utilized datasets of pictures and their sentence

Table 1 Related literature studies

References (et al.)	Year	Image encoder	Language model
Mao [9]	2015	AlexNet, VGGNet	RNN
Chen [10]	2015	VGGNet	RNN
Karpathy [1]	2015	VGGNet	RNN
Vinyals [6]	2015	GoogleNet	LSTM
Xu [7]	2015	AlexNet	LSTM
Jin [11]	2015	VGGNet	LSTM
Wang [12]	2016	VGGNet	LSTM
Ma [13]	2016	AlexNet	LSTM
Yao [14]	2017	GoogleNet	LSTM
Park [15]	2017	ResNet	LSTM
Aneja [5]	2018	VGGNet	Language CNN
Wang [16]	2019	ResNet	RNN
Gao [8]	2019	ResNet	LSTM

depictions. Their work portrayed a multimodal recurrent neural detail that uses the gathered colinear structuring of highlights in order to be advised the best approach to produce novel portrayals of pictures. Aneja et al. [5] on the other hand proposed a convolutional neural network model for AI and restrictive image generation. Vinyals et al. [6] introduced a generative model comprising of a deep learning structure that use AI to produce coherent descriptions of an image by guaranteeing most elevated likelihood of the created sentence to precisely depict the objective picture. Xu et al. [7] presented an attention-based model that figured out how to clarify the image features naturally. The model was prepared utilizing generalized backpropagation strategies by amplifying a variable bound. The model automatically learnt to distinguish object limits while at the same time create an exact caption of the image. Gao et al. [8] on the other hand noticed that the models with attention-based mechanism as decoders evaluated the non-visual words ("a," "an," "the") using the visual signals, which could be easily prepared using natural language processing and therefore proposed a hierarchical LSTM with adaptive attention mechanism to enhance overall performance Table 1.

3 Dataset

In order to tackle problems based on image classification and caption generation, several open-source benchmark datasets were available. Statistics of a few such datasets are given in Table 2.

Table 2 Open-source benchmark datasets

Name of dataset	Dataset split		
	Training	Validation	Test
Flickr8K	6000	1000	1000
Flickr30K	28,000	1000	1000
MSCOCO [17]	82,783	40,504	40,775

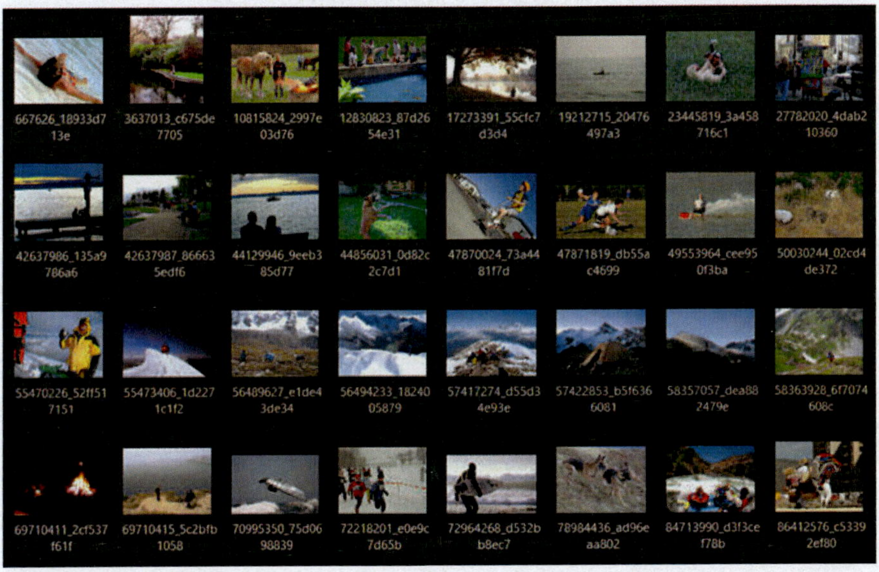

Fig. 2 Dataset images

We found the Flickr 8k dataset as the most suitable considering the computational limitations we faced. Flickr8K, as the name suggests, contains 8000 images, with each image linked to five human-generated captions, each having different scale of relevance to the image. Figure 2 depicts the snapshot of the dataset images.

4 The Proposed CapGen Model

The model consists of the following phases:

4.1 Automatic Feature Extraction

To represent images, each image is translated into a vector of fixed length structure using the convolutional neural network model known as ResNet50 image encoder, consisting of 50 deep layers. ResNet50 model inputs images in the dimensions (224, 224, 3) where the last unit depicts the number of channels, red, blue, and green [15], to extract an output of 2048 fixed length vector representation of the image which will later be passed onto the LSTM layer. By removing the last softmax dense layer from the ResNet model which performs the 1000-class classification, we are able to generate a fixed length feature vector for each image that is then fed into the deep learning model, thereby extracting a 2048 length vector. This process of vector generation is also known as automatic feature engineering.

4.2 Caption Preprocessor

In order to create word embeddings for the words in the vocabulary, created from the captions given in the dataset, every word is mapped to a vector of the 50 fixed length, using a pretrained GloVe model from Stanford NLP. GloVe, which stands for Global Vectors for generating word embeddings, can be defined as an unsupervised learning algorithm for translating words into a defined vector representation. It permits one to take a corpus of text and instinctively translates each word in that corpus into a vector mapping in a high-dimensional space, implying that all homogeneous words will be put together.

4.3 Data Preparation

Data preparation takes place in such a manner so as to make the give data fit to be fed as inputs to the proposed deep learning architecture. For preparing the data for the given model, generator functions were used, which primarily act as iterator of code, in the deep learning model. Considering the basics of deep learning, in order to prepare any proposed neural network on a given dataset considered for training, we employ a stochastic gradient descent (SGD), such as AdaGrad and Adam. While using a stochastic gradient descent, the loss is not computed on the whole training set in order to find new values, i.e., updating the gradients. Instead of that, instead loss is computed on a subset of the training data in each iteration. These batches are usually of the powers of 2, frequently used are 64, 128, 256, 512, etc. This suggests that it is not mandatory to store the complete dataset at once, thereby implying that employing just the current batch of datasets in order to update the values of the gradients is sufficient to solve the purpose. Figure 3 illustrates the summary of the model.

Model: "model_2"

Layer (type)	Output Shape	Param #	Connected to
input_3 (InputLayer)	(None, 35)	0	
input_2 (InputLayer)	(None, 2048)	0	
embedding_1 (Embedding)	(None, 35, 50)	92400	input_3[0][0]
dropout_1 (Dropout)	(None, 2048)	0	input_2[0][0]
dropout_2 (Dropout)	(None, 35, 50)	0	embedding_1[0][0]
dense_1 (Dense)	(None, 256)	524544	dropout_1[0][0]
lstm_1 (LSTM)	(None, 256)	314368	dropout_2[0][0]
add_17 (Add)	(None, 256)	0	dense_1[0][0] lstm_1[0][0]
dense_2 (Dense)	(None, 256)	65792	add_17[0][0]
dense_3 (Dense)	(None, 1848)	474936	dense_2[0][0]

Total params: 1,472,040
Trainable params: 1,472,040
Non-trainable params: 0

Fig. 3 Model summary

4.4 Decoder

Decoder is the final phase of the model which consisted of a feed-forward network [18] where the outputs from the image model as well as the caption model were fed into a layer consisting of 256 neurons. The dense layer at the end has a softmax function so as to predict the target word, out of the 1848 unique possible outcomes stored in the vocabulary prepared from the partial captions in the dataset during the caption preprocessing phase

Since the words are shortlisted based on their maximum probability distribution of occurrence, this process can also be defined as maximum likelihood estimation (MLE), i.e., the word shortlisted is which is without a doubt as demonstrated by the model for the given data.

4.5 Deployment

After the training, the models were stored in the local directory, so as to be locally hosted in a chatbot deployed on a localhost server, so as to enable us to input any image into the recurrent network model and evaluate a description of the image using Flask microweb framework. This recurrent network model was integrated with the speech synthesis framework, using Windows text-to-speech application program interface for speech synthesis, to convert the generated caption from the image given as input in the deployed website into audio samples.

5 Implementation

The model was implemented in a Python environment using Jupyter notebook. The latest version of the Keras framework enabled training of the deep neural networks with TensorFlow as backend. The ResNet convolution neural network was used for the object identification and conversion of the dataset of images into a fixed length vector. Google developed a deep learning library called TensorFlow which provides heterogeneous platform for execution of algorithms. The model architecture is shown in Fig. 4.

Pretrained embedded layer requires us to freeze that layer before training the model, so as to prevent overwriting during the backpropagation. During training phase, an input image along with its assigned captions as fed to the image captioning model. The ResNet convolution network, trained to recognize every single imaginable item, extracts all the features from the image and store it in a fixed length image vector.

The next step is to input these vectors in the deep learning model, which uses long short-term memory and generator functions to assess the target words based upon

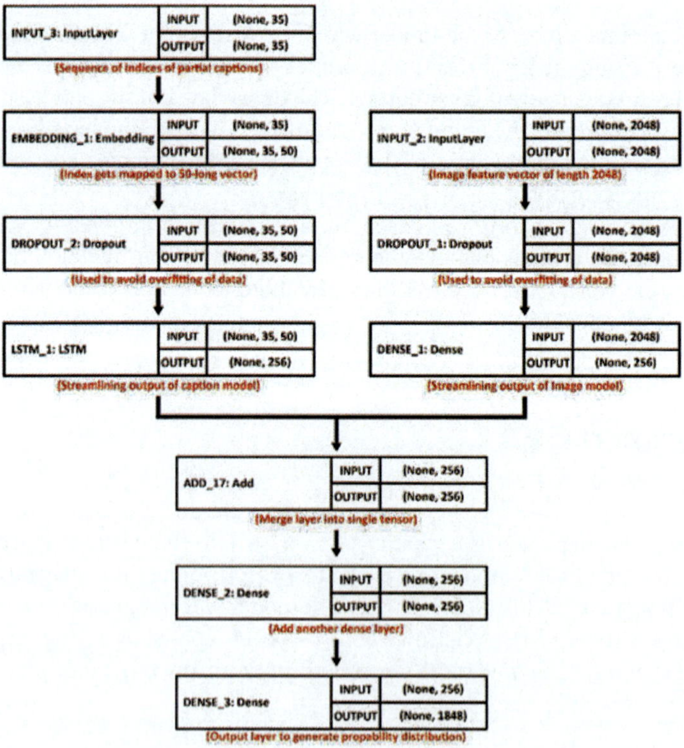

Fig. 4 Model architecture

the image vector of the input image and every predicted word prior to this. Also, two extra tokens were added to each caption to mark head and tail of the string literal. At whatever point stop word was encountered, the model quits the process of the generating sentence further and marks it as end of string. Initially, comparing the model with other studies, the proposed CapGen model was trained for 20 epochs at the learning rate of 0.001, which resulted in feeble outputs. Based on human evaluation, the deep neural network was then trained for 50 epochs with the initial learning rate and pictures per batch as 0.001 and 3, respectively. The deep neural network proposed in this work was trained on the Nvidia GeForce 1050Ti Max-Q graphics processing unit.

6 Results

The execution of the deep neural network model resulted in creating practically identical descriptions in contrast with human produced captions in the dataset. For evaluation of this, we used the BLEU metric score [18], which stands for bilingual evaluation understudy score. The candidate sentence picked and compared with reference sentences in order to evaluate the BLEU Score, in n-grams. Unigrams are considered while drawing in a contrast between the candidate and reference sentences for BLEU-1, while for evaluation of BLEU-2 metric score, bigrams are used for matching, and so on. In order to obtain the best correlation with human-generated descriptions, an upper cap order of 4 is empirically determined. Considering BLEU metric score analysis, the BLEU score evaluated with unigrams accounts for the adequacy, while all the higher *n*-gram scores account for the fluency of the model.

$$p_n = \frac{\sum_{C \in \{\text{Candidates}\}} \sum_{n\text{gram} \in C} \text{Count}_{\text{dip}}(\text{ngram})}{\sum_{C\prime \in \{\text{Candidates}\}} \sum_{n\text{gram}\prime \in C\prime} \text{Count}(\text{ngram}\prime)} \tag{1}$$

BLEU is notable in light of the fact that it is one of the prime metrics in evaluation of machine deciphered substance and has a reasonable association with human choices of significant worth.

The ResNet model first converts the image into fixed length vectors so as to send it as input into the deep learning model, along with the word embeddings from the partial captions generated from the pretrained GloVe model, using the generator function. This is then input into the LSTM cells, which then processes the vectors to form comprehensible sentences using the input vectors. Figure 5 shows the sample input image with its generated caption output. When the given image was input into the deep learning model, the generated caption was, *"Two dogs are playing with each other in the grass."*

The captions in the dataset however were as follows (Fig. 6).

This resulted in the BLEU score [19] of 0.61 for this image. Figure 7 illustrates the comparison with the existing works.

Fig. 5 Caption generation output

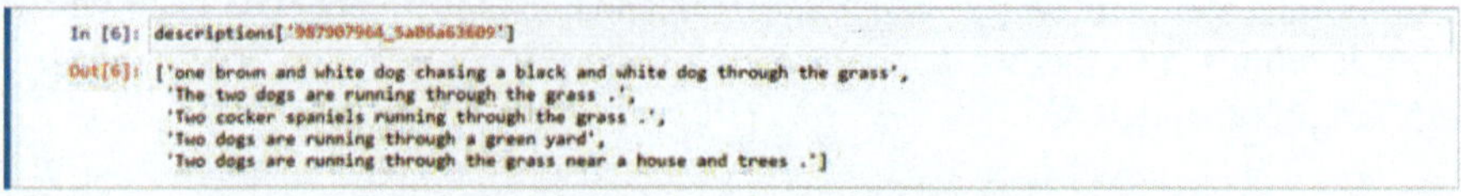

Fig. 6 Human-generated captions in dataset

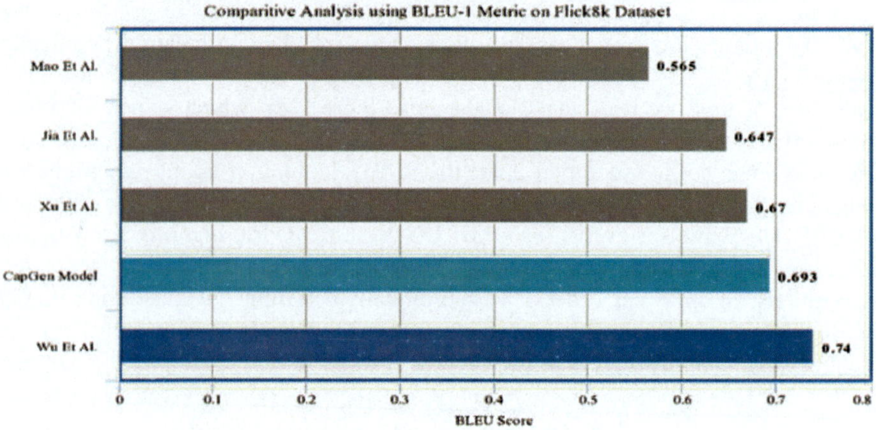

Fig. 7 Comparative analysis

7 Conclusion

With this endeavor, we successfully manage to study the deep recurrent neural network architecture approach for the captioning of images. The use Keras Library to create Merge Models, and utilizing TensorFlow as a backend to train the deep neural network model resulted in translation of images to descriptive human comprehensible sentences which were then translated to audio samples on generation. With this approach, we were able to achieve a cumulative BLEU score of 0.693 for our model, which was validated using 200 pictures from the validation and test dataset. The evaluation metric considered in this undertaking, BLEU score, is evaluated comparing a generated sentence to reference sentences in the dataset. A perfect match where the candidate sentence matches with all the reference sentences achieves a score of absolute 1.0, while in a situation where the candidate sentence does not match with any of the reference sentences results in a score of 0.0.

This study can have many future road tracks: *Aid to the blind*—A guidance system which inputs real-time feed from a camera and relays direction to the physically disabled. *Reviewing CCTV footage*—can generate relevant information which can be retrieved to detect suspicious activities faster. *Automatic captioning*—can enhance Google Image Search by magnifying the search capabilities. However, authors are also planning to use edge detections to extract features from the images and also planning to word by reducing the channels in the convolutional neural network model to reduce the processing time and increase processing speed. This model was hypertuned based on the computational limitations faced; however, different configurations can be considered for assessment of improved execution.

References

1. Andrej K, Li F-F Deep visual-semantic alignments for generating image descriptions. https://cs.stanford.edu/people/karpathy/cvpr2015.pdf
2. Kumar A, Goel S (2017) A survey of evolution of image captioning techniques. Int J Hybrid Intell Syst 14(3):123–139
3. Karpathy A Github repository: "neuraltalk2"; https://github.com/karpathy/neuraltalk2
4. Krizhevsky A, Sutskever I, Hinton G (2017) ImageNet classification with deep convolutional neural networks. ACM 60(6):84–90
5. Aneja J, Deshpande A, Schwing AG (2018) Convolutional image captioning. In: Proceedings of the IEEE conference on computer vision and pattern recognition, pp 5561–5570
6. Vinyals O, Toshev A, Bengio S, Erhan D (2015) Show and tell: a neural image caption generator. In: Proceedings of the IEEE conference on computer vision and pattern recognition, pp 3156–3164
7. Xu K, Ba J, Kiros R, Cho K, Courville A, Salakhudinov R, Zemel R, Bengio Y (2015, June) Show, attend and tell: neural image caption generation with visual attention. In: International conference on machine learning, pp 2048–2057
8. Gao L, Li X, Song J, Shen HT (2019) Hierarchical LSTMs with adaptive attention for visual captioning. IEEE Trans. Pattern Anal. Mach, Intell
9. Mao J, Huang J, Toshev A, Camburu O, Yuille AL, Murphy K (2016) Generation and comprehension of unambiguous object descriptions. In: Proceedings of the IEEE conference on computer vision and pattern recognition, pp 11–20
10. Chen X, Zitnick CL (2015) Mind's eye: a recurrent visual representation for image caption generation. In: 2015 IEEE conference on computer vision and pattern recognition (CVPR), pp 2422–2431
11. Jin J, Fu K, Cui R, Sha F, Zhang C (2015) Aligning where to see and what to tell: image caption with region-based attention and scene factorization. arXiv preprint arXiv:1506.06272
12. Wang M, Song L, Yang X, Luo C (2016) A parallel-fusion RNN-LSTM architecture for image caption generation. In: 2016 IEEE international conference on image processing (ICIP), Phoenix, AZ, pp 4448–4452. https://doi.org/10.1109/icip.2016.7533201
13. Ma S, Han Y (2016) Describing images by feeding LSTM with structural words. In: 2016 IEEE international conference on multimedia and expo (ICME), Seattle, WA, pp 1–6. https://doi.org/10.1109/icme.2016.7552883
14. Yao T, Pan Y, Li Y, Qiu Z, Mei T (2016) Boosting image captioning with attributes. arXiv 2016. arXiv preprint arXiv:1611.01646

15. Chunseong Park C, Kim B, Kim G (2017). Attend to you: personalized image captioning with context sequence memory networks. In: Proceedings of the IEEE conference on computer vision and pattern recognition, pp 895–903
16. Wang W, Hu H (2019) Image captioning using region-based attention joint with time-varying attention. Neural Process Lett 1–13
17. Lin TY, Maire M, Belongie S, Hays J, Perona P, Ramanan D, Dollár P, Zitnick CL (2014) Microsoft COCO: common objects in context. arXiv 2014. arXiv preprint arXiv:1405.0312
18. Hossain MZ, Sohel F, Shiratuddin MF, Laga H (2019) A comprehensive survey of deep learning for image captioning. ACM Comput Surv (CSUR) 51(6):1–36
19. Papineni WJ (2002) Bleu: a method for automatic evaluation of machine translation. In: Proceedings of the 40th annual meeting of the association for computational linguistics. Assoc Comput Linguist 311–318

An Improved Method for Denoising of Electrocardiogram Signals

Nisha Raheja and Amit Kumar Manocha

Abstract Today, the prospect of human's life becomes limited because of cardiovascular diseases (CVDs). The CVDs is the one of leading cause of morality. Electrocardiogram (ECG) is the only tool which measures the electrical activity of the human heart variations in the form of signal. During recording, ECG signal contains various type of noises. So for analysis of ECG signal, noise must be removed. There are different type of noises exist in ECG signal that is baseline wander, power line interference, and EMG. In this paper, an improved method as combination of median filter, Savitzky–Golay (SG) filtering, and wavelet transform is presented for the reduction of noises from the ECG signal. The proposed method is validated on standard database of MIT-BIH for different records and measured in the form of signal-to-noise ratio (SNR) and compared these results with the existing works. The results show that proposed method archived better SNR than that reported in other literature.

Keywords Electrocardiogram · Baseline wander · Power line interference · EMG · Signal-to-noise ratio · Savitzky–Golay filter · Wavelet transform

1 Introduction

The electrocardiogram (ECG) signal is used for the representation of electrical activity of heart. ECG is mostly used because of its low cost and gives us a large amount of diagnosis information. The ECG signal is produced by depolarization and re-polarization of ventricular and atrial muscles of the human heart. The ECG signal consists of P wave due to depolarization of atrial, QRS complex is because of atrial re-polarization and depolarization of ventricular, and T wave is due to re-polarization of ventricular [1]. A typical ECG waveform is shown in Fig. 1. It is also

N. Raheja (✉) · A. K. Manocha
Department of Electrical Engineering, Maharaja Ranjit Singh Punjab Technical University, Bathinda Punjab, India
e-mail: nisha.5165@gmail.com

A. K. Manocha
e-mail: manocha82@gmail.com

A. Khanna et al. (eds.), *Data Analytics and Management*, Lecture Notes on Data Engineering and Communications Technologies 54,
https://doi.org/10.1007/978-981-15-8335-3_47

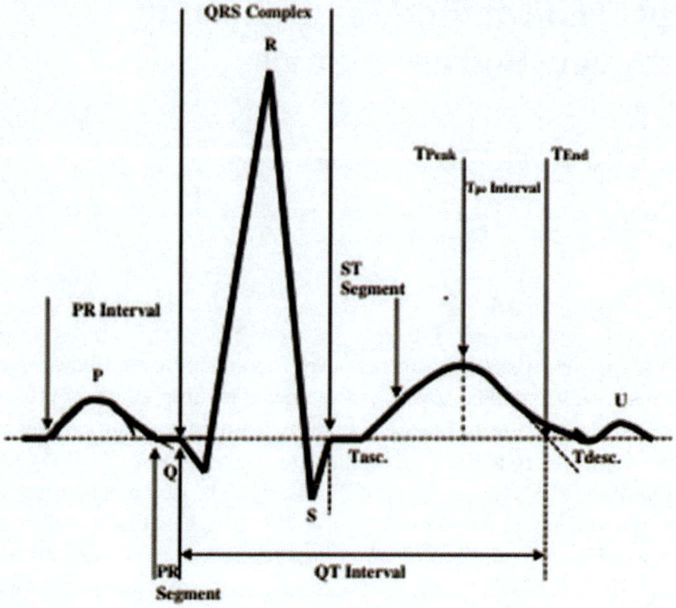

Fig. 1 Components of ECG waveform

used for recording purposes. During recording, it gets affected by various types of noises such as baseline wander, power line interference, and EMG. The analysis of the ECG signal takes place in two steps: preprocessing and feature extraction. In the preprocessing step, there is the removal of noise from the ECG signal; and in feature extraction, the information related to ECG is extracted [2]. In [3], a method based on adaptive filter is proposed for removal of baseline wander noise. In [4], an analog integrated circuit based on telecardiology system is proposed for wireless ECG system. In [5], a method based on adaptive filter is proposed for removal of noise and strengthening the ECG signal for real-time monitoring system. For noisy input, a bias compensated least mean square algorithm is proposed in [6]. In [7], a least mean square algorithm is presented for removal of power line interference. For removal of artifacts from ECG signal, an S-Median thresholding technique is presented in [8]. For the reduction of additive white Gaussion noise in ECG signal, a non-local wavelet transform is proposed in [9]. For continuously monitoring the heart activity, a new biotelemetry-based system is presented in [10]. For rural population, a method based on telecardiology is proposed for the detection of heart disease in real time [11].

In [12], a discrete wavelet-based method is presented in this paper for removal of various artifacts, i.e., baseline wander, power line interference, and EMG. In [13], hybrid IIR/FIR filter and wavelet transform are proposed in this paper for signal denoising and reduction of various noises. In [14], the dual-tree complex wavelet transforms, and logistic regression is proposed for hearing loss identification and for detection of heart diseases. This paper covers the research gap, and these research gaps motivate us to do work in this direction and to achieve the objectives. These are the research gap which is recently viewed by the author according to literature. The proposed work in this paper represents a novel technique to solve the problem of smoothing the noised records of ECG signal. This paper explains the flow diagram for removal of baseline wander and high frequency noise. In this flow diagram, a new filter Savitzky–Golay (SG) and wavelet are used for removal of artifacts present in ECG signal. The baseline wander is a very low frequency noise that is generated from the movement of body during recording. So, it displaced the level of ECG from zero up and down that makes very difficult to estimate the low frequency. So, this is saturated by using median filter and if ECG signal not comes to zero level, it shows that there is a presence of low frequency, so this is removed by Savitzky–Golay (SG) filter that completely smooths the noised waveform. The proposed method is validated on standard database of MIT-BIH for different records and measured in the form of signal-to-noise ratio (SNR) and compared these results with the existing works. The results show that proposed method archived better SNR than that reported in other literature.

The paper has been divided into five sections; first section is introduction to ECG and also explains the noises in ECG signal, second section includes materials and methods, third section is simulated results and discussion, fourth section makes the comparison with existing methods, and last section includes conclusions.

1.1 Noises in ECG Signal

1.1.1 Baseline Wander

This noise is mainly due to electrode which is not placed properly on the chest of human, movement of patient body, and respiration. The frequency range of this noise is 0.5–0.6 Hz. It is a very low frequency, so it is difficult to detect the baseline wander noise [15].

1.1.2 Power Line Interference

This type of noise occurs when the ECG machine is not properly equipped with a proper filter; then, a thick line takes place to get overlapping the ECG waveform components. This noise makes the analysis of ECG signal very difficult for doctors. Power line interference (PLI) is having the frequency range of 50/60 Hz. So, it is very important to remove this noise [15].

1.1.3 Electromyographic (EMG) Noise

This noise is caused by muscles activities. It creates the problem in ECG signal during recording. It overlaps the P, QRS complex, and T component which creates the problem for analysis of ECG waveform components [15].

2 Materials and Methods

2.1 ECG Database

Various type of database are available in PhysioBank that consists of various signals recordings, and these recordings are used by biomedical research communities. In this paper, MIT-BIH database is used. This type of database comprises of 48 signals and each signal of half an hour duration and sampled at frequency 360 Hz. In this database, the number of beats is 110,007 [16].

2.2 Method

The analysis of biomedical signals can be done both in time domain and in time-frequency domain using Fourier transform and Wavelet transform. The Fourier transform is used for the analysis of a stationary signal in frequency domain. It is not suitable for analysis of non-stationary data, so this problem can be resolved by wavelet transform. The wavelet transform is used for the analysis of signal both in time and frequency domain. It is a tool used for minimization of noise and to analyze the components of a non-stationary signal. The wavelet function is defined as:

$$W_s f(x) = f(x)\Psi(x) = \frac{1}{s}\int f(t)\Psi\left(\frac{x-t}{s}\right)\mathrm{d}t \tag{1}$$

where s is a scaling factor and $\Psi_s(x) = \frac{1}{s}\Psi\left(\frac{x}{s}\right)$ is pre-dilation of basic wavelet (x) with scaling factor s.

For $s = 2^j$, the wavelet transform is called digital wavelet transform (DWT). The wavelet transform of a digital signal is calculated by mallet algorithm as given in Eqs. (2) and (3)

$$s_2^j f(n) = \sum h_k s_2^{j-1} f\left(n - 2^{j-1}k\right) \tag{2}$$

$$W_2^j f(n) = \sum g_k W_2^{j-1} f\left(n - 2^{j-1}k\right) \tag{3}$$

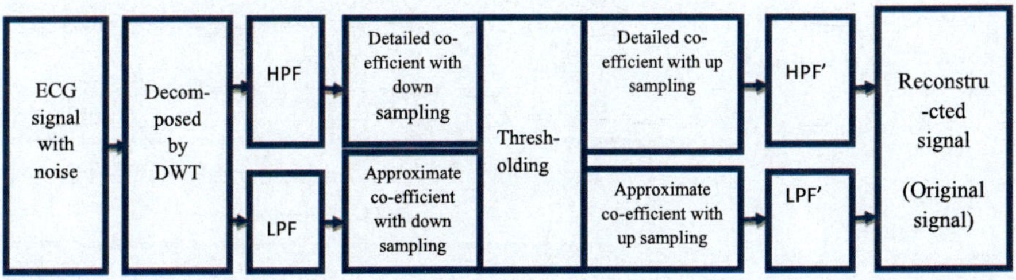

Fig. 2 Wavelet decomposition and reconstruction process

Where s_2^j is a Smoothing operator and $s_2^0 f(n) = d(n)$, where $d(n)$ is digital signal to be and $s_2^j f(n)$ is the wavelet transform of the digital Signal $f(n)$. $\sum h_k$ and $\sum g_k$ are the coefficients of a low pass filter *H(w)* and high pass filter *G(w)* respectively.

2.3 *Discrete Wavelet Transform*

In this section, there is a usage of discrete wavelet transforms (DWT), and the ECG signal is divided into two coefficient. First is the series of approximation coefficient, and second is detail coefficient. In each step, the approximation and detail coefficient contain the information required to reconstruct the signal.

This is known as Mallat tree decomposition which is shown in Fig. 2, and the corresponding flowchart described the systematic steps for removal of noises in ECG signal as shown in Fig. 3. It can be seen that the denoising of the ECG signal is carried out in the various steps such as removal of baseline wander and high frequency noise using median filter, Savitzky–Golay (SG) filter, and wavelet function. The SNR is calculated using MIT-BIH database, and the results are compared with [19] and found good agreement.

2.4 *Savitzky–Golay (SG) Filter*

The least-square approximation polynomial is the basic principle of Savitzky–Golay filter. Savitzky and Golay present a technique for smoothing the data. The various input samples are required to calculate the results in the form of polynomial at a single point. Savitzky–Golay filter also acts as a digital filter for smoothing the noisy data using least-square approximate polynomial and also used to increase the signal-to-noise ratio.

Fig. 3 Flowchart for removal of artifacts in ECG

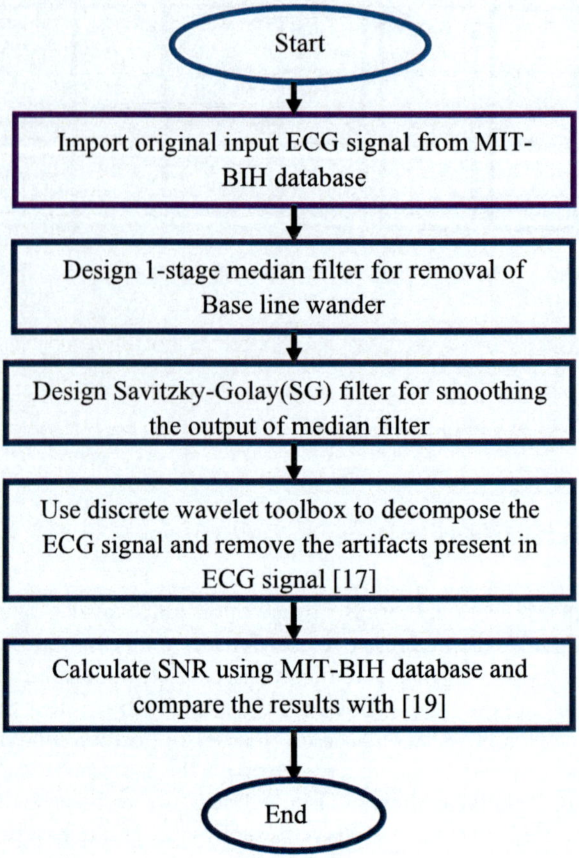

3 Simulated Results and Discussion

3.1 Removal of Baseline Wander

This noise mainly occurs due to improper placement of electrode on the chest of human, movement of patient body, and respiration. The frequency range of this noise is 0.5–0.6 Hz. It is a very low frequency, so it is difficult to detect the baseline wander noise. The baseline signal as shown in Fig. 4a, b is a low frequency signal that shift the DC level up and down.

For removal of DC component, the median of ECG signal has taken and subtracted from the original signal, and the output is given to Savitzky–Golay filter for smoothing the noised waveform [18] and we get free baseline ECG signal. In this proposed work, the noise is removed using MIT-BIH database for records 109, 111 as shown in Fig. 4a, b. These figures show baseline ECG signal, detected baseline, and baseline-free ECG signal.

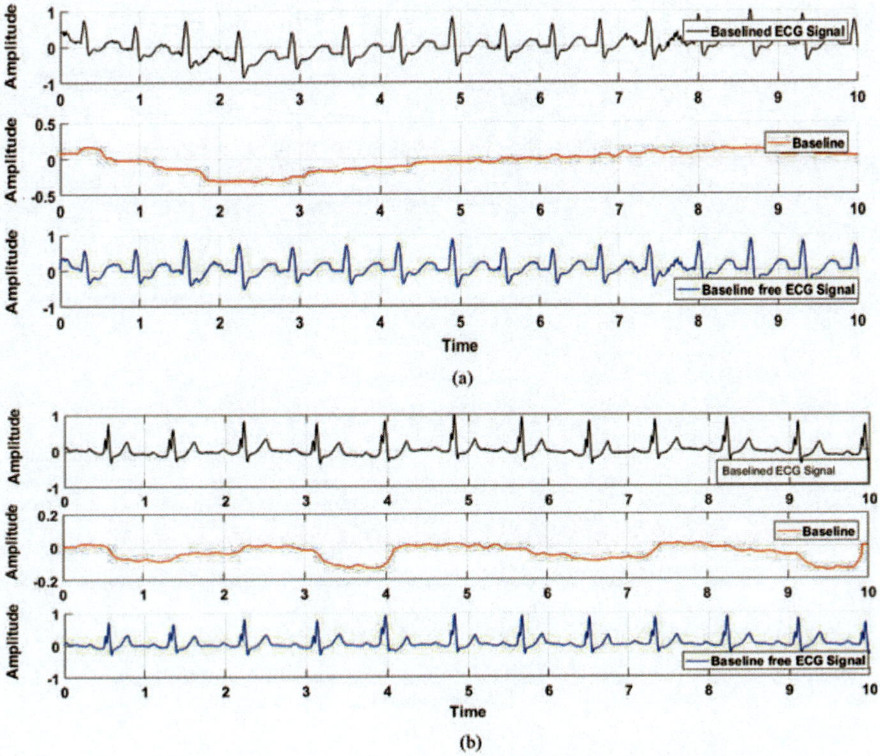

Fig. 4 Results for removal of baseline wander for MIT-BIH database (a)109 record (b)111 record

3.2 *Removal of High Frequency Noise*

During recording, high frequency noise completely overlaps the ECG signal components that make it difficult for the specialist to analyze ECG signal. The range of high frequency noise is between 100 and 150 Hz. This type of noise cannot be efficiently removed using a proper band pass filter. This noise can be removed by using wavelet function. The removal of baseline signal as shown in Fig. 4a, b is a low frequency signal that shifts the DC level up and down. For removal of this DC component, the median of ECG signal has taken and subtracted from the original signal, and this output is passed to Savitzky–Golay filter for smoothing the waveform, and we get baseline-free ECG signal. After that wavelet transform is used for the decomposition of ECG signal up to tenth level, (Cd1, Cd2 … Cd10) and (Ca1, Ca2 … Ca10). Cd1, Cd2 (detail co-efficients) contain high frequency components [18]. So using wavelet, the co-efficient Cd1, Cd2 are discarded which contain high frequency noise. So in this proposed work this noise is removed using MIT-BIH database for records 109, 111 as shown in Fig. 5a, b. These figures show that high noised ECG signal detected high noise and high noise-free ECG signal.

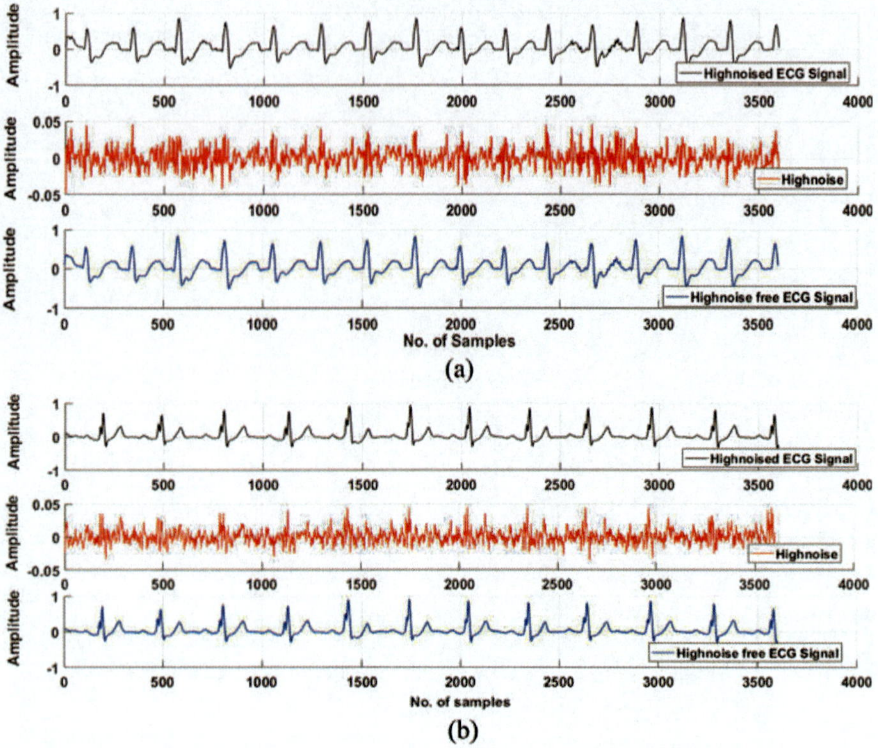

Fig. 5 Results for removal of power line interference for MIT-BIH database **a** 109 record **b** 111 record

4 Comparison with Existing Method

The various techniques have been used for removal of baseline wander, power line interference (PLI), and muscle tremors have been explained in literature. In [16], authors described a technique based on optimal selection of wavelet functions for removal of various artifacts and calculated percentage root mean square difference (PRD). The value of PRD is smallest, i.e., 0.3980. In [17], authors proposed a method based on multi-resolution wavelet along with window for removal of various artifacts present in an ECG signal, and percentage root mean square difference (PRD) has been calculated. The method is having the smallest value of PRD for removal of noises. In [19], wavelet transform method is used for removal of noises and calculated the results in term of signal-to-noise ratio. In this proposed work, the comparison of SNR is done for MIT-BIH database for records, i.e., 109, 111, 112, 114, 115, 116, with ref [19] that provides better results as given in Table 1. The proposed method is having large value of SNR for different records of MIT-BIH database when compared with existing work, i.e., [19] using db4 wavelet function.

Table 1 Comparison of the signal-to-noise ratio

Sr. no.	MIT-BIH database	Wavelet transform with IIR filter for MIT-BIH database [19]	Wavelet transform +median filter + SG filter median filter for MIT-BIH database (proposed)
1	109	−1.8472	25.5315
2	111	−2.7015	24.1984
3	112	−0.1926	23.6709
4	114	−0.6912	19.1777
5	115	−10.997	28.5704
6	116	−5.7531	28.9598

5 Conclusion

In this paper, median filter, SG filter, and wavelet transform have been used to decompose the ECG signal for removal of various artifacts. Wavelet transform provides the information in low frequency and high frequency decomposed ECG signal. The simulated results shown that proposed method achieved better SNR for MIT-BIH database compared to the existing work. Hence, this method can be employed for analysis and diagnosis of CVDs.

References

1. Pater C (2005) Methodological considerations in the design of trial for safety assessment of new drugs and chemical entities. Trials 6(1):1–13
2. Priya MS (2015) MATLAB based ECG signal noise removal and its analysis. In: International conference on recent advances in engineering & computational sciences, IEEE
3. Jane R, Laguna P et al (1992) Adaptive baseline wander removal in the ECG: Comparative analysis with cubic spline technique. In: Proceedings of computers in cardiology conference, USA, pp 143–146
4. Tsai T-H, Hong J-H, Wang L-H, Lee S-Y (2012) Low-power analog integrated circuits for wireless ECG acquisition systems. IEEE Trans Inf Technol Biomed 16(5):907–912
5. Rahman MZU, Karthik GVS, Fathima SY, Lay-Ekuakille A (2013) An efficient cardiac signal enhancement using time–frequency realization of leaky adaptive noise cancelers for remote health monitoring systems. Measurement 46(10):3815–3835
6. Kang B, Yoo J, Park P (2013) Bias-compensated normalised LMS algorithm with noisy input. Electron Lett 49(8):538–539
7. Taralunga DD, Gussi I, Strungaru R (2015) Fetal ECG enhancement: adaptive power line interference cancellation based on Hilbert Huang transform. Biomed Signal Process Control 19:77–84
8. Awal MA, Mostafa SS, Ahmad M, Rashid MA (2014) Anadaptive level dependent wavelet thresholding for ECG denoising. Biocybern Biomed Eng 34(4):238–249
9. Yadav SK, Sinha R, Bora PK (2015) Electrocardiogram signal denoising using non-local wavelet transform domain filtering. IET Signal Process 9(1):88–96
10. Wang L-H, Chen T-Y, Lin K-H, Fang Q, Lee S-Y (2015) Implementation of a wireless ECG acquisition. IEEE J Biomed Health Informan 19(1):247–255

11. Raheja N, Manoacha AK (2020) A study of telecardiology-based methods for detection of cardiovascular diseases. Adv Intell Syst Comput Book Ser 1124
12. Hadji S, Salleh M, Rohani M, Kamat M (2016) Wavelet-based performance in denoising ECG signal. In: Proceedings of the 8th international conference on signal processing systems, pp 148–153
13. Eminaga Y, Coskun A, Kale I (2018) Hybrid IIR/FIR wavelet filter banks for ECG signal denoising. In: IEEE biomedical circuits and systems conference (BioCAS), pp 1–4
14. Wang SH, Zhang YD, Yang M, Liu B, Ramirez J, Gorriz JM (2019) Unilateral sensorineural hearing loss identification based on double-density dual-tree complex wavelet transform and multinomial logistic regression. Integr Comput Aided Eng 1–16
15. Rahul K (2019) Signal processing techniques for removing noise from ECG SIGNALS. J Biomed Eng Res 3:101
16. Kumar A, Singh M (2015) Optimal selection of wavelet function and decomposition level for removal of ECG signal artifacts. J Med Imaging Health Inf 5:138–146
17. Kumar A, Singh M (2016) Robust multiresolution wavelet analysis and window search based approach for electrocardiogram features delineation. J Med Imaging Health Informatics 6:146–156
18. Kumar A, Singh M (2016) Statistical analysis of ST segments in ECG signals for detection of ischaemic episodes. Trans Inst Meas Control 1–12
19. Sangaiah AK, Arumugam M, Bian G-B (2019) An intelligent learning approach for improving signal classification and Arrhythmia analysis. Artificial Intelligence Med 101788

Product Recommendation Platform Based on Natural Language Processing

Vanita Jain, Mankirat Singh, and Arpit Bharti

Abstract Online Videos command a majority portion of the worlds internet traffic. In 2019, consumer Internet video traffic had accounted for 80% of all consumer Internet traffic. At the forefront of this are platforms like YouTube. As the consumption of the online videos increases, it is imperative for various platforms to be able to monetize these videos using relevant advertisements. To maximize revenue from ads, we propose a method of ranking and sorting products based on relevancy to a given video. Using the metadata and captions, the system aims to generate product recommendations that a user may find useful. The system uses Rapid Automatic Keyword Extraction to find the words that best summarize the video and proposes recommendations using similarity measures. The system is then evaluated on a database. Finally we proceeded to conduct an extensive User Study to judge the accuracy of our system. The results conclude that the proposed model is capable of recommending contextually relevant products for a wide variety of Online videos.

Keywords Recommendation System · Natural Language Processing · Keyword Extraction · online videos

1 Introduction

As online videos become more popular across various social networks, especially specialized platforms like YouTube, it has become important to be able to monetize them effectively. Online Advertisement has not only become critical for the revenue generation of these platforms but also as part of the User Interface. Thus, in this paper

V. Jain (✉) · M. Singh · A. Bharti
Bharati Vidyapeeth's College of Engineering, Paschim Vihar, New Delhi 110063, India
e-mail: vanita.jain@bharatividyapeeth.edu

M. Singh
e-mail: mankirat.work@gmail.com

A. Bharti
e-mail: arpitbharti73@gmail.com

A. Khanna et al. (eds.), *Data Analytics and Management*, Lecture Notes on Data Engineering and Communications Technologies 54,
https://doi.org/10.1007/978-981-15-8335-3_48

we are proposing a system for recommending ads that proposes relevant products for a given source video, based on Natural Language Processing.

There are two main perspectives that are key to an online advertisement, user and advertiser. The latter is assessed online through metrics such as scrolling across patterns and electronic snap polling. Previous research suggests that user experience can help understand the negative impacts of advertisement on users [10]. In more studies by McCoy et al. [6, 7] on the topic it has been revealed that irrelevant advertisements have a negative effect on the user's perception of the brand.

1.1 Literature Review

The given topic is thoroughly investigated in previous literature. In a proposal by Mei et al. [8], ad recommendation was proposed by relying on global textual relevance such as tags and queries along with local visible features like colour and audio. Likewise, [16] Chose relevant ads using text based information, and improved the rankings using image based information. Sengamedu et al. presented [13] an approach to find the correct timing of displaying an ad by analysing the video in chunks and run analysis to identify faces and objects that are used for ad recommendation. In [14] video frames were converted to captions by computer vision and based on the captions video advertisements were recommended. Vedula et al. [15] proposed that we can add other parameters such as sentiments and previous browsing data [3], biddings [18] for ad recommendation.

We also studied different recommendation systems implemented in previous work, Rana et al. [9] proposes a recommendation that uses Collaborative filtering with Jaccard Similarity to give more accurate book recommendations. Further research and various implementations of Jaccard similarity have been conducted in [1, 2, 17]. We have analysed these approaches to design the system. We also studied the comparative analysis of various similarities [12].

While the use of keyword extraction and text similarity algorithms is widespread we haven't observed these being used for product recommendations for Online Videos.

2 Proposed Methodology

The system has three parts. In the first part, a Youtube video is submitted as input in our system via a YouTube link. We use the YouTube Data API to extract audio captions and video specific metadata (description, title etc.). The second part of the algorithm uses Natural Language Processing to first pre-process the textual data, extract keywords and generate a list of recommended products which are germane to the given video, using our dataset of products, PRODUCTS50. In the third step we measure the semantic similarity between the video and products based on the

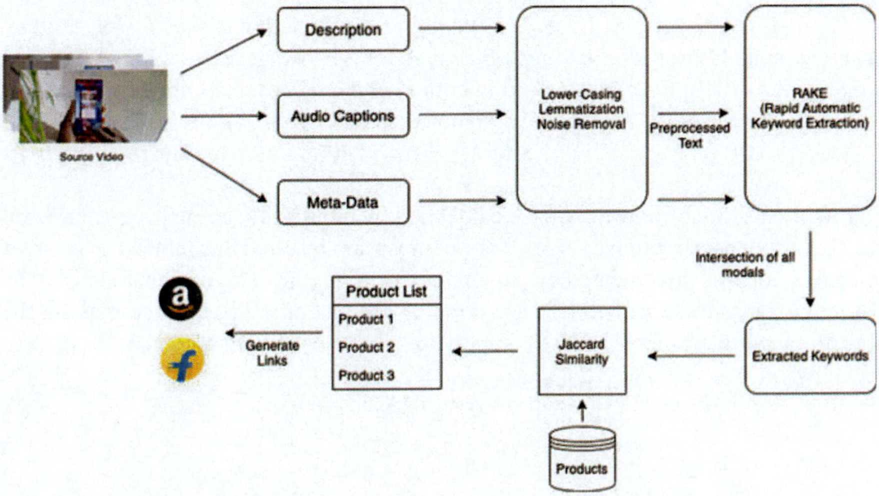

Fig. 1 Architecture of the system

similarity score between the keywords from video and product description. Figure 1 describes the architecture of the system.

2.1 Collecting Data for Analysis

First, we collect the data needed for analysis. The Youtube API is used to extract captions and metadata of a given video. A combined text, *source_description*, is formed by concatenating the speech and metadata above. This *source_description* is then used to propose relevant products and is passed to the next step.

2.2 Recommending Relevant Products

In this step of the model, a list of germane products is recommended. The products are selected from our PRODUCTS50 database which includes 50 products from various domains. The model uses the *source_description* from the first step and makes a comparison with the description of all products recursively.

Let $D = \{d_1, d_2, d_3, \ldots, d_n\}$ and $P = \{p_1, p_2, p_3, \ldots, p_n\}$ symbolize the *source_description* string and the *product_description* string. First, the strings D and P are converted to tokens followed by lower casing and lemmatization of their words. After the initial pre-processing, a bag of words, $bow = \{w_1, w_2, \ldots, w_n\}$, is generated for each string. We then applied RAKE (Rapid Automatic Keyword Extraction) [11] to extract the keywords from the *bow*.

The list of keywords extracted in the above step are denoted as k_s for *source_description* and k_p for *product_description*. After this, we generate a list of lists that contains the synonyms of each word in both k_s and k_p. These lists are symbolized as S_v and S_p for k_s and k_p respectively. Denoted as $S_v = \{\{v_{11}, v_{12}, \ldots, v_{1x}\}, \{v_{21}, v_{22}, \ldots, v_{2y}\}, \ldots \{v_{n1}, v_{n2}, \ldots, v_{nz}\}\}$ and $S_p = \{\{p_{11}, p_{12}, \ldots, p_{1x}\}, \{p_{21}, p_{22}, \ldots, p_{2y}\}, \ldots, \{p_{n1}, p_{n2}, \ldots, p_{nz}\}\}$.

Finally a score of relevancy s_c is calculated by recursively comparing each word v_{mn} in S_v with each word p_{mn} in S_p. If the comparison resulted in a match the score of relevancy, s_c, was incremented by one. After a word v_{mn} in S_v found a match in S_p, the list iterator was incremented by one, skipping the rest of the list, hence making the algorithm more efficient. Mathematically,for each *product_description* P_k, s_c was given by,

$$\begin{aligned} \forall i, j \quad \mathit{If}\ v_{ij} \text{ in } S_p : \\ s_c = s_c + 1 \\ i = i + 1 \end{aligned} \tag{1}$$

The products with a s_c greater than or equal to the threshold, length of K_s, were recommended as relevant products, symbolized by *product_list*. This *product_list* was then sent to the next step for a semantic comparison.

$$\begin{aligned} \mathit{If}\ s_r >= \text{len}(K_s) : \\ \mathit{product\_list.append}\,(s_r) \end{aligned} \tag{2}$$

2.3 Semantic Comparison of Descriptions

In this step of the system, each product description in *product_list* is vectorized and a semantic comparison is done between it and the *source_description*. First we encode each description into a dimensional vector of size 4000 based on [4]. It is similar to the approach taken in [5]. As this GRU based encoder doesn't limit the system to a particular category or genre, it is a good fit for our system.

The *product_list* is represented as $P = \{P_1, P_2, \ldots, P_n\}$ in which P_i symbolizes a single product description. Each product description P_i is vectorized to form $V(P_i)$. $V(P_i)$ can be denoted as $V(P_i) = [p_1, p_2, p_3, \ldots, p_{4000}]$ in which p_i is a decimal number. Similarly, the *source_description* from the first step is encoded into a vector $V(S_v)$ where S_v is the *source_description* from the first step.

Semantic similarity between S_v and P_i was calculated using Jaccard Similarity, $J(x, y)$

$$J(X, Y) = \frac{|V(P_i) \cap V(S_v)|}{|V(P_i) \cup V(S_v)|} = \frac{|V(P_i) \cap V(S_v)|}{|V(P_i)| + |V(S_v)| + |V(P_i) \cap V(S_v)|} \tag{3}$$

Fig. 2 User interface for the system

Many different similarity measures were tried, such as Manhattan Distance, Cosine Similarity, Jaccard Similarity, Euclidean Distance. The proposed system uses Jaccard Similarity because it is a great fit when dealing with keyword comparison, as we have done in step 2. This step further filters the list of recommended products.

Finally, the top three products having maximum similarity measure were recommended from the *product_list*. The implementation of the entire system can be seen in Fig. 2.

3 PRODUCTS50 Database

We created a Database of products for testing the recommendation platform. Due to the unique nature of the specifications, ex. specific metadata for the products, we decided to create a limited product collection, PRODUCTS50. This collection included 50 products from a variety of different brands and domains like cooking, smartphones, laptops, programming tutorials, shoe reviews, travel guides etc. Finally, we ended up with a JSON file containing a list of products from different domains. Each product is an object with some properties that describe various feature of a product. The object contained the product_name, product_description, link, domain as well as the brand the product belonged to. These properties of the products were then compared with that of the source video as detailed above.

4 Results

We tested our system with 20 YouTube videos that belonged different categories, to test product recommendation across a multitude of domains. The products and videos were chosen independently of one another to prevent any bias in our analysis. The PRODUCTS50 database was used as a products database. We reviewed the performance of our system using both a Qualitative Analysis and a User Study.

Table 1 Some results of our user study

Video title and ID	Products recommended (Top 3)	User study (%)	Rating
Title: Samsung Galaxy S20 review: better than the Ultra ID: xJ3e0RDbT1w	Samsung Galaxy S20 Redmi 8A Mobile Phone Microsoft surface pro laptop	95 42 23	4.1
Title: The Most Common Natural Ingredients for glowing skin ID: 0VwpOJp6CGY	Garnier BB cream Lakmé Liquid foundation Nivea total face cleanup	75 76 73	4.0
Title: Stop Buying the MacBook Air ID: _VotJ1IZPSw	Apple macbook air laptop Apple macbook pro laptop Microsoft surface pro laptop	80 67 48	4.3
Title: Best running shoes |stability, Cheap, cushioned, long distance ID: CMVFNp5zRgQ	Nike shoes Adidas shoes Zemic furniture—shoe rack	86 78 40	4.0
Title: 15 Travel Essentials for Men | What to Pack ID: 34fj-M4bcew	GoTrippin travel adapter Boat Bassheads 100 headphones Canon EOS 1500D Camera	88 61 48	4.3
Title: Ricoh GR III—The Best Camera you should have with you ID: 5msrk6yNRpk	Nikon D3400 Camera Canon EOS 1500D camera Sony alpha camera	90 82 76	4.2
Title: how to make butter chicken At Home | restaurant style recipe The bombay chef ID: a03U45jFxOI	Fortune Soyabean Oil Tata salt Everest masala	97 94 90	4.6
@@Title: Top 3 Programming Languages to Learn in 2019 ID: Pb3AAfz5Yjg	Object-Oriented with C++ Core Python Programming Eloquent Javascript	96 94 94	4.3
Title: Graphic Design For Newbies (Canva Tutorial 2019 - Canva 2.0) ID: hiBAn1exlmc	Canva Creative Software Afinity Designer Snapseed	95 85 85	4.2
Title: Xbox One X Review ID: 4AVra_cGnxI	Xbox One S 1TB Console PS4 Gaming Console Microsoft SurfacePro Laptop	100 84 49	4.2
Title: How are there SO MANY good photographers ID: 7-Y34BT6EXA	Canon EOS 1500D Camera Nikon D3400 Camera Boat Headphones	80 78 20	4.0

4.1 Qualitative Analysis

In Table 1 we can see the results of 11 videos chosen out of a set of 20 videos. The first video is a video reviewing a mobile device results in recommendation of a mobile phone as a top choice, the second and third choices are also tech products and relevant. The second video is about skin care, our algorithm extracted words such as "skin care", "natural ingredients" from the source video and proposed relevant products related to Skin Care. Similar to the first, the third video is about a laptop, our algorithm again accurately proposes 3 similar Laptop products for the user. The fourth video belongs in the apparels category, it is a review of shoes, in this our system not only proposes two shoe products but also a shoe rack. In the fifth video, we have a travel guide and our system proposes important travel items. In the sixth video we have a camera review and our algorithm proposes 3 cameras from our data set. The seventh video is a cooking tutorial and again our system accurately proposes products not only related to cooking but also to the specific recipe. The eighth video is an educational learning video about programming, model proposes books about programming. Similarly in the ninth, tenth and eleventh video the model accurately proposes software for graphic design, gaming console and cameras accurately. Hence, we can see that the results contain videos covering different domains of interests and observe that the recommendations are also part of the same or a closely related domain.

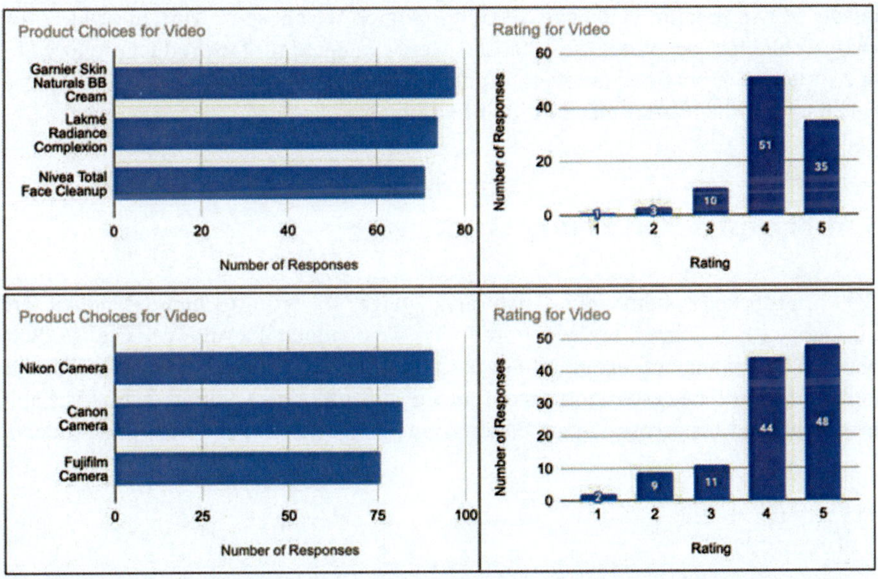

Fig. 3 Results for two specific video with title: the most common natural ingredients for glowing skin and title: Ricoh GR III —the best camera you should have with you

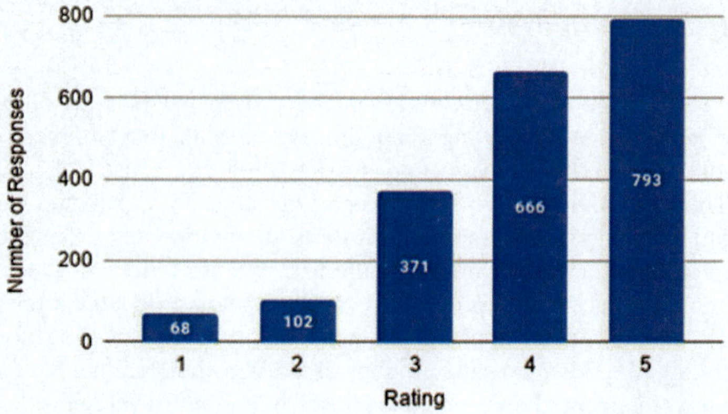

Fig. 4 Cumulative rating for all 20 videos by 100 users

4.2 User Study

We created a survey to run a user study to analyse the relevancy of our 100 adults aged 15–60 years were chosen and asked to answer two questions related to every video. First the users were asked to select all products that they thought was relevant to the video. Second, they rated the relevance of the proposed products on a Scale of 1–5 with higher rating meaning more relevance of products to the source video. The results of two specific video can be seen in Fig. 3. We chose 20 YouTube videos for analysis and for each video we 100 users were evaluated for the product proposed by our system. We obtained an overall rating of 4.0/5 as observed in Fig. 4 which was calculated as the overall mean rating of all videos.

5 Conclusion and Future Work

For this proposed system we used natural language processing techniques and Jaccard Similarity to recommend relevant products. We conducted a survey, a user study to check the relevancy of our system in the real world. In the future, we would like to build our prototype system into an optimized and automatic solution that will be able to recommend ads across a larger database and wider domain of videos and products.

References

1. Ayub, M., Ghazanfar, M.A., Maqsood, M., Saleem, A.: A jaccard base similarity measure to improve performance of cf based recommender systems. In: 2018 International Conference on Information Networking (ICOIN), pp. 1–6 (2018)

2. Bag S, Kumar SK, Tiwari MK (2019) An efficient recommendation generation using relevant jaccard similarity. Information Sciences 483:53–64
3. De Bock K, Van den Poel D (2010) Predicting website audience demographics for web advertising targeting using multi-website clickstream data. FUNDAMENTA INFORMATICAE 98:49–70
4. Hou J, Kang J, Qi N (2010) On vocabulary size in bag-of-visual-words representation. In: Qiu G, Lam KM, Kiya H, Xue XY, Kuo CCJ, Lew MS (eds) Advances in Multimedia Information Processing - PCM 2010. Springer, Berlin Heidelberg, pp 414–424
5. Kiros R, Zhu Y, Salakhutdinov R, Zemel RS, Torralba A, Urtasun R, Fidler S (2015) Skip-thought vectors
6. Li H, Edwards SM, Lee JH (2002) Measuring the intrusiveness of advertisements: scale development and validation. J Advertising 31(2):37–47
7. McCoy S, Everard A, Polak P, Galletta DF (2007) The effects of online advertising. Commun ACM 50(3):84–88
8. Mei T, Hua XS, Yang L, Li S (2007) Videosense: towards effective online video advertising. In: Proceedings of the 15th ACM international conference on multimedia, MM '07. ACM, pp 1075–1084 (2007)
9. Rana A, Deeba K (2019) Online book recommendation system using collaborative filtering (with jaccard similarity). J Phys Conf Ser 1362, 012,130 (2019). https://doi.org/10.1088/1742-6596/1362/1/012130
10. Rohrer C, Boyd J (2004) The rise of intrusive online advertising and the response of user experience research at yahoo! In: CHI '04 extended abstracts on human factors in computing systems, CHI EA '04. ACM, pp 1085–1086. https://doi.org/10.1145/985921.985992
11. Rose S, Engel D, Cramer N, Cowley W (2010) Automatic keyword extraction from individual documents, pp 1–20. https://doi.org/10.1002/9780470689646.ch1
12. Saad SM, Kamarudin SS (2013) Comparative analysis of similarity measures for sentence level semantic measurement of text. In: 2013 IEEE international conference on control system, computing and engineering, pp 90–94
13. Sengamedu SH, Sawant N, Wadhwa S (2007) vadeo: Video advertising system. In: Proceedings of the 15th ACM international conference on multimedia, MM '07. ACM, pp 455–456
14. Singh M, Lamba R (2020) Proposing contextually relevant advertisements for online videos. In: Thampi SM, Trajkovic L, Li KC, Das S, Wozniak M, Berretti S (eds) Machine learning and metaheuristics algorithms, and applications. Springer Singapore, Singapore, pp 218–224
15. Vedula N, Sun W, Lee H, Gupta H, Ogihara M, Johnson J, Ren G, Parthasarathy S (2017) Multimodal content analysis for effective advertisements on youtube. In: 2017 IEEE international conference on data mining (ICDM)
16. Xiang C, Nguyen TV, Kankanhalli M (2015) Salad: a multimodal approach for contextual video advertising. In: 2015 IEEE international symposium on multimedia (ISM), pp 211–216
17. Yan H, Tang Y (2019) Collaborative filtering based on gaussian mixture model and improved jaccard similarity. IEEE Access 7:118690–118701
18. Zhang W, Yuan S, Wang J (2014) Optimal real-time bidding for display advertising. In: Proceedings of the 20th ACM SIGKDD international conference on knowledge discovery and data mining, KDD '14. ACM, pp 1077–1086

A Secured Supply Chain Network for Route Optimization and Product Traceability Using Blockchain in Internet of Things

Poonam Rani, Vibha Jain, Mansi Joshi, Muskan Khandelwal, and Shivani Rao

Abstract The conventional method of implementing supply chain management suffers several limitations like cost, unpredictability, complicated environment, and wide-open access to sensitive information. To defeat these issues, the supply chains should be progressively more intelligent. To build a large-scale smart infrastructure of physical objects, products, and all the other supply chain-related processes, supply chain management can be integrated with the Internet of things (IoT). Open interrelation communication opens several threats in the supply chain environment. In the proposed work, we construct a supply chain network (SCN) secured with blockchain to find the optimum route of a product from a manufacturer to customer and to trace back a product to its origin. A genetic algorithm-based scheme is used to find the optimum route for a product and approach to trace back the origin of a product is proposed, which is especially beneficial in cases like a product turning out to be infected and the need to track the rest of the products of that batch. Simulation results are compared using two different consensuses for verification of transactions: PoW and proof of elapsed time. Graphical results show that proof of elapsed time outperforms PoW in terms of cost, processing power, and transaction verify per second.

Keywords IoT · Blockchain · Supply chain · Route optimization

P. Rani (✉) · V. Jain · M. Joshi · M. Khandelwal · S. Rao
Netaji Subhas University of Technology, Dwarka, New Delhi, Delhi, India
e-mail: poonamrani2017.nsit@gmail.com; Poonam.rani@nsut.ac.in

V. Jain
e-mail: vibha.jain.cs19@nsut.ac.in

M. Joshi
e-mail: mansij.co.17@nsit.net.in

M. Khandelwal
e-mail: muskankhandelwal369@gmail.com

S. Rao
e-mail: shivanirao1910@gmail.com

A. Khanna et al. (eds.), *Data Analytics and Management*, Lecture Notes on Data Engineering and Communications Technologies 54,
https://doi.org/10.1007/978-981-15-8335-3_49

1 Introduction

Internet of things (IoT) describes a network of devices like smart locks, smart meters, smart homes, smart parking, watches, IP Cameras, etc., to connect them to the Internet and share information. This network connectivity would allow accessing and controlling the components of the network. IoT in a broader sense means all these devices connected through a defined network and this connectivity helps to gather data information, analyses it, and perform an action. The device or the thing in the IoT network could be any device embedded with electronics or sensors like a smart air conditioner, a smart refrigerator, etc. This would also help in reducing human intervention and making the network more integrated. Applications of IoT devices are everywhere like in the medical field in detection of skin cancer [1], in industrial field in image compression for data transmission in IHoT [2], in oppIOT [3] network, etc. One more applications of IoT is supply chain management. According to the supply chain operations reference model (SCOR) [4], the supply chain is an embedded set of different entities including manufactures, customers, suppliers, distributors, retailers, wholesalers, etc., which are intrigued by each other to achieve an ultimate end goal [4]. Supply chain management signifies that all the products must be available to the customer within sufficient time in correct condition and correct price. A supply chain network is used to simulate the behavior of logistics over time. With an effective supply chain management, organizations can plan their operations well in advance to reduce costs and enhance profits. This discipline has been gaining importance in recent years because of increased competition as an outcome of globalization [5].

A problem frequently arises in the supply chain is that data may be isolated with suppliers and procurement officers, which can be tampered by any unauthorized entity. [6] describes how even a single point of failure in an IoT network can bring a disastrous impact on human lives. A centralized point is vulnerable to attack by hackers; therefore, the system needs to be decentralized using blockchain. A blockchain can be defined as an immutable record of time-stamped data, which is managed by a distributed collection of nodes. Being transparent, blockchain makes it easier to trace a product's journey to its point of origin [7]. The benefit of being able to trace a product's history, location, and other details is increased consumer trust in the network. They can be made aware of the transformations brought about in a product throughout the supply chain.

Another major problem in supply chain management is to optimize the route for transportation of goods originated for a customer to maximize the profits of the participants of the supply chain. We have solved this problem by proposing a genetic algorithm for the same. Genetic algorithms have been very useful in solving optimization problems and are gaining significance lately. The major contributions of this paper are:

- Comparing various consensus algorithms for the blockchain securing a supply chain network.

- Proposing a genetic algorithm for finding the optimum route for a product from a manufacturer to a retailer in terms of cost and time.
- Proposing a method to trace back the path of a product using its unique identification.

This paper is organized as follows: Section 2 comprises the related work done in this field. Section 3 describes the proposed work in detail. Section 4 includes the results of the simulation conducted along the lines of the proposed work. Section 5concludes the paper and describes the future scope of the work.

2 Related Work

Blockchain has brought about increased traceability of the products in the supply chain network of various organizations like Walmart, Project Provenance Ltd, British Airways, etc. [8]. Small businesses can trust firms that might be remote to them initially. For instance, a small farmer might be able to trust a transportation firm to deliver his farm product to cold storage that might provide him better prices for his harvest than involving a middleman in the process.

Hence, a blockchain can add trust to the supply chain network and help the participating firms find the optimal path for their product to the customer to maximize their profits. In the field of IoT, the genetic algorithms are widely used to solve the optimization problem like finding the optimal path, as its convergence rate is high compared to other methods. Algorithms like Dijkstra's can find an optimized route but they do not guarantee that constraints such as that of time, cost, and capacity will be met [9]. This paper devised an experiment based on car navigation systems that prove the superiority of genetic algorithms over Dijkstra. Arafat et al. [10] present one such algorithm in a single-objective supply chain network design problem. However, it has a single objective (i.e., to minimize the cost) whereas in reality, there are multiple objectives to be considered and trade-offs need to be done. As an improvement, we have formulated a 2-point objective to be optimized.

A major problem while working with decentralized networks is the Byzantine Generals problem. Since there is no controlling authority, we need to establish consensus amongst all peers so as to validate the transactions. Choosing the right parameters for the consensus algorithm and blockchain parameters like block size is vital for setting up the blockchain network. The paper [11] has analyzed the consensus algorithm proof of elapsed time, and compared the blockchain throughput with increasing network size. However, no analysis is done with respect to different block parameters like block size.

3 Proposed Work

3.1 Blockchain

Blockchain is an apt solution to secure the decentralized network. In a blockchain, the various transactions are stored as blocks. There can be one or more transactions in a single block. Because there can be multiple blocks, each block has a unique ID. We have used the Python SHA-256 hashing function, to add a digital footprint to all the blocks of the data contained in that block. These blocks are then chained together. Storing the hash of the previous block in the current block does this. Therefore, any change made in the previous block invalidates the entire chain. The first block is called the genesis block and it can be generated either manually or through some unique logic.

Blockchain can be broadly categorized as public blockchain and private blockchain. A public blockchain has an open access (both read as well as write). Anybody can join the network in an unrestricted manner. They can even participate in consensus. A private blockchain, on the other hand, is accessed only within a single organization or enterprise, which has the authority to decide who will participate in consensus, thus preventing any malicious activities. In the case of supply chain management, a public blockchain has proved to be an effective solution [12]. Information updates are available to components thereby preventing any possible miscommunication and errors.

3.2 Consensus Algorithms

3.2.1 Proof of Work

In this consensus algorithm, a constraint, i.e., the hash must start with "n leading zeroes" where n is any positive integer, is added to every block of the blockchain. This makes it difficult for any generic person to add a new block to the blockchain since computational work is needed to calculate a valid hash for the current block.

Table 1 describes the block parameters, node parameters, and the network parameters of various blockchain networks like Bitcoin, Litecoin, Dogecoin, helped us understand the various parameters of the blockchain network [13].

Table 1 Blockchains based on PoW

Parameter	Bitcoin	Litecoin	Dogecoin
Number of nodes	6000	800	600
Block interval	10 min	2 min 30 s	1 min
Block size	534 KiB	6.11 KiB	8 KiB

3.2.2 Proof of Elapsed Time

Proof of elapsed time is a consensus algorithm developed by Intel, which aims to put an end to the waste of energy that takes place during mining activity in the proof of work consensus algorithm. This algorithm allows more number of transactions to be validated in less time; therefore, increasing the throughput of the network. In this, every validator node waits for a random time that is decided by a code executed inside a "trusted execution environment" (TEE) and as soon as this time ends, the first node with the minimum waiting time broadcasts the new block to the network. It has been implemented in the open-source Hyperledger Sawtooth blockchain framework [11]. Factors that determine the suitability of a consensus algorithm are:

1. Performance measured as transaction per second or transaction throughput and average transaction latency
2. Scalability is the measure of the efficiency of a network as the number of nodes increases [9].

Proof of work is the most scalable; however, it gives the minimum throughput. Proof of elapsed time is considerably scalable and gives good throughput. We used both the consensus while implementing our supply chain model.

3.2.3 Tracing the Product Back to Its Origin

Every transaction is uniquely identified by a product identity, which indicates the product that has been received or sent. A block stores these transactions such that any transaction dealing with a specific product can be found in minimal time within the block. This enables the quickest retrieval of all the transactions related to a specific product in the blockchain (Fig. 1).

3.3 Route Optimization

The proposed supply chain network comprises three types of nodes: manufacturers, wholesalers, and retailers. Every node is determined by:

- The latitude and longitude values of the node location.
- Its maximum capacity of goods.
- Its trust score, defined by how old the node is in the blockchain network and if any invalid transaction has been reported by the node.

The objectives for route optimization via genetic algorithm are to minimize the cost of transportation and the time needed to deliver the product. The constraint is that we can only send a product to a node, which has not yet reached its capacity. Each chromosome is encoded as a 3-member list indicating the manufacturer, wholesaler, and retailer who provide the optimized route. Let x be the cost of transportation in a

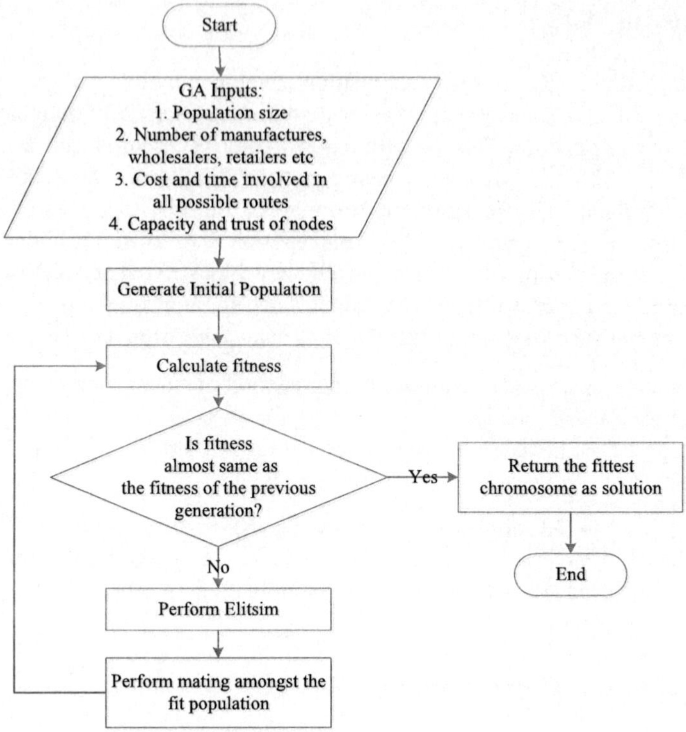

Fig. 1 Flowchart showing the genetic algorithm

path and y be the time consumed by that path in hours. The fitness of a chromosome is defined as 1000 * capacity/$(1 + x + y)$. 10% of the fittest population passes on to the next generation while mating is done between the fittest 50% to obtain the remainder of the 90% of the population. The mutation rate is kept at 2%. The terminating condition is for the average fitness of the population to become almost constant.

4 Simulation and Results

The metrics used to measure a blockchain's performance is throughput and finality. Throughput is measured in terms of transactions per second(TPS). Higher the TPS, the better, it signifies lesser blockage in the supply chain. The simulation was done in a macOS computer with a 2.5 GHz Intel Core i5 processor. Table 2 shows the different simulation parameters used while implementing the proposed model. To measure the mining time of the blockchain, transactions of all types (manufacturer to wholesaler and wholesaler to the retailer) were simulated. Different numbers of

Table 2 Simulation parameters

Parameter	Value
Number of simulations	10
Number of transactions	4500
Range of throughput (transactions per second)	0–20 tps
Block size (in the average number of transactions)	10,000 transactions

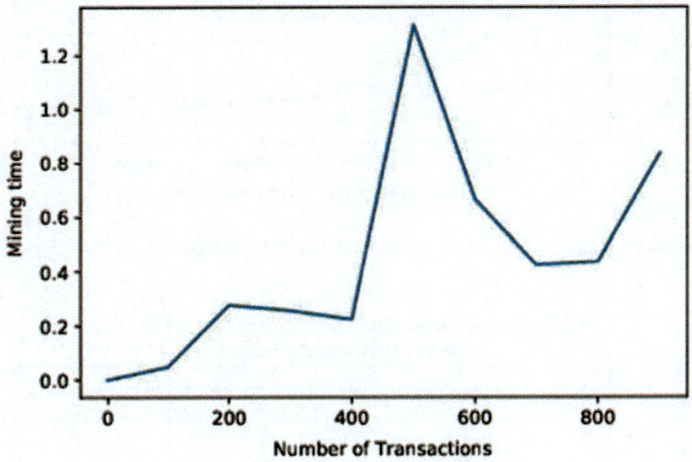

Fig. 2 Mining time versus number of transactions using PoW consensus

transactions were conducted over ten simulations. The total transaction count was 4500.

We use two different consensus algorithms while implementing blockchain: Proof of work (PoW) and proof of elapsed time. Figure 2 shows the variation of mining time (in seconds) with the number of transactions using PoW consensus, when difficulty is chosen 2. While the graph of the number of transactions per second versus various mining times (in seconds) is shown in Fig. 3.

Figure 4 shows the number of transaction per second concerning various elapsed time with the average number of transaction in each block is 10,000 and 5000, respectively. The average fitness of the population after every generation keeps increasing as depicted by Fig. 5. We were able to minimize the time and cost involved in the transport of goods. Its impact over different generations can be seen in Fig. 6 (for time) and Fig. 7 (for cost). The time is measured in hours and the cost is measured in hundred rupees

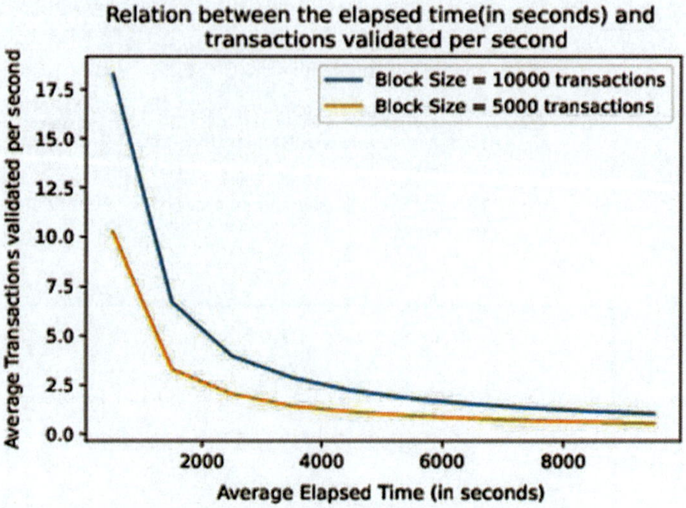

Fig. 3 Relation between mining time and transactions validated per second

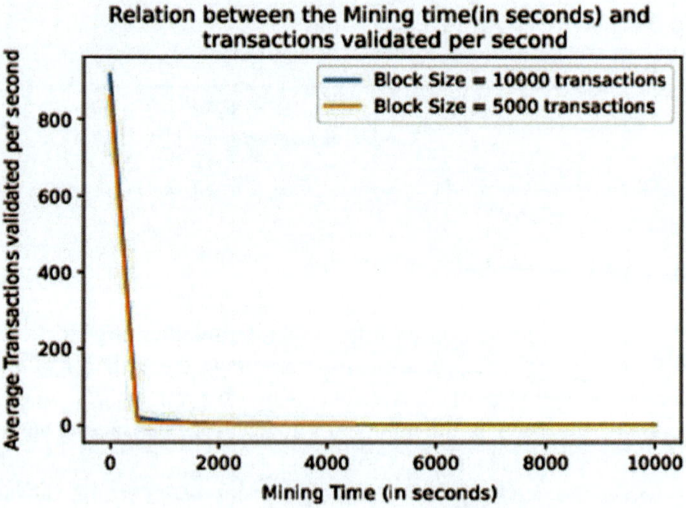

Fig. 4 Number of transaction per second versus various elapsed time average with 5000 transactions per block and with 10,000 transactions per block

5 Conclusions and Future Scope

We have successfully integrated blockchain with supply chain network. Future work includes further extension and development of the simulator. A front-end application

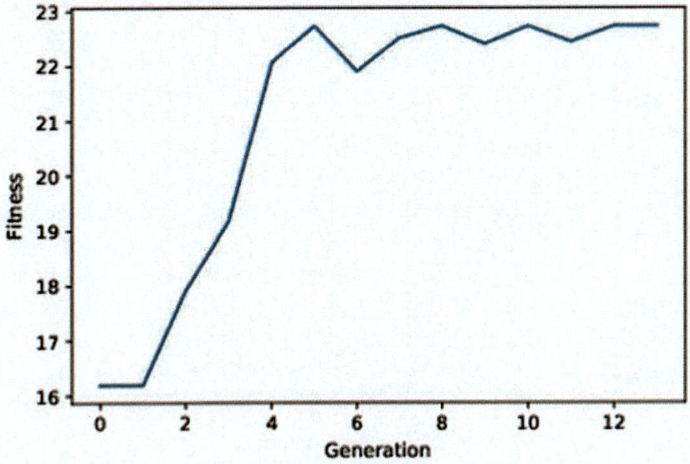

Fig. 5 Fitness of the population over different generations

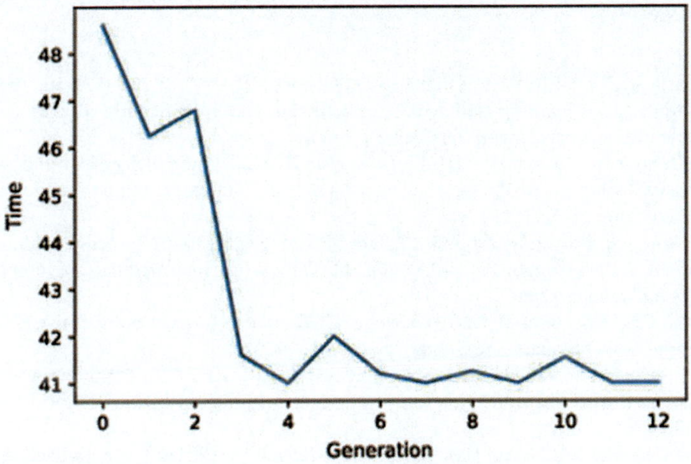

Fig. 6 Time needed for transportation over different generations

can be developed for easy access by a user to trace their products and for manufacturers for finding the optimal path. We can also create various attacker nodes to measure the security aspects of the blockchain. IoT can also be used to monitor the storage conditions of a product and enhance the quality aspects of the supply chain management system. We can also apply this approach to make secure social networks as discussed in different aspects, issues, and challenges explored in paper [14].

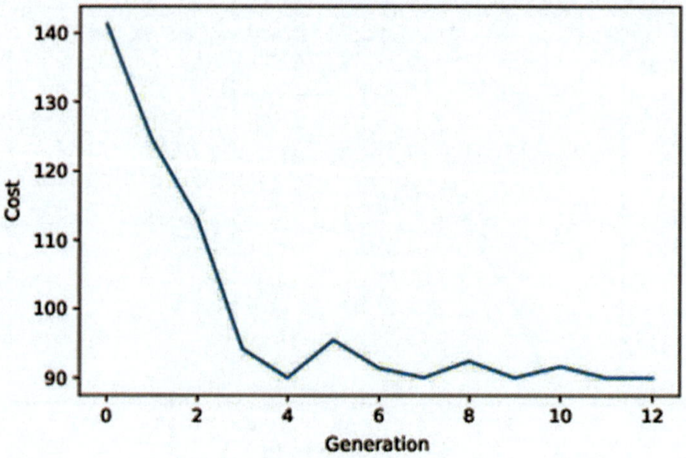

Fig. 7 Cost Needed for transportation over different generation

References

1. Khamparia A, Singh PK, Rani P, Samanta D, Khanna A, Bharat B (2020) An internet of health things-driven deep learning framework for detection and classification of skin cancer using transfer learning. Trans Emerg Tel Tech, pp 1–11
2. Sujitha B, Subbiah V, Laxmi PE, Rani P, Polkowski Z (2020) Optimal deep learning based image compression technique for data transmission on industrial Internet of things applications. Trans Emerg Telecommun Tech 1–13
3. Rani P, Singh A, Balyan PP, Shokeen J, Jain V, Sangwan S (2020) A secure epidemic routing using blockchain in opportunistic Internet of Things. In: International conference on data analytics and management
4. Delipinar GE, Kocaoglu B (2016) Using SCOR model to gain competitive advantage: a literature review. Procedia Soc Behav Sci 229:398–406
5. Sangaiah AK, Hosseinabadi AAR, Shareh MB, Bozorgi Rad SY, Zolfagharian A, Chilamkurti N (2020) IoT resource allocation and optimization based on heuristic algorithm. Sensors 20(2):539
6. Waheed U, Khan MS, Awan SM, Khan MA, Mansoor Y (2019) Decentralized approach to secure IoT based networks using blockchain technology. 3C Tecnol. May 2019, pp 182–205
7. Wamba SF, Queiroz MM (2020) Blockchain in the operations and supply chain management: benefits, challenges and future research opportunities. Int J Inf Manage 52:102064
8. Awwad MA, Obispo SL, Kalluru SR, Airpulli VK, Zambre MS (2018, 2019) Blockchain technology for efficient management of supply chain blockchain technology for efficient management of supply chain
9. Kanoh H (2007) Dynamic route planning for car navigation systems using virus genetic algorithms, vol 11, pp 65–78
10. Arafat H, Alam MR, Zahid N, Zoarder Z, Haque A, Marjia Route optimization in supply chain network. Imp J Inter Discip Res
11. Corso A (2019) Performance analysis of proof-of-elapsed-time (PoET) consensus in the sawtooth blockchain framework
12. Hanebeck HC, Hewett N, McKay PA (2019) Inclusive deployment of blockchain for supply chains: Part 3—public or private blockchains—which one is right for you ?

13. Aoki Y, Otsuki K, Kaneko T, Banno R, Shudo K (2019) SimBlock: a blockchain network simulator. In: IEEE INFOCOM 2019—IEEE conference on comput commun work (INFOCOM WKSHPS), pp 325–329
14. Rani P, Tayal DK, Bhatia MPS (2018) Different Aspects, challenges, and impact of social networks with a mathematical analysis of teaching learning process. JARDCS 14:1576–1590

Enhancing Image Resolution and Denoising Using Autoencoder

A. S. Keerthi Nayani, Ch. Sekhar, M. Srinivasa Rao, and K. Venkata Rao

Abstract Nowadays, in real time, image processing is involved in various sectors like security, health care, banking, and face recognition. While capturing an image, there is more chance of noise engaged with multiple aspects of the surroundings. To improve the quality of the image and to get better classification results, we need to clean the picture, which is called pre-processing of the image. For the past 30 years, there is tremendous research happening on image processing by many researchers. Deep learning-based autoencoders are producing better results with minimum loss. Image denoising can be achieved with autoencoder architecture. The denoised image is taken as input to the next level to improve the resolution. In this paper, we have considered the popular dataset fashion mnist to denoising the image, which includes the noise. We used back-to-back autoencoders to perform both image denoising and resolution enhancement. In this approach, we can do the pre-processing stage once on the dataset for both image denoising and enhancement of image resolution. We have used binary cross-entropy as loss function to evaluate the performance of the model, and later, we have focussed on improving the resolution to the image. Denoising of an image followed by resolution enhancements in the same process minimizes the time and pre-processing steps separately.

Keywords Image denoising · Noise · Autoencoder · CNN · Super resolution · Deep learning

A. S. Keerthi Nayani
Department of ECE, Matrusri Engineering College, Hyderabad, India
e-mail: naskeerthi@gmail.com

Ch. Sekhar (✉) · M. Srinivasa Rao · K. Venkata Rao
Department of CSE, Vignan's Institute of Information Technology, Visakhapatnam, India
e-mail: sekhar1203@gmail.com

M. Srinivasa Rao
e-mail: srinivas.marada22@gmail.com

K. Venkata Rao
e-mail: vrkoduganti@gmail.com

A. Khanna et al. (eds.), *Data Analytics and Management*, Lecture Notes on Data Engineering and Communications Technologies 54,
https://doi.org/10.1007/978-981-15-8335-3_50

1 Introduction

The image denoising plays a crucial role in day-to-day life in applications such as remote sensing, satellite TV, individual recognition. Once the image was captured with a hardware device, the quality of an image may degrade due to lighting effect and noise. We have to reconstruct the image from noise; it can be achieved with the help of image denoising technique [1, 2].

The technique of image denoising involves extracting of latent clean image from the noisy image. Image denoising is similar to operations such as reduction of a blur and removal of watermarks. This method is used during pre-processing operation of computer vision tasks. The noise in the image is of type white noise and color noise based on correlation, additive, and multiplicative noise based on nature, quantization, and photon noise based on the source. If the image x is corrupted by additive white Gaussian noise (AWGN) model, it can be formulated as [2, 3].

$$y = x + \varepsilon \tag{1}$$

where: y belongs to R^N, i.e., observed noisy image

x belongs to R^N, i.e., observed clean image
ϵ is denoted the Gaussian noise vector with zero mean and covariance matrix.

Image resolution is normally depicted in PPI, which alludes to what numbers of pixels are shown per inch of a picture. Higher goals imply that there more pixels per inch (PPI), bringing about more pixel data and making an excellent, fresh picture. Image improvement is the name of the system that can be used to overhaul the perception or by the day's end the interpretability of data information in pictures which is accessible in the pixels for making a not too bad quality for human watchers and also better "input" delivered for other robotized or modified techniques for picture getting ready. The standard objective of picture improvement or development is to modify the properties of an image in order to grow its quality with the objective that it is better sensible for a particular task and moreover for the specific watcher or watcher. For this improvement to be happened, a bit of the characteristics of the images are should have been changed or improved by using certain procedures, for instance, the spatial region methodology is where it works by dealing with the pixels of the image. For obtaining the improvement, rough calculation of the pixels is controlled. Second is the recurrent region frameworks, and the image record should have been changed over into rehash space before going after it. Despite whatever else, image is made due with the Fourier change [4, 5].

- The central theme of this paper is to eliminate the noise during the capturing of image, process to handle the noisy image, and removing the noise in an image with the support of autoencoders. Once the noise was removed from the captured image, the next step is to increase the resolution of the picture [13].

- Usually, image denoising and improving the resolution of the image are done separately with autoencoders. But here, we are merging the two operations as one operation by applying back to back autoencoders.
- With this approach, the processing time will be reduced, as we do the pre-processing step only once on the dataset.

2 Literature Review

Image denoising concept can be implemented with a deep learning approach. We have given precise information on various researchers who worked on deep learning for image denoising.

Chunwei Tian et al. (2019) made a comparison of various DL methods toward image denoising in his research. They did CNNs for additive noise, deep CNNs for real noise, blind denoising, and noisy hybrid images. They worked with public data sets to implement deep learning methods and made comparison analysis. Based on their review, it is found that Gaussian noisy image denoising method is giving better results. They suggested that the use of the high-level hardware device while capturing a high-quality image to minimize the noise [3].

Junyuan Xie et al. (2012) proposed denoising autoencoder approach for low-level vision issues images which is more comparable to that of KSVD sparse coding method. Autoencoder is actually designed to learn an approximation to the unlabeled data. They showed experimental results with efficient accuracy in handling the image denoising. And also proposed a training model for denoising autoencoder for both denoise and in paint images with a unified framework. The future scope is to explore image super-resolution and missing data completion [1].

Gondara Lovedeep et al. (2016) proposed a model that plays a significant role at the pre-processing stage of denoising in medical image processing. Performing deep learning methods on an image requires large training samples and high-performance CPUs/GPUs. They have used autoencoders applying convolution layers to handle the small sample sizes for useful denoising of health-related images. To improve denoising performance, the sample size can increase by mixing heterogeneous images with 300 samples used for the training dataset and achieved excellent results [6].

Sudipta Singha Roy et al. (2018) to handle the images significant issues with the noise present in the image and decrease the performance of classifiers. They proposed a model that trains the DAE with low-level noise injected images and a CNN with noiseless native images independently. They arranged the trained model with different combinations like CNN, DAE-CNN, and DAE-DAE-CNN and then classified the images corrupted with zero, regular, and massive noises, respectively. This approach is implemented on MNIST handwritten numeral dataset with various noise levels. They got better results than the individual trained models [7].

Chao Dong et al. (2015), implemented end-to-end model to achieve the super-resolution for the output image. Mapping is done between the low-resolution image and high-resolution image with deep convolution neural network. Old methods use each component individually, and their model combines all layers with lightweight design. The main objective is the faster usage in online applications. They implemented with three color channels in parallel with good quality of reconstructed image [8].

3 Autoencoder

Data points without any label to the group are considered as a common set. These can be grouped based on the similarity or distance measure between the points such as learning called unsupervised machine learning. Autoencoder is a type of unsupervised learning with feed-forward neural network architecture. The network takes the input and produces the output which is same as the input. Autoencoder architecture is shown in Fig. 1, with significant layers encoder, hidden layer, and decoder. Neural network design is like bottleneck fashion between encoder and decoder layers. Bottleneck purposes the function by compressing the knowledge representation of the input data points. The compressed input again uncompressed at output form.

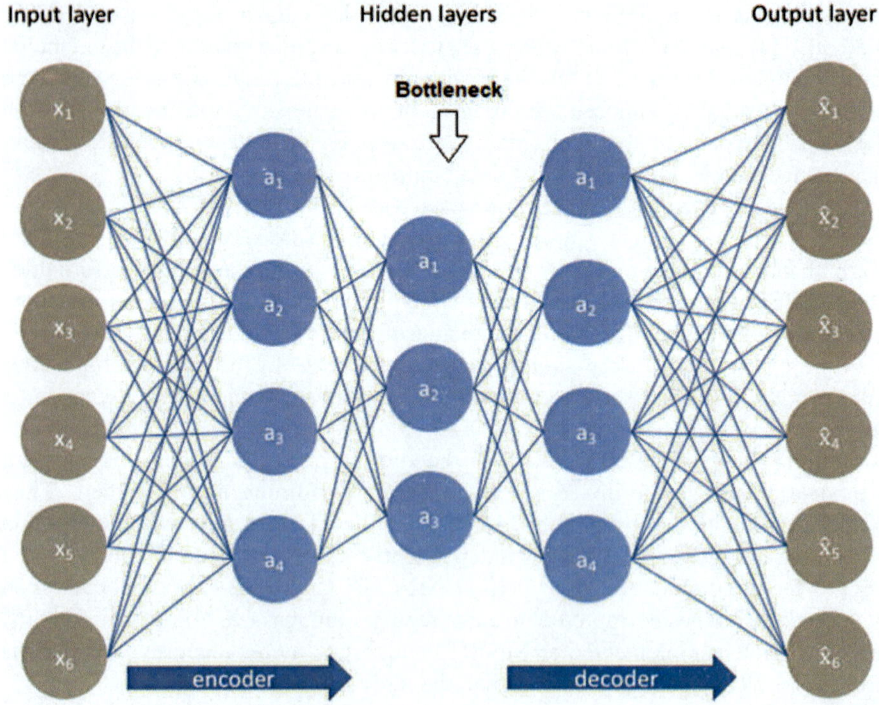

Fig. 1 Autoencoder process mechanism [9]

Compression of the input data can be treated as latent space representation [9]. The compression mechanism in autoencoder is well suited for dimensionality reduction.

Figure 1 shows the input features fed to the encoder layer; it was fully connected neural network the given output. The decoder layer also like full connected neural network and produce the code. The main objective is to get the output the same as the input. The major part of an autoencoder is to set up network architecture with the following hyper parameters

1. **Nodes**: Number of perceptron nodes in the middle layer. To get high compression results, it is better to use less number of nodes.
2. **Layers**: We can place as many layers in encoder and decoder blocks to get deep, Input and output need not be considering.
3. **Number of nodes per layer**: Number of nodes in the encoder block layers decreases with subsequent layers, whereas a number of nodes in the decoder block increase.
4. **Loss function**: The main objective of the loss function is the performance metric of decoder, which shows the similarity of the output of the decoder how to original input data. Mean squared error (MSE) or binary cross-entropy is used as the loss function.

$$\text{Encoder: } h(x) = \text{sigmoid}(W * x + b) \tag{2}$$

$$\text{Decoder: } x = \text{sigmoid}(W * h(x) + c) \tag{3}$$

where

x denotes the inputs
b denotes the bias each layer.

The various applications of autoencoders are like dimensionality reduction, image processing, anomaly detection, information retrieval, drug, discovery, and machine translation. The initial autoencoders are meant for majorly dimensionality reduction and information retrieval only. As trend change in the modern model, design of autoencoders is well implemented in many applications mentioned above.

4 Data Set

Dataset is an essential parameter while working with machine/deep learning methods. Han Xiao et al. [10] designed a fashion MNIST dataset containing 70,000 fashion-related images of type grayscale. The images broadly categorized 10 labels with 28 × 28 grayscale images. The dataset split into two parts, training and testing datasets. The training dataset consists of 60,000 images and test dataset with 10,000 images (Fig. 2; Table 1).

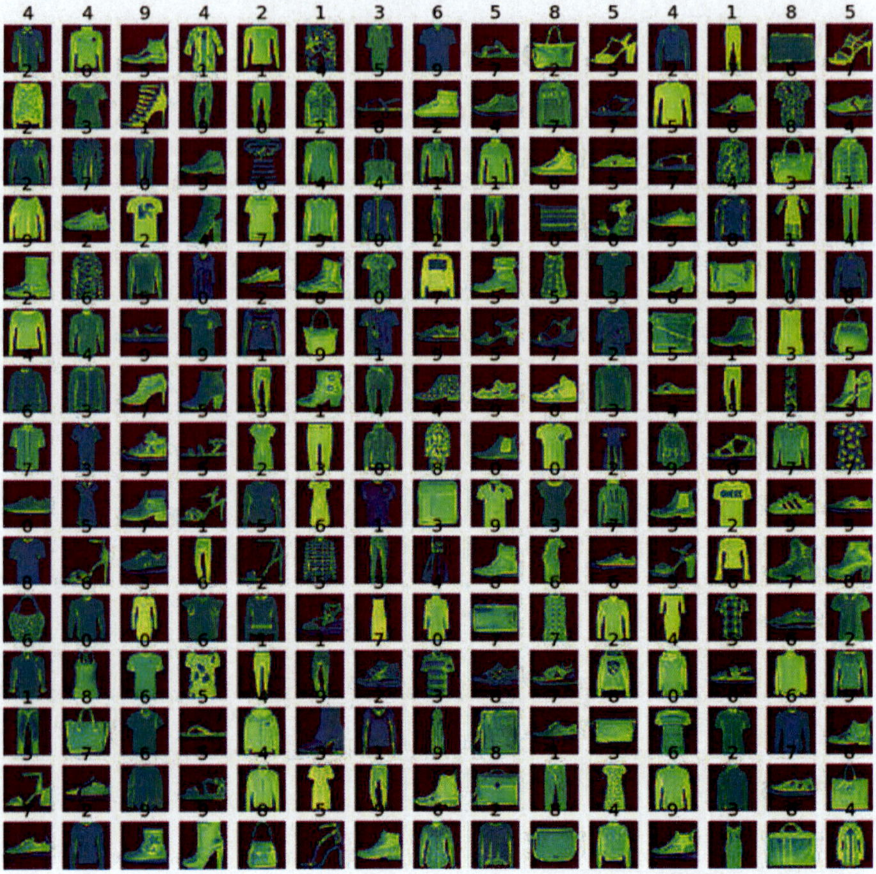

Fig. 2 Sample training images from fashion MNIST dataset [10]

Table 1 MNIST image labels [10]

Label	Description
0	T-shirt/top
1	Trouser
2	Pullover
3	Dress
4	Coat
5	Sandal
6	Shirt
7	Sneaker
8	Bag
9	Ankle boot

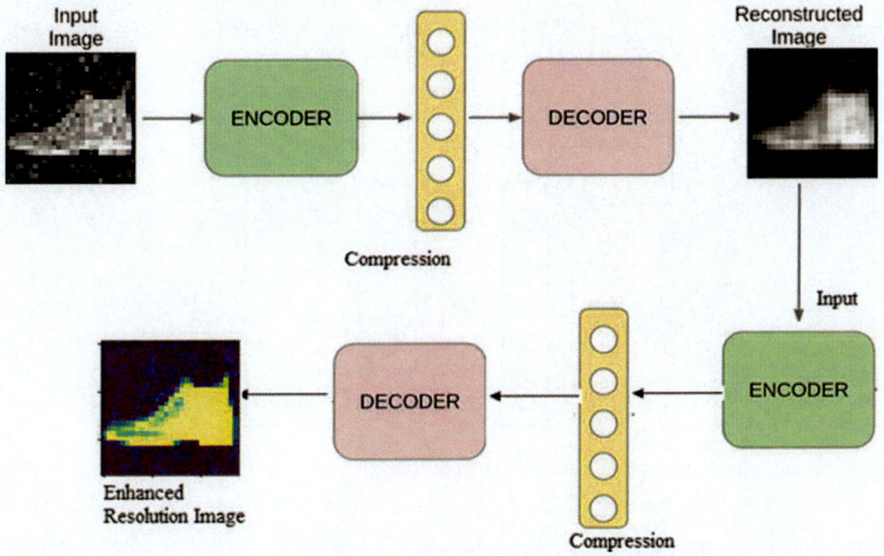

Fig. 3 Two-level autoencoder model for image denoising and enhancement of resolution of the image

5 Implementation and Results

In this paper, we have implemented the image denoising and super-resolution of images with autoencoders. Autoencoders are a type of artificial neural networks (ANN) that are used to perform a task of data encoding. In Fig. 1 shown, the basic architecture of autoencoder consists of encoder, hidden layer, and decoder layers. Bottleneck is implemented between encoder and decoders. This bottleneck advantage is to create compressed data of the input data. Autoencoders work well if correlations existing between input [9, 11]. The main aim of the autoencoder is to minimize the reconstructed error rate. The error is the difference between the input and output.

Figure 3 shows the combination of two autoencoder models, the first model does the image denoising process, and the second model enhances the image resolution. The denoise image is taken as input to the next level to improve the resolution with the combined architecture. Advantage of combined architecture avoids the pre-processing steps separately and hence better performance with optimal time.

6 Image Denoising Stage

Image denoising with autoencoder architecture is shown below in Fig. 4, we will feed noisy image from the dataset of fashion mnist as input, and the output is a clean image. The steps involved to achieve the task are as follows

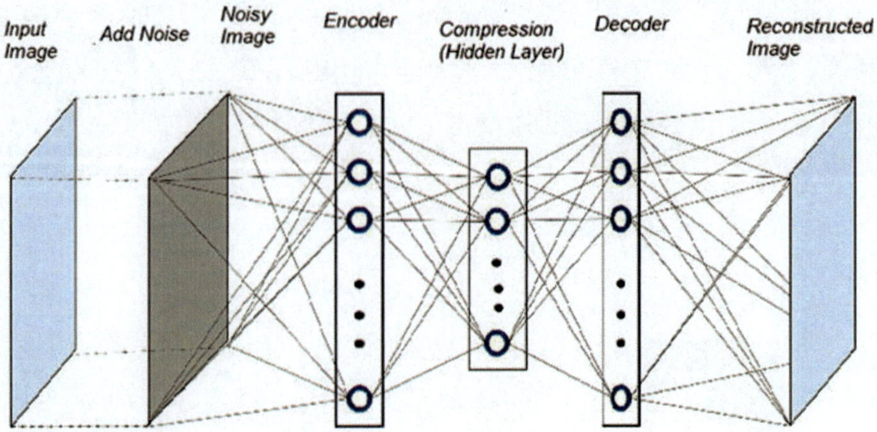

Fig. 4 Image denoising with autoencoder architecture [7]

Step 1: Import the required Python libraries.
Step 2: Load the Fashion MNIST dataset and split the dataset into train and test sets.
Step 3: Apply pre-processing methods on train and test sets.
Step 4: Add random noise to the input image.
Step 5: Build and train the autoencoder model with convolution layers.
Step 6: Evaluate trained model performance by calculating the loss function.

At step 5, we are going to build the encoder, compression, and decoder layers. The input image of size 28 × 28 is encoded into 14 × 14 and further into 7 × 7 in the convolution layers [12]. In the process, we were doing the compression of the input image. Again the reconstruction of image is done with the decoder layer, as shown in Fig. 4 experimentally.

The experimental results we can observe in the above output Fig. 5, top images showing the images with noise and image denoising results shows in the bottom.

Figure 6 shows that the autoencoder model that runs for ten epochs, visualizing each epoch loss value. After the second epoch, there is no notable improvement in the loss.

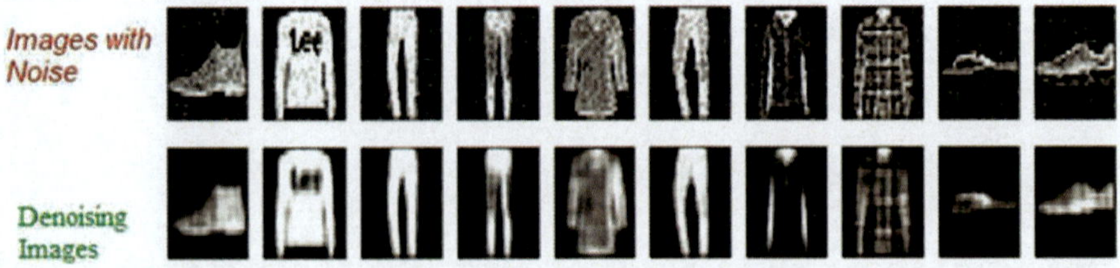

Fig. 5 Image denoising stage output at the end of autoencoder model

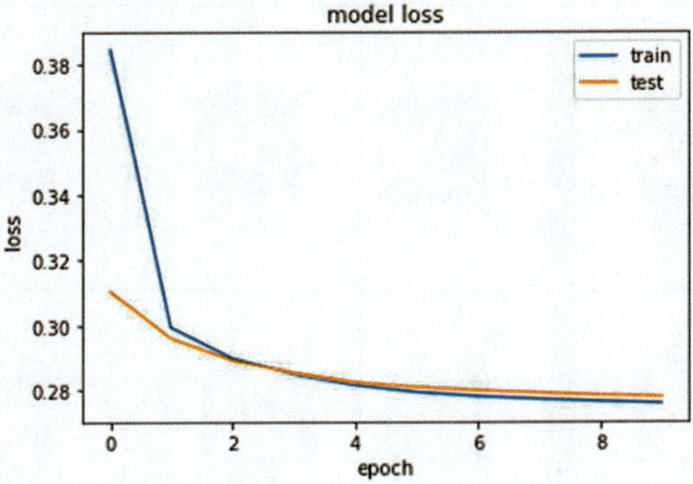

Fig. 6 Prediction loss value at each stage of an epoch running on training and test sets

7 Enhancement of Image Resolution

Step 1: Output of the Previous Stage take as Input.
Step 2: Design an autoencoder model with Sequential Layers

Encoder: Input Layer and Flatten Layer.
Decoder: Input Layer and Dense Layer.

Step 3: Build and compile the autoencoder model.
Step 4: Calculate the loss value of the reconstructed image.

In the image enhancement stage, the image size is $32 \times 32 \times 3$ (width: 32, height: 32, color channels: 3). Autoencoder builds with sequential layer architecture, and each layer feeds into the next layers. It consists of flatten layer and dense layers. The role of flatten layer is to flatten the image of size $32 \times 32 \times 3$ matrix size into a one-dimensional array and the dense is the neural network layer to find the optimal parameters to achieve better results.

Figure 7 shows those results of enhancement of image stage with two sample images (Shoe and T-shirt from fashion mnist dataset). Left-side output is an image with 32 pixels; it's not looking proper resolution in the output due to loss of data as it was with very less number of pixels outcome after the compression stage of the autoencoder. Right-side image with 1000 pixels, it shows that the image is much clear compared to the left sid [12, 13]

Figure 8 shows that the autoencoder model runs for 20 epochs, visualizing each epoch loss value. After the third epoch, there is no notable improvement in the loss.

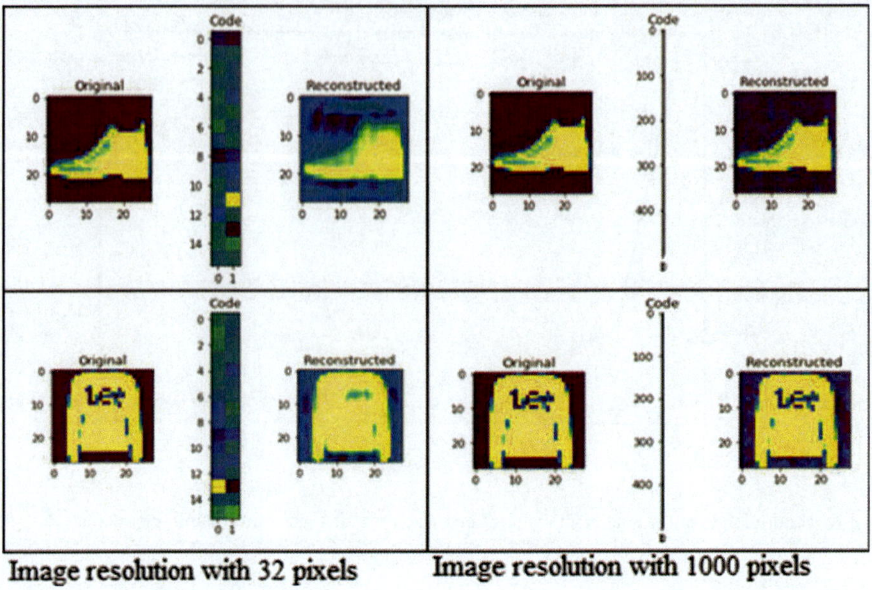

Fig. 7 Image resolution enhancement with autoencoder

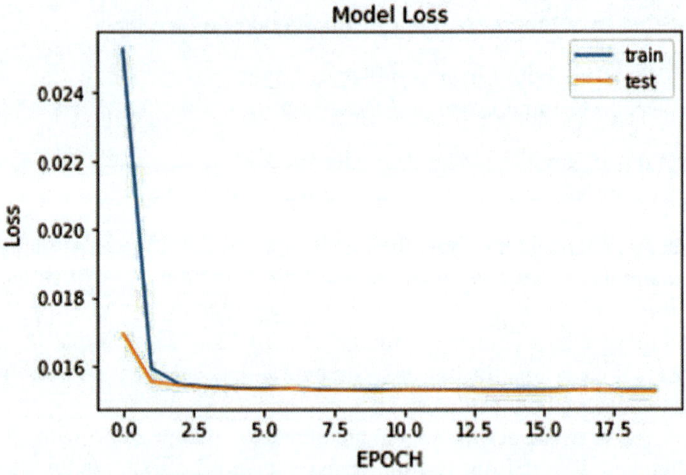

Fig. 8 Prediction loss value at each stage of epoch during. Image resolution enhancement process

8 Conclusion

The unsupervised learning method, i.e., the autoencoder model, plays a vital role in the image processing related tasks. In real time while captured the image with a hardware device that usually contains noise. In this paper, we have taken the popular dataset fashion mnist to denoising the image, which includes the noise. We used binary cross-entropy as loss function to evaluate the performance of the model. We combined the image denoising and resolution enhancement as combined architecture. Advantage of combined architecture avoids the pre-processing steps separately and better performance with optimal time.

References

1. Tian C, Fei L, Zheng W, Xu Y, Zuo W, Lin C-W (2019) Deep learning on image denoising: an overview
2. Xie J, Xu L, Chen E (2012) Image denoising and inpainting with deep neural networks. Adv Neural Inf Process Syst
3. Gondara L (2016) Medical image denoising using convolutional denoising autoencoders 241–246. https://doi.org/10.1109/icdmw.2016.0041
4. Roy SS et al (2018) A robust system for noisy image classification combining denoising autoencoder and convolutional neural network. Int J Adv Comput Sci Appl 9
5. Arden D, Applied deep learning: autoencoders. https://towardsdatascience.com/applied-deep-learning-part-3-autoencoders-1c083af4d798
6. Jordan J, Introduction to autoencoders
7. Han X, Kashif R, Roland V (2002) Fashion-MNIST: a novel image dataset for benchmarking machine learning algorithms
8. Thakur RS, Yadav RN, Gupta L (2019) State-of-art analysis of image denoising methods using convolutional neural networks. IET Image Process 13(13):2367–2380
9. Dong C, Loy CC, He K, Tang X (2016) Image super-resolution using deep convolutional networks. IEEE Trans Pattern Anal Maschine Intell 38(2):295–307. 1 Feb 2016. https://doi.org/10.1109/tpami.2015.2439281
10. Bhabatosh C, Dwijest Dutta M (2002) Digital image processing and analysis
11. Weeks RW Jr (2006) Fundamental of electronic image processing. Spie Press, Bellingham
12. Nishio M, Nagashima C, Hirabayashi S, Ohnishi A, Sasaki K, Sagawa T, Hamada M, Yamashita T (2017) Convolutional auto-encoder for image denoising of ultra-low-dose CT. Heliyon 3(8):e00393. ISSN 2405-8440. https://doi.org/10.1016/j.heliyon.2017.e00393
13. Deep learning based image super-resolution to enhance photos. https://cv-tricks.com/deep-learning-2/image-super-resolution-to-enhance-photos/

Detecting Organic Audience Involvement on Social Media Platforms for Better Influencer Marketing and Trust-Based E-Commerce Experience

Ayushi Dewan

Abstract This study addresses a major problem of online advertisement. Usage of Internet and specifically social media platforms for promoting businesses and services has seen a drastic increase in the recent times. Influencer-based marketing is a major area for promotion. With its major dominance, comes its downsides. This influencer-based model alternatively being referred to as influencer marketing has a major drawback which concerns one of the foremost aspects of digital media, organic and inorganic results. Many of the social media platforms involve inorganic audience which includes fake popularity for the concerned account. This heavily impacts the advertisements business being run on social media and other platforms. This could lead to wrong audience impact involving losses for the business as well as the customers being targeted. In this study, a trust-based model is designed in which the social media platform, Instagram, is checked for fake audience involvement and their consecutive detection for better future impressions. For the detection of inorganic audience involvement, firstly, a study is done regarding the parameters that could assist the detection process. These parameters are further used to generate the nodes of a decision tree. Second of all, the dataset is prepared for the detection process. Machine learning is used to make the predictions on the dataset as with minimum human intervention, it can help automate the entire process and make it much faster than manual detection. Decision tree is then applied to the dataset for building a prediction based model, which is used to predict the occurrence of fake accounts in test dataset. This is followed by the concluding of the result in which prediction has been done. Also, the accuracy of the same has been checked in the conclusion with the help of confusion matrix which turns out to have a good accuracy, thus, successfully concluding the study.

Keywords Inorganic audience · Machine learning · Decision tree · Influencer marketing · Social media platforms · Confusion matrix

---

A. Dewan (✉)
USICT, GGSIPU, Sector 16 C, Dwarka, Delhi 110078, India
e-mail: ayushidewan@gmail.com

A. Khanna et al. (eds.), *Data Analytics and Management*, Lecture Notes on Data Engineering and Communications Technologies 54,
https://doi.org/10.1007/978-981-15-8335-3_51

1 Introduction

Social media networks are widely used nowadays by individuals and they surely have seen a huge rise in the number of users from the past few years. Some of the most used social media networks are Facebook, Twitter, Snapchat, Instagram, YouTube and LinkedIn. Apart from being used for sharing content, these platforms are also widely used by brands to sell their products to the audience.

- A marketing model that is widely being used on these platforms is influencer-based marketing. Influencer-based marketing refers to the marketing strategy which is an amalgam of longstanding and new marketing tools. It basically forms its base similar to celebrity endorsements. Influencer marketing is the collaboration between brands and influencers. Rather than directly marketing to large number of audience, instead, influencers are paid to give out the message. Influencers have a good reach on various social media platforms through which they promote a brands idea, message, advertise products so that the audience reach the product easily. Brands can create huge revenues with a well-structured campaign. These influencers are expected to increase the reach of the word through their channels or accounts on Facebook, Twitter, YouTube, Instagram, LinkedIn, etc. The influencers basically create content on their social media platforms for the brand. Influencer campaigns mainly involve two kind of marketing strategies: Social media marketing and Content-based marketing.
- Instagram has come across as an idyllic fit for e-commerce marketing in various industries. The platform is often used as the place to make big sales in businesses. With over a billion users, it is widely used, and according to a survey, 90% of accounts follow at least one business on Instagram. Brands want to invest more on the marketing action here. Eighty-three percent of Instagram users ascertain new merchandises and services on the Instagram. This number is increasing exponentially. Instagram mainly provides businesses the prospect to start a business page which is free of cost and is used by them to promote their products as well as increase reach of their brand. Businesses along with a free profile also have all the access to impression and reach metrics along with free engagement. As quoted by official Instagram's account, more than 1 million advertisers use Instagram worldwide to share their stories and get better business results, and also, about 2 million monthly advertisers and users like 4.2 billion posts daily come across Instagram. Additionally, 60% of people are said to discover new products through Instagram. According to a study by 'Mediakix,' Instagram is the most preferred platform for influencer-based marketing. Instagram is preferred 89%, followed by YouTube at 70%. Facebook is at 45%. The top three industries that have collaborated with Instagram influencers are 91% luxury brands, 84% sports clothing brands and 83% are of beauty industry. These statistics are based on a survey conducted by sprouts social.
- With such huge user base and ads funding, fake engagement and inorganic traffic generation is becoming a major concern. Scammers attempt to exploit this for personal gains and trick the businesses to collaborate with them. The problem

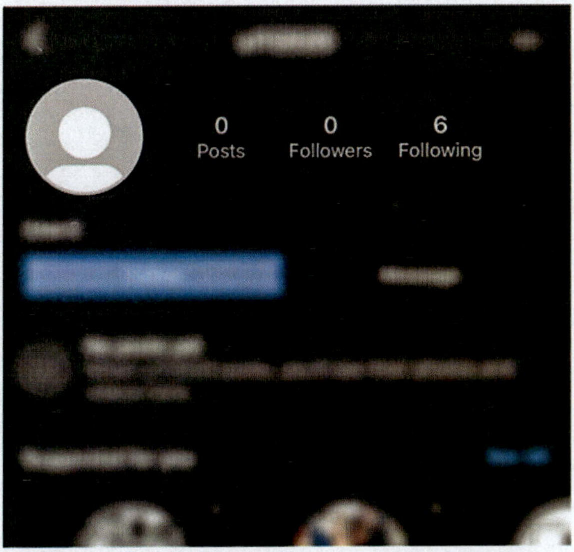

Fig. 1 Fake account example

of the fake impressions on these platforms is emerging. Influencer is using fake metrics by buying followers, likes and comments so as to trick the brands to collaborate with them. Thus, one of the major problems in influencer-based marketing nowadays is the building of large number of fake profiles. Fake accounts are basically accounts that are created to gain some personal use against the rules. The focus lies on how to detect this fake activity involvement. This is the ultimate problem as one needs to consider various factors for this.

A lot of parameters can be considered. Usually considered ones are:

1. User follower to following count ratio
2. Number of likes to comments ratio on media being uploaded
3. Number of media uploaded
4. User biography
5. Username and number of characters and digits in it, etc.

Therefore, it is crucial to preserve the healthy environment in such an important social media platform. An example of a fake account on Instagram has been shown in Fig. 1. Also, an example of a real Netflix India account is shown in Fig. 2.

2 Literature Review

Removing fake accounts has attracted the attention of many researchers. Extensive researches have been carried out on the identification of fake accounts in social networks. Different approaches are proposed and to find fake accounts based on

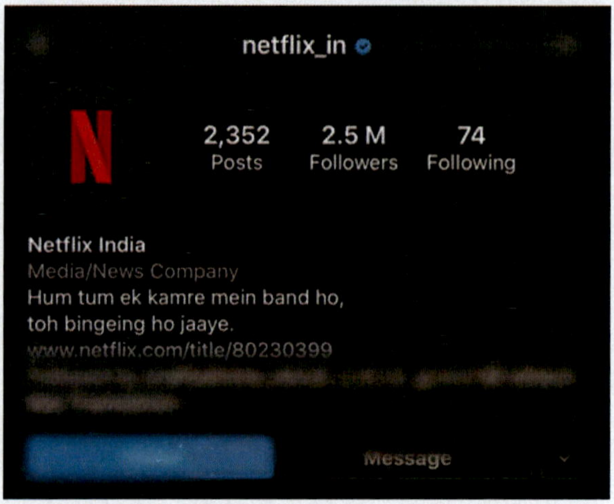

Fig. 2 Real netflix india account

attribute similarity, similarity of friend networks, and profile analysis for a time interval, etc. Now, this particular approach can be further expanded to various other areas considering the wide usage of this detection algorithm. Some of the researches and past areas in the literature that have been included and studied for the establishment of a reliable prediction model for fake and real accounts analysis in this research have been implemented on various other social media networks like Facebook, Twitter, LinkedIn, YouTube, etc. Some of these are, to detect the Twitter-based fake accounts, have been done in [1] using support vector machines and logistic regression. Now, similar Twitter-based analysis is done in [2] using graph-based methods, and also in [3] using the joint usage of naive Bayes classifier with entropy minimization discretization. [4] mainly is implemented to detect fake accounts using NLP and machine learning tools but also proposes some security architectures and focuses on Facebook side by side.

Some of these researches and previously done works also involve Instagram, which is our major area of concern in this study. In the recent times, gradually, Instagram has been ruling in terms of social media involvement and activity being posted and users accessing this platform. Some of the studies that show and focus on Instagram mainly are ‘worth its Weight in Likes: Towards Detecting Fake Likes on Instagram’ [5]. Now considering this one specifically, it focuses on examining the likes of media and posts on Instagram for a user and it also identifies the legitimate likes to minimize the impact of fake likes on influencer marketing. In this study, the main concern is to estimate what is the probability that a user can like the post of another user based on the network closeness, interest overlap, liking frequency, influencer effect and link farming hashtag effect. They have plied a simple feed-forward neural network multi-layer perceptron (MLP) which got a precision of about 83%. [6] focuses on the detection of not real accounts in social media networks. This is a literature review plied to detect fake social media profiles classified to the different approaches focusing on analyzing individual profiles. Thus, our proposed system is an approach involving the platform chosen, i.e., Instagram and the algorithms used like decision tree for classification. Through all of these readings, it has been observed that there is not much study that has been done considering the organic and inorganic traffic detection for Instagram influencer marketing with widely accessible information. Study on real and fake profile detection for Instagram using machine learning algorithms is done in this work and the algorithm implemented has been discussed.

3 Methodology

The adopted methodology has been explained in detail step by step. In the start, characteristics of the dataset have been explained. This includes all the parameters that were taken to analyze the result, from where the dataset was extracted and what kind of actions were performed for preprocessing and preparing the train and test data. This is followed by the actual implementation of the model designed. All the steps regarding this system have been given in detail along with the supervised machine learning algorithm that has been used which is decision tree algorithm. The prediction compares the characteristics of the given factors for organic as well as inorganic audience. After comparing, decision tree is used to detect and build the system. Decision tree classification algorithm is very simple which assumes that the classification attributes are independent and they do not have any correlation between them. Decision tree is implemented in various areas to solve real-world problems.

The given Fig. 3 is recommended by the author in which a decision tree basic structure has been shown which is top-down induction algorithm.

Figure 4 is recommended by the author and it illustrates the process of how machine learning is implemented. This is the machine learning pipeline. The same methodology is followed here to build the required model. We import and store the

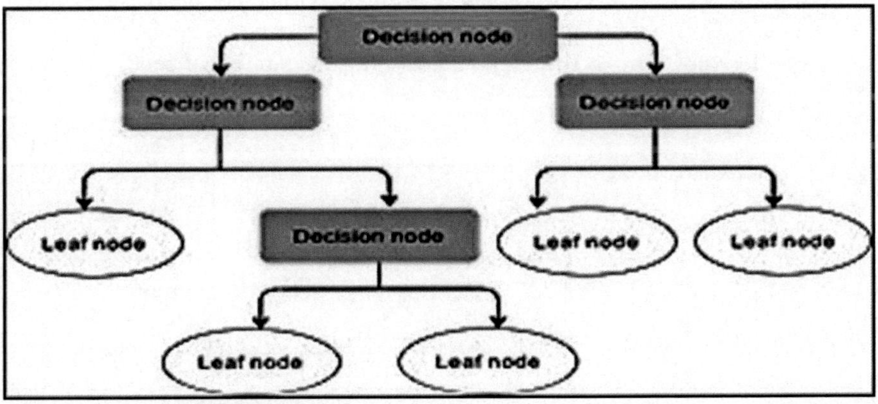

Fig. 3 Decision tree structure

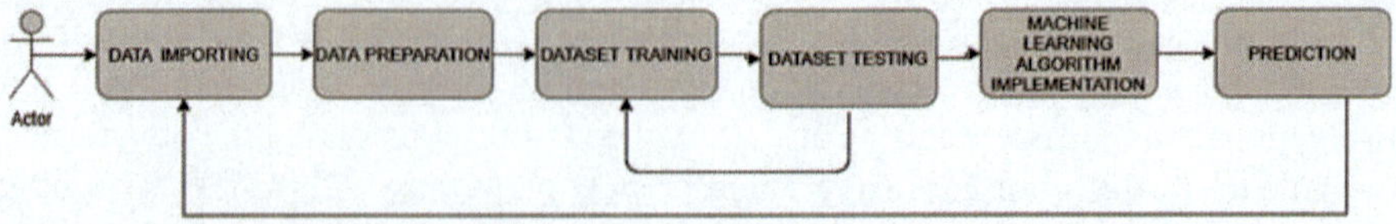

Fig. 4 Machine learning components

data. Then, we prepare, train and test our system, and finally, we deploy this model. With new data being generated, repetition of this process occurs.

3.1 *Characteristics of Dataset*

The dataset takes into consideration, some accounts with organic audience involvement and some accounts with inorganic traffic involved. The dataset consists of accounts of fake and real people. In total, 1196 rows are present, i.e., data of 1196 accounts have been considered. Out of this number, 995 accounts are organic and the remaining 201 are inorganic. This dataset has been taken from https://github.com/fcakyon/instafake-dataset. Instagram dataset can be collected by web scraping. Instagram has JavaScript JSON data in their HTML source. One can parse this by any scripting language and then get JSON data of Instagram. Now this particular dataset contains of mainly two files namely, 'fakeaccounts.json' and 'realaccounts.json'. These files were shuffled into a single file named 'mixed sample data.' This final file has in total nine columns. Each of these column names denotes a parameter considered for detecting if the account is fake or real. The below Fig. 5 is recommended by the author and it shows how the two datasets were merged to get a dataset containing all the values of real and fake accounts (Table 1).

Following are the description of each of the parameter considered for this study.

1. **userFollowerCount**—This column gives the total number of followers of the account.
2. **userFollowingCount**—This column gives the total number of people following the account.

Fig. 5 Dataset creation

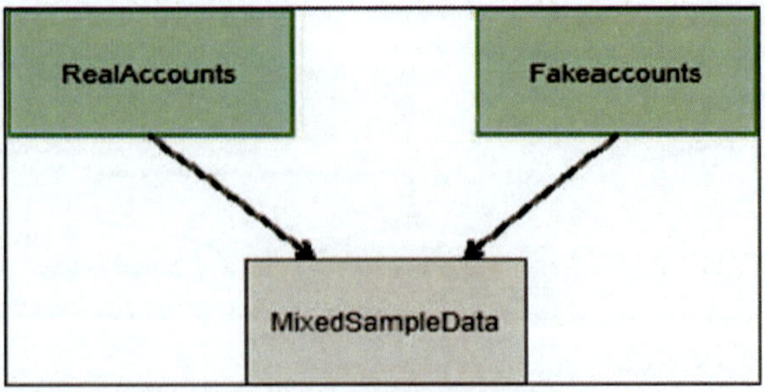

Table 1 Dataset real and fake followers count

Fake and real accounts data	Rows	Columns
Real accounts	995	9
Fake accounts	201	9

3. **userBiographyLength**—Here in this column, the length of the biography of the account is given in numbers.
4. **userMediaCount**—In this column, the number of media uploaded by the account, be it video or image, is denoted by a count.
5. **userHasProfilPic**—This column tells if the account has a profile picture uploaded or not. Given by binary values 0 and 1. 1 if the user has a profile picture and 0 if no profile picture has been uploaded. It is a common trend in fake profiles not to upload a profile picture.
6. **userIsPrivate**—Instagram has two kind of privacy settings. One is public profile, in which the content is visible to all and the other one is private account in which only the people following the account can view the content of the account. This parameter taken binary values where 1 denotes that the account is private and 0 denotes that the account is not private, i.e., public.
7. **usernameDigitCount**—This column gives the count of digits in the username of the account. The fake profiles have been observed to have a lot of digits in their username.
8. **usernameLength**—This column gives the count of length of the username of the particular account.
9. **isFake**—This is the parameter which denotes by binary values if the account is fake or not. This parameter will be used to check the accuracy of the result predicted by decision tree algorithm. It gives a value of 1 if the account is fake and a value of 0 if the account is real.

3.2 Analysis

The data is imported from an external source as mentioned above. R has been used to perform the analysis on the dataset. R is one of the most powerful machine learning platforms and is used by the top data scientists in the world. R is an excellent choice if data analytics is at the core of the study being carried out. It enables rapid prototyping and functioning with datasets to develop machine learning systems.

Figure 6 is recommended by the author and it portrays the value of isFake column that is if the account is fake or not on x-axis with respect to the number of followers of the account on the y-axis. The number of followers has been extensively used as a measure to detect manually, the reach of the account in consideration. The result shows that most of the accounts with more than 1000 followers are real. But, if we observe the values in detail, we see that some accounts are still there that are fake and still have more than 1000 followers. Thus, there is a need to have other parameters that should be considered for achieving a better result that is more accurate and

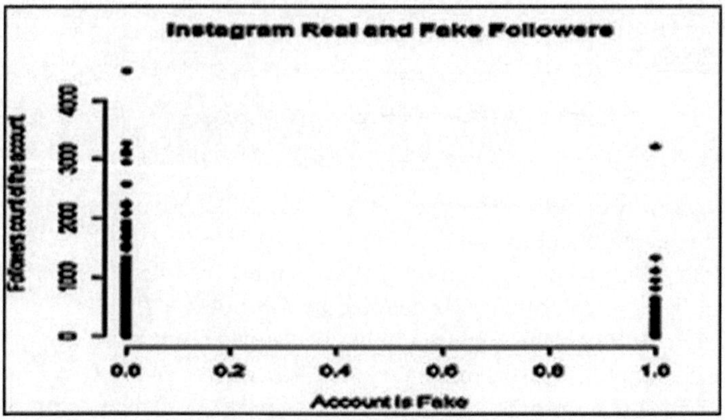

Fig. 6 Plot of fake accounts with respect to number of followers

Table 2 Test/train split

Test and train data	Rows	Columns
Train	835	9
Test	359	9

reliable to use as automated detection systems. We further move to our next step that is applying a machine learning algorithm for an automated result. To prepare the dataset for building the model, a train/test split has to be generated. The model is trained with the help of this 'train' dataset. And 'test' is used to predict the result that is if the account in consideration is a fake account or not. The splitting of 'train' and 'test' data is performed randomly; here, we have introduced a split which contains 70% of 'train' data and 30% of 'test' data. This 70% of data is used to train the model and the remaining 30% of the data is used to make predictions as show in Table 2.

Using the library rpart() and rpart.plot() to fit the model and the value of the result is stored. The rpart () function has the arguments, formula, data and method. The 'formula' refers to the function that is to be predicted. The 'data' specifies the data frame and 'method' for a classification tree.

4 Results

The model has performed appropriately as required. This section mainly includes the decision tree that has been created and its outcomes explained in detail. This is followed by the prediction and testing of the dataset 'test' which actually predicts if the account is fake or not using the experience gained from the training. Starting at the root node, the overall probability of an account being fake or not. '0' in this tree denotes that the account is real and '1' denotes that the account is fake. By default, Gini impurity measure is taken to split the nodes by rpart () function. With higher value of Gini coefficient, more different instances are there in the node. In this decision tree:

- At the top of the tree, the overall probability of an account being fake or real is shown. Left side consists of 688 accounts which is basically the first split in the tree.
- This node basically splits on the basis of the number of followers of the particular account. If userFollowerCount >=90, we follow the left child node. 81% of the population taken have followers >=90. The rest 19% are denoted by the right child node which denotes that account is fake.
- Next split is done on the basis of the number of people being followed by this particular account. If the userFollowingCount < 630, we follow the lefty child node which denotes that account is real. In case the follower count is less than that, we follow the other nodes in the tree.
- Similarly, we split the tree on other levels as well like UserMediaCount, UserNameLength, etc. to get better and more precise results.
- Finally, the last level containing the termination nodes has the result with '0' denoting the percentage of real accounts in our dataset and '1' denoting the fake accounts in the dataset considered.

The above Fig. 7 of decision tree is recommended by the author and using the above result, the next step is to test the model that is built in this study a prediction is made. Now, a table is created to count how many of the accounts taken in consideration are classified as fake and are accurate according to the classification. The output of this step is a table that predicts that 299 accounts are real but it also classified seven real accounts as fake. Similarly, the model correctly classified 41 accounts as fake. But, it also classified 12 accounts as real which were actually fake. The result of this has been shown in table.

Here, in the above table, '0' denotes if an account is real and '1' denotes the account is fake. By performing the analysis, it is concluded that in this model, the detection of organic and inorganic accounts has been performed successfully. This study has been carried out as this problem leads to a lot of issues for the brands and the public in the influencer marketing sector. To reduce this problem to a certain extent, supervised machine learning algorithm decision tree has been implemented. Machine learning has been used to automate the process of fake accounts detection. The contributions that have been done in this study are, to our knowledge, proposing the features that can be used to detect fake and real accounts detection, proposing a technique to predict if the account is fake or real using classification method. To measure the performance so as to conclude this work, an accuracy measure for classification task with the confusion matrix has been computed. This has been chosen as it is a good choice to count the number of times true instance has been classified as false, that is, how many times real accounts have been classified as fake and vice versa. As shown in Table 3, each row in the confusion matrix denotes an actual target whereas each column a predicted target. According to the result of this study, the first row of the matrix considers the real accounts, that is, the 'false' class in which 299 are

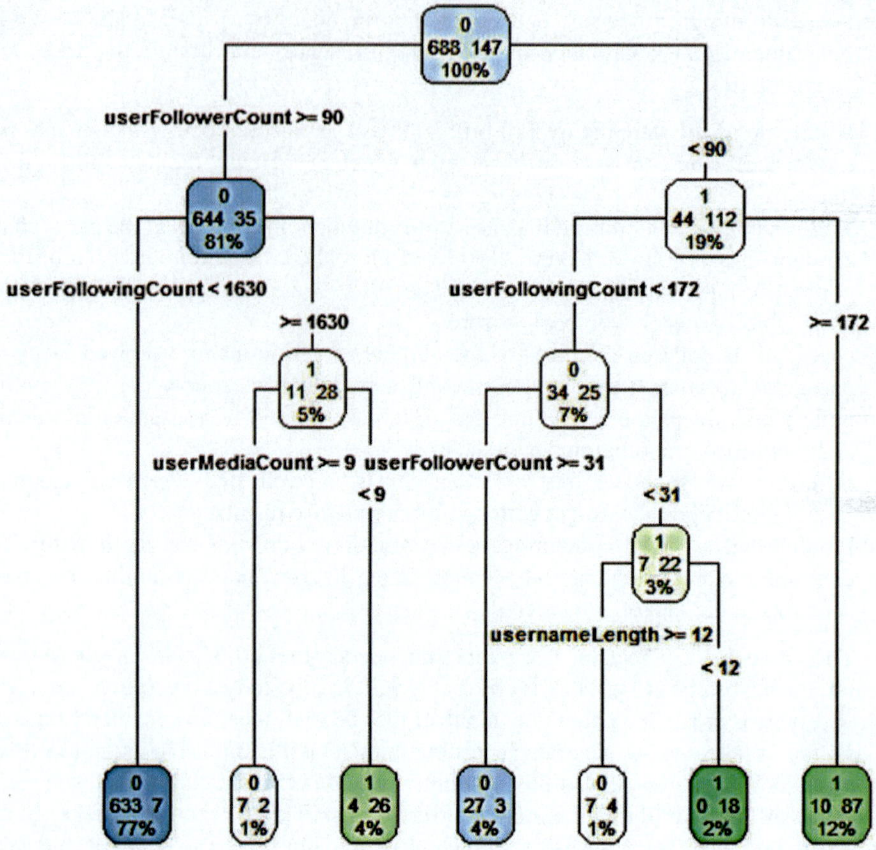

Fig. 7 Resulting decision tree

Table 3 Prediction results

Actual/predicted result	0 (False)	1 (True)
0 (FALSE)	299 (TN)	7 (FP)
1 (TRUE)	12 (FN)	41 (TP)

classified correctly, that is, 'true negative,' and the remaining seven as 'false positive.' Coming to the second row, which denotes the fake followers count, the positive class consisted of 41 which is 'true positive' and 12 as 'false negative.' The accuracy test can be computed from the confusion matrix by using the Eq. 1.

$$\text{Accuracy} = \frac{(41(\text{True Positive}) + 299(\text{True Negative}))}{(41(\text{True Positive}) + 299(\text{True Negatuve}) + 7(\text{False positive}) + 12(\text{False negative}))} \quad (1)$$

Fig. 8 Confusion matrix visualization

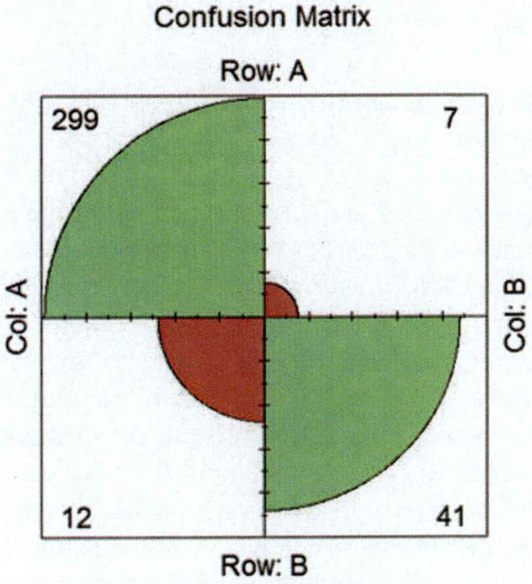

Accuracy of result.

This is the ratio of true positive and true negative with respect to the sum of the matrix. By performing the above calculation in R, the accuracy for the test comes out to be '0.9470752'. This score is of 94% for the dataset. This score should be of higher than 0.78 for the model to perform well. Thus, it has been concluded that an automated model has been built which performs adequately and is a reliable, easy to use and an accurate one. The visualization of confusion matrix has been shown in Fig. 8. This figure has been recommended by the author.

5 Conclusions

On the basis of the result drawn, it is concluded that Instagram can be used for influencer-based marketing by the brands. It is a safe platform but as it is discussed in this paper that many influencers try to scam the businesses by creating a fake metrics, so for this issue, a prediction-based model has been designed in this paper using supervised machine learning algorithm, decision tree. Using this model will help keep brands aware of where they are investing their money and if it will prove beneficial for them. As seen in literature review, some researches have been done on many social media platforms. In this model, the accuracy rate achieved is 94.70% which is a good score as compared to previous works.

6 Future Work

This study can be further implemented to various other fields in E-commerce. An analysis can be done on the review-based model of E-commerce websites. As most of the customers rely on review-based model to trust a particular product, this model can be tested and an agent can be designed for the detection of fake reviews being made to fool the customers into buying these products and services. Further, various other algorithms can be used to detect this inorganic result problem, and thus, a better trust-based system can be designed for usage. E-commerce being one of the major platforms of buying and selling goods can be analyzed based on the user reviews. Websites like Amazon, Flipkart, Myntra, and Paytm have huge number of customers who solely rely on the customer reviews being shown. Some of these reviews are fake and are published just to increase the recommendations ranking and create revenue. This is a scam which can be avoided by detecting the number of fake reviews being posted based on the certain important factors such as active usage of account, display picture, buying patterns and buying counts, user name. Many such factors can be taken into account.

References

1. Efthimion PG, Payne S (2018) Supervised machine learning bot detection techniques to identify social twitter bots. SMU Data Sci Rev 1(2):5
2. Mohammadreza Mohammadrezaei M (2018) Identifying fake accounts on social networks based on graph analysis and classification algorithms. Secur Commun Netw 2018
3. Er B, Sahin, ÖA (2017) Twitter fake account detection. In: International conference on computer science and engineering (UBMK). IEEE, sf, pp 388–392
4. Raturi R (2018) Machine learning implementation for identifying fake accounts in social network. Int J Pure Appl Math 118(20):4785–4797
5. Sen AA (2018) Worth its weight in likes: towards detecting fake likes on instagram. Web Sci 205–209
6. Romanov A, Semenov A (2017) Detection of fake profiles in social media. In: 13th International conference on web information systems and technologies
7. Ananya Dey, HR (2019) Detection Of fake accounts in instagram using machine LearnING. Int J Comput Sci Inf Technol (IJCSIT), 11(5)
8. arvix (2019) Retrieved from arvix: https://arxiv.org/pdf/1910.03090.pdf
9. Bhaskar N. Patel, SG (2012) Efficient classification of data using decision tree. Bonfring Int J Data Mining, 2(1)
10. Caelen O (2017) A Bayesian interpretation of the confusion matrix. Ann Math Artif Intell
11. Chen J (2020) Instagram stats. Retrieved from sprouts social: https://sproutsocial.com/insights/instagram-stats/
12. Cagatay Akyon F, Kalfaoglu E(2019) Instagram fake and automated account detection. IEEE
13. Fcakyon (n.d.) Dataset for intagram fake and automated account detection. Retrieved from github: https://github.com/fcakyon/instafake-dataset
14. Hub IM (2018) Influencer fraud. Retrieved from Influencer Marketing Hub. www.influencermarketinghub.com
15. Instagram (2020) Netflix India Official. Retrieved from Instagram. https://www.instagram.com/netflix_in/?hl=en

16. Kumar PT (2017) chapter 11 conclusion. In: Ubiquitous machine learning and its applications, p 217. IGI
17. Pentreath N (2015) The components of a data driven machine learning system. In: MS Rajdeep dua, MAchine learning with spark. Packt Publishing. Retrieved from Subscription packtpub: https://subscription.packtpub.com/book/big_data_and_business_intelligence/9781785889936/3/ch03lvl1sec33/the-components-of-a-data-driven-machine-learning-system
18. R Decision Trees (n.d.) Retrieved from guru99. https://www.guru99.com/r-decision-trees.html

A Framework for Sandboxing of Pandemic Spread

Siddharth Swarup Rautaray, Manjusha Pandey, and Hrushikesha Mohanty

Abstract Pandemic disease, like Corona spreading by social contacts, needs "lock-down", a measure to limit the virus spread. But the measure is too expensive for a nation for its adverse impact on national economy. Sandboxing followed in system security is a proactive and resilient mechanism that allows a system to function either in full or partial capacity without compromising its security. Similarly, in order to limit a community spread with resilience, a proactive mechanism is required to predict and safeguard the area that is the most vulnerable to a pandemic disease infection and has the potential of a super spreader resulting to a community spread. The early care of the region may protect it from a community spread of an infection. Social analytics on immunity and connectivity are proposed in this research paper to predict the vulnerable regions. Based on this idea, a tool is under development and this paper presents a framework of the tool.

1 Introduction

The social network is generated by the interactions any human being does in his day-to-day life to ideally fit the study of spread of any pandemic and generated some precautionary model this social network would be of great help. The model thus developed would be analogous to directed graphs in computation theory and can utilize by information theoretic spread and control predictions. The major problem in establishing this analogy is the complexity of the interactions any individual has in

S. S. Rautaray (✉) · M. Pandey
School of Computer Engineering, KIIT Deemed to be University, Bhubaneswar, India
e-mail: siddharthfcs@kiit.ac.in

M. Pandey
e-mail: manjushafcs@kiit.ac.in

H. Mohanty
School of Computer & Information Sciences University of Hyderabad, Hyderabad & KIIT Deemed to be University, Bhubaneswar, India
e-mail: h.mohanty@kiit.ac.in

A. Khanna et al. (eds.), *Data Analytics and Management*, Lecture Notes on Data Engineering and Communications Technologies 54,
https://doi.org/10.1007/978-981-15-8335-3_52

his lifetime to model the same correlation dimensions can be added to the world graph generated following the correlations rules of personal, professional and livelihood interactions of any individual.

To model the social network, many researchers have proposed many models as has been done by Liu et al. [1] as they propose C-RBFNN an improved Radical Basis Function (RBF) neural network that is an combination of fuzzy theory and neural network algorithm for prediction of user behaviour and its role in spread of any topic. The C-RBFNN has been advocated to be effective in simulating the nonlinear relationships among complex behaviours of users as compared to traditional neural networks as it possesses the advantages of fast convergence and local approximation to eigenvalues when dealing with large-scale network topic data and does not suffer from the problems posed by local minimum. Besides, time-decay function is introduced to make RBF neural network can adapt to dynamic changes of various influence factors in the social networks.

To achieve further depth of AI and spread, we have gone through a systematic review presented by Arji et al. [2] on emergence of infectious diseases, their spread and impact on health and economics with evaluation and classification usage of fuzzy logic methods in diseases spread predictions. The authors propose that the findings of their evaluation present that the fuzzy logic methods can be used for diagnosis of epidemic diseases and the key fuzzy logic based methods used for the infectious disease included fuzzy inference system; rule-based fuzzy logic, fuzzy cognitive maps and adaptive neuro-fuzzy inference system. The fuzzy logic-based methods are able to model the epidemic spread but to automate the process deep learning and big data methods would be playing a major role to have a flavour of these we referred work done by Chae et al. [3] on prediction of infectious disease using deep learning and Big Data In their work for Korea Centre for Disease Control (KCDC) made use of the deep neural network (DNN) and long-short term memory (LSTM) learning models also comparing with the autoregressive integrated moving average (ARIMA) when predicting three infectious diseases one week into the future. To model the carrier predictors and detecting infectious disease using big data such as Internet search queries has been advocated by the authors as the Internet search data can be gathered and processed at a speed that is close to real time. For analysis, the authors applied a lag of seven days between the input variables (optimal variable combination) and their associated output variable (disease occurrence).

The proposed paper provides following major contributions

- Novel framework of social data collection and social network generation.
- The modelling of social network.
- The modelling of spread in the social network.
- Correlation of social model based on their professional, personal as well as the livelihood network.

2 Proposed System Framework

Here, a comprehensive view of the system is presented for a quick introduction to the proposed research and the deliverable. The proposed research not only intends to discover new ideas on prediction of infection spreads particularly driven by social contacts but also to develop a system implementing the research ideas and algorithms. The proposed software system has four primary modules to perform the activities viz. social data collection, world graph generation, immunity prediction and spread prediction as shown in Fig. 1. It also has a module for visualization of region spread. The visualization other than demarking the region into different categories based on infection status categories should also help in tracing of people of contact and home quarantine.

A higher level description of the system dynamic behaviour is presented in Fig. 2. For a given population, data are collected both offline and online. At the beginning of a population, random sample-based data from different regions and strata of life are collected. It can also have online data feed from cameras and IoT sensors. The collected are undergone a pre-processing stage at module social_data_collection, making those input ready for the next module that generates world graph for different worlds a person has. As assumed, there are mainly three worlds a person deals with. Those are personal world, professional and livelihood world. All these worlds are represented by graphs following the associations each has. Next network of networks is generated by execution of the module world_graph_generation. On the network, immunity of each entity of a region is established and assigned to each node of personal world graph. This is shown in Fig. 2 as an activity succeeding to world graph generation activity. Then spread prediction based on the world model of a

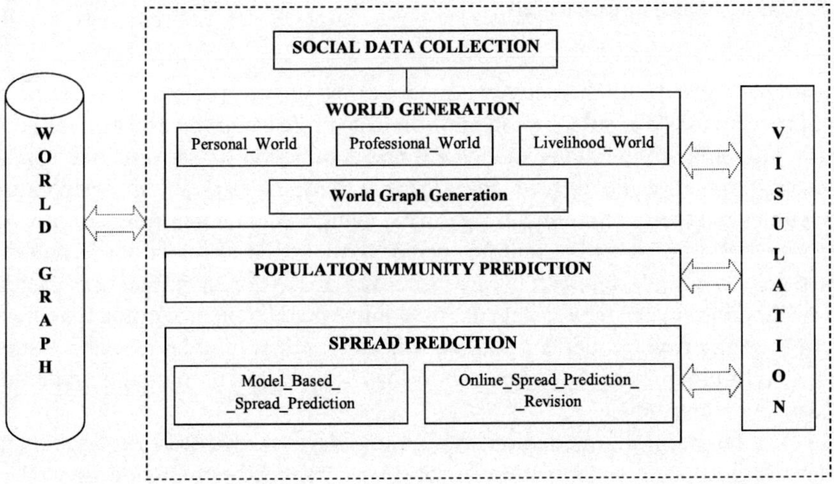

Fig. 1 Proposed system framework

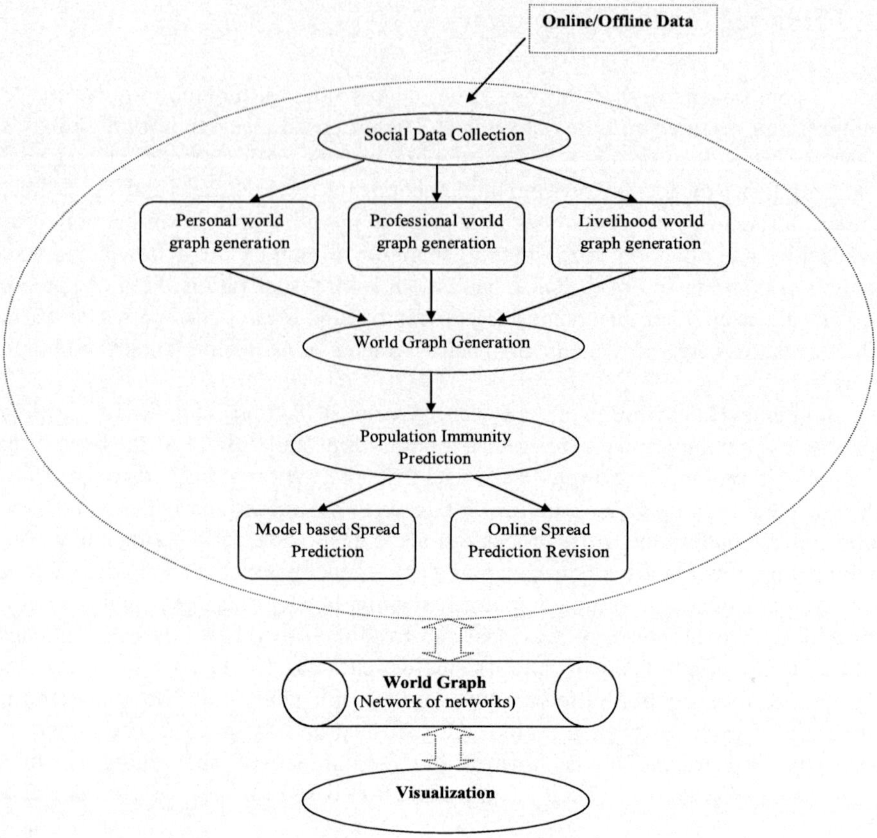

Fig. 2 Working of the proposed model

population is carried out. This activity is model-based spread prediction that performs bulk of transactions a population can perform in its professional as well as livelihood graphs. The bulk is essentially model a spreader function to perform like spread waves on a population. For this, we propose to use Petrinets for its suitability to model transactions and their conditional triggering semantics and temporal sensitivity. An offline prediction essentially identifies persons vulnerable to infections. Based on them, tentative spreads of infections are visualized. In case of an outbreak, as ground data on infection appear, then system goes for online prediction on learning the spread pattern evolving from ground social data collected online. In order to make system responsive to ground situation, we wish to focus our research on online prediction and learning algorithms.

On corroboration of a prediction with ground data on reported infection, social memory is generated so the pattern is later used for ready prediction as well as reminders to the officers of the centre for disease control. The proposed system is a generic one to implement for a region for twenty-four hours' service on infection spread prediction helping health workers and administration for ground work. Such systems can be networked across the nation for integrated intervention while allowing a model at a place to care for its location specific ingenuity among the population.

Next before positioning our research proposal, we will put a comprehensive picture of current research in related areas.

3 Current Research Status

With respect to the stated framework Fig. 1, a survey on related works is presented here. The survey has been divided into five divisions viz. epidemic model, social structure and infection spread, immunity prediction and artificial intelligence. The survey in no way claims to be complete but indicative to show the role of social data in early prediction and spread of an infection.

3.1 Social Structures

The existing social contacts and interaction patterns among individuals play a major role in encoding infectious diseases transmission route and further result in the realistic characterization as well as modelling of epidemics spread. The major hurdle to this modelling is accessibility of contact patterns in human populations and mixing patterns among groups based on differently characteristics like professional interactions, economic class, etc. In the same line, Laura Fumanelli et al. [4] proposed individual mixing patterns estimation using parameterized virtual populations based on demographic and census data and generating synthetic contact matrices. This method would be helpful to provide the ability of mixing patterns in population, with highlight common features as well as country-specific features and provided a numerical approach for social structure generation improved accuracy and mathematical model even in case of non-availability of specific social structure data. The authors have reported average age to be determinant of attack rates in different countries and also the synthetic contact matrices as a mathematical model for predictions and support of public health decisions. Also, the authors have given suggestions based on routine data collected data for social structure modelling.

To determine the contact patterns between susceptible and infectious people, Kathy Leung et al. [5] in their work of modelling social structure with the help pf contact between individuals for respiratory disease faced the lack of any published data to quantify the social contact patterns; hence, they conduct a survey using online questionnaires and tried to present a social structure using prolonged and frequent contacts, and contacts at home, school and work were more likely to involve physical contacts that decreased over age but increased with household size, years of education and income level. Prolonged and frequent contacts, and contacts at home, school and work were more likely to involve physical contacts. The authors suggest to improve the design of future social contact surveys and parameterize transmission models for generation of analysis-based contact network of a detailed virtual society, and compute matrices of contacts for social modelling and establishing correlations between epidemiological parameters and socio-demographic features of the populations.

Carlo Piccardi [6] in his work proposed a new framework to map the traditional epidemic spread on social network model that are based on social contacts with an assumption of homogeneous interactions among the population. But the authors even after more than ten years of extensive research have found their work extended in various directions adding to the complexity of the epidemic spread as well as the social network model. The author advocates inclusion of more realistic description of features to include associated linking/unlinking of connections in the social network due to birth death of individuals and their mobility through adaptive networks.

Increased modern-day connectivity and easy availability of advanced hardware as well softwares (social networking websites and apps) have infused additional velocity to the diffusion of contagions like diseases and rumours as well as computer viruses. This increased infusion leads to many detrimental to human societies like security compromises, economic loss and also threat to life in order to mitigate these identification of social networks as well as epidemic spread network model are essential for timely quarantining and breaking of spread of epidemics. Syed Shafat Alia et al. [7] in their work on infection source identification propose to use classical graph centrality measures with a combination of eccentricity and closeness (EC + CC) to understand the increase in infection size. The authors have theoretically justified the effect of density on source identification using k-regular trees and suggest focusing on classical graph centrality measures improvement.

3.2 Spread Modelling

Various methods proposed, developed and implemented for social network modelling as discussed in section above would be fruitful only if the spread modelling can be mapped with the social network models developed; hence, in this section, we would describe some of the work done in spread modelling.

Kautz [8] in his work of modelling spread of disease from social interactions proposes usage of fine-grained modelling infectious spread mapped to the real-world model of social network specifically concentrating of social ties and interactions between specific individuals that contribute to contagion. The authors used health-related messages from public Twitter data to identify sick individuals from the content of online communication and try to develop social ties to infected, symptomatic people and other co-location model the disease spread. Also, the authors try to provide qquantifiable estimates of disease characteristics contributing to transmission on large scale using SVM model.

The authors have proposed for an extension of their work to focus on larger geographical regions including airplane travel with fine-grained location for modelling and prediction of global epidemics based on daily interactions of individuals.

Guclu et al. [9] in their disease spread modelling work communicated to letters to nature make use of differential equations with uniform mixing assumptions similar to ad hoc models for modelling of the contact process detailing it further with use of dynamic bipartite graphs that can model contact patterns of physical interactions resulting from individual's movements. The graphs thus based on urban traffic of actual census, land-use social-mobility data were found to be small-world-like graph that were strongly connected and had well-defined scale of degree distribution. The graphs were suggested by the authors to be followed for placement of sensors to generate highly efficient epidemic spread detection. The limitation of the proposed model was its dependency on census data and sensor data both of which could be compromised. The authors have also developed a tool foe spread modelling called EpiSims that is an agent-based simulation tool which works with the help of TRANSIMS: Transportation Analysis and Simulation System that was developed by Los Alamos National Laboratory.

4 Model Functionalities

The proposed model functionalities include social data collection and social world graph generation, the collected social data would be translated into graph-oriented database workloads for social analytical big data tasks and social network analysis from the world graphs thus generated. The working of the proposed model requires personal world graph generation, professional world graph generation, livelihood generation and as it is evident from the above literature survey that the problem of social network modelling with epidemic spread modelling and inclusion of immunity as well based on human mobility because of his personal and professional worlds is a multi-dimensional problem that requires the help of computing efficiency for analysis and prediction; hence, AI has a major role to play in network model generation and the detections as well as predictions of epidemic spread [10–13].

The social world graphs are unique in nature as these graphs the nodes do not have structures inside but the structure lies in the way nodes the nodes are connected to each other that are structural correlations. These structural correlations have been modelled based on the three characteristics of the social structures namely: (1) personal interaction structural correlations for personal world graph generation, (2) professional interaction structural correlations for professional world graph generation, and (3) livelihood interaction structural correlations for livelihood world graph generation. These world graph generations are done based on a scalable and structure-correlated social graph with the help of graph generating algorithm1.

4.1 Graph Model

Graph generating algorithm initially creates a random social graph with non-uniform value distributions and random structural correlations, and the initial generated graph is then scaled for collected data of personal, professional and livelihood interactions. We thus propose the decomposition of correlated social world graph generation in three different passes where each pass focuses on each individual correlation dimension of personal interaction, professional interaction and pivelihood. The graph model thus generated would be the social world graph model *G* (*N*, *V*, *P*, *C*) where *N* is a set of nodes representing individuals in a geographical location, *V* is a set of edges, *P* is a set of properties and *C* is a set of classes a node is correlated with. Any labelled edge of the social graph will have two nodes and the edge represents the property through which one node belongs to an object class and the other node. The edge property is then a literal property or a relation property, respectively. One node can have many edges with the same edge property and there is no edge connecting two literal nodes. As an example, the set of nodes $N = \{P1, P2, P3, P4\}$ of the individuals of the specific group. $V = \{R1, R2, R3\}$ may be relating properties between these individuals. The set of properties P contains relation properties such as <is>, <belong T o>, <works with> and literal properties such as <name>, <liveAt>, <studyAt>. Accordingly, there are edges between object nodes whose labels are those of the relation properties (e.g. (<person>, <belongT o>, <place>)), and edges between an object node and a literal node whose labels are those of literal properties (e.g. (*P*4, <works at>, "place")).

Initialize

$$
\begin{aligned}
&G(N, V, P, C)\\
&N = \bigcup_{r \in C} \mathrm{Or} \bigcup_{r \in C} L\\
&P = \{P_{L(x)} | \forall_x \in C\} U \{P_{R(x,y)} | \forall_{x,y} \in C\}\\
&E = \{(n_1, n_2, P) | \{n_1 \in O_x \wedge n_2 \in L \wedge P \in p_{2(x)}\}\\
&\quad \cup \{n_1 \in O_x \wedge n_2 \in O_y \wedge p \in p_{r(x,y)}\}\}
\end{aligned}
$$

Here

O_r object of a class *r* in *C*

$p_{L(x)}$ is set of properties

$pR(x, y)$ is the set of relation properties

4.2 Correlation Dimensions

To model the correlated dimensions in the three types of data, correlations are addressed in proposed model of social world graph generation, including the correlations between property values (e.g. the value of property <name> is correlated with values of property <liveAt>), the correlations between property values and graph structures (e.g. the availability of edges labelled <knows> between nodes is correlated with values of properties <studyAt>, <liveAt>), and the correlations between the graph structures themselves (e.g. the existing of edge (<P2>, < knows>, <P3>) is correlated with the existing of edges (<P1>, <knows>, <P2>), (<P1>, <knows>, <P3>)).

In order to generate the correlated graph data, our proposed framework is built on the notions of correlation dimensions of personal world, professional world and livelihood that represent the underlying causes of (some) correlations among individuals to be depicted in the social world graphs generated. The correlation dimensions are actually real-time relations that are property value, like the location name for the property <Workat>—data collected and designed dictionaries along with parameterized frequency distribution would be used for generating these property values. The included correlation dimensions have influence on (i) property values of node by changing the frequency distributions of the value in dictionaries that generate these property values, and (ii) the connectedness, i.e. edges between pairs of two nodes in the graph, thus generating the correlations among correlations and network of networks for the social graphs [14, 15].

4.3 Correlation Rules

The correlation rules have to be defined based on the three dimensions of correlation to be modelled in the social world graph like a correlation rule can be $P4 \rightarrow (R1, f(A))$, meaning that the values assigned to $R1$ are distributed according to a distribution function f built on the set of the values of the correlations defined by a function A. The correlation rules followed by the social world graphs can be divided into three major categories:

Category1: The "correlation dimensions-property values" correlation rules

$$CR_{CP} = \{A \rightarrow (B, f(A)) | A \in S_c B \in S_c \cup P\} \tag{1}$$

Category2: The "correlation dimensions-graph structure" correlation rules

$$CR_{CE} = \{A \rightarrow (B, f(A)) | (A \in S_c \cup P) \wedge B \in E\} \tag{2}$$

Category3: The "graph structure-graph structure" correlation rules

$$CR_{EE} = \{A \rightarrow (B, f(A)) | A \in E \wedge B \in E\} \tag{3}$$

S_c	Set of correlation dimensions
$CRcp$	Contains rules for distributing the property value of each node given the values of correlation dimensions
CR_{CE}	Contains the rules between the correlation dimensions and the graph structure
CR_{EE}	Contains the correlation rules between the graph structures which specify the probability of creating a graph structure in considering generated graph structures

4.4 Population Immunity Prediction

This section intends to provide a gist of similar work done and techniques used social model generation and usage of AI techniques in immunity modelling and spread infection predictions. To attain the summarized total, world graph generation can be done using graph model as mentioned in Sect. 4.1 and the social world graph has been generated as a weighted graph $G(N, V, P, C)$ and the connectivity in the graph has been represented by correlation dimensions mention in Sect. 4.2 following the correlation rules as detailed in Sect. 4.3. The same graph with assumed vertex order represented by a function $F(V)$ the weighted connectivity has been modelled using a weighted matrix $W_{i,j}$ with $w(i,j)$ representing the edge weight between i and j the particular ordering and flow of information between all the edges i and j would be represented by a probability function P

$$P_{i,j} = \frac{\alpha_i \ W_{ij}}{\sum_k W_{ik}} \tag{4}$$

when $\sum_{k \in V} W_{ik} > 0$ and $P_{ij} = 0$ otherwise. Here, $\alpha_i \in (0, 1]$ for all i.

Based on the above, the probability function of P and data analytics approaches would be of use while performing the classifications clustering and predictions of epidemic spread in their original formats or they may be modified for specific diseases based on the features identified and network modelled. Once the immunity and spread predictability is achieved on the network networked world graph, the results could be visualized using data analytics visualization tools to provide the users a better understand ability of the results generated [16].

5 Conclusion

The recent encounter with Covid-19 is painful but in a way eye-opening to humanity to take a look and review the ways it interfaces with world revolving around it for sustainable life in the peace of Nature. Remaining alert is the first principle of security. An alert system avoids not only the dangers but also its possibility. While hotspot is found a beneficial mechanism in breaking the chain of spread, sandboxing is a proactive chasing technique to stop an infection spread ensuring operational resiliency in social system. The research paper aims to develop data analytics-based social network modelling and immunity prediction techniques using social analytics on social, professional, livelihood management networks of a population.

The novelty of our approach is to derive a graphical view of social world graph generated by the random graphs generated and model the information flow graphs for modelling and view of immunity from social data analytics techniques, to devise spread prediction algorithms and to suggest probable sandboxing approaches for a person or region based on medical advises and guidelines suggested by the authorities. The current research paper focuses on proposal of the framework for social network modelling and generation of world graphs from the models generated followed by spread and immunity modelling and the future work would be generation of visualization to the spread and immunity models along with the world graphs generated. The visualization thus generated can help in accurate demarcation of the regions into different categories or zones based on infection and immunity of local residents modelled and this would be of great help for the contact tracing of individuals and home quarantine suggestions.

References

1. Liu Y, Zhao J, Xiao Y (2018) C-RBFNN: a user retweet behavior prediction method for hotspot topics based on improved RBF neural network. Neurocomputing 275:733–746
2. Arji G, Ahmadi H, Nilashi M, Rashid TA, Ahmed OH, Aljojo N, Zainol A (2019) Fuzzy logic approach for infectious disease diagnosis: a methodical evaluation, literature and classification. Biocybernetics Biomedical Eng
3. Chae S, Kwon S, Lee D (2018) Predicting infectious disease using deep learning and big data. Inter J Environ Res Public Health 15(8):1596
4. Laura F et al (2012) Inferring the structure of social contacts from demographic data in the analysis of infectious diseases spread. PLoS Comput Biol 8(9)
5. Leung K et al (2017) Social contact patterns relevant to the spread of respiratory infectious diseases in Hong Kong. Sci Reports 7(1):(1–12
6. Piccardi C (2013) Social networks and the spread of epidemics. Lettera Matematica 1(3):119–126
7. Ali, SS, Anwar T, Rizvi SAM (2020) A revisit to the infection source identification problem under classical graph centrality measures. Online Soc Netw Media 100061
8. Adam S, Kautz H, Silenzio V (2012) Modeling spread of disease from social interactions. In: Sixth international AAAI conference on weblogs and social media
9. Eubank S, Guclu H, Kumar VA, Marathe MV, Srinivasan A, Toroczkai Z, Wang N (2004) Modelling disease outbreaks in realistic urban social networks. Nature 429(6988):180–184

10. Mohanty H, Venkataswamy M, Ramaswamy S Shyamasundar R (2009). Translating security policy to executable code for sandboxing linux kernel. 124–129. https://doi.org/10.1109/ems.2009.42
11. Oommen TK (2007) Knowledge and society: situating sociology and social anthropology. Oxford University Press, USA
12. Kapadia A, Outcalt PA, PG Shealy (2020) Replication control among redundant data centers. U.S. Patent No. 10,599,676. 24 Mar. 2020
13. wiki: https://en.wikipedia.org/wiki/Sandbox_(computer_security)
14. Yang L-X et al (2013) Epidemics of computer viruses: a complex-network approach. Appl Math Comput 219(16): 8705–8717
15. Bin S, Sun G, Chen C-C (2019) Spread of infectious disease modeling and analysis of different factors on spread of infectious disease based on cellular automata. Int J Environ Res Public Health 16(23):4683
16. Cosma G, Brown D, Archer M, Khan M, Pockley AG (2017) A survey on computational intelligence approaches for predictive modeling in prostate cancer. Expert Syst Appl 70:1–19

Hybrid Recommender System Using Artificial Bee Colony Based on Graph Database

Rohit Beniwal, Kanishk Debnath, Deobrata Jha, and Manmeet Singh

Abstract Recommender systems hoard appropriate suggestions guided by the interactions and previous choices of the users. In a world of the ever-growing amount of data, we are overpopulated with undesired information, which is making it difficult to operate or choose. Hence, we require a recommender system that is enough capable to deduce desired suggestions which can be valuable for the users. Thus, they are rising in popularity and becoming part and parcel of day-to-day activities in our general life. Here, in this paper, we implement the recommender system, which is conceptualized on a hybrid filtering algorithm that helps in dealing with limitations of both content and collaborative filtering reinforced with artificial bee colony optimization along with k- nearest neighbor for better performance. For this purpose, we used MovieLens dataset, which contains information regarding users, movies, and ratings given by the users. Here, we gathered a pre-filled user project scoring matrix and have compared multiple models of recommender systems for their precision and recall factor. We are using a recommender system based on graph database which uses graph traversal and pathfinding algorithms to establish relations and hence is more robust and faster in implementation. The thickness of the edges connecting movie nodes and the indegree of a movie node specifies the recommendation limit of the movie. The experiment results on MovieLens dataset establishes scope for future scalable models and delivers competent outcomes brought in comparison with traditional systems. Preliminary results show improvement of 9% in precision and 3% improvement in recall over traditional systems.

R. Beniwal · K. Debnath (✉) · D. Jha · M. Singh
Department of Computer Science and Engineering, Delhi Technological University, New Delhi 110042, India
e-mail: kanishk.coe@gmail.com

R. Beniwal
e-mail: rohitbeniwal@yahoo.co.in

D. Jha
e-mail: deobratajha@gmail.com

M. Singh
e-mail: 97singh.manmeet@gmail.com

A. Khanna et al. (eds.), *Data Analytics and Management*, Lecture Notes on Data Engineering and Communications Technologies 54,
https://doi.org/10.1007/978-981-15-8335-3_53

Keywords Artificial bee colony · Hybrid recommender systems · Recommender system · Graph database

1 Introduction

Enlargement of network technology space is evident with development in the computational technologies for better user experience. Today, billions of rows of data are being added to the network every day. So, it became extremely crucial to develop methods to extrapolate useful information from the ocean of irrelevant data. Recommender systems are an effective tool to handle online data and information overload which is carefully filtered by using different machine learning models.

The recommender system automatically computes the user's preference and synthesize recommendations. The mainstream recommender systems use a collaborative and contend-based approach, however, a hybrid system using the results of both can overcome their collective limitations. For effective recommendations, we employed optimization through artificial bee colony (ABC) and k-means. Present approaches are highly memory-driven or based on computational models. Model-based techniques have shown favorable results for sparse data. ABC is an optimization procedure based on a swarm meta-heuristic algorithm inspired by the rummaging behavior of honey bee colonies. ABC applied in various domains such as industrial data clustering, protein structure study, computer and sensor networks, image processing, and mechanical engineering. For the dataset, we used a graph database that uses graph data structures querying semantics with nodes, weighted-edges, and properties to represent and store data. Relationships in a graph database are established using nodes and their edges. Data is stored in the linkages created during graph construction and are highly efficient for retrieval of data. Due to heavy interconnections in the data, graphs traversal algorithms are an effective way to query relationships. This has improved our results by 9% in precision and 3% in recall when existing models were compared.

Fields like movies, music, electronic commerce, and social networking use recommender systems to enhance their user experience. We usually have a habit to read reviews and scores of any movie before watching it hence it helps in developing an association between user interaction and ratings. Using these ratings, we can generate a score for recommendation and as per the user's available data, we can recommend a movie to them.

The rest of the paper is organized as follows: Sect. 2 discusses the related work; Sect. 3 demonstrates the research approach of the hybrid recommender system; Sect. 4 discusses implementation of recommender system followed by Sect. 5, which concludes the research paper and provides direction for future work.

2 Related Work

The related work is divided into three sub-sections namely (1) introduction of recommender systems, (2) artificial bee colony and (3) graph database which are as follows:

2.1 Introduction of Recommender Systems

Recommender system uses users' preference, items' information and users' intere-actions with application of knowledge filtering algorithm to produce suggestions of items to the users. Figure 1 shown below is the working diagram of a generic recommender system.

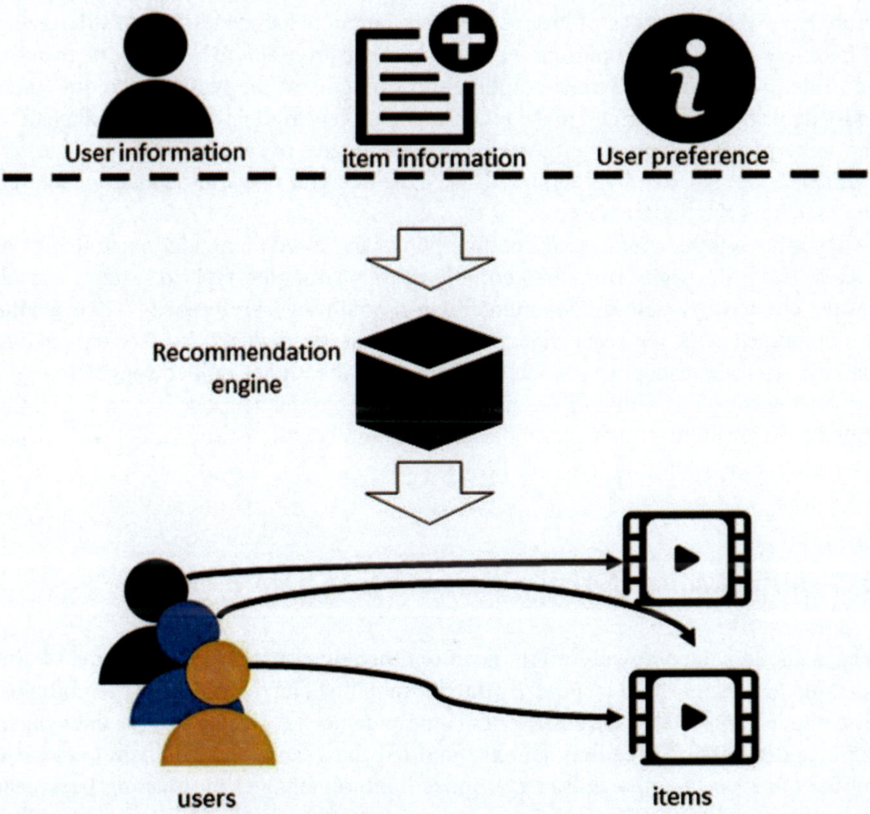

Fig. 1 Conceptual diagram of the recommendation system

The recommendation system is considered as a black box. The data source is the accepted input. Generally, the recommendation engine needs data sources which include metadata of content, for example, user's preferences for items, global rating associated to the item, features of similar items, user general data such as location, age, sex, region, etc. The likeliness information of these users can be either explicit or implicit feedback by users [1].

There are three kinds of recommender engines developed so far which include content-based filtering, collaborative filtering, and hybrid filtering [2]. Content-based filtering and collaborative filtering have their advantages and disadvantages. The content-based approach has three specific problems which are (1) lack of content description due to privacy issues, (2) over-specialization of user's preferences, and (3) difficulties in differentiating between subjective knowledge such as humor and point of view. Collaborative filtering systems introduce defects of its own such as primal score problem where the initial recommendations are likely to be wrong because of (1) lack of participating user interactions (cold start), (2) existence of the number of items exceeding the amount a user is able/willing to delve into creates sparsity problem, (3) the existence of groups with overlapping characteristics but still having differences in individual opinions lead to gray sheep problems [3]. Therefore, to make accurate predictions, one must combine the strengths of the two to overcome their individual shortcomings and make recommendations based on the weighted mean of the content- and collaborative-based results. The degree of each item being suggested could be a metric for the weight. In this way, the maximum recommendation factored item receives the highest score.

Hybrid recommender systems couple two or more recommendation strategies in diverse ways to advance from their complementary strengths. Hybrid systems use the results obtained by both user-item interactions and user's preferences. Their results are combined with the respective weights, which are deduced by their respective rankings in their respective models [4]. This way, the higher-ranked suggestion gets upvoted and hence results in the more generic recommendation. Beyond that, the final selection model then picks the desired number of recommendations as per user's preference.

2.2 Introduction of Artificial Bee Colony Algorithm

The artificial bee colony algorithm is an optimization procedure based on a swarm meta-heuristic algorithm inspired by the rummaging behavior of honey bee colonies. The model consists of three essential components (1) engaged and unengaged rummaging bees, (2) feeding sources, and (3) third component is being searched by the other two components, i.e., engaged and unengaged rummaging bees near their colony. The model also discusses two preeminent types of behavior which are crucial for self-regulating and cumulative knowledge: the backing of rummagers to affluent feeding source resulting in positive feedback and dereliction of poor source of feeding source resulting in negative feedback [5].

To beat the shortcomings of the individual models used during the pre-combination stage, we implemented “a hybrid cluster and optimization-based technique” to improve the accuracy rate [6]. We used k-means and ABC as optimization for a better and competent recommender system. Firstly, the difference between the user’s preference score and the centroid of clusters was computed. These clusters are selected randomly initially and refined in further steps. Now, user with the lowest distance is assigned to the particular cluster. However, this cluster is not the final accurate cluster at this instant and furthermore computations are required for better accuracy. From this stage, iterative repositioning would take place and would continue to reposition until no further changes are discovered. After this point, when no relocation will take place, we would mark that as completion of the clustering process [7].

The next k-means procedure is optimized by the ABC algorithm. We have developed a function to fit the user-centroid distances that helps in achieving more accurate recommendations in the following stages. This fitness function would transform the previous centroids for a limited iteration count and then, it helps in classifying the users again by computing the cosine similarity with the centroids resulting in minimum centroid differences [8].

2.3 Graph Database

In our further investigation, we came across graph databases which can be utilized as the relationships are stored in the form of nodes and edges. All relational queries are then translated into graph traversal techniques hence making the process even more efficient [9].

Graph database is a “NoSQL database based on graph theory. The basic elements of a graph are nodes and edges. An obvious difference between a graph and relational database is the use of relations to connect nodes instead of foreign key” [1]. The searching procedure is highly dependent on these foreign keys. These operations are memory and time heavy and hence graph databases are employed to simplify it. Graph databases are able to organize the data in a more simple and expressive manner than the traditional relational database and another NoSQL database [9]. Based on the above improvements of the database, we have used the Neo4j graph database.

3 Research Approach of the Hybrid Recommender System

This section is divided into six sub-sections, namely preprocessing of data, content-based filtering, collaborative filtering, k-means and artificial bee colony algorithm, cosine similarity, and combining results, which are as follows:

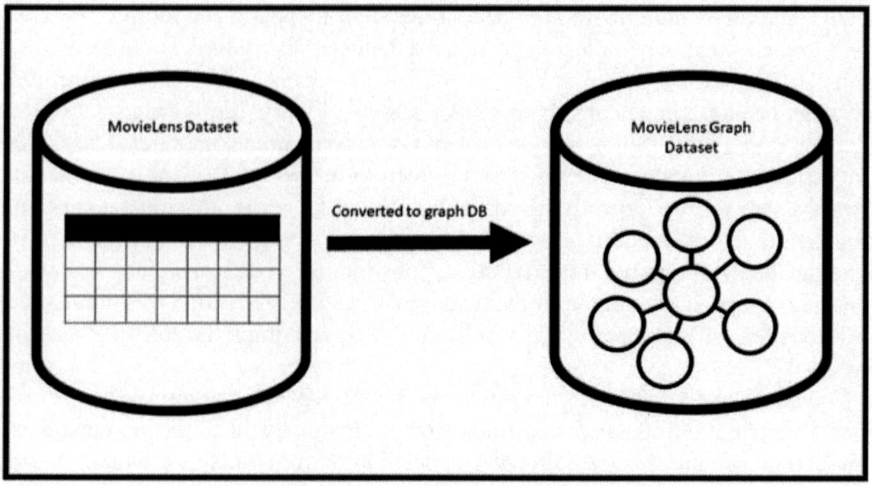

Fig. 2 Preprocessing of relational data into graph data

3.1 Preprocessing of Data

Relation-based data is first converted into graph-based data using Neo4j graph database. The following Fig. 2 shows the preprocessing of data.

Three kinds of nodes in the graph DB are: (i) User, (ii) Movie, (iii) Genre

Two kinds of edges are:

1. Genre: If movie M_j is of genre G_i, then G_j is exclusive with G_i with a relationship of type "genre".
2. Rated: If user U_i has rated movie M_j, then U_i exclusive with M_j with relationship type "rated". The rating $r(M_j, U_i)$ is recorded as the degree of edge.

3.2 Content-Based Filtering

In content-based filtering, the items or movies are clustered based on the features of the movies using modified k-means and artificial bee colony algorithm. The content-based filtering recommends items based on the objects already rated by the users. The items already present in the clusters will be recommended to the target user. This way, the content-based filtering technique will recommend target user the items already appreciated by him. The cosine similarity will be used to rank the items selected to form the cluster.

3.3 Collaborative Filtering

In collaboration filtering, the users will be clustered using the clustering algorithm (k-means and artificial bee colony algorithm). The users will be clustered based on the appreciation they have given to movies. The users in the same cluster will have the same taste and the items seen or rated by the users in the same cluster can be used to recommend items to the target user.

3.4 K-Means and Artificial Bee Colony Algorithm

Artificial bee colony algorithm is a swarm-based optimization algorithm. Following Fig. 3 is the flowchart of the aforementioned algorithm.

3.5 Cosine Similarity

The cosine similarity will be used to select similar items in the cluster. In content-based filtering, items will be clustered using the k-means ABC technique. Items similar to the movies rated by target user will be selected based on cosine similarity. Let S_i be an item with parameters as a genre g_j and average rating r_i. The similarity between the two items will be computed as

$$\mathrm{sim}(S_i, S_j) = \vec{S}_i \cdot \vec{S}_j \tag{1}$$

where $S_i = g_i G + r_i R$ and $S_j = g_j G + r_j R$

In collaborative filtering, let n be the number of movies shared by users in a cluster. The similarity between the two uses $\vec{U}_i$ and $\vec{U}_j$ will be calculated as

$$sim(U_i, U_j) = \vec{U}_i \bullet \vec{U}_j \tag{2}$$

where $\vec{U}_i = r_{1i} A_1 + r_{2i} A_2 + \cdots + r_{ni} A_n$ and $A_1, A_2, \ldots, A_n$ are movies and $r_{1i}, r_{2i}, \ldots, r_{ni}$ are their ratings by user $\vec{U}_i$.

3.6 Combining Results

The final recommendations will be the combinations from both the filters (Content as well as collaborative). The final results will be selected by assigning weights to the item from the two filters equal to their average ratings in the database. The top N recommendations with the highest weights will be given to the targeted user.

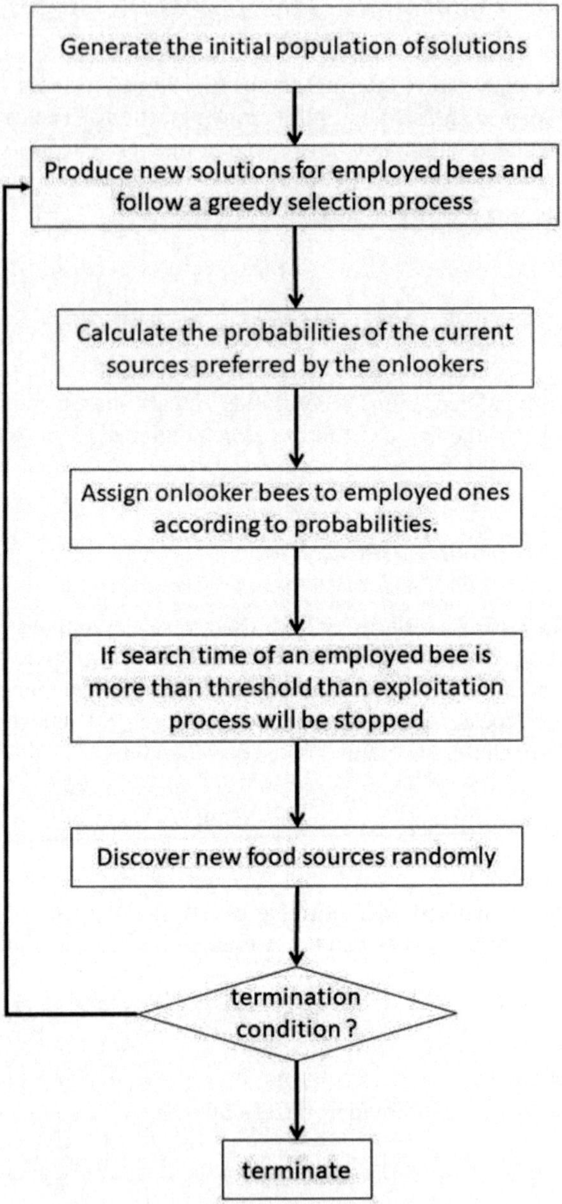

Fig. 3 Flowchart of the artificial bee colony algorithm

4 Implementation of Recommender System

We implemented our architecture of the recommender system in python language. Based on the model, we create CSV files which store the following information about the dataset:

1. Clusters of Users-It specifies the cluster of similar users.
2. Clusters of Movies-It specifies the cluster of similar movies.
3. Users-It contains information about each UserId and their corresponding User Cluster.
4. Movies-It contains information about each Movie and their corresponding Movie Cluster.
5. Genre-It stores the information about the genre of the movies in the dataset.
6. Relationship between user and movie-It specifies the average rating given by the users in cluster u_i to movies in cluster c_i. The thickness specifies the magnitude of the rating.
7. Relationship between movie and genre-If movie cluster mi is linked to g_i, it specifies that the movie in the cluster c_i belong to the genre g_i.

The following Fig. 4 explains the flow of the recommendation system from the preprocessing stage to result combination and final generation of recommendations.

We created the graph database in Neo4j by importing the CSV files using Cypher Language. We use Cypher Language to query the database, which outputs recommendation for a user based on query. We used Graphlytic to render the graphs and represent our outputs in the form of a graph.

The figures below show the recommendations for the user (userID = 1). Figure 5 shows the list of movies recommended for the specified user. Figure 6 shows the graphical representation of the recommended movies for the user. The movie with the highest recommendation rating is represented with a yellow edge. The width of the edge represents the degree of the recommendation of the movie for the user.

We made use of MovieLens dataset which is publicly available. We use our proposed architecture to build the recommender using MovieLens. To compute the performance of the recommender engine, we compute precision and recall. For comparing our model with existing models, we use precision and recall values and compare it with our model. We measure precision and recall values given by Equations.

$$\text{Precision} = \frac{|\text{Interested} \cap \text{TopN}|}{N} \tag{3}$$

$$\text{Recall} = \frac{|\text{Interested} \cap \text{TopN}|}{|\text{Interested}|} \tag{4}$$

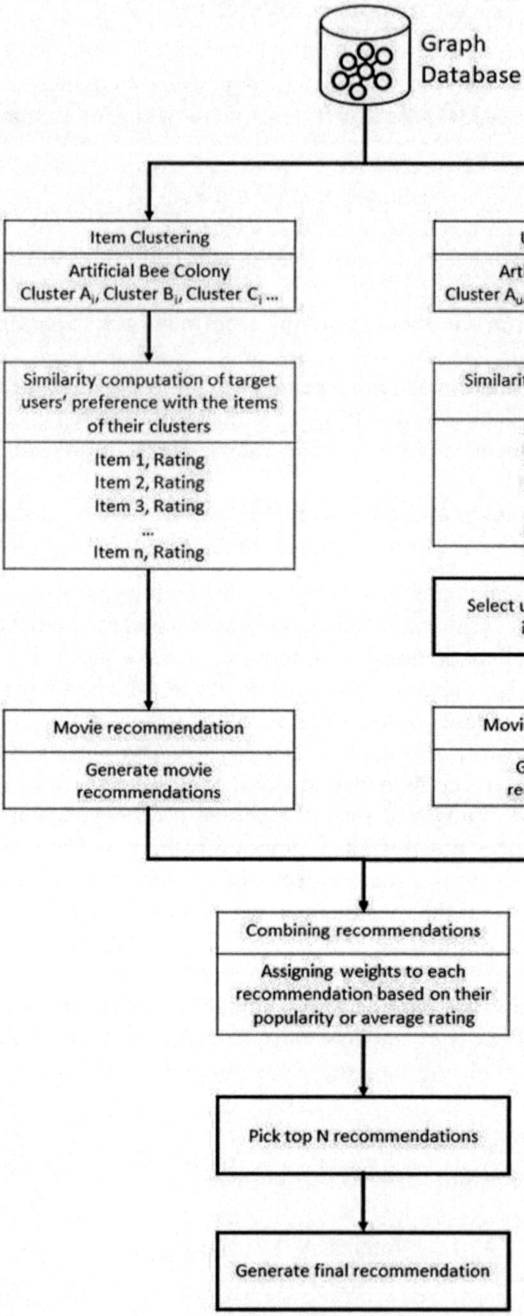

Fig. 4 Flow diagram of the recommendation engine

```
$ MATCH (u1:Users)-[:WATCHED]→(m3:Movies) WHERE u1.userId
```

"userId"	"title"	"rating"
"1"	"Fight Club (1999)"	"4.228310618821568"
"1"	"Paths of Glory (1957)"	"4.199341795956746"
"1"	"Touch of Evil (1958)"	"4.141198424217293"
"1"	"Decalogue, The (Dekalog) (1989)"	"4.137667304015296"
"1"	"Lawrence of Arabia (1962)"	"4.136035830866166"

Fig. 5 Top 5 recommendations for user with UserID 1

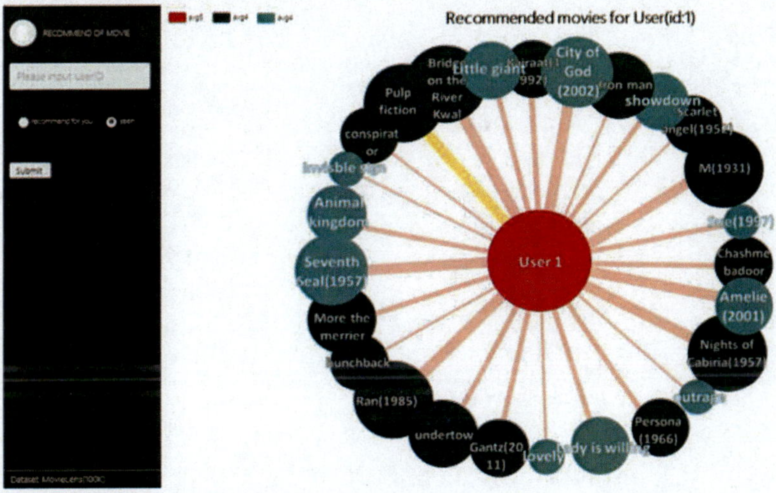

Fig. 6 Graphical representation of recommendations for user with UserID 1

In Fig. 7, we have shown the precision values of our system and compared with existing approaches. Precision metrics of our proposed method are clearly superior than the previously developed methods with recall methods being on par with existing methods.

In Fig. 8, we have shown the recall values of our system and compared with existing approaches. The recall values of our proposed method is better than various traditional models for specific cluster count.

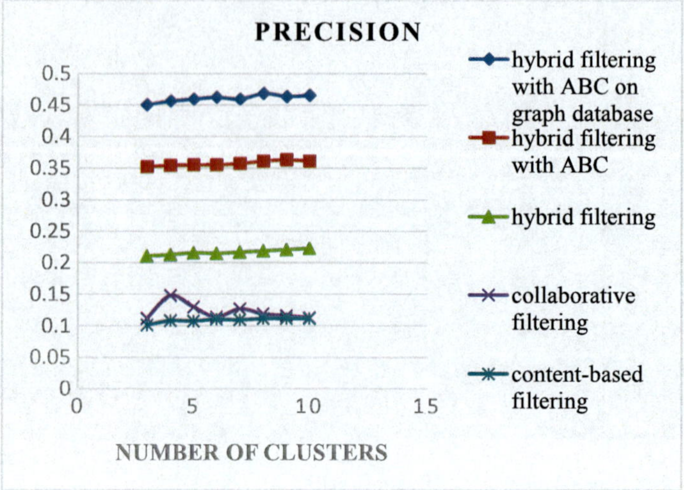

Fig. 7 Comparison of precision with previously developed methods

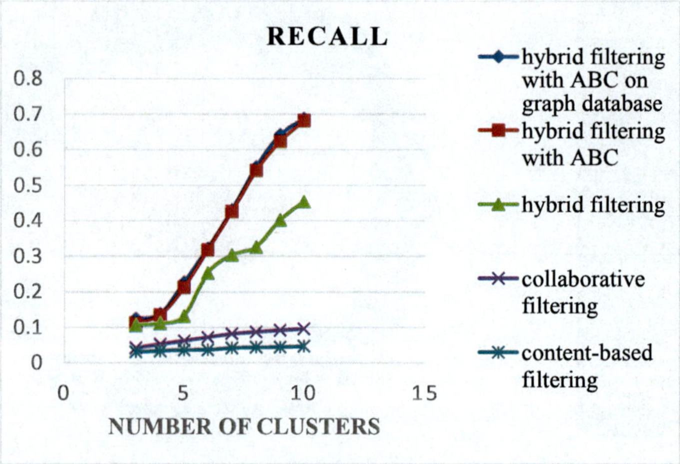

Fig. 8 Comparison of recall with previously developed methods

5 Conclusion and Directions for Future Work

In this paper, we proposed a hybrid recommender system that uses meta-heuristic artificial bee colony optimization. We used the graph database to facilitate the processing of the large dataset efficiently. We used MovieLens dataset to build our recommender system and used precision, recall, and accuracy measures to gauge the performance of our model and compare it with existing approaches. We found that our model did really well when it came to precision measures and outperformed other models in recall measures for the specific number of clusters. Results shows improvement of 9% in precision and 3% improvement in recall over traditional systems. For future work, we would like to try to evaluate our model on MovieLens on a high configuration machine. We would also like to evaluate our model based on other characteristics such as privacy concerns and the social network of the user. The experiment results on MovieLens dataset establish scope for future scalable models.

References

1. Yi N, Li C, Feng X, Shi M (2017) Design and implementation of movie recommender system based on graph database. In: 2017 14th Web information systems and applications conference (WISA). IEEE, pp. 132–135
2. Gedikli F, Jannach D, Ge M (2014) How should I explain? A comparison of different explanation types for recommender systems. Int J Human-Comput Stud 72(4):367–382
3. Vozalis E, Margaritis KG (2003) Analysis of recommender systems algorithms. In: The 6th Hellenic European conference on computer mathematics and its applications, pp 732–745
4. Yuan J, Kanagal B, Josifovski V, Garcia-Pueyo L, Ahmed A, Pandey S (2002) Focused matrix factorization for audience selection in display advertising. Interaction 12(4):331–370
5. Karaboga D, Ozturk C (2011) A novel clustering approach: artificial bee colony (ABC) algorithm. Appl Soft Comput 11
6. Daraboga D, Gorkemli B, Ozturk C (2012) A comprehensivesurvey: artificial bee colony (ABC) algorithm andapplications. Artif Intell Rev
7. Karaboga D, Basturk B (2007) Artificial bee colony (ABC) optimization algorithm for solving constrained optimization problems. In: International fuzzy systems association world congress. Springer, Berlin, Heidelberg, pp 789–798
8. Katarya R (2018) Movie recommender system with metaheuristic artificial bee. Neural Comput Appl 30(6):1983–1990
9. Webber J (2012) A programmatic introduction to neo4j. In: Proceedings of the 3rd annual conference on systems, programming, and applications: software for humanity, pp 217–218

NPMREC: NPM Packages and Similar Projects Recommendation System

Rohit Beniwal, Sonika Dahiya, Deepak Kumar, Deepak Yadav, and Deepanshu Pal

Abstract Node.js has a default package manager called Node Package Manager (NPM). There exists a command line client, called NPM, and an online database of public and paid-for private packages, known as the NPM registry. The registry is accessed via the user, and the available packages can be browsed and searched through the NPM Web site. Given a new project description, it is crucial to determine the most favorable NPM packages that can be used for the overall success of the project because of the reusable nature of these packages for rapid development. Though the hurdle faced by most of the developers is to select the right one from the vastly present number of NPM packages. Thus, to solve this issue, we propose a method called NPMREC known as NPM package and Similar Projects Recommendation System. It takes a project description as an input and gives a ranked list of NPM packages as the output that can then be used to implement the project with better efficiency. We used custom-built datasets for our approach using libraries.io Web site. The training dataset contains two datasets, firstly, the past project dataset with 589 NPM projects/NPM modules with information about their dependencies/NPM packages; secondly, the NPM package dataset with 759 NPM packages containing the detailed information about the dependencies of these 589 NPM projects/NPM modules. The test dataset contains 105 NPM projects/NPM modules along with information about their dependencies.

R. Beniwal · S. Dahiya · D. Kumar (✉) · D. Yadav · D. Pal
Department of Computer Science and Engineering, Delhi Technological University, New Delhi 110042, India
e-mail: deepak.k25101997@gmail.com

R. Beniwal
e-mail: rohitbeniwal@yahoo.co.in

S. Dahiya
e-mail: sonika.dahiya11@gmail.com

D. Yadav
e-mail: d.yadav.yadav75@gmail.com

D. Pal
e-mail: deepanshupal000@gmail.com

A. Khanna et al. (eds.), *Data Analytics and Management*, Lecture Notes on Data Engineering and Communications Technologies 54,
https://doi.org/10.1007/978-981-15-8335-3_54

Keywords Node package manager · NPM · NPM packages · Recommender system

1 Introduction

Developing a well-defined project is not a simple task, as clients generally request numerous features to be implemented. Therefore, to ease out the development work, programmers frequently utilize outside libraries and packages that give relevant functionalities [1]. Libraries and packages can be used to provide functionalities that can be reused by the programmer to develop the software. The use of a package helps the programmer to concentrate on important tasks rather than focusing on rediscovering the pre-existing code. Henceforth, the use of correct packages in the project is a smart way to implement the features of the project in a faster way. The advantage of this approach lies in the well designed and tested nature of these packages (in terms of functionalities) by numerous customer applications. Finding the correct package, however, is not as simple as it might appears. A vast number of packages have been created to be used for different purposes, and programmers often do not have sufficient knowledge about the presence of such packages, which may be suitable for some part of the feature of the project being developed by them. Although some packages are notable, however, in most of the cases, packages do not enjoy such comfort [2].

In this research paper, we present a new methodology NPMREC for recommending NPM packages to new project along with the similar projects to this new project, based on its descriptions. We considered the project description along with relevant keywords as the textual description of the requirements to be fulfilled. There is no need for source code by the user because the user can always use these NPM packages once they have installed the NPM library in their project.

Given a new project textual description, our methodology suggests NPM packages and similar projects by exploring past project dataset and the NPM package dataset. There are two phases used in the NPMREC, namely Training phase and Deployment phase that will be explained in detail in the later section.

To give an idea about the relevance of our strategy, let us consider a situation in which a software developer decides to start a new project. The issue that will be faced by him will be the absence of his knowledge about the NPM packages that can be used to develop the project effectively. For this, he will spend numerous hours looking, as well as trying out the appropriate packages. Despite this, he may not get what he is looking for. For such a problem, we created NPMREC, a recommendation system that can help in easing the exertion faced by a software developer to locate the privilege NPM packages.

We assess the viability of our methodology with the help of the Hit@N metric [3]. Our experiments show that our methodology accomplishes Hit@5, Hit@10 scores of 0.552, 0.742 individually. The Hit@10 score infers that for 74.2% of the projects,

NPMREC can effectively return in any event, one right NPM package in top-10 packages, which can be used to implement the project.

The rest of the paper is organized as follows: Sec. 2 discusses the related work; Sec. 3 explains the prerequisites for NPMREC; Sec. 4 expounds the research approach for NPMREC followed by Sec. 5 that shows implementation work along with the discourse on results and analysis; at last, Sec. 6 concludes the research paper and provides direction for future work.

2 Related Work

In Ferdian et al. [4], WebAPIRec uses text mining and does not need details about APIs that are, or are going to be used in a project. Thus, WebAPIRec can be used in the initial stage of development, where only a project requirement is understood. WebAPIRec uses a "personalized ranking model," which ranks APIs particular to a project. Based on Web API use history data, WebAPIRec learns a model that decreases the wrong ordering of Web APIs, i.e., when a utilized API is classified below an unutilized Web API.

The work done by the authors in [4], is with respect to the recommendation of Web APIs. Compared to this, our approach is based on the recommendation of NPM packages instead of Web APIs. We have used a cosine similarity-based approach for recommending NPM packages and similar projects. We obtained a Hit@10 score of 0.742. In the future, we would like to further extend the datasets and test our datasets on the "personalized ranking" approach used by the authors in [4].

3 Prerequisites

The prerequisites for NPM packages and similar projects recommendation system are divided into two sub-sections which are as follows:

3.1 Libraries.Io Web site

Libraries.io is an open-source Web administration that lists software development project dependencies and alarms developers about the latest version of those software libraries they are utilizing. Libraries.io is composed by Andrew Nesbitt, who has additionally utilized the code as the reason for DependencyCI, an assistance that tests project dependencies. A key feature is that the service checks for software license compliance. It collects the data by utilizing the package manager for each programming language. The site sorts out them by programming language, package manager, permit, and by catchphrase.

Table 1 A sample NPM package profile

Bootstrap package	
URL	https://libraries.io/npm/bootstrap
Description	The most popular front-end framework for developing responsive, mobile first projects on the Web.
Summary	Front-end framework
Category	NPM
Keywords	CSS, sass, mobile-first, responsive, front-end, framework, Web, bootstrap, css-framework, html, javascript, scss

Table 2 A sample project profile

Bootstrap package	
URL	https://libraries.io/npm/bootstrap-social
Title	Bootstrap-social
Description	Front-end framework
NPM packages used	Bootstrap, font-awesome
Keywords	Bootstrap, css, social-buttons

The structure of an NPM package in our dataset can be seen in Table 1. The structure of an NPM package contains its URL, Name, Summary, Description, and Keywords. Moreover, we represent each NPM package by its textual description, which is produced by merging the description and keyword of the NPM package.

Libraries.io contains a great number of NPM projects. The structure of a project in our dataset can be seen in Table 2. The structure of the project consists of a few snippets of data including URL, Description, Title, NPM packages used, Keywords.

3.2 Information Retrieval & Natural Language Processing Techniques Used in NPMREC

It comprises of text preprocessing, cosine similarity, TF-IDF measure, and vector space model (VSM). Now, we describe these techniques in detail.

Text preprocessing. In this stage, we break information into an increasingly reasonable portrayal that in later phases can be changed into an information retrieval (IR) model. Additionally, since content information is frequently noisy (i.e., it contains various unimportant words, solidly related words that are in different tenses, etc.), extra preprocessing steps are required. These now, pre-processing steps are:

- A procedure of removing words that appear multiple times and as such, it does not help in separating one document from another [5], known as stop word removal. Cases of these stop words include “I,” “you,” “are,” etc.

- A procedure of breaking information into word tokens, known as tokenization [6]. Delimiters, like accentuation stamps and blank areas, are used as cutoff points between single word token and another. After completion of this process, information is represented by a collection of word tokens.
- A procedure of changing a word to its base structure, usually by removing the postfix form that word has, known as stemming. To better understand this, let us suppose we apply stemming to words "running" and "ran." These will eventually be changed over to "run." In the absence of stemming, these words are taken as dissimilar altogether. To make words to their stemmed structure, we utilize the Porter stemming strategy [7].

Vector space model. Significant words are not distinguished from unimportant ones in the "bag of words" representation. To reflect on the relative importance of words, vector space model (VSM) was proposed by IR researchers, which uses a vector of weights [8] to represent a textual document. The general significance of a word is observed with the help of the weight assigned to it.

Although various weighting schemes can be utilized to induce the significance of a word, we use the well-known, term frequency-inverse document frequency (tf-idf) scheme [9] in this work.

Cosine Similarity. It is used to measure the similarity between two documents. After obtaining the VSM representations of the document, cosine similarity [10] can be used to compare the two vectors of weights.

4 Research Approach for Npmrec

Figure 1 represents the overall architecture of NPMREC. The input to NPMREC consists of a textual description of the new project, NPM packages, and past projects. From the past project dataset, for each project NPMREC takes its textual description and NPM packages that were used by them. Furthermore, for each NPM package, NPMREC takes its textual description and eventually, after analyzing these inputs, it generates a decreasing list of NPM packages and similar projects based on their similarity score, to be recommended for the new project.

Training and deployment are the two phases used in NPMREC. In the training phase, to train the recommendation model, NPMREC takes the textual description of all past projects (training data), eliminates stop words, break the remaining words, and form a tf-idf bag of words vector form out of it. A similar procedure will be performed on the NPM package dataset.

In the deployment phase, NPMREC converts the new project description into its vector form using tf-idf and then, it performs two operations. Firstly, it computes the cosine similarity between the vector form of new project and vector form of past project dataset (as obtained in training phase). Secondly, it computes cosine similarity between the vector form of new project and vector form of NPM package dataset (as obtained in training phase). Now, two lists are obtained, one with the NPM

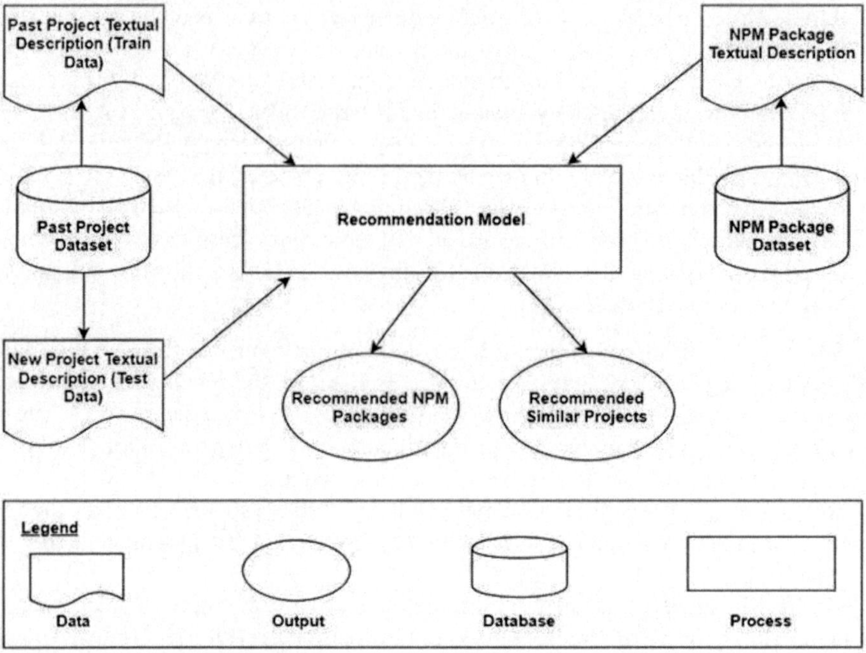

Fig. 1 Architecture of NPMREC

packages used by similar past projects and the other is the NPM packages from the NPM package dataset. Now, we combine both the lists and sort it according to the similarity scores. After that, we present the top-k NPM packages to the user. While combining the two lists, NPMREC handles the case of repetition by only taking into consideration of the NPM package with a higher score. Also, top-p similar projects are shown to the user.

NPMREC uses similarity-based recommendation model to solve the NPM package recommendation problem, which is divided into the following two subsections. Our goal is to produce list of NPM package according to its similarity to each project.

4.1 Recommendation Model

The similarity score between project *p* and an NPM package is calculated by our similarity-based ranking model.

$P(p_i, \{n_{1pi}, n_{2pi}, \ldots\ldots, n_{kpi}\}, S_{pi})$
$N(n_i, S_{ni})$
Where,
p_i is the project

```
        {n1pi, n2pi, ......, nkpi} list of packages that belong to project i
        Spi is similarity score of project i
        ni is the ith NPM package
  Final Packages list = {nf1, nf2, ....}
        Where,
        nf1 is final first package with highest similarity score
        {nf1, nf2, ....}is decreasing list based on similarity score
        {nf1, nf2, ....} =npi, if Spi>= Sni
                                          ni, otherwise
```

4.2 *Training and Testing*

We trained our model with 589 Projects and 759 NPM packages and the remaining 105 projects were used for testing purpose.

4.3 *Updating Data Set from User Feedback (Insertion of New Projects)*

In our approach, after providing a project description as the input, our system recommends the top 10 NPM packages as well as similar projects in their decreasing order of similarity score. Now, the user develops the project and after this, we allow him/her to send us a feedback as to which packages among the recommended ones were actually used in the project. We then add that project along with the packages it used in our datasets. This way the system learns by itself with expansion in the dataset and, therefore, overall efficiency of the system increases with time. Ultimately, it results in better recommendations.

5 Implementation, Results and Analysis

This section is divided into four sub-sections, namely dataset used, evaluation metric, experimental setting, and results and analysis, which are as follows:

5.1 *Dataset Used*

We used custom-built datasets for our approach using libraries.io Web site. Libraries.io is an open-source Web administration that lists software development project dependencies and alarms developers about the latest version of those software libraries they are utilizing. The training dataset contains two datasets: firstly, the project dataset with 589 NPM projects/NPM modules with information about their dependencies/NPM packages and secondly, the NPM package dataset with 759 NPM packages containing the detailed information about the dependencies of these 589 projects/NPM modules. The test dataset contains 105 NPM projects/NPM modules along with information about their dependencies/NPM packages. Our research goal is to investigate whether NPMREC can return appropriate NPM packages provided a new project's description. A project's ground truth NPM packages are the NPM packages/dependencies that appear on the library.io Web site in the project's page. In our initial research with a few projects on library.io, we noticed the absence of their descriptions; therefore, we searched about the project on its homepage which was generally present on Github.

5.2 Evaluation Metric

We used the assessment metric Hit@N to test our approach. It counts the fraction of ranked lists formed while recommending NPM packages to projects, whereas a minimum, one right NPM package is present in the top N results. Two values of N are used in this paper, namely 5 and 10.

5.3 Experimental Setting

We used 589 projects and 759 NPM packages for training and then 105 projects for testing our approach. The machine used for conducting experiments had an Intel(R) Core (TM) i3-2100 @3.10 GHz PC processor with windows 7 ultimate operating system.

5.4 Results and Analysis

In testing, for each test project, we will be having its description, keywords, and NPM packages used by it. Our approach takes its textual description as input and recommends top-10 NPM packages along with top-10 similar projects. Table 3 shows the recommendation example by NPMREC.

Our methodology achieves ratings of 0.552 and 0.742, respectively, at Hit@5 and Hit@10. This means that not all NPM packages are relevant for most cases in the top-N results, but, at least one of them is relevant. Based on the results of Hit@10, we discovered that a valid NPM package used to execute a project is among the top 10 NPM packages returned by NPMREC for 74.2% of the projects.

Table 3 Example recommendation

Koa-router	
Description	Router middleware for koa. Provides RESTful resource routing.
Keywords	Koa, middleware, route, router
NPM packages used	Debug, http-errors, koa-compose, methods, path-to-regexp, urijs
Recommended NPM packages	Path-to-regexp, ip, opn, express, workbox-routing, change-case, got, hash.js, tunnel, yargs, debug
Recommended similar projects title	Easy-router, saucelabs, connect, morgan, cors, react-native-easy-router, webpack-dev-middleware, hapi-stateless-notifications, phatom-pool, redux-logger

6 Conclusion and Future Work

We proposed an approach NPMREC that took the new project textual description as input and recommended NPM packages as well as similar projects. We assessed our methodology on 759 NPM packages and 694 projects (train and test dataset combined) from libraries.io. NPMREC achieved Hit@5, Hit@10 of 0.552 and 0.742, respectively. NPMREC can thus effectively advise correct NPM packages in top-10 positions for 74.2% of the projects.

The efficiency of NPMREC decreased when a project had used a single NPM package, because the probability of existence of that package was quite low in our dataset. Thus, we would like to address this issue by exploring more NPM packages and more projects from other data sources outside libraries.io in the future. We would like to introduce a mechanism through which the system learns form past user experience. The clustering algorithms such as k-mean clustering [11] can also be applied to recommend NPM packages and similar projects.

References

1. Raemaekers S, van Deursen A, Visser J (2012) An analysis of dependence on third-party libraries in open source and proprietary systems. In: Sixth international workshop on software quality and maintainability, SQM, vol 12, pp 64–67
2. Yu S, Woodard CJ (2008) Innovation in the programmable web: characterizing the mashup ecosystem. In: International conference on service-oriented computing. Springer, Berlin, Heidelberg, pp 136–147
3. Avazpour I, Pitakrat T, Grunske L, Grundy J (2014) Dimensions and metrics for evaluating recommendation systems. In Recommendation systems in software engineering. Springer, Berlin, Heidelberg, pp 245–273
4. Thung F, Oentaryo RJ, Lo, D, Tian Y (2017) Webapirec: recommending web apis to software projects via personalized ranking. IEEE Trans Emerg Topics Comput Intell 1(3):145–156

5. Silva C, Ribeiro B (2003) The importance of stop word removal on recall values in text categorization. In: Proceedings of the international joint conference on neural networks, vol 3. IEEE, pp 1661–1666
6. Méndez JR, Iglesias EL, Fdez-Riverola F, Díaz F, Corchado JM (2005) Tokenising, stemming and stopword removal on anti-spam filtering domain. In: Conference of the Spanish association for artificial intelligence. Springer, Berlin, Heidelberg, pp 449–458
7. Porter MF (1997) An algorithm for suffix stripping. Readings in information retrieval
8. Schütze H, Manning CD, Raghavan P (2008) Introduction to information retrieval, vol 39. Cambridge University Press, Cambridge, pp 1041–4347
9. Rajaraman A, Ullman JD (2011) Mining of massive datasets. Cambridge University Press
10. Nguyen HV, Bai L (2010) Cosine similarity metric learning for face verification. In: Asian conference on computer vision. Springer, Berlin, Heidelberg, pp. 709–720
11. Likas A, Vlassis N, Verbeek JJ (2003) The global k-means clustering algorithm. Pattern Recogn 36(2):451–461

Electronic Wallet Payment System in Malaysia

Md Arif Hassan, Zarina Shukur, and Mohammad Kamrul Hasan

Abstract Today's Internet has dramatically reshaped how people make payments and transfer money. Electronic transactions are playing a vital role now in these days to shop, pay bills and money transfer, and many more. Among the electronic payment system nowadays, E-wallet is one of the most famous payment systems. Many countries have already implemented the use of electronic wallets as part of their daily purchase transaction options for their customers. The goal of the study is to recognize the overview study of E-wallet and the features of theirs and to determine the different level of use and learn the best E-wallet and what the reason for practicing. Specifically, the research investigation focuses on Malaysian public university students. To achieve this, we conducted a survey to collect data from 120 participants from the university campus. Statistics were evaluated by using SPSS software and have been documented in this survey. The results show that, among 120 respondents, received 120 valid responses. Among the 120 valid responses, about 101 (84.2%) of them are using the E-wallet payment system. The following respondents who never use e-wallet payment 19 which achieve a rate of 15.8%. The study also reveals that the new financial service has a huge effect on the use of e-wallet among bachelor students. While analyzing gender well the usage of e-wallet, man has larger satisfaction levels and also requires e-wallet often than female students. In addition, students' perceptions of their perceived use, practice, and reason had a significant impact on their confidence. For practical involvement, the use of E-wallets by most respondents confirms that there is a great potential for this payment spread in Malaysia.

M. A. Hassan (✉) · Z. Shukur · M. K. Hasan
Center for Cyber Security, Faculty of Information Technology, National University Malaysia (UKM), UKM, Bangi 43600, Selangor, Malaysia
e-mail: arifhassane72@gmail.com

Z. Shukur
e-mail: zarinashukur@ukm.edu.my

M. K. Hasan
e-mail: mkhasan@ukm.edu.my

A. Khanna et al. (eds.), *Data Analytics and Management*, Lecture Notes on Data Engineering and Communications Technologies 54,
https://doi.org/10.1007/978-981-15-8335-3_55

Keywords Electronic payments system · E-wallet · Boost · M-money · Touch n go · Lazada · AEON · Grab pay · Vcash, WeChat · Alipay

1 Introduction

Today's Internet has dramatically reshaped how people make payments and transfer money. Electronic transactions are playing a vital role now in these days to shop, pay bills and money transfer and many more [1, 2]. Electronic payments are a financial exchange that takes place online between buyers and sellers. These exchange materials are usually some form of digital financial instruments. In the last two decades, electronic payment systems have attracted much attention because of their vital role in modern electronic commerce [3]. The electronic payment system is referred to as a mode of payments over an electronic network [4] which a person can make online payments for his purchase of goods and services [5]. There are several e-payment services that have been developed within the payment system around the world, [6, 7] In their study they classified electronic payment systems into four categories such as e-cash, online banking, debit–credit and e-wallet. Among the electronic payment system nowadays, E-wallet is one of the most famous payment systems. An electronic wallet, sometimes called "digital wallet" or "e-wallet," is an electronic adaptation of a payment card that is approved for you to direct exchanges. In certain countries such as Japan, digital payment systems, or otherwise known as e-wallet, had been introduced to popular usage since as early as 2004 [8, 9]. These wallets are mostly on a smartphone; however, desktops and laptops can likewise be kept electronic. E-wallet might be utilized with mobile payment methods that enable buyers to cover their shopping using smartphones [10].

Electronic wallets save time for customers as they scan their card or e-wallet app rather than just queuing to look for cash and purchase tickets. Many countries have already implemented the use of electronic wallets as part of their daily purchase transaction options for their customers. Malaysia has taken seriously the development of cashless societies in particular. Based on the financial blueprint 20–20, the Central Bank of Malaysia aims to move towards a new cashless society with the goal of increasing financial sector efficiency and effectiveness. This may be realized by the attempt of the Central Bank of Malaysia to reduce the fee for an instant e-payment and increase the check fee up to RM 5000 to reflect higher processing costs [11]. In 2018, almost 50% of Malaysian cashless payments are going on with most debits and credits. A survey performed by Visa discovered that nearly half of Malaysians might survive with no cash money [12]. It reveals that Malaysia is prepared for development in a cashless society. Payment methods in Malaysia are split into two categories namely electronic-based and paper-based which is shown in Fig. 1.

Paper-based payments or Cash payment may be the earliest, most typical payment process that is popular and is probably the ideal means for little payments since it requires absolutely no credit. Paper-based payments are in the cheques, need drafts,

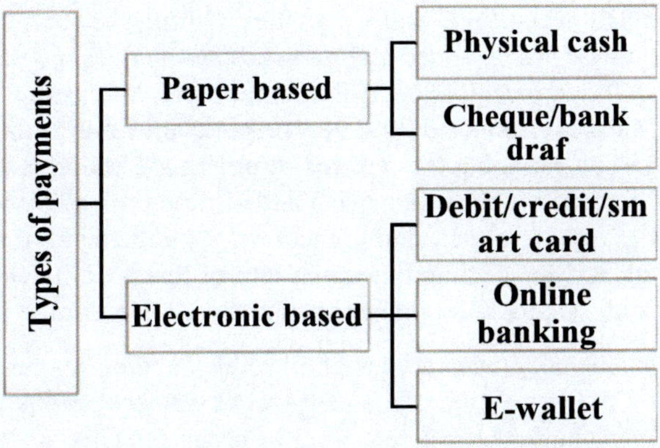

Fig. 1 Payment types of Malaysia [13]

transaction orders, banker's cheques, refund orders, warrants, etc. In Malaysia, electronic retail payments are being continually created to change or even bring down paper-based payments [14, 15]. Many brands of recent payment service came into existence in the recent years; these include E-wallet debit cards, recognition cards, electric money transfers, direct debits, direct credits, and internet banking and e-commerce payment systems. There are 42 e-money certificates has issued by Central Bank of Malaysia, including 5 banks and 37 non-banks [16] which shown in Tables 1 and 2. The Central Bank of Malaysia (BNM) has established the goal of it has to flip Malaysia right into a cashless society. Despite cyber-security concerns, people are using e-wallets for bills, food, tolls, petrol, grocery and merchandising spending. Based on Bank Negara Malaysia's information, e-money transactions amounted to 1.4 zillion and RM10.6 billion coming from January to August 2019 [17]. There has been a positive adoption of E-wallets amongst Malaysians, and 2.9 Malaysians are using an e-wallet [18]. E-wallet prospect is of a current and broad interest in Malaysia since it is going to affect the method of company, financial markets as well as payment process in Malaysia. Malaysia is one of several promising nations being involved

Table 1 Demographic characteristics

Statistics

		Gender	Age	Status	Education	Usage	Practice	Wallet	Reason
N	Valid	120	120	120	120	120	120	120	120
	Missing	0	0	0	0	0	0	0	0

Table 2 Total count of gender responders

	Frequency	Percent	Valid percent	Cumulative percent	
Valid	Male	76	63.3	63.3	63.3
	Female	44	36.7	36.7	100.0
	Total	120	100.0	100.0	

much deeper in e-wallet activities. During the recent past in Malaysia, technology literacy has exploded after tremendous changes in federal policy. It provides specifically an incentive of RM30 on the E-wallet customers that had been a part of the Malaysia Budget 2020 [19]. The government of Malaysia is committed to applying and encouraging the rewards of utilizing e-wallet uses therefore the device can be accepted by most Malaysian and also be utilized by almost all levels of communities to relieve their daily functions. With the hand of Malaysia's financial blueprint 20–20, we identified the chief aim of this paper. Hence, based on the needs of the study, we took the following objectives to determine the focus of the study:

- To study the overview of E-wallet and their features
- To identify the usage of E-wallet services among the university students and to evaluate the different level of usage e-wallet among various levels of students, and also to discover most using an e-wallet, and determine the reason behind the practice of e-wallet services.

We organize this paper as follows. Section 2 discusses the scope of the study, Literature Review, and research methodology. Section 3 discuss the research results and debate based on the analysis objectives. Section 4 discusses the accomplishments of the findings and conclusions are probably available in Sect. 5.

1.1 Scope of the Study

This paper provides a focus on E-wallet services, which is a modern pattern in electronic payments. We attempt a considerable study of students' use and understanding of the present mechanisms. We restrict the geographical proper care of the research to the university campus. The respondents in the research are the students of the various faculty. This paper offers a guide to the present arrangement of electronic payments and highlights electronic wallets including new technology and its effectiveness. It also helps to understand the various services offered in E-wallet.

2 Literature Review

E-wallets are one of the best inventions of the twenty-first century. They are quick and easy, allowing people to pay with a single tap. Digital wallets are a growing trend in Malaysia. Bank Negara Malaysia (BNM) has set its goal to turn Malaysia into a cashless society. Despite cyber-security concerns, more and more people are using digital wallets for bills, food, tolls, petrol, grocery, and retail spending. According to Bank Negara Malaysia data, e-money transactions amounted to 1.4 million and RM10.6 billion from January to August 2019 [17]. The payment methods used by consumers will have great effects on the future of the financial system and the business model of a country. As E-wallet is getting even more famous and broad nowadays, more shops and business is offering rewards to customers or consumers who use E-wallet to complete their payment. This section will summarize some related work. In Malaysia, Amin [20] performed e-wallet adoption investigations in Malaysia in Sabah. The survey questionnaire was definitely employed to get information from the primary respondents are bank's clients in Sabah. Likewise, the study goal to learn the elements affecting the adoption of bank clients towards E-wallet with the extension of using a TAM concept in the research and also make a much better representation of the adoption of an e-wallet in Malaysia. Perform the same research by Karim et al. [21], In their study, they conduct a technology acceptance model (TAM) based survey among Malaysian Young Adults. After conducting a Survey concerns to 330 individuals, the results from this analysis show that perceived convenience, perceived ease of use and privacy and security measures have a positive and significant relationship with a behavioral aim to use the e-wallet. Do a similar extended model of technology acceptance survey was conduct by [22]. They conducted a survey on college students in Semarang, Indonesia. The results of this study show that the perceived value of e-payment security and the buying intentions of customers is better than mediated. Perceived ease of use has a significant indirect impact on e-commerce purchasing intentions through e-payment protection. The results also give a heightened awareness of security in electronic transactions. In electronic payment systems such as digital wallets, security is an immense problem [23], and without secure commercial information exchange and secure electronic financial transactions through networks, nobody will trust this facility/service. Users play a key role in the system, as they connect different devices to the network using the application provider.

As mobile payments gained popularity, the attention also turned to cybercriminals, as it is clear, that the nature of e-wallet mobile apps requires stronger protection than other smartphone apps [24, 25]. It is also a major challenge that what kind of e-wallet design customers wants to use [26] because it requires a usable interface and support for all financial transactions that a user may wish to perform. Security Concepts based survey was conducted by [7]. The study was discussed, explore, and raise awareness of different concepts related to EPS, including its advantages, challenges and safety aspects. The study carried out has shown that while the usage of electronic payment systems can pose a variety of challenges, they are seen as a significant step towards a nation's economic growth. Argimbayeva et al [27], they conducted a survey on customer satisfaction in Abu Dhabi National Oil Company (ADNOC). The research would investigate the consumer service implications of ADNOC Distribution's e-wallet. It aims to educate ADNOC Delivery regarding possibilities and differences in e-wallet adoption. The work would also include guidelines to advise the introduction of technology through ADNOC Delivery to enable potential transfer payments. The Sunway University students conducted a survey in Malaysia [28] have been undertaking the efficacy of the e-wallet study. The study showed a significant effect on consumers' purchasing behavior by the usage of the E-wallet on their convenience, cost savings and health.

3 Materials and Methods

3.1 Methodology

The research methodology can be considered as the key factor for researchers to achieve their goals as it shows the direction of how the goal goes forward and what tools and techniques are used to achieve it. Different researchers utilized different methodologies for acquiring their objectives. For instance, [29] used the methodology determined by both qualitative and quantitative approaches, which was conducted in the urban area of Bangalore city. In this combined methodology, the researcher used open-and closed-ended questions and observations to carry out the qualitative and quantitative analysis. This section summarizes the methodology focused on how the research was carried forward. The method and techniques were used to achieve the outcomes.

3.1.1 Data Collection

In this paper, we have identified research approaches, and it involves both primary data and secondary data. Involves qualitative methods of data gathering strategy interviews or data examination, where the information used to finish the study are primarily discussions, bulletin, market study reports and articles [30]. In this paper, we gather primary data through surveys and secondary data through literature, articles, journals, conferences, bulletin and news.

- Primary Data

Primary information is information that is newly produced from primary sources concluded questionnaires, interviews, or maybe observations to find solutions associated with a particular research project [31]. The interview questions are open-ended and semi-structured, specifically designed to help students explore their experiences of e-wallets. However, the cost of distributing questionnaires and interviews will be easily reduced by using an advanced technology called the Internet. For the current study, have been collected through live conversation. Furthermore, it helps us monitor the data collection process. Before the survey began, the researcher gave an introduction and explained the purpose of the study. Respondents answered questions anonymously and were assured that their responses would be treated confidentially. Each respondent was carefully requested to complete the questionnaire. Altogether, 120 valid responses were collected from interviewers.

- Secondary Data

Secondary data is raw information, which has already been selected by another person, for some general information purposes, for example, the official census or any other government purpose or even for a particular research project. According to Collis and Hussey (2014), "Secondary information is research information that is collected from existing sources such as publications, databases or internal records and may be available in hard copy form or on the Internet" [32]. In this study, we planned secondary research with literature reviews. Secondary data have been collected from various sources including websites, newspapers, various published and unpublished articles. The purpose of the literature review is to select the most relevant content of the literature related to the research objectives.

3.1.2 Sampling Design

A multi-stage sampling technique was adopted to select respondents for the study. We selected students from the Bangi Campus of the National University of Malaysia in four terms. There are about five public research universities in Malaysia, UKM and UPM are among the five public research universities in the state of Selangor, only the UKM Bangi Campus was selected for study between the two universities.

Twelve hostels are available on the UKM campus. Sample size—In the present study, 120 local Malaysian students were taken as sample sizes from twelve hostels. There are 10 students selected in each hostel. The questionnaires were distributed among 120 respondents, but we received 120 valid responses. We show the procedure of sampling design in Fig. 2.

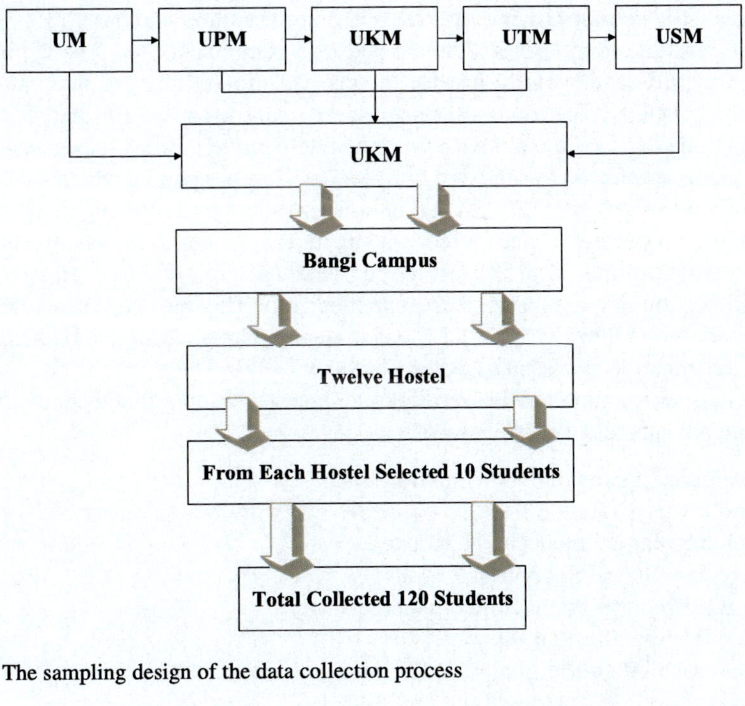

Fig. 2 The sampling design of the data collection process

3.1.3 Statistical Tools and Techniques

We conducted an analysis of the information while using software SPSS by IBM, while the raw information is pre-edited in Microsoft Excel. Toward analyze the composed data, numerous statistical methods and tools for example averages, frequency distribution tables, histogram charts, and regular distribution were used.

4 Results and Discussion

4.1 Overview of E-Wallet in Malaysia

E-wallet is the invention of finance technology that makes our transaction payment easily and fast. Online purchases can be made automatically through an electronic device, using an E-wallet. Electronic wallets are simple and fast, enabling individuals to pay with one tap. In 2019, the "market research future" in their article report that, the global e-wallet market is expected to grow at a CAGR of 15% and estimated to reach a market size of approximately USD 2,100 billion through the end of estimate period 2017–2023 [33]. An E-wallet aims to eliminate the need for carrying physical cash. E-wallet is a virtual storage method [34], which may record your distinctiveness and numerical IDs and compromise to an electronic gadget or web-based service that provides a person to commit online purchase, [35–37]. An E-wallet consumes similarly a software and info component [36]. The application offers encryption and security just for the private info and just for the real online transaction. Usually, electronic wallets are saved on the prospect side and are readily self-maintained and companionable with many e-commerce sites. A server-side E-wallet, likewise referred to such as a tinny wallet [38], is a single, which a company makes for and keeps on its servers. Server-side E-wallets are becoming more popular between traders because of the protection, effectiveness, and additional effectiveness it offers to the customer, which boosts the enjoyment of theirs of the general purchase. The info portion is a database of user-inputted info. This info comprises delivery, delivery address, billing address, payment strategies, namely containing charge card figures, termination dates, and security numbers [39]. There are several benefits of E-wallets to examine if you are thinking about embracing this technology. The following advantages of E-wallets are:

- Transmit and get payments within Malaysia.
- Boundless transfers.
- Simple regular payments and handover.
- Online and Offline Shopping
- Accomplish our account from our smartphone.
- Authentication after transactions
- Appeal cash into our digital wallet from at all bank account.
- Receive wired funds straight into our e-wallet.

- Transmission money from wallet to wallet deprived of allocation personal account numbers.

There are three types of E-wallets [37] such as open E-wallet, closed E-wallet and semi-open E-wallet. An open wallet refers to the wallets that have been issued by the banks. There are multiple services, open wallets. Open E-wallets can be utilized to buying goods or services from merchants, containing commercial services, such as transferring funds to merchant locations or receiving points at point-of-sale (POS) terminals and cash withdrawals at ATMs or business communicators (i.e.) Merchantrade Money. Note: Both types of E-wallets issue e-money and can be open-loop (Fig. 3).

A semi-Closed wallet may purchase goods and facilities, comprising financial services at the site or organization of the identified merchant. They have a specific agreement with E-Wallet for accepting payments. It does not allow the removal or redemption of cash by the E-wallet holder. One example of a semi-closed E-wallet is a boost. This can be seen in Fig. 4.

This is especially popular among e-commerce companies. Companies to buy products and services from them only issue closed wallets. These tools do not allow for cash withdrawals or discounts. Most of these companies act as bank accounts where the money is deposited, and the refund is because of the withdrawal or return of a product or service. This can be seen in Fig. 5.

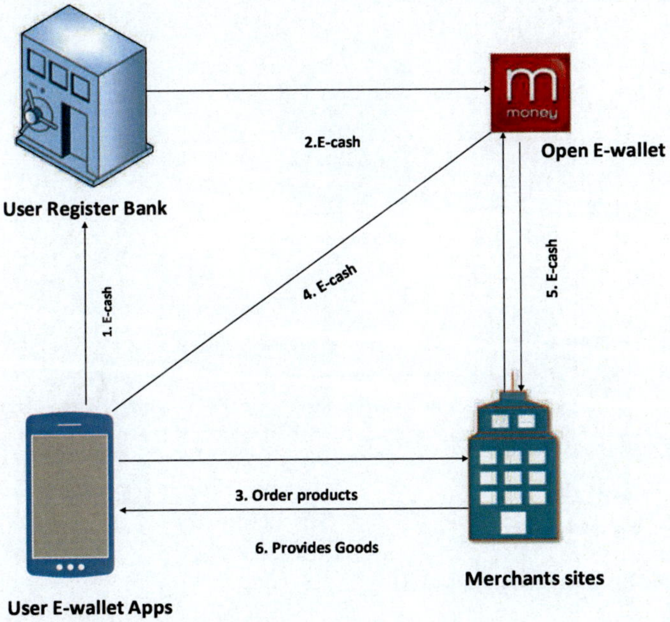

Fig. 3 Open E-wallet in Malaysia [37]

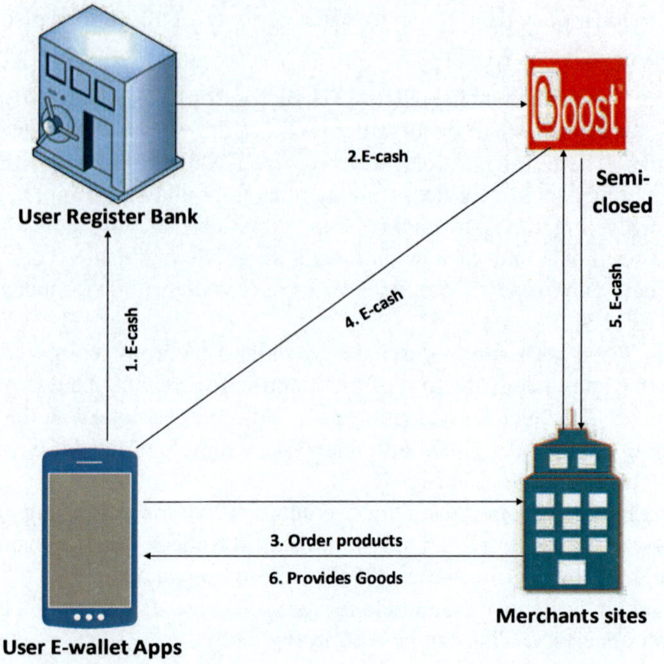

Fig. 4 Semi-closed E-wallet in Malaysia [37]

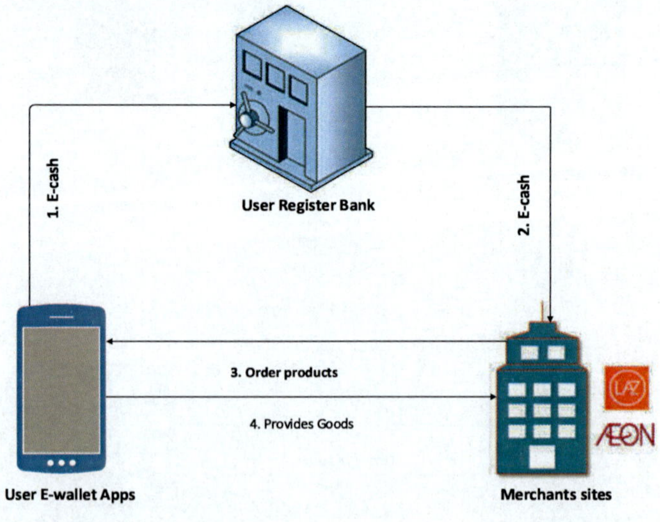

Fig. 5 Closed E-wallet in Malaysia [37]

4.2 The Technology Used in E-Wallet

E-wallet method is a revolutionary breakthrough mobile commerce solution, which provides the comforts of cashless shopping and making remote payments via any smartphone, in which financial, as well as merchandising transactions, are completed with the contact of the fingertips [40]. Traditionally, E-wallets have been saved on the desktops of individual computers. By the early 2000s, new E-wallets were appropriate with wireless along with other mobile devices and were much more often saved on a central server owned by an E-wallet vendor Internet service provider. In this section, we are going to discuss the technology used in E-Wallet.

4.2.1 Contactless Payment

There are numerous systems using and containing NFC technologies. The qualities of NFC enabled many solutions, which helped the company to think of recent ideas, and solutions, which provide the industry customers. The customary smart card banking services now are going to smartphones. This change is facilitated by Near Field Communications (NFC) technological innovation that allows a smartphone to emulate a credit card using E-wallets. This imitation is usually delivered through a hardware Secure Element (SE) Host Card Emulation (HCE) [25, 35, 36]. People be able to wave their Smartphone near an NFC supported terminal, and the transaction moves concluded. To apply this, each merchant point of sale made (POS) terminal and the phone of yours needs to have the technology. If the phone of yours does not have, some E-wallet companies use stickers you can follow your telephone and tap it on an NFC enable POS machine. The contactless based payment in Malaysia is Samsung Pay and BigPay.

4.2.2 Tracing Based Store

The E-wallet app is going to consumption GPS on the smartphone of yours and find a store near you, which receives finances expenses. This will probably come available as just several stores have E-wallet acceptance infrastructure. Besides finding a proximate store, which accepts wallets, it will furthermore alert you of offers out there in case you spend going with an E-wallet. It will in addition display channels nearby exactly where you can load money in the E-wallet (i.e. Setel). This is a distinctive payment procedure and one that is sadly a closed environment. It bound the product to just a tiny number of Petronas stations at the second. On another hand, it is helpful for individuals who would rather not swipe the credit cards of theirs at gasoline stations.

4.2.3 Virtual Cards Payment

Despite the title, it is actually not a card, though a pair of numbers, which is produced by a financial service provider to allow a customer-making internet, purchases (very similar to a recognition credit card). It is also referred to as Controlled Payment Numbers. Approximately E-wallet suppliers enable you to produce a simulated card together with the cash in the wallet. The card is in partnership with Mastercard or Visa. The virtual card is going to have a 16-digit card number, expiry day, and CVV quantity like every other credit or maybe a debit card and you can make use of it on any internet platform. In order to get the card, you have to perform an onetime registering. Some apps immediately produce a simulated card whenever you burden the cash (i.e. razer pay, mPay).

4.2.4 Quick Response (QR) Code

QR Code is a kind of 2D bar code that was developed by Denso Wave, a Japanese automated details record gear business (Denso 2009), within 1994 [26]. It is clear by reasonably equipped mobile cell phones with cameras and QR scanners. A QR code is effective at keeping 7089 numeric characters, 4296 alphanumeric characters, 2953 binary bytes, 1817 Kanji figures or maybe a combination of them [41–51]. Recently, QR codes grew to be commonly used in many disciplines like advertising, document verification, item tracking, social networking and electrical payment system. A QR code is a mechanism-readable bar code, which has details of the shoes you purchase. The client has to browse the code to your smartphone's camera to create a payment. Initially, the client will have to sign in on the app which enables the client to scan the QR code. Banks along with Fintech businesses provide the transaction process. In recently QR is the most famous payment system in the payment industry. There are the number of QR code-based payment systems in Malaysia (i.e. boost, grab pay, touch n go, aeon) etc.

4.3 E-Wallet Companies in Malaysia

The e-wallet is akin to the wallet, though it has electronic cash rather than physical money or perhaps currencies. Thus, the e-wallet facilitates the consumer to do convenient and quick modes of transactions. You will find various e-wallet companies in Malaysia and also provide unique solutions to the people. This section will provide some e-wallet companies.

4.3.1 Boost

Boost is free to use E-wallets that allow users to pay for all merchants taking part on its platform, such as the app through 99 Speedmart, Steam, Shopify, and more. To do this, users need to top-up the digital wallet using credit, debit card or online banking. Boost's primary market is small merchants, which are just experimenting with cashless payments [42]. Boost provides protected purchases by reloading wallet with seventeen reliable bank associates as well as credit/debit card suppliers [43]. Boost E-wallet provides the following features: Add Currency, Scan & Pay, Supplier's QR Code, consume QR Code, send and request money, buy vouchers and send as a gift, manage Postpaid/Prepaid bills, pay utility bills, pay for movie tickets, pay for an event, donation, online shopping, order food distribution, and book transport [43].

4.3.2 Merchant Money

Merchantrade Money is an E-wallet that has several uses, including foreign currency exchange. Merchantrade Money is the prepaid card & e-wallet for the jet-setting traveler. Buy, store & sell multiple foreign currencies and manage them within the app. There are a number of a feature on this e-wallet including Add Cash, Scan & Pay, Merchant's QR Code, consume QR Code, send money, buy vouchers and conduct as a gift, manage Postpaid/Prepaid bills, online shopping [44].

4.3.3 Touch n Go

Touch 'n Go's E-wallet was primarily released, making cost payments convenient. It is, in fact, among the simplest ways of paying for tolls in Malaysia. The new e-wallet scheme is a joint scheme between Touch 'n Go and Ant Financial (Alibaba). TNG has just recently declared exceeding one million people and they have an advantage in advertising and also accessibility in public transportation use cases including cost, parking, bus transaction, etc. The features are provided by TNG are following Movie Tickets, Mobile Reload, Transfer Money, Shopping, Flight Tickets, Bill Payment, Petrol Station, Beauty & Healthcare, Pay Direct & RFID, Food & Beverages and App Store [45].

4.3.4 Lazada

Lazada Wallet is among the several E-wallets to offer cashback on most purchases. This comes at the enormous cost of just being in a position to make use of it on Lazada's very own online shop and also getting cashback credits that expire after several months. This is probably the most limited of the e-wallets considering the scope. It locks owners into the Lazada environment; therefore, folks must understand what they are getting themselves into.

4.3.5 Aeon

Aeon is among the biggest list groups in Malaysia with countless current Aeon members in its RM12-per-year paid loyalty program [46]. Aeon Credit Service creates the brand new e-wallet, which ties to some Visa Prepaid Card. They are vigorously recruiting users via all the brick-and-mortar outlets. Self-checkout at retailers using Aeon Wallet might stay in the pipeline as well.

4.3.6 Grab Pay

GrabPay, among Malaysia's foremost E-wallet providers, provides you with the comfort of having to pay for day services like bills, prepaid reloads, rides services, food, groceries, and much more all within a single app [47]. For GrabPay, absolutely no pin is necessary to do transactions. Users are enabled by each transaction to make points, that may be utilized to have totally free Grab trips and deals on GrabFood. A minimum of RM10 along with a maximum of RM1 or RM500, 500 is reloaded into the Grab wallet. Money packed into the wallet cannot be withdrawn at any price.

4.3.7 Vcash

Vcsash is a mobile wallet driven by Dig.com Video. Nevertheless, users will make use of the wallet with virtually any Malaysian telecommunications business so long as they have a Malaysian phone number. To use Vcash, users have to download the application and credit money in their e-wallet through Digi Stores, online banking system or Jompay. Then they can scan the QR code and make a payment. For visas, it requires a six-digit PIN or fingerprint to perform the transaction. It allows Bill splitting to be paid to the group. Depending on the wallet, it can reload the minimum RM10 or maximum RM500 on the wallet [48] Exceptional Wallet users can extract cash from the account of theirs at an RM1 cost per withdrawal.

4.3.8 Wechat Pay

WeChat said that in Malaysia there are 20 million WeChat users, with Malaysia the first country to be e-money licensed outside China [46]. Unlike Alibaba, who wants JV with local players? WeChat Pay is a multiplicity payment method Quick Pay, a QR code enables users to send funds individually or grouped to withdraw the e-wallet balance from any bank in Malaysia, and cross-border [49]. WeChat payment is based upon multiple payment methods: Quick Pay, QR code, In-app web or Native Payment in-app.

4.3.9 Alipay

China's indigenous mobile wallet AliPay had set foot in Malaysia early last year. AliPay is sponsored on over 80 markets around the country [50] by over 110,000 retail stores. It allows users to enjoy paying for their taxi ride, hotel bookings, movie tickets and even doctor appointments. Once the user have the app downloaded, sign up using the user mobile number. User can load up their wallet by linking their bankcard and entering their credentials. The features of Use AliPay are users can online and offline purchases: 7-Eleven outlets, Berjaya Group merchants, Starbucks outlets, etc.

4.4 E-Wallet Usage Among Students

4.4.1 Demographic Profile of the Respondents

Data is collected from the pre-schedule questionnaire form. The data collections and analysis has done in an iterative process it consists of 44 females and 76 males. The participants were between 20 and 45 years old and were conducted via live conversation. The common of the respondents were between 20–25 and 25–30 years old. The authors would like to statement the sharing of surveys through live interviews in different hostel canteens as a possible cause of this situation. To get a better understanding of the distribution of our subjects, we show the table and Fig. 6.

Based on Table 1, it shows that the frequency of demographic characteristics and using electronic payment as one of the demographic information. There are

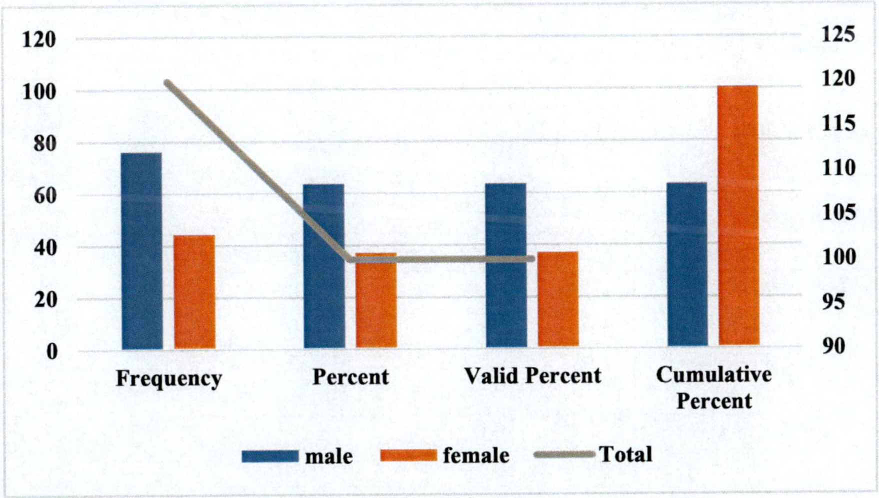

Fig. 6 Demographic characteristics for gender

gathered with 120 respondents who have been chosen randomly for this research study purpose. Among 120 respondents, received 120 valid responses.

Based on the information above, the size of male respondents is greater than female respondents. Among all of the respondents, male respondents have gathered 76 people, which is 63.3% for this research study. Other than that, there are a total of 44 female respondents which represent 36.7% are being conducted for this research (Fig. 7).

From Table 3, the age of 20–25 years, old is the highest percentage among the age group. The 20–25 years old had 50.8% while the age of 25–30 years old consist of 35.0%. The 30–35 years old is 6.7%. The age 35–40 years old 2.5%, are the lowest among the age group. The age 40–45 years old having 5.0% as the second-lowest percentage among the age group (Fig. 8).

From Table 4, the marriage status of single tends to get the highest percentage of

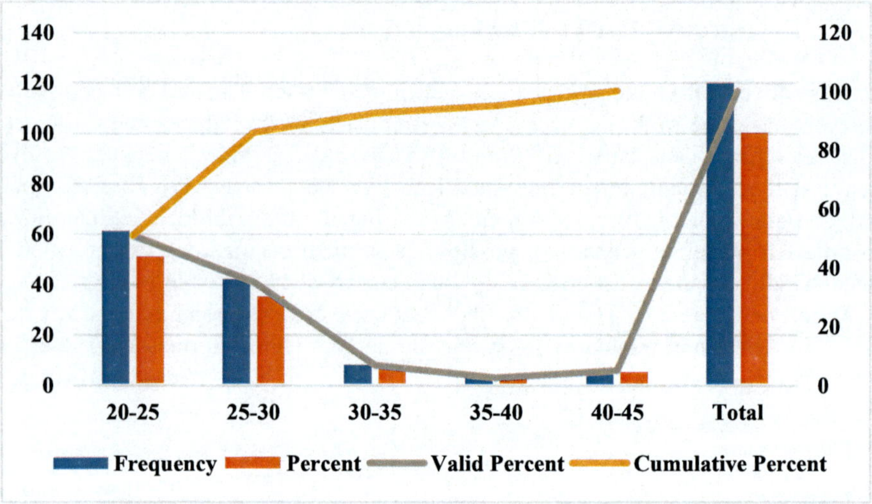

Fig. 7 Total count of gender responder

Table 3 Total of total responder's age

	Frequency	Percent	Valid percent	Cumulative percent	
Valid	20–25	61	50.8	50.8	50.8
	25–30	42	35.0	35.0	85.8
	30–35	8	6.7	6.7	92.5
	35–40	3	2.5	2.5	95.0
	40–45	6	5.0	5.0	100.0
	Total	120	100.0	100.0	

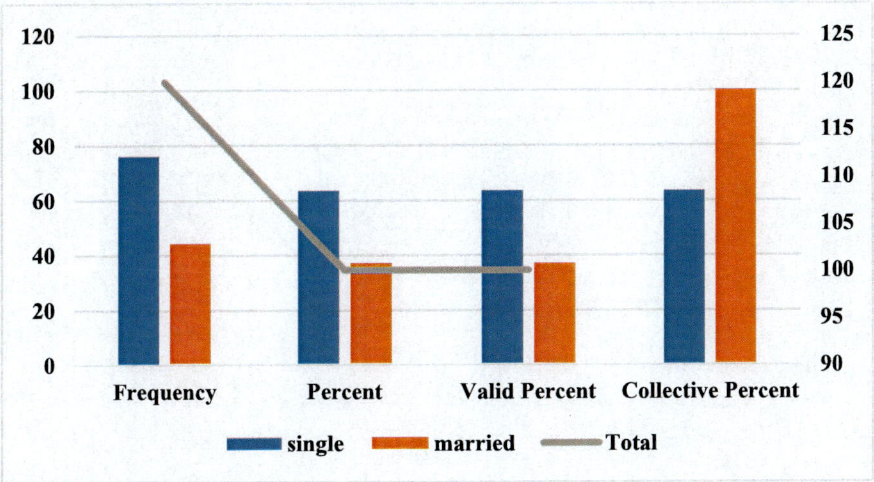

Fig. 8 Material status of responders

Table 4 Material status of responders

	Frequency	Percent	Valid percent	Collective percent	
Valid	Single	76	63.3	63.3	63.3
	Married	44	36.7	36.7	100.0
	Total	120	100.0	100.0	

63.3% among the other. The married having the second and last percentages, which are 36.7% (Fig. 9).

According to Table 5, the amounts of respondents 66 (55.0%) respondents have Bachelor's Degree, and 37 (30.8%) respondents have Master's Degree. For the amounts of respondents who have PhD, there are only 17 (14.2%) (Fig. 10).

Tables 6, shows the frequency of respondents using E-wallet payment. Among the 120 selected respondents, about 101 (84.2%) of them are using the e-wallet payment system. The following respondents who never use e-wallet payment 19 which achieve a rate of 15.8%. After the analysis of Table 8, we can say that, by using innovative technology, the majority of the students decide to pay each online payment. Because of they can easily pay for almost anything with e-wallets, flight tickets, clothes, including groceries, gadgets and many other things However, who never uses are from a rural area who does not know about the e-wallet payment system (Fig. 11; Table 7).

Table 7, shows the frequency of respondents using electronic wallet payments. Among the 120 selected respondents, the most frequent e-wallet used are boost 23 (19.2%). An increase is a separate e-wallet that enables people to make payments without the hassle of holding physical cash. It may be worn nearly anywhere, maybe even in night markets. Lazada is the second-highest position with 21 (17.5%). The

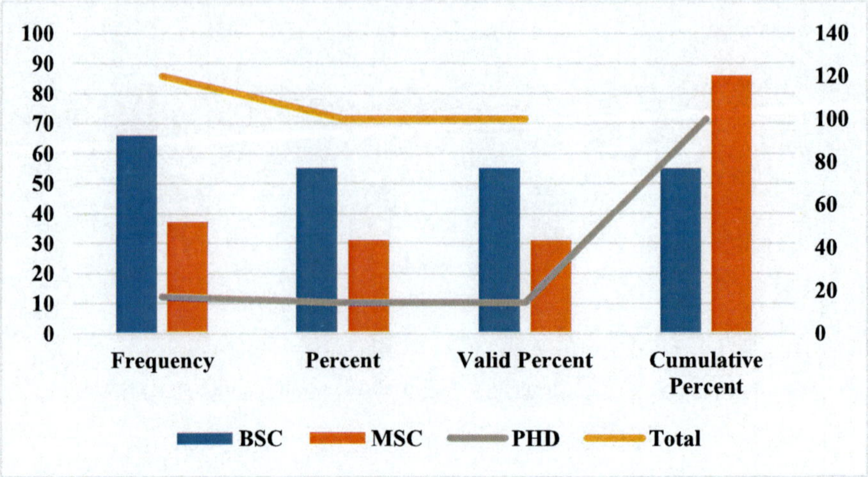

Fig. 9 The education level of the responders

Table 5 The education level of the responders

	Frequency	Percent	Valid percent	Cumulative percent	
Valid	BSC	66	55.0	55.0	55.0
	MSC	37	30.8	30.8	85.8
	PHD	17	14.2	14.2	100.0
	Total	120	100.0	100.0	

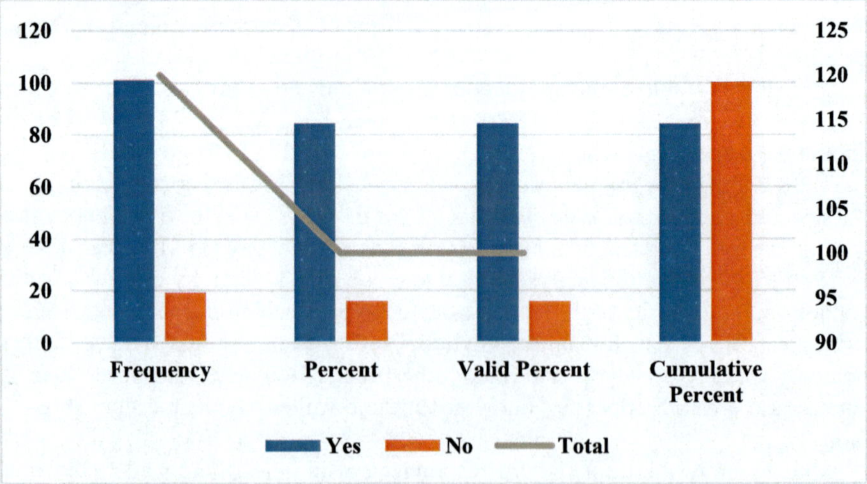

Fig. 10 Responder's response to e-wallet practice

Table 6 Responder's response to e-wallet practice

	Frequency	Percent	Valid percent	Cumulative percent	
Valid	Yes	101	84.2	84.2	84.2
	No	19	15.8	15.8	100.0
	Total	120	100.0	100.0	

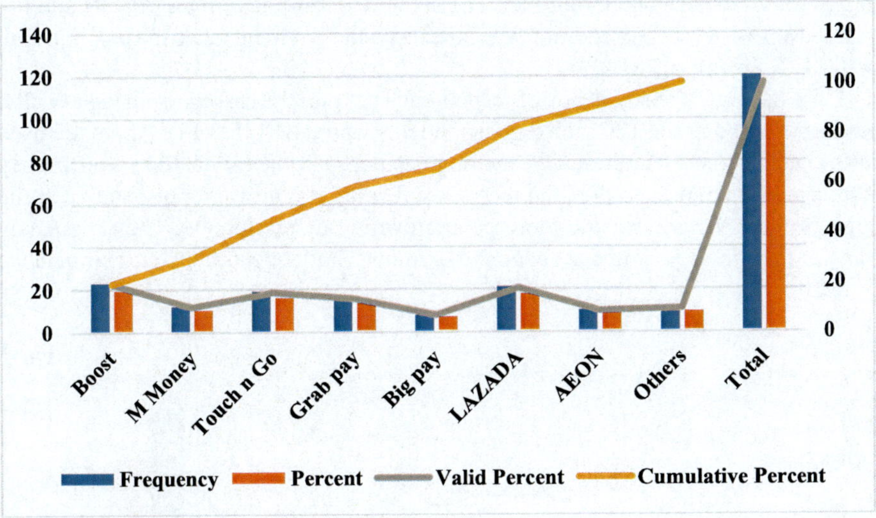

Fig. 11 Responder's response to e-wallet usage

Table 7 Responder's response to e-wallet usage

	Frequency	Percent	Valid percent	Cumulative percent	
Valid	Boost	23	19.2	19.2	19.2
	M Money	12	10.0	10.0	29.2
	Touch n Go	19	15.8	15.8	45.0
	Grab pay	16	13.3	13.3	58.3
	Big pay	8	6.7	6.7	65.0
	LAZADA	21	17.5	17.5	82.5
	AEON	10	8.3	8.3	90.8
	Others	11	9.2	9.2	100.0
	Total	120	100.0	100.0	

reason for utilizing Lazada is, it is the recognized e-commerce side in Malaysia and student can purchase anything because of this side furthermore it provides strong customer care. They also provide a bonus voucher every week. Touch 'n Go is the third position with 19 (15.8%), It allows seamless payments for the RFID toll system and it doesn't need an RFID tag, the user can pay PayDirect which allows deducting toll charges from the wallet instead of using the physical card's balance. Students also pay in different types of public transport. Grab pay is the fourth position with 16 (13.3%). Grab pay is yet another most widely used e-wallet for a drive and in addition provide food. Because a student can readily reserve a ride for going away, provide a meal online. A student could top up their Grab e-wallet and begin having to pay for Grab trips and GrabFood. The six and seven-position are AEON and big Pay 10 (8.3%), 8 (6.7). In addition, the other e-wallet payment takes the position fifth with 11 (9.2%) (Fig. 12).

Table 8, shows the frequency of respondents beyond the reason for using e-wallet payments. Among the 120 selected respondents, about 31 (25.8%) of them are used for transfer money. Traditionally, sending cash abroad meant handing over money at an agent location, and then the receiver collected it out of a representative at the opposite end. With more and more people owning a smartphone. They can easily transfer their money from one account to another. Student can transfer their balance

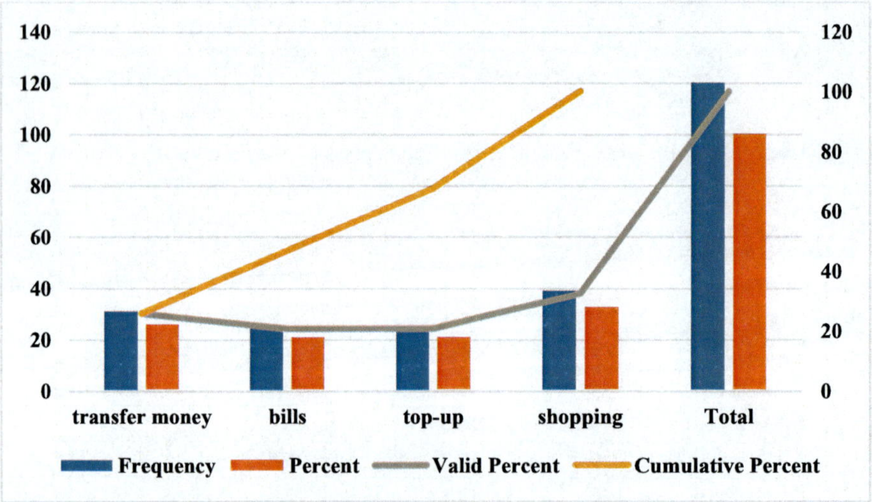

Fig. 12 Responder's Response for utilizing the e-wallet

Table 8 Response to utilizing the purpose of e-wallet

Reason	Frequency	Percent	Valid percent	Cumulative percent	
Valid	Transfer money	31	25.8	25.8	25.8
	Bills	25	20.8	20.8	46.7
	Top-up	25	20.8	20.8	67.5
	Shopping	39	32.5	32.5	100.0
	Total	120	100.0	100.0	

easily anywhere where they need. Shopping which the first place in the group with 39 (32.5). Students like shopping online from all around the world flock to the web making purchases. It has turned into a much simpler method to shop considering it does not even leave the house. The most apparent advantage of online shopping is the reality that it enables individuals to stay away from long lines and crowds. They could look for precisely what they would like. It is simple to compare reviews and prices. The following respondents who using e-wallet regarding bills payment and top-up money are the same values with 25 (20.8%), 20 (20.8%).

5 Discussion

It has been analyzed that E-wallet payments have become more aware and responsive to digital payments and Malaysia is contributing to some or alternative paths around the growth and achievement of virtual creation. The achievement of the following survey is; the maximum (63.3%) of the respondents are male and (36.7%). The majority (50.8%) of the respondent's age is between 20–25 years old. The majority (63.3%) of the responder's material status is single. The majority (55.0%) of the respondents were from the bachelor degrees, whereas the next maximum (30.8%) from the master's degree. Among all the respondents maximum (84.2%) were having-wallet applications where (15.8%) do not use electronic wallets. The majority (85.0%) of the respondents are having practice always were (15.0%) of the respondents were using electronic wallets occasionally. The majority (19.2%) of the respondents were using a boost e-wallet. Most of the participants assumed that e-wallet protected their time and (32.5%) of the respondents using e-wallet in order to online shopping. The maximum (25.8%) of the respondents were using the electronic wallet for transferring money. (20.8%) of the responders use the electronic wallet for bills and mobile top-up. The following recommendations may be useful for electronic wallet services and for promoting their services among the supplying customers. Recognition must be produced among the student because (15.8%) do not because they have come for the rural area, recognition was the main likely reason behind non-practicing E-wallet. Advertisement must be created within the social media networks that will record people that are young to enter the use. They must make a discount and offer apparent as advertising resources for all the students (Fig. 13).

6 Conclusion

E-wallet is now a major component of the electronic payment industry. It provides users with greater flexibility to cover the financial activities of theirs and make buying in obsolete locations and anytime on the day. Concisely, this study examined E-wallet service among UKM students within the Bangi campus to recognize

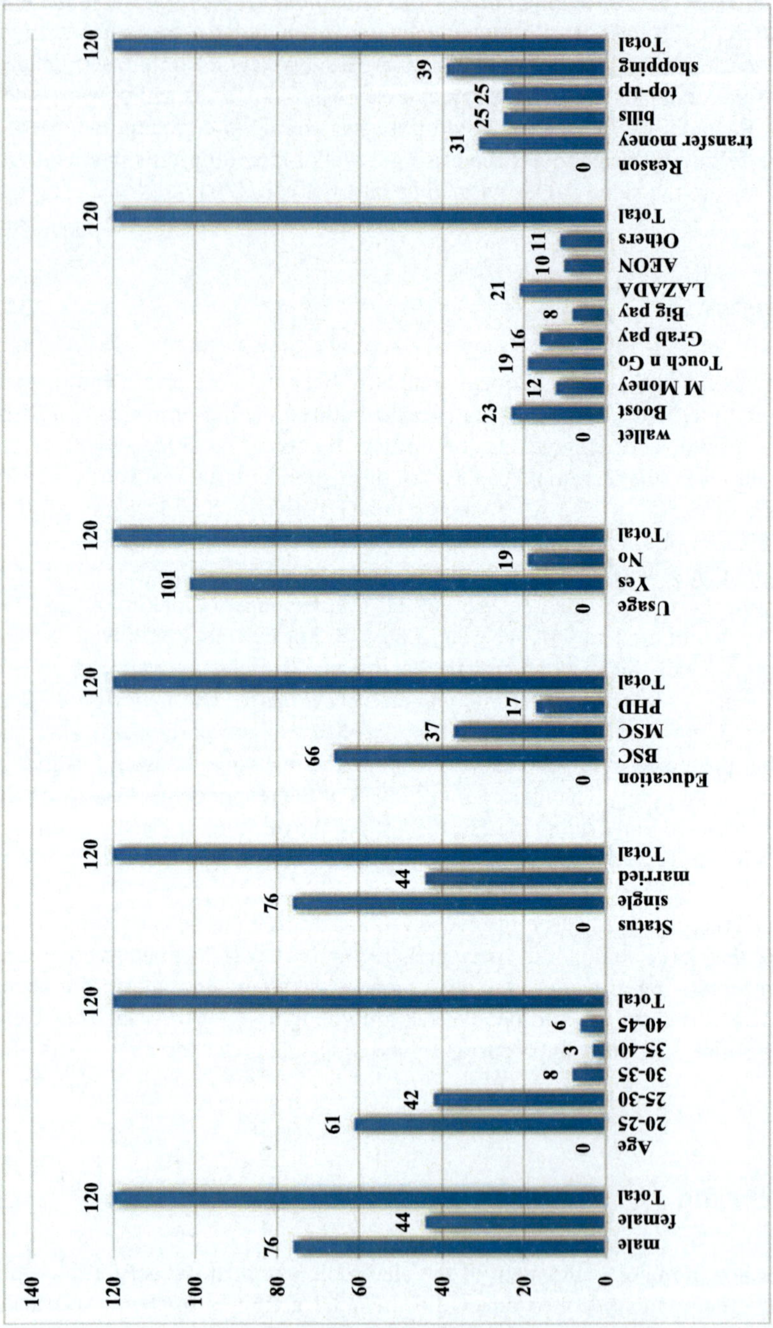

Fig. 13 Overvier of the responder's response

the various aspects of E-wallet services. The findings of the research work facilitated the overview study of the E-wallet adoption and practice of electronic payments among students. The primary goal of this paper is usually to determine the different level of use and learn the best E-wallet and what the reason for practicing. Moreover, the financial institutions should offer improved hi-tech safety methods and well advertisements to adopted electronic payments among consumers that they can more enjoy adopting E-wallet. This study covers select students of Malaysian public universities and was purely based on primary and secondary data. We also limited the sample size for the study was also to 120 students. We might perform future studies with a larger sample size so that outcomes might be widespread.

Acknowledgements The authors would like to thank the anonymous reviewers for their helpful feedback. This research was funded by a research grant code from Ya-Tas Ismail - University Kebangsan Malaysia EP-2018-012.

Appendix A

NO	Issuers (Non-Banks)	Wallet name	NO	Issuers (Non-Banks)	Wallet name
1	AEON	AEON e-wallet	19	MOL AccessPortal	Razer Pay e-wallet
2	Alipay Malaysia	Alipay	20	MRuncit Commerce	MCash e-wallet
3	Axiata Digital eCode	Boost e-wallet	21	Numoni DFS	NAPP- Money Transfer App
4	Bandar Utama City Centre	Onecard	22	PayPal Pte.Ltd.	PayPal
5	BigPay alaysia	BigPay e-wallet	23	Petron Feul Interternational	Petron Fleet Card
6	Celcom eCommerce	N/A	24	Presto Pay	Presto e-wallet
7	DIV Services	Card for School	25	Raffcomm.	HotWallet e-wallet
8	Fass Payment Solutions	Fasspay e-wallet	26	Shell Malaysia Trading	Shell Fleet Card
9	Fiexus Cards	Kedah Pay	27	SiliconNet Technologies	Sarawak Pay e-wallet
10	Fullrich Malaysia	TaPay e-wallet	28	Silverlake Global Payments	UTAR Silver Card
11	GPay Network (M)	GrabPay e-wallet	29	SMJ Teratar	eWang e-wallet

(continued)

(continued)

NO	Issuers (Non-Banks)	Wallet name	NO	Issuers (Non-Banks)	Wallet name
12	IPay88 (M)	White label e-wallet	30	Touch 'n Go	Touch 'n Go Card
13	I-serve Payment Gateway	Upay	31	TNG Digital	Touch 'n Go eWallet
14	JuruQuest Consulting	QBpay e-wallet	32	Valyou	Vcash e-wallet
15	ManagePay Services	Maxis Pay	33	WeChat Pay Malaysia	WeChat Pay e-wallet
16	Merchantrade Asia	Merchantrade Money e-wallet	34	Webonline Dot Com	KiplePay e-wallet
17	Mobile Money International	Mobile Money	35	XOX Com	MyXOX
18	Mobility One	One Pay	36	qBayar	qBayar
			37	Valyou	Valyou

No	Banks	Products
1	AmBank (M) Berhad	Prepaid Card (MasterCard)
2	Bank of China (M) Berhad	Prepaid Card (China Union Pay)
3	CIMB Bank Berhad	Prepaid Card (MasterCard) CIMB Pay
4	Malayan Banking Berhad	QR Pay
5	RHB Bank Berhad	Prepaid Card (Visa)

References

1. Yusoff H, Alomari MA, Latiff NAA, Alomari MS (2019) Effect of e-commerce platforms towards increasing merchant's income in Malaysia. Int J Adv Comput Sci Appl 10:466–470. https://doi.org/10.14569/ijacsa.2019.0100860
2. Alam SS, Ali MH, Omar NA, Hussain WMHW (2020) Customer satisfaction in online shopping in growing markets: an empirical study. Int J Asian Bus Inf Manag 11:78–91. https://doi.org/10.4018/IJABIM.2020010105
3. Faiesal Z, Mohamad B, Hamzah A, Abdul B (2018) En hancing e-commerce to sustain economy in Malaysia. Short Review 4750:718–724. https://doi.org/10.5281/zenodo.1493057
4. Kabir MA, Saidin SZ, Ahmi A (2015) Adoption of e-payment systems : a review of literature. In: Proceeding of Conference on E-Commerce, pp 112–120
5. Singh PR (2013) Issues and challenges of electronic payment systems. Int J Res Manag Pharmacy(IJRMP) 2:25–30
6. Aigbe P, Akpojaro J (2014) Analysis of security issues in electronic payment systems. Int J Comput Appl 108:10–14. https://doi.org/10.5120/18946-9993

7. Masihuddin M, Islam Khan BU, Islam Mattoo MMU, Olanrewaju RF (2017) A survey on e-payment systems: elements, adoption, architecture, challenges and security concepts. Indian J Sci Technol 10:1–19. https://doi.org/10.17485/ijst/2017/v10i19/113930
8. Balan RK, Ramasubbu N, Tayi GK, Digital wallet : requirements and challenges 2 the Singapore factor. Challenges 1–6
9. Karim NA, Shukur Z (2015) Review of user authentication methods in online examination. Asian J Inf Technol 14:166–175. https://doi.org/10.3923/ajit.2015.166-175
10. Investopedia digital wallet available online: https://www.investopedia.com/terms/d/digital-wallet.asp. Accessed on Dec 7 2018
11. Online TS, Towards a cashless Malaysia available online: https://www.thestar.com.my/news/nation/2019/10/13/towards-a-cashless-malaysia. Accessed on Feb 28 2020
12. Kadar HHB; Sameon SSB, Din MBM, Rafee P, Amirah BA (2019) Malaysia towards cashless society. Lect Notes Electr Eng 565:34–42. https://doi.org/10.1007/978-3-030-20717-5_5
13. Malaysia CB of payment systems in Malaysia available online: https://www.bnm.gov.my/index.php?ch=ps&pg=ps_mps_type&ac=177&lang=en. Accessed on May 5 2020
14. Satar NSM, Dastane O, Ma'arif MY (2019) Customer value proposition for E-Commerce: a case study approach. Int J Adv Comput Sci Appl 10:454–458 https://doi.org/10.14569/ijacsa.2019.0100259
15. Alih PM, Sarmidi T, Shaari AH, Said FF (2018) Investigating the relationship between financial innovation and money demand in Malaysia : an Ardl Approach To Co-Integration. Int J Accounting, Financ Bus 66–85
16. Tech.netonboard E-wallet growth in Malaysia available online: https://tech.netonboard.com/e-wallet-growth-in-malaysia/. Accessed on Nov 1 2019
17. Reserve TM, The best e-wallets in Malaysia, as ranked by users available online: https://themalaysianreserve.com/2019/10/29/the-best-e-wallets-in-malaysia-as-ranked-by-users/. Accessed on Dec 26 2019
18. Oppotus navigating the e-wallet landscape of Malaysial oppotus available online: https://www.oppotus.com/e-wallet-landscape-malaysia/. accessed on Apr 19, 2020
19. Teoh Teng Tenk M, Hoo Chin Yew LTH (2020) E-wallet adoption: a case in Malaysia. Int J Res Commer It Manag 2:135–3
20. Amin H (2009) Mobile wallet acceptance in Sabah: an empirical analysis. Labu Bull Int Bus Financ 7:33–52
21. Karim MW, Haque A, Ulfy MA, Hossain MA, Anis MZ (2020) Factors influencing the use of e-wallet as a payment method among Malaysian young adults. J Int Bus Manag 3:1–11. https://doi.org/10.37227/jibm-2020-2-21
22. Noor Ardiansah M, Chariri A, Rahardja S (2020) Udin the effect of electronic payments security on e-commerce consumer perception: An extended model of technology acceptance. Manag Sci Lett 10:1473–1480. https://doi.org/10.5267/j.msl.2019.12.020
23. Isaac JT, Sherali Z (2014) Secure mobile payment systems. J Enterp Inf Manag 22:317–345. http://doi.ieeecomputersociety.org/10.1109/MITP.2014.40
24. Alam SS, Islam MR, Mokhbul ZKM, Makmor NB (2018) Factors affecting intention to use online dating sites in Malaysia. Int J Eng Technol 7:192–198. https://doi.org/10.14419/ijet.v7i4.28.22578
25. Kaur R, Li Y, Iqbal J, Gonzalez H, Stakhanova N (2018) A security assessment of HCE-NFC enabled e-wallet banking android apps. Proc Int Comput Softw Appl Conf 2:492–497. https://doi.org/10.1109/compsac.2018.10282
26. Shetty S, Shetty T, Amle R (2014) QR-Code based digital wallet. Int J Adv Res Comput Sci 5:105–110. https://doi.org/10.26483/ijarcs.v5i7.2264
27. Argimbayeva G, Menasria I, Elshareif E, Lewis A (2020) The impact of e-wallet on ADNOC' s customer satisfaction. IC4E 364–368
28. Nizam F, Hwang HJ, Valaei N (2019) Measuring the effectiveness of e-wallet in malaysia. Springer International Publishing, 2019; vol 786. ISBN 978-3-319-96802-5
29. Nayana N, KPV (2019) Choosing e-wallets is compatible ? 8:103–113

30. Saunders Scientific research Available online: https://www.scirp.org/(S(czeh2tfqyw2orz553k1w0r45))/reference/ReferencesPapers.aspx?ReferenceID=1571834. Accessed on Nov 12 2019
31. Blaikie N (2019) Analyzing quantitative data. From description to explanation. Sage Publications, Thousand Oaks, 353. References—Scientific Research Publishing Available online: https://www.scirp.org/(S(vtj3fa45qm1ean45vvffcz55))/reference/ReferencesPapers.aspx?ReferenceID=2001417. Accessed on Nov 12 2019
32. Collis H (2014) A Practical Guide for Undergraduate and Postgraduate Students. Qual Mark Res An Int J 5:12–14. https://doi.org/10.1108/qmr.2002.21605daa.001
33. Future (2019) Market research E-wallet market research report—global forecast to 2023
34. Batra R, Kalra N (2016) Are digital wallets the new currency? vol 11
35. Olsen M, Hedman J, Vatrapu R (2011) E-wallet properties. In: Proceedings of 2011 10th International conference on mobile business, ICMB 2011, 158–165. https://doi.org/10.1109/icmb.2011.48
36. Hassan MA, Shukur Z (2019) Review of digital wallet requirements. In: 2019 International Conference on cybersecurity, ICoCSec 2019 2019, 43–48. https://doi.org/10.1109/icocsec47621.2019.8970996
37. Singh J (2017) Offline transactions functionality in ewallets. Indian J Sci Technol 10:1–4. https://doi.org/10.17485/ijst/2017/v10i16/110781
38. Akpan, AG, Mmeah S, Polytechnic KS, Baah B (2018) E-Wallet system implementation : impact on small scale business in Nigeria. Int J Adv Res Comput Sci Softw Eng
39. Nair A, Dahiya M, Gupta N (2016) Educating consumers about digital wallets. Int J Res. Available 2016 03
40. Lee ZW, Daniel KPT (2018) Transforming mobile phones into e-wallets in Malaysia. Bank Negara Malaysia, pp 36–43
41. Alhothaily A, Alrawais A, Song T, Lin B, Cheng X (2017) Quickcash: secure transfer payment systems. Sensors (Switzerland) 17:1–20. https://doi.org/10.3390/s17061376
42. Gazi F (2019) Most popular e-wallet comparison Available online: https://www.imoney.my/articles/choosing-e-wallet. Accessed on Dec 26 2019
43. Boost BoostTM eWallet| Malaysia's Homegrown Lifestyle Mobile Payment App Available online: https://www.myboost.com.my/. Accessed on Dec 25 2019
44. Merchantrade Merchantrade Money| How It Works Available online: https://www.merchantrademoney.com/how-it-works/. Accessed on Dec 26 2019
45. Digital T (2019) Home—TNG digital—Touch'n Go Ewallet Available online: https://www.tngdigital.com.my/. Accessed on Dec 26 2019
46. Oh A (2019) The e-wallet infinity war in Malaysia—everything you need to know about e-wallet starts here Available online: https://www.ecinsider.my/2018/12/malaysia-ewallet-battle-landscape-analysis.html. Accessed on Dec 26 2019
47. GrabPay GrabPay—mobile wallet payment solution Available online: https://www.grab.com/my/pay/. Accessed on Dec 26 2019
48. Vcash vcash—Malaysian eWallet App| pay with your mobile phone available online: https://vcash.my/. Accessed on Dec 26 2019
49. Team G (2019) 8 Popular E-wallets trending in malaysia right now| GoBear Malaysia Available online: https://www.gobear.com/my/blog/8-popular-e-wallets-trending-in-malaysia-right-now. Accessed on Oct 28 2019
50. Alipay-Trust makes it simple Available online: https://intl.alipay.com/. Accessed on Dec 26 2019
51. Hasan MK, Ismail AF, Islam S, Hashim W, Ahmed MM, Memon I (2019) A novel HGBBDSA-CTI approach for subcarrier allocation in heterogeneous network. Telecommun Syst 70(2):245–262

Implementation of Violence Detection System using Soft Computing Approach

Snehil G. Jaiswal and Sharad W. Mohod

Abstract Numerous techniques have been evolved for the detection of violence in human beings. Prior detection of human action can help to prevent and control suspicious and criminal activities. The offline video processing system has been used for post-action analysis. We address the violence detection trouble of humans in real-time visual surveillance such as punching, fighting. The present research work proposes a novel framework that processes real-time video data received from fixed cameras installed area of interest under surveillance. To determine the security level, we developed a new algorithm based on the decision-making classifier to recognize the violent situation in real time. In the view of human violence detection, the proposed work is simple and unique. The transition effects observed during violence detection are deliberated in detail. It has wide applications in the area of visual indexing, biometrics, telehealth, and human–computer interaction.

Keywords Action recognition · Surveillance · Computer vision · Violence detection · Feature extraction · Histogram · Surveillance camera · Classifier · Human–computer interaction

1 Introduction

Action recognition of human beings is one of the most demanding applications in surveillance system. Violence recognition is not merely to determine the various action performed by the human body but it is the prediction of human thoughts and intention. Violence detection is as the detection of violence by using a computer system that analyzes the video sequence to determine violence present in the video.

S. G. Jaiswal (✉)
Research Scholar, Sant Gadge Baba Amravati University, Amravati, India
e-mail: Mr.Snehil.Jaiswal@gmail.com

S. W. Mohod
Professor, Department of Electronics & Telecommunication, Prof Ram Meghe Institute of Technology & Research, Amravati, India
e-mail: sharadmohod@rediffmail.com

A. Khanna et al. (eds.), *Data Analytics and Management*, Lecture Notes on Data Engineering and Communications Technologies 54,
https://doi.org/10.1007/978-981-15-8335-3_56

Since the past decade machine learning and human action recognition is a complex and challenging area that has received much attention in the computer vision communities. The closed-circuit television (CCTV) cameras almost easily exist in most of the places. A visual surveillance system (VSS) has a wide range of security applications used for surveillance that may need to monitored. VSS consists of multiple cameras to capture video frames with human operator monitoring the frames to identify and track the suspicious frames from the video, but it is difficult to capture effective and discriminative features as a result of variation of human body.

The most challenge that occurs in the field of violence detection is that these events are rare and found infrequently. The aim of the present research work is to design an intelligent surveillance system that is able to learn what normal behavior is and is able to distinguish between normal and abnormal activities or action performed by the human body [1–6].

Mohannad et al. [1] proposed a method to detect the suspicious behavior of people by obtaining 3D object-level information present in the video sequence by using blob matching technique in real time. Depends upon the chronological properties of this technique, violent scenes are identified by employing object movement features. Chen et al. [2] proposed an additional feature to detect violence with the aid of bloody frame detection. The task of bloody frame detection was determined with the help of defining blood color cluster decision boundaries for RGB color space but these techniques are not suited for violence detection in real-time videos. The soft computing techniques can also be efficiently applied to usability prediction [7, 8], identification systems [9], and file synchronization [10].

The complexities occurred above should be solved in order to identify violence detection in multimedia content. The generalized model of violence detection system is as shown in Fig 1.

The paper proposes a novel algorithm that can detect the violent scene in a real-time video sequence to prevent violent action in spite of the number of human

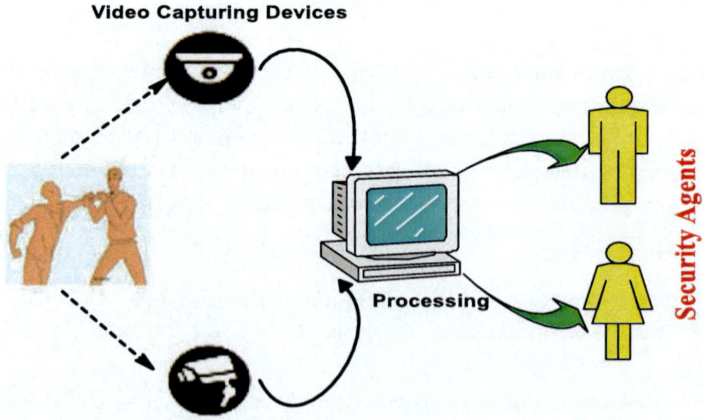

Fig. 1 Generalized model of violence detection system

being concerned. The rest of the paper is divided into four main sections. Section 2 discusses the research methodology based on the literature review. Section 3 presents the proposed technique used to detect the violent scene. In Section 4, experimental results in view of transition effects are discussed in detail. Section 5 in the end summarizes the research work following the conclusion.

2 Video Surveillance

In video surveillance, different techniques are used for human action recognition in the last few years. Since there is the uncertainty definition of violence, a novel violence detection framework in videos is proposed. The current violence detection framework can be classified into three different stages. On the basis of visual in videos, the problem of violence scene detection in a human being like kicking, hitting and fighting is addressed. The second stage is based on an audio aspect to identify the clear audio present in the videos. The third one is based on the synthesis of both audio and video features.

Efficient use of motion intensity, motion angle, blood analysis and shot and explosion present in the video is done for detecting violence present in video scene sequence. In addition to this hidden information support vector machine (SVM) is used to improve computer efficiency and time [4].

A visual-based approach to extract relevant visual features classified as local and global features is implemented in [5]. The local features such as position, shape, veins, color and global features such as average speed, relative position variations and the relationship between objects and background are extracted. The use of K self-organizing map to identify blood and skin pixels present in every frame and violent actions are detected using motion intensity analysis [5].

The semantic complete scene framework of video is utilized by decomposing the task as a violent sequence and wounded frame are use to extract the violent scene [6]. The motion analysis finds the point in the frame where some thing is moving to determine the magnitude and direction of motion of every point in the frame. The technique used violence detection features like average motion intensity, camera motion ratio, shot cut off frequency and are fed to the classifier. Figure 2 shows the detection of motion intensity feature.

A hierarchical approach based on the Gaussian mixture model and hidden Markov model to identify violence by recognizing gunshot, car-braking and explosion in audio is mentioned [11]. The method uses frame-level audio approach as input to SVM classifier together as of time and incidence field, which results into classification of violent scene from a normal one in video [12]. Zaigham Zaheer et al. mentioned a machine learning approach based on audio identification. The input to the system is MFCC features after interpolation. By the use of self-recorded scream datasets

Fig. 2 Detection of motion intensity

and with forbidden premeditated parameters, the proposed system achieved 100% accuracy [13].

Nam et al.[14] proposed a novel technique for violence identification in visual sequences based on flame and blood pixels and imprisoning the degree of action and signal sounds of violent videos. Figure 3 shows the video frames containing blood. Based on audio-video information [15] proposed a technique for detection of violence in movie scenes that utilize a dataset of average motion, acoustic characteristics and action orientation characteristics in visual sequence combined in a K-nearest neighbor classifier to detect the video is violent or not Gong et al. [15]. Proposed an action detection technique using high-level audio effects, low-level visual and auditory features identifying potential violence scenes in movies.

Thi Thi Zin et al. proposed a novel framework of video scene investigation for human interaction in which pretense depiction is based on the image features. This method basically uses the different poses of persons to classify the action and interaction present in a scene video. Punching, kicking, pushing, hugging and handshaking are the classes of interaction. Depending on the interactions again, video will be

Fig. 3 Blood frames

classified as violent and non-violent visual scenes so that safety precautions would be taken [16].

Daniel et al.[17] came up with an approach to designing a computer-aided solution for violence detection. For violence action identification, it was a content depiction technique having chronological strong features that very fast grasp the video sequences, and without human intervention classify them as violent or non-violent. The method also promises speedy and valuable classification of further identification tasks. For identification of the violence, primarily human gestures and body motion which can be extended to emotion recognition from the face and movement of hand are used in majority of the cases [18].

3 Proposed Technique

In this section, we present the methodology of soft computing approach is used for violence detection system. The proposed system working resides of three main steps: video data acquisition, feature detection and classification of human violence detection. In the initial step, the continuous data is acquired to extract the cutting edge of the object of the video sequences in each frame. In the next step, the process of extraction of features is to be done from the acquired video obtained in the initial step. The last step is to segregate features extracted from the previous stage by implementing an adequate classification application. The main flowchart of the proposed violence detection system is shown in Fig. 4.

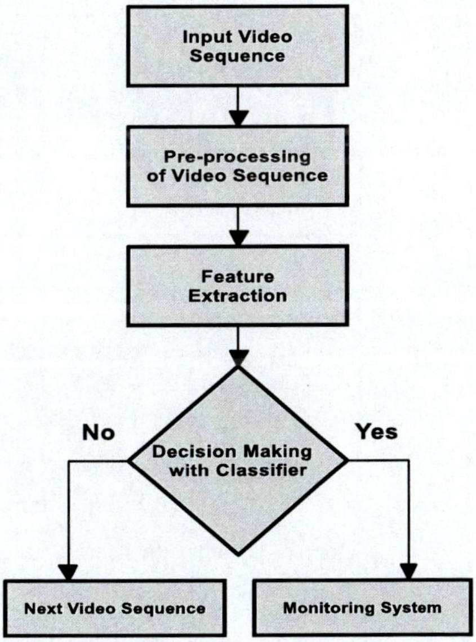

Fig. 4 Proposed model of violence detection from a video sequence

In the initial stage, continuous video raw data is acquired with the help of various closed-circuit television cameras (CCTV) fixed at the region of interest under surveillance. The analysis of actual time images for tracking and object representation in case of violent activities. The problem that occurs with these videos is that it does not have audio and another contextual source of information. Under such circumstances, the quality of the video may be degraded because of limited transmission bandwidth and low storage capacity. Thus, before the feature extraction process, preprocessing operation is required on raw video data. To extract more accurate information, preprocessing stage enhances and improves the image quality. Figure 5 describes all the step-by-step operations performed in proposed system.

The output produced by the preprocessing stage is given to the feature extraction stage as an input. The main function of this extraction stage is to fetch the primary descriptive data and make it available in an understandable form which can be treated as input by the classifier stage. The task of the feature extraction stage in the field of violence detection is in demand on account of system effectiveness, storage requirement and operating time.

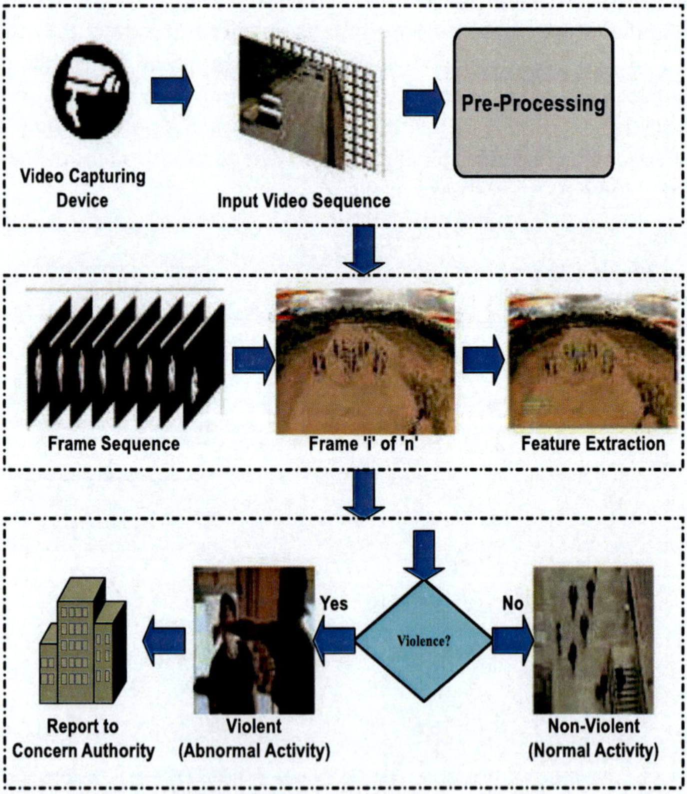

Fig. 5 Step-wise operations performed in proposed system

In the detection of violence, the extracted features of every distinguished human being in every frame are equated with the data set available by training and learning method. Training and learning framework helps to judge whether the extracted features are falls in the category of violent or non-violent video. By observing normal video sequences, a minimum threshold can be determined to differentiate violent videos from normal videos using a suitable classifier. When violence is detected, the violence detection system automatically sends an alert SMS to the respective authority so that preventive steps can be taken.

4 Experimental Results

To evaluate the efficiency and accuracy of the proposed model extracting the features, the following images are taken from different video sequences filmed between a group of persons as shown in Fig 6.

Such a suspicious frame of a video scene is then processed for feature extraction as shown in Fig 7.

The result of violence detection in the video taken for the experiment is provided in Table 1. The features like cut, dissolve, fade and wipe effects are used to prove the results more clearly.

Fig. 6 The suspicious scene

Fig. 7 The suspicious scene with feature extraction

Table 1 The result of violence detection

	Present	Correctly detected	Missed detected	Error detected	Recall rate	Precision rate
Cut	45	38	7	8	95	84.44
Dissolve	NIL	NIL	NIL	NIL	NIL	NIL
Fade	1	1	0	1	98.9	99
Wipe	0	0	0	0	0	0

A shot is termed as sequence of frames produced during continuous camera operation and imitate a progressive action in time and space. Video editing and processing results in abrupt and gradual shot transition. A gradual change may occur over various frames and it is the result of fade ins, fade outs, dissolves or cut.

It is extremely useful to detect shots and the types of transitions used to link them. This study gives us an insight if inter-shot relationships which the very contributory for high level interpretation. The terms fade in and fade out are used to describe a transition to and from a blank image. Figure8 shows the fade in fade out effect.

Figure9 shows the cut effect which is referred to as a process of partitioning an

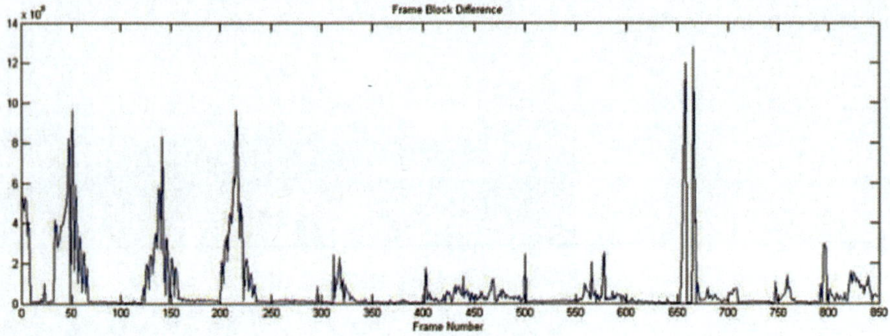

Fig. 8 Fade in fade out effect

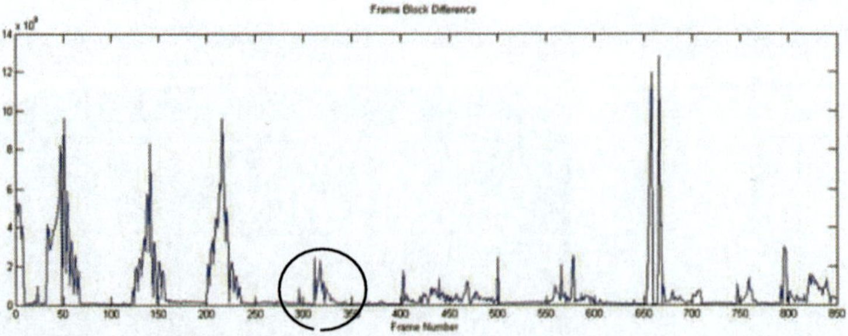

Fig. 9 Abrupt cut for violence detection

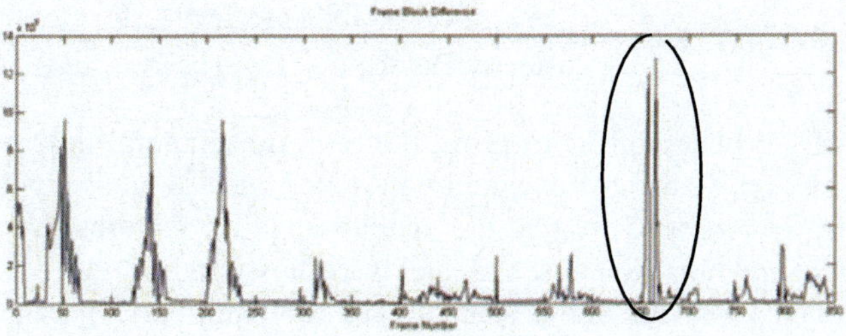

(a) Wipe effect

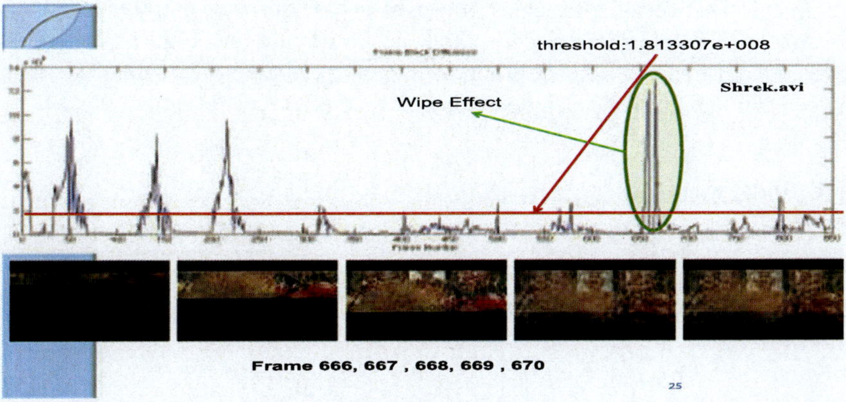

(b) Wipe effect (Shrek Trailer)

Fig. 10 Wipe effect

image into different segments. The purpose of cut effect is to simplify the image and represent it into something that is easier for analysis.

A wipe effect is a type of transition where one screen scene replaces by another by traveling from one side to another side of the frame. This effect is shown in Fig. 10a. The cut and wipe effect in the area of interest is shown by a black circle. Wipe effect for the video scène from the famous movie Sherk 3 is also shown in Fig. 10b.

The performance measures like precision rate and recall rate give the highest efficiency.

The precision rate [14] is the ratio of correct experimental detections over the number of all experimental detections as defined with equation (1).

$$\text{Precision Rate} = \frac{\text{Correctly Detected}}{(\text{Correctly Detected} + \text{Missed Detceted})} \tag{1}$$

The recall rate [8] is defined as the ratio of correct experimental detections over the number of all true detections as defined with equation (2).

$$\text{Recall Rate} = \frac{\text{Correctly Detected}}{(\text{Correctly Detected} + \text{Errorly Detected})} \quad (2)$$

The aim of the histogram is to provide long term temporal description of the surrounding motion. To create histogram flow, fields must be created within images. The histogram is proposed to measure spatial-temporal features that distinguish between violent and non-violent scenes and is as shown in Fig. 11.

Scatter plots plays a vital role in statistics and estimation as they can describe the extent of correlation, if any, between the values of quantities or phenomena under observation. The points will appear randomly scattered on coordinate plane if there is no correlation between the variables. However, in case of strong correlation between the variables, the points tend to concentrate near a straight line. Figure 10 shows the scatter plot of perceived violence in annotations and prediction. The natural clustering between annotations and predictions can be observed on scatter plot in Fig 12.

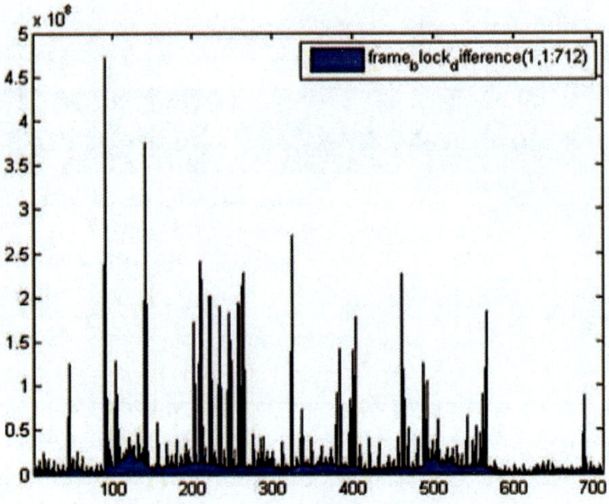

Fig. 11 Histogram of sample video

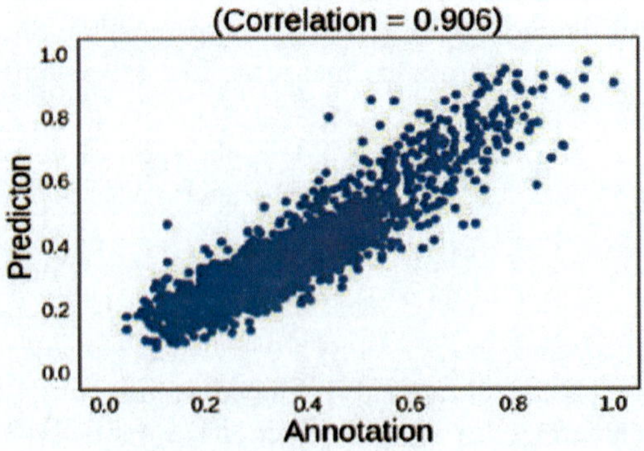

Fig. 12 Scatter plot for violence

5 Conclusion

In this paper, a novel framework is proposed that processes raw data of video streams receiving through the surveillance system. The violence scenes are detected by the process of feature extraction and decision making classifier to judge, whether the extracted features fall in the category of violent video or not. The experimental results analytically examine in contrast the distinctness between a fade, cut and wipe effect to choose precision improving. The results of the proposed experiment have several applications in the field of visual surveillance, security for indoor/outdoor environment and tagging of video content. The results also show that the proposed framework is effective in the field of violence detection.

References

1. Elhamod M, Levine MD (2013) Automated real-time detection of potentially suspicious behavior in public transport areas. IEEE Trans Intell Transport Syst 14(2):688–699. https://doi.org/10.1109/tits.2012.2228640
2. Chen L, Hsu H, Wang L, Su C (2011) Violence detection in movies. In: 2011 Eighth international conference computer graphics, imaging and visualization, Singapore, pp 119–124. https://doi.org/10.1109/cgiv.2011.14
3. Donahue JJ et al (2017) Long-Term recurrent convolutional networks for visual recognition and description. IEEE Trans Pattern Anal Mach Intell 39(4):677–691. https://doi.org/10.1109/tpami.2016.2599174
4. Ji X, Wu O, Wang C, Yang J (2014) Visual feature-based violent video detection. In: 2014 IEEE 3rd International conference on cloud computing and intelligence systems, Shenzhen, pp 619–623. https://doi.org/10.1109/ccis.2014.7175809
5. Clarin C, Dionisio J, Echavez M, Naval P (2005) DOVE: detection of movie violence using motion intensity analysis on skin and blood. Technical report, University of the Philippines
6. Chen L, Su C, Hsu H, Violent scene detection in movies. Int J Pattern Recogn Artif Intell 25:1161–1172
7. D Gupta, A Ahlawat (2017) Usability prediction of live auction using multistage fuzzy system. Int J Artif Intell Appl Smart Devices 5(1)
8. D Gupta, A Ahlawat (2016) Usability determination using multistage fuzzy system. Procedia Comput Sci. Elsevier, Scopus. https://doi.org/10.1016/j.procs.2016.02.042
9. Patnaik A, Gupta D (2010) Unique Identification System. Int J Comput Appl
10. Gupta D, Sagar K (2010) Remote file synchronization single-round algorithms. Int J Comput Appl
11. Cheng W, Chu W, Ling J (2003) Semantic context detection based on hierarchical audio models. In: Proceedings of the ACM SIGMM workshop on multimedia information retrieval, pp 109–115
12. Giannakopoulos T, Pikrakis A, Theodoridis S (2006) Violence content classification using audio features. In: Proceedings of the 4th Hellenic conference on artificial intelligence, Crete, Greece, pp 502–507
13. Zaheer MZ, Kim JY, Kim H-G, Na SY (2015) A preliminary study on deep-leaning based screaming sound detection. In: Proceedings of 5th international conference on IT convergence and security (ICITCS). IEEE, pp 1–4. https://doi.org/10.1109/icitcs.2015.7292925
14. Giannakopoulos T, Makris A, Kosmopoulos D, Perantonis S, Theodoridis S (2010) Audio-Visual fusion for detecting violent scenes in videos. In: Proceeding of 6th hellenic conference on AI, Lecture Notes in Computer Science, vol 6040. Springer, Berlin, Heidelberg, pp 91–100

15. Gong Y, Wang W, Jiang S, Huang Q, Wen Gao (2008) Detecting violent scenes in movies by auditory and visual cues. In: Proceedings of the 9th pacific rim conference on multimedia. Lecture Notes in Computer Science, vol 5353. Springer, Berlin, Heidelberg, pp 317–326
16. Zin TT, Kurohane J (2015) Visual analysis framework for two-person interaction. In: 2015 IEEE 4th global conference on consumer electronics (GCCE), Osaka, pp 519–520. https://doi.org/10.1109/gcce.2015.7398694
17. Moreira D et al (2017) Temporal robust features for violence detection. In: 2017 IEEE winter conference on applications of computer vision (WACV), Santa Rosa, CA, pp 391–399. https://doi.org/10.1109/wacv.2017.50
18. Jaiswal SG, Mohod SW (2020) Recapitulating the violence detection systems. In: Kumar A, Mozar S (eds) ICCCE 2019. Lecture Notes in Electrical Engineering, vol 570. Springer, Singapore

An Algorithm to Design a Scalable Control Layer for a Software-Defined Network

Shailender Kumar, Divtej Singh Sethi, Kanchan Kispotta, and Deepanshu Verma

Abstract Software defined networking (SDN) architecture streamlines the contemporary networks by separating the data forwarding capabilities of the data plane from the routing capabilities of the control plane that were previously carried out in the network nodes itself. Network changes in the data plane are propagated to the control plane through an interface existing between a switch and its controller and also among the controllers. Majority of the relevant research work focuses on building a control layer that aims to minimize communication delay between a switch and its controller. Such a control layer may compromise on the data loss that occurs as a result of a link break in the network. We propose an algorithm that aims to cut down the data loss as a result of a link break, resulting in a control layer which is more failure resilient. In addition to this, we consider each controller's individual capacity to handle requests thereby, assigning it as many switches as it is capable of handling.

1 Introduction

Conventionally, all the three planes of a network architecture were enforced in the firmware of the switches making the refinement of the network topology intricate. However, the segregation of the control plane [1] and the data plane in the software-defined network (SDN) [2] paradigm is what imparted flexible node administration, configurable traffic control mechanisms and extensive capabilities to the control plane.

S. Kumar · D. Singh Sethi (✉) · K. Kispotta · D. Verma
Delhi Technological University, New Delhi, Delhi 110042, India
e-mail: divtej97@gmail.com

S. Kumar
e-mail: shailenderkumar@dce.ac.in

K. Kispotta
e-mail: kanchankispotta@gmail.com

D. Verma
e-mail: singhdeep12370@gmail.com

A. Khanna et al. (eds.), *Data Analytics and Management*, Lecture Notes on Data Engineering and Communications Technologies 54,
https://doi.org/10.1007/978-981-15-8335-3_57

The SDN's control plane is an overlay network, that consists of a set of controllers, and is realized above its fundamental physical network, where each node transmits requests to the controller node by means of its very own control path. The architecture follows the classic "master-slave" concept where the master refers to the controllers and the slave refers to the nodes in the network. Every single controller therein supervises a group of nodes. The distance over the two, the controllers and the nodes, is termed as control paths employed to communicate and trade control information.

The primary function of a controller node is to set up flows over the network. The flow establishment can be proactive or reactive in nature. The switches in a proactive flow establishment are aware of the measures to be taken when the packet flow is received. On the contrary, in a reactive flow establishment, the switches acquire flows that correspond to none of the entries in the table. Thus, the switches advance the flow to the controller node to gain further information. The total time to establish a flow is undermined by the propagation delay in wireless networks.

To respond to incoming flow requests from switches i.e. to direct the incoming packet to other switches and eventually the destination, the controllers need to agree on a common network state. Any change in the data layer (for example, the addition or removal of a switch) may result in a change in one or more routing tables. Such a change needs to be propagated to the control layer. A link break prevents the particular switch from conveying any change in its sub-network to the control layer. This acts as an obstacle for the controllers leading them to agree to a faulty state in the network. Different from the previously published studies that have tried to place the multi-controllers with the aim of minimizing propagation delay in wireless networks, our work:

i. focuses on building a failure resistant control layer by minimizing data loss that occurs as a result of a link break in the controller's network of switches that it manages.
ii. constrains the propagation delay between a switch and a controller to make the response time feasible for real time events.
iii. uses the concept of a pivot node which allows us to manage the network with the least number of controllers possible.
iv. considers the network to be homogeneous in nature where every possible controller location in the network topology graph may have a different request handling capacity. This makes sure that none of the controllers is overloaded and appoints new controllers if necessary.

We evaluate our algorithm based on the following characteristics—(i) average depth of the control tree, and (ii) data loss incurred as a result of a link break. We also compare our algorithm against an already proposed k-center algorithm [16] which builds a control layer that focuses on minimizing propagation delay.

2 Related Work

The scalability of a controller is determined by its potential to deal with a resolute flow rate and the time it takes to respond to any change in the network. These determinants decide the no. of controllers essential and their deployment in the network. To address the scalability problem in SDN, the right placement of controllers is imperative to (i) reduce the communication time amongst the controllers and switches and, (ii) balance the load among the controllers to avoid network congestion. The two ultimate solutions to the controller placement problem are (i) the installation of the controller nodes and, (ii) their association to the switches in the network grid. Placement of controllers to cover a Software-defined network is referred to as Controller Placement Problem (CPP). A CPP solution can be either static or dynamic in nature. Earlier, the researchers assumed that the association among the switches and controllers never underwent any change and remained the same throughout the experiment. Later, they comprehended that a static controller placement does not ensure optimum efficiency at all times due to inconsistent network traffic for small-scale networks.

2.1 Static CPP Solutions

The controller placement problem (CPP) [3] was initiated and established by Heller et al. It explains the impact of placing the controllers across the network and examines the average latency as well as worst-case latency therein. References [4, 5] took notice of the reliability and flexibility facets of the control paths. The researchers in [6], deemed load balancing on controllers apart from propagation latency to introduce a heuristic method. Sallahi and St-Hilaire [7] proposed a statistical method that realized costs for controller installation and the association between switches. Zhoa et al. [8] put forward a variation of affinity propagation clustering algorithm [9] to find optimum controllers, and thereafter used it to create clusters and compute the number of controllers. However, a variation of the algorithm was proposed because it used Euclidean distance to assess the node similarity. Euclidean distance cannot be used in SDN since nodes are not usually accessible right away in an actual network topology. Wang et al. [10] based their approach on incorporation of various latencies (propagation latency, queuing latency, etc.) as a measure of performance for the placement of controllers. Here, a clustering method was used to divide the network into subnets so as to allocate controllers to one of these subnets on the basis of node concentration. They exercise Dijkstra's algorithm [11] to estimate the shortest route among the nodes and pick cluster centers at random so as to simplify the conjecture that the shortest route includes the latencies through and through.

2.2 Dynamic CPP Solutions

The dynamism in SDN is addressed by redistributing the flows. Flow redistribution may be acquired by per-flow redistribution, redistribution of a proportion of flows, or redistribution of all the flows of a switch. Redistributing all the flows of a switch essentially translates to relocating a switch to another controller. In [12], a proportion of the flows is dispensed to different controllers in lieu of balancing the load per flow. A multiple-mapping approach pertaining to the network traffic is used to associate a switch to several controllers. Such a method provides more resistance to a switch with respect to controller breakdown. Regardless, it's major shortcoming is that the switches have to stay informed about the flows and allocate the flows to the controllers appropriately. In [13], the authors present the problem of load oscillation where intended controllers accustomed to take some load off the overloading controllers, themselves turned out to overload rapidly. They state that it happens because it takes only the load of the switches into consideration and not the network state. Gao et al. [14] addressed the issue of traffic load imbalance in devolved controllers [15] by articulating it as an issue to balance the traffic load amidst *m* partitions.

3 Principles Involved

We elaborate on the various principles leading to the algorithm discussed.

3.1 Tree Robustness

To minimize data loss, we conclude the feasibility of a configuration in accordance with tree robustness and homogeneous index. Robustness of a network tree is related to the amount of data-loss when a link or a switch in the tree fails. Furthermore, a network tree is inherently homogeneous if the number of switches across various branches is well-balanced. Our algorithm seeks to generate a set of trees where each switch is managed by a controller node which is the root of that particular tree. The network tree presents itself with maximum data-loss when the linkage from the controller node with most nodes in relevant subtree breaks. In contrast with the k-Median and k-Center algorithms which aim to minimize propagation delay, our algorithm seeks to construct trees which have higher degree of root and the leaf nodes being not too far from the root in terms of number of edges. This results in trees with more switches closer to the controller and spread across more branches, thereby minimizing data loss when a link to the controller breaks.

Illustratively, to demonstrate that our algorithm gives better results than the two earlier suggested algorithms, we consider a network in Fig. 1 and generate trees using the three aforementioned algorithms. We observe maximum loss in the k-center

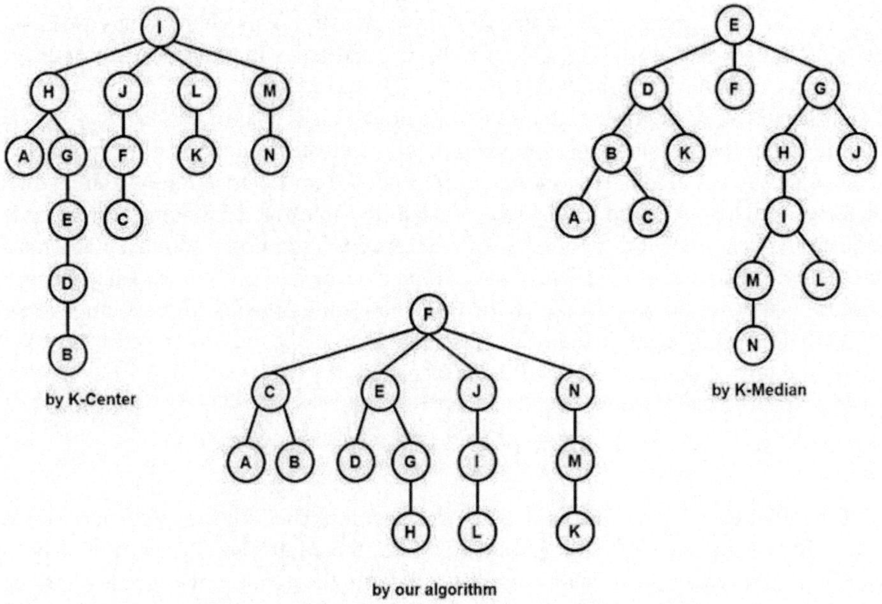

Fig. 1 Data loss comparison between our algorithm, k-center and k-median algorithm

approach where the data associated with 6 nodes (when link I–H breaks) is lost whilst in k-median it is that of 7 nodes (when link E–G breaks) and the least being in our algorithm with 4 nodes (when link F–E breaks). Therefore, this property makes the system extremely reliable and operational if any link fails.

3.2 *Request Handling Capacity of the Controller*

We assume that the possible controllers in the network are heterogeneous in nature and there is some factor associated with each node if it becomes a controller. This is termed as the request handling capacity of a node and is measured by the number of forwarding elements it can handle or packets it can address per second. A placement solution that only tries to minimize data loss in the control layer may result in a placement that is not in accordance with some controller's request handling capacity.

3.3 *Latency Requirements and Possible Controller Location*

One requirement for building a software defined network is having a low convergence time. It is associated with how quickly a change in the network is sensed by

a controller and reaches other controllers in a network. This change may result in certain changes in the routing table which may perhaps affect the shortest path for certain nodes in the network.
The first part of our algorithm deals with a mathematical quantity called T_{req}. T_{req} is the maximum permissible communication delay between a node and its controller. It is a quantity set by the network operator as per the traffic estimations and latency demands. It is constrained to make the system feasible for responding to real time requests. We estimate the "goodness" of a certain configuration through a mathematical quantity called theta (θ). We construct the control tree by exploring a node as a controller that serves as the root of the tree and establish paths to those uncovered nodes that have propagation delay less than T_{req}. Hence, the maximum delay between a root and any of its leaf nodes will be less than T_{req}.

$$\theta = \frac{\text{Deg}}{N_{\text{Total}} - N_i} + \frac{R_C}{R_{\text{Max}}} \tag{1}$$

The parameter (θ) in (1) examines the tree accomplished for the relevant possible controller on the basis of node associativity and the controller's inherent ability to handle packet requests. Node associativity relates to the number of edges leaving the root (degree) and the number of nodes the controller covers at a maximum distance (in terms of number of edges) i from it where i can be referred to as the number of hops and is decided by the network operator. Here, N_i is the number of nodes in the tree that are less than or equal to i distance from the root, Deg is the number of edges leaving the root and N_{Total} is the number of nodes in the control tree. R_C denotes the request handling capacity of controller C and R_{Max} denotes the maximum request handling capacity among all the candidates. The purpose of our algorithm is to construct trees that end near to the controller and are broad so that maximum nodes are closer to the controller. For two controllers having the same data loss for the network, we prefer the one with more request handling capacity.

3.4 Notion of Pivot Node

Out of all the nodes which are not yet managed by any controller, the node which has the maximum average distance to other uncovered nodes is elected as the pivot node, C_n. All uncovered nodes in the network which have a delay to the pivot node less than T_{req}, serve as the candidate nodes and have the ability to transform into a controller node. This idea of deciding controller locations based on the pivot node produces the least number of controllers needed to cover the entire network. Initially, deciding a controller position to manage a node other than the pivot node may still require a separate controller for the pivot node as it is farthest from others in the uncovered network.

4 Algorithm

We consider a given network graph of N_{Total} nodes. Initially, all the nodes in the network will be considered as a switch and each node in the network graph may later become a controller in the network to satisfy the condition that every switch must be part of a cluster controlled by some controller.
Considering a wireless LAN as the underlying network, propagation delay between any two nodes will be known to us. We refer to the controller's inherent limit on handling requests as R_C, measured in packets per second or by the number of switches that the controller can handle.
We would be known with R_C for each probable controller position and T_{req}. As an initial pre-computation for the algorithm, a square matrix of dimensions N_{Total} X N_{Total} is created which has the minimum propagation delay between any 2 nodes of the network.
From the set of uncontrolled switches, we find the node that is furthest to other nodes in this set and label it as pivot node (C_n). Then, figure out the candidates capable of managing C_n based on T_{req}. For each candidate controller, we construct its control tree using the shortest path to uncovered switches with propagation delay less than T_{req}, evaluate the function θ and select the controller that gives the maximum value. Then remove nodes (apart from the pivot node) from the controller if they exceed the controller's capacity. We first consider switches that are farthest from the controller in terms of number of hops. If two switches are equidistant from the root in terms of number of hops, then we remove the one with higher propagation delay to the root. This is repeated until the controller can handle all its underlying nodes in the worst case too.
The above process is carried out until each node v in the network is either a controller itself or is managed by a controller C_k with a communication delay between them $T(v, C_k)$ to be less than T_{req}.
After every node is under a controller, we try to minimize propagation delay for switches if possible. For every controller with its current capacity less than its maximum capacity, migrate switches from its original controller to the controller that gives a lesser propagation delay. This should be done only when the nodes that are in the shortest path to the particular switch from this new controller are already managed by this controller. This is repeated until there are no changes in the network.
The steps for the algorithm are as follows (considering request handling capacity in terms of number of switches that can be handled by the controller):

1. Select the pivot node i.e., the one having maximum average distance to other uncovered nodes.
2. Find the possible candidates that can act as the controller for this pivot node.
3. Select the candidate having the maximum θ as the controller. Label it C.
4. Remove nodes from C's control tree till the packets it receives per second or the number of switches that are assigned to it are more than its maximum capacity. To minimize data loss, first prefer removing switches that are more links away

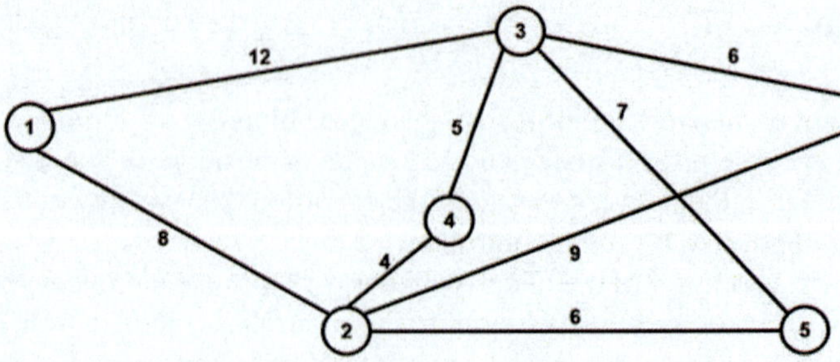

Fig. 2 Example network topology graph

from the controller. Then, among those that are equal links away, remove them in the order of decreasing propagation delay.

5. Repeat steps 1–4 till the set of uncovered nodes is not empty
6. After covering entire network, for every controller C' that still has capacity to manage a switch, check for every switch S', if the nodes that lead to the shortest path from C' to S' are present in the control tree of C' and propagation delay decreases for S' when shifted to C', then shift S' from its original controller to C'.
7. Repeat step 6 till there is no change in the network.

We demonstrate the above algorithm through an example.
Consider the network topology (edge length refers to the propagation delay in ms) shown in Fig. 2. We consider that each probable controller at a given location in the graph has a maximum capacity of handling 3 switches. We also consider T_{req} to be equal to 10 ms.

1. For the network given, as node 1 is farthest from the rest of the network (calculated using average distance to the rest of the nodes to which it has a link), it is the pivot node.
2. Candidate to become its controller is only node 2 as its propagation delay is less than T_{req}. Its control tree is shown in Fig. 3. Its

$$\theta = \frac{4}{6-4} + \frac{1}{1} = 3 \tag{2}$$

3. Since node 2 has a maximum capacity of handling three switches, we need to remove two switches from nodes 1, 4, 5, 6, 3. Since, node 3 is at two links from the root, while others are at one link distance, we remove node 3 to minimize data loss when the link 2–4 breaks. Now, another node needs to be removed from 1, 4, 5, 6. Since all the nodes are at 1 link away from the controller, we remove the one with maximum propagation delay i.e. node 6. The final control tree for node 2 is shown in Fig. 4.

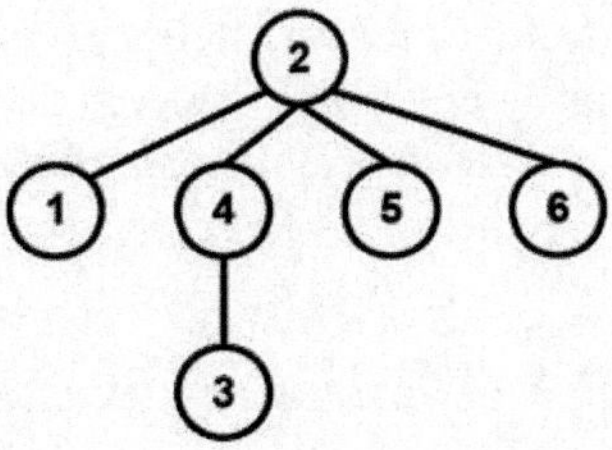

Fig. 3 Node 2 control tree

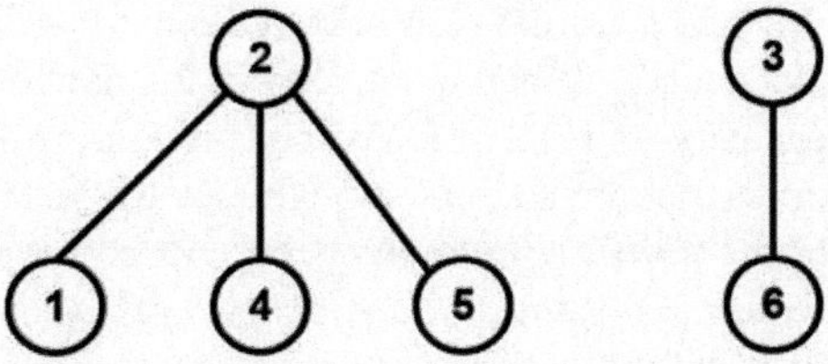

Fig. 4 Final control layer of the network

4. Now, nodes left without any cluster are nodes 3, 6. Any one of them can become the controller due to equal delay along both directions. Considering node 3 to act as a controller for node 6. Its

$$\theta = \frac{1}{6-1} + 1 = 1.2 \tag{3}$$

5. Since, node 3 can accommodate more switches, we can consider if any switch when shifted from controller 2 to controller 3 offers a lesser propagation delay. Since switches 1, 4 and 5 have lesser propagation delay to controller 2 as compared to controller 3, they are left with controller 2.
6. The final control layer thus consists of controllers 2 and 3 managing the entire network as shown in Fig. 4.

5 Simulation Results

In this section, we evaluate the performance of our algorithm and compare it with previously proposed solutions. Based on the location of all the nodes in the graph, every single node has an alternative to remain as a switch or mediate as a controller depending on the requirement of the algorithm. We assess our algorithm upon sparse, moderate and dense network grids. Networks with nodes having neighbors between 1 and 10 are considered sparse in nature, between 20 and 30 are medium in nature and between 40 and 50 are dense in nature. For comparative analysis, we generated 100 networks, each with 100 nodes for each density type. The results obtained were averaged for each density type. The code for the proposed algorithm was implemented on Intel Core i7-7500 (4 core) processor with 8GB RAM and written in C++. We analyze the results using MS Excel.

5.1 Controller's Packet Processing Equality

Preceding studies that have attempted to balance the load have taken into account that each and every controller in the network graph has equivalent request handling capacity and shift the switches dynamically based on the difference in the packet flows that are addressed by the controller. Our work takes into account the capacity of a controller in terms of the number of switches it can handle or the number of requests it can respond to per second. As a result, this modus operandi offers better distribution of switches across controllers. Majority of the network grids that we have executed our algorithm on, present us with a switch equality rate of 0.9 while the k-center and k-mean algorithms provide a switch equality rate of 0.55.

5.2 Average Depth of the Control Tree

We evaluate and compare the average depth of the control tree obtained using the k-center algorithm and our proposed algorithm.

Figure 5 shows us that the average depth of the control tree for our proposed algorithm is lower than that of the k-center algorithm. This is in accordance with the premise that we seek to construct a control layer having trees which end near to the controller with leaf nodes being not too far from the root in terms of the number of links along the path. For both the algorithms, the average depth of the tree decreases with the increase in density. This is due to the fact that dense networks being more connected will have lesser diameter. The maximum distance between any two nodes will be lesser for a dense network as compared to a sparse network for the same number of nodes.

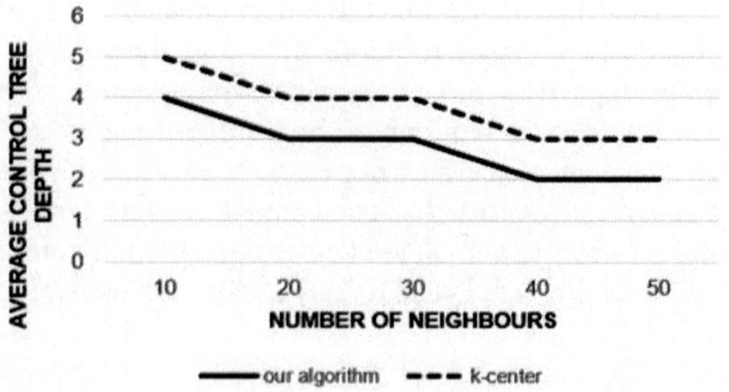

Fig. 5 Comparison of the average control tree depth between our algorithm and k-Center algorithm

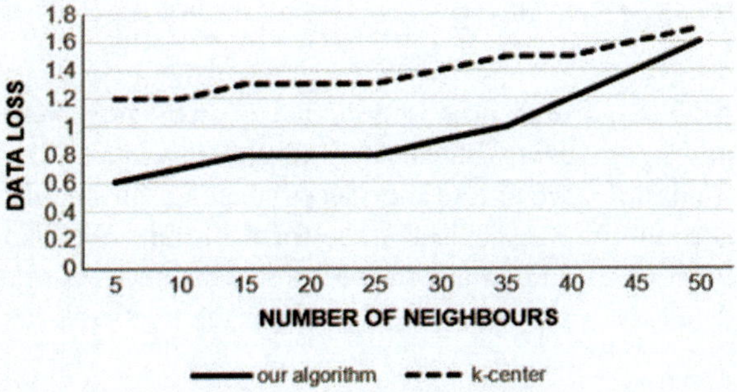

Fig. 6 Data loss comparison between our algorithm and k-Center algorithm

5.3 *Anticipated Data Loss*

For a controller managed software defined network, data loss refers to the breaking of any linkage in the network which thereby forbids the switch to convey any modifications in the network to their respective controller as it loses the connection to the former. Whenever a link fails, all the nodes in the link's subtree are unable to communicate any information to the controller. Generally, with the increase in density of networks, the data loss increases due to the closeness and high degree of the nodes in the network. The algorithm we present in this paper chooses from all the probable candidate nodes, a controller- based on the wideness and shortness of the tree created so as to minimize data loss. A switch attached to their current controller is detached to another controller only if their packet producing rate exceeds the controller's capacity to process it. We give primary importance to scaling down data-loss, contrary to the k-center and k-median algorithms which intend to minimize the propagation delay.

We compute data loss for a node as the number of nodes in the subtree divided by the total number of nodes in the network. This measure is summed over all the nodes in the network for the aggregate network data loss. Figure 6 outlines a comparative analysis of data loss in our algorithm against the k-center algorithm along increasing network density. The curve obtained for our algorithm is lower as against the k-center algorithm for different densities.

6 Conclusion

In this paper, we provide an algorithm which aims to place controllers to cover a software defined network in an efficient way. The criteria we followed was to minimize data loss giving due respect to the network topology in question. Figures 5 and 6 confirm that our initial criteria of constructing shorter control trees where the root node has a higher degree leads to a robust control layer thereby, minimizing data loss. Along with this, we also take into account the fact that each controller has an inherent limit on its ability to handle switches or process packets.

7 Future Work

In the algorithm proposed we have considered data loss as the primary factor for designing a control layer. After minimizing data loss, we minimize propagation delay if possible. We aim to introduce balancing of load dynamically in our algorithm to account for changing flow rate for the future. We also intend to improvise the theta function by adding few factors to give the network operator an option of minimizing data loss or propagation delay as per the requirements.

References

1. Yeganeh SH, Tootoonchian A, Ganjali Y (2013) On scalability of software-defined networking. IEEE Commun Mag 51(2):136–141
2. Feamster N, Rexford J, Zegura E (2014) The road to SDN: an intellectual history of programmable networks. ACM SIGCOMM Comput Commun Rev 44(2):87–98
3. Heller B, Sherwood R, McKeown N (2012) The controller placement problem. ACM SIGCOMM Comput Commun Rev 42(4):473–478
4. Guo M, Bhattacharya P (2013) Controller placement for improving resilience of software-defined networks. In: 2013 Fourth international conference on networking and distributed computing, IEEE, pp 23-27
5. Hu Y, Wang W, Gong X, Que X, Cheng S (2014) On reliability-optimized controller placement for software-defined networks. China Commun 11(2):38–54
6. Yao G, Bi J, Li Y, Guo L (2014) On the capacitated controller placement problem in software defined networks. IEEE Commun Lett 18(8):1339–1342
7. Sallahi A, St-Hilaire M (2016) Expansion model for the controller placement problem in software defined networks. IEEE Commun Lett 21(2):274–277
8. Zhao J, Qu H, Zhao J, Luan Z, Guo Y (2017) Towards controller placement problem for software-defined network using affinity propagation. Electron Lett 53(14):928–929
9. Dueck D (2009) Affinity propagation: clustering data by passing messages. University of Toronto, Toronto, p 144
10. Wang G, Zhao Y, Huang J, Wu Y (2017) An effective approach to controller placement in software defined wide area networks. IEEE Trans Netw Serv Manag 15(1):344–355
11. Johnson DB (1973) A note on Dijkstra's shortest path algorithm. J ACM (JACM) 20(3):385–388
12. Sridharan V, Gurusamy M, Truong-Huu T (2017) On multiple controller mapping in software defined networks with resilience constraints. IEEE Commun Lett 21(8):1763–1766
13. Zhou Y, Wang Y, Yu J, Ba J, Zhang S (2017) Load balancing for multiple controllers in SDN based on switches group. In 2017 19th Asia-Pacific network operations and management symposium (APNOMS), IEEE, pp 227–230
14. Gao X, Kong L, Li W, Liang W, Chen Y, Chen G (2016) Traffic load balancing schemes for devolved controllers in mega data centers. IEEE Trans Parallel Distrib Syst 28(2):572–585
15. Liang W, Gao X, Wu F, Clien G, Wei W (2014) Balancing traffic load for devolved controllers in data center networks. In: 2014 IEEE global communications conference, IEEE, pp 2258–2263
16. Khuller S, Sussmann YJ (2000) The capacitated k-center problem. SIAM J Discrete Math 13(3):403–418

Hybrid Model with Word2vector in Information Retrieval Ranking

Shweta Pandey, Iti Mathur, and Nisheeth Joshi

Abstract People have realized the importance of finding and archiving information with the computer advents for thousands of years, and storing of large amount of information became possible. It is actually not related to the fetching of the documents, it informs the user on the whereabouts and existence of the documents. In this paper, hybrid model has been used in which the document is classified using the support vector machine (SVM) classifier, and after the condition is applied, if it is satisfied, the extraction of the matched paragraph and the sentence is responsible for the generation of relevant answer. The knowledge base gets updated if condition does not match, and new updated answer will be generated. Finally, the best answer is displayed after ranking by using the PSO optimization. Word2vector is applied for feature extraction. In this paper, comparison of RankSVM, RankPSO and RankHSVM + PSO for the implementation of IR ranking is considered. Here, first SVM is used as a classifier for dividing most relevant and non-relevant results, and afterward PSO is used for the optimization of the result means extraction of the best answer or document. Selection of appropriate parameters is difficult in case of simple SVM, but for the ranking of the answers it gives potential solutions. PSO is used for optimization which has global search capability and is easy to implement and thus to optimize the ranking of document retrieval. We propose the RankHSVM + PSO model to find the fitness function. This technique improves the performance of the system as comparative to other techniques. The result shows that the algorithm applied here improves the value of performance evaluation by 4–5%. TREC 2004 QA DATA dataset is used which contains my datasets. It has a question answering track since 1999. The task was defined in each track. Retrieval of true equivalent test

S. Pandey (✉) · I. Mathur · N. Joshi
Department of Computer Science, Banasthali Vidyapith, Jaipur, India
e-mail: shweta.dubey12@gmail.com

I. Mathur
e-mail: mathur.iti@rediffmail.com

N. Joshi
e-mail: nisheeth.joshi@rediffmail.com

A. Khanna et al. (eds.), *Data Analytics and Management*, Lecture Notes on Data Engineering and Communications Technologies 54,
https://doi.org/10.1007/978-981-15-8335-3_58

collection for standard retrieval is an open problem. In a retrieval test collection, the unit that is judged the document has a unique identifier.

Keywords Information retrieval · Ranking · PSO · SVM · Machine learning

1 Introduction

Information retrieval system [1] is the system that finds the relevant information for a document, then review the content extracted and finally all the unstructured documents that is relevant to the information is collected from an achieve of information resources. Information retrieval system helps the end users to search the required information which are more relevant to the query fired by the user (Fig. 1).

Indexing: For efficient retrieval the documents crawled by the search engine are stored in an index. To understand the different processes that is implemented let's take an example from dataset:

Take a query-How long was Lincoln's legal career?

Query preprocessing
Tokenize the query:['How','long','was',"Lincoln's",'legal','career?']
Lower case:["how long was lincoln's legal career?"]
Punctuation removal:['how long was lincolns legal career']
Stop word removal: ['lincolns legal career']
Stemming: ['lincoln legal career']
Extract key words:['lincoln','legal','career']

1.1 Ranking

The basic problem of the information retrieval system is query ranking [2]. The main problems associated with the query ranking with the group of documents are: The documents are to be sorted in terms of relevant information that matches to the phrases of the query. Ranked retrieval is based on the score of statistical ranking. The

Fig. 1 Classic search model

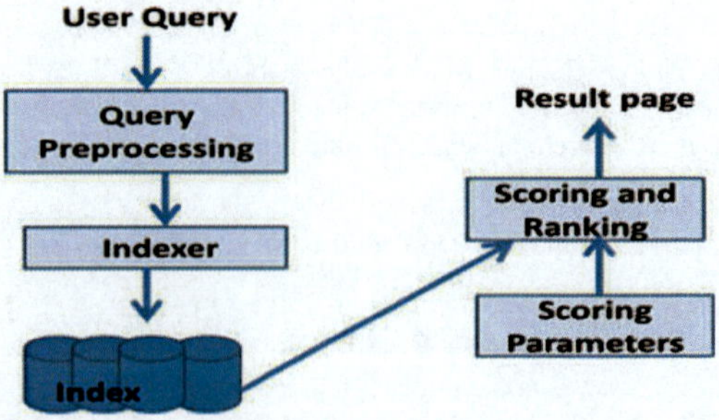

document having the highest score is assigned the first rank and like this the process continues. In ranking, the score of better result is calculated by using the weighted or simple match.

There are two types of ranking models: query ranking against individual documents and query ranking against the group of related documents. Based on these types of ranking models, the categorization of ranked models is as follows:

Boolean model: The simplest model of IR is Boolean model, that is based on set theory [3]. Partial matched documents are not recovered because it's supported binary concept, just those documents are often retrieved that matched exactly. This model requires expertise to create the queries to retrieve the more relevant document.

Vector Based Model: To overcome the matter of Boolean model that fetches the documents that are completely matched vector model [4] is introduced that is based on weight rather than binary values, tf (term frequency) and idf (inverse document frequency). The combination of tf-idf results the term weights that are approximately calculated. Together these factors are tf-idf measure.

Probabilistic Model: Probability of relevancy of documents is the base of this model [5]. In this, the probability of documents that are estimated as relevant for the query is calculated. The representation of document and query is the base of probability calculation.

1.2 Ranking-Related Problems

To extract related documents from number of documents, ranking is the main research area now days. But there are many challenges in this area as discussed below:

1. To form one ranking function, combination of all the ranking functions is extremely tough.
2. The considered ranking factors that treated as ranking function are page content, structure, link, etc. As the number of ranking factors is many, so the speed becomes the most challenging task.
3. Mean average precision (MAP) and Normalized Discounted Cumulative Gain (NDCG) are non-convex measures for the algorithms based on ranking in IR. So optimization of conventional measures becomes difficult for optimization tool.
4. Retrieval of irrelevant information that is bulk in number is the main problem in ranking.
5. With same fingerprints, encoding schemes and kinds of document are different.

1.3 Evaluation Measures

Calculation of performance or efficiency of a system, evaluation measures [6] are used, i.e., it tells about how the system reached the expectation level of the user's information. The four evaluation measures are:

Precision: It's the ratio of the retrieved documents to the relevant documents.

$$\text{P} = \text{True Positive}/\text{True Positive} + \text{False Positive} \quad (1)$$

Precision = retrieval of relevant documents/the retrieved documents that are both relevant and non-relevant

Recall: It's the ratio of relevant documents in collection to the retrieved documents.

$$\text{Recall} : \text{R} = \text{True Positive}/\text{True Positive} + \text{False Negative} \quad (2)$$

Recall = retrieval of relevant document/retrieval of relevant documents + the documents that are relevant but not retrieved)

F-measure: The weighted mean of precision and recall is calculated that results into F-measure. It is commonly denoted by F1 or F.

F1-Score:

$$\text{F1} = 2 * \text{Precision} \times \text{Recall}/\text{Precision} + \text{Recall} \quad (3)$$

Accuracy: In machine learning classification work, accuracy may be a commonly used evaluation measure, but in IR it's not a really useful measure. The rationale behind this is often that altogether the conditions the info is extremely asymmetric.

Accuracy is that the fraction of the right classifications.

$$\text{Accuracy} : \text{ACC} = \text{TruePositive} + \text{TrueNegative}/\text{TruePositive} + \text{TrueNegative} + \text{FalsePositive} + \text{FalseNegative} \quad (4)$$

1.4 PSO

Best solution can be represented in an n-dimensional space as a point or surface for dealing with problems by particle swarm optimization that is a global optimization algorithm. With their own velocity vectors particle vectors $V_j = [v\text{j}_1, v_{j2},..,v_{j\text{D}}]$ and position vector $P_j = [p_{j1}, p_{j2}, ..., p_{jD}]$ to point its current state. In iterations, consistently particle is updated with two "best" values. One is that is local called pBestj, the other is the global best called gBest.

1.5 SVM

Isolated hyperplane is the characteristic of support vector machine (SVM) and may be a discriminative classifier. Because it has labeled training data it is often defined as a supervised learning. SVM gives a perfect hyperplane which sorts new models. The hyperplane of this classifier separates a plane in two sections in two-dimensional spaces. It very well may be utilized for both binary and multiclass classification. Classifying the hypothesis of SVMs can get extremely specialized. This probably will not be the most definite classification, yet it is a decent beginning stage.

1.6 Word Embedding Using Word2Vector

The reason and helpfulness of Word2vector is to bunch the vectors of comparable words in vector space. That is, it identifies similitude scientifically. Word2vector makes vectors that are circulated numerical portrayals of word highlights, highlights, for example, the setting of individual words. It does as such without human intercession.

1.7 Objectives of Current Research

Information retrieval ranking is now the very active research area. Evaluation of IR system for cross-lingual information retrieval system is quite not feasible for this experiment. So we have restricted our experiment on monolingual IR system for English-English language. Our major objective is to improve the performance of IR system ranking. In our experiment, we have compared our IR system ranking using the RankHSVM + PSO machine learning technique with the concept of word2vector model in place of tf-idf. So, we propose to address the following objectives to remove the shortcoming of above papers:

- Extracts the features from monolingual documents.
- SVM (support vector machine) and PSO (particle swarm optimization) combination is used to implement a system which assigns the rank to the relevant information.
- Our work gives best-performance evaluation result compared with other machine learning techniques.

2 Related Works

In 2016, Kakde et al. [7] proposed a recognition system for recognizing the Devnagari script by using the combination of PSO and SVM algorithms.

In 2017, Liu et al. [8] designed an approach which includes the two steps: question selection and question diversification. They used probabilistic retrieval model for question selection. They also introduced a recurrent neural network (RNN) encoder–decoder.

A problem occurs in the feature extraction from information retrieval was addressed by Gaurav Pandey et al. [9]. The proposal of linear feature extraction was also given by them and also called as Life Rank algorithm for ranking. Kehinde Agbele et al. [10] proposed predictive ranking model by using DROPT technique to measure the individual search in their domain of knowledge for the document. They used the query context to determine the relevance of information retrieved.

Qinglan Fan et al. [11] proposed a credit scoring model based on the combination of support vector machine (SVM) and adaptive mutation partial swarm algorithm (APSO) for optimization. Initially, to simplify the credit scoring data they used principal component analysis (PCA) method and normalization method both. Secondly, they used SVM in order to search optimal parameters values, and they also presented APSO algorithm by means of introducing premature convergence indicator and adaptive mutation operator, which can solve the premature convergence in traditional PSO (Fig. 2).

3 Proposed Methodology

The proposed approach consists the following steps: Initially, the preprocessing is occurred to preprocess the raw document by stemming, removal of stop words, etc. After that the processed data has been sent to knowledge base.

In the similar manner, the preprocessing of the query will take place in which the query will be splitted to extract keywords after that word to vector conversion will occur. In the next step, the splitted query and the knowledge base is provided for clustering. The SVM classifier is used to classify the document into relevant and non-relevant documents. Among all the relevant documents, the optimization algorithm (PSO) will be applied for the more relevant answer among all the answers. If the answer is not relevant then the knowledge base will get updated and new answer will be generated.

At last with the help of PSO, the answers are ranked and best answer is displayed. PSO ranking score is generated by using fitness value. PSO fitness is,

$$F_i^d(d) = \sum_{j \in k\text{best},\, j \neq 1} r_1 \cdot F_{ij}^d(t), \tag{5}$$

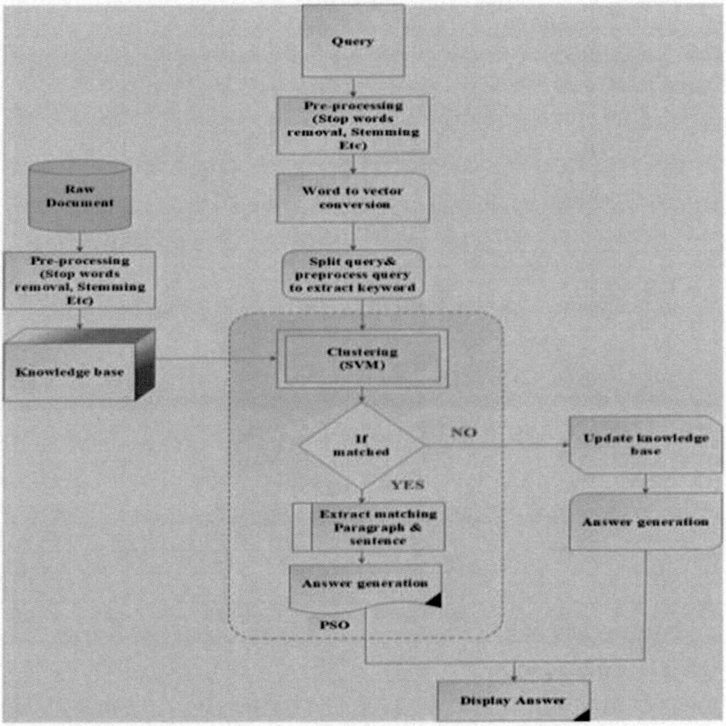

Fig. 2 Structure of hybrid model (HSVM + PSO)

where *kbest* is the initial best value, *r* is the uniform random variable in the interval [0, 1], *Fdij* is the word to vector of matching sentences. Fitness is calculated for each matching sentence. Also the sentence which has high fitness is selected as the correct answer. Finally, the score and the answer are displayed.

4 Results and Discussion

This section includes the experimental results of the proposed approach. The TREC 2004 QA DATA [12] dataset is taken for the experimental purpose, and Spider (python 3.7) is used for implementation. Let's take an example from the dataset [12] for experimental result:

Take a query-How long was Lincoln's legal career?
Tokenize the query:['How','long','was',"Lincoln's",'legal','career?']
that has been given above:-
Take a query-How long was Lincoln's legal career?
Tokenize the query: ['How','long','was',"Lincoln's",'legal','career?']

Lower case: ["how long was lincoln's legal career?"]
Punctuation removal: ['how long was lincolns legal career']
Stop word removal: ['lincolns legal career']
Stemming: ['lincoln legal career']
Extract key words: ['lincoln','legal','career']

Output or the result of the query:

```
Word to Vector for Question key word
 [[0, 0, 0, 0, 0, 0, 0, 0, 0, 0, 0, 0, 0, 0, 0, 2, 0, 0, 0, 0, 0, 0, 5, 318, 21, 1, 16, 2, 0, 0, 0, 0,
0, 0, 0, 0, 0], [0, 0, 0, 0, 1, 1, 2, 0, 0, 0, 3, 2, 0, 3, 5, 0, 2, 2, 6, 0, 4, 1, 0, 6, 1, 0, 0, 2, 0,
0, 0, 0, 0, 0, 0, 0, 0], [0, 0, 0, 0, 0, 0, 0, 0, 0, 0, 0, 0, 0, 0, 0, 0, 0, 0, 0, 0, 1, 0, 0, 4, 3, 0,
1, 0, 0, 0, 0, 0, 0, 1, 1, 0, 0]]


Matching Sentence is:

1 => ['fillmor', 'refus', 'join', 'republican', 'parti', 'whig', 'includ', 'abraham', 'lincoln',
'found', 'refuge']
2 => ['ï»¿abraham', 'lincoln', 'abraham', 'lincoln', 'februari', '12', '1809', 'â\x80\x93', 'april',
'15', '1865', 'sixteenth', 'presid', 'unit', 'states']
3 => ['dori', 'kearn', 'goodwin', 'team', 'rival', 'polit', 'geniu', 'abraham', 'lincoln', '2005', 'p']
4 => ['donald', '1995', '15051', 'lincoln', 'involv', '5100', 'case', 'illinoi', '23year', 'legal',
'career']
```

```
The Correct Answer is:
====================
['donald 1995 15051 lincoln involv 5100 case illinoi 23year legal career', 0.4799855634696781]
```

The value 0.479985 … is the best score value created by our model.

5 Experimental Evaluation

In this section, we evaluate the performance of these algorithms: RankSVM, RankPSO and RankHSVM + PSO based on single-word queries and multiple words queries is evaluated and compared. For a given user query, the learned ranking function is applied to rank the documents. The performance of these algorithms is evaluated and compared based on the categorization of questions in the dataset as easy, medium and hard questions.

The proposed algorithm [13] collects the user queries and their relevant retrieved documents. Human interpretation is used for the more significant judgment. Based on the relevancy of the answer, the ranking is assigned to each document. If answer is more relevant corresponding to the document, then it will contains the more score. The document which scores the highest ranking will be ranked at the top. For this experiment, the corpus of 1000 documents is taken, and 70% of the data has been taken for training purpose and for testing purpose, rest 30% of the data will be used. Here, a query set of 300 queries will be considered, 150 queries will be considered as single word queries and rest as multiple word queries. The proposed algorithm

Table 1 Single-term queries

Techniques/parameters	RankSVM			RankPSO			RankHSVM + PSO		
	Types of questions								
	Easy	Medium	Hard	Easy	Medium	Hard	Easy	Medium	Hard
Precision	81.33	89.34	86.66	92.0	94.67	89.33	98.66	98.67	94.66
Recall	82.43	79.76	89.04	88.46	91.02	93.05	94.87	97.36	97.26
F1-score	81.87	84.27	87.83	90.19	92.81	91.15	96.73	98.01	95.95
Accuracy	82.0	83.33	88.0	90.0	92.67	91.33	96.67	98.0	96.0

Table 2 Multi-term queries

Techniques/parameters	RankSVM			RankPSO			RankHSVM + PSO		
	Types of questions								
	Easy	Medium	Hard	Easy	Medium	Hard	Easy	Medium	Hard
Precision	86.66	92.0	88.0	93.33	96.0	90.54	97.33	97.34	94.66
Recall	80.24	80.23	89.18	90.90	93.50	93.05	94.80	98.64	97.26
F1-score	83.34	85.71	88.59	92.10	94.73	91.78	96.05	97.98	95.94
Accuracy	82.67	84.67	88.67	92.0	94.67	92.0	96.0	98.0	96.0

provides the more feasible rank of the document in both, single-word queries and multiple-word queries. It also provides the better response time within an acceptable memory cost. The comparison of the algorithms are depicted in Table 1 (for single-term queries) and Table 2 (for multiple-term queries).

The graphs are created based on the values of confusion matrix as follows:

Confusion Matrix for single-term queries:

	True Positive	False Positive
False Negative	[71]	[4]
True Negative	[1]	[74]

Confusion matrix for multiple-term queries:

	True Positive	False Positive
False Negative	[71]	[4]
True Negative	[2]	[73]

Confusion matrix is created for each type of question (easy, medium and hard) for both single-term queries and multiple-term queries.

The performance measures have been calculated based on the following:

$$\text{Recall} = \text{TP}/(\text{TP} + \text{FN}) * 100$$

It is true positive rate (TPR)

$$\text{Precision} = \text{TP}/(\text{TP} + \text{FP}) * 100$$

It is positive predictive value (PPV)

$$\text{F1score} = 2 * ((\text{precision1} * \text{recall1})/(\text{precision1} + \text{recall1}))$$

$$\text{Accuracy} = (\text{TP} + \text{TN})/(\text{TP} + \text{FP} + \text{FN} + \text{TN}) * 100$$
$$\text{Accuracy} = \text{np.around}(\text{Accuracy}, 2)$$

The graphs have been created based on the values of Tables 1 and 2 (Figs. 3, 4, 5, 6, 7, 8, 9, and 10).

Fig. 3 Recall for single-term questions

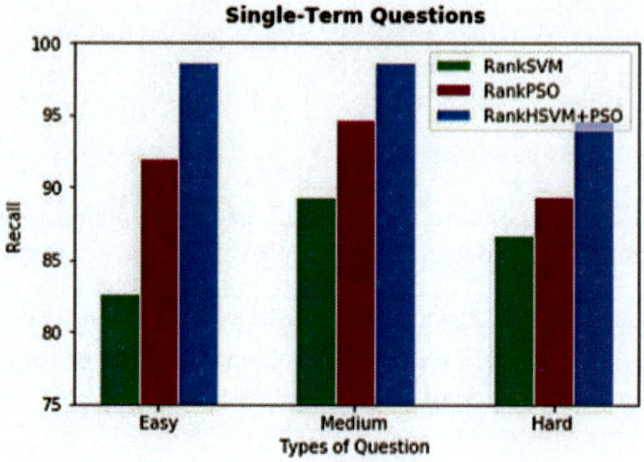

Fig. 4 Precision for single-term questions

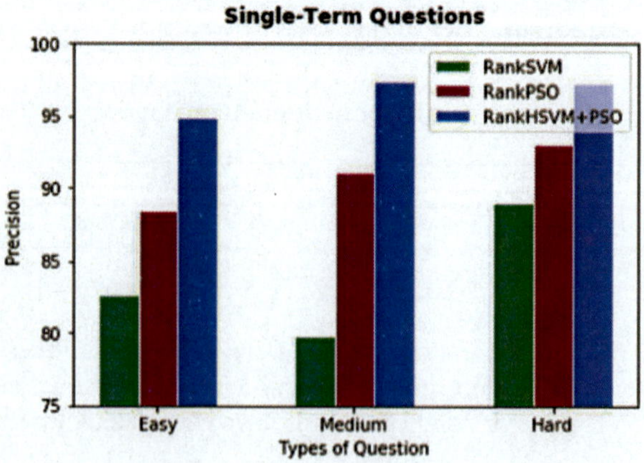

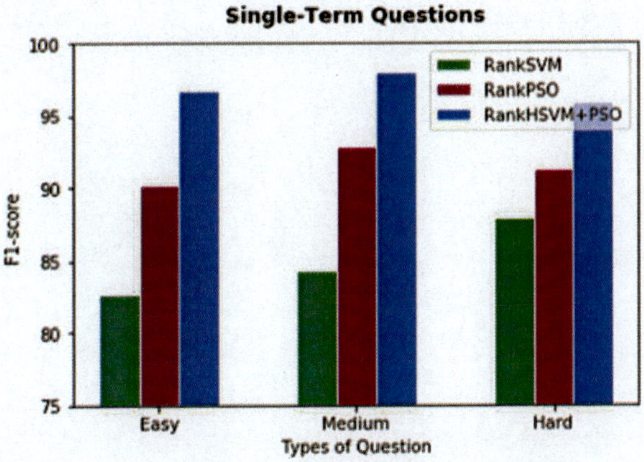

Fig. 5 F1-score for single-term questions

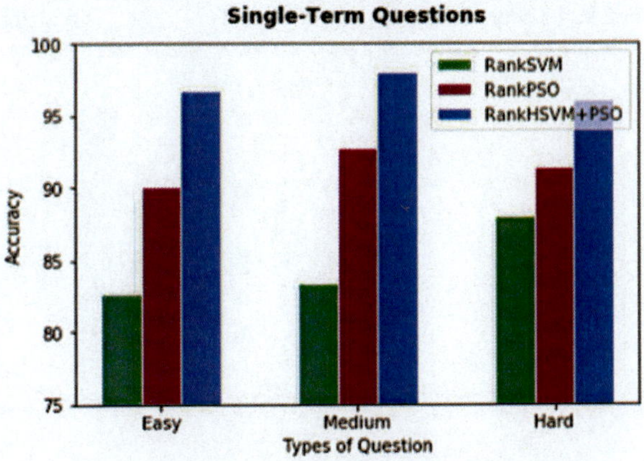

Fig. 6 Accuracy for single-term questions

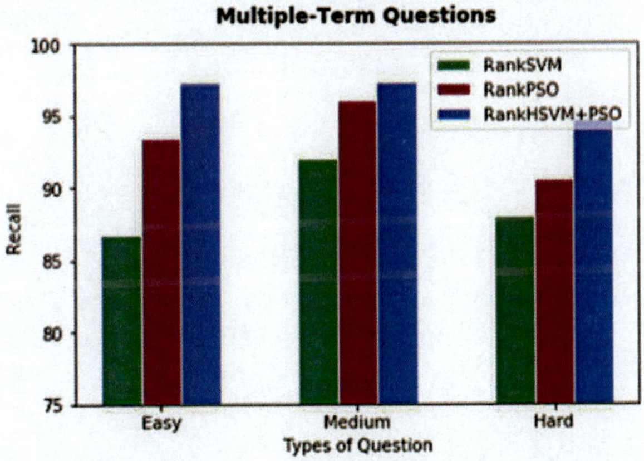

Fig. 7 Recall for multiple-term questions

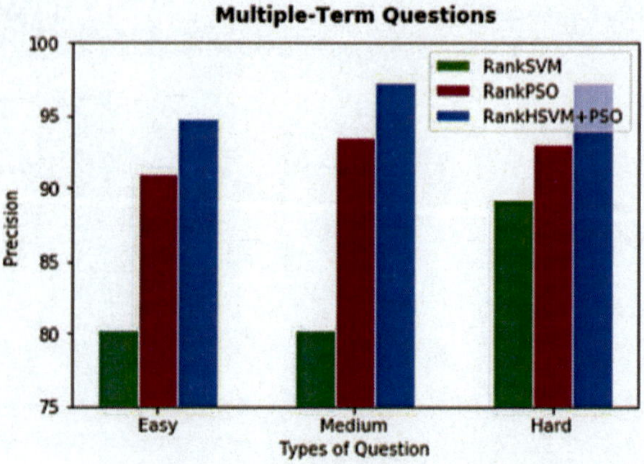

Fig. 8 Precision for multiple-term questions

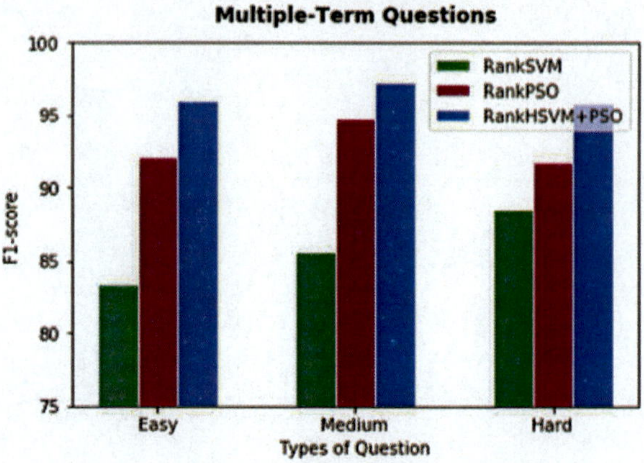

Fig. 9 F1-score for multiple-term questions

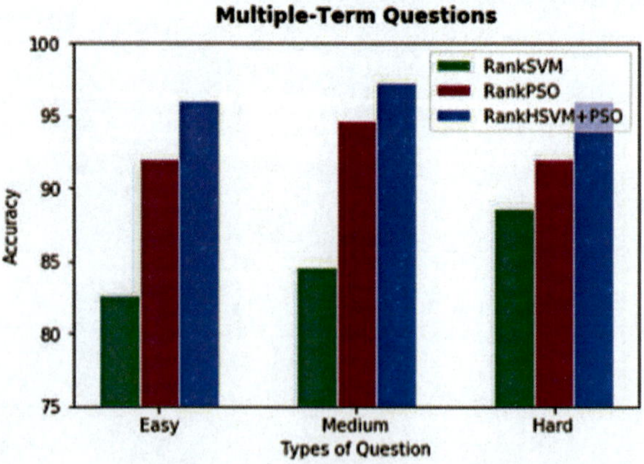

Fig. 10 Accuracy for multiple-term question

6 Conclusion and Future Scope

It is very difficult to specify the information required for the information retrieval system. Sometimes the users also do not know their need of information precisely. This system combines the PSO and SVM algorithms. It improves the performance of the existing ranking system, and it also reduces the previous problems associated with information retrieval ranking. This paper uses the SVM and PSO machine learning algorithms with word2vector as a feature extraction from the documents to contain the ranking system for monolingual only. This research can further be carried for real-time information retrieval system and for cross-lingual.

References

1. Telkapalli MK, Fanta H, Appalabatla S (2015) Trends and techniques in information retrieval system (IRS): a comparative study. Int J Adv Res Comput Sci Softw Eng
2. Cao Y et al. (2006) Adapting ranking SVM to document retrieval. In: Proceedings of the 29th annual international ACM SIGIR conference on research and development in information retrieval. ACM, 2006
3. Lee JH (1994) Properties of extended Boolean models in information retrieval. In: Proceedings of the 17th annual international ACM SIGIR conference on research and development in information retrieval. Springer, New York, Inc.
4. Lee DL, Chuang H, Seamons K (1997) Document ranking and the vector-space model. IEEE Softw 14(2):67–75
5. Jones KS, Steve W, Stephen ER (2000) A probabilistic model of information retrieval: development and comparative experiments: Part 2. Inf Process Manag 36(6):809–840
6. Kalervo J, Kekäläinen J (2000) IR evaluation methods for retrieving highly relevant documents. In: Proceedings of the 23rd annual international ACM SIGIR conference on research and development in information retrieval. ACM
7. Kakde PM, Dr. Gulhane SM (2004) A comparative analysis of particle swarm optimization and support vector machines for devnagri character recognition: an android application, Procedia Computer 1877–0509 © 2016 The Authors. Published by Elsevier B.V. This is an open access article under the CC BY-NC-ND license Peer-review under responsibility of the Organizing Committee of ICCCV 2016 https://doi.org/10.1016/j.procs.2016.03.044 Science Direct 7th International Conference on Communication, Computing and Virtualization 2016. Proceedings of the 27th annual international ACM SIGIR conference on Research and development in information retrieval, pages 64-71, 2004
8. Liu T-Y, Xu J, Qin T, Xiong W, Li H, LETOR: benchmark dataset for research on learning to rank for information retrieval
9. Pandey G, Ren Z, Wang Z, Vaijilainen J, De Rijke M (2018) Linear feature extraction for ranking. Inf Retrieval J Springer link 21(6):481–506
10. Agbele K, Ayetiran E, BabalolaO (2018) A context-adaptive ranking model for effective information retrieval system. Copyright © 2018 The Author(s). Published by Scientific & Academic Publishing
11. Fan Q, Liu X, Zhang Y, Bao F, Li S (2018) Adaptive mutation PSO based SVM model for credit scoring. CSAE '18, October 22–24, 2018, Hohhot, China © 2018 Association for Computing Machinery. ACM ISBN 978-1-4503-6512-3/18/10...$15.00 https://doi.org/10.1145/3207677.3278014
12. https://trec.nist.gov/data/qa/t2004_qadata.html
13. Pandey S, Mathur I, Joshi N (2019) Information retrieval ranking using machine learning techniques. In: Amity international conference on artificial intelligence (AICAI), IEEE

Heuristic Approach Towards COVID-19: Big Data Analytics and Classification with Natural Language Processing

Sabyasachi Mohanty, Ritika Sharma, Mohit Saxena, and Ankur Saxena

Abstract Data has tremendously incorporated our lifestyle. With advancements in technology and reduced Internet cost, data usage has increased many folds resulting in generation of huge heaps of unstructured data called as big data. This unstructured big data is difficult to handle using existing database management technology. We observed that genetic information related to coronavirus is tremendously increasing everyday. With implementation of big data analytics, these databases will be easily manageable leading to advancements in COVID-19 research. In this article, we have used HDFS system for efficient data management. In our work, we classified gene classes present in complete sequence so as to quickly detect mutation in no time. To achieve this, we predicted machine learning models to classify gene sequences faster in-class with libraries like matplotlib to construct detailed graph of the data. We choose three different sequences to classify gene sequence using natural language processing technique of Sklearn library and tested our results using logical regression.

Keywords COVID-19 · Matplotlib · Sklearn · Natural language processing, classification · Logistic regression

S. Mohanty · R. Sharma · A. Saxena (✉)
Amity University, Noida, Uttar Pradesh, India
e-mail: asaxena1@amity.edu

S. Mohanty
e-mail: sabya.mohanty12@gmail.com

R. Sharma
e-mail: ritika.sharma787878@gmail.com

M. Saxena
Sup'Biotech, Paris, 94800 Villejuif, France
e-mail: mohitsaxena1106@gmail.com

A. Khanna et al. (eds.), *Data Analytics and Management*, Lecture Notes on Data Engineering and Communications Technologies 54,
https://doi.org/10.1007/978-981-15-8335-3_59

1 Introduction

The coronavirus outbreak hit millions of people's lives and came across with numerous deaths. The threat of this virus increases every day with new cases [1]. However, the countries affected by this virus are taking different measures using artificial intelligence and big data technologies [2].

The available data is now processed by machine learning software to recognize the pattern and after this formed algorithmic models [3]. After processing, these models used to predict the number of currently infected cases [4]. Disease control centres use the big data from several years and forecasting tools to predict the models from different multiple sources to contain the coronavirus. According to the World Health Organization (WHO), artificial intelligence and big data play a very important role in response to COVID-19. Many countries use big data analytics to minimize the further risk of coronavirus [5]. The outbreak of novel coronavirus has spread rapidly from its origin in Wuhan, Hubei Province of China in late December of 2019. A case of pneumonia with unknown ethology was reported to the WHO country office in China on 31 December 2019. Early investigations suspect that the cause of COVID outbreak to humans could be the Wuhan South China Seafood and wildlife market where all varieties of seafood and other kinds of meat were sold altogether pointing towards a single torrent crisis from an animal reservoir. Previous investigations of publicly available sequence data have presented phylogenetic estimates of SARS-CoV-2 time of most recent common ancestor (TMRCA) and growth rates using Bayesian phylogenetic methods [6]. Further studies revealed that many clinical, epidemiological and radiological features of COVID-19 are similar to those of severe acute respiratory syndrome coronavirus (SARS) with pangolins or bats as the most likely animal reservoir which outbreak in 2003, Middle East respiratory syndrome (MERS) outspread in 2012. Immediately, it was found that the virus transmitted between human and human in hospital, family setting and other gatherings.

From the Fig 1, gene expression analysis of big data generated based on the research done on the corona virus, we can see that the genetic information has been increasing exponentially everyday, so we can draw a conclusion that over the next years, we need more accurate fast machine learning models which can be applied to big data in order to solve or cure such pandemic cases like COVID-19 case right now, and seeing the increasing amount of the data, we worked on machine learning model on big data that can classify the genes in sequence faster as compared to earlier models. GAN has already been applied to predict COVID-19 [7]. These three viruses belong to the same genera beta coronavirus (Table 1).

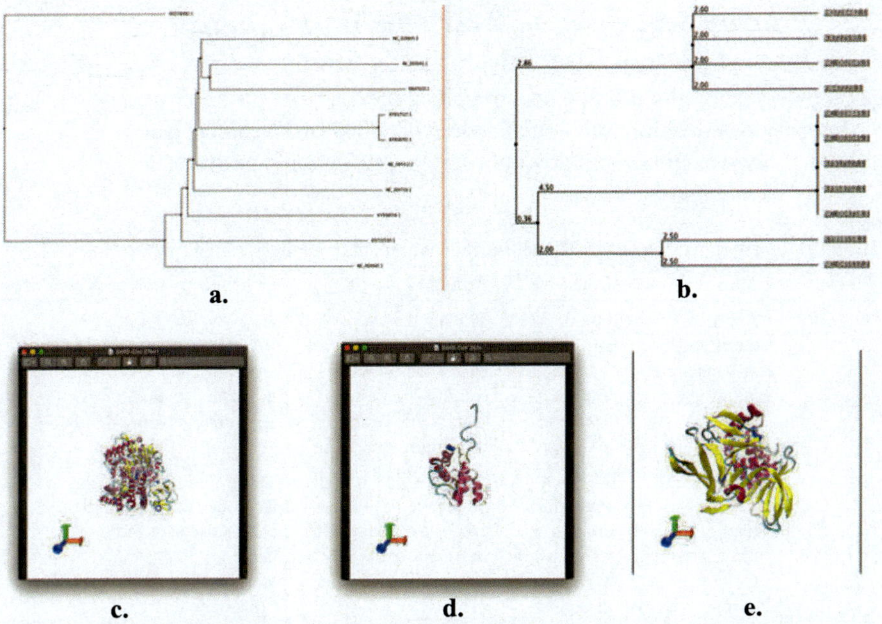

Fig. 1 Increased genetic big data expression of coronavirus over the past years **a** phylogenetic tree using neighbouring joining using BLOSUM62 on region from original of MAFFT multiple sequence alignment of coronavirus disease host retrieved from Uniprot, **b** phylogenetic tree using average distance between coronavirus disease hosts as mentioned in Fig. 2 using BLOSUM62 on region from original of MAFT by JalView v2.5.10, **c**, **d**, **e** VMD predicted cartoon structure of SARS-CoV-2, SARS-CoV and MERS, respectively, obtained from crystal structures available on Protein Data Bank

2 Methodology

2.1 About the Dataset

We have collected this dataset through means of using Google.com, then we found the genetic information about the coronavirus. The data was collected from NCBI by doing an advanced search by doing main search as (COVID-19, CORONAVIRUS, HUMAN BEINGS, SARS, MERS, GENOMICS, GENES). We further refined the search results by selecting the information available of past 30 years only so the data can be said that it is latest or up to date.

Upon finding out the proper gene sequences, further processing was done that involved converting the fast a sequence format data to text format and then to csv format of the data as it is easy to handle with help of pandas, aligning of the sequences together assigning the gene classes of gene sequence file in a structured format and assigning gene classes to the gene sequence. The dataset processing was instrumental to correct unambiguous presentation and seamless execution of ML tools on the data.

2.2 Primary Sequences on Which the Work is Based

https://www.ncbi.nlm.nih.gov/nuccore/NC_045512
https://www.ncbi.nlm.nih.gov/nuccore/NC_004718.3?report=fasta
https://www.ncbi.nlm.nih.gov/nuccore/NC_019843.3?report=fasta

Table 1 Recent works on artificial intelligence and machine learning for COVID-19

S. No.	Title	Objective	Results	Conclusion	References
1	Artificial intelligence and machine learning to fight COVID-19	Use of AI and machine learning for biological processes	The big data is managed and processed with AI resulting in better therapeutics and prevention operation and diagnosis	A cyberinfrastructure algorithm is necessary to deal with this pandemic	[8]
2	Deep learning-based drug screening for novel coronavirus 2019-nCov	Potential drugs to be tested using ML including big data on nCoV protease	Sequence has been processed and results in proper dock match in 18 patients	Modelled structure can be used for the deep learning-based drug screening	[9]
3	Mapping the landscape of Artificial Intelligence applications against COVID-19	Use of AI, to study COVID-19 crisis at different scales including molecular, clinical and applications	Research highlighted the angiotensin-converting enzyme 2 (ACE2) protein, facilitates the virus' entry into host cells	Applications of AI were used to better understand the proteins involved in SARS-Cov-2 infection	[10]
4	Leveraging deep learning to simulate coronavirus spike proteins has the potential to predict future zoonotic sequences	Use of deep learning in order to control the spike protein formation to predict the future zoonotic sequence	With test set of 100 simulated sequences, all had best BLAST matches to spike proteins in searches against NCBI nonredundant dataset	Simulated sequences from neural network may guide us in future with prospective targets for vaccine discovery	[11]

(continued)

Table 1 (continued)

S. No.	Title	Objective	Results	Conclusion	References
5	COVIDier: a deep-learning tool for coronaviruses genome and virulence proteins classification	Deep learning algorithm that can predict virus family from its genome	Deep learning tool that uses scikit-learn packages to classify between Alpha and Beta coronavirus, MERS, SARS-CoV-1, SARS-CoV-2 and bronchitis-CoV genomes can give an accuracy above 90% which can be used	A tool like COVIDier has potential to replace blast	[12]
6	Artificial intelligence against COVID-19: an early review	Role of AI in the fight against COVID-19 I	AI models do not have enough open datasets and models to work on, problems of big data hubris, non-adjustment of algorithms, outlier data	Innovations in AI's impact is limited, and any innovation further in this domain may be a result of this pandemic situation	[13]
7	COVID-19 epidemic analysis using machine learning and deep learning algorithms	Measure everyday exponential behaviour along with the prediction of future reachability	Number of cases were scaled using minmax scaler to fit the LSTM model; predicted cases were rescaled to original range	Early prediction of transmission can help to take necessary actions	[14]
8	Review of big data, artificial intelligence and nature-inspired computing models for performance improvement towards detection of COVID-19 pandemic case and contact tracing	Computing models that can be adopted to enhance the performance of detecting and predicting the COVID-19 pandemic cases	Identified a blend of CNN and whale optimization algorithm for COVID-19 diagnosis and prediction for patient response to treatment	NIC and big data Preprint analytics tools can be adopted in contact tracing of COVID-19 pandemic case to identify 'hot spots', and to alert people	[15]

2.3 Data Storage

The role of data storage is very crucial in this analysis experiment as to easily keep the data separated and for fast and effective workflow of the analysis process (although the work can be done without any data storage format but we preferred the data storage format in order to understand these viruses more and to easily handle the data).

Without data storage format, it would have taken a very long time to do the analysis as there would not have been a proper structure.

The data downloaded was stored in a MapReduce format.

- First, we divide the input into three splits as shown in the figure. This would help to spread the work among all the map nodes.
- Then, we assign a value.
- Now, a list of primary value pairs will be created where the value is nothing but the individual words and value which is one. So, for the first line (COVID-19 SARS MERS), we have three key value pairs—COVID, 1; SARS, 1; MERS, 1. The mapping process remains the same on all the nodes.
- After the mapping phase, a partition process took place where we did the sorting and shuffling process so that all the tuples with the same value are sent to the corresponding reducer.
- Now, each reducer has a unique value and a list of values corresponding to that very key. For example, MERS, SARS, etc.
- Then, the Reducer counts the values which are present in that list of values. Reducer got a list of values like SARS (bats) (1, 1, 1). Then, it counts the number of ones in the very list and gives the final output as—Bats, 3.
- Finally, result is given.

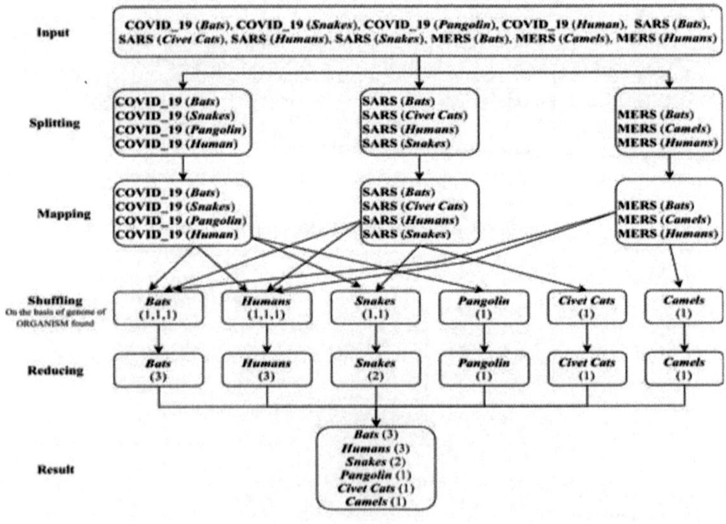

Fig. 2 Data organizing in MapReduce format

2.4 Environment Setup

We started the work by installing Anaconda, as it allows the library installation process to be pretty easier, it is used with Python version 3.7.

We used Jupyter Notebook as our primary IDE as it is said to be one of the most precious standard IDE used in the field of machine learning as it has a very simple user interface and very user friendly as well.

2.5 Starting

In our work, we used advanced machine learning techniques like natural language processing with most popular libraries like Sklearn in our work. The data was imported in comma separated values (CSV) format (Figs. 3, 4 and 5).

Fig. 3 Dataset of COVID-19

	sequence	class
0	attaaaggtttataccttcccaggtaacaaaccaaccaactttcga...	1
1	ttctgcaggctgcttacggtttcgtccgtgttgcagccgatcatca...	1
2	agactccgtggaggaggtcttatcagaggcacgtcaacatcttaaa...	1
3	cgaaataccagtggcttaccgcaaggttcttcttcgtaagaacggt...	0
4	atgcactttgtccgaacaactggactttattgacactaagaggggt...	2

Fig. 4 Dataset of SARS

	sequence	class
0	ATATTAGGTTTTTACCTACCCAGGAAAAGCCAACCAACCTCGATCT...	1
1	TGCAGTCGATCATCAGCATACCTAGGTTTCGTCCGGGTGTGACCGA...	2
2	CCTTAAGCACCAATCACGGCCACAAGGTCGTTGAGCTGGTTGCAGA...	0
3	GTGACGAGCTTGGCACTGATCCCATTGAAGATTATGAACAAAACTG...	1
4	CCGAACAACTTGATTACATCGAGTCGAAGAGAGGTGTCTACTGCTG...	0

Fig. 5 Dataset of MERS

	sequence	class
0	GATTTAAGTGAATAGCTTGGCTATCTCACTTCCCCTCGTTCTCTTG...	1
1	TCGTGTCTCTTGTACGTCTCGGTCACAATACACGGTTTCGTCCGGT...	0
2	GGTTCATGGATGGCGAAAATGCCTATGAAGTGGTGAAGGCCATGTT...	2
3	ATTGCTTGTGAAAATCCATTCATGGTTAACCAATTGGCTTATAGCT...	2
4	ACCTCTTGCCCTGAGTGGATGGACGATTTTGAGGCGGATCCTAAAG...	3

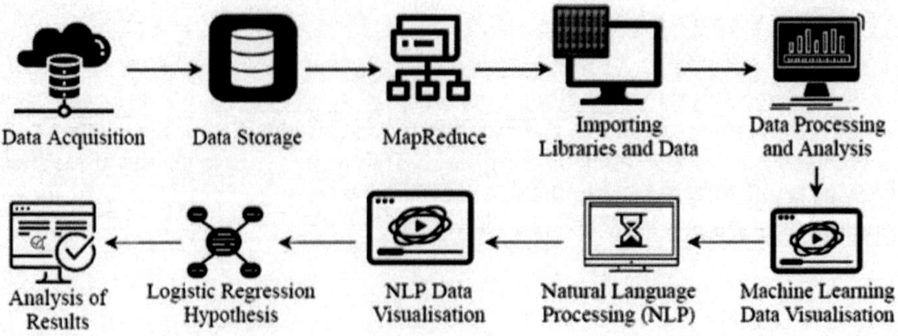

Fig. 6 Workflow of study

2.6 *The Dataset Looks like*

.head() function of pandas shows the first 5 rows of the dataset.

The complete process can be summarized as (Fig. 6).

After we installed the Jupyter Notebook, which is an integrated development environment, on our PCs, we utilized the pre-installed libraries in the IDE for further structuring like matplotlib for plotting graphs, Sklearn for classification training and natural language processing, pandas, numpy for mathematical operations and then pand as library to import our sequence datasets into Jupyter. We performed a natural language processing technique for training classification (DNA sequencing) and then visualized the data by plotting graphs. Then, we compared the individual sequences of SARS, MERS and COVID-19 by comparing their nucleotides against each other.

3 Results

3.1 *K-mer Formation of Sequence or Natural Language Processing of the Data*

The dataset above is not in vector form or in uniform length, which is a necessity for data to which is to be classified or a regression algorithm.

DNA and protein sequence can be said to be the transcript of life. This transcript codes information for molecules that are found in all life forms. This transcript is written as the (book) genome, gene class or family (sentence or chapter), kmers → motifs (words), and nucleotide bases and amino acids (alphabets or characters).

First, the biological sequences were broken to k-mer length overlapping 'words'. Like 'ATGCATGTCA' becomes 'ATGCAT', 'TGC ATG', 'GCATGC' and 'CATGCA'.

Therefore, the above example is broken into four parts.

Fig. 7 K-mer formation of COVID-19

	class	words
0	1	[atatta, tattag, attagg, ttaggt, taggtt, aggtt...
1	2	[tgcagt, gcagtc, cagtcg, agtcga, gtcgat, tcgat...
2	0	[ccttaa, cttaag, ttaagc, taagca, aagcac, agcac...
3	1	[gtgacg, tgacga, gacgag, acgagc, cgagct, gagct...
4	0	[ccgaac, cgaaca, gaacaa, aacaac, acaact, caact...

Fig. 8 K-mer formation of SARS

	class	words
0	1	[attaaa, ttaaag, taaagg, aaaggt, aaggtt, aggtt...
1	1	[ttctgc, tctgca, ctgcag, tgcagg, gcaggc, caggc...
2	1	[agactc, gactcc, actccg, ctccgt, tccgtg, ccgtg...
3	0	[cgaaat, gaaata, aaatac, aatacc, atacca, tacca...
4	2	[atgcac, tgcact, gcactt, cacttt, actttg, ctttg...

Fig. 9 K-mer formation of MERS

	class	words
0	1	[gattta, atttaa, tttaag, ttaagt, taagtg, aagtg...
1	0	[tcgtgt, cgtgtc, gtgtct, tgtctc, gtctct, tctct...
2	2	[ggttca, gttcat, ttcatg, tcatgg, catgga, atgga...
3	2	[attgct, ttgctt, tgcttg, gcttgt, cttgtg, ttgtg...
4	3	[acctct, cctctt, ctcttg, tcttgc, cttgcc, ttgcc...

We used six character words which can be changed according to the user or parameter setter. The length of the words and amount of overlaps are needed to be figured out for any application.

In the field of genomics, this sort of data manipulation technique is called k-mer counting or counting the number of each k-mer sequence. We used Python's natural language processing as it is very easy. After k-mer formation, the dataset was further converted to words (Figs. 7, 8 and 9).

3.2 Individual Class Comparison of Dataset in Form of Graphs or Histograms Which is Very Important for Data Analysis

This step is a necessary step which is done in order to compare the amount of characterized gene classes in a particular viral sequence.

From Fig. 10, we can conclude that COVID-19 dataset has the most second class

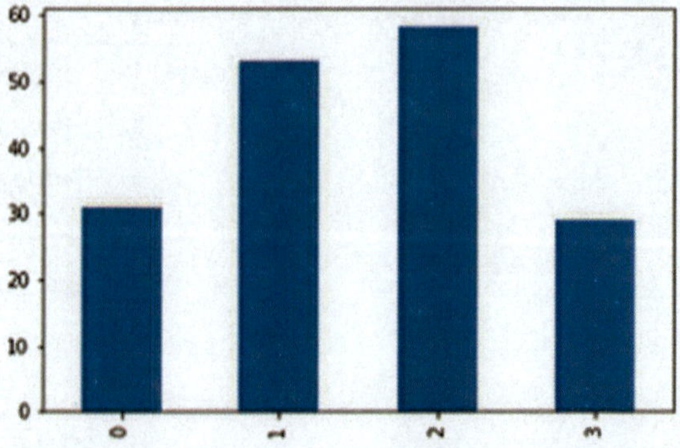

Fig. 10 Individual parameter study of classes in form of graph of COVID-19 dataset

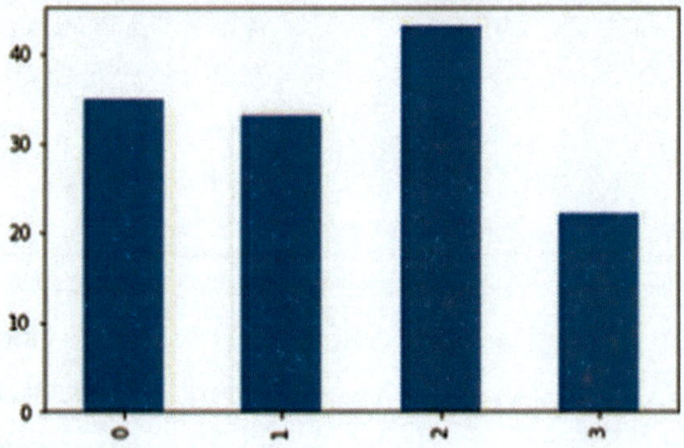

Fig. 11 Individual parameter study of classes in form of graph of SARS dataset

of genes at maximum and the third class has the least amount.

$$2 > 1 > 0 > 3$$

From Fig. 11, we can conclude that the SARS dataset has the most second class of genes at maximum and the third class is of the least amount.

$$2 > 0 > 1 > 3$$

From Fig. 12, we can conclude that the MERS dataset has the most second class of genes at maximum and the zeroth class is of the least amount.

$$2 > 1 > 3 > 0$$

From all three, we can conclude that second class of gene is maximum in all three sequences.

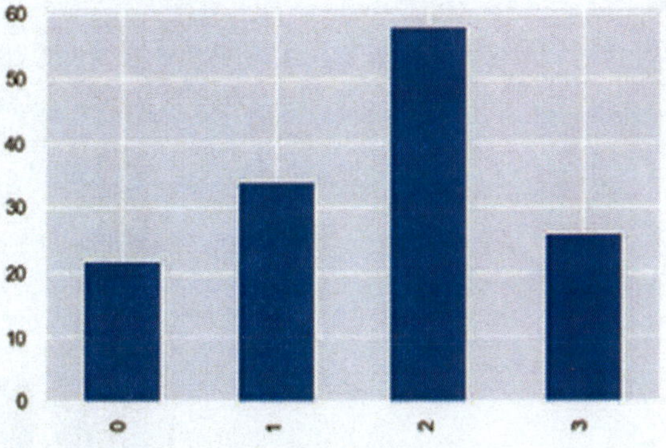

Fig. 12 Individual parameter study of classes in form of graph of MERS dataset

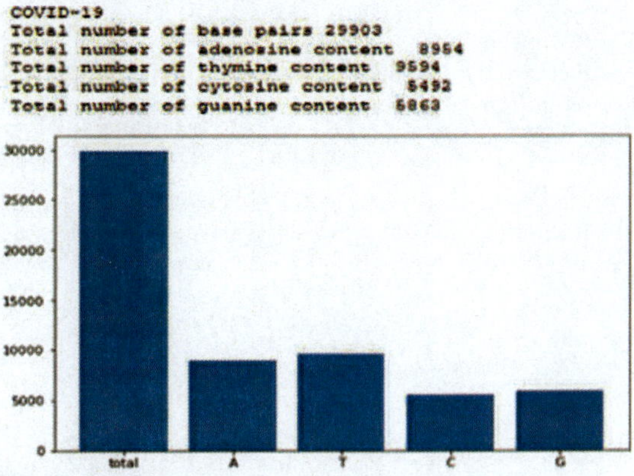

Fig. 13 Nucleotide parameter study of COVID-19 dataset in form of graph

3.3 Individual Nucleotide Comparison of Dataset in Form of Graphs or Histograms Which is Very Important for Data Analysis

From the above three analysis, we can conclude that all sequences are in format from maximum amount to minimum $T > A > G > C$ (Figs. 13, 14 and 15).

3.4 For Better Analysis We Did a 1 to 1 Nucleotide Comparison of Data

See Fig. 16.

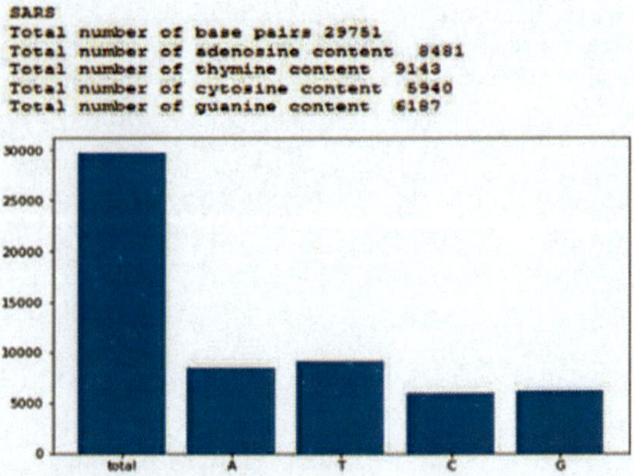

Fig. 14 Nucleotide parameter study of SARS dataset in form of graph

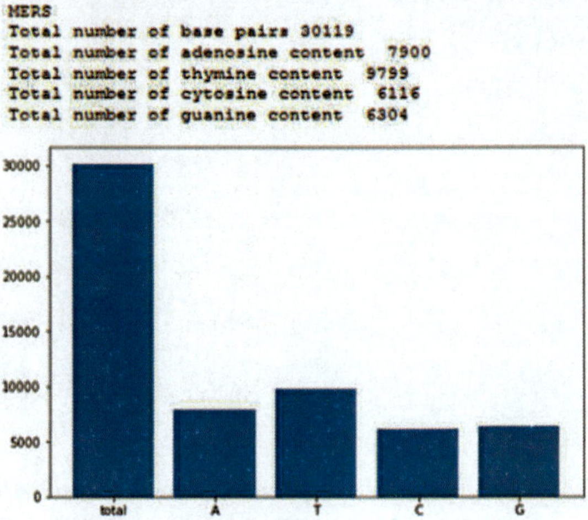

Fig. 15 Nucleotide parameter study of MERS dataset in form of graph

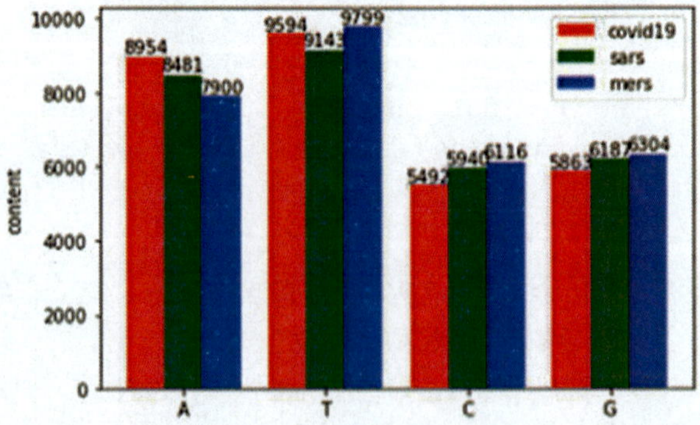

Fig. 16 Graph shows the comparison of nucleotide comparison of all three sequences

3.5 Interpretation from the Above Plot

The above graph shows the ATGC content of COVID-19, SARS and MERS which represents the comparison between these three sequences. The red colour rectangle represents COVID 19, the green one shows SARS, and the blue rectangle shows MERS.

3.6 Logistic Regression Applied in Order to Classify the Three Sequences

Logistic regression was applied after the natural language processing as if we had done a normal base pair sequence, the outcome will not predict anything, whereas the result after performing the natural language processing, the result predicts the potential direction of gene mutation.

Logistic regression is a statistical technique used to model a binary dependent variable. It is another technique by machine learning in the field of statistics. Logistic regression is named as logistic function which is also called sigmoid function. Mathematically, it is dependent on two class values '0' and '1', and it is an S-shaped curve.

Formula: $1/(1 + e\hat{}\text{-value})$ where e is the base of the natural algorithms.

Figure 17 After we did the Kmers formation using natural language processing technique, we applied the logistic regression equation to the COVID-19 gene sequence in order to check the changes in the data and notice the patterns in it. Then, the result was plotted with help of matplotlib library.

Figure 18 Further, we applied and noticed the pattern in the SARS and COVID-19 sequence as they have been worked more upon and have suitable drug candidates for curing the ill patients or for testing first on research animals and then for human trials.

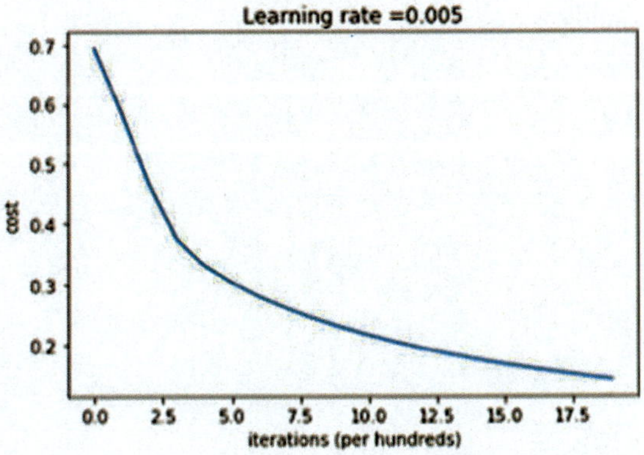

Fig. 17 Graph shows the logistic Regression equation applied on COVID-19 sequence

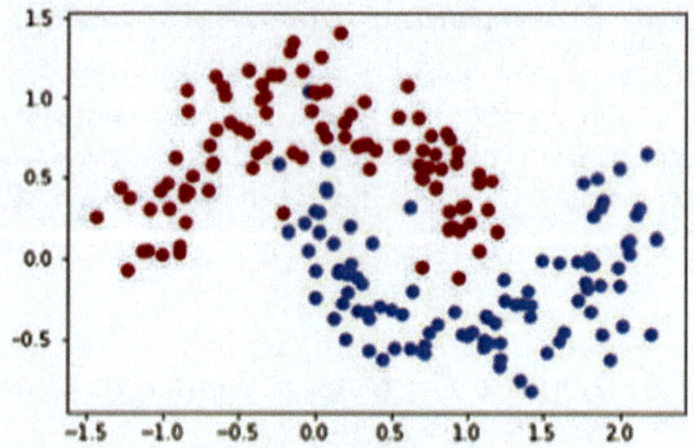

Fig. 18 Graph shows the logistic regression equation applied on COVID-19 sequence and SARS sequence

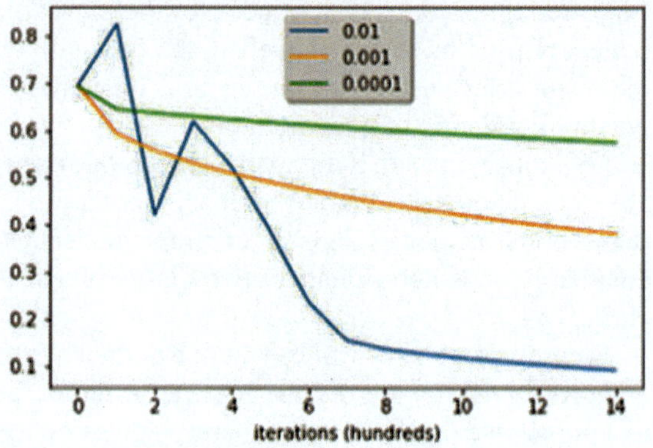

Fig. 19 Graph shows the logistic regression equation applied on COVID-19 sequence,MERS sequence and SARS sequence

Figure 19 Here, the orange line depicts COVID-19 gene sequence, blue line depicts SARS gene sequence, and green line depicts MERS gene sequence. The above result shows that the most active sequence according is MERS although it got eliminated very quickly and then followed by COVID-19 which is going on now and then SARS.

Thus, we can conclude that although all three viruses are of the same family as they follow the same pattern of sequence and genes but no two viruses of the corona family are the same.

4 Discussion

In this paper, we highlighted on how humongous amount data or also called as big data can be easily structured/organized in an efficient way and can be worked upon using modern era technique to handle the data and with machine learning training the personal computers to predict and analyse the genomic sequence data accurately without human intervention [8]. We used techniques such as natural language processing in order to structure the sequence data and then worked further upon it for increasing accuracy using different machine learning models. In the previous papers or studies, the work has been done on huge scale on bigger level such as classifying then virus but not concentrated to specific gene so taking inspiration from the recent studies on we developed a model to classify the genes present in the complete sequence effectively so that mutations can be noticed easily saving time and then we compared the sequences with each other drawing attention more towards the abnormalities present the Viral genomes which has many interpretation like one of those is that all the mutation are Natural and not Synthetically done.

Epidemiologists have estimated the worldwide spread of COVID-19 and different mutated morphological features using machine learning and deep learning algorithms [14]. Researchers have used nature-inspired computing models and artificial intelligence models along with big data analytics in order to draw attention towards COVID-19 detection prevention and diagnosis [15]. Scientists have been constantly tracking and predicting the spread using AI dashboards like for previous pandemic zika virus using models and deep neural networks; the same model is being trained with the large dataset generated due to COVID-19 [13]. With a strong file storage system such as HDFS file system of storage and machine learning techniques including deep learning, a faster pipeline for drug screening process can be established where all potential drug targets are tested against nCoV protease effectively [9]. The research institutions or researchers need more and more data each data for example 100 sequences dataset of corona virus using deep learning algorithms and multimodal neural networks could predict the sequence alignment more accurately than BLAST which is widely being used plus the use of this system works more effectively and is more memory efficient with low memory usage as compared to BLAST and it can help in with suitable target selection for vaccine test or discovery in a novel zonnois case [11]. The researchers have drawn attention towards the role of angiotensin-converting enzyme 2 (ACE2) protein the receptor that makes the COVID virus enters easily into the human cells or other carrier organisms [10]. The scientist has also been able to prepare multiple deep learning algorithms that can relate the family of viruses from its genome [12].

5 Conclusion and Future Scope

Thus, we can conclude that from all the big data generated in the past days needs a very robust system such as MapReduce for storage and processing of the data, and the above analysis also shows that that although all three viruses are of same family as they follow the same pattern of sequence and genes but no two viruses of corona family are same, so no same treatment can be done for the viruses, whereas similar sort of molecules used in drugs used to treat any virus can be used, and with help of machine learning, we can analyse its effect on the virus virtual without spending any big amount of money in experimenting in highly expensive lab setups. This model requires better, more robust entries that are accurate and curated. We will work on the other zoonotic sequences too. Since the diagnosis is not specific so it cannot be analysed with just a few parameters as more information is needed to be analysed due to difference in multiple ailments. The data should be 98% accurate for it to be acceptable in real-time diagnostic tool development. The dataset is required to be trained rigorously to make the analysis more efficient. Also, the future work may involve deep learning and neural networks like BERT and other better algorithms after an improvized dataset is formed.

Acknowledgements We would like to thank Amity Institute of Biotechnology and our family, without their support throughout the process, this paper would have not been accomplished. We would also like to thank Amity University for giving us this great opportunity.

References

1. Volz E, Baguelin M, Bhatia S, Boonyasiri A, Cori A, Cucunubá Z, et al. Report 5: Phylogenetic analysis of SARS-CoV-2
2. Agarwal A, Saxena A (2019) Comparing machine learning algorithms topredict diabetes inwomen and visualize factors affecting it the most—a step toward better healthcare for women. In: International conference on innovative computing and communications. https://doi.org/10.1007/978-981-15-1286-5_29
3. Saxena A, Kaushik N, Chaurasia A, Kaushik N (2019) Predicting the outcome of an election results using sentiment analysis of machine learning. In: International conference on innovative computing and communications, https://doi.org/10.1007/978-981-15-1286-5_43
4. Agarwal A, Saxena A (2020) Comparing machine learning algorithms to predict diabetes in women and visualize factors affecting it the most—a step toward better health care for women. In: International conference on innovative computing and communications. Springer
5. Ying S, Li F, Geng X, Li Z, Du X, Chen H et al (2020) Spread and control of COVID-19 in China and their associations with population movement, public health emergency measures, and medical resources. medRxiv
6. Zhou Y, Hou Y, Shen J, Huang Y, Martin W, Cheng F (2020) Network-based drug repurposing for novel coronavirus 2019-nCoV/SARS-CoV-2. Cell Discovery 6(1):1–18
7. Waheed A, Goyal M, Gupta D, Khanna A, Al-Turjman F, Pinheiro PR CovidGAN: data augmentation using auxiliary classifier GAN for improved Covid-19 detection. IEEE Access. https://doi.org/10.1109/ACCESS.2020.2994762

8. Alimadadi A, Aryal S, Manandhar I, Munroe PB, Joe B, Cheng X (2020) Artificial intelligence and machine learning to fight Covid-19
9. Zhang H, Saravanan KM, Yang Y, Hossain MT, Li J, Ren X, Wei Y (2020) Deep learning based drug screening for novel coronavirus 2019-nCov
10. Bullock J, Pham KH, Lam CSN, Luengo-Oroz M (2020) Mapping the landscape of artificial intelligence applications against COVID-19. https://arXiv.org/2003.11336
11. Crossman LC (2020) Leverging deep learning to simulate coronavirus spike proteins has the potential to predict future zoonotic sequences. bioRxiv
12. Habib P, Alsamman AM, Saber-Ayad M, Hassanein SE, Hamwieh A (2020) COVIDier: a deep-learning tool for coronaviruses genome and virulence proteins classification. bioRxiv
13. Naudé W (2020) Artificial Intelligence against COVID-19: an early review
14. Punn NS, Sonbhadra SK, Agarwal S (2020) COVID-19 epidemic analysis using machine learning and deep learning algorithms. medRxiv
15. Agbehadji IE, Awuzie BO, Ngowi AB, Millham RC Review of big data, artificial intelligence and nature-inspired computing models for performance improvement towards detection of COVID-19 pandemic case and contact tracing

SSDA: Sleep-Scheduled Data Aggregation in Wireless Sensor Network-Based Internet of Things

Rachit Manchanda and Kanika Sharma

Abstract In Wireless Sensor Network (WSN)-based Internet of Things (IoT), the data transmissions from the nodes to the sink consume a lot of energy of the sensor nodes. To add to this, the process of data aggregation in cluster-based network spends the energy resources substantially. In this paper, the process of sleep scheduling is introduced among cluster member nodes in a way that based on the pre-defined threshold value of the distance, the nodes are made to act or sleep. Eventually, the active nodes transmit their data to the Cluster Head (CH) which performs data aggregation before forwarding it to the sink. Further, the CH selection is performed by considering energy, distance, and number of neighbor nodes. It is observed from the simulation analysis that the SSDA improves the stability and survival period of the network with a prodigious magnitude of 47% and 48%, as compared to the EFTA and DAFA protocols, respectively.

Keywords Wireless sensor network · Data aggregation · Cluster head · Sleep scheduling

1 Introduction

Wireless Sensor Network (WSN) comprises of various wireless sensor nodes that deployed randomly in the remote areas, where the sensed information is processed and forwarded to the sink in the structured manner [1]. WSN suffers from the various limitation, and the most dominant one is the limited energy resources of the sensor nodes employed in the network [2]. Various attempts have been reported that addresses the limited battery stock of the sensor nodes. One of the highly used approach is clustering in the WSN [3]. Clustering is the grouping of sensor nodes

R. Manchanda (✉) · K. Sharma
NITTTR, Chandigarh, India
e-mail: rachit.tech2@gmail.com

K. Sharma
e-mail: kanikasharma80@yahoo.com

A. Khanna et al. (eds.), *Data Analytics and Management*, Lecture Notes on Data Engineering and Communications Technologies 54,
https://doi.org/10.1007/978-981-15-8335-3_60

in a way that one of the node is selected as Cluster Head (CH) which is assigned the role of data collecting from the cluster member nodes and aggregates it before sending it to the sink [4].

When in the network various sensor nodes tend to follow the data transmission, the energy consumption occurs based on the number of transmissions. Further, sometimes due to the random deployment of the nodes, the nodes are likely to be placed adjacent to each other [5]. While doing so, they sense the same data, and the redundant data is forwarded to the sink which causes heavy damage to the energy resources of the WSN.

Therefore, it becomes highly significant to reduce the number of unnecessary transmissions in the network [6].

Data aggregation is the process in which the CH node removes the redundant data before forwarding it to the sink [7]. While doing so, it consumes significant amount of energy. Therefore, it becomes essential to utilize the number of data packets from the cluster member nodes in a way that the energy consumption of the network is reduced in the context of data aggregation [8].

There could be two solutions to perform this task of reducing the number of transmissions. One of the data packets removal and other is packets compression. In this work, the removal of data packets is done by considering the closely located nodes will share the same sensed data with the sink.

The other factor that is given due importance is the CH selection [9]. There has been a plethora of research on the selection of CH in the network. Various researchers have explored the variety of parameters that are essentials to be included for the CH selection [10]. Most of them have used energy or the distance parameter as CH selection. Therefore, it leaves a scope for the improvement in a way that if the additional parameters can be included for the selection of CH; the network can be optimized to the great level.

In this work, the protocol DAFA [11] and EFTA [12] are taken into consideration for performance evaluation of the SSDA protocol.

To address aforesaid problems, in this paper, we have attempted to make data aggregation more energy efficient by limiting the number of data transmission in the network.

The major contributions that we have reported in this manuscript are stated as follow.

- The SSDA is proposed that reduces the energy consumption gigantically, and the data aggregations is performed based on the data received from the active nodes for particular time interval.
- The CH selection is performed with use of residual energy, separation factor among nodes, i.e., distance, and the number of neighboring nodes.
- It is the first ever work that makes use of the sleep scheduling along with the energy-efficient CH selection for the WSN.
- The performance evaluation is done against the competitive data aggregation techniques, based on the essential metrics of performance check, namely stability and survival period of network, etc.

The organization of the rest of the paper is done as follow. Section 2 reports about the background overview, and Sect. 3 explains the operation of SSDA, and description of proposed SSDA algorithm is also given. The simulation results are explained in the Sect. 4. Finally, the conclusion is done in Sect. 5. Hence, the references are reported.

2 Background Overview

Since the development of the WSN, the energy efficiency has been the bottleneck for the progressive growth of sensing technology. The traditional approach in networking followed the address centric approach where the data packets have to be communicated. However, the authors [7] in his work gave the concept of data aggregation that implied the data centric approach for the communication of data packets in WSN. Since then, a huge number of research works are reported in the field of data aggregation.

The authors in [13] produced very informative literature review to cover various research methods proposed in the field of data aggregation. The authors in [12] proposed mobile agent paradigm to perform data aggregation in WSN. Despite of sending the data to the sink, the nodes are made to move for their data transmission. The essential and novel thing the authors have proposed in this work is the fault tolerance. It is observed that the algorithm has outperformed the traditional ones. The authors in [14] exploited uniform an Bernouli sampling to perform data aggregation, and while doing so, the authors have targeted the maximum value and query of dataset.

The authors in [15] presented a secure data aggregation algorithm named as CSDA that gave the good flexibility and also made use of the slice assemble technology. The fragments will change according to the network size being employed. Hence, it helps in the energy saving by reducing the communication overheads.

The authors in [16] proposed adaptive neuro-fuzzy inference system (ANFIS)-based model to perform data aggregation that made use of the high-energy nodes. The data transmission from the CH was done to the two aggregating nodes that helped in the deletion of the duplicate data.

Further, weighted compressive data aggregation (WCDA) method was reported by the authors in [17] that allotted the weight factor to each node, based on the energy level of that node. Hence, the node that has the highest weight value and is given the task of aggregating the data.

Apart from data aggregation, the some of the researcher focused on introducing the sleep scheduling in the WSN. However, they did not consider the CH selection with the energy-efficient parameters that reduced the optimal performance of the network. The authors in [18] proposed least spanning algorithm that focused on the data collection based on the similarity index of the information. The authors in [19] presented a method that rendered the multi-level sleep scheduling that gave the information about the difference in the data arrival in WSN.

The authors in [20] reported Energy-efficient Sleep Scheduling Mechanism (ESSM) in which they distributed the nodes in various clusters. Based on fuzzy theory, the similarity is checked, and hence, the nodes are made to be in the sleep mode.

In nutshell, either the presented techniques have focused on the data aggregation or they have totally emphasized on the CH selection. In this work, the both factors are considered for the evaluation of the SSDA algorithm. The operational functioning of SSDA is discussed in next section.

3 Operational Functioning of SSDA

The SSDA is operated through the following stages. As soon as the network is deployed with the sensor nodes and the sink is located in the middle of the network, the process of clustering is initiated. As explained in the conventional cluster-based protocols, the setup and steady state phase are encountered. The following section explains the setup, steady state phase, and the process of sleep scheduling among the cluster nodes.

3.1 *Setup Phase*

The clustering process is done with the initiation of the messages from the sink to the nodes in the network. The primarily CH selection is done based on some parameters. The whole operation of the SSDA is done in terms of rounds. A round is an iteration in which the sensor nodes send data to the sink for one time. The proposed approach is a distributed approach in which the nodes compare themselves with the following factors for CH selection.

(a) **Remaining Energy**
In every round, the energy of the nodes is updated. Therefore, it is quite necessary to check the status of the node in terms of energy for selecting as a CH. Higher the remaining energy (R_{enr}), more is the chances for a node to be selected as CH.

(b) **Separation among nodes, i.e., Distance**
The most influencing factor for the CH selection is the distance (D_{ist}), as it decides how far the data has to be transmitted. Higher is the distance from the sink, lesser it will be preferred as a CH.

(c) **Number of neighboring nodes**
To select a node as a CH and to ensure least distance of the nodes from the CH, a node should be selected with more number of neighboring nodes (N_{nb}).

These aforesaid parameters are considered in the threshold formula as given in Eq. (1) that is the conventional method for CH selection. The threshold calculations are performed for each node; simultaneously, a random number is generated for each node. It is checked if the random number is less than the threshold value for any node, the node is said to be CH; otherwise, the node stays as cluster member node. P_{opt} is the optimum pre-defined number of CHs in the network.

$$T_{ld(i)} = \left(P_{\text{opt}} \times R_{\text{enr}} \times N_{nb}\right)/\left(D_{\text{ist}} \times \left(1 - P_{\text{opt}}\left(r \times \text{mod}\left(1/P_{\text{opt}}\right)\right)\right)\right) \quad (1)$$

The factors remaining energy is computed instantly. However, the other factors, namely distance and number of neighboring nodes are computed through Euclidean distance as performed in [10].

3.2 Steady State Phase

In this phase, the data transmissions are performed. The cluster members are assigned Time Division Multiple Access (TDMA) slot for sending their data packets to the CH. The nodes send their data to the CH and thereafter CH forward the collected data after performing aggregation to the sink. CHs are assigned Code Division Multiple Access (CDMA) slots to send their data to the sink.

3.3 Sleep Scheduling

It is one of the most prominent aspect of the reported work. In this work, the participating nodes are reduced to decrease the energy consumption of the cluster. The algorithm for the sleep scheduling is given as below.

Algorithm 1 Sleep Scheduling

Input: *n*, $CH_{selected}$,

Output: Z= {N_{active}, N_{sleep} };

for i=1: C_{size}
  for j=1: C_{size}
  compute distance D_{ist} for each node from the other

$$D_{ist} = \sqrt{(n(i).x - n(j).x)^2 + (n(i).y - n(j).x)^2}$$

if $D_{ist(i-j)} > D_{thrld}$ **then**
  if $E_{(i)} > E_{(j)}$ **then**
    N(i)→ *active_mode*
    N(j)→ *sleep_mode*
    N(i)→ $CH_{selected}$
  else
    N(j)→ *active_mode*
    N(i)→ *sleep_mode*
    N(j)→ $CH_{selected}$
  end if
else
  [N(i) N(j)]→ *TDMA_slot*
  N(i)→ $CH_{selected}$
  N(j)→ $CH_{selected}$
end if
end for
return {N_{active}, N_{sleep} }

The functioning of SSDA is illustrated in Fig. 1 as it shows the various steps that are followed. It is noted that the energy consumption of the nodes is done through the radio energy model as exploited in various cluster-based algorithms [21].

3.4 Network Assumptions

There are some network assumptions that are considered while implementing SSDA.

(a) The whole network is static including the sink. The nodes are energy limited, and sink is having no constraint of energy.
(b) The security aspects are not considered. The other physical factors like reflection, refraction of signals are not taken into consideration.
(c) The nodes are assumed homogeneous in every aspect.

Once the nodes are dead, there is no provision for replacing the batteries of the nodes.

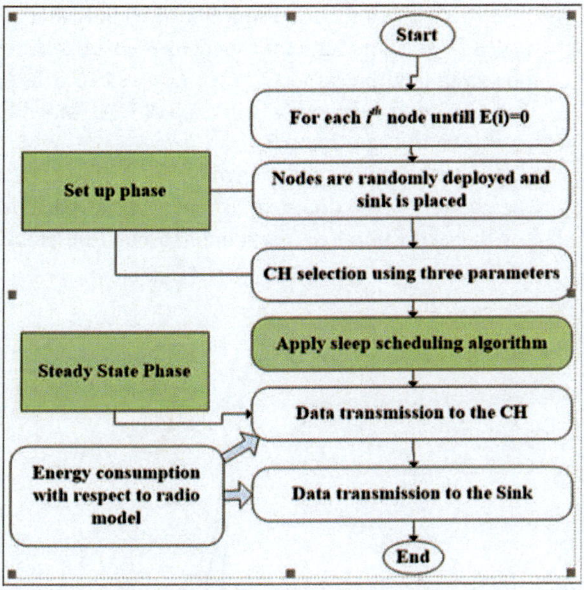

Fig. 1 Operational functioning of SSDA

4 Simulations Results and Discussion

The SSDA is simulated in MATLAB Software using the simulation parameters as given in Table 1. The network area is taken as 100 m × 100 m^2. The number of nodes deployed is 100, and all are homogeneous with initial energy of 1 J. The sink is placed at the middle of the network.

4.1 Performance Metrics

There are various performance metrics that are used to evaluate the performance of SSDA. These metrics are explored from their simulation analysis given as below.

(a) **Stability of Network**

Table 1 Simulation table

Parameter	Value
Network area	100 * 100 m^2
Total nodes	100
Energy of nodes	1 J
Threshold distance	5 m
Sink location	(50, 50)

"It is defined as the total number of rounds that are completed when first node is dead." It is very crucial aspect as it ensures the reliability of the network. It is observed that the stability of the network of SSDA is 1671 rounds whereas it is just 1127 and 1136 rounds in case of DAFA and EFTA protocols, respectively, as shown in Figs. 2 and 3.

The reason behind such improvement is the sleep scheduling the sensor nodes of the cluster; while doing so, the number of participating nodes is kept reserved for the extra number of rounds. Hence, the stability period of the network is enhanced.

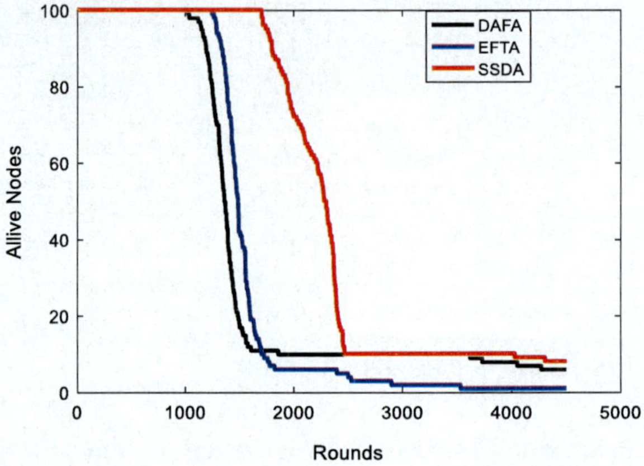

Fig. 2 Alive nodes comparison

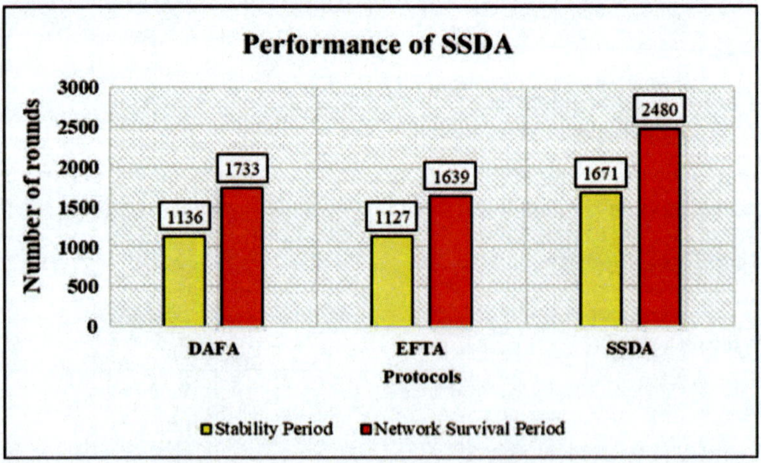

Fig. 3 Performance comparison

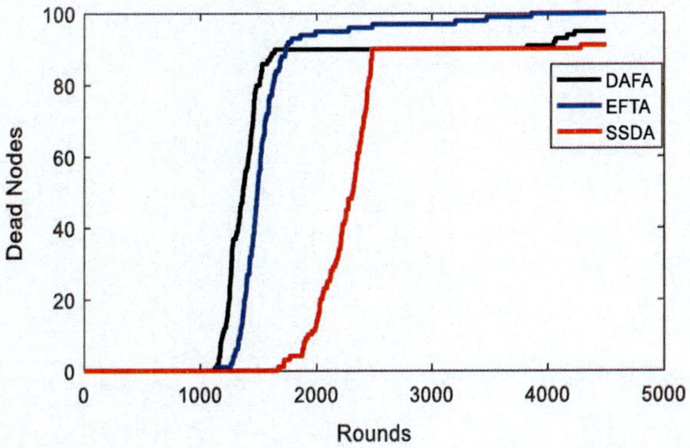

Fig. 4 Dead nodes comparison

(b) **Survival Period of the Network**
"It is defined as the total number of rounds completed when the 90% of all nodes of the network are dead." It is observed that SSDA acquires network survival period equal to 2480 rounds whereas it is just 1733 rounds and 1639 rounds in case of DAFA and EFTA protocols, respectively, a shown in Fig. 3
The reason behind such improvement in the network survival period is the energy-efficient CH selection that considers the crucial parameters including energy, distance, and number of neighboring nodes.

(c) **Alive nodes versus rounds**
As the data transmission is process, the energy of the network decreases correspondingly. As shown in Fig. 4 the alive nodes versus rounds of SSDA is compared with the competitive protocols. It is observed that SSDA comprehensively improves the status of alive nodes versus rounds due to the energy balancing in the network acquired through the sleep scheduling and energy-efficient CH selection.

(d) **Throughput**
It is defined as the number of packets successfully sent over the rounds. The throughput of SSDA is higher than EFTA and DAFA algorithms as shown in Fig. 5.

5 Conclusion

A lot of research efforts are reported that deal with the process of data aggregation to combat the ever growing energy expenditure of the sensor nodes. In this work, we have proposed Sleep Scheduled Data Aggregation (SSDA) that incorporates the process of sleep scheduling among the cluster member nodes as they tend to get into sleep or active mode based on the distance from each other. The decisive factor is the pre-defined threshold distance that decides which node will be active or sleep. It is the first ever work reported that considers distance-based sleep scheduling for optimizing data aggregation. The simulations are performed in MATLAB, and results

Fig. 5 Throughput comparison

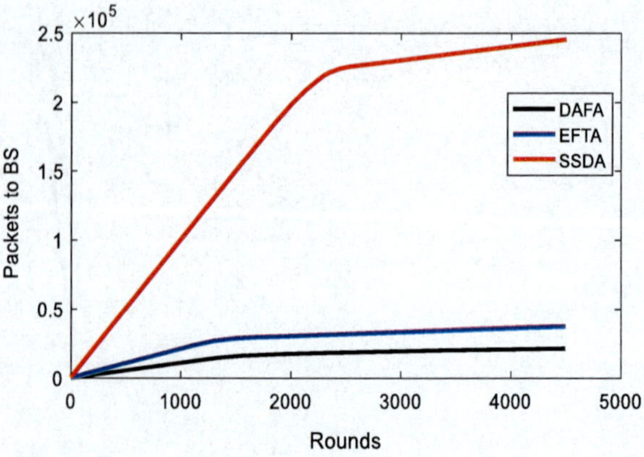

show that the SSDA has outperformed the competitive protocols in the context of performance metrics, namely stability, survival period, etc. The proposed SSDA is a promising approach pertaining to the various WSN-based IoT applications where the magnitude of the data has to be reduced.

References

1. Akyildiz IF, Su W, Sankarasubramaniam Y, Cayirci E (2002) Wireless sensor networks: a survey. Comput Netw 38:393–422
2. Verma S, Sood N, Sharma AK (2019) Genetic algorithm-based optimized cluster head selection for single and multiple data sinks in heterogeneous wireless sensor network. Appl Soft Comput 105788
3. Verma S, Sood N, Sharma AK (2019) A novelistic approach for energy efficient routing using single and multiple data sinks in heterogeneous wireless sensor network. Peer-to-Peer Netw Appl 12:1110–1136
4. Pant D, Verma S, Dhuliya P (2017) A study on disaster detection and management using WSN in Himalayan region of Uttarakhand. In: 2017 3rd international conference on advances in computing, communication & automation (ICACCA) (Fall). IEEE, pp 1–6
5. Meng J-T, Yuan J-R, Feng S-Z, Wei Y-J (2013) An energy efficient clustering scheme for data aggregation in wireless sensor networks. J Comput Sci Technol 28:564–573
6. Wang C, Zhang Y, Wang X, Zhang Z (2018) Hybrid multihop partition-based clustering routing protocol for WSNs. IEEE Sens Lett 2:1–4. https://doi.org/10.1109/LSENS.2018.2803086
7. Krishnamachari L, Estrin D, Wicker S (2002) The impact of data aggregation in wireless sensor networks. In: 2002 Proceedings of 22nd international conference on distributed computing systems workshops. IEEE, pp 575–578
8. Petrovic D, Shah RC, Ramchandran K, Rabaey J (2003) Data funneling: routing with aggregation and compression for wireless sensor networks. In: 2003 Proceedings of the first IEEE international workshop on sensor network protocols and applications. IEEE, pp 156–162
9. Verma S, Sood N, Sharma AK (2018) Design of a novel routing architecture for harsh environment monitoring in heterogeneous WSN. IET Wirel Sens Syst 8:284–294
10. Verma S, Sood N, Sharma AK (2019) QoS provisioning-based routing protocols using multiple data sink in IoT-based WSN. Mod Phys Lett A 34:1950235

11. Mosavvar I, Ghaffari A (2019) Data aggregation in wireless sensor networks using firefly algorithm. Wirel Personal Commun 104:307–324
12. El Fissaoui M, Beni-Hssane A, Saadi M (2019) Energy efficient and fault tolerant distributed algorithm for data aggregation in wireless sensor networks. J Ambient Intell Humanized Comput 10:569–578
13. Rajagopalan R, Varshney PK (2006) Data aggregation techniques in sensor networks: a survey
14. Li J, Siddula M, Cheng X, et al (2020) Approximate data aggregation in sensor equipped IoT networks. Tsinghua Sci Technol 25:44–55. https://doi.org/10.26599/TST.2019.9010023
15. Fang H, Cui X, Liu Q (2005) Review of the localization problem in wireless sensor networks. Comput Inform Technol 6
16. Acharya S, Tripathy CR (2018) An ANFIS estimator based data aggregation scheme for fault tolerant wireless sensor networks. J King Saud Univ Comput Inform Sci 30:334–348
17. Abbasi-Daresari S, Abouei J (2016) Toward cluster-based weighted compressive data aggregation in wireless sensor networks. Ad Hoc Netw 36:368–385
18. Chauhan R, Gupta V (2012) Energy efficient sleep scheduled clustering & spanning tree based data aggregation in wireless sensor network. In: 2012 1st international conference on recent advances in information technology (RAIT). IEEE, pp 536–541
19. Hassan S, Nisar MS, Jiang H-B (2017) Multi-level sleep scheduling for heterogeneous wireless sensor networks. In: Proceedings of the 2016 international conference on computer science, technology and application (CSTA2016). World Scientific, pp 227–233
20. Wan R, Xiong N (2018) An energy-efficient sleep scheduling mechanism with similarity measure for wireless sensor networks. Human-centric Comput Inform Sci 8:18
21. Verma S, Mehta R, Sharma D, Sharma K (2013) Wireless sensor network and hierarchical routing protocols: a review. Int J Comput Trends Technol (IJCTT) 4:2411–2416

Prediction Using Machine Learning in Sports: A Case Study

Megha Kasera and Rahul Johari

Abstract Analysis of data is important to extract information. Algorithms are designed based on this analysis. Machine Learning is an important subset of artificial intelligence. Based on the data trends and the relationship among them, algorithms are designed. Random forest and logistic regression are among the two most widely used algorithms of machine learning. Logistic regression is a probability based algorithm. It consists of response and predictor variables. The hypothesis is based on Bernoulli's distribution. Random forest is a collection of trees. Each tree is dependent upon a random vector sampled independently. Features are randomly selected. Entropy should decrease, and information gain should increase with each split. It is more robust with respect to noise. The objective of the paper is to forecast the winner of the Cricket World Cup by taking into deliberation the several factors that are necessary for deciding the result of the game. Random forest and logistic regression algorithms are applied on R to predict the final winner of the tournament. The paper concludes that logistic regression predicts outcome more 8.825% more accurately than random forest.

1 Introduction

Science provides ideas which enable man to manipulate the data and extract information from the records so that the knowledge evolves [1]. There are many common sense task. They provide knowledge about the events and their effect. Common sense task can be solved using mathematical logic [2]. Through mathematical science, it is possible to deduce all the important properties from the assumptions [3]. These assumptions lead to heuristics. Heuristic programming is used for problem solving. There are five main elements of heuristic programming that include search, pattern

M. Kasera (✉) · R. Johari
USICT, GGSIPU, New Delhi, India
e-mail: kaseramegha07@gmail.com

R. Johari
e-mail: rahuljohari@hotmail.com

A. Khanna et al. (eds.), *Data Analytics and Management*, Lecture Notes on Data Engineering and Communications Technologies 54,
https://doi.org/10.1007/978-981-15-8335-3_61

recognition, learning, planning, and induction. A machine does what it is instructed to do. When solution to a problem is not known, then the entire data can be searched for solution using pattern recognition. Pattern recognition can be combined with learning to gain experience [4]. This experience can be used to plan the record that needs to be searched. The main question arises can machine learn or think? Alan Turing coined a concept known as Turing Test [5]. This test is used to determine whether a machine is able to learn or not. In Turing Test, imitation game is played. There are three players: first is the machine, second is interrogator, and third is a man. The interrogator is in a separate room from them. He questions the machine and the man. The machine should be able to make the interrogator believe that it is the man. If the interrogator is not able to identify between the man and the machine, it can be said the machine can think like a human. Machines that learn to modify their behavior require algorithms to represent changes. The study and development of algorithms to transform raw data into intelligent actions are called Machine learning. A computer program is said to learn from experience E with respect to some class of tasks T and performance measure P, if its performance a tasks in T, as measured by P, improves with experience E [6].

2 Literature Survey

Artificial intelligence has evolved as the broad and exciting field in computer science over the past decade. In this the machine performs a task that a human may do. Machine learning gives solution to questions to build a machine that improves its performance based on its experiences. It is a combination of computer science and statistics [7]. As technology has improved, the capability of a machine to learn has increased. It learns through past experiences. It uses the available data to train and learn. Data mining techniques are used to extract information. It searches for hidden patterns [8]. There are two types of machine learning algorithms. Supervised learning consists of inputs that are already labeled [9]. In unsupervised learning on training data is present. It is not labeled. It is used to find patterns. Random forest and logistic regression are supervised learning algorithms. Random forest algorithm was given by Leo Beriman [10]. It is based on bagging concept given by Beriman. It is an efficient method that can operate over large data very fast. There are two main aspects of random forest bagging [11] and boosting. In boosting, different learners are created by reweighing the instances of the training dataset. A vote is cast depending on its performance. A subset of data is selected from the training dataset, and then, it is fed to the learner. For every tree, a different subset is selected. This concept is known as bagging. There is no literature as to how many trees should be used in a random forest. Increasing the number of trees does not guarantee improvement in performance. There is a threshold after which even if number of trees is increased there is no significant change in the performance [12]. There are two methods for classification of tree known as boosting and bagging [11]. In boosting, successive trees give extra weights to points that are incorrectly predicted. Weighted vote is taken in the end. In bagging, successive tree is independent of the previous trees. In the end, majority vote is considered. Random forest adds an additional layer to bagging [10]. It adds additional randomness to bagging. In addition to being independent of the previous trees, it also calculates split by randomly choosing features. This is different from normal classification as not all variables are selected. It is robust against overfitting [10]. It only requires two parameters, first the number of trees and second the number of features. In statistics, datasets that contain one or more independent variables for them logistic regression is used to predict the outcome. It is a popular method for binary and multinomial outcomes. Predictor variables are categorical in nature. When all the predictor variables are highly correlated this may lead to unstable parameters [13]. Response variables are not required to be multicollinear. This model can also provide the percentage of correct classification. Deep learning can be applied to predict pneumonia [14] and neuromuscular disorder [15].

3 Application of Machine Learning: Cricket as Case Study

The ICC Cricket World Cup is an international sporting event that started in 1975. It takes place every four years. There are preliminary qualification rounds that lead the teams up to the finals. A sphere ball and a bat are used to play the game. What sets it aside from other games is the set of rules that are different for other balling games. The game has evolved over the years. First test matches were played; then, one-day matches came into trend, and now, T20 cricket has taken a lot of attention. Despite so many matches being played every year, the ICC Cricket World Cup is the tournament that each team waits to participate in. It is a form of limited overs match (50 overs). Physical demand of wicket keeper, batsman, and bowler for different formats of the game [16, 17] depends upon their physiology [18]. Studies have been conducted to calculate the performance of individual players by calculating their impact [19], but no prediction model has been created to study the team as a whole. The development of a prediction model in sports could be one of the solutions that will help in predicting the match result. The developed model will be helpful to the team captain, coaches, and team managers to make different tactics during the half time or for the upcoming matches.

4 Problem Statement

Cricket is a popular game played worldwide. Sports prediction leads to improvement in performance. It can help team and management to model various batting and bowling strategy. It can also help with business such obtaining television contract, merchandise sales, sponsorship. This project aims to predict the outcome of the ICC World Cup being held in 2019 using various machine learning algorithms.

5 Materials and Methods

5.1 Decision Tree and Random Forest

To make difficult decision, problem can be divided into series of smaller but more specific choices. Based on these choices and features, rules can be made and applied to solve the problem. In decision tree, the model is represented in the form of a tree structure. The decision tree becomes complex with many more nodes, branches, and leaves. It would be challenging to identify the feature to split. To overcome this challenge, entropy is calculated. Once entropy has been calculated, the data is split. Split in data is based on information gain. Information gain is the measure of decrease in entropy after the dataset is split. Random forest is a method that operates by constructing multiple decision trees during training phase. The decision of the majority of the trees is chosen by the random forest as the final decision. It creates a forest with multiple trees. More the number of trees in the forest more accurate results. In random forest, the splitting of nodes happens randomly. There are two methods for classification of tree known as boosting and bagging. It calculates split by randomly choosing features. It is robust against overfitting. It only requires two parameters, first the number of trees and second the number of features.

5.2 Algorithm

Notation used in Algorithm

t = number of tress
l = leaf node
k = feature
m = total number of features
n = number of features selected o ut of m
$E(k_i)$ = entropy
c = number of classes
p_j = proportion of values falling into class level j
d = node of tree
D_A, D_B = daughter nodes of d.
Algorithm

1) while(!t)
2) while(!l)
3) Randomly select $k_n \in \{k_1, k_2, k_3, \ldots k_m\}$ where $n < m$
4) for $i \in [1, n]$
5) $E(k_i) = \sum_{j=1}^{c} -p_j \log_2 p_j$
6) $x = \min(E(k_i))$

7) x- > [D_A, D_B] for node d

5.3 Logistic Regression

Logistic regression is a classification method. It gets its name from the logistic function of statistics. It is used to assign observations to different classes. It is used to predict based on probability. For prediction, sigmoid function is used. It is a S-shaped curve that can take real valued numbers and map it into value between 0 and 1. It models the probability of classes. Logistic regression is different from linear regression. Linear regression is used to predict dependant continuous values whereas logistic regression is used to predict dependant discrete categorical values. Logistic regression has the word regression but it a classification algorithm. Logistic regression has a S-shaped curve whereas linear regression has a straight line graph. In linear regression, least square estimation method is used for accuracy. In logistic regression, maximum likelihood estimation is used for measuring accuracy. The hypothesis for linear regression can have large range of values whereas logistic regression can only have values between 0 and 1. Hypothesis for logistic regression is:

$$h(x) = \frac{1}{1 + e^{-\theta^T x}}$$

5.4 Experimental Setup

R is a programming language that was developed by Ross Ihaka and Robert Gentleman [20]. It is used for statistical computing and graphics. It consists of various computing and graphical techniques, linear and non linear modeling, time series analysis, clustering, classical statistical testing, classification. The packages used are SDMTools, dplyr, and random forest.

6 Results

Prediction for winner of the tournament is a classification problem. Random forest and logistic regression have been used to predict the outcome for the 2019 cricket tournament. The dataset contains details about all the matches that were conducted between 2007 and 2018. Data related to Afghanistan is not considered as it started participating few years back, and if used for prediction all matches played by Afghanistan will predict its lose for all the testing data. It consists of two teams: team A played the first innings, and team B played the second innings. This dataset is used as the training data. Data related to matches conducted in 2019 are used as the testing data. There are two classes for prediction. First class is true when the match is won by team A, and second class is true when team B wins a match that is team A lost the match. Team A won is the response variable; if team A wins the value is set to 1 indicating first class else the value is set to 0 indicating second class (Fig. 1). Data analysis is performed on the data by means of pivot table. Since last two years, 2017 and 2018, India and England have been playing very well. They are trending on the top. In 2011, the final match was conducted between India and Sri Lanka. Even though Australia had been ranking higher and more chances of going to the final but Australia lost to India in the second quarter finals. Sri Lanka played against New Zealand in the semi finals and won by five wickets. New Zealand emerged to a competition to Australia in 2012 and 2014 matches. For 2019 world cup India, New Zealand, England, and South Africa are the strongest contenders. After data analysis, random forest model is applied on the training data. When random forest is applied on the training data the error rate is high. Based on Fig. 2, the confusion matrix false positive is high. The number of variable used for prediction in total

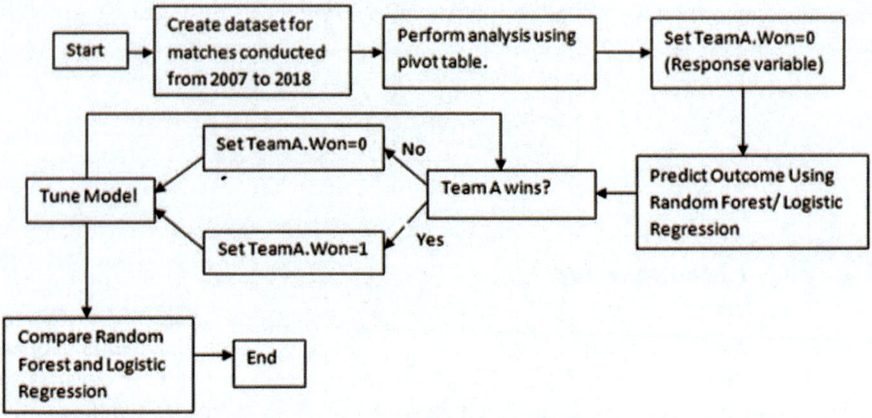

Fig. 1 Working model

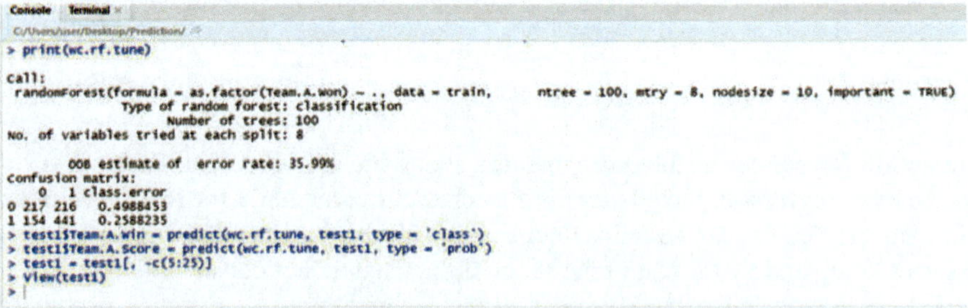

Fig. 2 Confusion matrix and error rate

Fig. 3 Out of bag error

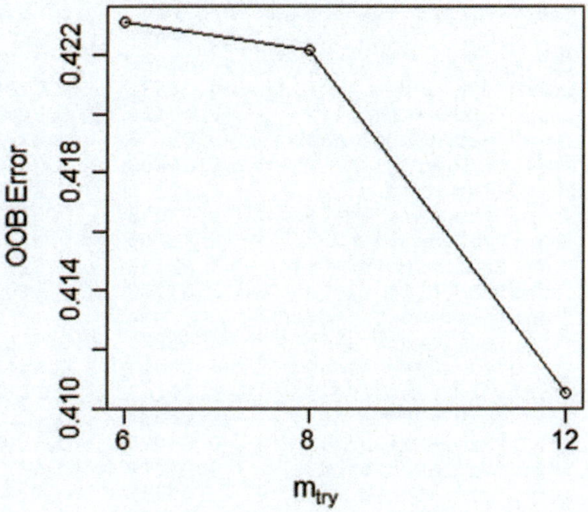

was 6. According to out of bag error (Fig. 3), if the number of variables used for prediction is greater than 12 then the error rate can be decreased. But the data does not consist of 12 variables. The model applied on training dataset is then tuned. Even after tuning the error rate is high. The model is applied to the testing data. It predicts the correct winner, but the probability of winning is only 18% (Table 1). A second model, logistic regression is applied on the training dataset. The accuracy of prediction is higher than that of random forest model. The model is able to predict the correct winner with a probability of 75.35% (Table 2). The output of random forest and logistic regression are compared for all the matches that are being conducted for cricket tournament in 2019. Random forest predicts 11 wrong outcomes whereas logistic regression predicts only 6 wrong outcomes (Fig. 4).

Table 1 Prediction for random forest

Month	Year	Team A	Team B	Team A win	Team A score
14-Jul	2019	England	New Zealand	1	0.18

Table 2 Prediction for logistic regression

Monrh	Year	Team A	Team B	Team A win	Predict logit
14-Jul	2019	England	New Zealand	1	0.7533276

```
Coefficients: (2 not defined because of singularities)
                         Estimate Std. Error z value Pr(>|z|)
(Intercept)             -52.44057   37.90727  -1.383 0.166545
Year                      0.02634    0.01883   1.398 0.161997
Trim.Team.AAustralia      1.43418    0.30674   4.676 2.93e-06 ***
Trim.Team.ABangladesh    -0.09173    0.32710  -0.280 0.779141
Trim.Team.AEngland        1.06863    0.30281   3.529 0.000417 ***
Trim.Team.AIndia          1.20279    0.29706   4.049 5.14e-05 ***
Trim.Team.ANew.Zealand    1.13608    0.31239   3.637 0.000276 ***
Trim.Team.APakistan       0.41705    0.32335   1.290 0.197123
Trim.Team.ASouth.Africa   1.20453    0.32711   3.682 0.000231 ***
Trim.Team.ASri.Lanka      0.63961    0.30697   2.084 0.037192 *
Trim.Team.AWest.Indies         NA         NA      NA       NA
Trim.Team.BAustralia     -1.56277    0.31849  -4.907 9.26e-07 ***
Trim.Team.BBangladesh     0.30018    0.46477   0.646 0.518362
Trim.Team.BEngland       -1.27143    0.32570  -3.904 9.47e-05 ***
Trim.Team.BIndia         -1.65087    0.32277  -5.115 3.14e-07 ***
Trim.Team.BNew.Zealand   -0.68495    0.33922  -2.019 0.043467 *
Trim.Team.BPakistan      -0.95779    0.32231  -2.972 0.002962 **
Trim.Team.BSouth.Africa  -1.49333    0.32708  -4.566 4.98e-06 ***
Trim.Team.BSri.Lanka     -1.05889    0.30695  -3.450 0.000561 ***
Trim.Team.BWest.Indies         NA         NA      NA       NA
---
Signif. codes:  0 '***' 0.001 '**' 0.01 '*' 0.05 '.' 0.1 ' ' 1

(Dispersion parameter for binomial family taken to be 1)

    Null deviance: 1399.5  on 1027  degrees of freedom
Residual deviance: 1274.6  on 1010  degrees of freedom
AIC: 1310.6
```

Fig. 4 Significant data and deviation

7 Conclusion

Random forest algorithm had an error rate of 35.61% which is very high. The prediction made by random forest algorithm is not accurate. Logistic regression has better outcomes as compared to random forest algorithm. It predicted that England would win. The percentage of winning was 75.33% which can be considered as near to accurate. Comparison between both the algorithms predicted that logistic regression has more accurate outcome as compared to random forest. Random forest algorithm needs to be tuned.

References

1. Bush V (1996) As we may Think. Interactions 3(2):35–46
2. McCarthy J (1989) Artificialintelligence, logicandformalizingcommonsense. In Philosophical logic and artificial intelligence, Springer, pp 161–190

3. McCarthy J (1993) Towards a mathematical science of computation. Program Verification. Springer, Dordrecht, pp 35–56
4. Minsky M (1961) Steps toward artificial intelligence. Proc IRE 49(1):8–30
5. Turing AM (1950) Mind. Mind 59(236):433–460
6. Mitchell TM et al. (19997) Machine learning. Burr Ridge, IL: McGraw Hill, 45(37):870–877
7. Jordan MI, Mitchell TM (2015) Machine learning: trends, perspectives, and prospects. Science 349(6245):255–260
8. Teng X, Gong Y (2018) Research on application of machine learning in data mining. In: IOP conference series: materials science and engineering, vol 392. IOP Publishing, pp 062202
9. Supervised Machine Learning (2007) A review of classification techniques, sb kotsiantis. Informatica 31:249–268
10. Breiman L (2001) Random forests. Mach Learn 45(1):5–32
11. Breiman L (1996) Bagging predictors. Mach Learn 24(2):123–140
12. Oshiro TM, Perez PS, Baranauskas JA (2012) How many trees in a random forest? In International workshop on machine learning and data mining in pattern recognition, Springer, pp 154–168
13. James L (2015) Assumptions of logistic regression, statistics solutions
14. Jaiswal AK, Tiwari P, Kumar S, Gupta D, Khanna A, Rodrigues JJPC (2019) Identifying pneumonia in chest X-Rays: a deep learning approach. Measurement (Elsevier). 145:511–518
15. Khamparia A, Singh A, Anand D, Gupta D, Khanna A, Arun Kumar N, Tan J (2018) A novel deep learning based multi-model ensemble methods for prediction of neuromuscular disorders. Neural Comput Appl (Springer). https://doi.org/10.1007/s00521-018-3896-0
16. Amritashish B, Shiny R (2015) Anthropometric and physical variables as predictors of off-spin performance in cricket: a multiple regression study. Int J Sports Sci Fit 5:314–322
17. Christie CJA, King GA. Heart rate and perceived strain during batting in a warm and cool environment. Int J Fitness, 4(1)
18. Noakes TD, Durandt JJ (2000) Physiological requirements of cricket. J Sports Sci 18(12):919–929
19. Petersen C, Pyne DB, Portus MJ, Dawson B (2008) Analysis of twenty/20 cricket performance during the 2008 indian premier league. Int J Performance Analysis Sport 8(3):63–69
20. Ihaka R, Robert Gentleman R (1996) A language for data analysis and graphics. J Comput Graph Stat 5(3):299–314

Analysis of Vehicle Collision Prediction Algorithms Using CNN

Tanya Jain, Garima Aggarwal, and Sumita Gupta

Abstract In today's driven world, vehicle collision (VC) is one of the primary causes of injuries and fatalities on the road. The recent advances in technology help us to predict and potentially avoid such incidents for a safer and smarter traveling experience. Thus, there is also a need to evaluate, compare and improve on these technologies. This paper includes analysis of 108 convolution neural networks (CNN) created with different permutations of configurations (config.): Gaussian mixture model, Kaiming weights and biases, average or max pooling, dropout and additional fully connected layer, negative log likelihood loss, cross-entropy loss or multi-class hinge loss, stochastic gradient descent or Adam's optimizer and padding in convolution layers. The detection of VCs is performed upon 8284 data points using CNN. The analysis of best and worst performing CNNs has also been presented to understand the nature of the prediction resulting due to certain pairings. The major contribution of this paper involves the proposal of a collision detection system which is highly efficient, accurate and lossless with low computation cost in memory and time, making it implementable in applications requiring less infrastructure. It also analyzes the different config. that work for this task of detecting collisions.

1 Introduction

Introduction of autonomous vehicles (AV) can reduce maneuvers of conventional vehicle (CV) drivers like 'right of way violation' [1] and accidents with pedestrian by 90% [2]. Yet, AVs increase CV drivers' maneuvers of 'following too closely' and 'unsafe speed' [1]. Though AVs perform better in structured environments over

T. Jain (✉) · G. Aggarwal · S. Gupta
Department of CSE, Amity University, Sector-125, Noida, Uttar Pradesh, India
e-mail: tanyaacjain@tanya-jain.xyz

G. Aggarwal
e-mail: gmehta@amity.edu

S. Gupta
e-mail: sgupta4@amity.edu

A. Khanna et al. (eds.), *Data Analytics and Management*, Lecture Notes on Data Engineering and Communications Technologies 54,
https://doi.org/10.1007/978-981-15-8335-3_62

complex ones [3], AVs can fail in right decision making despite testing in highly controlled settings. To prevent unforeseen accidents and improve the AVs' efficiency and safety, research and development are required on VC and its impact.

Traditional machine learning algorithms [4] have been used to detect the accident via computer vision techniques such as random forest classifiers (RFC) [5], support vector machines (SVM) and artificial neural networks (ANNs). Traditional feature extraction methods like local binary pattern [6], histogram of gradients [7], maximally stable extremal regions [8] and speeded-up robust features [9] have been used on the RFC to extract image features in a single matrix. The proposed model (PM) is inspired from using max pooling (MaxPool) to select the local maxima from multiple input layers and generate the desired features in the output layer [10]. The ResNet50 model was used in [11] for collision detection. Other much extensive works include using the LSTM architecture integrated along with augmented context mining (ACM) into the faster R-CNN detector to complement the accuracy for small pedestrian detection [12].

The major challenge faced was the system's ability to identify vehicles' accidents. Accidents result in deformity of more than one vehicles when in contact. While the non-accident database may contain any image other than accident, such as pedestrians, birds and trees, there are attributes unique to every accident image. This problem of object detection and analysis being further essential has been met by making the accident dataset more concentrated than the non-accident dataset.

The major contributions of this paper are as follows:

1. A CNN collision detection system that delivers 100% accuracy with full score in precision, recall and f1-score, which is lossless in nature.
2. A low computation and run time system feasible for real-world applications.
3. Result and analysis of the different config. of 108 CNNs and their trends.

The rest of this paper is organized as follows: Sect. 2 presents the preliminaries, Sect. 3 provides information on the database, Sect. 4 describes the PM, Sect. 5 includes information on methodologies, Sect. 6 presents the analysis' results, and finally, conclusion is mentioned in Sect. 7.

2 Preliminaries

Gaussian mixture model (GMM) [13] is used for better feature extraction.
Kaiming weights and biases (W, B) [14] work efficiently with nonlinear activations, especially ReLU, to help the model converge much easily.
Dropout on first fully connected layer and a second fully connected layer (DFc) does not let extreme or rare data values affect model's results by being biased.
Average pooling (AvgPool) helps in detection of smoother features, in comparison with **MaxPool** that detects edges and corners more efficiently.
Padding = 0 on 1st convolution layer (C1P(0)) reduces image dimensions and saves memory.

Negative log likelihood loss (NLLLoss) function considers the uncertainty of the prediction based on deviation from the actual class, represented as follows:

$$L(y) = -\log(y) \tag{1}$$

Multi-class hinge loss (hinge) is widely used for classifications with SVMs.
Cross-entropy loss function is the average difference between the true and predicted probability distributions which minimizes nearing to zero (ideal score).
Stochastic gradient descent (SGD) optimizer is cheaper, noisier and takes more steps to reach the minima, unlike gradient descent. SGD is used with either no momentum (**SGD(0)**) or with momentum = 0.9 (**SGD(0.9)**).
Adam optimizer is a first-order gradient that requires little memory. Individual adaptive rates are calculated for varying parameters from first and second gradient moments' estimates [15].

3 Database Used

While certain laws prohibit data retrieval for detecting vehicle collisions, minimal to no data is available on open data platforms as accidents are rare events [16] and adequate amount of cameras are not being used on devices and roads [17]. Dataset for this work has been collected via web scraping from Google and Accident Images Analysis Dataset [10].

4 The Proposed Model

In the PM as visualized in Fig. 1, image processing techniques have been performed on input images, which are changed into grayscale color composition and transformed into tensors of constant dimensions ($28 \times 28 \times 1$). Grayscale aids in feature extraction while consuming less memory with one color channel as compared to three color channels of RGB. The analysis has been performed on 8284 datapoints.

Before our system is subjected to classification, some of the config. involve using GMM on the input tensors before being fed to the PM. The PMs are fit to the needs with the below described config. and trained on 30 epoch iterations with the train loader divided into batches of 20.

- This CNN comprises four convolution layers with a 3×3 kernel and padding = 1 each. First convolution layer of some config. has no padding.
- The output channels of the convolution layers which are 16, 32, 64, 128, respectively, undergo 2D batch normalization, ReLU activation and then pooling.
- Pooling is either MaxPool or AvgPool, of kernel size 2×2 and stride 2.

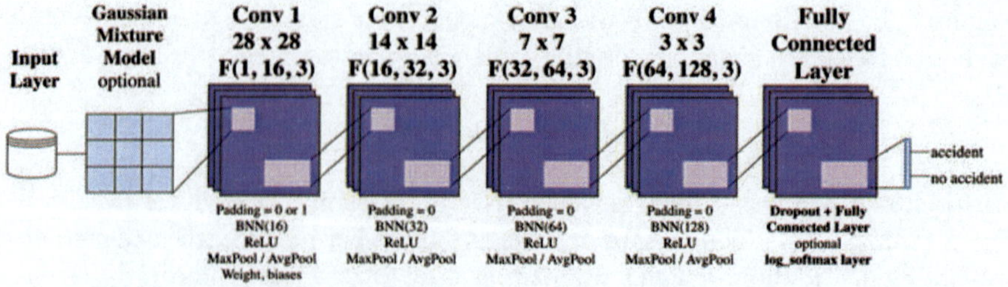

Fig. 1 Proposed CNN with config. to classify the images as 'with accident' or 'without accident'. Denotions used are *Conv*: convolution layer, *BNN*: 2D batch normalization, *ReLU*

- The CNN has one fully connected layer and a log softmax layer to convert all of the final output channels into a distribution of class scores.
- Some config. of the CNN adds DFc before the log softmax layer.

5 The Process

The dataset is split into a 4:1 ratio by keeping 80% of the data as a training set, split into batches of 20 and 20% of data as a testing set, split into batches of 22.

In Fig. 2, the test performance of CNN 1 of Table 2 is visualized on a batch of test images. The classifier's aim is to detect accidents in images. A 'True' label signifies that an accident has taken place; whereas, a 'False' label signifies otherwise.

Table 1 depicts the baseline CNNs created on which further configurations were made with addition and removal of: GMM, [W, B], DFc, [AvgPool or MaxPool], C1P(0), [NLLLoss, cross-entropy, hinge] loss, [SGD(0), (SGD(0.9), Adam] optimizer. These resulting 108 CNNs were categorized into the best performing model (illustrated in Table 2) and the worst performing models (illustrated in Table 3) based on the following parameters:

Accuracy, Precision, Recall, F1-Score [18] The PM has results with a full score in classification report (CR) which signifies an accuracy of 100% with precision, recall and f1-score to be 1.0/1.0 in score, the ideal score.
Loss of 0.00 is the ideal score for the system in which the PM has achieved.

Fig. 2 Test performance of CNN 1 of Table 2: W, B + Hinge loss + Adam's optimizer. Every image is captioned in the format of *predicted class (actual class)*

Table 1 Base models

	CNN configuration	Loss function	Optimize function	Accuracy			Precision		Recall		F1-score		Loss	Training time
				0	1	Overall	0	1	0	1	0	1		
1	W, B + MaxPool	Hinge	SGD(0)	99	99	99.71	1.0	1.0	1.0	0.99	1.0	1.0	0.006895	2:29.33
2	W, B + MaxPool	Cross-Entropy	SGD(0)	99	99	99.67	1.0	0.99	1.0	0.99	1.0	0.99	0.006895	2:15.35
3	MaxPool	Cross-entropy	SGD(0)	92	96	93.92	0.98	0.85	0.93	0.96	0.96	0.90	0.170758	3:40.21

Table 2 Best performing models

	CNN configuration	Loss function	Optimize function	Accuracy			Precision		Recall		F1-score		Loss	Training time
				0	1	Overall	0	1	0	1	0	1		
1	W, B + MaxPool	Hinge	Adam	100	100	100	1.0	1.0	1.0	1.0	1.0	1.0	0.0	2:31.95
2	GMM + W, B + MaxPool	Hinge	SGD(0.9)	100	100	100	1.0	1.0	1.0	1.0	1.0	1.0	0.0	2:25.48
3	GMM + W, B + MaxPool	Hinge	Adam	100	100	100	1.0	1.0	1.0	1.0	1.0	1.0	0.0	2:33.49
4	W, B + MaxPool + DFc	Hinge	SGD(0.9)	100	100	100	1.0	1.0	1.0	1.0	1.0	1.0	0.000119	3:50.20
5	GMM + W, B + MaxPool + DFc	Hinge	SGD(0.9)	100	100	100	1.0	1.0	1.0	1.0	1.0	1.0	0.000119	3:51.20
6	W, B + MaxPool	Hinge	SGD(0.9)	100	100	100	1.0	1.0	1.0	1.0	1.0	1.0	0.00014	2:31.76
7	W, B + C1P(0) + AvgPool	Hinge	SGD(0.9)	100	100	100	1.0	1.0	1.0	1.0	1.0	1.0	0.000297	2:34.28
8	GMM + W, B + MaxPool	Cross-entropy	SGD(0.9)	100	100	100	1.0	1.0	1.0	1.0	1.0	1.0	0.000346	2:16.82
9	AvgPool	Cross-entropy	SGD(0.9)	100	99	99.99	1.0	1.0	1.0	1.0	1.0	1.0	0.0019	2:34.80
10	GMM + W, B + MaxPool	Hinge	SGD	99	100	99.95	1.0	1.0	1.0	1.0	1.0	1.0	0.0012	2:34.14

Training time and test time are more preferable with less values as they help ensure the system is more robust and capable of immediate action.

6 Results and Discussion

The presented computations were performed using a system with 16 GB RAM, a core i7 process and an Intel series GPU, and all operations were primarily performed on the terminal using Python scripting language.

6.1 The Base Models

Table 1 illustrates the significant baseline models created for this analysis, which are further configured resulting into 108 permutations of CNNs.

These are

- The CNN 3 has cross-entropy loss function and SGD(0) optimizer.
- The CNN 2 improves drastically by 5.75% on addition of W, B.
- The CNN 1 sees a further improvement with the use of hinge loss function.

6.2 Best Performing CNNs

- In Table 2, CNNs with W, B + MaxPool + Hinge loss perform well.
  - CNN 1, 2 and 3: *0.00 loss* is achieved with three of the seven CNNs which give *full score in CR*. The CNNs with run time lower than these three CNNs have a negative effect on their performance. Hence, CNN 1 with Adam runs in the *fastest testing time* of 108 CNNs, given it performs well.
  - CNN 6 gives loss = 0.00014 with SGD(0.9).
  - CNN 4 with DFc gives loss of 0.000119, and CNN 5 is similar with +GMM.
  - CNN 7 with AvgPool + C1P(0), the loss decreases to a slight, 0.000297.
- CNN 9: With AvgPool + Cross-Entropy + SGD(0.9), yields a 1.0 recall, precision and f1 score and 99.99% of accuracy with 99% accuracy of the accident data and 100% accuracy of the non-accident data.
- GMM + W, B + Hinge loss has resulted in one of the best performing CNNs.
  - Only CNN 10, with SGD(0) has accuracy of non-accident dataset greater than accuracy of accident dataset even though it is smaller in size.
  - CNN 2 with SGD(0.9) and CNN 3 with Adams yield *full score in CR and 0.00 loss* and differ in training and testing time.

  - CNN 8: With cross-entropy and SGD(0.9) gives *full score in CR*, loss = 0.000346.
- CNNs with NLLLoss have not made it to the well performing CNNs.

6.3 Worst performing CNNs

- In Table 3, CNN 13: With W, B + C1P(0) + AvgPool + Cross-Entropy + SGD(0.9) performs considerably well with an overall accuracy of 99.79%, a CR tending to 1.0, and a loss = 0.009001. Yet, its second highest training time (7:24 min) attribute makes it among the worst performing CNNs.
- NLLLoss + SGD(0) is the most common duo to yield CNNs with one of the poorest performances when paired with W, B + DFc + MaxPool (CNN 11), AvgPool (CNN 10), GMM + MaxPool (CNN 2) and MaxPool (CNN 1):
  - CNN 10, with AvgPool, runs with the third slowest training speed (6:15 min) and has the slowest testing speed (5.1658 s).
  - CNN 1, with MaxPool, performs among the worst with 100% accuracy of the accident data and 0% accuracy of the non-accident data.
- The CNN with W, B + DFc yielded three of the worst performing CNNs:
  - With NLLLoss + Adam, CNN 12 performs better than CNN 3 for using MaxPool and not AvgPool.
  - CNN 8, with Hinge + Adam, trains in the slowest time (over 9 s).
- Unlike CNN 9 of Table 2, CNNs using AvgPool over MaxPool have not performed well when used with Hinge + SGD(0) (CNN 3), Hinge + SGD(0.9) (CNN 6), Hinge + Adam (CNN 9) and NLLLoss + SGD(0.9) (CNN 10).
- Like the CNN 10 of Table 2, CNN 10 with AvgPool + NLLLoss + SGD(0.9) and CNN 4 with NLLLoss + Adam of Table 3 have yielded a higher accuracy for non-accident dataset than the accident dataset given that the dataset of accidents was much larger than that without accident images.

6.4 Trends

- NLLLoss has considerably affected the CR and loss *negatively*.
- CNNs with W, B have significantly better performance than baseline CNNs.
- GMM + W, B tends to *enhance the CNN's performance further*.
- Hinge loss function improved the CNN's performance, especially when complemented with SGD(0.9).
- SGD(0.9) has shown *better results* as compared to SGD(0).
- DFc increased the training time for most CNNs; hence, its effect is subjective.

Table 3 Worst performing models

	CNN configuration	Loss function	Optimize function	Accuracy			Precision		Recall		F1-score		Loss	Training time
				0	1	Overall	0	1	0	1	0	1		
1	MaxPool	NLLLoss	SGD(0)	100	0	70.63	0.71	0.00	1.0	0.00	0.83	0.00	0.580795	0:06.50
2	GMM + MaxPool	NLLLoss	SGD(0)	98	5	70.71	0.71	0.63	0.99	0.05	0.83	1.00	0.006895	0:02.78
3	AvgPool	Hinge	SGD(0)	95	60	84.75	0.85	0.85	0.95	0.6	0.9	0.71	0.3496	2:29.05
4	MaxPool	NLLLoss	Adam	79	98	85.07	0.99	0.67	0.79	0.99	0.88	0.8	0.672155	3:22.70
5	GMM + AvgPool	Hinge	SGD(0)	90	82	88.49	0.93	0.79	0.91	0.83	0.92	0.81	0.2996	2:31.56
6	AvgPool	Hinge	SGD(0.9)	89	88	89.21	0.95	0.78	0.9	0.88	0.92	0.83	0.2637	2:32.20
7	MaxPool	Hinge	SGD(0)	99	65	89.5	0.87	0.99	1.0	0.66	0.93	0.79	0.266	3:35.20
8	W, B + MaxPool + DFc	Hinge	Adam	99	75	92.29	0.91	1.0	1.0	0.75	0.95	0.86	0.2777	9:04.75
9	AvgPool	Hinge	Adam	99	77	92.63	0.91	0.98	0.99	0.77	0.95	0.86	0.3754	2:43.98
10	AvgPool	NLLLoss	SGD(0)	93	94	94.16	0.98	0.86	0.94	0.95	0.96	0.9	0.1677	6:15.65
11	W, B + MaxPool + DFc	NLLLoss	SGD(0)	95	91	94.6	0.97	0.9	0.96	0.92	0.96	0.91	0.16909	3:56.25
12	W, B + MaxPool + DFc	NLLLoss	Adam	98	89	96.13	0.96	0.97	0.99	0.9	0.97	0.94	0.3038	3:44.28
13	W, B + C1P(0) + AvgPool	Cross-entropy	SGD(0.9)	100	99	99.79	1.0	1.0	1.0	0.99	1.0	1.0	0.009001	7:24.04

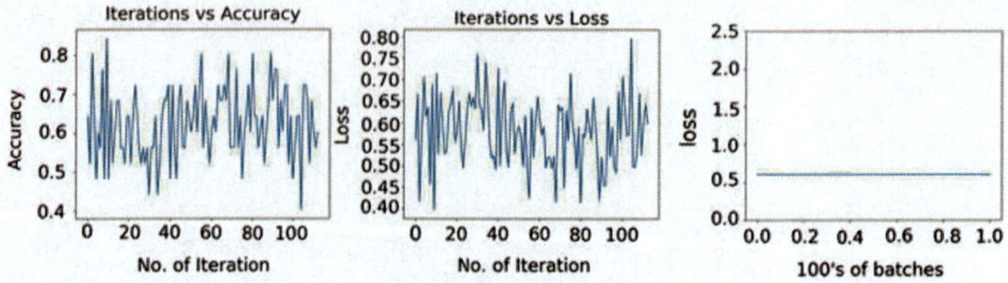

Fig. 3 Worst performing network—Table 3 network 1 uses NLLLoss

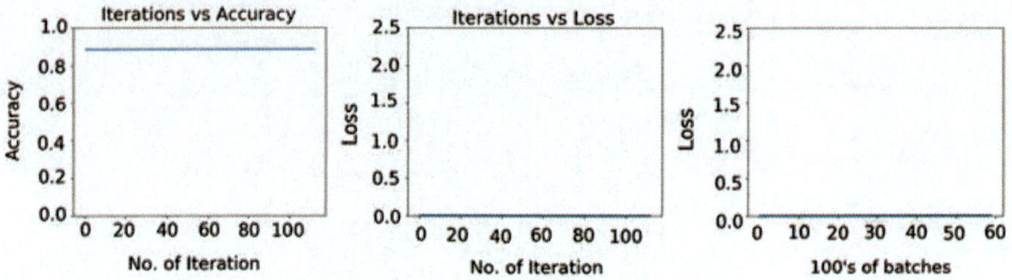

Fig. 4 Best performing network—Table 2 network 1 uses hinge loss

Figures 3 and 4 show the changes in three graphs: The first one determines the number of iterations on which the model is trained versus accuracy, second depicts iteration versus loss, and the third depicts changes in loss with every 100 batch.

7 Conclusion

In autonomous driving, detecting vehicle collisions minimizes the risk of accidents in real-world implementable scenarios. This paper proposes a collision detection system by forming a baseline CNN on which further models are derived with different configurations resulting in analysis of 108 CNNs for the possible methods to use for vehicle accident classification on collision image dataset such that there is maximum feature extraction. Additionally, the proposed models yield full score in CR. Hence, the PMs are efficient, highly accurate and lossless in nature. The results have been achieved in minimal computation cost, that is, by consuming less memory and time making the entire system feasible for application in real-world scenarios.

References

1. Petrović D, Mijailović R, Pešić D (2020) Traffic accidents with autonomous vehicles: type of collisions, manoeuvres and errors of conventional vehicles' drivers. Transp Res Procedia 45:161–168
2. Combs S, Sandt S, Clamann P, Mcdonald C (2019) Automated vehicles and pedestrian safety: exploring the promise and limits of pedestrian detection. Am J Prev Med 56:1–7
3. Guo J, Kurup U, Shah M (2018) Is it safe to drive? an overview of factors, challenges, and datasets for driveability assessment in autonomous driving. ArXiv, abs/1811.11277
4. Dogru N, Subasi A (2012) Traffic accident detection by using machine learning methods
5. Dogru N, Subasi A (2018) Traffic accident detection using random forest classifier. In: 2018 15th learning and technology conference (L&T). IEEE, Jeddah, pp 40–45
6. Ahonen T, Hadid A, Pietikainen M (2006) Face description with local binary patterns: application to face recognition. IEEE Trans Pattern Anal Mach Intell 28(12):2037–2041
7. Wang X, Han TX, Yan S (2009) An HOG-LBP human detector with partial occlusion handling. In: 2009 IEEE 12th international conference on computer vision. Kyoto, pp 32–39
8. Chen H, Tsai SS, Schroth G, Chen DM, Grzeszczuk R, Girod B (2011) Robust text detection in natural images with edge-enhanced Maximally Stable Extremal Regions. In: 2011 18th IEEE international conference on image processing. IEEE, Brussels, pp 2609–2612
9. Bay H, Tuytelaars T, Van Gool L (2006) SURF: speeded up robust features. In: Leonardis A, Bischof H, Pinz A (eds) Computer vision—ECCV 2006. ECCV 2006. Lecture notes in computer science, vol 3951. Springer, Berlin, Heidelberg
10. Pashaei A, Ghatee M, Sajedi H (2019) Convolution neural network joint with mixture of extreme learning machines for feature extraction and classification of accident images. J Real-Time Image Process 17:1051–1066
11. He K, Zhang X, Ren S, Sun J (2016) Deep residual learning for image recognition. In: 2016 IEEE conference on computer vision and pattern recognition (CVPR). IEEE, Las Vegas, NV, pp 770–778
12. Shah AP, Lamare J, Nguyen-Anh T, Hauptmann AG (2018) CADP: a novel dataset for CCTV traffic camera based accident analysis. In: 2018 15th IEEE international conference on advanced video and signal based surveillance (AVSS). Auckland, pp 1–9
13. Reynolds DA (2009) Gaussian mixture models. In: Encyclopedia of biometrics. Springer, Boston, MA
14. He K, Zhang X, Ren S, Sun J (2015) Delving deep into rectifiers: surpassing human-level performance on ImageNet classification. In: 2015 IEEE international conference on computer vision (ICCV) (ICCV '15). IEEE Computer Society, USA, pp 1026–1034
15. Kingma DP, Ba J (2014) Adam: a method for stochastic optimization. In: International conference on learning representations
16. Chan FH, Chen YT. Xiang Y. Sun M (2017) Anticipating accidents in dashcam videos. In: Lai SH, Lepetit V, Nishino K, Sato Y (eds) Computer vision—ACCV 2016. ACCV 2016. Lecture notes in computer science, vol 10114. Springer, Cham, pp 136–153
17. Shah AP, Lamare J, Nguyen-Anh T, Hauptmann AG (2018) Accident forecasting in CCTV traffic camera videos. ArXiv, abs/1809.05782
18. Tharwat A (2018) Classification assessment methods. Appl Comput Inform

Multimodal Deep Learning Architecture for Identifying Victims of Online Death Games

Anshu Malhotra and Rajni Jindal

Abstract Online death games are a fairly recent public health concern of the modern technology-driven world. Various dangerous online games like Blue Whale Challenge and MOMO challenge have grown popular through social networking sites where players or victims engage in self-harming activities, often leading to death. This problem domain has not been studied in depth till date and no known technology-based solutions exist to prevent the spread of such dangerous challenges. The prime objective of our research is to explore the use of deep learning and transfer learning techniques for content analysis of user-generated posts over various social networking sites and design an early warning system which can be used by healthcare authorities for timely identification of victims of these games so as to avoid any fatalities. In this paper, we first discuss in detail the numerous challenges in building required technology-driven solutions for this domain. Next we propose a multimodal deep learning-based system for identifying victims of online death games, using state-of-the-art feature generation techniques for two modalities in user's social media posts: image and text. To the best of our knowledge, our proposed system is the first technology-driven public healthcare administration tool for this this domain.

Keywords Multimodal deep learning · Transfer learning · Online death games · Online social networks · User-generated content

1 Introduction

In the past decade, with smartphone penetration and easy cheap access to high-speed Internet, various social networking sites became popular and so did online gaming. Playing of online games through one's social network and competing with "friends"

A. Malhotra (✉) · R. Jindal
Delhi Technological University, Delhi, India
e-mail: anshumalhotra5@gmail.com

R. Jindal
e-mail: rajnijindal@dce.ac.in

A. Khanna et al. (eds.), *Data Analytics and Management*, Lecture Notes on Data Engineering and Communications Technologies 54,
https://doi.org/10.1007/978-981-15-8335-3_63

in virtual space for fun and entertainment has become a trend in teenagers and young adults. Many gaming applications are also available on mobile's play store or application store. In addition to this came the trend of playing healthy "online challenges" for fun and posting them as one's post or status on social networking site, e.g. Planking, Bottle Flipping challenge.

An online suicide game, popularly known as "The Blue Whale Challenge", is believed to have originated in Russia on their social media site named "VKontakte", back in 2013 and supposedly is the cause of over 150 plus teenage suicides worldwide [3, 4]. Numerous teenage suicides were reported in Russia around November and December 2015 having a lot of commonalities, the most important of them being all of them having a whale engraved on their arm with a sharp object and having cuts and injuries on their bodies. Further investigation of these suicides led to the disclosure of the online death game, which was named "The Blue Whale Challenge"; this death game was secretly getting popular among teenagers through various online social networking sites, through messages and private chat rooms and forums.[1]

This online suicide game eventually spread to other parts of the world, and many teenage suicides were reported in India as well[2] [5]. Seeing teenage suicides linked to Blue Whale Challenge on rise, the Government of India through MeiTY asked the global big social media and Internet companies to detect, remove and report any content and links related to the Blue Whale Challenge which may have led to more teenagers participating in this life-taking game [3]. The law enforcement agencies of other countries were also working to trace the moderators of these games [6].

This game has 50 challenges that are to be competed over a period of 50 days, with tasks becoming more difficult, dangerous and deadly with each passing day, the last one being taking your own life to finally win the game (Fig. 1). The player is given one challenge every day by the moderator or curator, and the player has to post a selfie indicating the task completion to go to the next level. Majorly, all the tasks are of self-harming nature where the victim has to induce some form of pain or fear on himself, e.g. carving one's body, cutting her lips, standing on a cliff or a roof top. The final level to successfully win the game is to take one's own life.

The administrator/moderator/healer or curator, as they are popularly called, initiate conversations with the "selected" victims by contacting them online over social networking sites; they identify and lure potential weak and vulnerable teenagers who can be psychologically manipulated to play this deadly and life-taking game on the pretext of attaining happiness.[3] It has been mentioned in various studies [5, 7] that the players suffered from depression and low self-esteem, which led them to play this game with the aim to boost their self-confidence, attain a sense of gratification and win approval and attention of society [6]. As discussed by Ramamurthy [7], the tasks at initial levels are designed to reduce the fear of death by increasing the

[1]https://www.bbc.com/news/blogs-trending-46505722.

[2]https://timesofindia.indiatimes.com/city/mumbai/boys-suicide-turns-focus-on-the-blue-whale/articleshow/59853004.cms.

[3]https://timesofindia.indiatimes.com/india/meet-the-22-year-old-creator-of-the-blue-whale-death-game/articleshow/59860662.cms.

tolerance levels for bearing physical pain by repeatedly experiencing painful stimuli for 49 days consecutively, before attempting suicide on the day 50.

The nature of Blue Whale Challenge itself makes it difficult to discover and protect the victims from predatory curators of this game. The victims are coerced to keep the Blue Whale Challenge strictly confidential and not share the details about the game even with their friends or family. Due to the anonymity and confidentiality involved, it is difficult for cybercrime investigators to detect the moderators of the game. Since this game is neither available as a downloadable mobile application from the "mobile app store", nor it is an Internet-based game [4]; the government or law enforcement agencies do not have an option to ban this game by blocking the URL or blocking the app to get downloaded from the play store so as to prevent children from participating in it. Hence, as also mentioned by [8], other alternative technology-driven solutions must be explored to prevent the spread of such games. Currently, no technology-based solution is known to exist that can detect and remove the content related to the Blue Whale Challenge or help the law enforcement personnel to track the social networking accounts of the victims or predators.

Research Contributions: Online death games are a fairly recent public health concern of the modern technology-driven world, and this domain has not been studied in depth, neither from public health and psychology perspective, nor from the perspective of finding technology-driven solutions to prevent the same. In this paper, we first discuss various concerns related to the Blue Whale online death game in Sect. 1; then in Sect. 2, we elaborate the related literature survey and also study the technology research done to curb the spread of online death games; next in Sect. 3, we elaborate the difficulties and roadblocks in developing public health monitoring systems for preventing fatalities from online death games; and finally in Sect. 4, we discuss in detail our proposed deep learning and transfer learning-based detection and warning system to curb the spread of online death games before they become popular and claim innocent lives. The key research contributions of our work are enlisted below:

- This problem domain has not been studied in depth till date and no known technology-based solutions exist; we first study and discuss in detail the numerous roadblocks in building technology-driven solutions for the said domain.
- We discuss the use of state-of-the-art deep learning-based feature generation techniques for the two most common modalities present in user's posts on online social networks, i.e. textual sentences and images; that are also used by players of Blue Whale Challenge to post status updates about level/task completion.
- To the best of our knowledge, our proposed architecture is the first technology-driven public healthcare monitoring system for mitigating the spread of online death games by using novel multimodal deep learning techniques.

2 Related Literature Survey

Not much research has been done around developing technology-driven solutions to curb the spread of online death games and timely identifying the victims. Most of the related literature work in this domain have either done a secondary research study to analyse the impact of online death games and the public sentiment around them, or discuss preparedness of society at large to address the concerns arising due to growing popularity of such deadly online games.

The closest research to our work is by Baghel et al. [9] who have proposed VGG-16-based deep learning model trained using transfer learning to identify dangerous "KiKi Challenge" videos, where the users have to dance besides their moving car on a song. They have contributed in collection and annotation of data set for this challenge and released their data set publicly for the purpose of future research. Their data set comprises 2000 videos related to "KiKi Challenge", out of which 220 have been annotated as dangerous. Their model trained over this data set gives an accuracy of 87%. The positive data set used for training is small and unimodal in nature. It has not been explained how their technique can be used for real-time identification of similar new challenges. The other features sets like the text posts and hashtags have not been utilized for this task of classifying dangerous "KiKi Challenge" videos.

Similar to our work, Sumner et al. [10] also emphasize that social media and web posts can be used to develop a public health monitoring and awareness system to detect and prevent teenagers from participating in online death games or indulge in any form of self-harming online activity. They have studied the temporal, geographical and linguistics characteristics of social media and web posts related to the Blue Whale Challenge. They manually collected 95,555 publicly available user posts from various sources like Twitter, YouTube, Reddit, Tumblr, public blogs and forums during the period of 1 January 2013 to 30 June 2017. They used a semi-supervised machine learning approach to classify the collected Blue Whale Challenge-related posts as "pro" or "anti", where "pro" posts indicate curiosity or willingness to participate in the game. They trained a classifier using only 155 manually annotated posts and used that to classify all the other collected posts as "pro" or "anti". They validated the performance of their classifier by using a random sample of 200 posts, which gave an accuracy of 86%. Apart from our work, this is the only related research study where machine learning techniques have been used to classify user posts related to the Blue Whale Challenge. However, there are numerous limitations: the authors have not mentioned the classification algorithms used and have not compared the performance of various classifiers on their data set; the data set used is extremely small, i.e. 155, the number of samples in each class has not been mentioned, the prepossessing techniques used and feature generation from the collected data set are also not explained in their research paper, and lastly, news articles have not been removed from the data set before classification task of "pro" and "anti" sentiment.

Khatter et al. [11] have done detailed analysis of social media posts from: Instagram, Twitter and VK (Russian Social Networking website); around 2700 posts were collected from these social networking sites. They have done temporal analysis to understand user behaviour, their demographics and spread of the game; next they have done network analysis to understand the communication between curators, player and the users in their network; and at last, they have done content analysis to understand the characteristics of collected social media posts. Their data set was collected using the APIs of these networking sites through the use of popular hashtags and terms associated with the game.

Pai et al. [12] studied the sentiment of Twitter users regarding the impact of social media games, chat rooms and related aggressive and self-harming activities. Similar content analysis of sentiments reflected in news articles published about Blue Whale Challenge in leading Indian dailies has been done by Pramod and Natrayan [13]. Both these studies are secondary analysis regarding overall public sentiment about the Blue Whale game. Neither do they provide any analysis or insights about the Blue Whale game itself, nor they have discussed any tech solutions to curb the menace of such online death games.

The research article by Mukherjee and Shukla [5] tries to understand various sociological and psychological reasons, e.g. conflict theory, post-modern and structural/functional perspectives of society as to why teenagers would attempt suicides and specifically through death games and online self-harming challenges like Blue Whale.

A survey of front line medical respondents working in the domain of clinical psychology and paediatrics was done by Mahadevaiah and Nayak [8] to gauge their awareness about the dangerous online games and their preparedness to detect the same in children. The authors have conducted a descriptive cross-sectional study with 54 respondents to assess their perceptions about the subject. Based on the analysis of the results of semi-structured open-ended survey, it was derived that there is gap in understanding and the lack of awareness in the medical fraternity about the subject. This further necessitates developing integrated technology-driven automated solutions to prevent the fatalities from dangerous online death games. Their study defined and classified previous online challenges into two categories: harmless and harmful; "Harmless challenges are those which are played for fun and entertainment and do not involve the use of body parts to complete the challenge", e.g. Planking, Bottle flipping challenge; "Harmful challenges are defined as those involving inflicting some form of force or injury on any body part", e.g. Salt & Ice challenge, Cinnamon challenge [8].

As can be seen from the background literature study, there is scarcity of research done for the domain of online death games. Also, we could not find any related research work that proposes technological solutions for detecting and controlling the spread of such games and timely identification of victims.

3 Research Challenges

The timely detection of online death games and identification of its victims through automated techniques is challenging due to various reasons such as dynamic and confidential nature of these games, e.g. new private groups and version of the game may emerge with modified language or new tasks; lack of availability of clinical and past data and tools to study the characteristics of these games, privacy-enabled structure of popular social networks like Facebook, Instagram and others where all user posts are not publicly available for analysis and monitoring [10]. Lack of publicly available large data sets to study the characteristics of such online death games remains a major roadblock for conducting novel research in this domain. Deep learning-based techniques like one-shot learning and transfer learning which automatically learn the feature representations from a very small data set or by transferring the feature leanings from another domain, respectively, can be useful in developing the required technology solutions.

Developing machine learning and deep learning-based systems for continuous content analysis of user posts across social networking sites and other web sources can help discover trends related to popular online death games and identify curators, victims and related posts. However, the development of required technology systems can only be done by the large Internet and social media platforms themselves, due to privacy concerns of sharing user's data with third-party applications. It is also not feasible for parents and teachers to install such third-party applications (if made available) on mobiles and laptops of teenagers, due to obvious resistance and confidence issues from the teenage child. Hence, the content analysis of user posts must be done via an integrated and platformized approach on real-time basis. In the following section, we discuss our proposed framework which can be adopted by global Internet technology companies as a fundamental layer on their platform to screen user posts on real-time basis. As per our knowledge, no such system currently exists on any social networking platform for detecting posts related to online death games and identifying the victims.

4 Proposed Architecture

4.1 Algorithm and Framework

The list of 50 challenges or tasks of the Blue Whale Challenge as known from various sources is shown in Fig. 1. Since the player of the game is required to share or post an image clearly indicating the task completion, the same can be used to extract the unique feature sets related to each tasks; these learned unique features can in turn be used as a cue to screen other user's social media messages and posts and identify the victims of online death games. Children curious to participate in these games also post status like "Curator Find Me" [8]. As can be seen from the list, most of the tasks have some distinctive text or image-related characteristics that can be used to identify the victims through automated techniques and alert the concerned public health authorities or family and friends in the user's social network. Keeping in mind the above design requirements, characteristics of the problem domain and challenges

1. Carve with a razor "f57" on your hand, send a photo to the curator.
2. Wake up at 4.20 am and watch psychedelic and scary videos that curator sends you.
3. Cut your arm with a razor along your veins, but not too deep, only 3 cuts, send a photo to the curator.
4. Draw a whale on a sheet of paper, send a photo to curator.
5. If you are read to "become a whale", carve "YES" on your leg. If not, cut yourself many times (punish yourself).
6. Task with a cipher.
7. Carve "f40" on your hand, send a photo to curator.
8. Type "#i_am_whale" in your VKontakte status.
9. You have to overcome your fear.
10. Wake up at 4.20 am and go to a roof (the higher the better).
11. Carve a whale on your hand with a razor, send a photo to curator.
12. Watch psychedelic and horror videos all day.
13. Listen to music that "they" (curators) send you.
14. Cut your lip.
15. Poke your hand with a needle many times.
16. Do something painful to yourself, make yourself sick.
17. Go to the highest roof you can find, stand on the edge for some time.
18. Go to a bridge, stand on the edge.
19. Climb up a crane or at least try to do it.
20. The curator checks if you are trustworthy.
21. Have a talk "with a whale" (with another player like you or with a curator) in Skype.
22. Go to a roof and sit on the edge with your legs dangling.
23. Another task with a cipher.
24. Secret task.
25. Have a meeting with a "whale".
26. The curator tell you the date of your death and you have to accept it.
27. Wake up at 4.20 am and go to rails (visit any railroad that you can find).
28. Don't talk to anyone all day.
29. Make a vow that "you're a whale".

30 – 49. Everyday you wake up at 4.20 am, watch horror videos, listen to music that "they" send you, make 1 cut on your body per day, talk "to a whale".

50. Jump off a high building. Take you life.

Fig. 1 Blue Whale Challenge levels and tasks [1, 2]

listed in previous section, we describe our proposed architecture in the following subsections for detecting the victims of Blue Whale game using deep learning and transfer learning techniques. Flow chart enumerating the key processing steps in our proposed deep learning-based architecture is depicted in Fig. 2.

4.2 Processing Images

Sample images corresponding to various tasks/levels of the online suicide game—Blue Whale—that were posted by teenagers on their different social media profiles are shown in Fig. 3. The images in the first row correspond to the task where the victim has to sit on the edge of a roof with his legs dangling (level 22) or stand on the edge of the roof of a tall building (level 17). The images in the second row are of level 11 wherein the victim has to carve a whale on his arm with a sharp object. The state of the art, deep learning algorithm for object detection, localization and classification: faster R-CNN [16] can be used to detect and flag images with objects that are related to various tasks or levels of any death game. For example, in our present study of Blue Whale, faster R-CNN can detect objects like feet, roof of a building, edge of a building, deep red-coloured outline of a whale, the characters "f57" or "f40" carved on hands with sharp objects, cuts, injuries and needle pricks, crane, railroads and others as per the list of tasks discussed earlier (Fig. 4).

For identifying a victim, the foremost task is to detect these object related to the game in images from user's social network posts. In order to achieve this, the next

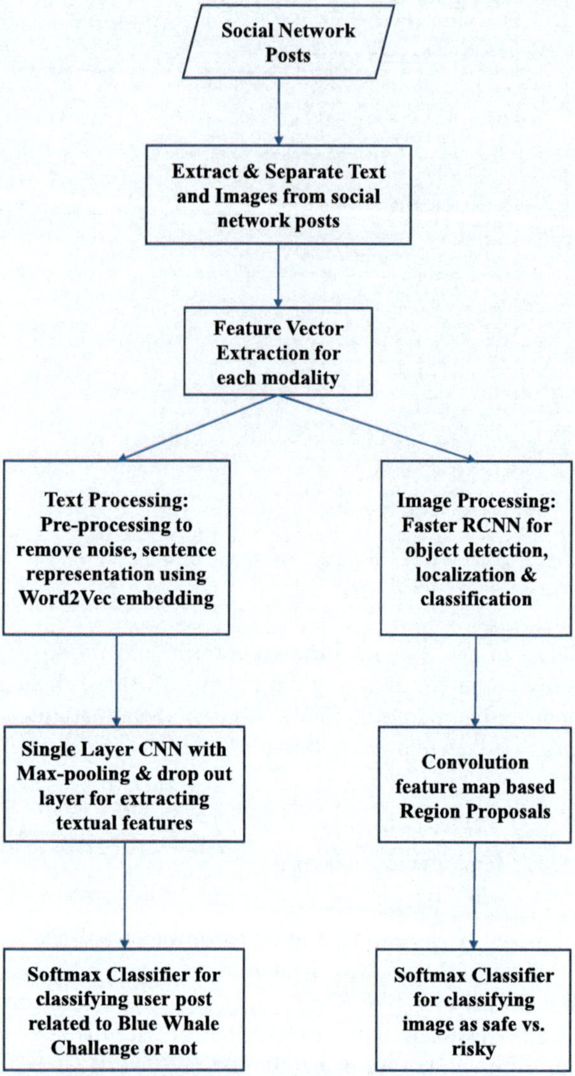

Fig. 2 Flow chart enumerating the key processing steps in our proposed deep learning-based architecture

step is to localize various objects and people in the image; and to finally predict if the objects related to the game are present or not and whether the person is in a dangerous situation. To achieve all these tasks, a tweak in faster R-CNN architecture would be sufficient and can be used on each image from user's posts in real time to detect, localize and predict the dangerous objects and circumstances.

R-CNN family of algorithms was developed by Ross Girshick et al. where different regions of interest from the image are passed to pre-trained CNN models to detect, localize and classify the objects present in an image. R-CNN uses selective search algorithm to extract 2000 region proposals from an image, post which CNN is used

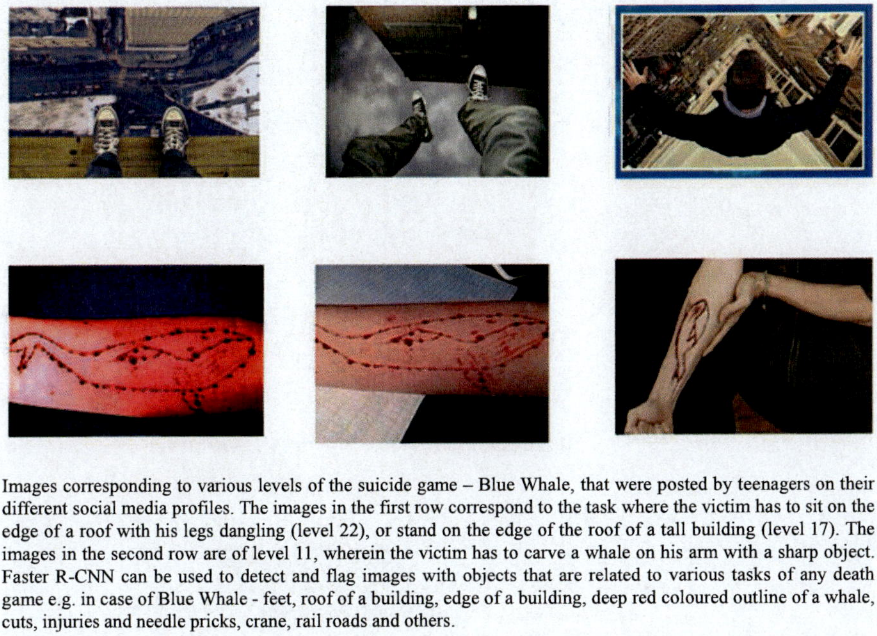

Fig. 3 Sample images shared by victims on their social network accounts as a proof of completing Blue Whale Challenges [14, 15]

for features extraction from these region proposals; and at the last step SVM classifier is used for predicting the presence of object of interest in the image. *Fast R-CNN* was an optimization over the previous algorithm achieved by generating convolution feature map only once per image instead of 2000 convolution operations of the previous R-CNN technique. The regions of proposal are then generated from the convoluted feature map instead of using selective search technique [16–18].

Ren et al. [16] proposed *faster R-CNN*, which is a development over *R-CNN* and *fast R-CNN*, and is the most optimized and state-of-the-art technique of this family. *Faster R-CNN* uses a convolution neural network over the entire image to extract a feature map for the whole of the image at once. This gives faster R-CNN computational advantage over R-CNN, where feature maps are generated for each of the 2000 proposals by multiple convolution operators. However, instead of using selective search like fast R-CNN, faster R-CNN allows the network to learn the region proposals over CNN feature maps. Hence, faster R-CNN is even faster than fast R-CNN due to computation speed up achieved by eliminating selective search technique for making region proposals [16–19].

For generating the feature maps, either ZF-net or VGG architecture can be used. The architecture can be pre-trained either with Pascal VOC data set or ImageNet data set, post which the faster R-CNN model can be trained on images associated with tasks of various levels of Blue Whale Challenge. This concept is known as

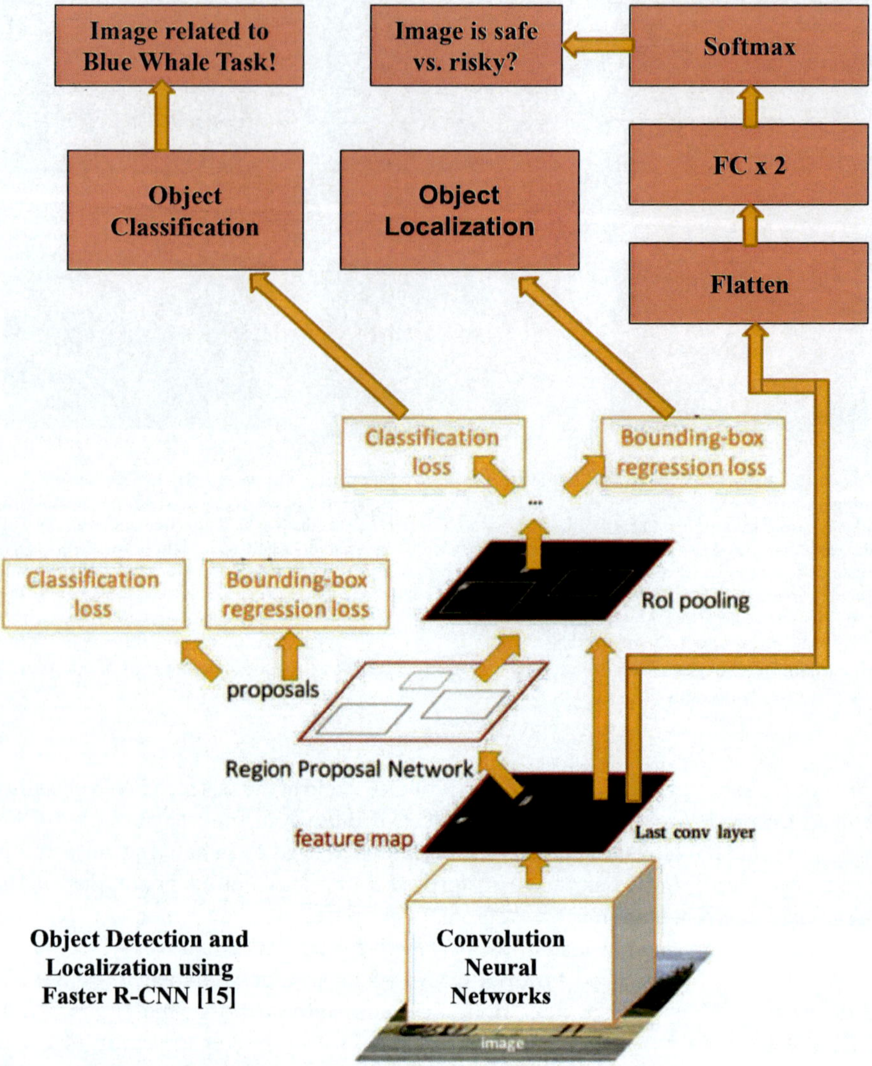

Fig. 4 Proposed deep learning-based architecture for detecting known objects related to Blue Whale Challenge tasks in users' posts using faster R-CNN [16]

transfer learning. Transfer learning is a popular technique of using pre-trained deep learning models for image and text predictive modelling or classification problems of a new domain where training data is scarce. By fine-tuning the model weights, a pre-trained model is adapted for another task domain where training instances are limited (as in the case of online death games, where initially not many training samples are available). This technique is extremely useful for our research problem, where new dangerous online challenges and trends surface get viral. The pre-trained CNN models, e.g. to detect injuries on one's body parts in images posted online or CNN-based text classifiers to detect posts with words like "challenge", "cut", "kill" learnt from Blue Whale Game can be used to detect social media posts related to some other dangerous self-harming challenge which may become popular in future.

As explained before, the features maps serve as input towards three different tasks. The first input is to the Region Proposal Network (RPN), where the proposals are created, even if these proposals may not contain an object. The second input is to the ROI pooling layer, which further refines the proposals, does object localization followed by classification layer to predict the presence of objects of interest. Lastly, the proposed tweak takes as input the feature map generated in previous step. The feature map is flattened to vectorize it, followed by two fully connected (FCs) layers and a softmax layer. The third branch predicts the class probabilities for the whole image to be safe or risky at an overall level, which means it is the input image indicative of any dangerous or risky situation a person may be in, e.g. a person standing on railway tracks or an edge of a cliff. This prediction will help in discovering dangerous and risky circumstances, self-harming activities and behaviours that people may be engaging in, and those may be associated with other online death games about which not much is known at the moment.

4.3 Processing Text

User's social networking posts and status messages are most commonly comprised of two modalities: text and images. In addition to analysing images shared by users for detecting if the user is a victim or player of online death games, the textual content in the user's post can also be leveraged to find cues for the same. Usually, it is seen that social media posts and status updates are short in length, with the presence of hashtags. Some of the literature studies have shown that posts related to Blue Whale challenge have presence of particular words, hashtags and are also short in length, for example: "I am Whale", "Curator Find Me". We propose the use of single-layer CNN architecture for sentence classification as developed by Kim [20] for this task, the same is illustrated in Fig. 5. First, sentence representations obtained using pre-trained Word2Vec embedding, which are then used to train a simple single-layer CNN-based architecture for sentence classification task, i.e. in our cases classifying user posts having short sentences related to the Blue Whale Challenge as discussed above.

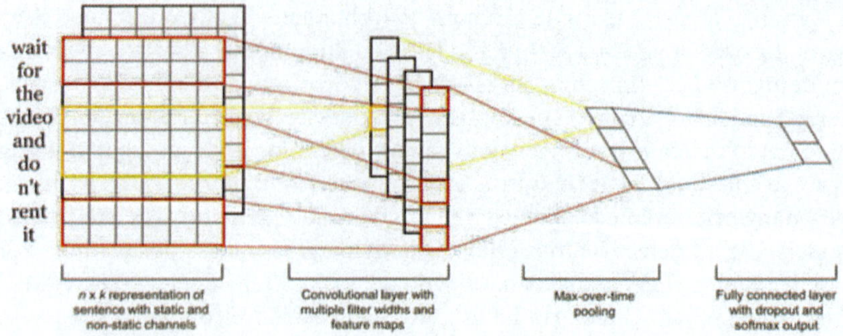

The deep learning model proposed by Yoon Kim is known to give state of the art results for short sentence classification. A single layer of Convolution Neural Network is trained over pre trained Word2Vec vector representations obtained for sentences in users' social media posts obtained after common natural language preprocessing. As very minimal hyper parameter tuning is required for this architecture, we use the same in our proposed framework. The word embeddings like the one used here Word2Vec are learnt by pre training on enormous size of text corpus, hence they offer universal features for natural language text and hence out perform in any text classification task where enough training data from the domain may not be available. This is also referred as transfer learning as explained previously.

Fig. 5 Identifying users' posts related to the Blue Whale Challenge using CNN architecture for sentence classification proposed by Kim [20]

The preprocessing of user's social media posts is done through standard NLP pipeline operations comprised of case conversion, removing punctuations and stop words, spell correction, tokenization followed by either stemming or lemmatization. Word embedding for sentences is obtained through any of the state-of-the-art pre-trained deep learning-based methods like Word2Vec, Glove, FastText, CBOW. Word embedding is a commonly used technique to represent words, sentences or documents using dense vectors, where position of any word in that vector is derived from the occurrence of that word among other words in natural language as it is used. The convolution functions with varying kernel size are applied, followed by max pooling layer to extract textual features and a dropout layer for regularization. The sentence classification is then done using the final softmax layer. This architecture is known to give the best results for sentence classification tasks across various data sets and requires minimal hyper parameter tuning [21].

4.4 Data Set and System Integration

Mukhra et al. [4] have rightly mentioned that the Blue Whale challenge is not a game but rather an unethical and immoral crime of the modern society. The Internet and social media platforms must develop state-of-the-art technology-driven systems to curb the spread of such online death games by monitoring both public and private activity of their users. As has been described above, the proposed system analyses user's social media posts using deep learning techniques to detect textual and image-based features and characteristics related to various tasks of Blue Whale Challenges. Many a times, task completion posts are also shared as private messages in community or group chats. In both the scenarios, the content or message exchange is happening over these online user-generated content delivery platforms; hence, it is possible to screen the content of these message exchanges. If frameworks like the one proposed by us are integrated by these websites, it can help in large-scale public health administration for timely detection of victims of online death games and other self-harming challenges before they become a popular trend.

However, due the private structure of most of these popular online social networks and lack of availability of user posts and data in public domain, the support and cooperation from these online platforms are essential to develop and integrate such systems. The system must screen and flag user posts if classified as risky based on the classification outcome of our proposed system. The immediate next step should be to reach out to the user through available channels and automated chat bots as the user could be a potential victim of the death game. The other users in his network marked as his family or close friends should also be alerted via the online platform and informed that the user could potentially be at risk. In addition to this, the victim or user's post should not be published or broadcast in order to prevent the spread of such dangerous challenges.

Additionally, a large data set is also required for training the faster R-CNN model for images and single-layer CNN model for text in order to extract feature representations for posts that may be related to dangerous online games. Since no known publicly available data set exists for this domain (except for a small data set related to KiKi Challenge [9] as mentioned in Sect. 2), hence, for each known task of an online death game, the data set needs to be artificially generated through data augmentation techniques like generative adversarial network (GANs) and so on. For example, images of bridges, rooftops, cranes, railroads, injury, needle prick marks and cuts on human body can be collected from other sources. Similarly, the corpus of text can be built for words related to the game, e.g. whale, death, rails, "f57", "f40" and others. The data set needs to be periodically augmented and enriched as and when new tasks related to online challenges are known. For this, metadata of user posts like the hashtags must be also be monitored.

4.5 Limitations and Future Scope

Due to the limitation of availability of data sets for this domain, the proposed framework could not be evaluated as no know publicly available data set exists to study the posts and messages related to dangerous online death games. However, we have proposed a novel state-of-the-art approach for achieving the task of identifying victims of online death games. As next steps, first we plan to collect and curate a data set for this domain using artificial data augmentation techniques like GANs as described above. Secondly, we shall evaluate our proposed algorithm and architecture on the artificially generated data set related to the Blue Whale Challenge. As a part of our future research, we shall also evaluate other deep learning architectures and study the applicability of techniques like zero-shot learning and one-shot learning to enhance the accuracy of our proposed framework. Since new variants of self-harming or online death games and challenges may come up in future, hence, we plan to build a generic framework detecting dangerous and risky situations from a user's social media posts which may indicate self-harm in any form or any other risk to his life.

5 Discussion

The popular trend of various forms of online challenges on different social networking sites is here to stay; often they are even endorsed by celebrities. What is important is to have a framework to distinguish between the challenges played for fun and entertainment versus those which are dangerous, risky and can be fatal. Such self-harming online challenge games where people even go to the extent of taking their lives is clearly a public health issues of the modern world. The concerned authorities often learn about these risky online challenges after a significant time has elapsed and some fatalities are reported by the traditional media. However, if novel data sources and techniques are used, such public health concerns may be alerted much earlier [10]. We believe using the multimodal deep learning and transfer learning-based system as proposed by us in this research, timely detection of user posts with injured body parts, or with them in dangerous backgrounds or physical spaces can easily be detected. This will help discover trends about dangerous online challenge before they become popular and help curb their spread by screening off the posts. To the best of our knowledge, this is the first such system that has been proposed to identify the victims of online death games, and no such known system exists on any social networking sites currently. Hopefully, with the development of such AI and ML-driven platforms, the "Blue Whale Challenge" would become the first and last such online death game where the victims lost their life to manipulative predators with nefarious psychology.

References

1. https://www.nationalheraldindia.com/national/blue-whale-suicide-game-now-available-under-different-names-unesco-advisory
2. https://indianexpress.com/article/what-is/what-is-the-blue-whale-challenge/
3. Siddiqui SA (2017) Cyberbullying and cyber-victimization: from online suicide groups to 'Blue Whale' menace. Indian Pediatr 54(12):1056
4. Mukhra R, Baryah N, Krishan K, Kanchan T (2019) 'Blue Whale Challenge': a game or crime? Sci Engi Ethics 25(1):285–291
5. Mukherjee A, Shukla SK (2018) Decoding the Blue Whale Challenge from sociological perspectives. J Natl Dev 31(1):112–119
6. https://www.safecommunitiesportugal.com/wp-content/uploads/2015/02/The-Blue-Whale-Challenge.pdf

7. Ramamurthy P (2017) The Blue Whale Challenge: why do people commit suicide for an online game? J Curr Res Sci Med 3(2):137
8. Mahadevaiah M, Nayak RB (2018) Blue whale challenge: perceptions of first responders in medical profession. Indian J Psychol Med 40(2):178
9. Baghel N, Kumar Y, Nanda P, Shah RR, Mahata D, Zimmermann R (2018) Kiki kills: identifying dangerous challenge videos from social media. arXiv preprint arXiv:1812.00399
10. Sumner SA, Galik S, Mathieu J, Ward M, Kiley T, Bartholow B, Mork P (2019) Temporal and geographic patterns of social media posts about an emerging suicide game. J Adolesc Health 65(1):94–100
11. Khattar A, Dabas K, Gupta K, Chopra S, Kumaraguru P (2018) White or Blue, the Whale gets its vengeance: a social media analysis of the Blue Whale Challenge. arXiv preprint arXiv:1801.05588
12. Pai RR, Alathur S (2020) Social media games: insights from Twitter analytics. Int J Web Based Commun 16(1):34–50
13. Pramod S, Natrayan B (2018) Content analysis and sentiment analysis of Blue Whale Challenge issue reported in news websites
14. https://www.india.com/viral/the-blue-whale-game-a-silent-house-a-sea-of-whale-wake-me-up-at-2-40am-and-other-internet-suicide-games-are-in-ban-list-of-schools-2097218/
15. https://www.india.com/viral/blue-whale-game-rules-facts-how-to-identify-the-50-day-challenge-that-pushes-kids-to-commit-suicide-2442613/
16. Ren S, He K, Girshick R, Sun J (2015) Faster R-CNN: towards real-time object detection with region proposal networks. In: Advances in neural information processing systems, pp 91–99
17. https://www.analyticsvidhya.com/blog/2018/11/implementation-faster-r-cnn-python-object-detection/
18. https://towardsdatascience.com/r-cnn-fast-r-cnn-faster-r-cnn-yolo-object-detection-algorithms-36d53571365e
19. http://www.agv.iitkgp.ac.in/blog/?p=59
20. Kim Y (2014) Convolutional neural networks for sentence classification. arXiv preprint arXiv:1408.5882
21. http://www.wildml.com/2015/11/understanding-convolutional-neural-networks-for-nlp/

TS-GAN with Policy Gradient for Text Summarization

Nobel Dang, Ashish Khanna, and Viswanatha Reddy Allugunti

Abstract Text summarization is a much evolving task, especially since neural networks were introduced. Similarly, generative adversarial networks (GANs) can be used to perform this task due to their ability to produce features or learn the whole sample distribution and produce correlated sample points. Thus, in paper, the authors exploited the characteristics of generative adversarial networks (GANs) for the abstractive text summarization task. The proposed generative adversarial model has three components: a generator which encodes the input sentences into much shorter representations; a discriminator which enforces generator to create understandable summaries; and a second discriminator which exerts upon generator to curb the output co-related to the input. The generator is optimized using policy gradient method, converting the problem into reinforcement learning. The ROUGE scores achieved by the model are as follows: **R-1: 41.52, R-2: 16.20, R-L 37.21**.

Keywords Abstractive text summarization · Generative adversarial network · Convolutional neural network

1 Introduction

Text summarization is the task of representing the whole given or available document with fewer semantically correct words. While doing so, the key information must be preserved. Text summarization is of two types: extractive and abstractive. Extractive text summarization involves extracting the right words or sentences from the

N. Dang (✉)
Maharaja Agrasen Institute of Technology, Delhi, India
e-mail: nobeldang@gmail.com

A. Khanna · V. R. Allugunti
Arohak Inc, Monmouth Junction, USA
e-mail: ashishkhanna@mait.ac.in

V. R. Allugunti
e-mail: itvissu@gmail.com

A. Khanna et al. (eds.), *Data Analytics and Management*, Lecture Notes on Data Engineering and Communications Technologies 54,
https://doi.org/10.1007/978-981-15-8335-3_64

corpus itself which sustains the meaning of the document, and the survey of various techniques used is done in [1]. The right words or phrases are then stacked to create the summaries. Whereas, in the abstractive text summarization, new sentences or words are formed which may not be originally present in the corpus to form semantically related summaries. Clearly, the abstractive text summarization task is much more appealing and indulging as it is related to the way in which humans write summaries. But abstractive text summarization is an exhaustive and difficult task to perform than extractive. Another problem which arises in carrying out text summarization is the varying lengthy input sentences. If the input corpus and its sentences are lengthy, then it becomes difficult to capture the context and find semantically related keywords or phrases. Extractive summaries often give better results than abstractive summaries because of the fact that abstractive summarization methods face various problems such as semantic representation, inference and natural language generation which are relatively harder than data-driven approaches which generally involves just targeting and extracting appropriate words or phrases. These two tasks, however, have escalated since the introduction of neural networks.

Text summarization is an important task because there is plenty of redundant information available in the form of text which has to be skimmed through in order to obtain the important gist. Also, the data contributed through the Internet has exploded within the past decade. The main advantage of text summarization is the reduction in the reading time of the user. A noteworthy text summary system should reproduce the entitled theme and essence of the document while keeping the repetition to minimum. There are various applications of text summarization, ranging from creating applications for newspaper articles to the meeting's minutes. The key structure used in text summarization; recurrent neural network (RNN); and convolutional neural network (CNN) also have many applications separately like generating captions as done in [2].

The authors of the paper perform the abstractive text summarization task on the CNN/Daily Mail dataset using a custom built generative adversarial network (GAN) model which contains one generator and two discriminators. The generator is responsible for encoding the input sentences in shorter lengths. Whereas, out of the two discriminators, one is responsible for creating human-understandable words and the other to provide feedback and restrain the generator to produce input words related output. Also, during the training, policy gradient method is also used to propagate the loss back to the generator from the discriminators. Finally, the relevance of summaries is evaluated by the ROGUE-n score, described in [3], and a comparison with that of state-of-the-art is done. The proposed architecture is derived from [4] but with subtle changes in the architecture of the generator and hyperparameters.

2 Dataset

The dataset used to train the TS-GAN is CNN/Daily Mail dataset which originally contained **287,113** training examples, **13,368** validation, and **11,490** testing examples and was initially introduced in [5]. The average token length in the long-text articles is about 785, whereas the corresponding summaries averages to be around 56. The dataset is preprocessed as done similarly in [4], and the output of the generator is limited to 80 tokens. Further, the different four classes for the generator's output were obtained as follows:

(1) Similar class: The original summaries, s^i, were used with text sentences x^i.
(2) Incomplete class: Some of the key features or words were removed from summaries, s^i, resulting in $s^i_{\text{incomplete}}$ as additional summaries.
(3) Redundant class: The text sentences from x^i were chosen and added to the respective, s^i, summary.
(4) Irrelevant class: Random summaries, s^j $(i! = j)$, were chosen and added in the summaries of x^i, resulting in $s^i_{\text{incomplete}}$ summaries.

3 Literature Review

For text summarization a lot of work has been done, especially in the past five years due to escalation in variants of neural and deep learning models. The first text summarizer was introduced in [6] where the authors used attention neural mechanisms to produce summaries on the input sentences. Their attention-based model gained remarkable performance in the DUC-2004 shared task compared with several strong baselines. Later on many advancements have been done to improve the proficiency and results with different variants like encoder-decoder models, etc. However, the model built in [6] performed well only on short sentences.

In [7], the authors performed abstractive text summarization using attention-based encoder-decoder, sequence-to-sequence recurrent neural network (RNNs) on larger text sentences and corpus as introduced and described in [8, 9]. The authors also introduced a new dataset, CNN/Daily Mail, which consists of multi-sentence summaries.

The authors in [5] performed summarization with pointer-generator networks which helped in increasing the state-of-the-art ROGUE score by two points. They proposed a novel architecture that changed the standard attention-based sequence-to-sequence model in two orthogonal ways. First, they used a hybrid pointer-generator network to accurately reproduce information by copying words from the source text via pointing, while retaining the ability to produce novel words through the generator. Second, they discouraged repetition by using coverage to keep track of what has already been summarized.

The ROGUE-L score in the state-of-the-art is further improved in [10] in which for the first time generative adversarial networks (GANs) were used with policy gradient for text summarization. They modeled the generator as a stochastic policy in reinforcement learning, and whose goal is to generate the summaries. Whereas the discriminator aimed to calculate and predict the probability that the training set created the summary rather than the generator.

A new intra-attention-based model is introduced by authors in [11]. The model is a deep-reinforced model. They combined the supervised learning with reinforcement learning to throw the "exposure bias" caused due to the supervised learning. The authors generated more readable summaries by amalgamating the standard word prediction with global sequence prediction training of the reinforcement learning.

The idea of creating human-like and human-readable summaries was also exploited in [12]. The authors of [12] proposed an auto-encoder model. The model consists of an encoder to encode the input sentences to shorter length sentences, and a reconstructor to produce generator input from the generator's output. To make the generator's output human readable, a discriminator is introduced to restrain the output to resemble human-like summaries.

The authors of [13] introduced a GAN model with reinforcement learning on the CNN/Daily Mail dataset. They simultaneously trained both, the generator and discriminator. They showed a qualitative comparison of more abstractive, readable, and diverse summaries by achieving one of the best ROGUE scores, R-2 to be 17.65.

The authors of [4] built a GAN [14] trained with a policy gradient algorithm, simulating [10], but they modified the architecture with one generator and two discriminators. This helped them increase the ROGUE-1 score to 40.11 and be the state-of-the-art.

4 Proposed Work

4.1 Ts-Gan

The proposed model, TS-GAN, to do text summarization consists of majorly three components.

4.1.1 Generator

First component is a generator which takes the long sentences as input and then encodes it to shorter length text. The generator can either output words directly from the text or make a prediction from the vocabulary. The generator takes an ith sentence as $x^i = \{x^i_1, x^i_2 \ldots, x^i_N\}$, where x^i_n is the nth word in the ith sentence. This will produce a distribution of words from the vocabulary which is then sampled to produce $y^i = \{y^i_1, y^i_2 \ldots, y^i_T\}$. Thus, the loss function of generator is:

$$L_G = \frac{1}{N}\sum_{i=1}^{N} -\alpha C\left(x^i, y^i\right) - \beta D\left(y^i\right) \tag{1}$$

where $C(x^i, y^i)$ is the output of the similarity discriminator, and $D(y^i)$ is the output of the readability discriminator, and α, β are two hyperparameters as shown in Eq. (1). After this, the generator is optimized using policy gradient which is discussed in [4, 15, 16]. The generator consists of both the encoder and decoder, both of which are two-layer LSTM-based models with 128 and 256 hidden states, respectively, as projected the use in [17]. As the output of the generator is a four-dimensional classifier tensor, the loss used is cross-entropy with the human-written summaries as ground truth values.

4.1.2 Similarity Discriminator

The second component is the similarity discriminator which is a four-class text classifier. The four classes are: Similar, Incomplete, Redundant, and Irrelevant class and are as follow:

(a) Similar class: The short text is much related to the long text.
(b) Incomplete class: Some of features present in long text are missing from the short text.
(c) Redundant class: The repetitive or redundant information present in the short text thus resulting in inappropriate summaries.
(d) Irrelevant class: The short text not related to the long text.

There are two encoders in the similarity discriminator which are CNN-based models with convolution, max pooling, and activation layers in the suggested order. One encoder is for short text whereas the other for long text. The weights of CNN are initialized with xavier initialization. The results of the two encoders, e_l and e_s, are then concatenated, multiplied element wise, and the absolute difference is taken, which is then fed to the fully connected layers and later on the softmax for probability distribution whose final output is a one-hot tensor with four dimensions. The architecture of the similarity discriminator is shown in Fig. 1.

The encoder in the similarity discriminator uses convolutional layers with filters of window sizes 3, 5, and 7 with a total of 256 kernels and is pre-trained as mentioned in [18]. The resulting 256 activation maps were then connected to a fully connected layer with 512 neurons as in [4]. The optimizer used for training the similarity generator is SGD with a learning rate of 0.06 with batch size of 128.

4.1.3 Readability Discriminator

The third component present in our proposed TS-GAN is a readability discriminator which is another CNN-based model. It tells whether the summary is generated by the generator or human. Thus, it is essentially a binary classifier whose architecture consists of a convolutional layer, then max-pool, and then finally a softmax classifier for predicting the probability distribution of the two classes. It is pre-trained by following the methods explored in [18]. The class depicting human-written

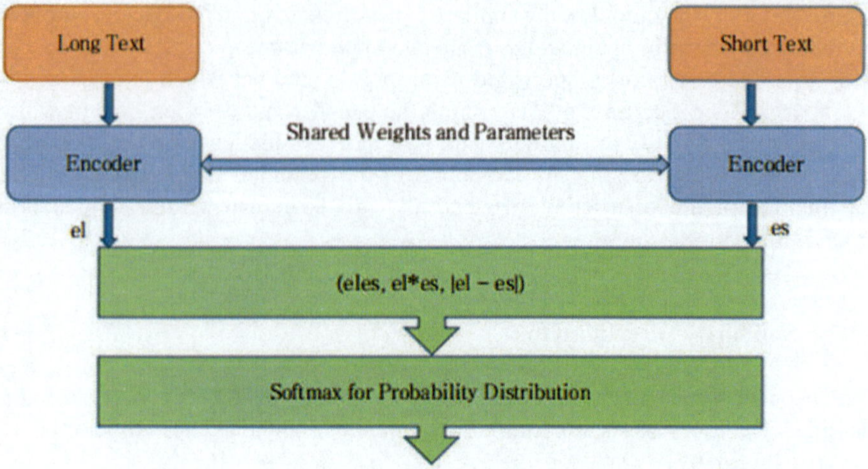

Fig. 1 Architecture of similarity discriminator

summaries are treated as "positive" whereas the summaries generated by the generator as "negative." The loss function used in the binary classifier of the readability discriminator is for positive and negative sample inputs is described in Eq. (2).

$$L_D = \frac{1}{N}\sum_{i=1}^{N} -\log\big(1 - D\big(x_-^i\big)\big) - \log\big(D\big(x_+^i\big)\big) \tag{2}$$

The following readability discriminator is also a CNN model with kernels of window sizes 3, 4, and 5. Each filter uses and results in 128 activation maps. Before training it with the generator and similarity discriminator, the readability discriminator is pre-trained with the pre-trained generator only. The outputs of the generator were treated as "negative" samples whereas the original text summaries as the "positive" instances. The architecture of the TS-GAN is shown in Fig. 2.

4.2 *Policy Gradient*

The output of the generator is not directly fed into the two discriminators; rather the sampled discrete sequences are used to feed the two discriminators because of the non-differentiability problem. The task of learning of the TS-GAN is converted to a reinforcement problem by using SeqGAN from [15, 16]. The reward from the similarity discriminator is given when the generator produces semantically related summaries. Similarly, the reward from the readability discriminator is given when the summary produced by the generator is human understandable. The policy gradient algorithm for training the proposed architecture's generator is discussed in [4]. Thus,

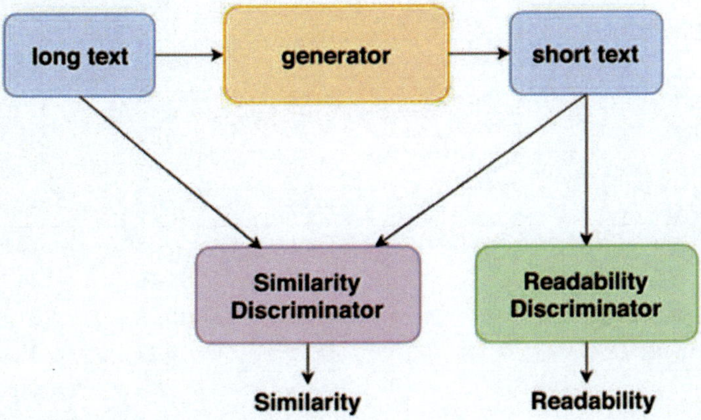

Fig. 2 Architecture of TS-GAN [4]

policy gradient helps train the generative adversarial network with reinforcement learning and eliminating the "exposure bias" introduced while supervised training as discussed in [11].

5 Results

The proposed TS-GAN model by Dang N. Khanna A. and Allugunti R. V. was able to be the state-of-the-art when it comes to overlapping of unigram word, i.e., R-1, and the longest common subsequence, i.e., R-L, between the systems and their generated summaries, respectively. The output summaries by the generator of TS-GAN achieved the ROGUE scores, introduced by the authors of [3], as follows: **R-1: 41.52, R-2: 16.20** and **R-L: 37.21**. The two-layer LSTM architecture with different hyperparameters from the [4] resulted in better summaries, and the comparison of result is done in Table 1.

As mentioned in Table 1, the proposed model, TS-GAN, achieved better R-1 and R-L scores. However, the model lags behind in R-2 score or the overlapping of bi-grams between article's text and their generated summaries. This may be the result of losing the context with bi-grams intact, which can be further improved by using pre-trained word embeddings enabling learning of the bi-grams or n-grams suitably.

6 Conclusion

The proposed generative model, TS-GAN, by Dang N., Khanna A. and Allugunti R. V., was able to generate the summaries for the given long article or corpus

Table 1 Comparison of ROGUE metric

Method	R-1	R-2	R-L
Pointer-generator [5]	39.53	17.28	36.38
SummaRuNNer [7]	39.6	16.2	35.3
GAN + policy gradient [10]	37.87	15.71	39.20
Deep-RL [11]	39.87	15.82	36.90
WGAN adversarial reinforcement [12]	35.51	9.38	20.98
GAN [13]	39.92	**17.65**	36.71
Zhuang H. and Zhang W. [4]	40.11	16.23	37.18
Proposed model TS-GAN	**41.52**	16.27	**39.21**

with unprecedented ROGUE values, R-1 and R-L. The model reported R-1 score to be **41.52** and R-L to be **39.21.** The GAN model which comprises three components, generator, similarity discriminator, and readability discriminator, is pre-trained individually firstly and then jointly using reinforcement learning's policy gradient algorithm [16]. The model's architecture is similar to the architecture described by the authors in [4], but the generator and the hyperparameters have been modified accordingly producing better results. However, in future these results can be further improved or noted by using Bert Embeddings as introduced by the authors in [19]. These are the pre-trained k (mostly $k \Rightarrow 768$) dimensional word vectors for words in the vocabulary which can be used via transfer learning or can be custom trained too. These pre-trained word embeddings can help further improve the ROGUE metric.

References

1. Moratanch N, Gopalan C (2017) A survey on extractive text summarization. In: International Conference on Computer, Communication and Signal Processing (ICCCSP)
2. Khamparia A, Pander B, Tiwari S, Gupta D, Khanna A, Rodrigues J (2019) An integrated hybrid cnn-rnn model for visual description and generation of captions. In: Circuits, Systems and Signal Processing, Springer
3. Yew CL (2004) ROUGE: a package for automatic evaluation of summaries. In: Proceedings of the Workshop on Text Summarization Branches Out (WAS 2004), Barcelona, Spain
4. Zhuang H, Zhang W (2019) Generating semantically similar and human-readable summaries with generative adversarial networks. IEEE Access 7:169426–169433
5. See A, Liu JP, Manning DC (2017) Get to the point: summarization with pointer-generator networks. In Proceedings 55th Annual Meeting Assoc. Comput. Linguistics, vol 1, pp 1073–1083
6. Rush MA, Chopra S, Weston J (2015) A neural attention model for abstractive sentence summarization. In: Proceedings Conference Empirical Methods Natural Lang. Process, pp 379–389
7. Nallapati R, Zhai F, Zhou B (2017) SummaRuNNer: a recurrent neural network based sequence model for extractive summarization of documents, *AAAI*, pp 3075–3081
8. Sutskever I, Vinyals O, Le VQ (2014) Sequence to sequence learning with neural networks. Adv Neural Inf Process Syst (NIPS) 27:3104–3112

9. Ramachandran P, Liu P, Le VQ (2017) Unsupervised pretraining for sequence to sequence learning. In: Proceedings of Conference on Empirical Methods in Natural Language Processing, 2017, pp 383–391
10. Rekabdar B, Mousas C, Gupta B (2019) Generative adversarial network with policy gradient for text summarization. In: Proceeding IEEE 13th International Conference Semantic Computer (ICSC), pp. 204–207
11. Paulus R, Xiong C, Socher R (2018) A deep reinforced model for abstractive summarization. In: International Conference on Learning Representations
12. Wang Y, Lee H-Y (2018) Learning to encode text as human-readable summaries using generative adversarial networks. In: Proceeding Conference Empirical Methods Natural Language Process, Association for Computational Linguistics, Stroudsburg, PA, USA, pp. 4187–4195
13. Liu L, Lu Y, Yang M, Qu Q, Zhu J, Li H (2017) Generative adversarial network for abstractive text summarization
14. Goodfellow I, Pouget-Abadie J, Mirza M, Xu B, Warde-Farley D, Ozair S (2014) Generative adversarial nets. In: Processing Advances in Neural Information Processes System, pp. 2672–2680
15. Yu L, Zhang W, Wang J, Yu Y (2017) Seqgan: sequence generative adversarial nets with policy gradient. In: Proceeding 21st AAAI Conference Artificial Intelligence, pp. 2852–2858
16. Sutton R, McAllester D, Singh S, Mansour Y (2000) Policy gradient methods for reinforcement learning with function approximation. In: Advances in Neural Information Processing Systems, pp 1057–1063
17. Hochreiter S, Schmidhuber J (1997) Long short-term memory. Neural Comput 9(8):1735–1780
18. Zhang H, Xu J, Wang J (2019) Pre Training-based natural language generation for text summarization. arXiv:1902.09243
19. Devlin J, Chang WM, Lee K, Toutanova K (2019) BERT: pre-training of deep bidirectional transformers for language understanding. NAACL-HLT 2019

Energy-Efficient Routing Protocols for Cluster-Based Heterogeneous Wireless Sensor Network (HetWSN)—Strategies and Challenges: A Review

Preeti Gupta, Sachin Tripathi, and Samayveer Singh

Abstract Regardless of the advancement in wireless sensor network (WSN), efficient energy utilization is still essential to increase network lifetime. In real-time applications, due to battery constraints of sensor nodes, the network lifetime depreciates. To ameliorate the energy depletion problem, cluster-based heterogeneous WSN (HetWSN) integrated with optimization techniques is one of the solutions. Technologies like IoT, machine learning, and neural network could be of great importance for optimization. In this paper, we address comprehensive literature review of cluster-based routing protocols along with their pros and cons for HetWSN, covering period of 2009–2019. In addition, we also briefly compare energy-based and hybrid clustering algorithm for static and mobile HetWSN on the basis of various clustering attributes. As an outcome of our review, we present a statistical study of the survey which will give researchers a direction to propose novel energy-efficient protocol in future. Finally, open issues in WSNs followed by some discussion and conclusion are presented in the paper.

Keywords Heterogeneous WSN · Clustering · Mobility · Machine learning · Internet of things

P. Gupta (✉) · S. Tripathi
Indian Institute of Technology (ISM) Dhanbad, Jharkhand, India
e-mail: preetigupta216@gmail.com

S. Tripathi
e-mail: var_1285@yahoo.com

S. Singh
National Institute of Technology, Jalandhar, Punjab, India
e-mail: samayveersingh@gmail.com

A. Khanna et al. (eds.), *Data Analytics and Management*, Lecture Notes on Data Engineering and Communications Technologies 54,
https://doi.org/10.1007/978-981-15-8335-3_65

1 Introduction

WSN has become one of the most attracted topics of research due to its ability to detect and monitor physical or dynamic environmental conditions over a time frame. How vital has wireless sensor network become nowadays can be deciphered from the high volume of research being carried out related to it in past decade. Many sensor nodes combined to form a WSN. The sensor node in WSN has ability to gather information through sensing unit, process the collected data through computing unit, and perform communication. Any physical or dynamic activity in the environment can be sensed by a sensor node in which they are installed. A sensor node encompasses components like a battery (acts as power source), sensing unit (sensors and analog-to-digital converters), transceiver, and processing unit (storage and processor). Other possible parts may be generator for energy, mobilizer to move sensor node whenever required, and a system for finding a location with high accuracy. Source of power must be used proficiently in the sensor nodes, because of extensive network structure the replacement of the embedded batteries in sensors is a quiet tough procedure once they have been deployed. In cluster-based network, the collected data is transferred to the cluster head (CH) and then to the node from where data can certainly be sent to user either through a cables or wireless medium. In clustered WSN, cluster head makes the data ready to be used at the base station and further to be used by users. The setup of sensor node in cluster-based network can be done using static or mobile element.

Several applications in WSN such as environment monitoring [1], control system for pollution, operations of military, vehicle motion controlling, finding earthquake, target tracking system and surveillance system, monitoring system in medical [2] are of great significance. Efficiency of cluster-based HetWSN can be examined through some parameters like sensor node deployment, clustering properties, CH selection, mobility of nodes, network scalability, energy efficiency, connectivity between network, data delivery ratio at network layer, etc.

1.1 Preliminaries

Clustering is the method in which samples are apportioned into sets with alike members, and these groups are called clusters. It falls under one of the divisions of unsupervised learning. Cluster-based HetWSNs can be classified according to the clustering techniques, clustering algorithm, mobility, and node heterogeneity. The purpose of the grouping nodes in a cluster is to ameliorate energy efficiency. However, grouping node in one cluster is a quiet challenging issue. Classification of cluster-based HetWSN is presented in Fig. 1.

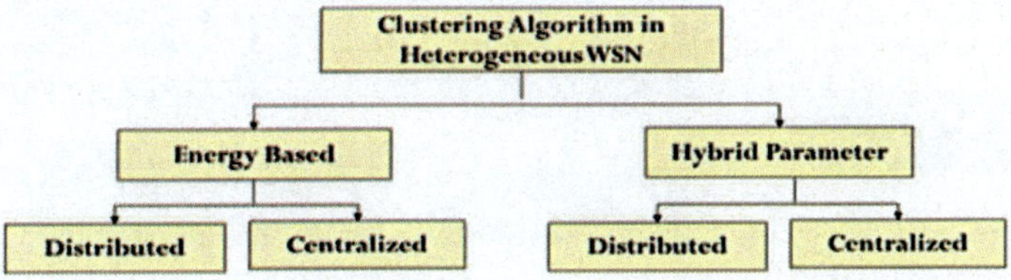

Fig. 1 Classification of clustering algorithm

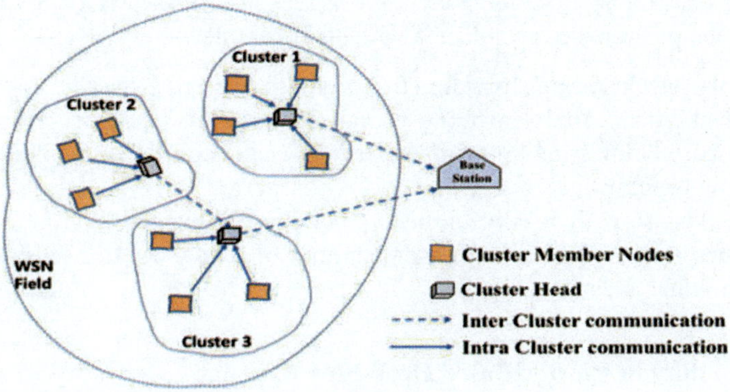

Fig. 2 Architectural design of cluster-based HetWSN

1.1.1 Clustering Techniques

To differentiate the cluster-based networking protocols, there are several parameters employing clustering techniques for grouping nodes in a cluster depending on network structure, data processing, and node deployment in HetWSN. Nodes placed in network area may have similar or different energy levels, processing capability, or linking potential. Figure 2 shows the two-level hierarchy of cluster member nodes. Furthermore, networks can be established on basis of hop (single/multiple) and number of sinks in the network. Cluster head nodes send the gathered data to CH, and collected aggregated data is sent to the sink (base station), respectively. The challenges for obtaining the best result in clustering come out from the process of clustering. The process of clustering includes cluster formation and cluster head (CH) selection algorithms like distributed, hybrid, and centralized. Incorporating machine learning techniques, neural network, or IoT could be effective in optimizing the clustering result [3–5].

1.1.2 Clustering Algorithm in HetWSN

Sensing devices placed in network field may have similar or different energy levels, processing capability, or linking potential. The nodes with higher energy level have

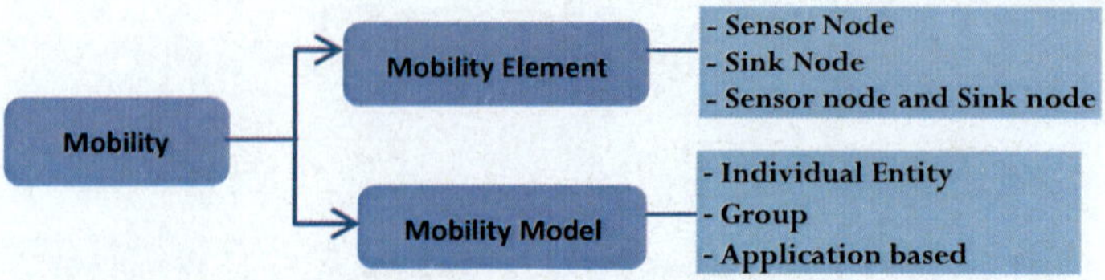

Fig. 3 Classification of mobility for stationary and moving element

the greater probability to become a cluster head. Cluster-based HetWSN protocols are either energy-based or established with other parameters.

- Energy-based clustering algorithm: In this type of clustering, the election of cluster head is based on total remaining energy of nodes and the energy consumed. Energy-based clustering protocols in HetWSN can be further classified based on clustering topology.
- Hybrid clustering algorithm: In this type of clustering algorithm, the communication between base station and member node of cluster decides selection of the head node in a cluster.

1.1.3 Mobility in Cluster-Based HetWSN

Mobility is one of the parameters that is also helpful in numerous of application-based WSN. Mobility is significant generally in application-based network and remarkably, for the IoT. The invention of the Internet of things (IoT) has accelerated avant-garde protocols which ameliorate the lifetime of a network [6]. Mobility in WSN can be categorized on the basis of physical as well as architectural aspects. Figure 3 shows categorization of mobility in a network for stationary or moving element.

1.1.4 Challenges in Clustering

The procedure of clustering includes cluster formation and election of a head node of a cluster. Difficulties in forming a cluster and selecting a cluster head are mentioned below

a. Challenges in formation of cluster are total number of clusters, type of communication (inter/intra), load balancing, and cost.
b. Challenges in selecting cluster head (CH) are energy efficiency, connectivity, load balance among all sensor devices and node distance.

Though various methods have been established for the appropriate grouping of node, but the sensors are clustered suitably or not still do not meet the accountability for answering that the clustering done is best or not.

1.2 Outline

Section 2 explains the work done on different clustering routing protocols in the direction of achieving elongated lifetime for HetWSNs. Contributions and corresponding discussions which contain node deployment, cluster formation, cluster head selection based on several parameters, and node heterogeneity of various protocols are comprehensively discussed in Sect. 3. Section 4 presents statistical analysis of protocols in this review paper. Section 5 discusses the persisting problems in this area. Conclusion of entire paper is reported in Sect. 6. A little direction to the future work is also given along with the conclusion. Thereafter, the references in the proposed work are listed.

2 Literature Review

A protocol, low-energy adaptive clustering hierarchy (LEACH) for WSNs, efficiently utilized energy using clustering approach which in turn increases network lifetime, W. R Heinzelman et al. [7]. The issues associated due to random selection of CH in LEACH protocol are (i) CH nodes are not permanent and (ii) CHs distributed disproportionately in the zone where they are installed. These two issues can be solved by uniformly distributing the CHs throughout the network. The same issues have been investigated and solved suing LEACH-C protocol and a fixed LEACH protocol. Simulation and analysis of two-level hierarchical network (HWSN) done by Smaragdakis et al. [8] solve the problem of uneven CH uneven energy distribution. Election of CHs based on the leftover energy of every sensor node follows the procedures of weighted election probabilities per node. It gives reduced performance for multi-level HWSN.

A major effort for multi-level hierarchical network (HWSN) is investigated by Qing et al. [9]. They proposed clustering algorithm, distributed energy-efficient clustering (DEEC), which provides high performance by saving energy of nodes and thus prolonged network life, Lindsey and Raghavendra [10]. Another extension of LEACH protocol is hybrid energy-efficient distributed (HEED) clustering protocol [11], which uses primary constraints and secondary constraint to choose the CHs. Overhead is less in relations to processing and exchange of message. Also, it does not presume any distribution of the nodes or location awareness.

Yazid et al. [12] analyzed the routing protocols in an extensive manner based on clustering method. Comparing all those architectures, the design of TSEP and SEP was efficient than that of ZSEP and ESEP. Therefore, in order to enhance network lifetime and to save energy, TSEP architecture can be a decent preference for heterogeneous WSN. Several reviews of cluster-based routing protocols in heterogeneous WSN over the period of time are summarized in Table 1.

From the survey of literature in Table 1, it can be seen that hierarchy of sensor nodes, consumption of energy, redundancy in data, SNs size, consistency, safety,

Table 1 Several reviews of cluster-based routing protocols in heterogeneous WSN

Year and author	Objective	Pros	Cons
2009 GM et al. [13]	Classification based on energy efficiency, minimize bandwidth, and latency	Efficacy and weak point of every protocol described	Comprehensive tabular analogy is not performed
2010 Vivek et al. [14]	Research on different clustering algorithm for HetWSN. Protocols are categorized based on energy efficiency and system stability	Incorporated different heterogeneity types	Few parameters selected for comparison
2010 Shio et al. [15]	Survey of energy-efficient protocol based on category and system stability	Grouping on the basis of various categories of routing protocol	Tabular classification was missing
2011 Chunjuan et al. [16]	Design of accuracy-based 3D network and duty cycled WSN	Underlined problems in clustering	Tabular classification was missing
2012 Kewei et al. [17]	Articulation of problems based on cluster size optimization and choice of proper communication mode	Qualitative review of protocols	Heterogeneity not considered while comparison
2012 Sahoo et al. [18]	Layout of protocols for multi-path infrastructure, energy efficiency, design, and challenges in WSN	Inclusion of the concept of MAC protocol	No tabular comparison
2013 Sanjeev et al. [19]	Selection of optimal routing path, energy efficiency, network lifetime cluster head selection	Tabular approach to explain categorization of routing protocols	All heterogeneities not considered
2013 Tyagi et al. [20]	Grouping according to cluster head selection, security and load balancing	Pictorial classification of each categories of routing protocols	Types of heterogeneity not considered
2013 Zahariah et al. [21]	Aim to project protocols based on centralized topology management and network lifetime	Strategy of designing routing protocol along with performance evaluation encompassed	Focused only on hierarchical routing protocols

(continued)

Table 1 (continued)

Year and author	Objective	Pros	Cons
2015 Sudeep et al. [22]	Comprehensive discussion on various protocols, to highlight the pros and cons with respect to some performance evaluation parameters	Good and extensive research on heterogeneous protocols	Focused only on energy heterogeneity
2015 Singh J et al. [23]	To describe the operations of WSN algorithms and to assess them based of several clustering features	Brief review of protocols	Mobility of node in HetWSN is not discussed
2016 Jing et al. [24]	Categorize and brief discussion on persisting routing protocols into homogeneous or heterogeneous WSNs	Tabular method for comparison of protocols	Covered less protocols for comparison
2018 Ali et al. [25]	Classification of clustering algorithms that are stability oriented and energy efficient according to operation model and network architecture	Good and extensive research on heterogeneous protocols	Not discussed much about mobility in heterogeneous wireless sensor network
2019 Bhagya Shri et al. [26]	Current researches performed to obtain the result for less energy utilization and better network lifespan	Good and extensive research on clustering protocols	Detailed description missing. Types of heterogeneity not taken into consideration

and fault tolerance is some features exhibited by many of routing protocols, very few persisting researches have contemplated the node heterogeneity with respect to various parameters. In our work, we have defined classification of various heterogeneous protocols as well and focused more static and mobile heterogeneous WSN.

3 Classification of HetWSN

HetWSN consists of sensor devices having dissimilar level of energy. The processing, computing, and linking capacity vary in the WSN area. Further classification in HetWSN could be done on the basis of mobile or static HetWSN. For clustering in

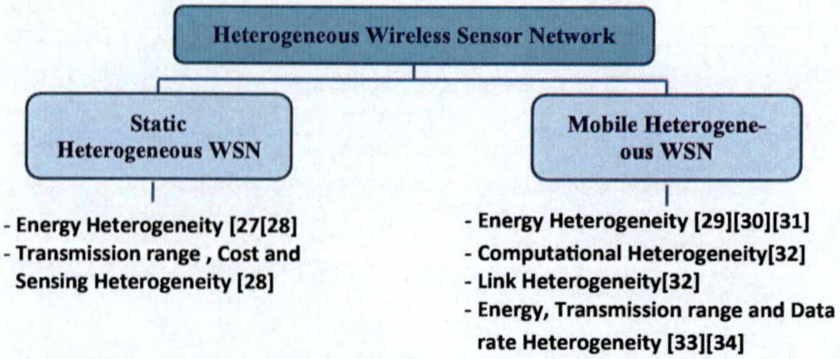

Fig. 4 Classification of energy-efficient heterogeneous routing protocols based on static and mobile environment

HetWSN, various cluster-based protocols can be distinguished based upon centralized, distributed, or hybrid clustering algorithms. Division of routing protocols on basis of mobility of node for heterogeneous wireless sensor network is illustrated in Fig. 4.

A comprehensive explanation of cluster-based protocols for static and mobile heterogeneous WSNs (HetWSN) centered on energy heterogeneity is provided in this paper. Employing node heterogeneity in the design of cluster-based WSNs reduces the latency and enhance the data throughput transfer to the destination from the sink. Different types of heterogeneity are energy heterogeneity, computational heterogeneity, link heterogeneity, and hybrid category. Using heterogeneity, concept in cluster-based HetWSN can boost the energy efficiency of the network. Mobile nodes are useful to minimize the number of hops traveled by a data packet in a network using single or else multi-hop communication.

3.1 Static Heterogeneous WSN (HetWSN)

A network made up of the sensing devices in which the nodes are static in nature but have different processing capabilities. Sensor nodes with hierarchy of nodes supporting different level of energy help gaining improved network lifespan.

3.1.1 Energy-Based CH Selection Algorithms

ECDC [34] protocol used in coverage of area and a particular point in heterogeneous WSNs for improving efficiency in network lifetime. This protocol has three various divisions for sensor nodes in terms of the energy that are cluster head (CH), plain node, and a cluster member. CH is selected based on coverage and residual energy which leads to even distribution of cluster size. Additionally, the lifetime of CH is decided from the starting time to the period when more than thirty percent of sensors are dead. The simulation yields depict the decreased energy consumption and improved coverage performance when compared with HEED and LEACH. This protocol is used where nodes are uniformly deployed.

Chand et al. proposed a protocol heterogeneous HEED [35] for HetWSN in which three-level hierarchy of sensor nodes is considered based on energy heterogeneity. Authors also incorporated distance factor also followed by fuzzy logic implementation to elect the CHs. Incorporating fuzzy logic system improved the network lifespan by many folds because a greater number of packets are forwarded to the sink with reduction in energy consumption. The simulation work of heterogeneous HEED protocol shows that there is an increase in number of active sensor devices on increasing the energy heterogeneity level. Hence, the network lifetime is improved at excessive level when higher energy heterogeneity is used.

A protocol multi-level HEED proposed by S. Singh [36] considered six-level heterogeneity in the experiment. The simulation result shows improved throughput and network lifetime with reduced aggregate delay. CH selection is based on total energy remaining in a node and node density. ECCR [37] is cluster-based energy-centric protocol proposed by A. S. M. Sanwar Hosen et al. and the selection of cluster head is based on residual energy, distance from member devices.

Verma et al. proposed improved dual hop routing (IDHR) and multiple data sink-based energy-efficient cluster-based routing (MEEC) protocols [26] in HetWSN. Selecting a cluster head in IDHR and MEEC depends on parameters, i.e., energy, node density, and distance between the sink and a node. There are three types of node namely super, advance, and normal nodes. Although network structure for MEEC and IDHR is different, cluster head selection and the network functioning are same for both the protocols. Probability of each node being alive is calculated once it is checked that residual energy of node is greater than threshold energy. Thereafter, a random number is generated using following equation: $\mathrm{ARN}_0 = \frac{N-D_N}{N} \times R_N$.

If ARN_0 generated is greater in value than the threshold value for advance, super, and normal node then respective node becomes a cluster head. MEEC reduces the energy-hole problem thereby enhancing the network lifespan. From the simulation results, it is concluded that IDHR and MEEC individually outperform state-of-the-art protocols.

Xiaoqiang et al. worked on routing protocol in HetWSN using bio-inspired optimization technique to propose a new protocol using modified grey wolf optimizer (GWO) [38]. Cluster head is selected on the basis of remaining energy of a node and distance. Cluster is formed using various fitness functions for the nodes in HetWSN. The experimental results show signification improvement in network energy. Energy-based clustering protocol has been compared in Table 2.

3.1.2 Hybrid Parameters-Based Clustering Algorithm

The authors of lifetime extended multi-levels heterogeneous routing (LEMHR) showed that improvement of energy of a network initially may not assure an improvement of energy for higher level nodes in the starting when compared with the nodes at

Table 2 Brief comparison of energy-based clustering protocols for static HetWSN

Protocol and reference	Cluster head selection constraints	Cluster structure	Data transmission (inter–intra-cluster)	Cluster optimization technique/model	Node hierarchy	Deployment strategy	Stability	Load balancing	Reliability
2014 ECDC [34]	Residual energy and coverage	Distributed	Single hop–single hop	Radio energy model	Three levels (CH, plain node, and cluster member)	Uniform in square filed	–	–	–
2014 Heterogeneous HEED [35]	Residual energy and neighbor density of nodes	Distributed	Single hop–single hop	Fuzzy logic	Three levels	Random (square filed)	–	Yes	Yes
2017 ML HEED [36]	Residual energy and node density	Distributed	Single hop–single hop	Radio energy model	Six levels (n level)	Random (square area)	–	Yes	–
2018 ECCR [37]	Residual energy, no. of member nodes distance from base station	Hierarchical	Multi-hop–single hop	First-order radio dissipation model	One level	Uniform and random	Good	–	–

(continued)

Table 2 (continued)

Protocol and reference	Cluster head selection constraints	Cluster structure	Data transmission (inter–intra-cluster)	Cluster optimization technique/model	Node hierarchy	Deployment strategy	Stability	Load balancing	Reliability
2019 MEEC [26]	Residual energy, distance factor of node from sink, node density	Hierarchical	Single hop–single hop	Radio energy model	Three levels (normal advance and supernode)	Uniform and random	–	–	Yes
2019 IDHR [26]	Residual energy, distance factor of node from sink, node density	Hierarchical	Dual hop–single hop	Radio energy model	Three levels (normal, advance, and supernode)	Uniform and random	–	–	Yes
2020 HMGWO [38]	Residual energy, distance between the node and the BS	Hierarchical	Single hop	Bio-inspired	Multi-level	Random	Very good	Yes	Yes

lower level. EEMHR [39] uses k levels of vertical energy heterogeneity but LEMHR [40] uses k levels of horizontal energy heterogeneity. Because of energy heterogeneity, it is noticed that the lifetime of a network using LEMHR nearly doubles up compared to EEMHR. LEMHR can be used in all the applications using smart grid.

S Singh et al. proposed a cluster-based three-level heterogeneous DEEC protocol [41]. HetDEEC works on energy dissipation model. This protocol is location unaware because nodes are not equipped with any GPS. HetDEEC—this cluster head is based on weighted election probabilities and threshold function. There is not any cluster optimization technique used in the projected protocol. However, the comparison of one level, two levels, and three levels of DEEC with one level, two levels, and three levels of HetDEEC respectively shows that there is improvement in network energy using HetDEEC protocol. Authors also proposed three-level HetSEP protocol [42] and compared the network energy with three-level SEP protocol. There was 100% increase in network energy as compared to stable election protocol.

Sahoo et al. [43] proposed enhanced stable routing algorithm (ESRA) for HetWSN. Two-level node hierarchy is employed which not only increases the cluster head selection but improves the operation of node. For normal and advance nodes, a cluster head is selected using below equation: $P_{\text{nrm}} = \frac{P_{\text{opt}}}{(1+\text{am})\times E_i n}$; $P_{\text{adv}} = \frac{P_{\text{opt}}}{(1+\text{am})\times E_i n}(1+a)$.

Threshold value of normal node and advance node is calculated and then is compared with the random number generated by the respective node. If threshold value is greater than the generated random number of respective node, then that particular normal and advance node will be chosen as cluster head otherwise it will remain as a normal node or member node, respectively. However, cluster head selection has been categorized as NP-hard problem in their work.

A nature-inspired method GSA-DEEC in WSNs is introduced by Samayaveer Singh [36] which is effective in terms of communication and operating cost of the systems. An optimization technique gravitational search algorithm inspired from nature is used for CH selection. GSA-DEEC offers balance stability among all the nodes deployed randomly in an area to perform for both HetWSNs as well as homogeneous WSN thereby making it efficient selection for CH.

A protocol [27] ECRCP proposed by M Zeng et al. is a cluster-based heterogeneous protocol in which the cluster selection is based on maximum coverage ratio. The nodes deployed at fixed position and are equipped with global positioning system. Nodes have different transmission range and data range. The energy model is established before clustering and then according to maximum coverage ratio cluster head is elected. To optimize cluster head selection, it uses objective function in coverage control algorithm. It is observed from the experimental observations that ECRCP is energy efficient in terms of network energy. This protocol can be used for the applications where base station is located in the center of the WSN.

Raji pal et al. proposed a cluster-based energy-efficient weighted clustering method [44]. This protocol uses genetic algorithm technique to optimize cluster head selection and cluster formation. The objective function to elect cluster head is modified on the basis of density of nodes, distance between nodes, and the number of CH.

This protocol can be used in application of image segmentation, simulation, image compression, medical image processing, etc. A cluster base dynamic energy-aware routing protocol gateway clustering energy-efficient centroid (GCEEC) proposed by K. N. Qureshi et al. The cluster head selection is on the basis of centroid position of a sensor node in a cluster [45]. The data load of cluster head is reduced due to positioning of gateway node between cluster head and the base station. It is observed from the simulation that GCEEC outperform the state-of-the-art protocols and can be used in applications based on humidity or temperature monitoring and illumination in farming area.

Table 3 gives a brief description of several protocols highlighting static heterogeneous WSN where the cluster head selection is based on parameters like weighted election probabilities (WEP), threshold function, distance factor, coverage ratio, and centroid position.

3.2 *Mobile Heterogeneous WSN (HetWSN)*

A network is made up of the sensing devices in which the nodes are mobile in nature and also have different processing capabilities. A mobile wireless sensor network and heterogeneity may improve the network's efficiency.

In cluster-based HetWSN, sink or cluster head can be moving element. A mobile node is used to collect the sensed information by moving around the network. A mobile node can track various kinds of mobility designs in the sensor deployed area, such as controlled mobility, random mobility, and fixed path (TO-FRO) or predictable mobility. Sink mobility plays significant role for data collection and energy efficiency strategies.

Figure 5 shows mobility scenario in heterogeneous WSN where sink is a mobile element. It is moving in to and fro motion (fixed mobility pattern) with the purpose to balance load among all the nodes in network. Sink mobility method is categorized depending on the mobility patterns like fixed/predictable, controlled modalities, and random. The path selection for the sink mobility can be optimized using techniques like bio-inspired protocols and mobility pattern.

3.2.1 Energy-Based Clustering Algorithm

A cluster-based hierarchical adaptive and reliable routing protocol proposed by F. J Atero et al. [46]. In this two-level nodes are deployed in the network field where cluster head and member nodes are at one level whereas sink is at another level. The cluster head selection is based on residual energy. Another protocol s-Harp is introduced by the authors to optimize the cluster head selection. It uses novel threshold value which reduces the chance of a node to become CH which are at far distant from the base station and have low energy. Thus, protocol supports in maintaining efficient link and providing fault tolerance.

Table 3 Brief comparison of hybrid parameter clustering protocols for static HetWSN

Protocol and reference	Heterogeneity/WSN type	Cluster head selection	Clustering structures	Optimization technique/model	(Inter–intra-cluster) data transmission	Node hierarchy	Deployment strategy	Stability	Load balancing	Reliability
2014 EEMHR [39]	Energy, HetWSN	WEP	Hierarchical	Radio model	Multi-hop–single hop	Multi-level (normal and advance nodes)	Random	Good	–	–
2015 LEMHR [40]	Energy, HetWSN	WEP	Hierarchical	Radio free space/multi-path model	Multi-hop–single hop	Multi-level normal node	Uniform	Good	–	Yes
2017 hetDEEC [41]	Energy, HetWSN	WEP and threshold function	Distributed	Radio dissipation model	Single hop or multi-hop	Three levels	Random (square area)	–	Yes	–
2017 hetSEP [42]	Energy, HetWSN	WEP and threshold function	Distributed	Radio dissipation model	Single hop or multi-hop	Three levels	Random (square field area)	–	Yes	Yes
2019 ESRA [43]	Energy, HetWSN	Distance factor of node from sink	Hierarchical	Radio energy consumption model	Multi-hop–single hop	Two levels (normal and advance node)	Random	Very good	Yes	–
2019 GSA-DEEC [36]	Energy, HetWSN/homogeneous	Fitness function, distance factor of node from base station, and node density	Distributed	Bio-inspired	Multiple hop or/and single hop	Multi-level for HetWSN	Random (square filed)	Very good	Yes	Yes

(continued)

Table 3 (continued)

Protocol and reference	Heterogeneity/WSN type	Cluster head selection	Clustering structures	Optimization technique/model	(Inter–intra-cluster) data transmission	Node hierarchy	Deployment strategy	Stability	Load balancing	Reliability
2019 E-CRCP [27]	Data and range transmission, HetWSN	Maximum coverage ratio	Centralized	Energy model	Single hop or multi-hop	Multi-level	Random	Good	Yes	–
2020 EEWC [44]	Energy, HetWSN	Weighted fitness function	Distributed	Genetic algorithm	Single hop–multi-hop	Two levels (normal and advance node)	Random (square filed)	Very good	Yes	–
2020 GCEEC [45]	Data and range transmission, HetWSN	Centroid position	Centralized	Energy model	Multi-hop–single hop	Multi-level–single hop	Random	Good	Yes	Yes

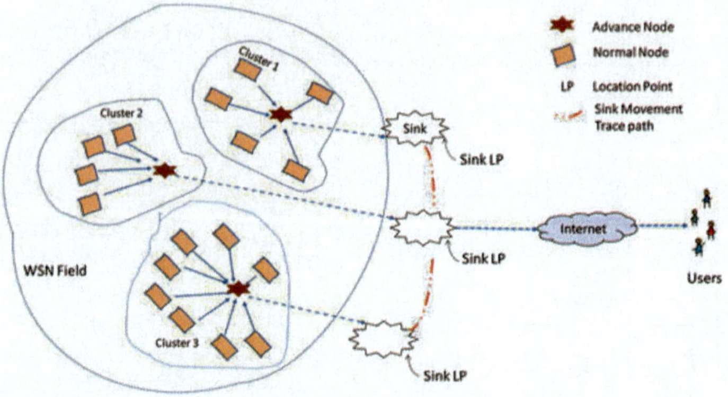

Fig. 5 Mobility scenario in heterogeneous wireless sensor network

A. M. Krishnan and P. G. Kumar proposed an effective clustering approach with data aggregation using multiple mobile sinks for HetWSN. With the help of mobile sink, CHs are elected. And the selection is based on residual energy and a threshold value. Mobile sink trajectory is a fixed predefined straight line inside the heterogeneous WSN field. [47].

In protocol energy-efficient cluster-based dynamic routes adjustment approach (EECDRA) [32], the sensor devices in the network field are distributed in a circular area uniformly and then partitioned in identical cluster heads and clusters. There are two movable sinks placed at the edge of the deployed area. Mobility of sink is fixed either clockwise or anticlockwise. It reduces the updating of path thereby enhancing the power efficiency and lifespan of a network. It also gives accountability for load balancing.

A nature-inspired protocol using multiple mobile sink is proposed by R. Vijayashree et al. [48]. The protocol multiple mobile sink using artificial bee colony (MMABC) can be used in applications with large WSN area. The cluster head selection is based on total remaining energy of nodes, threshold value, and distance factor. For mobility path optimization, it uses bio-inspired artificial colony algorithm. Also, mobile sink powered with more energy is compared to other nodes in the sensor network field. It is observed from the experimental analysis that MMABC not only optimizes the sink path but also stabilizes the cluster head selection.

S. Zafar et al. proposed two cluster-based and mobility-aware protocols mobility-aware centralized clustering algorithm (MCCA) and mobility-aware centralized clustering algorithm (MHCA) [5]. Nodes are deployed randomly in network area and have different data rate processing data, energy, and transmission range. MCCA algorithm uses centralized clustering, whereas MHCA uses hybrid clustering. The cluster head selection in both the protocol is based on residual energy and node velocity. Authors employed first-order radio model for energy consumption in data transfer. Experimental results show that both the protocols help in reducing the loss and are energy-efficient protocols. Table 4 gives a brief description of several protocols highlighting energy-based mobile HetWSN.

Table 4 Brief comparison of energy-based clustering protocols for mobile HetWSN

Protocol and study reference	Heterogeneity/WSN type	Mobility node	Mobile trajectory	Cluster structure	Cluster head selection	Data transmission	Node hierarchy	Deployment strategy	Stability	Load balancing	Reliability
2011s-HARP [46]	Energy, HetWSN	Mobile cluster head	Random	Hierarchical	Residual energy	Multi-hop	Two levels (cluster head and normal node)	Random	Good	Yes	Yes
2015 [47]	Energy, HetWSN	Multiple mobile sink	Fixed and predefined	Distributed	Residual energy, threshold	Single hop	Two levels	Uniform	Good	Yes	Yes
2017 EECDRA [32]	Data rate and transmission range HetWSN	Mobile sink	Clockwise or anticlockwise to network boundary	Distributed	Residual energy	Single Hop	Two levels	Uniform (circular region)	–	Yes	–
2019 MMABC [48]	Energy, homogeneous/HetWSN	Multiple mobile	Random	Distributed	Residual energy, threshold, distance factor	Multi- hop	Two levels	Random (square filed)	Good	Yes	Yes
2019 MCCA [5]	Energy, data rate and transmission range, HetWSN	Mobile nodes, static base station	Random	Centralized	Residual energy, velocity	Single hop	Three levels	Random (square filed)	Good	–	Yes

(continued)

Table 4 (continued)

Protocol and study reference	Heterogeneity/WSN type	Mobility node	Mobile trajectory	Cluster structure	Cluster head selection	Data transmission	Node hierarchy	Deployment strategy	Stability	Load balancing	Reliability
2019 MHCA [5]	Energy, data rate and transmission range, HetWSN	Mobile nodes, static base station	Random	Hybrid	Residual energy, velocity	Single hop	Three levels	Random (square filed)	Good	–	Yes

3.2.2 Hybrid Parameter-Based Clustering Algorithm

A mobile HetWSN protocol RAHMoN [49] distinguishes sensing devices as mobile and static. Cluster head is elected on the basis of distance to the sink, mobility level, and energy of node. Sink mobility is based on random waypoint model. There is not any optimization technique used for clustering. But, the simulation work and performance calculation show that it is effective and adaptable in any kind of environment.

A mobile sink clustered heterogeneous sensor network (HSN) protocol was proposed by Sudarmani et al. [50]. Network deployment consists of three-level hierarchical structure of nodes. HSN follows particle swarm optimization for movement of sink. It also demonstrated that the loss of data incurs when mobility in sink increases. Also, it is appropriate for large-scale wireless sensor network.

G Hie et al. proposed cluster-based heuristic tour planning algorithm in which cluster head is selected on the basis of threshold function. Mobility elements are the mobile nodes, mobile vehicle, or a mobile robot powered with adequate energy. The energy consumption is based on radio energy consumption model. It uses minimum spanning tree to solve the traveling salesman problem. The energy dissipation model is represented using below equations

$$\mathrm{ET}_x(m, d) = E_{\mathrm{elec}} * m + \varepsilon * m * d^n; \quad \mathrm{ER}_x(m) = E_{\mathrm{elec}} * m$$

where $\mathrm{ET}_x(m; d)$ is the energy transmission, $\mathrm{ER}_x(m)$ is the energy dissipation, E_{elec} and ε are the constants, 'm' is the number of bits, and 'd' is the distance [51]. EC-PSO is an improved routing schema with special clustering proposed by J wang et al. [52]. Cluster head selection is based on node energy.

Table 5 gives a brief description of several clustering protocols where CH selection is based on hybrid parameters for mobile HetWSN.

4 Statistical Analysis

In this section, we discuss the statistical study of recent research topics for routing protocols that are cluster-based in heterogeneous WSN. Comparative analysis of energy base and other clustering protocols is shown in the graph (Fig. 6).

We find that most of the researchers focus on energy heterogeneity. In contrast to it, less research has been carried out by considering transmission range and transmitting data rate. Static HetWSN protocols have been employed more in large area WSN. However, mobile HetWSN outperforms in application-specific protocols. Probabilistic parameter is most commonly used in the election of head node of a cluster. Residual energy and distance are next two most used parameters in CH selection.

Transmission and data rate heterogeneity are least explored area as compared to energy heterogeneity. In our survey, 76% of protocols are static in cluster-based HetWSN which is 26% more than mobile HetWSN protocols. Altogether, protocols

Table 5 Brief comparison of hybrid-based clustering protocols for mobile HetWSN

Protocol and study reference	Heterogeneity/WSN type	Mobility node	Mobile trajectory	Clustering structure	Cluster head selection	Data transmission	Node hierarchy	Deployment strategy	Stability	Load balancing	Reliability
2012 RAHMoN [49]	Energy, HetWSN	Mobile sink, cluster head, static nodes	Random	Hybrid	Distance to the sink, mobility level, and energy of node	Multi-hop	Two levels (mobile and static node)	Uniform	–	–	Yes
2013 HSN [50]	Energy, data rate, transmission range, HetWSN	Mobile sink	Random way point	Hierarchical	High energy-level node	Single hop	Three levels (h-sensor, l-sensor, sink)	Random and uniform	–	Yes	–
2016 Heuristic tour planning algorithm [51]	Transmission, energy, HetWSN	One mobile sink, static nodes	Random	Hybrid	Threshold function	Single Hop	Two levels	Random	–	Yes	–
2019 EC-PSO [52]	Energy, transmission range, HetWSN	Mobile data collector, static nodes	Determined	Centralized	Node energy	Multi-hop	Two levels	Random	Good	Yes	Yes

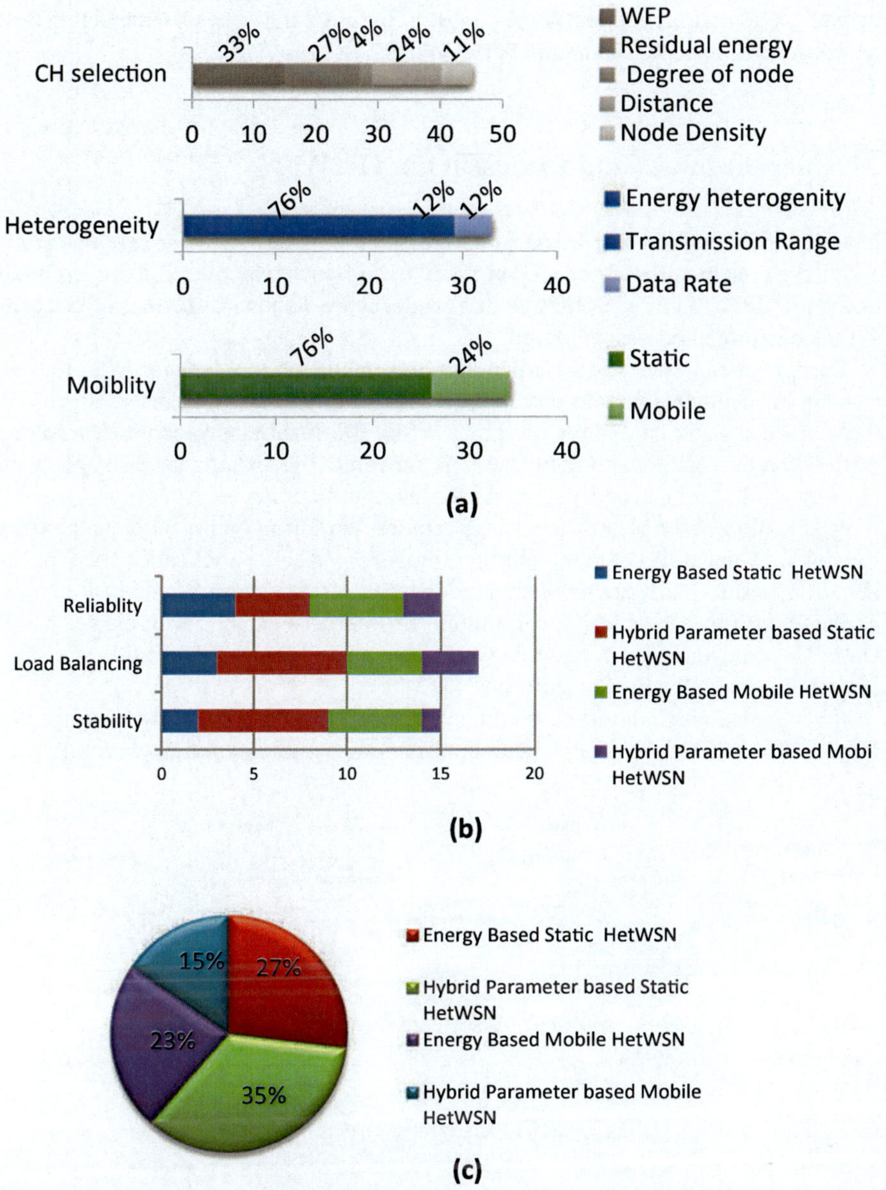

Fig. 6 **a** Factors considered in protocols of cluster-based heterogeneous WSN. **b** Static and mobile HetWSN comparative analysis on basis of stability, load balance, and reliability. **c** Total percent of cluster-based static and mobile HetWSN algorithms considered in survey

in both static and mobile HetWSN based on hybrid parameter provide better load balancing and stability and are more reliable.

5 Current Issues and Discussion in HetWSN

From the review and comparison of several protocols designed for heterogeneous WSN, we conclude that there are yet many open issues which need to be explored and a new direction of work may be designed. Figure 7 shows directions of research yet to be explored or least explored.

Energy efficiency is yet a challenging and important problem for WSNs. New scholars or beginners can do research on computational and link heterogeneity as these are also significant in the design of WSN. Bio-inspired protocols are also one of the effective methods to inculcate sink movement. This reduces energy loss in election of cluster head and also improves in target tracking.

Application-based clustered network requires exploring optimization techniques for sink movement. As of date, routing protocol for specific applications also needs attention beside hardware implementation. The state-of-the-art protocols can be explored with invent of IoT. Incorporation of Internet of things with cluster-based HetNET could be great for topology management, ensuring reliability and high availability in application-specific WSN [53].

Machine learning algorithms could be of great importance in sensor readings, for large-scale clustering, data aggregation, target tracing, and to optimize routing path

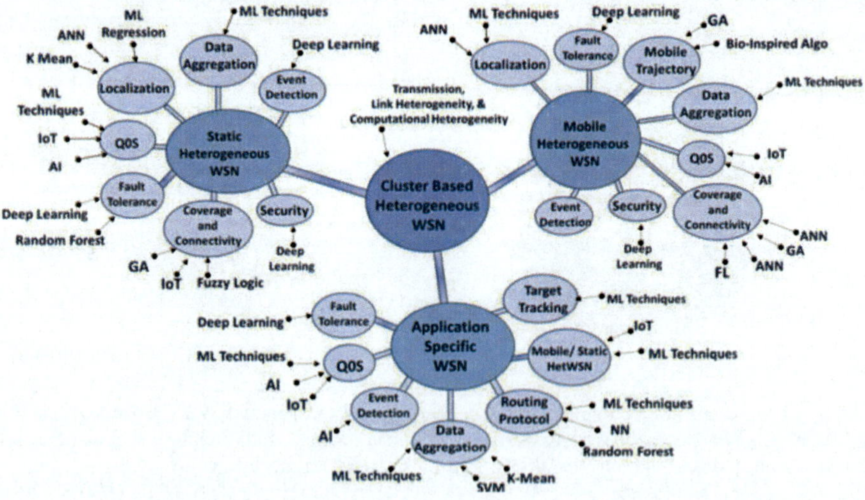

Fig. 7 Open issues in cluster-based heterogeneous WSN. AI—artificial intelligence; ML—machine learning techniques; IoT—Internet of things; GA—genetic algorithm; FL—fuzzy logic, and ANN—artificial neural network

for a cluster-based routing algorithms [54]. Hybrid parameter clustering for mobile HetWSN is also less explored. Neural network can be useful as an effective tool in decreasing duty cycling, data driven, and mobile-oriented approach in cluster-based heterogeneous wireless sensor network.

6 Conclusion

We have focused on cluster-based routing protocols that are energy efficient for mobile and static heterogeneous wireless sensor networks (HetWSN) in this paper. Clustering protocols based on hybrid parameters are more reliable and provide better load balancing which improves the network stability. Mobility-based protocols also outperform in many scenarios and ameliorate energy consumption in clustering process. Mobility plays a significant role in network longevity in most of the application-based networks. We found that cluster-based network incorporated with optimization techniques outperforms the state-of-the-art protocols in some cases. Hence, incorporation of intelligent techniques with the clustering protocol is recommended for a cluster-based HetWSN. Several directions in future work may include implementation of an efficient protocol using optimized clustering and incorporating node mobility for heterogeneous WSN.

References

1. Peizhe L, Muqing W, Wenxing L, Min Z (2017) A game-theoretic and energy-efficient algorithm in an improved software-defined wireless sensor network. IEEE Access 5:13430–13445. https://doi.org/10.1109/access.2017.2727139
2. Chi Y (2013) Chang H (2013) An energy-aware grid-based routing scheme for wireless sensor networks. Telecommun Syst 54:405–415. https://doi.org/10.1007/s11235-013-9742-x
3. Singh S, Chand S, Kumar B (2016) Energy efficient clustering protocol using fuzzy logic for heterogeneous WSNs. Wireless Pers Commun 86:451–475. https://doi.org/10.1007/s11277-015-2939-4
4. Atero J, Vinagre J, Ramiro-Bargueno J, Wilby M (2011) A low energy and adaptive routing architecture for efficient field monitoring in heterogeneous wireless sensor networks. 449–455. https://doi.org/10.1109/hpcsim.2011.5999859
5. Zafar S, Bashir A, Chaudhry SA (2019) Mobility-aware hierarchical clustering in mobile wireless sensor networks. IEEE Access 7:20394–20403. https://doi.org/10.1109/access.2019.2896938
6. Silva R, Sa Silva J, Boavida F (2014) Mobility in wireless sensor networks—survey and proposal. Comput Commun 52:1–20. https://doi.org/10.1016/j.comcom.2014.05.008
7. Heinzelman WR, Chandrakasan A, Balakrishnan H (2000) Energy-efficient communication protocol for wireless microsensor networks. In: Proceedings of the 33rd Annual Hawaii International Conference on System Sciences, vol 2, Maui, HI, USA, pp 10. https://doi.org/10.1109/hicss.2000.926982
8. Chen Y, Edward C, Song H (2005) Energy efficient multipath routing in large scale sensor networks with multiple sink nodes. In: Advanced Parallel Processing Technologies, Springer Berlin Heidelberg, pp 390–399

9. Jain TK, Saini DS, Bhooshan SV (2015) Lifetime optimization of a multiple sink wireless sensor network through energy balancing. Hindawi Publishing Corporation Journal of Sensors, vol 921250. https://doi.org/10.1155/2015/921250
10. Lindsey S, Raghavendra CS (2002) PEGASIS: power-efficient gathering in sensor information systems, computer systems research department, the aerospace corporation. http://ceng.usc.edu/~raghu/pegasisrev.pdf
11. Younis O, Fahmy S (2004) HEED: a hybrid, energy-efficient, distributed clustering approach for ad hoc sensor networks. IEEE Trans Mob Comput 3(4):366–379. https://doi.org/10.1109/tmc.2004.41
12. Yazid Y, Ez-zazi I, Salhaoui M, Arioua M, González A (2019) Extensive analysis of clustered routing protocols for heteregeneous sensor networks. In: Third International Conference on Computing and Wireless Communication Systems, ICCWCS 2019. European Alliance for Innovation (EAI). https://doi.org/10.4108/eai.24-4-2019.2284208
13. Shafiullah M, Gyasi-Agyei A, Wolfs PJ (2008) A survey of energy-efficient and qos-aware routing protocols for wireless sensor networks. In: Sobh T, Elleithy K, Mahmood A, Karim MA (eds) Novel Algorithms and Techniques In Telecommunications, Automation and Industrial Electronics. Springer, Dordrecht, ISBN: 978-1-4020-8736-3, pp 352–357. https://doi.org/10.1007/978-1-4020-8737-0_63
14. Vivek K, Narottam C, Surendra S (2010) Clustering algorithm for wireless sensor network: a survey. Int J Appl Eng Res 1(2):273–287
15. Singh S, Singh M, Singh, D (2010) Routing protocols in wireless sensor networks—a survey. Int J Comput Sci Eng Surv. https://doi.org/10.5121/ijcses.20https://doi.org/10.1206
16. Wei C, Yang J, Gao Y, Zhang Z (2011) Cluster-based routing protocols in wireless sensor networks: a survey. In: International Conference on Computer Science and Network Technology, IEEE conference, pp 1659–1663
17. Sha K, Gehlot J, Greve R (2013) Multipath routing techniques in wireless sensor networks: a survey. Wireless Pers Commun 70(2):807–829
18. Sahoo BPS, Rath S, Puthal D (2012) Energy efficient protocols for wireless sensor networks: a survey and approach. Int J Comput Appl (0975 8887), 44(18)
19. Sanjeev kumar G, Neeraj J, Poonam S (2013) clustering protocols in wireless sensor network: a survey. Int J Appl Inf Syst 5(2):41–50
20. Manap Z, Ali BM, Ng CK, Noordin NK, Sali A (2013) A review on hierarchical routing protocols for wireless sensor networks. Wireless Pers Commun 1–28
21. Tanwar S, Kumar N, Rodrigues JJPC (2015) A systematic review on heterogeneous routing protocols for wireless sensor network. J Netw Comput Appl Elsevier 53:39–56. https://doi.org/10.1016/j.jnca.2015.03.004
22. Singh J, Kumar R, Mishra AK (2015) Clustering algorithms for wireless sensor networks: a review. In: 2015 2nd International Conference on Computing for Sustainable Global Development (INDIACom), pp 11–13, Electronic ISBN: 978-9-3805-4416-8
23. Yan JJ, Zhou M, Ding Z Recent advances in energy-efficient routing protocols for wireless sensor networks: a review, IEEE. https://doi.org/10.1109/access.2016.2598719
24. Rostami AS, Badkoobe M, Mohanna F, Keshavarz H, Hosseinabadi AAR, Sangaiah AK (2018) Survey on clustering in heterogeneous and homogeneous wireless sensor networks. J Supercomput 74(1):277–323
25. Julme B, Patil P (2020) A review on clustering algorithms in wireless sensor networks for optimal energy utilisation. In: Balaji S, Rocha Á, Chung YN (eds) Intelligent Communication Technologies and Virtual Mobile Networks. ICICV 2019. Lecture Notes on Data Engineering and Communications Technologies, vol 33. Springer, Cham, ISBN: 978-3-030-28363-6, 2019. https://doi.org/10.1007/978-3-030-28364-3_66
26. Verma S, Sood N, Sharma AK (2019) A novelistic approach for energy efficient routing using single and multiple data sinks in heterogeneous wireless sensor network. Peer-to-Peer Netw. Appl 12:1110–1136. https://doi.org/10.1007/s12083-019-00777-5
27. Zeng M, Huang X, Zheng B, Fan X (2019) A heterogeneous energy wireless sensor network clustering protocol. Wireless Commun Mob Comput 7367281:11. https://doi.org/10.1155/2019/7367281

28. Kumar D, Aseri TC, Patel RB (2011) Multi-hop communication routing (MCR) protocol for heterogeneous wireless sensor networks. Int J Inform Technol Commun Convergence 1(2):130–145. https://doi.org/10.1504/ijitcc.2011.039281
29. Guney E, Aras N, Altınel IK, Ersoy C (2012) Efficient solution techniques for the integrated coverage, sink location and routing problem in wireless sensor networks. Comput Oper Res 39(7):1530–1539
30. https://doi.org/10.1016/j.cor.2011.09.002
31. Amgoth T, Jana PK (2015) Energy-aware routing algorithm for wireless sensor networks. Comput Elect Eng 41:357–367. https://doi.org/10.1016/j.compeleceng.2014.07.010
32. Wang J, Cao J, Sai J, Park JH (2017) Energy-efficient cluster-based dynamic routes adjustment approach for wireless sensor networks with mobile sinks. Springer Science + Business Media New York 2017. https://doi.org/10.1007/s11227-016-1947-9
33. Vijayashree R, Suresh Ghana Dhas C (2019) Energy efficient data collection with multiple mobile sink using artificial bee colony algorithm in large-scale WSN. J Control Meas Electron Comput Commun. ISSN: 0005-1144 (Print) 1848-3380. https://doi.org/10.1080/00051144.2019.1666548
34. Gu X, Yu JG, Yu DX, Wang GH, Lv YH (2014) ECDC: an energy and coverage-aware distributed clustering protocol for wireless sensor networks. J Comput electr Eng 40(2):384–398
35. Chand S, Singh S, Kumar B (2014) Heterogeneous HEED protocol for wireless sensor networks. Wireless Pers Commun 77(3):2117–2139
36. Singh S (2017) Energy efficient multilevel network model for heterogeneous WSNs. Eng Sci Technol Int J 20(1):105–115
37. Sanwar Hosen SM, Cho GH (2018) An energy centric cluster-based routing protocol for wireless sensor networks. Sensors 18(5):1520. https://doi.org/10.3390/s18051520
38. Zhao X, Ren S, Quan H, Gao Q (2020) Routing Protocol for Heterogeneous Wireless Sensor Networks Based on a Modified Grey Wolf Optimizer. Sensors 20:820. https://doi.org/10.3390/s20030820
39. Sudeep T, Neeraj K, Niu JW (2014) EEMHR: Energy-efficient multilevel heterogeneous routing protocol for wireless sensor networks. Int J Commun Syst 27(9):1289–1318. https://doi.org/10.1002/dac.2780
40. Tyagi S, Tanwar S, Gupta SK, Kumar N, Rodrigues JJPC (2015) A lifetime extended multi-levels heterogeneous routing protocol for wireless sensor networks. J Telecommun Syst 59(1):43–62. https://doi.org/10.1007/s11235-014-9884-5
41. Singh S, Malik A (2017) hetDEEC: Heterogeneous DEEC protocol for prolonging lifetime in wireless sensor networks. J Inf Opt Sci 38(5):699–720
42. Singh Samayveer, Malik Aruna (2017) hetSEP: heterogeneous SEP protocol for increasing lifetime in WSNs. J Inf Opt Sci 38:721–743. https://doi.org/10.1080/02522667.2016.1220093
43. Sahoo BM, Gupta AD, Yadav SA, Gupta S (2019) ESRA: enhanced stable routing algorithm for heterogeneous wireless sensor networks. In: 2019 International Conference on Automation, Computational and Technology Management (ICACTM), London, United Kingdom, pp. 148–152. https://doi.org/10.1109/icactm.2019.8776740
44. Pal R, Yadav S, Karnwal R et al. (2020) EEWC: energy-efficient weighted clustering method based on genetic algorithm for HWSNs. Complex Intell Syst
45. Qureshi KN, Bashir MU, Lloret J, Leon A (2020) Optimized cluster-based dynamic energy-aware routing protocol for wireless sensor networks in agriculture precision. J Sens Appl Agric Environ Monit 9040395:19. https://doi.org/10.1155/2020/9040395
46. Atero FJ, Vinagre JJ, Ramiro J, Wilby M (2011) A low energy and adaptive routing architecture for efficient field monitoring in heterogeneous wireless sensor networks. In: Processing of IEEE International Conference High Performance Computer Sim, pp. 449–455
47. Muthu Krishnan A, Ganesh Kumar P (2016) An effective clustering approach with data aggregation using multiple mobile sinks for heterogeneous WSN. Wireless Pers Commun 90:423–434. https://doi.org/10.1007/s11277-015-2998-6

48. Vijayashree R, Suresh Ghana Dhas C (2019) Energy efficient data collection with multiple mobile sink using artificial bee colony algorithm in large-scale WSN. J Cont Meas Electron Comput Commun. ISSN: 0005-1144 (Print) 1848-3380. https://doi.org/10.1080/00051144.2019.1666548
49. Vilela MA, Araujo RB (2012) RAHMoN: routing algorithm for heterogeneous mobile networks. In: 2012 Second Brazilian Conference on Critical Embedded Systems, Campinas, pp. 24–29. https://doi.org/10.1109/cbsec.2012.19
50. Sudarmani R, Kumar KR (2013) Particle swarm optimization-based routing protocol for clustered heterogeneous sensor networks with mobile sink. American J Appl Sci 10(3):259–269. https://doi.org/10.3844/ajassp.2013.259.269
51. Xie G, Pan F (2016) Cluster-based routing for the mobile sink in wireless sensor networks with obstacles. IEEE Access 4:2019–2028. https://doi.org/10.1109/access.2016.2558196
52. Wang J, Gao Y, Liu W, Sangaiah AK, Kim H-J (2019) An improved routing schema with special clustering using pso algorithm for heterogeneous wireless sensor network. Sensors 19:671
53. Mainetti L. Patrono L, Vilei A (2011) Evolution of wireless sensor networks towards the internet of things: a survey. In: SoftCOM 2011, 19th International Conference on Software, Telecommunications and Computer Networks, Split, 2011, pp. 1–6
54. Praveen Kumar D, Amgoth T, Annavarapu CSR (2019) Machine learning algorithms for wireless sensor networks: a survey. Information Fusion 49:1–25. https://doi.org/10.1016/j.inffus.2018.09.013

Voice-Based Gender Identification Using qPSO Neural Network

Ruchi Jha, Anvita Saxena, Jodh Singh, Ashish Khanna, Deepak Gupta, Prerna Jain, and Viswanatha Reddy Allugunti

Abstract Gender identification is a classic problem. A considerable amount of research has been carried out in this field. Gender identification has many applications in various fields, especially in marketing, online research and media industry. This work proposes a novel method of quantum-inspired Particle Swarm Optimised neural network for voice-based gender identification. To the best of our knowledge, the above-mentioned approach has been presented for the first time in this paper. In the presented work, statistical extracted features are fed into a qPSO optimised feed-forward four-layered neural network. The data set was cleaned and preprocessed using principal component analysis and was then fed into the network, which gave us an optimal accuracy of 91.15%. The proposed approach performed significantly better than the classical Particle Swarm Optimisation approach.

R. Jha (✉) · A. Saxena · J. Singh · A. Khanna · D. Gupta · P. Jain
Maharaja Agrasen Institute of Technology, Plot No. 1, Rohini Sector-22, Delhi 110086, India
e-mail: ruchijha28@gmail.com

A. Saxena
e-mail: anvanvita2207@gmail.com

J. Singh
e-mail: jetjodh@gmail.com

A. Khanna
e-mail: ashishkhanna@mait.ac.in

D. Gupta
e-mail: deepakgupta@mait.ac.in

P. Jain
e-mail: prernaten@gmail.com

V. Reddy Allugunti
Arohak Inc., 4105 US-1 #16, Monmouth Junction, NJ 08852, USA
e-mail: itvissu@gmail.com

A. Khanna et al. (eds.), *Data Analytics and Management*, Lecture Notes on Data Engineering and Communications Technologies 54,
https://doi.org/10.1007/978-981-15-8335-3_66

Keywords Gender identification · Gender detection · Principal component analysis (PCA) · Quantum computing · Quantum inspired algorithms · Particle swarm optimisation (PSO) · Quantum inspired particle swarm optimisation (QPSO) · Neural networks

1 Introduction

Gender identification and detection is a very broad and classic problem. Subcategorisation of large datasets helps in narrowing down the search space and increases efficiency. Hence, dividing a particular dataset into categories like male, female, etc., significantly increases the efficiency and reduces the computation time. A significant amount of research has been carried out in the same field over the years. Fingerprint-based gender detection, facial gender identification, frequency-domain gender identification and voice-based gender identification are some of the popular subfields. It has many practical applications. For example, it is extensively used in multimedia-related fields to process a variety of speech conditions [1]; it is also used for gender identification of the names present on web pages [2], gender identification of authors on the Internet [3], gender identification using images provides a means which helps in monitoring people in any social place, company or organisation better [4]. In this work, we present a novel method of voice-based gender identification and detection using qPSO optimised neural network.

The voice signals of male and female have different acoustic and physiological features. Difference in stress and pitch of the voice helps the human brain in identifying the voice as male or female [5]. The aim of voice-based gender identification systems is to make the model identify these differences with optimal accuracy. The preprocessing of voice signals is also a strenuous task [6]. Voice-based gender identification and detection systems have potential applications in automatic speech recognition systems. Models which are gender specific have higher accuracy than the gender-independent systems [7].

In this work, we have used the gender recognition by voice data set [8], which consists of 3168 recorded voice samples of male and female speakers. The voice samples are pre-processed by acoustic analysis in R using the seewave and tuneR packages, with an analysed frequency range of 0–280 Hz (which is the human vocal range). The data set consists of extracted statistical features such as the mean frequency, median frequency, mode frequency, skew, etc. First, a minimum number of features were selected randomly to determine the highest accuracy. Next, the features were fed into a feed-forward three-layer (two hidden layers) neural network which was optimised using the qPSO algorithm. The objective of this work is to introduce and assess a novel gender identification and detection method with an optimal accuracy and minimum errors.

Section 2 discusses the forerunner algorithms, Sect. 3 discusses the modified algorithm, dataset and classification steps, Sect. 4 concludes the results of the paper, Sect. 5 is the conclusion followed by references.

2 Background

2.1 Conventional Particle Swarm Optimisation Algorithm

Particle Swarm Optimisation (PSO) is a versatile evolutionary algorithm based on swarm intelligence. This algorithm mimics the behaviour of a bird or fish flock. The two main features of the algorithm are self-organisation and division of labour [9]. Self-organisation is the personal experiences of a particle and division of labour is the learning through the experiences of other particles. Every particle in the swarm amounts to a solution, the local best of every particle is termed as pbest and the global best is termed as gbest [10]. The activity of every particle is decided by the velocity of the particle. The velocity again is governed by the local best and global best values. The constant updation of the particles on the basis of optimisation factors creates solution sets. Because of the versatility, simplicity and easy implementation, PSO is employed to solve problems such as function optimisation, model classification, neural network training, signal processing, etc. [9].

Every solution in the set is considered as a point. If there are m points in a group, the location of the j_{th} point ($j = 1, 2, 3, \ldots, m$) can be denoted as P_j, the best location is represented as P_{best}, and the global best of the entire group can be represented as G_{best}. The velocity can be represented as X_j.

X_j is a function of the values of P_{best} and G_{best} such that-

$$X_j = wX_j + C_o \times \text{rand}() \times P^j_{\text{best}} + C_1 \times \text{rand}() \times G_{\text{best}} \tag{1}$$

$$P_j = P_j + X_j \tag{2}$$

A particular and problem specific PSO is determined by w, C_o, C_1, m and n, where

w = weight of inertia,
C_o, C_1 = constants,
m = number of points,
n = number of iterations.

2.2 Quantum Optimisation

Quantum bit is the fundamental unit of quantum computations. The existence of a quantum bit is a super-positioned state of co-existing 0s and 1s. It is represented as (Fig. 1):

$$|\phi\rangle = \alpha|0\rangle + \beta|1\rangle \tag{3}$$

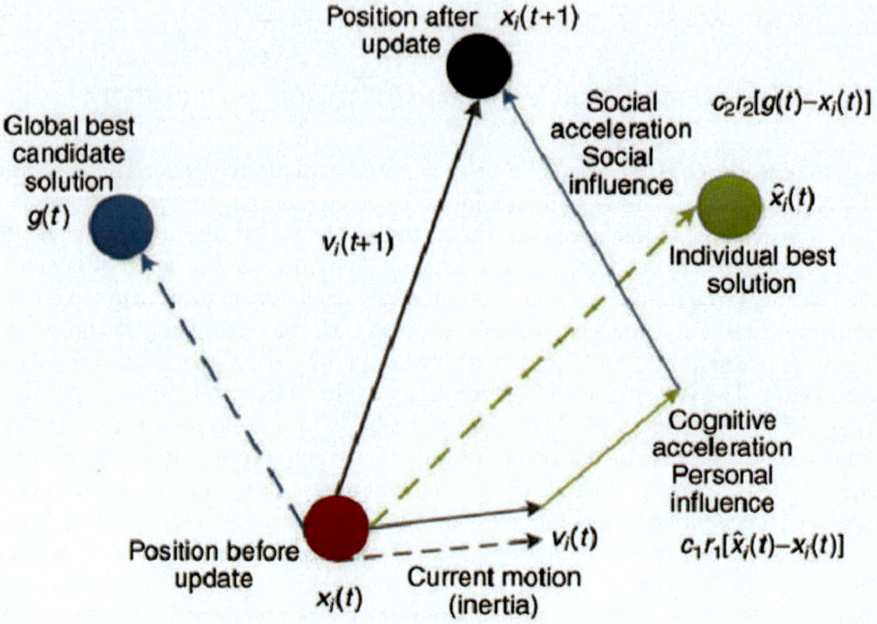

Fig. 1 Pictorial representation of Particle Swarm Optimisation

where

α = real number which denotes quantum bit probability values,
β = real number which denotes quantum bit probability values.

And,

$$|\alpha|^2 + |\beta|^2 = 1[12]. \tag{4}$$

A quantum bit is often conceptualised as a complex number of the form $e_{i\theta}$, where,

$\alpha = \cos\theta$,
$\beta = \sin\theta$,
Real part = probability amplitude for $|0\rangle$,
Imaginary part = probability amplitude for $|1\rangle$.

The computations and results in quantum computing are usually represented as matrices. For example, a quantum bit is represented as

$$\begin{bmatrix} \alpha \\ \beta \end{bmatrix} \tag{5}$$

We have principally used the rotation and inverse operations in our implementation [13]. These are usually performed using rotation gate $R(\theta)$ and NOT gate. These can be denoted as

$$R(\Delta\theta) = \begin{bmatrix} \cos(\Delta\theta) & -\sin(\Delta\theta) \\ \sin(\Delta\theta) & \cos(\Delta\theta) \end{bmatrix} \quad (6)$$

$$\text{NOT Gate} = \begin{bmatrix} 0 & 1 \\ 1 & 0 \end{bmatrix} \quad (7)$$

After performing the above two operations, rotation changes the number into a complex number with a rotated angle, i.e. $e^{(\theta+\Delta\theta)}$ and NOT gate interchanges the real and imaginary parts resulting into a complex number, i.e. $e^{i(\frac{\pi}{2}-\theta)}$.

2.3 *Optimised Evolutionary Algorithm*

Evolutionary algorithms are genetic algorithms which are generally used to optimise biological problems [14–17]. Quantum-behaved Particle Swarm Optimisation (qPSO) is fairly a new method for optimising problems. It has earlier been used for optimising composite structures [18], image enhancing and processing [19, 20], face detection [21], etc. Our work is a novel approach of classifying gender using a quantum optimised Particle Swarm Optimisation algorithm.

3 Methodology

3.1 *Dataset*

The name of the dataset used is "Gender Recognition by Voice" [8] which consists of 3168 voice samples of males and females. The dataset has been pre-processed using the seewave and tuneR packages. An acoustic analysis of the voice samples is done in R using the above-mentioned packages, with an analysed frequency range of 0–280 Hz. Table 1 enlists the 21 features present in the dataset.

3.2 *Preprocessing*

The data was cleaned before feeding it into the neural network. First, feature scaling was performed and later, and principal component analysis was employed for dimensionality reduction. PCA is an efficient feature selection technique which finds the best fitting line by producing an orthogonal basis of one line perpendicular to another

Algorithm 1: qPSO-based neural network

Input: Array of selected statistical features obtained after cleaning
Output: Predicted binary valued array
1. Initialise all the coefficients and parameters such as 'a' (lower limit of weight) and 'b' (upper limit of weights used) required for the problem and define the population size and set the initial count of total iterations.
2. Initialise the population size in a range of

$$-1/\sqrt{n} \ to \ 1/\sqrt{n}$$

where n is the number of layers, and multiply them with 2.
3. Generate complex numbers using following equation:

$$(X_i)_j = \frac{1}{2}(b_i \times (1 + e^{\phi^{ij}i})) + (a_i(1 - e^{\phi^{ij}i}))$$

4. Calculate the cost function and compare the fitness function with existing $f_{p_{\text{best}}}$.
5. Update the p_{best} and g_{best}, if they are providing a lesser minimum.
6. For chosen number of iterations run (7–8).
7. For each particle update velocity vector using differences from p_{best} and local value and g_{best} and local value.

$$(\Delta\phi_{ji})_{op} = \begin{bmatrix} 2\pi + (\phi_{ji})_{op} - \phi_{ji}, & (\phi_{ji})_{op} - \phi_{ji} < -\pi \\ (\phi_{ji})_{op} - \phi_{ji}, & -\pi < (\phi_{ji})_{op} - \phi_{ji} < \pi \\ (\phi_{ji})_{op} - \phi_{ji} - 2\pi, & (\phi_{ji})_{op} - \phi_{ji} > \pi \end{bmatrix}$$

$$(\Delta\phi_{ji})_{g} = \begin{bmatrix} 2\pi + (\phi_{ji})_{g} - \phi_{ji}, & (\phi_{ji})_{g} - \phi_{ji} < -\pi \\ (\phi_{ji})_{g} - \phi_{ji}, & -\pi < (\phi_{ji})_{g} - \phi_{ji} < \pi \\ (\phi_{ji})_{g} - \phi_{ji} - 2\pi, & (\phi_{ji})_{g} - \phi_{ji} > \pi \end{bmatrix}$$

Here $(\Delta\phi_{ji})_{op}$ is difference between the local best and local value of population and $(\Delta\phi_{ji})_g$ is the difference between the global best and the local value of the population. ω, c1, c2, r1 and r2 are varying parameters of the given particle swarm optimisation technique in (1).
8. Evaluate the fitness function and update the global best and local best if required.
9. The final global best matrices define the final weights to be used in our neural network.
End

[11]. In our approach, PCA was employed for reducing the number of features from 21 to 10. These ten features were fed into the qPSO optimised neural network.

3.3 *Classification*

3.3.1 Designing the Network

The feed-forward neural network used with quantum optimisation is based on a logistic regression framework. The fitness function has been defined by error function

Table 1 The list of extracted features from the voice signals

Mean frequency (in kHz)	Standard deviation	Median frequency (in kHz)
First quantile (Q25) (in kHz)	Third quantile (Q75) (in kHz)	Interquartile range (in kHz)
Skew	Kurt	Spectral entropy
Spectral flatness	Mode frequency	Centroid
Average of fundamental frequency measured across acoustic signal	Peak frequency (frequency with highest energy)	Minimum fundamental frequency measured across acoustic signal
Maximum fundamental frequency measured across acoustic signal	Average of dominant frequency measured across acoustic signal	Minimum of dominant frequency measured across acoustic signal
Maximum of dominant frequency measured across acoustic signal	Range of dominant frequency measured across acoustic signal	Modulation index. Calculated as the accumulated absolute difference between adjacent measurements of fundamental frequencies divided by the frequency range

given by (5) which is the summation of individual logistic losses of all training variables.

$$J(\theta) = -\frac{1}{m}\sum [y^{(i)} \log(h\theta(x(i))) + (1 - y^{(i)}) \log(1 - h\theta(x(i)))] \quad (8)$$

A four-layered neural network was designed with a reduced subset having ten features. The input layer with 10 units and a bias were fed to the neural network. The hidden layers had 512, 128 and 32 units and 1 output neuron totalling 75,457 parameters. The population size consists of 50 particles which set updated values of p_{best} and $f_{p_{\text{best}}}$.

3.3.2 Training

The input dataset is in the form of an $N \times 10$ matrix, which is fed to the input layer. The desired output is predicted to be 0 if the voice is that of a 'male' and 1 for 'female'. Each of the 50 particles is evaluated in every iteration, and the local best of the particle is stored in the p_{best} matrix. Global best of all the layers are stored in matrices, and each one has different g_{best} to generate an output. An improved working learning algorithm is discussed in Algorithm 1.

Population matrices are updated with every iteration regardless of update in p_{best} and g_{best}. With each iteration, both the sine and cosine values of the randomly generated angles are taken into consideration and the minimum of two are compared with existing best values till then. The final usable weights are given by individual g_{best} matrices, and the minimum cost is stored in g_{best} (Figs. 2, 3, 4 and 5).

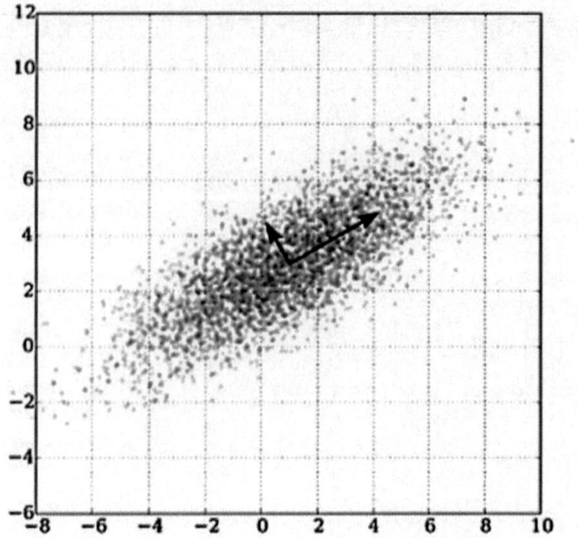

Fig. 2 Pictorial representation of principal component analysis (PCA)

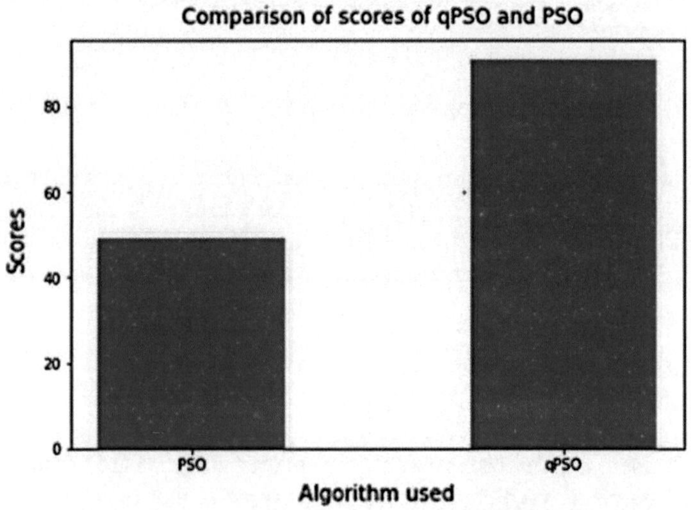

Fig. 3 Comparison of accuracy of qPSO and PSO

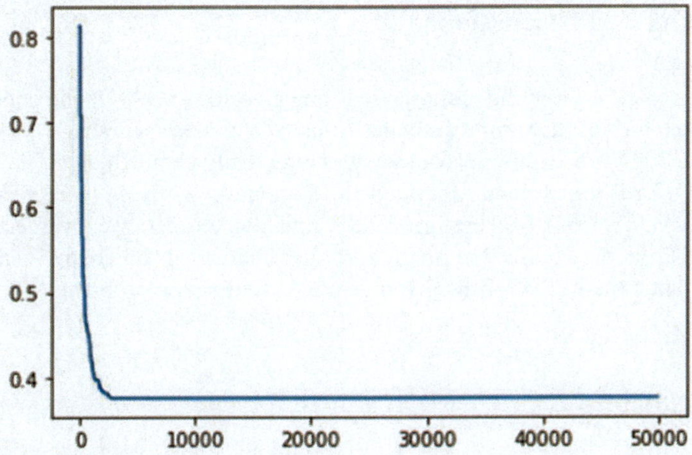

Fig. 4 Depiction of error rate versus number of iterations

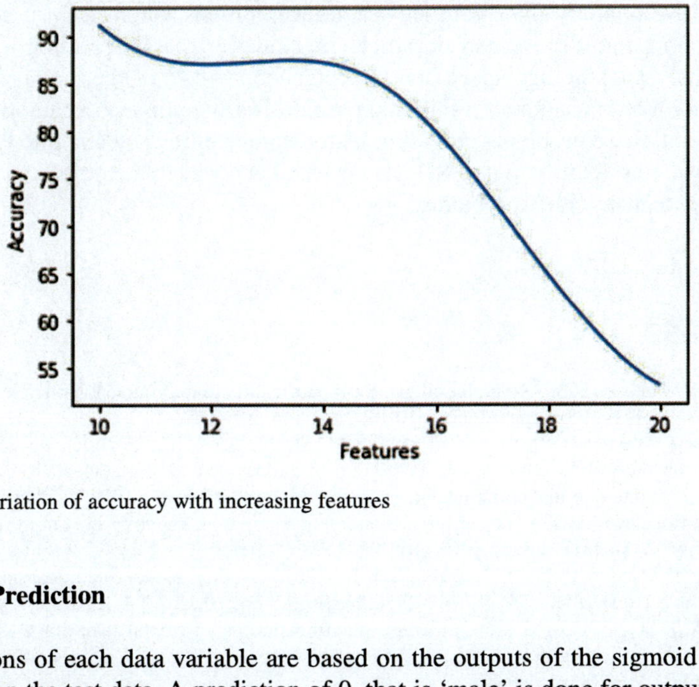

Fig. 5 Variation of accuracy with increasing features

3.3.3 Prediction

Predictions of each data variable are based on the outputs of the sigmoid function applied on the test data. A prediction of 0, that is 'male' is done for outputs valued lesser than 0.5 and 1, that is 'female', for outputs greater than 0.5. Final accuracy is calculated based on the predicted and actual values.

4 Results and Discussions

The comparison between the proposed learning algorithm (QPSO) and general PSO is shown in Fig. 2. qPSO significantly performs with its best accuracy 91.15% against 49.09% for PSO. An optimum accuracy was effectively obtained after 50 iterations. The range of random values generated for population matrices worked well with $(-1/\sqrt{n}, 1/\sqrt{n})$. The global best gradually slows and ultimately converges to global minimum. From the accuracy of predictions, it is confirmed that the qPSO algorithm performs better than PSO for the given gender classification problem.

5 Conclusion

The aim of this work was to develop a novel quantum computing inspired method for the problem of voice-based gender identification. Moreover, traditional and conventional evolutionary algorithms are versatile algorithms which can be used for feature selection, optimisation and classification. However, these algorithms have some limitations. Large storage space and convergence are two of the major problems. The quantum optimisation of traditional particle swarm optimisation algorithms has countered all these problems providing better storage efficiency and improved time cost. The results confirm that qPSO is a powerful and versatile algorithm and can be used to solve many other problems.

References

1. Harb H, Chen L (2005) Voice-based gender identification in multimedia applications. J Intell Inf Syst 24:179–198. https://doi.org/10.1007/s10844-005-0322-8
2. Karimi F, Wagner C, Lemmerich F, Jadidi M, Strohmaier M (2016) Inferring gender from names on the web: a comparative evaluation of gender detection methods. In: Proceedings of the 25th international conference companion on world wide web (WWW '16 companion). International world wide web conferences steering committee, Republic and Canton of Geneva, CHE, pp 53–54. https://doi.org/10.1145/2872518.2889385
3. Cheng N, Chandramouli R, Subbalakshmi KP (2011) Author gender identification from text. Digit Invest 8(1):78–88. ISSN 1742-2876. https://doi.org/10.1016/j.diin.2011.04.002
4. Jain A, Huang J, Fang S (2005) Gender identification using frontal facial images. In: 2005 IEEE international conference on multimedia and expo, Amsterdam, 2005, p 4. https://doi.org/10.1109/ICME.2005.1521613
5. Price PJ (1989) Male and female voice source characteristics: inverse filtering results. Speech Commun 8(3):261–277. ISSN 0167-6393. https://doi.org/10.1016/0167-6393(89)90005-8
6. Keerio A, Mitra BK, Birch P, Young R, Chatwin C (2009) On preprocessing of speech signals. Int J Signal Process 5(3):216–222. ISSN 2070-397X
7. Harb H (2003) Gender identification using a general audio classifier. https://doi.org/10.1109/ICME.2003.1221721
8. Gender Recognition by Voice (identify a voice as male or female). Kaggle. https://www.kaggle.com/primaryobjects/voicegender#voice.csv

9. Qinghai B (2010) Comput Inf Sci 3(1). www.ccse.net.org/cis
10. Sahni S, Aggarwal V, Khanna A, Gupta D, Bhattacharyya S (2020) Diagnosis of Parkinson's disease using a neural network based on QPSO. In: Khanna A, Gupta D, Bhattacharyya S, Snasel V, Platos J, Hassanien A (eds) International conference on innovative computing and communications. Advances in intelligent systems and computing, vol 1087. Springer, Singapore
11. Ringnér M (2008) What is principal component analysis? Nat Biotechnol 26:303–304. https://doi.org/10.1038/nbt0308-303
12. Akama S (2015) Elements of quantum computing. Springer, Berlin
13. Nielsen MA, Chung IL (2000) Quantum computation and quantum information. Cambridge University Press, Cambridge, UK
14. Gupta D, Rodrigues JJPC, Sundaram S, Khanna A, Korotaev V, Albuquerque VHC (2018) Usability feature extraction using modified crow search algorithm: a novel approach. Neural Comput 1–11
15. Gupta D, Sundaram S, Khanna A, Hassanien AE, de Albuquerque VHC (2018) Improved diagnosis of Parkinson's disease based on optimized crow search algorithm. Comput Electr Eng 68:412–424
16. Sharma P, Sundaram S, Sharma M, Sharma A, Gupta D (2018) Diagnosis of Parkinson's disease using modified grey wolf optimization. Cogn Syst Res 52:36–48
17. Gupta D, Julka A, Jain S, Aggarwal T, Khanna A, Albuquerque VHCD (2018) Optimized cuttlefish algorithm for diagnosis of Parkinson's disease. Cogn Syst Res 52:36–48
18. Omkar SN, Khandelwal R, Ananth TVS, Narayana Naik G, Gopalakrishnan S (2009) Quantum behaved Particle Swarm Optimization (QPSO) for multi-objective design optimization of composite structures. Expert Syst Appl 36(8):11312–11322. ISSN 0957-4174. https://doi.org/10.1016/j.eswa.2009.03.006
19. Xu X, Shan D, Wang G, Jiang X (2016) Multimodal medical image fusion using PCNN optimized by the QPSO algorithm. Appl Soft Comput 46:588–595. ISSN 1568-4946. https://doi.org/10.1016/j.asoc.2016.03.028
20. Huang Y, Wang S (2008) Multilevel thresholding methods for image segmentation with Otsu based on QPSO. In: Congress on image and signal processing, Sanya, Hainan, 2008, pp 701–705. https://doi.org/10.1109/CISP.2008.76
21. Li S, Wang R, Hu W, Sun J (2007) A new QPSO based BP neural network for face detection. In: Cao BY (ed) Fuzzy information and engineering. Advances in soft computing, vol 40. Springer, Berlin, Heidelberg

A Machine Learning Approach for the Classification of the Buddha Statues of Borobudur (Indonesia)

Shivanee, Nikhil Kumar Rajput, and Ajay Jaiswal

Abstract The displacement of artwork from one part of the world to another is very common in highly connected world. In this paper, our study is limited to the images of Buddha. The statues of Buddha in different part of the world are having some distinct features, which can be used to classify them. We have made a sincere attempt to recognize and classify the Buddha statues of Borobudur, Indonesia, owing to its remarkable and significant features. A machine learning-based approach has been applied to do the same. A face recognition system is also being used, which extracts the features of a statue using machine learning libraries. A two-step approach has been performed to identify the face: face encoding using deep neural networks and classification using support vector machines.

Keywords Buddha statues · Borobudur · Machine learning · Classification · Support vector machine · Deep neural network · OpenCV · Face recognition

1 Introduction

Borobudur is a world-famous Buddhist monument located in Kedu Plains of Central Java, Indonesia. This monument is being dated to eighth and ninth century CE which is assumed to be built during the reign of rulers of Sailendra dynasty in Central Java. The structure of the monument seems to be designed in three levels: a pyramid base with five clustered square terraces, a vortex closet with three circular platforms, and a colossal stupa on the upper edge [1]. The vertical division into base, body, and

Shivanee (✉)
Center for Historical Studies, JNU, New Delhi, India
e-mail: shivaneejha@gmail.com

N. K. Rajput
Department of Computer Science, Ramanujan College, DU, New Delhi, India
e-mail: nikhilrajput@gmail.com

A. Jaiswal
Department of Computer Science, SSCBS, DU, New Delhi, India
e-mail: a_ajayjaiswal@yahoo.com

A. Khanna et al. (eds.), *Data Analytics and Management*, Lecture Notes on Data Engineering and Communications Technologies 54,
https://doi.org/10.1007/978-981-15-8335-3_67

superstructure of the Borobudur temple ideally correlates with the universe creation in Buddhist cosmology. The walls and balustrades have beautiful narrative reliefs which cover a total area of 2500 m^2. There are seventy-two openwork stupas, each housing a Buddha image, all across the circular platforms. Borobudur was gone astray in dormancy for a longer period. It was, however, rediscovered and restored in many stages since 1814, though finally in the 1970s with the support of UNESCO.

The sacred landscape of Borobudur temple has been built in the shape of a lotus and represents an odd combination of the very core concept of indigenous ancestral worship and the Buddhist concept of Nirvana. The ten mounting terraces of the whole structure lead to the next phases to be completed by the Bodhisattva before achieving the Buddhahood.

There are about five hundred four Buddha statues at the monumental structure of Borobudur, all of which are sculpted on a red stone *sabongkah.* These statues, explicitly illustrating the Dhyani Buddhas, are spread to different levels of the monument which basically have been put into three categories according to the Buddhist cosmology, as well as the levels going forward up until the top of the monument [2]. The three levels are *kamadhatu* (sphere of desires), *rupadhatu* (sphere of forms), and *arupadhatu* (sphere of formlessness), from the base to the top of the monument.

The Dhyani Buddha is distinct from the original Buddha. They are not enlightened earthly beings, but transcendental savers, sitting in meditation posture with half-closed eyes on their lotuses, in eternal veneration and peace. The simplicity of the Dhyani Buddha is perhaps the most remarkable aspect of him. He is depicted as a Buddha in a monk's garb with folds falling or covering his body as "wet clothes." Only his wrists, neck, and ankles are shown through his cloak. A point on his forehead, right between the eyebrows, is another striking physical feature. This simplicity in the Buddhist image blends integrity with compassion and makes this statue elegant. Each statue has its own characteristics, even though there are resemblances. Each statue shows the same thing in addition to its apparent innate persona.

At the outer side of the wall, on the top, the statue is mounted in a line. At the rupadhatu level, there are a total of 432 niches, and they are positioned as follows: while the first and second ledge has 104 niches each, the third has eighty-eight, at the fourth seventy-two, and the fifth has sixty-four niches. This arrangement seems to be following the concept of decreasing the size of the level while moving upwards through the monument.

The Buddha statues are usually placed in hollow stupas which in turn arranged in three concentric circles level wise. There are thirty-two stupas in the first level of the circle, twenty-four on the second level, and sixteen on the third level. Thus, there are seventy-two stupas in total at three levels that seem to suit the Javanese tradition of saying "seventy-two guarded principles in a cage" in Borobudur.

The Buddhas have been depicted in six mudras in the statue at Borobudur. These mudras are (i) *Bhumisparsa mudra* (earth touching), (ii) *Abhaya mudra* (protection, blessing), (iii) *Varamudra* (charity, compassion), (iv) *Dhyana mudra* (meditation), (v) *Vitarka mudra* (gesture of judgment), and (vi) *Dharmacakra mudra* (teaching preaching). Research indicates that the Buddha statues in Borobudur vary significantly in their posture. While the left hand of the statue is usually seen with the palm facing upward and placed in the lap, the right-hand often portrayed in the specific mudra (gesture) related to particular historical happenings in the life of the Buddha.

The Buddhas of the very first four balustrades are represented with unique mudras, where each one on four sides of structure possesses distinctive characteristics. The statue on the east has the same mudra as other statues in the west, north, and south sides. The Buddhas at the top of the fifth balustrade have the same mudra in each direction. They all have one position at the circle level but are different from mudra at the lower levels. It equals the seventy-two Buddhas on the circle level.

2 Literature Review

Various available researches are using the content specifically of paintings focused on the extraction of hand-made features [3–6]. These basic features, such as the identification of author by brushwork decomposition using wavelets, were unique to their application [4]. Besides, SIFT features approached the more significant task of the classification of paintings in a much more traditional way [3, 7].

Naturally, this has been extended to highly effective use of in-depth visual features [8–12]. The pre-trained automatic classification networks were first used. Then, fine-tuned networks showed improved performance [13, 14]. The linear combination of multimedia information in the form of joint visual and textual models or using graph modeling for semantical painting analysis has been introduced in recent approaches [15]. Style analysis was also analyzed for time and visual characteristics [16]. Additional alternatives include exploring object/face domain transfer detection and acknowledgement [17, 18].

These techniques are primarily aimed at capturing the visual quality of paintings, and very well-cured datasets. Paintings are, however, somewhat different from Buddha statues, in that statues are 3D objects, made with strict laws. We also want to study the history of art, not only the visual appearance but also its historical, material, and artistic context. In this piece, we discuss different incarnations for various classifications of Buddha statues from the Ancient Tibetan rules up to modern visuals, as well as graphics and graphs.

We can also research recent works, such as an examination of East Asian statues that are close to our application domain [19, 20]. However, one previous work obtained recognition of the Thai statue through the use of handcrafted facial features [21]. Further works are concerned with the 3D acquisition and structural analysis of statues, sometimes also with classification objectives. The introduction of the Buddhist techniques to test and recover damaged parts should also be emphasized [22].

Since 3D scanning does not apply to the large volume of statues, we are analyzing 2D images of 3D statues, which are very carefully related to the Pornpanomchai et al. [21]. Besides the study of ancient dimensions, we also provide modern analytics on a minimal dataset which does not give any information for each class with visual, facial (which also involves a 3D analysis), or semantical features for multiple classification tasks.

3 Analysis of Buddha Statues Through the Application of Machine Learning

The creation and installation of Buddha statues were possibly directed by specific rules laid down in Buddhist canonical texts. However, the implementations of these rules differ widely over time and space. The goal of automatic art analysis is to identify and thus analyze these practical convolutions and challenges. In an attempt to use and compare between different deep learning characteristics for various classification tasks, in a small but rich dataset of Buddha statues, collected from specialists in the art of Buddhism. The contours defined by construction directives in the texts be effectively retrieved.

With the spread of Buddhism beyond the boundaries of its origin in Eastern India, multiple changes were introduced in the core Buddhist practices to serve a variety of communities across the territorial space. It was accommodated and assimilated according to the local needs and thus survived itself in the form of divergent sects in various geographical and cultural regions for many centuries. The indigenous people could develop Buddhism of their own as it penetrated new territories. They did not only follow traditional laws laid in the Buddhist canons but also tailored them to portray a culture of their own, leading to new types and styles of art related to the Buddhist cult.

The present research concentrates on Buddha, the core theme of Buddhist art and in particular, the Buddha statues. Statues are mostly 3D artifacts. Based on their construction, the Buddha statues have been grouped into several types. However, they are all subjected to crafting principles defined by canons. These canons constitute the very foundations of the Buddha representation by having several basic rules or principles. Although the canons were originally taught using language-based description, these rules have today been preserved and graphically transcribed into rule books [23]. An analysis of measurements of art objects or iconometry has been used to understand and examine the relationship and differences among different Buddha statues. As per dictionary definition, "iconometry is the art of estimating the distance or size of an object by the use of an iconometer" [24]. Iconometer, according to Merriam-Webster dictionary, is "an instrument that determines the proper objective to be used in taking a picture of a given size from a given standpoint and that consists of a diopter and an open rectangular frame sliding on a graduated rod."

The iconometry principle is focused on measuring the different parts of the body, such as head, torso, limbs, fingers, and trying to compare them to establish standards for uniform art codes. The dimensions are widely based on the various types of body, such as Buddha and other deities associated with the cult, as well as various human teachers related to Buddhism.

In Himalayan art, the first code of iconometry was from an Indian cultural esthetic. However, there are many Indian textual sources which contributed to the creation of new pan-Indian figurative esthetics. The Buddhists took this naturally and gradually made their way into the Himalayas and Central Asia. The Manjushri Mulakalpa, SamVarodaya, Krishna Yamari, and Kalakakratantras are examples of an earlier Buddhist text for the study of iconometric proportions.

The Himalayan, Tibetan, and Central Asian Buddhist artistic traditions believed in adopting body proportions directives in figurative art. Especially, in comparison to body forms such as Buddha, benevolent and wrathful deities and human teachers, measurements differ. Many variations exist. Although proportional systems exist, there is currently no uniform system in place for all and not even all artists use iconometric measurement systems. The dimensions and measurements vary significantly among the regions, cultural groups, schools of painting, and various art traditions. The Buddha images and Tantric deities were first exposed to early iconometry.

4 Automatic Art Analysis

Automatic art analysis aims at the classification and retrieval of artistic representations from an image set by computer vision and methods of machine learning [15]. Several researchers have proposed to enhance visual representation through contextual neural networks knowledge about art. While visual representations can capture information concerning the content and style of work, we also encode links between different artistic attractiveness such as the author, the college, or the historical era in our proposed context-conscious embedding systems. In automated art analysis, we design two separate methods for the use of meaning. The former incorporates contextual data through a multi-task learning model in order to identify visual associations between components. The former offers multiple attributes together. In the second method, an information graph encodes connections between artistic attributes creates meaning.

This work is an attempt to associate the historical artifacts, the statues of Buddha with their place of origin. In this preliminary attempt, we have tried to recognize the statues of Buddha belonging to Borobudur due to their unique characteristics as outlined before. Presently, we have used only the features reflected in the facial component for our classification model. Deep neural networks have been used to compute the required facial embedding, and a model based on support vector machines has been trained on these embedding. The confidence results obtained prove that the model is indeed promising and raises avenues for further deliberation in this area.

5 Proposed Model for Identification of the Buddha Statues of Borobudur

In this paper, we intend to present a model that can predict whether the 2D rendition of a particular statue of Buddha has its origins in Borobudur or not. For this task, classic machine learning techniques have been deployed that analyze the features of the face of the statue and make predictions on the basis of these articulates. The problem is defined as a binary classification task with the classes being Borobudur and other. The dataset comprises of sixty nine images of statues of Buddha from Borobudur and other places like Nalanda.

For accomplishing this task, OpenCV has been used to create a complete "pure" OpenCV face recognition pipeline which can be visualized in Fig. 1. The model responsible for actually quantifying each face in an image has been built from the OpenFace project, a Python and torch implementation of face recognition with deep learning.

The primary set of tasks involved in the process is outlined as follows:

- Detect faces of Buddha statues
  The image that represents a 2D rendition of the 3D statue of Buddha is input to the face recognition pipeline. A face detection scheme is applied to detect the presence and location of a face in the image. For this, OpenCV's deep learning-based face detector has been used to localize faces in the input image.
- Compute 128-d face embeddings to quantify the face and form the feature vector using deep neural networks
  After the face region of interest (ROI) is extracted, the blob constructed from it is passed to the face embedding model. FaceNet deep learning model is then used to compute a 128-d embedding that quantifies the face into the corresponding feature vector that can be used for training.
- Train a Support Vector Machine (SVM) on the embeddings and the labeled as Borobudur and Others

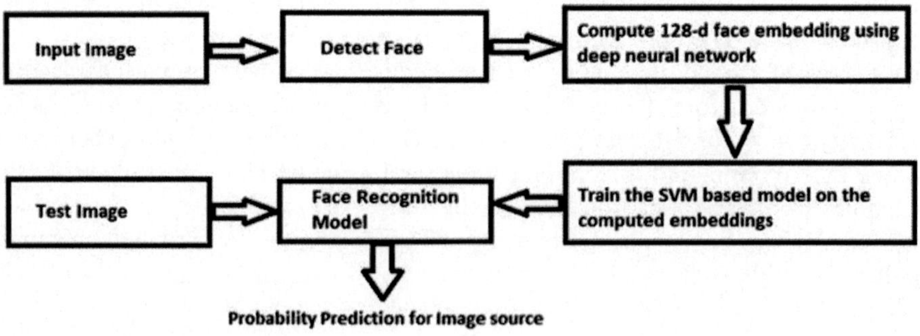

Fig. 1 OpenCV face recognition pipeline

These embeddings and their corresponding labels (Borobudur or other) are used to train a classification model based on the popularly used support vector machine to build the face recognition model.

- Use the model as a face recognition tool to identify the place of origin of the statues of Buddha

  This model can now be used to recognize and classify various unseen images of Buddha statues.

The model predicts whether the statue belongs to Borobudur or not and also outputs the probability or confidence associated with the prediction. Figure 2 presents the results for images that belong to Borobudur. It can be seen that the confidence interval lies in the range of [54.67–99.78%]. This shows that the model has been well able to identify distinguishing features of the statues from Borobudur.

Figure 3 shows some results when the model is deployed on images that belong to places other than Borobudur. Again, it can be seen that the confidence range lies between [60.33–97.24%] which is quite high.

Hence, the model has been able to produce satisfactorily good classification results.

Fig. 2 Buddha statues from borobudur

Fig. 3 Buddha statues belonging to places other than borobudur

6 Experimental Setup and Results

The dataset consists of total sixty-nine images. Out of sixty-nine images forty-eight belongs to Borobudur and rest are not from Borobudur. We used thirty-eight images from Borobudur and thirteen other images for the training of model. Total eighteen images ten from Borobudur and rest others are used for testing and obtained 94.44% accuracy (only one image was misclassified).

7 Conclusion

The spread of Buddhism has resulted in the displacement of artwork related to Buddha. Only a very few experts can identify the place where these artworks belong to. In this paper, a machine learning-based method is implemented for the identification of the origin of Buddha statues from the images, whether these statues belong to Borobudur or not. In the first step, a deep neural network has been utilized to extract the features from the face of the images of Buddha. Then, the classification of the Buddha images is carried out using the support vector machine. Images of Buddha statue from Borobudur as well as from some other places have been taken for training and testing purposes. The accuracy achieved is 94.44%.

The classification system can also be utilized further for the identification of archeological artifacts to place it in a proper class geographically and even historically.

References

1. Miksic J, Borobudur: golden Tales of the Buddhas; Tuttle Publishing; Reprint edition (February 28, 2017)
2. Kempers B (1976) 1976 ageless borobudur: Buddhist mystery in stone, decay and restoration, mendut and pawon. Folklife in Ancient Java, Servire
3. Carneiro G, Pinho da Silva N, Del Bue A, Paulo Costeira J (2012) Artistic image classification: an analysis on the printart database. In: ECCV
4. Johnson CR, Hendriks E, Berezhnoy IJ, Brevdo E, Hughes SM, Daubechies I, Li J, Postma E, Wang JZ (2008). Image processing for artist identification. IEEE Sig Process Magazine 25(4)
5. Khan FS, Beigpour S, Van de Weijer J, Felsberg M (2014) Painting-91: a large scale database for computational painting categorization. Mach Vision Appl
6. Shamir L, Macura T, Orlov N, Mark Eckley D, Goldberg IG (2010) Impressionism, expressionism, surrealism: automated recognition of painters and schools of art. ACM Trans Appl Percept 7(2):1–17
7. Mensink T, Van Gemert J (2014) The rijksmuseum challenge: museum-centered visual recognition. In: ICMR
8. Bar Y, Levy N, Wolf L (2014) Classification of artistic styles using binarized features derived from a deep neural network. In ECCV Workshops
9. Elgammal A, Mazzone M, Liu B, Kim D, Elhoseiny M (2018) The shape of art history in the eyes of the machine. arXivpreprint arXiv:1801.07729
10. Elgammal A, Saleh B (2015) Quantifying creativity in art networks.arXiv preprint arXiv: 1506.00711
11. Karayev S, Trentacoste M, Han H, Agarwala A, Darrell T, Hertzmann A, Winnemoeller H (2014) Recognizing image style. In: BMVC
12. Mao H, Cheung M, She J (2017) Deep Art: learning joint representations of visual arts. In: ACMMM
13. Chu W-T, Wu Y-L (2018) Image style classification based on learnt deep correlation features. IEEE Trans Multimedia 20(9):2491–2502
14. Strezoski G, Worring M (2017) Omni art: multi-task deep learning for artistic data analysis. arXiv preprint arXiv:1708.00684
15. Garcia N, Renoust B, Nakashima Y (2019) Context-aware embeddings for automatic art analysis. In: Proceedings of 2019 on International Conference on Multimedia Retrieval, ACM, pp 25–33
16. Garcia N, Vogiatzis G (2018) How to read paintings: semantic art understanding with multimodal retrieval. EECV Workshops
17. Crowley Elliot J, Zisserman Andrew (2016) The art of detection. Springer, ECCV
18. Crowley E, Zisserman A (2014) the state of the art: object retrieval in paintings using discriminative regions. BMVC
19. Bhaumik G, Samaddar SG, Samaddar AB (2018) Recognition techniques in buddhist iconography and challenges. In: 2018 International Conference on Advances in Computing, Communications and Informatics (ICACCI). IEEE, pp 1285–1289
20. Kamakura M, Oishi T, Takamatsu J,Ikeuchi K (2005). Classification of Bayon faces using 3D models. In: Virtual Systems and Multimedia, pp 751–760

21. Pornpanomchai C, Arpapong V, Iamvisetchai P, Pramanus N (2011) Thai Buddhist sculpture recognition system (tbusrs). Int J Eng Technol 3(4):342
22. Wang H, He Z, He Y, Chen D, Huang Y (2019) Average-face-based virtual inpainting for severely damaged statues of Dazu Rock Carvings. J Cult Heritage 36(2019):40–50
23. Renoust B, Oliveira Franca M, Chan J, Garcia N, Van L, Uesaka A, Nakashima Y, Nagahara H, Wang J, Fujioka Y (2019) Historical and modern features for Buddha statue classification. In: 1st Workshop on Structuring and Understanding of Multimedia heritage Contents (SUMAC'19), October 21, 2019, Nice, France. ACM, New York, NY, USA, pp 8. https://doi.org/10.1145/3347317.3357239
24. Merriam-Webster dictionary

Retrieval Mechanisms of Data Linked to Virtual Servers Using Metaheuristic Technique

Zdzisław Pólkowski, Suman Sourav Prasad, and Sambit Kumar Mishra

Abstract In the present situation, the large-scaled data supports the virtual environment specifically from small to medium ICT-based business. The implementation of virtual server, along with the associated technologies in general can reduce the processing costs, hardware costs as well as evaluation and optimization of large queries in multi query environments. As the processing capabilities during processing the queries are major issues, the process in such situation acquires more time as it cannot optimize the concerned data in the whole database in the stipulated time. Practically, it is difficult to process erroneous data with virtual link. To overcome this difficulty, canonicalization concept can be adopted to increase the performance of sustained data along with safe querying. Initially, it finds the similarities of the given query with the distinct database. After that it is essential to generalize the symbols linked with query stringing and enhances the system performance. This would result in better accuracy with limited query processing time and minimum computation cost. In this context, it has been thought of analysing the performance of query plans linked with the database associated with virtual servers using metaheuristic approach.

Keywords Query term · Metaheuristic · Canonical cover · Tuple · Pheromone · Semantic query · Cloud

---

Z. Pólkowski (✉)
Wroclaw University of Economics and Business, Wrocław, Poland
e-mail: zdzislaw.polkowski@ue.wroc.pl

S. S. Prasad
Ajay Binay Institute of Technology, Cuttack, affiliated to Biju Patnaik University of Technology, Rourkela, Odisha, India
e-mail: prasadsuman800@rediffmail.com

S. K. Mishra
Gandhi Institute for Education and Technology, Bhubaneswar, affiliated to Biju Patnaik University of Technology, Rourkela, Odisha, India
e-mail: sambitmishra@gietbbsr.com

A. Khanna et al. (eds.), *Data Analytics and Management*, Lecture Notes on Data Engineering and Communications Technologies 54,
https://doi.org/10.1007/978-981-15-8335-3_68

1 Introduction

Currently, small- and medium-sized enterprises represent two-thirds of global GDP. Only in Poland in present situation, having more than 2.1 million entities are registered, and almost 10 million people work in the entire sector. Increasingly more than ever before, small businesses are facing the pace of change and market transformation which in turn have more and more experienced customers expecting digital experience, the highest level of customer service or simple user interfaces.

With the increased use of cloud and web applications, a strong and secure network is the bloodstream of every small business. For this reason, many entrepreneurs from this sector are trying to use virtual technologies in their companies. The manifestations of these activities are attempting to implement databases in cloud computing. However, very often relational databases can be difficult to maintain, especially when a lot of data is stored and used by many users every day. Timeouts, deadlocks, long-running queries, breaks in operation are the main problems encountered by many users in the small business sector using cloud solutions, particularly in the usage of database system.

Thus, optimizing query along with query terms in heterogeneous cloud environment is sometimes a challenging task. The primary intention of this work is broadly to improve the performance of query processing in the cloud environment. Many techniques have already been proposed towards storing, processing and utilizing large amounts of data along with monitoring the processing time. In general, the voluminous data is preprocessed to eradicate the unnecessary information from the database. In that context, the metadata could be built towards data management, enhancement of business logic integration efficiency. Also the creation of metadata can be associated with the specific information to identify particular instances of data. In fact, the accuracy value of the metadata sometimes is useful towards query retrieval mechanisms. Many techniques have been applied to manage the retrieval mechanisms of data from cloud database in both consistent and inconsistent formats. The databases sometimes depend on the key value and scalable application of data sets. A specific methodology is implemented towards query plans in complex environments to accomplish data similarity and effective data placement. Also the expansion of query terms can be accumulated by evaluating the execution time of query and throughput. After that methodology can be adopted to process the framework of data to schedule the data in a cloud environment and to be performed by the cloud system.

In general, canonical cover is termed as simplified and reduced set of functional dependencies which can also be termed as irreducible set. In other words, it can be defined as a set of candidate functional dependencies with no redundant dependencies. Although it is difficult to identify the regular rules towards analysis of the structure accurately, some common features are easily recognized. It is advisable to classify the structural features, concerned patterns while encoding the databases. In this work, it has been projected on the query terms linked with the servers of virtual databases to measure the query response time. Also, it is essential to optimize the query terms within the data servers associated with the virtual database servers.

2 Review of the Literature

Mirajkar et al. [1] in their work have discussed the basic concepts of virtualization. The authors admit that the independent online utility sometimes allows the users to approach the Internet devices connected to remote as well as other connection. During their observation, they have also concentrated on viability of component-based design linked towards security of complex systems with heterogeneity. They opted for cloud security to project unified security architecture.

Kumar et al. [2] in their work have focused towards the provision of cloud service to manage hardware very efficiently and easily. Same as all the hardware can be utilized by all the computers linked to virtual servers, the operating expenditure may be reduced, and it is also required for any organization to accomplish the necessities of the clients.

Y. Huang et al. [3] in their article have projected the concept of outsourcing privacy. In such cases, the database owner may update the database linked to untrusted servers. It is presumed that, the clients of the database may not be able to focus on unauthorized access to the databases. Also the authors tried to focus on server-side indexing structure which could allow to be linked to database servers and efficiently use the multiple database clients.

E. Ayday et al. [4] in their work have discussed the privacy issues linked with genomic sequencing. It has also been focused towards research problems associated with several entities by collaborating with different organizations.

K. Lauter et al. [5] in their study have focused on homomorphic encryption along with privacy-preserving solution inclined towards the idea of computing over encrypted data without knowing the keys. In this regard, the data is encrypted along with suitable attribute specification in the form of virtual data.

A. Gholami et al. [6] in their work have implemented the security, cloud framework. It may somehow represent the proper storage and along with genomic data to be processed in a virtual environment. So, you can build the framework on the cloud privacy towards sequencing and processing the next-generation sequencing data.

S. Pearson et al. [7] in their work have described the basic concepts of A4Cloud. In such cases, the users may require to track the data usage and satisfy the utility and protection of data. In this regard, they suggested implementing suitable control and transparency over their data.

J. Nikolai et al. [8] in their article have focused towards cloud intrusion detection systems. The basic concept may be to enhance the efficiency of virtual machine and measure the performance of the data. Performance metrics may be directly involved with CPU utilization, which may be retrieved to analyze the collected data.

C. Klein et al. [9] in their work have improved the cloud service using the predefined load balancing mechanism. The primary intention in this case is to enhance the optional contents to provide a solution. The authors proposed a synchronization mechanism for distributed cloud accounting systems along with the set of accounting system requirements.

H. Takabi et al. [10] in their work have introduced the provision towards managing policies to utilize cloud resources. In this way, it solves the problems associated with authorization in access as well as specific applications.

Q. Zhang et al. [11] in their work focused towards sharing of virtual machines with the hardware as the resources in one virtual machine may not be transferred to another. In fact, sharing of resources in a virtual machine environment may be a challenging issue and difficult to be shifted to another.

J. Bhimani et al. [12] in their study evaluated the performance of tasks within the specific clusters in the virtual environments with support of physical machines. They made the comparison of clusters of fixed size.

S. Arnautov et al. [13] during their studies have focused towards internal mechanisms of kernel in an open source environment. They have also analyzed the security mechanisms along with execution support to protect the system.

R. Morabito et al. [14] in their work have evaluated the performance, implementing Internet of things. They have also compared the performance with Raspberry along with execution in the specific layers.

S. Singh et al. [15] in their study focused towards virtual machine along with the provision of services to users. In such cases, cloud service providers provide the services and allocate within the elastic resources as the resources may not have the specific allocation of scheduled tasks.

3 Performance Analysis and Representation

It is clear that a tuple may contain all the data for an individual record. While updating the cached tuples, the following steps may be essential.

Step 1: Accept the notion of query, q
Step 2: Display the result of the query, Qr
Step 3: Check all queries in Cache If Key[q] == Key [cache (j)]
Step 4: Update [cache (j)] = [cache (j)] + 1
Step 5: Update processing time

Sometimes semantic similarity may be a quantified lining with any notions. Therefore, modification towards semantic query is required to be matched with the query terms. The accuracy towards metadata is obtained from queries by evaluating the query response time.

3.1 Implementation of Metaheuristic Approach

Particularly, a metaheuristic approach is associated with search and optimization problems linked with one or more heuristics and will try to find a near-optimal solution. In general, it is iterative in nature, implementing stochastic operations in its search process to obtain solutions. Usually, the metaheuristic techniques can be motivated by natural, physical or biological principles along with maintaining criteria between exploration and exploitation. In such situation, exploration diversifies solutions in the search space having the global search behaviour. Accordingly, exploitation will utilize the information available from solutions from earlier iterations to obtain the optimality in the search space.

3.2 Application of Ant Colony Optimization Towards Query Optimization

Ant Colony Optimization (ACO) is defined as a set of probabilistic metaheuristics and an intelligent optimization algorithm, and implementation can be performed by observing the simulation of ant behaviour in the algorithm as follows.

3.3 Algorithm

Step 1: Identify the parameters associated with ant colony optimization
Step 2: Specify the maximum number of iterations, maxit = 100, queries in database,

n = 400;

Step 3: Identify the population size, p = 100, number of ants, a = 100, query plans in

query = 50;

Step 4: Evaluate the initial pheromone, query terms
Step 5: Initialize the exponential weight, heuristic weight and evaporation rate of

pheromone

Step 6: Evaluate the cost of pheromone matrix, i.e. query terms
Step 7: Regenerate the query terms based on optimality

```
for j = 1:maxIt
for q = 1:n
ant(q).queryterm = rand j([nVar]);
for j = 2:nVar
```

```
j = ant(q).queryterm(end);
Q = tau(j,:).^p.*eta(j,:).^a;
Q(ant(q).queryterm) = 0;
Q =/sum(Q);
ant(q).queryterm = [ant(q).j];
end
```

Step 8: Compare the cost of query with the query terms prioritizing the

```
Pheromone matrix
ant(q).cost = costquery(ant(q).Q);
if ant(q).Cost < optimalquery.cost
optimalquery = ant(q);
end
end
Step 9: Update Pheromones (Query terms)
for q = 1:n
optqry = ant(q).queryterm;
for l = 1:nVar
i = queryterm + l;
j = queryterm(l + 1);
taw(i,j) = taw(i,j) + Q/ant(q).cost;
end
end
```

3.4 *Experimental Analysis*

In general, the query terms associated with the virtual database servers can be measured as per their occurrence along with query response time. Therefore, classification and categorization of query terms are very much essential. In this regard, it is required to unify the activities within the data servers by applying the search engine to optimize the query terms. The weights associated with queries in the data servers depend on the link with each query term along with the retrieval mechanisms. But still the mechanisms cannot enhance the quality of data classifications. In such scenario, it is necessary to evaluate the cost of I/O and query response time along with the processing cost of queries. It has been observed that the I/O cost of queries along with the processing cost will obviously enhance with increasing number of queries associated with the data servers but significantly it may impact over the query response time (Table 1; Fig 1).

Table 1 Clustered index and associated processes data in server

Queries associated (server)	Estimated I/O cost	Estimated processing cost	Estimated number of executions	Query response time (ms)
1400	0.003127	0.0001593	11,554	0.46
1900	0.003439	0.0001597	11,579	0.37
2000	0.003467	0.0001599	11,584	0.34
2400	0.003472	0.0001603	11,593	0.31

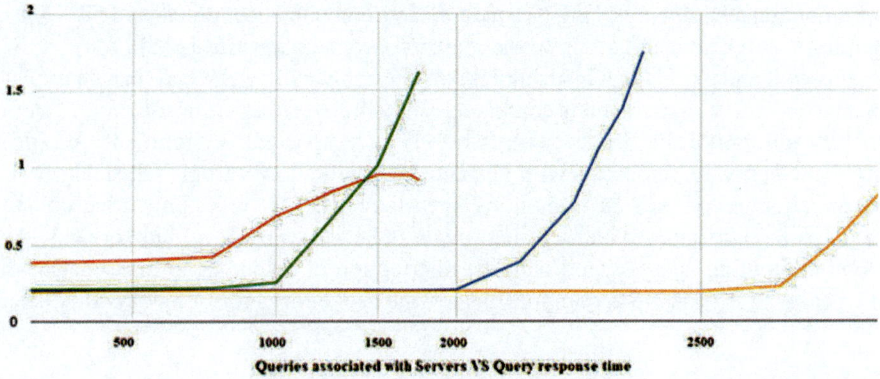

Fig 1 Queries associated with servers versus query response time

4 Optimization Criteria

Step 1: Minimization of scheduling tasks. The process may be initiated with the preceding tasks.
Step 2: Accumulation of completing time of all the tasks. In this case, to achieve optimization criteria, tasks may be scheduled to be executed according to occurrence.
Step 3: Waiting time, which is nothing but the task execution start time and the completion of the task.
Step 4: Turnaround time, which monitors how long it takes for a task to complete execution since its submission.

5 Discussion and Future Direction

An enterprise is worth as much as the data it has. It is generally understood as computerization and escaping towards large datasets that can be closed on a relatively small area and serve several employees and make SMEs more and more competitive.

There are still many problems in the analyzed area. The biggest of them seems to be data management, especially when it comes to differentiation of permitted levels. This is not always optimal in terms of performance, and there is also the issue of vulnerability to failure or attack. For these reasons, shared or virtual servers seem to be the best solution. Virtualization allows you to run multiple instances of virtual servers on one physical server. It optimizes resource allocation and minimizes the effects of failures. Unfortunately, the share of shared and virtual servers is still insufficient.

It is worth mentioning that one of the most heavily used, effective and important toolkits that improve the efficiency of a virtual environment are mathematical optimization techniques. Using the math in environmental economics and IT will be helpful to determine and describe the most important connections [16].

It is well understood that in virtualization, the application workloads can be implemented as virtual cases in many companies. In such cases, the virtual machine system will try to monitor the access towards resources in a virtual environment. Accordingly, the software associated with virtualization will allow a single physical server to run a number of separate computing environments. The server linked with cloud can also be associated with large data hubs linked to avail the cloud services. As cloud computing is associated with the distribution of resources to many users, in such cases, it is possible to combine data of one user or organization with other user.

6 Conclusion

Sometimes the concept of virtualization may be associated with a number of important utilities that make it a very strong tool to be used in a large number of applications. Of course, it is not to be confined to server consolidation, application sandboxing, access to different types of hardware and operating systems, debugging. By applying the other specific applications towards virtual machine, the performance of virtualization can be improved with time.

References

1. Mirajkar N et al. (2012). Implementation of private cloud uses eucalyptus and an open source operating system, arXiv preprint arXiv: 1207.3037
2. Smith Kumar R, Shilpi C (2015) An importance of using virtualization technology in cloud computing. Global J Comput Technol 1(2)
3. Huang Y, Goldberg I (2013). Outsourced private information retrieval. In: Proceedings of the 12th ACM Workshop on Workshop on Privacy in the Electronic Society, WPES'13, (New York, NY, USA), pp 119–130
4. Ayday E, Raisaro J, Hengartner U, Molyneaux A, Hubaux J-P (2014) Privacy-preserving processing of raw genomic data. In: Garcia-Alfaro J, Lioudakis G, Cuppens-Boulahia N, Foley S, Fitzgerald WM (eds) Data Privacy Management and Autonomous Spontaneous Security, vol. 8247 of Lecture Notes in Computer Science, pp 133–147

5. Lauter K, Lopez-Alt A, Naehrig M (2014) Private computation on encrypted genomic data, Tech Rep MSR-TR-2014–93
6. Gholami A, Laure E (2015) Advanced cloud privacy threat modeling. In: The Fourth International Conference on Software Engineering and Applications (SEAS-2015), to be published in Computer Science Conference Proceedings in Computer Science and Information Technology (CS/IT) Series
7. Pearson S, Tountopoulos V, Catteddu D, Sudholt M, Molva R, Reich C, Fischer-Hubner S, Millard C, Lotz V, Jaatun M, Leenes R, Rong C, Lopez J (2012) Accountability for cloud and other future internet services, in cloud computing technology and science (CloudCom). In: 2012 IEEE 4th International Conference, pp. 629–632
8. Nikolai J, Wang Y (2014) Hypervisor-based cloud intrusion detection system. In: Computing, Networking and Communications (ICNC), International Conference, pp 989–993
9. Klein C, Papadopoulos A, Dellkrantz M, Durango J, Maggio M, Arzen K-E, Hernandez Rodriguez F, Elmroth E (2014) Improving cloud service re silience using brownout-aware load-balancing, Reliable Distributed Systems (SRDS) In: 2014 IEEE 33rd International Symposium, pp 31–40
10. Takabi H, Joshi J (2012) Policy management as a service: An approach to manage policy heterogeneity in cloud computing environment, System Science (HICSS). In: 2012 45th Hawaii International Conference, pp 5500–5508
11. Zhang Q, Liu L, Su G, Iyengar Memflex AK (2017) A shared memory swapper for high performance vm execution. IEEE Trans Comput
12. Bhimani J, Yang Z, Leeser M, Mi N (2017) Accelerating big data applications using lightweight virtualization framework on enterprise cloud. In: High Performance Extreme Computing Conference (HPEC), IEEE, pp 1–7
13. Arnautov S, Trach B, Gregor F, Knauth T, Martin A, Priebe C, Lind J, Muthukumaran D, O'Keeffe D, Stillwell M, et al. Scone. (2016). Secure linux containers with intel sgx., OSDI, pp 689–703
14. Morabito R (2016) A performance evaluation of container technologies on internet of things devices, computer communications workshops (INFOCOM WKSHPS). In: 2016 IEEE Conference pp 999–1000
15. Singh S, Chana I (2016) A survey on resource scheduling in cloud computing: Issues and challenges. J Grid Comput
16. Vasilev J, Turygina V, Kosarev A, Nazarova Y (2016) Mathematical optimization in environmental economics, algorithm of gradient projection method. Int Multi Sci Geo Conf Surveying Geol Min Ecol Manage SGEM 3(2016):349–356

Machine Learning Approaches for Psychological Research Review

Marta R. Jabłońska and Zdzisław Pólkowski

Abstract With a broad adaptation of artificial intelligence (AI) tools in recent years, machine learning (ML) has encountered an increasing popularity in various aspects of business, science and individual life. We proposed a research question on possibilities and threat that machine learning tools may bring to psychological studies. We analyzed the current state of the art to provide contemporary adaptations and scope of implementation of ML into psychological research. The most popular ML languages and frameworks have been provided. Then, we defined core pros and cons of these deployment tools. Finally, we have presented a SWOT analysis for ML in psychology. Concerning our findings, we claim that ML seems to be a promising solution for improving psychological studies in areas of diagnosis, prognosis and treatment. These applications may be used for pattern recognition that are too complex for a human researcher. Still, ML is not a mature technology and still faces important challenges. It is not well-established, accessible and user-friendly.

Keywords Machine learning · Artificial intelligence · Psychology

1 Introduction

Psychology studies classically try to predict and understand behaviors as well as its roots, causes and determinants [1, 2]. Personality traits and emotions have been exposed as crucial components of intelligent behavior [1, 3]. Human behavior analysis is not limited to psychology and other sciences; it also focuses on this issue. One of the examples is computer science, which supports research using these emerging IT technologies, including ML [4, 5].

M. R. Jabłońska (✉)
Department of Computer Science in Economics, Faculty of Economics and Sociology, Institute of Logistics and Informatics, University of Lodz, Lodz, Poland
e-mail: marta.jablonska@uni.lodz.pl

Z. Pólkowski
Department of Technical Sciences, Jan Wyzykowski University, Polkowice, Poland
e-mail: z.polkowski@ujw.pl

A. Khanna et al. (eds.), *Data Analytics and Management*, Lecture Notes on Data Engineering and Communications Technologies 54,
https://doi.org/10.1007/978-981-15-8335-3_69

ML is an artificial intelligence method. Due to learning abilities gained by training using large amounts of data, there is no actual need to explicitly program them. ML algorithms supported by a set of data can learn and then perform analysis, such as classifying data due to unknown, hidden patterns or predict future values [6]. These models support decisions for all types of management level, and they can be used by governments, enterprises, society and single human beings [7, 8]. Studies prove that ML can be successfully implemented in psychology in areas such as diagnosing and predicting behavior, mental state and health as well as personality characteristics [6, 8, 9].

Experimental psychology seeks to clarify underlying processes of behavior as well as to predict it [2, 5]. Still, traditional methods have mainly focused on testing causal explanations rather than predictions [5]. On the opposite, ML is able to learn from multidimensional data and make outcomes about individuals that can be generalized. Thus, ML has an ability to advance decisions in psychology and psychiatry associated with the patient's analysis, prognosis and treatment [10]. Although the number of papers devoted to the machine learning application in this area is constantly growing, it is still a challenging issue with AI [4]. As traditional approach in psychology and psychiatry remains limited to diagnosis without an ability to generalize, ML could help these sciences to provide predictions and classifications able to provide an improved understanding of behavior [2].

This paper is aimed at presenting ML opportunities for psychological research, basing on a current state of the art. What is more, we have prepared a set of pros and cons of implementing these algorithms into psychological studies. The manuscript consists of the following sections. Section 2 presents the current state of the art in studies on ML approaches in psychology as well as its implementations and algorithms. Section 3 describes possible opportunities and advantages seen in ML deployment into psychological studies following by emerging difficulties and threats. The next part of the paper is a discussion summarizing machine learning approaches for psychological research.

We believe that our paper helps to clarify ML cons including the fact that this technology is not mature and well-established in psychology applications, nor accessible and easy to use. But, despite those limitations, it has a potential of analysis exceeding human capabilities, the ability to generalize and improving diagnoses, as well as patients' treatment.

2 Machine Learning Approaches for Psychology

2.1 State of the Art

This section aims to address some selected studies on ML in psychological research. We describe the current studies, utilized tools and employed measurements. We selected the studies basing on several criteria. All the manuscripts should be published in respected, psychology journals. To ensure this request, only papers indexed in the Web of Science database were included. Secondly, we considered the date of publication as only works that are perceived as up-to-date were enclosed. We chose papers from 2005 to 2019. And finally, we checked the content of the study whether it truly considered psychological research.

The current research describing ML approaches in psychology included especially studies on mental health, emotion and personality traits. The first problem domain—mental health—determines how a person thinks, experiences and deals with situations, so it is essential at every stage of life to preserve a healthy life balance. Mental health issues may cause: chronic stress, anxiety attacks, severe depression, addictions and personality disorders [6]. The second problem domain—emotions—has been determined as a core factor of intelligent behavior and cognitive processes [3]. Lastly, personality traits affect various aspects of an individual's behavior, including job performance, effectiveness, dominance, emotion, sentiment and deception [1]. To ensure a better transparency of collected data, we presented recent studies' juxtaposition in Table 1.

3 ML Tools and Frameworks

In this subsection, some broadly used tools and frameworks for accessing psychological data are discussed. We present ML algorithms, selected, supervised machine learning approaches and natural language processing features.

Recent studies suggest implementing ML such as support vector machines (SVM), decision trees, naive Bayes classification, k-nearest neighbor classification and logistic regression [6]. Algorithms that support supervised learning are perceived as broadly used [17].

Supervised ML uses labeled data for the purpose of learning an algorithm. Using historical data, the model tries to understand hidden patterns and then decides which output (label) should be given when a new data set is given. Among the most popular supervised ML approaches, the following should be mentioned: SVM, recurrent neural networks (RNNs), long short-term memory neural network (LSTM) and hidden Markov models (HMMs) as presented in Table 2.

These algorithms may be supported by natural language processing (NLP) which is used for studies on psychological traits, i.e., to analyze data from written medical documentation. Vast amounts of data collected from computer systems used in clinical psychology but also from online communities in cyberspace are parsed and used in research of psychological phenomena concerning mental disorders and health illnesses such as depression, anxiety, suicide attempts, lack of attention, crisis, schizophrenia, borderline syndrome, dementia and numerous disorders such as hyperactivity, bipolar, eating, obsessive–compulsive, seasonal affective and post-traumatic stress disorder [17–20]. To collect psycho-linguistic features from written data for NLP processing, Linguistic Inquiry and Word Count (LIWC), Linguistic Category Model (LCM), Sentiment Lexicons or CELEX are used [16, 17]. Future

Table 1 Recent studies on ML in psychological research

Authors	Year	Problem domain	Description
Devillers et al. [9]	2005	Emotions	Exploration of real-life non-basic emotion expression, representation and automatic detection of a subject's emotional state in speech
Kelly et al. [11]	2012	Mental health	Momentary cognitive–behavioral intervention delivery; maintenance of complex mental disorders and their mechanisms with a possibility of application to complex problems like suicide prevention
Chavan and Shylaja [12]	2015	Emotions	Extraction methods to disclose offensive comments toward those who are seen more negatively, causing cyberbullying
Ahmad and Adnan [14]	2015	Emotions	Exploring emotions, age and gender relations to cognitive skills
Halde et al. [13]	2016	Personality traits	Students' psychological traits prediction on failures during studies
Hu and Wang [22]	2016	Personality traits	A model that can alarm an online crime's event in advance by analyzing the Internet hate speech
Strömfelt et al. [3]	2017	Emotions	Considerations how emotions have been translated into ML
Borja-Borja et al. [4]	2017	Personality traits	Automated understanding and classification of the individual and group behavior
Ghai et al. [21]	2017	Emotions	Recognizing emotions in speech and classify them in output classes: anger, boredom, disgust, anxiety, happiness, sadness and neutral
Kshirsagar et al. [20]	2017	Personality traits	A model of for the detection of crisis in social media content. A modular approach to generating explanations similar to those produced by human annotators
Nath et al. [27]	2017	Mental health	Prediction of volatile substance abuse
Suryapranata et al. [28]	2017	Personality traits	Prediction of openness, conscientiousness and agreeableness personality traits of game players based on a game character design they choose before starting to play games
Stevens and Soh [15]	2018	Personality traits	Assessing factors of predicting similarity models of intertemporal choice

(continued)

Table 1 (continued)

Authors	Year	Problem domain	Description
Koul et al. [5]	2018	Personality traits	ML application enabling computational assessment of putative bio-behavioral markers for both prognosis and diagnosis. Dedicated to researchers with limited programming experience not only in the field of psychology, but also clinical neuroscience
Alharthi et al. [16]	2018	Personality traits	Automatic identification of needs and measurement of their satisfaction level
Cerasa et al. [23]	2018	Mental health	Identification of multivariate predictive patterns of personality profiles that could identify gambling disorder patients from healthy controls and assumption of an individual's risk of developing a gambling disorder
Arribas et al. [25]	2018	Mental health	Capturing the emerging relations between different elements of mood and diagnosing participants due to bipolar disorder and borderline personality disorder
Araújo et al. [26]	2018	Personality traits	Finding personality traits of young adults involved in a network
Großmann et al. [24]	2019	Personality traits	Using personality analysis to predict a result of a two partner's relationship

Table 2 Popular supervised ML models

Model	Description
Support vector machines (SVMs)	Aims to find a point in an N-dimensional space to properly classify data by calculating instances lengths to support vectors to calculate a dividing gap for classification and regression tasks [17]
Recurrent neural networks (RNNs)	RNN forms connections between nodes in a graph along a momentary sequence to display momentary dynamic behavior
Long short-term memory neural network (LSTM)	A subtype of RNN devoted to discovering long-term dependencies
Hidden Markov models (HMMs)	Probabilistic models, operate with unobservable states for modeling unseen events

research would investigate language used in order to detect the current emotional landscapes and personality traits [17].

Generally, among the most adopted ML languages, the following can be mentioned. Python is an universal language preferred for a high-quality, easy operating and a good structure and used often for prototyping models. For data science,

Table 3 Selected ML frameworks suitalble for psychology research

Framework	Description
Microsoft Azure Machine Learning	A cloud platform that may be used also by ML beginners not fluent in ML languages. May be used for building, teaching and implementing models. It is suitable for clinical and operational data research
Google TensorFlow	Similarly to the previous framework, proper for beginners as well as experts in ML. Offers easy and intuitive construction and deployment of models dedicated also for research experimentation
IBM Watson	A cloud service for ML and deep learning. It automates data preparation, simplifies model deployment and automates its retraining. Offers solutions dedicated to health care and life research
Amazon Machine Learning	It offers users numerous tools for automatization and simplification and speeding up creation of ML models. In psychology studies, it may be used in applications such as image and video analysis, text and document analysis, as well as predictions

a language and environment called R is probably the best choice. It allows not only statistical analysis but also graphical visualizations. Finally, a reliability, well-operating C++ is used, mainly for teaching purposes but also used in research as well. To build ML models more effortlessly and swiftly, various frameworks are gaining popularity. Frameworks may consist of interfaces, libraries or tools, and they are basically used to improve ML model performance and make it more user-friendly (Table 3).

The idea of using ML frameworks is to make the whole user experience faster, better and more user-friendly. According to Amazon, its framework offers three times increase in network throughput, 25% improvement in price and performance, 99.99% durability and unmatched availability, and increasing data queries four times.

4 Pros and Cons of Implementing Machine Learning Solutions

This part is mainly devoted to focus on defining advantages of implementing ML solutions into psychological research and secondly, on presenting some disadvantages and possible challenges that have to be faced and understood before deployment of these methods into studies.

To present the most vital pros of ML implementations into psychology, let us start with a glimpse on traditional methods of psychological research limitations. Methods that are typically assessed include mainly questionnaire surveys, self-report, personal interviews and social observation and are becoming to be currently considered as insufficient for large-scale identification and analysis [16]. Analytic methods in psychological studies based on long-term data are typically based on traditional testing of hypothesis [29]. Among their limitations, the following are usually mentioned: a difficulty to get results that are comparable within a large group, a huge time consumption, and correctness of the results depends on a model specification at the outset, providing reflection of only a small population's sample, vulnerability to outlying records and unlikely to provide complex decision boundaries, and possible information loss due to reduction of assessment measures [16, 29].

To overcome these limitations, ML solutions may be applied into psychological research. ML approach is important for future psychological studies as it has a potential to amplify diagnosis, individual or group predictions, and medical care [5, 10]. ML is adept at reviewing large volumes of clinical and biological data and identifying patterns and trends that might not be apparent to a human researcher. It also typically improves efficiency and accuracy over time, so it can generate better decisions and predictions while using it. ML tools reduce time consumption as they operate without the need for human intervention. ML techniques are able to adapt to changing conditions, identify suspicious cases, handle complex data in changing, ambiguous environments and gain experience over time. ML can model decisions, recommendations and diagnoses outcomes but also explain how they are made [15]. That is why, they can be useful for assessing decision strategies that are the main features of ML that makes it desirable for implementation into psychological research.

According to the current literature, ML tools may be used for creation of better psychological models, mental illness and future behavior prediction, as a monitoring tool for deviant behavior, and to be a chance to advance researchers understanding of personality [2, 6, 8, 26]. To sum up, ML may help psychology become a more predictive science [2].

Still, as any technology, ML has its flaws. As it is able to predict trends and patterns with ease, it also has a high level of error susceptibility as it needs good-quality data in order to learn. It can potentially be quite costly as it requires resources and massive data sets to bring results. Low quality or too poor training data may cause system's decisions and predictions biases. Poorly designed or learned ML algorithm can do more harm than good.

To present our findings in a more transparent and cohesive manner, we prepared a SWOT analysis concerning ML applications for psychology (Table 4).

Due to the current research on ML approaches in psychological studies, ML still needs continually lack settled, available and user-friendly solutions. There is still a limited number of ML implementation dedicated especially researchers with limited programming experience [5]. ML models are often perceived as black boxes as their algorithms may be too complex to understand by a human researcher. What is more, they can be independent to the researchers' conceptions, including knowledge, ontology, concepts, as well as aims [7]. There are even researchers that discredit basic features of ML, claiming that any contemporary form of ML is not real learning [30]. ML models should provide model prediction and its explanations, but they are still quite poor at this matter [7].

Table 4 ML for psychology research

Strenghts	Weaknesses
Easily identifies hidden patterns Highly automated Continuous retraining Wide area of applications in psychology sciences Able to analyze big data	Requires huge data sets Needs massive resources as well as enough time Perceived as a "black-box," hard to interpret High error susceptibility with biased training set
Opportunities	**Threats**
Automation of diagnosis Supporting decisions concerning treatment Discovering new relations concerning human behavior Analyzing tradicional, unstructured medical records	Requires psychology researchers to posses new skills in ML Higher risk of rejecting results due to hidden conclusion process (black-box) Data collected from questionnaires may be biased due to the limitations of this research method

To make ML more broadly used, they should be improved aiming at interpretability, transparency, trust and fairness of their solutions [7]. Perceiving is still as a rather fragmented field [17], ML frameworks for psychological studies certainly require further improvements.

5 Discussion and Conclusions

Machine learning tools are becoming widely implemented in various aspects of everyday life, business and science. The aim of this paper was to present possibilities of ML for psychological studies according to the current state of the art. We analyzed scientific papers from 2005 to 2019, indexed in the Web of Science database.

In Sect. 2, we presented contemporary studies, utilized tools and employed measurements concerning ML for psychology. ML methods may be more sensitive considering common univariate techniques [5]; thus, ML is perceived as a possibility of improvement of psychological studies. These tools may be used for diagnosis, prognosis and mental treatment as they are able to recognize patterns too complex for a human researcher. These findings, as well as a broad spectrum of ML applications, may be confirmed by various studies conducted on ML in psychology. Walsh, Ribeiro and Franklin [31] used ML to analyze routinely collected clinical data for predicting suicide risk among adolescents, while Poulin with a group of researchers conducted a similar study using unstructured clinical notes and ML to estimate the risk of suicide among US veterans [32]. ML was also implemented to assess generalized anxiety disorder, major depression and efficiency of medicines used to cure these disorders [33–36]. An interesting study was conducted by Healy, Donovan, Walsh and Zheng that used a video feed in real time to detect several emotions [37]. ML was also used to better understand the nature of subjective well-being and its relations [38]. ML was also successfully implemented in research on social media, including problematic smartphone use, social media addiction prediction [39, 40].

As artificial intelligence methods are becoming mature enough to be broadly adopted, ML advantages described in Sect. 3 seem to be quite promising for psychological research. Still, it is not a technology without challenges and possible threats as it was characterized in the same section among crucial flaws, and the fact of a lack of established and accessible implementations of ML tools for psychological research should be outlined. This state requires from researchers to possess advanced programming ML skills that makes adoption of these algorithms more difficult. Still, considering all pros and cons, ML tools seem to be rather promising solution for improving a quality of psychological research.

References

1. Gupta U, Chatterjee N (2013) Personality traits identification using rough sets based machine learning. In: 2013 International symposium on computational and business intelligence, New Delhi, pp 182–185.. https://doi.org/10.1109/iscbi.2013.44
2. Yarkoni T, Westfall J (2017) Choosing prediction over explanation in psychology: lessons from machine learning. Persp Psychol Sci 12(6):1100–1122. https://doi.org/10.1177/1745691617693393
3. Strömfelt H, Zhang Y, Schuller BW (2017) Emotion-augmented machine learning: overview of an emerging domain. In: 2017 seventh international conference on affective computing and intelligent interaction (ACII) San Antonio, TX, USA, pp 305–312. https://doi.org/10.1109/acii.2017.8273617
4. Borja-Borja LF, Saval-Calvo M, Azorin-Lopez J (2017) Machine learning methods from group to crowd behaviour analysis. In: Rojas I, Joya G, Catala A (eds) Advances in computational intelligence, IWANN 2017, Lecture notes in computer science, vol 10306. Springer, Cham, pp 294–305. https://doi.org/10.1007/978-3-319-59147-6_26
5. Koul A, Becchio C, Cavallo A (2018) PredPsych: a toolbox for predictive machine learning-based approach in experimental psychology research. Behav Res Methods 50(4):1657–1672. https://doi.org/10.3758/s13428-017-0987-2
6. Srividya M, Mohanavalli S, Bhalaji NJ (2018) Behavioral modeling for mental health using machine learning algorithms. J Med Syst 42:88. https://doi.org/10.1007/s10916-018-0934-5
7. Fabra-Boluda R, Ferri C, Hernández-Orallo J, Martínez-Plumed F, Ramírez-Quintana MJ (2018) Modelling machine learning models. In: Müller V (ed) Philosophy and theory of artificial intelligence 2017, PT-AI 2017, Studies in applied philosophy, epistemology and rational ethics, vol 44. Springer, Cham, pp. 175–186. https://doi.org/10.1007/978-3-319-96448-5_16
8. Bleidorn W, Hopwood CJ (2019) Using machine learning to advance personality assessment and theory. Pers Soc Psychol Rev 23(2):190–203. https://doi.org/10.1177/1088868318772990
9. Devillers L, Vidrascu L, Lamel L (2005) Challenges in real-life emotion annotation and machine learning based detection. Neural Netw 18(4):407–422. https://doi.org/10.1016/j.neunet.2005.03.007
10. Dwyer DB, Falkai P, Koutsouleris N (2018) Machine learning approaches for clinical psychology and psychiatry. Annual Rev Clin Psychol 7(14):91–118. https://doi.org/10.1146/annurev-clinpsy-032816-045037
11. Kelly J, Gooding P, Pratt D, Ainsworth J, Welford M, Tarrier N (2012) Intelligent real-time therapy: harnessing the power of machine learning to optimise the delivery of momentary cognitive–behavioural interventions. J Mental Health 21(4):404–414. https://doi.org/10.3109/09638237.2011.638001

12. Chavan VS, Shylaja SS (2015) Machine learning approach for detection of cyber-aggressive comments by peers on social media network. In: 2015 international conference on advances in computing, communications and informatics (ICACCI), Kochi, pp. 2354–2358. https://doi.org/10.1109/icacci.2015.7275970
13. Halde RR, Deshpande A, Mahajan A (2016) Psychology assisted prediction of academic performance using machine learning. In: 2016 IEEE international conference on recent trends in electronics, information & communication technology (RTEICT), Bangalore, pp. 431–435. https://doi.org/10.1109/rteict.2016.7807857
14. Ahmad S, Adnan A (2015) Machine learning based cognitive skills calculations for different emotional conditions. In: 2015 IEEE 14th international conference on cognitive informatics & cognitive computing (ICCI*CC), Beijing, pp. 162–168. https://doi.org/10.1109/icci-cc.2015.7259381
15. Stevens JR, Soh LK (2018) Predicting similarity judgments in intertemporal choice with machine learning. Psychon Bull Rev 25(2):627–635. https://doi.org/10.3758/s13423-017-1398-1
16. Alharthi R, Guthier B, El Saddik A (2018) Recognizing human needs during critical events using machine learning powered psychology-based framework. IEEE Access 6:58737–58753. https://doi.org/10.1109/ACCESS.2018.2874032
17. Johannßen D, Biemann C (2018) Between the lines: machine learning for prediction of psychological traits-a survey. In: Holzinger A, Kieseberg P, Tjoa A, Weippl E (eds) Machine learning and knowledge extraction. CD-MAKE 2018. Lecture notes in computer science, vol 11015. Springer, Cham, pp 192–211. https://doi.org/10.1007/978-3-319-99740-7_13
18. Reece AG, Reagan AJ, Lix KLM, Dodds PS, Danforth CM, Langer EJ (2017) Forecasting the onset and course of mental illness with twitter data. Nature Sci Rep 7(1):13006. https://doi.org/10.1038/s41598-017-12961-9
19. Morales M, Scherer S, Levitan R (2017) A cross-modal review of indicators for depression detection systems. In: Proceedings of the fourth workshop on computational linguistics and clinical psychology—from linguistic signal to clinical reality, Vancouver, BC, Canada, pp 1–12.. https://doi.org/10.18653/v1/w17-3101
20. Kshirsagar R, Morris R, Bowman S (2017) Detecting and explaining crisis. In: Proceedings of the fourth workshop on computational linguistics and clinical psychology—from linguistic signal to clinical reality, Vancouver, BC, Canada, pp 66–73. https://doi.org/10.18653/v1/w17-3108
21. Ghai M, Lal S, Duggal S, Manik S (2017) Emotion recognition on speech signals using machine learning. In: 2017 international conference on big data analytics and computational intelligence (ICBDAC), Chirala, pp 34–39. https://doi.org/10.1109/icbdaci.2017.8070805
22. Hu Y, Wang S (2016) Research on crime degree of internet speech based on machine learning and dictionary. In: 2016 3rd international conference on information science and control engineering (ICISCE), Beijing, pp 532–537. https://doi.org/10.1109/icisce.2016.120
23. Cerasa A, Lofaro D, Cavedini P, Martino I, Bruni A, Sarica A, Mauro D, Merante G, Rossomanno I, Rizzuto M, Palmacci A, Aquino B, De Fazio P, Perna GR, Vanni E, Olivadese G, Conforti D, Arabia G, Quattrone A (2018) Personality biomarkers of pathological gambling: a machine learning study. J Neurosci Methods 294:7–14. https://doi.org/10.1016/j.jneumeth.2017.10.023
24. Großmann I, Hottung A, Krohn-Grimberghe A (2019) Machine learning meets partner matching: predicting the future relationship quality based on personality traits. PLoS One 14(3):e0213569. https://doi.org/10.1371/journal.pone.0213569
25. Arribas IP, Goodwin GM, Geddes JR, Lyons T, Saunders KEA (2018) A signature-based machine learning model for distinguishing bipolar disorder and borderline personality disorder. Transl Psychiatry 8:274. https://doi.org/10.1038/s41398-018-0334-0
26. Araújo EFM, Simoski B, Klein M (2018) Applying machine learning algorithms for deriving personality traits in social network. In: Proceedings of the 33rd annual ACM symposium on applied computing, Pau, France, pp 346–349. https://doi.org/10.1145/3167132.3167377

27. Nath P, Kilam S, Swetapadma A (2017) A machine learning approach to predict volatile substance abuse for drug risk analysis. In: 2017 third international conference on research in computational intelligence and communication networks (ICRCICN), Kolkata, pp 255–258. https://doi.org/10.1109/icrcicn.2017.8234516
28. Suryapranata LKP, Kusuma GP, Heryadi Y, Abbas BS, Ahmad AS (2017) Personality trait prediction based on game character design using machine learning approach. In: 2017 international conference on innovative and creative information technology (ICITech), Salatiga, pp 1–5. https://doi.org/10.1109/innocit.2017.8319139
29. Askland KD, Garnaat S, Sibrava NJ, Boisseau CL, Strong D, Mancebo M, Greenberg B, Rasmussen S, Eisen J (2015) Prediction of remission in obsessive compulsive disorder using a novel machine learning strategy. Int J Methods Psychiatric Res 24:156–169. https://doi.org/10.1002/mpr.1463
30. Bringsjord S, Govindarajulu NS, Banerjee S, Hummel J (2018) Do machine-learning machines learn?. In: Müller V (eds) Philosophy and theory of artificial intelligence 2017. PT-AI 2017. Studies in applied philosophy, epistemology and rational ethics, vol 44. Springer, Cham, pp 136–157. https://doi.org/10.1007/978-3-319-96448-5_14
31. Walsh CG, Ribeiro JD, Franklin JC (2018) Predicting suicide attempts in adolescents with longitudinal clinical data and machine learning. J Child Psychol Psychiatry 59:1261–1270. https://doi.org/10.1111/jcpp.12916
32. Poulin C, Shiner B, Thompson P, Vepstas L, Young-Xu Y, et al (2014) Predicting the risk of suicide by analyzing the text of clinical notes. PLoS ONE 9(1):e85733. https://doi.org/10.1371/journal.pone.0085733
33. Hilbert K, Lueken U, Muehlhan M, Beesdo-Baum K (2017) Separating generalized anxiety disorder from major depression using clinical, hormonal, and structural MRI data: a multimodal machine learning study. Brain Behav 7:e00633. https://doi.org/10.1002/brb3.633
34. Chekroud AM, Zotti RJ, Shehzad Z, Gueorguieva R, Johnson MK, Trivedi MH, Cannon TD, Krystal JH, Corlett PR (2016) Cross-trial prediction of treatment outcome in depression: a machine learning approach. Lancet Psychiatry 3(3):243–250. https://doi.org/10.1016/S2215-0366(15)00471-X
35. Carpenter KLH, Sprechmann P, Calderbank R, Sapiro G, Egger HL (2016) Quantifying risk for anxiety disorders in preschool children: a machine learning approach. PLoS ONE 11(11):e0165524. https://doi.org/10.1371/journal.pone.0165524
36. McGinnis RS et al (2018) Wearable sensors and machine learning diagnose anxiety and depression in young children, In: 2018 IEEE EMBS international conference on biomedical & health informatics (BHI), Las Vegas, NV, pp 410–413. https://doi.org/10.1109/bhi.2018.8333455
37. Healy M, Donovan R, Walsh P, Zheng H (2018) A machine learning emotion detection platform to support affective well being. In: 2018 IEEE international conference on bioinformatics and biomedicine (BIBM), Madrid, Spain, pp 2694–2700. https://doi.org/10.1109/bibm.2018.8621562
38. Wilckens M, Margeret H (2015) Can well-being be predicted? a machine learning approach (February 8, 2015). Available at SSRN: https://ssrn.com/abstract=2562051 or http://dx.doi.org/10.2139/ssrn.2562051
39. Elhai JD, Yang H, Rozgonjuk D, Montag C (2020) Using machine learning to model problematic smartphone use severity: the significant role of fear of missing out. Addictive Behav 103:106261. https://doi.org/10.1016/j.addbeh.2019.106261
40. Valakunde N, Ravikumar S (2019) Prediction of addiction to social media. In: 2019 IEEE international conference on electrical, computer and communication technologies (ICECCT), Coimbatore, India, pp 1–6. https://doi.org/10.1109/icecct.2019.8869399

Transformation of Higher Educational Institutions from Distance Learning to the E-Learning 5.0: An Analysis

Tadeusz Kierzyk and Zdzisław Pólkowski

Abstract Considering the present pandemic situation, it is intended to focus on the issues associated with remote teaching supported by information telecommunication technology tools. In such a scenario, the introduction part represents the legal, organizational as well as technical aspects of remote teaching implementation at the Jan Wyzykowski University, Polkowice, Poland. Similarly, the portion of the literature review emphasizes specific e-learning methods and tools as well as e-learning solutions 3.0, 4.0, 5.0. The associated research queries are formulated in the methodological part, along with the proposed research method. The findings, as well as outcomes of the research, are basically the analysis of distance learning as well as the conceptual framework of the e-learning 5.0 system. Moreover, the overall conceptualizations and considerations are prioritized in the future work and conclusions.

Keywords Distance teaching · Distance learning · E-learning 3.0 · 4.0 · 5.0 · IoT

1 Introduction

1.1 Legal Aspects of Implementing Remote Learning

As a matter of fact, the issues linked to COVID-19 in the Polish legal system appeared in the Act of March 2, 2020, on special solutions toward prevention and eradication of COVID-19, other infectious diseases and crisis situations caused by them (Journal of Laws of 2020, No. 374). Pursuant to this Act, amendments to the Act on Higher Education were being introduced, giving the Minister of Science and Higher Education (Journal of Laws of 2020, item 85) the legal basis for introducing restrictions

---

T. Kierzyk (✉) · Z. Pólkowski
Jan Wyzykowski University, Polkowice, Poland
e-mail: t.kierzyk@ujw.pl

Z. Pólkowski
e-mail: z.polkowski@ujw.pl

A. Khanna et al. (eds.), *Data Analytics and Management*, Lecture Notes on Data Engineering and Communications Technologies 54,
https://doi.org/10.1007/978-981-15-8335-3_70

in the functioning of higher education in cases justified by extraordinary circumstances threatening the life or health of members of the community university. In these situations, the functionalities of the universities in the part of the country were temporarily suspended from anticipating the degree of danger in the area. Prioritizing the above-mentioned authorizations, the basic traditional learning systems, including classroom teaching methodologies to the students, have been suspended by the Minister of Science and Higher Education, Poland, following the Regulation of March 11, 2020, for the period from March 12, 2020. In such a scenario, the classes could only be implemented using distance learning methods and techniques, regardless of whether this was envisaged in the curriculum.

It is worth mentioning that distance learning is an outcome of distance education, where learners and teachers are separated by geographical and/or temporal distance, and a form of mediated learning can be achieved using a combination of technologies. Distance learning can be differentiated from e-learning, which may be undertaken at a distance or contiguously, or as a combination of both (blended learning) [1].

1.2 Organizational Aspects of Implementing Remote Teaching

Considering the present situation, at the Jan Wyżykowski University in Polkowice, as in the whole of Poland, education of students in all forms and education within the framework of implementing social projects financed by the local government have been suspended. The library activities and direct student service by the university administration have been stopped. All matters could be dealt with only remotely. In such a scenario, the rector issued an ordinance and specified all these issues in detail. On March 24, the rector sent a letter to all employees regarding the commencement of distance education. Rector emphasized about the awareness of the existing threats in this area, such as lack of knowledge of a never-used educational platform, concern for the efficiency of Internet connections as well as the fear of conducting classes without "face-to-face" contact. During the initiation of remote education, the rector obliged lecturers and students to use the Office 365 Teams/SharePoint platform and the university e-mail account. The choice of the platform has been preceded by detailed multi-criteria analysis. Classes were to be conducted by lecturers assigned to the subject in accordance with the schedule of classes in real time with the transmission on the Internet. The lecturers have been obliged to implement the learning outcomes provided for the given subject and to prepare an appropriate paper and electronic documentation.

1.3 Technical Aspects of Implementing Remote Teaching

As mentioned earlier, the selection of platforms took into account many criteria, including performance, ease of use and configuration, level of integration with the existing systems. The opinions of experts and persons experienced in this field were also taken into account. In this paper, the comparison among the most popular online cloud-based video conferencing solutions—Zoom, Microsoft Teams, and Google Meet—have been focused, as cited in Table 1.

The organization of the paper is shown in Fig. 1. The remainder of the paper is organized into five sections. Section 2 discusses the literature review explored in the paper. Section 3 discusses the methodology opted to carry out the research. Section 4 discusses the computed results. Section 5 discusses the future aspects of the paper. Finally, Sect. 6 concludes the paper.

2 Literature Review

The beginnings of distance learning are far away and more distant as compared to the Internet. Already in the eighteenth century, the first correspondence courses

Table 1 Comparative analysis of distance learning systems [2]

Zoom	Google meets	Microsoft teams
Pros:	*Pros:*	*Pros:*
• Free account available • Access from all devices • Host up to 100 people (free plan) and 500 people (paid plan) • Up to 49 persons on screen • Screen sparing and co-annotation • Meeting participants can work in a small-group discussion • Additional features	• Collaboration tools • Send files trough chat + instant messaging • Full integration with Google Apps • Host up to 250 persons for free until 30/09/2020 • No additional fee for call-in-participants • Access from all devices • Automatic captioning is available • No time limit on calls	• Collaboration tools • Full integration with Office 365 • Video conferencing with background blur • Recordings (in the cloud), file and screen sparing • Instant messaging • Access from all devices • Tabs to make winding information easy • Automatic captioning is available • No time limits on calls
Cons:	*Cons:*	*Cons:*
• Download the app • Time limit of 40 min • Security problems "zoom-bombing" • Purchase an audio conferencing plan for call ins	• No waiting room for participants that join early • The only person at a time can share his screen • Users privacy can be improved	• No waiting room for participants that join early • Users privacy can be improved • No grid view

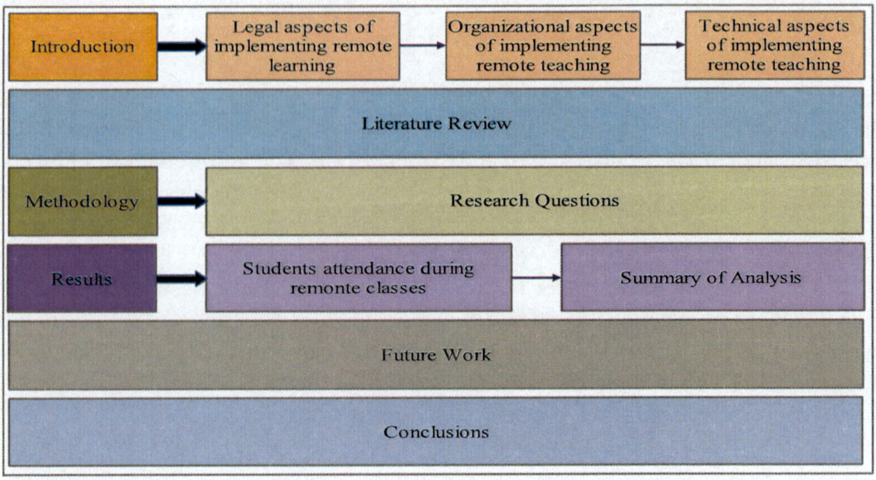

Fig. 1 Organization of the paper

were appeared to enable learning with the help of materials sent by post. Later, radio and television were being appeared as new communication channels. However, the biggest breakthrough was the discovery of the Internet [3]. The first e-learning system (PLATO, Programmed Logic for Automatic Teaching Operations) had been created around 1960, which was a CBT platform, i.e., computer-based learning. Hence, the largest number of studies on e-learning and information and communication technologies (ICT) in higher education began to appear after 1990. An often-cited study is the work "E-learning and the science of instruction: Proven guidelines for consumers and designers of multimedia learning [4]." E-learning trends were discussed by Poonam Gaur in "Research trends in E-learning" in 2015. During the analysis of various scientific studies on e-learning, a research gap related to the search for means and methods that would allow more to ensure the existence of relationships to occur during traditional classes has been noticed. Even fewer e-learning 4.0 publications can also be found [5, 6]. Several interesting studies in this field are available in German language. However, the biggest difficulty is finding studies on e-learning 5.0 studies [3, 7, 8, 9].

3 Methodology

The current situation in Poland has become the motivation to conduct the research, which obliged all teachers to reach for tools for remote work. This solution brings many benefits. First of all, it allows teachers and students to maintain workflow and avoid the accumulation of backlogs. But it is also a stimulus for many lecturers (especially for those who do not use educational applications, platforms, or online tools on a daily basis) to break down and gain new skills.

In this research, the case study method has been chosen. A case study is one of the qualitative research methods, where the main intention is to illustrate one case. It provides an in-depth analysis of a specific phenomenon containing the detailed analysis of the case, goals, assumptions, motives, actions. Nowadays, it is less and less used in research and science. In today's world, the emphasis is on mass phenomena rather than individuals. However, it should be emphasized that the case study is one of the ideal scientific materials. One analyzed material containing a large part of knowledge (both theoretical and practical) from the field of science is being studied.

This case study contains material developed on the basis of an analysis of the implementation of remote teaching at the UJW in Polkowice. For the purposes of the study, surveys were conducted among lecturers for distance learning. In addition, the author's personal experience in running a small IT company is an important source of valuable, reliable, and timely information on ICT in education.

The above-mentioned factors contributed to formulating the following research questions:

1. What impact did the use of the remote teaching method have on student attendance?
2. How were lecturers' behavior and fears noticeable during remote classes?
3. What model of distance learning system can be used in real conditions at universities?

4 Results

4.1 Students Attendance During Remote Classes

In accordance with the schedule of classes in force at the university, the first classes were being held on April 4 and 5, 2020. The dean of the technical department (where classes in the field of IT, mechatronics, logistics are taught) when assessing remote classes stated that the attendance at remote classes based on submitted reports is quite similar to that at direct contact classes, although it can be admitted that it is slightly better at remote classes in every technical field (Fig. 2).

It is noted that classes starting morning at 8.00 o'clock have much lower attendance than later classes. After 16.00 o'clock, attendance is similar to morning classes. In language and exercise classes, the attendance was very high, reaching 70–80%. The information provided by the dean shows that the vast majority use the TEAMS platform, although there were three cases where other solutions were used. In any case, this applies to lecturers, who are at least 70 years old. It is estimated that mentally, these people have failed in the new situation. These three people from the 50-person group constitute a small percentage of employees and generally do not affect the positive assessment. To sum up, according to the dean, remote learning

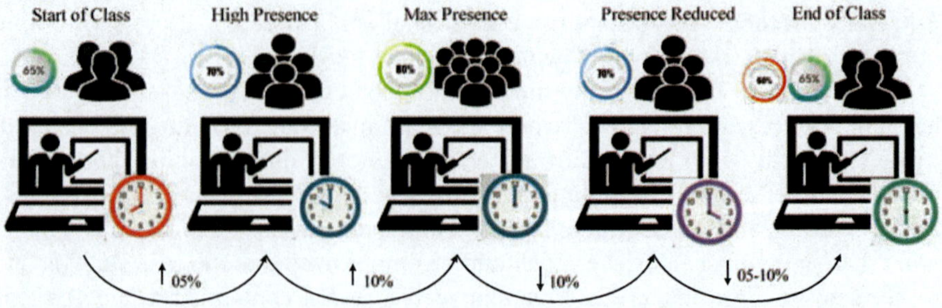

Fig. 2 Students' attendance during remote classes at technical faculty, own elaboration

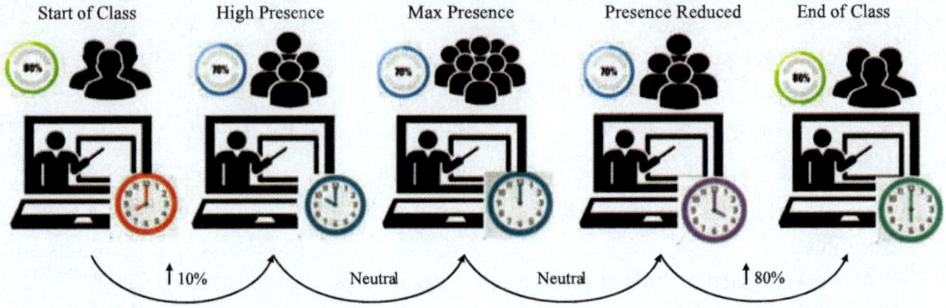

Fig. 3 Students' attendance during remote classes at Social Sciences faculty, own elaboration

cannot replace direct learning, especially in the case of exercises, laboratory, projects, etc. Lectures are best in this formula. One final note is that the preparation of materials for remote classes requires a lot of effort. At the Faculty of Social Sciences, in which education in the fields of Pedagogy and Administration is conducted, the presence of students in classes amounted to about 80%. The highest presence was at the beginning of the classes at 8 o'clock and end at 18 o'clock. The lowest was at lunchtime at around 13.00 o'clock. See Fig. 3.

At the branch of the university in Lubin, where education in the fields of Mining and Geology, Management and Production Engineering, Management and Law is conducted, the attendance in class is quite similar linked to Fig. 4.

An analysis of the presence of students in classes conducted entirely remotely shows that the form of these classes was accepted and enjoyed recognition. In individual cases, the participation of students who participated in classes using smartphones was observed. The conducted research showed that attendance at classes in all time intervals was higher than in the case of traditional classes conducted before the pandemic. Attitudes of lecturers have shown that the vast majority of lecturers, despite the previous concerns, have fulfilled their obligations. Almost all lecturers expressed fear of launching remote education at a fast, even express pace.

These concerns (Fig. 5) have appeared due to:

Fear 1. Lack of experience of using remote teaching methods,

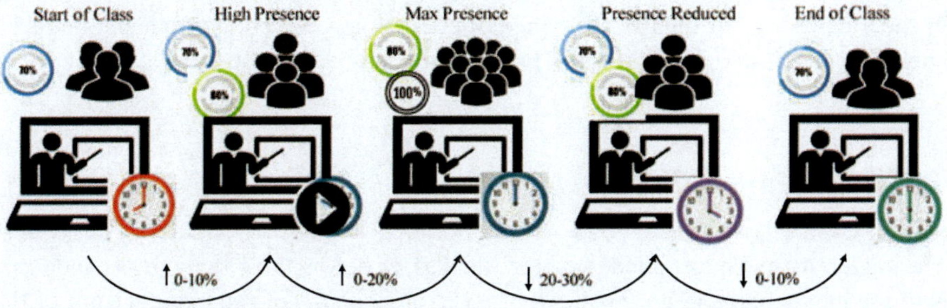

Fig. 4 Students' attendance during remote classes at technical faculty, own elaboration

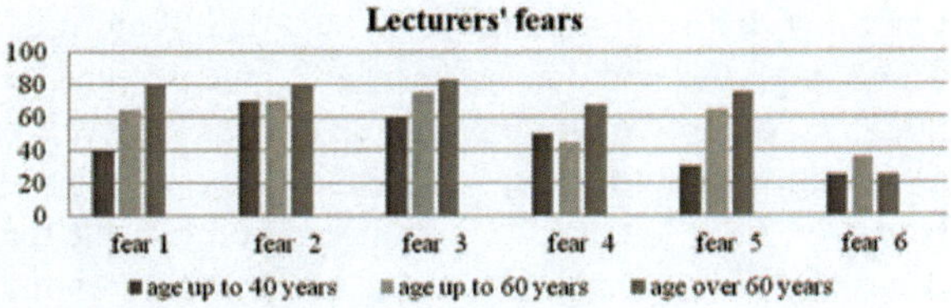

Fig. 5 Fears of lectures, own elaboration

Fear 2. Concerns about the efficiency of Internet connections,
Fear 3. Mental barriers resulting from the urgent need to speak to the computer and not to living people,
Fear 4. The need to apply self-limitation and self-discipline to a greater extent than in traditional forms,
Fear 5. Low self-assessment of technical competences related to system operation,
Fear 6. Other.

The analysis of the quality of completed classes shows that the most worried were the way of conducting laboratory and project classes, as well as the principles and mode of apprenticeships. Internships for students of the teaching specialization—pedagogy—were a serious problem. Due to blocking the functioning of schools, internships became impossible. In the month of May 2020, the regulations have been changed and the possibility of remote teaching practice has been allowed. Students were able to conduct classes with children remotely, check work, and perform other activities specified in the internship program.

However, it was not possible to take classes in laboratories—e.g. chemical. In such a scenario, the rector of the university has provided in his ordinance for the possibility of undergoing some classes at the university's premises, which is necessary with the prior consent of the dean and vice-rector for ensuring the quality of education. These were exceptions to the principle of distance learning. Lectures, seminars, and exercises were relatively easy. From April 14, 2020, the rector of the university introduced the obligation to start conducting diploma exams using distance communication devices. To conduct examinations, the Office 365 Teams/SharePoint platform was ordered. In individual, particularly justified cases, with the consent of the vice-rector for quality assurance of education, granted at the dean's request, another educational platform could be used to conduct the examination.

4.2 *Summary of the Analysis*

The university had never conducted remote classes before the coronavirus pandemic. Nobody had experience in this field. The rector's speech of the inauguration of the academic year in 2019 shows that the university is intended to launch remote education only from the new academic year 2020/2021. The vast majority of academic teachers did not know the TEAMS platform and never used it. Moreover, they did not know that such a platform exists at all. Among students who did not know the TEAMS platform and never heard of it, the percentage was even higher and reached 80–85%. After two months, remote classes received very high scores from students and lecturers. Both groups considered it a very good tool for conducting classes. However, both lecturers and students pointed out that the biggest discomfort was caused by the urgent need to speak to the computer and not to living people. Respondents also indicated a lack of emotional relationships between students and lecturers. The solution to this problem seems to be the further development of remote teaching systems through the implementation of modern technologies such as the Internet of things, big data, or virtual machines [10–17]. Taking into account the obtained research results, a conceptual model of e-learning 5.0 remote teaching has been developed to meet the demands of students and lecturers as reflected in Fig. 6.

5 Future Work

Gartner, Inc. forecasts that 4.9 billion connected things are in use since 2015, up to 30% from 2014, and will reach 25 billion by this 2020 year. The Internet of things (IoT) has become a dominant force for business transformation, and its disruptive impact will be felt across all industries and all areas of society. A similar situation is also expected in education.

The use of IoT solutions directly is expected in universities. Moreover, the use of hybrid solutions is assumed: implants, devices controlled by the Internet. So, we will deal with hybrid university (Fig. 7).

Visible to use IoT trends in education will influence shaping the IoT of the education model:

- Integration of devices or equipment,
- The use of video systems,
- Further, automation and the use of large-scale machines,
- Ability to control via the Internet from anywhere in the world.

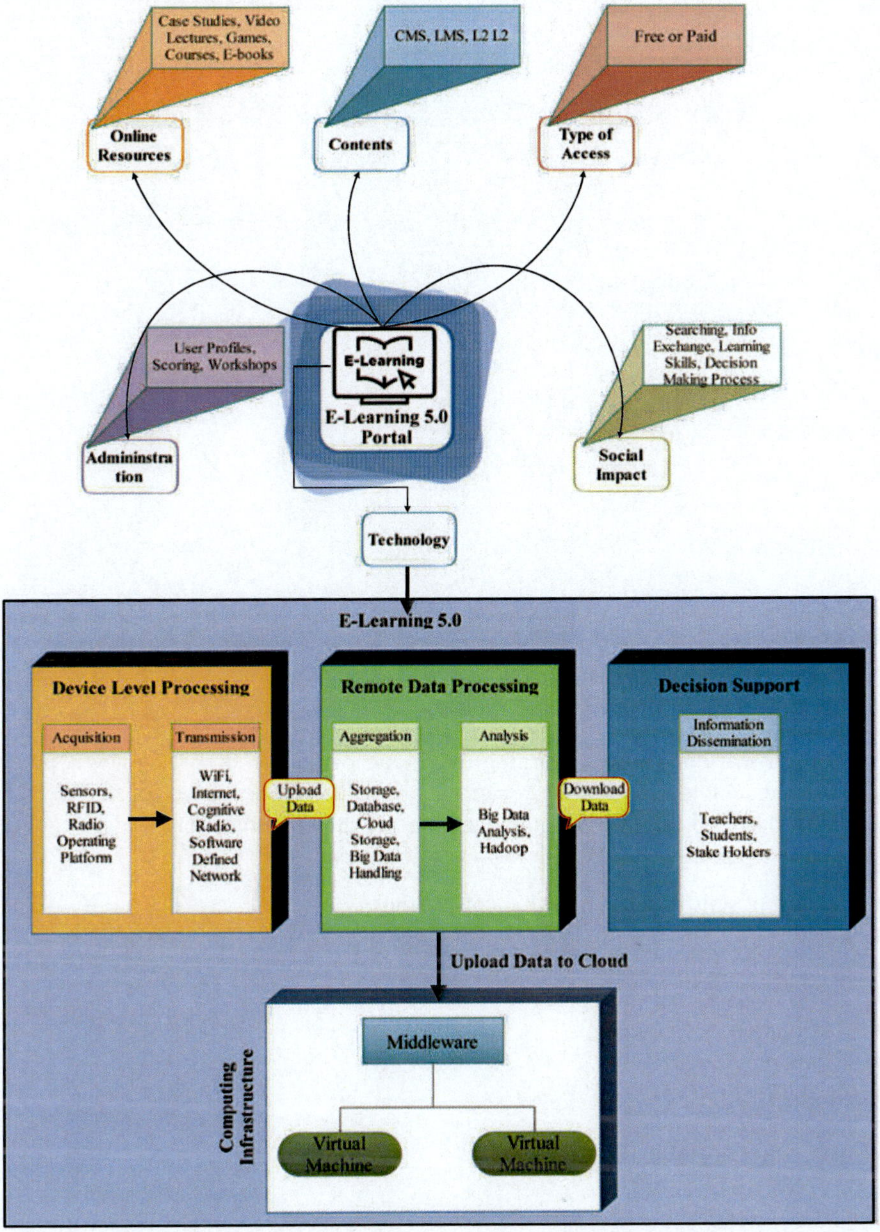

Fig. 6 Conceptual framework of e-learning 5.0 system, own elaboration

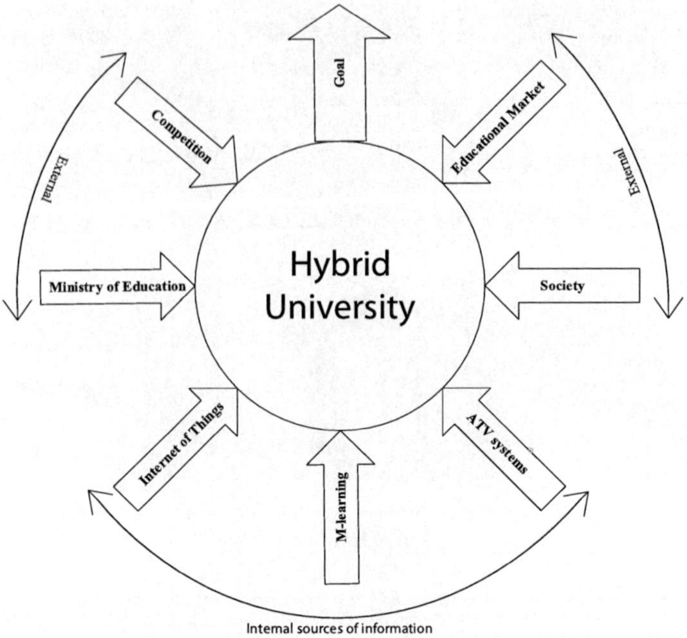

Fig. 7 Hybrid university, own elaboration

Also, M-learning (mobile learning) will appear in universities as a new type of electronic teaching using mobile devices. By using these, users have constant access to knowledge, and they can develop it and update their qualifications on an ongoing basis. Other possible technologies for use in universities are:

- Intelligent sensor (data-to-digital conversion),
- Smart robots and machines,
- Artificial intelligence,
- Three-dimensional (3D) printing technologies,
- Augmented reality (AR),
- Mixed reality (MR),
- Virtual reality (VR),
- Data analytics,
- Radio-frequency identification.

6 Conclusion and Future Scope

Based on the conducted research, it can be concluded that although, at the Jan Wyżykowski University until the outbreak of the COVID-19 pandemic, no remote classes were being conducted, the education continued smoothly. A surprisingly low level of awareness has been noted among staff and students about the existence of remote learning platforms, its functionality, and suitability for the remote education of students. Despite the low awareness, employees relatively quickly learned and adapted the platform as an effective and useful tool for distance learning. Lectures, seminars, projects, and exercises were being conducted very effectively. Inability to conduct remote laboratory classes, therefore, despite the pandemic, the rector envisaged in the ordinance a deviation from the ban and admission to traditional classes in the university building on an exceptional basis. There were difficulties in implementing internships in local government and educational institutions. These institutions were closed based on government decision for the public, and the implementation of practices became impossible. Remote engineering exams were efficient. Participants did not raise any objections. The lecturers' fears disappeared relatively quickly, although in few cases the classes were not done properly. The university will have to analyze these cases in detail and draw conclusions from it. Relatively, high attendance rates were noted higher than in traditional forms. After a relatively short period of conducting classes using the TEAMS platform, a very high degree of satisfaction was noted both in the group of lecturers and students. In the group of lecturers, it was 87% and in the group of students 84%. With all responsibility, it can be concluded that the awareness of the existence of very good IT tools for conducting remote education has increased at the UJW. This method will remain a permanent element of the education process. This is undoubtedly a positive aspect of this difficult situation, and education of such emerging types should be encouraged and developed. However, the research shows that distance learning should be complementary to traditional learning. The "master–student" relationship cannot be replaced by online contact. The atmosphere of the lecture, discussion, and exchange of thoughts—all this makes studying a unique period in the life of every student. One solution to this problem is the use of e-learning 5.0 (the sensory and emotive learning) which will be based on emotional interaction based on neurotechnology between people and computer devices. This study contains the design of the framework in this technology.

In the future, we will introduce e-learning 5.0 solutions by means of evolution, not revolution through the gradual expansion of the platforms.

References

1. (Online) Available at: https://link.springer.com/referenceworkentry/10.1007%2F978-1-4419-1428-6_432. Accessed 12 Apr 2020
2. (Online) Available at: https://www.fourcast.io/blog/comparing-zoom-microsoft-teams-and-google-meet. Accessed 12 Apr 2020
3. (Online) Available at: http://www.cel.agh.edu.pl/wp-content/uploads/2009/11/wprowadzenie_do_e-learningu.pdf. Accessed 12 Apr 2020
4. (Online) Available at: https://books.google.pl/books?hl=pl&lr=&id=v1uzCgAAQBAJ&oi=fnd&pg=PR17&dq=e-learning&ots=TMCEoJ9J8h&sig=P207Tm6-ZNbQwokjV0aBEupnSVY&redir_esc=y#v=onepage&q=e-learning&f=false. Accessed 12 Apr 2020

5. Digital Universities V.2 (2015) n. 2-3: International best practices and aplications. (Online) Available at: https://books.google.pl/books?hl=pl&lr=&id=GItVCwAAQBAJ&oi=fnd&pg=PA51&dq=e-learning+4.0&ots=MFjmHK08AF&sig=Z5571z7yOgWSlIqYE5__IRrCHhU&redir_esc=y#v=onepage&q=e-learning%204.0&f=false. Accessed 12 Apr 2020
6. Peter A, Henning, Learning 4.0, (Online) Available at: https://link.springer.com/chapter/10.1007/978-3-319-73546-7_17. Accessed 12 Apr 2020
7. Evolution of the weband E-Learning application. (Online) Available at: https://ieeexplore.ieee.org/abstract/document/7724435. Accessed 12 Apr 2020
8. European Internet Forum (2015) The digital world in 2030. (Online) Available at: https://www.eifonline.org/digitalworld2030.html. Accessed 12 Dec 2017
9. Van der Zwan R (2014) ICT trends 2014: what to expect in 2014 and beyond? (Online) Available at: http://blog.yenlo.com/nl/top-6-ict-trends-2014. Accessed 12 Dec 2017
10. Tanwar S, Patel P, Patel K, Tyagi S, Kumar N, Obaidat MS (2017) An advanced Internet of Thing based Security Alert System for Smart Home. In: 2017 International conference on computer, information and telecommunication systems (CITS), Dalian, pp 25–29. https://doi.org/10.1109/cits.2017.8035326
11. Kumari A, Tanwar S, Tyagi S, Kumar N, Maasberg M, Choo KKR (2018) Multimedia big data computing and Internet of things applications: a taxonomy and process model. J Netw Comput Appl 124:169–195
12. Tanwar S, Tyagi S, Kumar S (2018) The role of internet of things and smart grid for the development of a smart city. In: Hu YC, Tiwari S, Mishra K, Trivedi M (eds) Intelligent communication and computational technologies. Lecture Notes in Networks and Systems, vol 19. Springer, Singapore
13. Kumari A, Tanwar S, Tyagi S, Kumar N (2019) Verification and validation techniques for streaming big data analytics in internet of things environment. IET Netw 8(3):155–163
14. Kumari A, Tanwar S, Tyagi S, Kumar N, Parizi RM, Choo KKR (2019) Fog data analytics: a taxonomy and process model. J Netw Comput Appl 128:90–104
15. Gupta R, Tanwar S, Tyagi S, Kumar N (2020) Machine learning models for secure data analytics: a taxonomy and threat model. Comput Commun 153:406–440
16. Srivastava A, Singh SK, Tanwar S, Tyagi S (2017) Suitability of big data analytics in Indian banking sector to increase revenue and profitability. In: 2017 3rd International conference on advances in computing, communication & automation (ICACCA) (Fall), Dehradun, pp 1–6
17. Tanwar S, Ramani T, Tyagi S (2018) Dimensionality reduction using PCA and SVD in Big data: a comparative case study. In: Patel Z, Gupta S (eds) Future internet technologies and trends. ICFITT 2017. Lecture Notes of the Institute for Computer Sciences, Social Informatics and Telecommunications Engineering, vol 220. Springer, Cham
18. E-Learning 3.0 = E-Learning 2.0 + Web 3.0? Hussain, Fehmida. (Online) Available at: https://eric.ed.gov/?id=ED542649.. Accessed 12 Apr 2020
19. Rubens N, Kaplan D, Okamoto T, E-Learning 3.0: anyone, anywhere, anytime, and AI. (Online) Available at: https://link.springer.com/chapter/10.1007/978-3-662-43454-3_18. Accessed 12 Apr 2020

Author Index

A. Khanna et al. (eds.), *Data Analytics and Management*, Lecture Notes
on Data Engineering and Communications Technologies 54,
https://doi.org/10.1007/978-981-15-8335-3